Stability-Indicating
HPLC Methods for Drug Analysis

Third Edition

Notices

The inclusion in this book of any drug in respect to which patent or trademark rights may exist shall not be deemed, and is not intended as, a grant of or authority to exercise any right or privilege protected by such patent or trademark. All such rights or trademarks are vested in the patent or trademark owner, and no other person may exercise the same without express permission, authority, or license secured from such patent or trademark owner.

The inclusion of a brand name does not mean the authors or the publishers have any particular knowledge that the brand listed has properties different from other brands of the same drug, nor should its inclusion be interpreted as an endorsement by the authors or publishers. Similarly, the fact that a particular brand has not been included does not indicate the product has been judged to be in any way unsatisfactory or unacceptable. Further, no official support or endorsement of this book by any federal or state agency or pharmaceutical company is intended or inferred.

The nature of drug information is that it is constantly evolving because of ongoing research and clinical experience and is often subject to interpretation. Readers are advised that decisions regarding drug therapy must be based on the independent judgment of the clinician, changing information about a drug (e.g., as reflected in the literature and manufacturers' most current product information), and changing medical practices.

The authors, editors, and publishers have made every effort to ensure the accuracy and completeness of the information presented in this book. However, the authors, editors, and publishers cannot be held responsible for the continued currency of the information, any inadvertent errors or omissions, or the application of this information.

Therefore, the authors, editors, and publishers shall have no liability to any person or entity with regard to claims, loss, or damage caused, or alleged to be caused, directly or indirectly, by the use of information contained herein.

Stability-Indicating
HPLC Methods for Drug Analysis
Third Edition

Quanyun A. Xu, PhD

Research Scientist, Pharmacology Research Laboratory
Division of Pharmacy
University of Texas M.D. Anderson Cancer Center
Houston, Texas

Lawrence A. Trissel, FASHP

Research Consultant
TriPharma Research
Ponte Vedra Beach, Florida

American Pharmacists Association
Washington, D.C., USA

APhA

Pharmaceutical Press
London, UK

Acquiring Editor: Julian I. Graubart
Managing Editor: L. Luan Corrigan
Layout and Graphics: Kathryn A. Stromberger
Cover Design: Richard Muringer, APhA Creative Services

© 2008 by Quanyun A. Xu and Lawrence A. Trissel
Published by the American Pharmacists Association
1100 15th Street, N.W., Suite 400
Washington, DC 20005-1707, USA
and the
Pharmaceutical Press
1 Lambeth High Street
London SE1 7JN, United Kingdom and
100 South Atkinson Road, Suite 200
Grayslake, IL 60030-7820, USA

To comment on this book via e-mail, send your message to the publisher at aphabooks@aphanet.org

Library of Congress Cataloging-in-Publication Data

Xu, Quanyun A., 1961-
  Stability-indicating HPLC methods for drug analysis / Quanyun A. Xu, Lawrence A. Trissel. -- 3rd ed.
      p. ; cm.
  Includes bibliographical references and index.
  ISBN 978-1-58212-109-3 (APA) -- ISBN 978 085369 723 7 (PP)
  1. Drugs--Analysis. 2. High performance liquid chromatography. 3. Drug stability. I. Trissel, Lawrence A. II. American Pharmacists Association. III. Title.
  [DNLM: 1. Pharmaceutical Preparations--analysis. 2. Chromatography, High Pressure Liquid--methods. 3. Dosage Forms. 4. Drug Stability. QV 25 X85s 2008]

  RS189.5.H54X8 2008
  615'.1901--dc22
                                        2007041655

A catalog record for this book is available from the British Library.

To Hongqin, Michael, and Andy for love, joy, and support.

QAX

To those who recognize the value of pharmaceutical research,

And to Pam, for everything you do.

LAT

Contents

Preface

S tability-Indicating HPLC Methods for Drug Analysis, third edition, is a compilation of summaries of stability-indicating high-performance liquid chromatographic (HPLC) analytical methods that have appeared in the published literature. The information is presented as structured monographs on 572 different drug entities, including 96 new to this edition. These monographs have been prepared from 851 published references and present 1028 stability-indicating HPLC methods that have been described in the published literature. The work is unique in focusing only on stability-indicating HPLC methods, a specialized subset of all HPLC methodologies, that are useful to those involved with evaluating drug stability, dosage form development, quality control, and drug regulation.

The initial steps an analyst must take when beginning the process of analyzing an unfamiliar drug inevitably have included a search for previously published analytical methods for that drug. Often this takes the form of a computer search of the published literature. Usually being very general in nature, these searches often list a multiplicity of methods that are not relevant to the specific kind of application the analyst has in mind. This can lead to collecting and evaluating articles that, in the end, may or may not be useful. Too often the analytical methods tend to be too general in nature or specific for unrelated applications but not demonstrated to be useful for drug stability determinations. Because so much of the published literature describing analytical techniques focuses on determinations of drug concentrations in biological matrices, stability-indicating methods tend to be given less attention or even left out altogether.

For those analysts whose work involves the determination of the stability of pharmaceutical dosage forms, most published analytical methodologies have not been demonstrated to be suitable. Consequently, many methods will be of uncertain applicability at best when obtained from computer searches or published compilations.

It has always seemed to us that the best starting point in identifying a suitable stability-indicating analytical method is from methods that have previously been shown to be stability indicating in studies by other researchers. Of course, finding an analytical method in the literature that has been found to be stability indicating in a previous study does not preclude the need to evaluate the method in the analyst's hands. The analyst is still obligated to assure the adequacy of the method for specificity, precision, reproducibility, and sensitivity. Furthermore, the method must be shown to separate and quantify the intact drug in the presence of its decomposition products and other dosage form components. However, this task of finding a suitable method and documenting that it is both stability indicating and is functioning properly in one's own laboratory is aided when the analyst starts with a method that has been shown by others to be stability indicating.

This book is a result of the realization that finding suitable stability-indicating analytical methods would be greatly facilitated if a compilation limited to stability-indicating analytical methods was assembled. The compilation would be the first stop in the search to find a suitable methodology for a particular drug analysis application. Stability-indicating methods would be immediately at hand without lengthy literature searches. And if no stability-indicating HPLC method has been published on the subject drug, then wasting time in a futile search for a stability-indicating method could be avoided.

The introduction of high-performance liquid chromatography several decades ago revolutionized the determination of drug stability in dosage forms. HPLC rapidly replaced other techniques for most stability applications of drugs. Today, it is the most common stability-indicating methodology described in the published literature, almost to the exclusion of other techniques. Consequently, the methods we have described in this compilation are limited to HPLC analysis. By utilizing a textual format, we have provided a more-in-depth and extensive description of the HPLC analysis presented in the original publications than the more common sketchy and abbreviated outline formats that have appeared elsewhere. As an example, the summaries in this work include the manner of demonstrating the stability-indicating nature of the method when it is described in the original article.

The authors of this work acknowledge that there are undoubtedly numerous stability-indicating HPLC methods that have been developed but have never been published. The methods may have been relegated to dusty file cabinets if the analyst never found the time or had the inclination to publish the method. Or perhaps they are being held as the confidential and proprietary information of a commercial entity. In either case, they were simply unavailable for inclusion in this compilation.

Users of this work must recognize that it is not intended to be a primer or "how to" book of drug analysis. Rather, it is a compilation of stability-indicating methods from the published literature. It is intended for use by knowledgeable analysts skilled in HPLC analysis.

Stability-Indicating HPLC Methods for Drug Analysis, third edition, provides readily available and convenient access to the published literature on stability-indicating HPLC assays for the busy analyst who has no time to perform lengthy literature searches and spend tiresome hours in the library. It may also save analysts time, money, and frustration by helping them to identify suitable previously validated stability-indicating HPLC methods and avoid conducting expensive and time-consuming validations on methods of unknown applicability that may not prove to be stability indicating and may wind up being unsuitable for the purpose.

The authors would like to thank Mr. Julian Graubart, the American Pharmacists Association, and the Pharmaceutical Press for their continued encouragement and support.

Quanyun A. Xu
Lawrence A. Trissel
October 2007

Using This Book

Stability-Indicating HPLC Methods for Drug Analysis, third edition, is structured as a series of monographs on 572 separate drug entities and includes 1028 stability-indicating HPLC analytical methods. The information that is presented in the monographs has been gathered from the published literature and is summarized in a structured textual format composed of 30 elements.

Components

The first series of monograph elements were obtained from standard reference works.[1–4] This information is the **basic information** an analyst may find useful in analyzing various drug products. The information presented includes:

1. Chemical name(s)
2. Molecular structure
3. Other name(s)
4. Form(s) of the drug molecule available in drug products
5. Molecular formula
6. Molecular weight
7. CAS number
8. Appearance and description
9. Solubilities
10. pK$_a$ values

The basic information is followed by one or more summaries of previously published stability-indicating analytical methods. The summaries follow a highly structured text format. Each of the informational items that follow is included when presented in the original published articles. The reader should recognize that all too often one or more pieces of information were omitted in the original work and, consequently, do not appear in this compilation. However, the summaries have been made as complete as possible from the original published articles.

The **first section of the method summary** includes the following information:

11. Individual(s) who developed or published the method
12. Drug, salt form (if applicable), and presentation that was analyzed
13. Description of analytical system components
14. Description of column
15. Mobile phase composition and flow rate
16. Detection wavelength and sensitivity
17. Internal standard (if present)

The **second section of the method** includes:

18. Standard solution preparation
19. Sample preparation (dilution, extraction, derivatization, etc.)
20. Injection volume
21. Retention times for subject drug, secondary drug(s), and internal standard

The **third section** of the summary describes the demonstration of the stability-indicating nature of the method. The description includes:

22. The decomposition techniques and/or sample spiking that was used
23. Absence of interference of decomposition product with intact drug
24. Retention times for intact drug and degradation products

The **fourth section** of the method summary includes the following information:

25. Concentration range of the standard curve
26. Correlation coefficient and linearity

27. Coefficient of variation of the assay
28. Intraday and interday coefficients of variation of the samples
29. Limit of detection and limit of quantitation

And lastly, the **source** of the method is presented:

30. Reference(s)

The reference(s) from the published literature that utilized the particular method provides proper attribution and allows readers to more easily refer to the original work if that would be useful.

Limitations of the Literature

Unfortunately, not all of the published methods have included all of the informational items we think are appropriate for these kinds of studies. Too often editors (and authors) eliminate such details from journal articles in the interest of saving space. But for those who use such analytical methods in their work, this information is useful and needed. We have included all of the information as completely as possible from the original published articles.

Compared to the myriad number of HPLC assays that have been published for all purposes, the number of HPLC methods that have been demonstrated to be stability indicating in the published literature is relatively small. Also, one has to be cognizant of the number of stability-indicating methods that are held confiden-

tially by commercial concerns, which also contributes to this dearth. Even so, this compilation will enable the analyst to have ready access to stability-indicating assays for a large proportion of the drugs on the commercial market.

Analysts must also be aware that the presentation of a stability-indicating analytical method in the literature does not absolve them of the need to document the adequacy of the method in their own laboratory. The analyst still must determine fundamental performance parameters of the method including sensitivity, precision, reproducibility, and accuracy. Also, the stability-indicating ability of the method must be documented in the analyst's laboratory to demonstrate that the method is working properly. These are more likely to be easily achievable when the method has been shown to work in another analyst's hands

References

1. O'Neil MJ, ed., *The Merck Index, 14th ed.,* Whitehouse Station, NJ: Merck & Co., Inc., 2006.
2. *AHFS 07 Drug Information,* Bethesda, MD: American Society of Health-System Pharmacists, 2007.
3. Sweetman S., ed., *Martindale: The Complete Drug Reference, 35th ed.,* London: Pharmaceutical Press, 2007.
4. *Physicians' Desk Reference, 61st ed.,* Montvale, NJ: Medical Economics Co., 2007.

MONOGRAPHS

Aceclofenac

Chemical Name

[*o*-(2,6-Dichloroanilino)phenyl]acetate glycolic acid ester

Other Names

Aceclofar, Airtal, Preservex, Sanein

Form	Molecular Formula	MW	CAS
Aceclofenac	$C_{16}H_{13}C_{l2}NO_4$	354.2	89796-99-6

Appearance

Aceclofenac occurs as a white or almost white crystalline powder.

Solubility

Aceclofenac is practically insoluble in water. It is soluble in alcohol and freely soluble in acetone and dimethylformamide.

Method

Hasan et al. developed an HPLC method for the determination of aceclofenac in the presence of its major degradation product. A Shimadzu class LC-10 liquid chromatographic system was used with a Shimadzu SPD-10A diode-array detector. The stationary phase was a Zorbax C_{18} column (150 × 4.6 mm, 5-μm particle size). The mobile phase consisted of methanol and water (60:40, vol/vol). The flow rate was 1 mL/min. UV detection was performed at 275 nm.

Samples were prepared in mobile phase. The injection volume was 20 μL. Under these conditions, the retention time for aceclofenac was about 1.9 minutes.

The method was confirmed to be stability indicating by intentionally degrading the drug. Aceclofenac was refluxed in 1 N sodium hydroxide solution for 3 hours, cooled down, mixed with 2.0 N sulfuric acid, and heated to 90 °C. The degradation products did not interfere with the determination of aceclofenac. The retention time for the major degradation product, diclofenac, was about 4.6 minutes.

A calibration curve for aceclofenac was constructed from 1 to 50 μg/mL. The correlation coefficient was 0.9996.

Reference

Hasan NY, Abdel-Elkawy M, Elzeany BE, et al. Stability-indicating methods for the determination of aceclofenac. *Il Farmaco*. 2003; 58: 91–9.

Acetaminophen

Chemical Names
N-(4-Hydroxyphenyl)acetamide
4′-Hydroxyacetanilide

Other Names
Paracetamol, Tempra, Tylenol

Form	Molecular Formula	MW	CAS
Acetaminophen	$C_8H_9NO_2$	151.2	103-90-2

Appearance
Acetaminophen occurs as a white crystalline powder with a slightly bitter taste.

Solubility
Acetaminophen is soluble in boiling water and freely soluble in alcohol.

pK$_a$
Acetaminophen has a pK$_a$ of 9.51.

Method 1
Aukunuru et al. described the simultaneous determination of acetaminophen, salicylamide, phenyltoloxamine, and related products by HPLC. A Varian system consisting of a model 9010 pump, a model 9095 autosampler, a model 9050 UV absorbance detector, and a Rainin Dinamax MacIntegrator was used. The stationary phase was a Phenomenex Prodigy C$_8$ column (150 × 4.6 mm, 5-μm particle size). Mobile phase A was 0.1 M phosphate buffer (adjusted to pH 2.7 with phosphoric acid). Mobile phase B was acetonitrile. The mobile phase was linearly delivered from 5% B to 45% B in 17 minutes, followed by 10 minutes for equilibration. The flow rate was 1 mL/min. UV detection was performed at 220 nm. The injection volume was 50 μL. Under these conditions, retention times of acetaminophen, salicylamide, and phenyltoloxamine were 5.7, 10.9, and 15.9 minutes, respectively.

The stability-indicating nature of the assay was demonstrated by accelerated degradation of the drugs. Drug solutions were prepared in 1 N hydrochloric acid, 1 N sodium hydroxide solution, or 10% hydrogen peroxide and heated at 60 °C. No degradation products interfered with the analysis of drugs.

A standard curve for acetaminophen was generated from 0.06 to 300 μg/mL. The correlation coefficient was 0.999.

Reference
Aukunuru JV, Kompella UB, Betageri GV. Simultaneous high-performance liquid chromatographic analysis of acetaminophen, salicylamide, phenyltoloxamine, and related products. *J Liq Chromatogr Rel Technol*. 2000; 23: 565–78.

Method 2

Hewala described an HPLC method for paracetamol (acetaminophen), guaifenesin, sodium benzoate, and oxomemazine in the presence of degradation products. A Beckman Gold system consisting of a model 125 programmable pump, a model 166 programmable UV detector, and a Rheodyne 20-μL loop injector was used. The stationary phase was a stainless steel ODS C_{18} column (250 × 4.6 mm, 5-μm particle size) with a guard column (50 × 4.6 mm) of the same packing material. Mobile phase A was a mixture of methanol and water (18:82, vol/vol) adjusted to pH 3.9 with phosphoric acid, and mobile phase B was a mixture of methanol and water (80:20, vol/vol) adjusted to pH 3.9 with phosphoric acid. Mobile phase A was delivered from 0 to 12 minutes and mobile phase B from 12 to 22 minutes. The flow rate was 1.5 mL/min. UV detection was performed at 235 nm and 0.05 AUFS. Metronidazole was used as an internal standard.

A portion of the cough syrup (10 mL) was diluted to 100 mL with methanol, mixed with internal standard solution, and further diluted with methanol to 20 μg/mL of metronidazole in the final solution. Under these conditions, retention times for paracetamol (acetaminophen), metronidazole, sodium benzoate, guaifenesin, and oxomemazine were about 2.6, 5.5, 8.6, 10.7, and 16.8 minutes, respectively (estimated from the published chromatogram).

The method was evaluated to be stability indicating by assaying a mixture of active compounds and possible degradation products. Retention times for possible degradants 4-aminophenol and guaicol were about 3.7 and 7.0 minutes, respectively (estimated from the published chromatogram).

A standard curve for acetaminophen was constructed from 5.0 to 30.0 μg/mL. The correlation coefficient was 0.9998. Intraday and interday coefficients of variation were 1.06 and 1.86%, respectively.

Reference

Hewala II. Stability-indicating HPLC assay for paracetamol, guaiphenesin, sodium benzoate and oxomemazine in cough syrup. *Anal Lett*. 1994; 27: 71–93.

Method 3

Alvi and Castro reported a simultaneous analysis of acetaminophen and hydrocodone bitartrate in a tablet formulation by HPLC analysis. The liquid chromatograph consisted of a Waters model M6000A pump, a model 710B/712B WISP autosampler, a model 490 variable-wavelength detector, and a model 730 data module. The stationary phase was a Waters Nova-Pak C_{18} Radial-Pak cartridge column (100 × 8 mm, 4-μm particle size) with a Waters C_{18} guard column. The column temperature was 30 °C. The mobile phase consisted of an aqueous phosphate buffer and acetonitrile (84:16, vol/vol), where the phosphate buffer was 0.02 M monobasic potassium phosphate containing 0.2 mL of triethylamine and 0.2 mL of phosphoric acid and adjusted to pH 3.3 with 3 N phosphoric acid. The flow rate was 2.0 mL/min. UV detection was performed at 215 nm and 0.02 AUFS.

A portion of powder prepared from ground tablets equivalent to one tablet was weighed, transferred into a 100-mL volumetric flask, mixed with 50 mL of the diluent (0.5 mL of 3 N phosphoric acid in 1000 mL of water), and then shaken for 45 minutes. This mixture was diluted to volume with the diluent, mixed, filtered, and further diluted with the diluent. The injection volume was 10 μL. Under these conditions, the retention times of acetaminophen and hydrocodone bitartrate were 2.6 and 5.0 minutes, respectively.

The method was evaluated to be stability indicating by assaying a synthetic mixture of drugs and their known degradation products. Retention times for *p*-aminophenol, hydromorphone hydrochloride, codeine sulfate, and *p*-chloroacetanilide were 1.6, 2.3, 3.3, and 34.6 minutes, respectively. Samples were also refluxed in 1 N hydrochloric acid, 1 N sodium hydroxide solution, or 10% hydrogen peroxide for 4 hours. The decomposition products were separated from their intact drugs.

A standard curve for acetaminophen was obtained from 0.2998 to 0.6996 mg/mL. The correlation coefficient was 0.9995.

Reference

Alvi SU, Castro F. A stability-indicating simultaneous analysis of acetaminophen and hydrocodone bitartrate in a tablet formulation by HPLC. *J Liq Chromatogr.* 1987; 10: 3413–26.

Method 4

Sisco et al. developed an HPLC method for the simultaneous determination of acetaminophen, codeine phosphate, and sodium benzoate in an elixir formulation. The chromatographic system consisted of a DuPont model 850 high-pressure liquid chromatograph equipped with a DuPont automatic sampler with a 20-µL loop, a DuPont column oven, a DuPont 4100 integrator, and a Waters model 440 UV detector. A Hewlett-Packard model 1040A diode-array detector was used for specificity studies of stressed and unstressed samples. The stationary phase was a Waters µBondapak C_{18} column (300 × 3.9 mm). The mobile phase consisted of a buffer and methanol (80:20, vol/vol). The buffer contained 2.4 g of 1-butanesulfonic acid sodium salt (0.015 M), 2.04 g of monobasic potassium phosphate (0.015 M), and 2 mL of triethylamine per liter of water. The pH of this solution was adjusted to 4.8 ± 0.1 with dilute phosphoric acid. The flow rate was 2.0 mL/min. UV detection was performed at 214 nm.

Samples were diluted 1:40 with distilled water, and the injection volume was 20 µL. Under these conditions, retention times for acetaminophen, sodium benzoate, and codeine phosphate were 2.6, 3.8, and 4.7 minutes, respectively.

The HPLC method was determined to be stability indicating by accelerated decomposition of acetaminophen. An elixir sample was placed in a clear flint-glass bottle, capped, and stored in a 60 °C oven for 2 weeks. Also, a standard solution of acetaminophen and codeine phosphate was spiked with potential degradation products. Retention times for *p*-aminophenol, codeine *N*-oxide, and codeinone were 1.8, 6.4, and 8.8 minutes, respectively. Thus, degradation product peaks did not interfere with peaks of acetaminophen, sodium benzoate, and codeine phosphate.

A standard curve for acetaminophen was constructed from 0.3 to 0.9 mg/mL (estimated from the published figure). The correlation coefficient was 0.9999.

Reference

Sisco WR, Rittenhouse CT, Everhart LA, et al. Simultaneous high-performance liquid chromatographic stability-indicating analysis of acetaminophen, codeine phosphate, and sodium benzoate in elixirs. *J Chromatogr.* 1986; 354: 355–66.

Method 5

Wallo and D'Adamo developed an HPLC method for the simultaneous assay of hydrocodone bitartrate and acetaminophen in a tablet formulation. A Waters model ALC

204 liquid chromatograph was equipped with a Waters model 710B WISP autosampler, a Schoeffel SF 770 variable-wavelength UV detector, and a Waters data module. The stationary phase was a Waters μBondapak C_{18} reversed-phase column. The mobile phase consisted of 25% methanol and 75% of an aqueous solution containing 0.01 N monobasic potassium phosphate and 0.05 N potassium nitrate, adjusted to a pH of about 4.5 with 3 N phosphoric acid. The flow rate was about 1.1 mL/min. UV detection was performed at 283 nm and 2.0 AUFS. The injection volume was 13 μL.

The assay was determined to be stability indicating by spiking the sample with potential degradation products. Degradation product peaks did not interfere with the acetaminophen peak. Retention times for p-aminophenol, hydromorphone, acetaminophen, codeine, hydrocodone, and p-chloroacetanilide were 3.4, 5.2, 5.9, 7.3, 10.0, and 43.3 minutes, respectively.

Standard curves were constructed from 5 to 6.5 mg/mL. The correlation coefficient was 0.998.

Reference
Wallo WE, D'Adamo A. Simultaneous assay of hydrocodone bitartrate and acetaminophen in a tablet formulation. *J Pharm Sci*. 1982; 71: 1115–8.

Method 6
Sena et al. reported an assay of acetaminophen in an effervescent tablet by ion-pair HPLC. A Waters model 6000A pump was equipped with a Valco model CV-6-UHPa-N60 10-μL loop injector, a Waters model 440 fixed-wavelength UV detector, and a strip-chart recorder. The stationary phase was a Waters μBondapak phenyl column (300 × 4 mm, 10-μm particle size). The mobile phase was 15% (vol/vol) acetonitrile in 0.005 M tetrabutylammonium phosphate in distilled water, adjusted to pH 7.5 with phosphoric acid or sodium hydroxide. The flow rate was 3.0 mL/min. UV detection was performed at 254 nm and 0.2 AUFS.

An acetaminophen tablet was dissolved in 100 mL of a diluent of distilled water and acetonitrile (85:15, vol/vol), diluted by a factor of 200 with the diluent, and filtered. The injection volume was 10 μL. Under the described conditions, the retention time for acetaminophen was about 2 minutes (estimated from the published chromatogram).

The method was demonstrated to be stability indicating by spiking the acetaminophen solution with its degradation product, p-aminophenol. p-Aminophenol was well resolved from its parent compound.

A standard curve for acetaminophen was generated from 10 to 30 μg/mL and its correlation coefficient was 0.9998.

Reference
Sena FJ, Piechocki JT, Li KL. Stability-indicating assay of acetaminophen in an effervescent tablet by ion-pair high-performance liquid chromatography. *J Pharm Sci*. 1979; 68: 1465–6.

Acetazolamide

Chemical Names
N-[5-(Aminosulfonyl)-1,3,4-thiadiazol-2-yl]acetamide
5-Acetamido-1,3,4-thiadiazole-2-sulfonamide

Other Name
Diamox

Form	Molecular Formula	MW	CAS
Acetazolamide	$C_4H_6N_4O_3S_2$	222.3	59-66-5
Acetazolamide sodium	$C_4H_5N_4NaO_3S_2$	244.2	1424-27-7

Appearance
Acetazolamide is a white to faintly yellowish-white odorless crystalline powder. Acetazolamide sodium is a white solid.

Solubility
Acetazolamide is sparingly soluble in water and slightly soluble in alcohol. Acetazolamide sodium is freely soluble in water.

pK$_a$
Acetazolamide has pK$_a$ values of 7.4 and 9.1.

Method 1
Allen and Erickson studied the stability of acetazolamide in extemporaneously compounded oral liquids. A Hewlett-Packard series 1050 automated high-performance liquid chromatograph included a multisolvent mixing and pumping system, an auto-injector, a diode-array detector, and a computer with Chem Station software. The stationary phase was a Bakerbond C_{18} analytical column (250 × 4.6 mm). The mobile phase contained 950 mL of water, 20 mL of methanol, 30 mL of acetonitrile, and 4.1 g of anhydrous sodium acetate; the pH was adjusted to 4.0 with acetic acid. It was delivered isocratically at 2.0 mL/min. UV detection was performed at 254 nm.

Samples were diluted 1:100. The injection volume was 20 µL. The retention time for acetazolamide was 3.1 minutes.

The HPLC assay was determined to be stability indicating by degrading samples of acetazolamide in water and in various commercial oral vehicles using heat, acid, base, oxidizing agent, and light. A composite chromatogram of acetazolamide after accelerated decomposition showed that degradation product peaks did not interfere with the intact acetazolamide peak.

A standard curve was constructed from 50 to 500 µg/mL. The intraday and interday coefficients of variation were 1.1 and 1.5%, respectively.

Reference
Allen LV, Erickson MA. Stability of acetazolamide, allopurinol, azathioprine, clonazepam, and flucytosine in extemporaneously compounded oral liquids. *Am J Health Syst Pharm.* 1996; 53: 1944–9.

Method 2
Alexander et al. evaluated the stability of acetazolamide 25 mg/mL in a suspension using a stability-indicating HPLC analytical method. The system consisted of a Beckman 110B solvent-delivery system, a Beckman system organizer injector equipped with a 20-µL loop, a Beckman 420 system controller programmer, a Knauer 731.710 UV detector, and a Shimadzu C-R3A Chromatopac integrator. The stationary phase was a Whatman ODS-3 C_{18} Partisil column (250 × 4.6 mm, 10-µm particle size). The mobile phase was 65% 0.02 M phosphate buffer in deionized water and 35% methanol and was delivered isocratically at 1 mL/min. UV detection was performed at 254 nm and 0.04 AUFS. Theophylline 1 mg/mL in methanol was used as an internal standard.

Each sample was diluted with the mobile phase. The injection volume was 50 µL. Retention times were 4.5 and 5.8 minutes for acetazolamide and theophylline, respectively.

The HPLC method was determined to be stability indicating. The acetazolamide suspension was heated for 2 hours under acidic and alkaline conditions. Degradation product peaks did not interfere with either the acetazolamide or the theophylline peak.

A calibration curve was obtained by regression analysis of the ratio of acetazolamide peak area to the internal standard peak area versus the acetazolamide concentration from 10 to 100 µg per injection.

Reference
Alexander KS, Haribhakti RP, Parker GA. Stability of acetazolamide in suspension compounded from tablets. *Am J Hosp Pharm.* 1991; 48: 1241–4.

Method 3
Parasrampuria et al. studied the stability of acetazolamide sodium in 5% dextrose and 0.9% sodium chloride injections. A Waters ALC 202 high-performance liquid chromatograph was equipped with an Omniscribe recorder and a Schoeffel SF 770 multiple-wavelength detector. The stationary phase was a Waters µBondapak nonpolar C_{18} column. The mobile phase was an aqueous solution containing 0.02 M monobasic potassium phosphate, methanol, and acetonitrile (86:12:2, vol/vol/vol). The flow rate was 2.0 mL/min. UV detection was performed at 265 nm. Sulfamerazine was used as an internal standard for the purpose of calculating the acetazolamide concentrations. The sample injection volume was 20 µL. Retention times for acetazolamide and the internal standard were 4.2 and 6.8 minutes, respectively (estimated from the published chromatogram).

The HPLC method was determined to be stability indicating by accelerated decomposition of acetazolamide sodium. A 6-mL sample of the acetazolamide solution was mixed with 0.5 mL of 1 N sodium hydroxide and subjected to boiling for 25 minutes. The decomposition product peaks did not interfere with the intact drug peak. This result was confirmed by injecting a sample of the decomposed product onto the column.

Reference

Parasrampuria J, Gupta DV, Stewart KR. Stability of acetazolamide sodium in 5% dextrose or 0.9% sodium chloride injection. *Am J Hosp Pharm.* 1987; 44: 358–60.

Acetylcholine Chloride

Chemical Name

2-(Acetyloxy)-*N,N,N*-trimethylethanaminium chloride

Other Name

Miochol

Form	Molecular Formula	MW	CAS
Acetylcholine chloride	$C_7H_{16}ClNO_2$	181.7	60-31-1

Appearance

Acetylcholine chloride occurs as white or off-white crystals or crystalline powder.

Solubility

Acetylcholine chloride is very soluble in cold water and alcohol and practically insoluble in ether. It decomposes in hot water or alkalis.

Method

Tao et al. described a simple and rapid HPLC method for quantitating acetylcholine in a lyophilized preparation. The HPLC system consisted of a Waters 6000A pump, a Waters U6K universal loop injector, a Waters 401 refractive index detector, and a Spectra-Physics 4100 microprocessor-controlled data system. The stationary phase was a Waters μBondapak C_{18} column. The mobile phase was prepared by adding sodium 1-heptane-sulfonate (Waters PIC Reagent B-7) to 900 mL of water and adjusting the pH to 4.0 with 6 M ammonium hydroxide. After addition of 50 mL of acetonitrile, the resulting solution was brought to a volume of 1000 mL with additional water. The flow rate was 2.0 mL/min.

Samples were diluted with the mobile phase. The injection volume was 50 μL. Under these conditions, the retention time for acetylcholine ranged from 7.9 to 8.2 minutes.

The method was determined to be stability indicating by heat stressing the sample at 130 °C for 19 hours. Degradation product peaks did not interfere with the intact acetylcholine peak.

A standard curve was constructed from 50 to 150 µg of acetylcholine; the correlation coefficient was 0.9996.

Reference

Tao FT, Thurber JS, Dye DM. High-performance liquid chromatographic determination of acetylcholine in a pharmaceutical preparation. *J Pharm Sci.* 1984; 73: 1311–3.

Acetylcysteine

Chemical Names

N-Acetyl-L-cysteine

L-α-Acetamido-β-mercaptopropionic acid

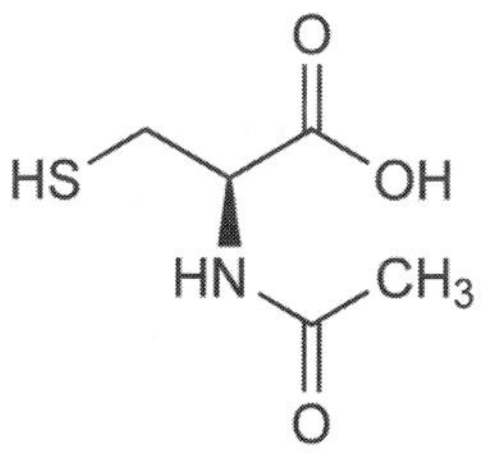

Other Name

Mucomyst

Form	Molecular Formula	MW	CAS
Acetylcysteine	$C_5H_9NO_3S$	163.2	616-91-1
Acetylcysteine sodium	$C_5H_8NNaO_3S$	185.2	19542-74-6

Appearance

Acetylcysteine is a white crystalline powder with a slight acetic odor.

Solubility

Acetylcysteine has an aqueous solubility of 1 in 5. In alcohol the solubility is 1 in 4. It is practically insoluble in chloroform and ether.

Method 1

Anaizi et al. used an HPLC method to determine the stability of acetylcysteine in an extemporaneously prepared ophthalmic solution. The HPLC system included a Perkin-Elmer model 250 binary LC pump, a model LC 290 UV detector, a model ISS 100 autoinjector, an Epson computer workstation, and a Perkin-Elmer model 0258-0195 reduced-activity Pecosphere C_{18} column (330 × 4.6 mm, 3-µm particle size). The mobile phase consisted of 50 mM phosphate buffer in HPLC-grade water (adjusted to pH 3.0 with phosphoric acid) and was delivered isocratically at 3.0 mL/min. UV detection was performed at 210 nm and 0.01 AUFS. DL-Phenylalanine 10 µg/mL in HPLC-grade water was used as an internal standard.

Samples were diluted with water and then the internal standard solution. The injection volume was 10 µL. Under these conditions, retention times for acetylcysteine and phenylalanine were 0.5 and 1.2 minutes, respectively.

To demonstrate the stability-indicating nature of the assay, solutions of acetylcysteine 100 mg/mL were adjusted to pH 0.9, 8.2, and 12.3 with 10 N sodium hydroxide and 10 N hydrochloric acid and were placed in a 60 °C oven for 7 days. The pH was then adjusted and solutions were diluted before injecting onto the column. Degradation product peaks did not interfere with the intact acetylcysteine peak.

A standard curve was generated from 5 to 40 µg/mL; the correlation coefficient was 1.000. The intraday and interday coefficients of variation were less than 1.88 and 1.22%, respectively.

Reference
Anaizi NH, Swenson CF, Dentinger PJ. Stability of acetylcysteine in an extemporaneously compounded ophthalmic solution. *Am J Health Syst Pharm.* 1997; 54: 549–53.

Method 2
Fawcett et al. investigated the stability of acetylcysteine 10% eyedrops stored in low-density polyethylene eyedrop bottles at various temperatures and under in-use conditions. The HPLC system included a Shimadzu LC-5A pump, a Rheodyne injector, a Shimadzu SPD-6AV variable-wavelength detector, and a Hitachi D-2500 chromatointegrator. The stationary phase was a Shandon Hypersil-ODS C_{18} column (100 × 4.6 mm, 5-µm particle size). The mobile phase was 50 mM potassium dihydrogen phosphate solution. The flow rate was 1.5 mL/min. UV detection was performed at 214 nm. DL-Phenylalanine was used as an internal standard.

Samples were diluted 1:200 with 0.05% sodium metabisulfite solution. The injection volume was 20 µL. Under these conditions, the retention times of acetylcysteine, the decomposition product (*N,N'*-diacetylcystine), and the internal standard were 1.4, 2.4, and 4.6 minutes, respectively.

The stability-indicating nature of the assay was demonstrated by the accelerated decomposition of the drug. Acetylcysteine was oxidized by hydrogen peroxide 30%. The decomposition product, *N,N'*-diacetylcystine, did not interfere with the analysis of the drug.

A standard curve for acetylcysteine was constructed from 0.25 to 0.75 mg/mL. The correlation coefficient of the curve was 0.999.

Reference
Fawcett JP, Woods DJ, Hayes P, et al. The stability of acetylcysteine eyedrops. *Aust J Hosp Pharm.* 1993; 23: 18–21.

Acyclovir

Chemical Names
2-Amino-1,9-dihydro-9-[(2-hydroxyethoxy)methyl]-6H-purin-6-one
9-[(2-Hydroxyethoxy)methyl]guanine

Other Names
Aciclovir, Zovirax

Form	Molecular Formula	MW	CAS
Acyclovir	$C_8H_{11}N_5O_3$	225.2	59277-89-3
Acyclovir sodium	$C_8H_{10}N_5NaO_3$	247.2	69657-51-8

Appearance
Acyclovir and acyclovir sodium occur as white crystalline powders.

Solubility
Acyclovir has solubilities of 1.3 mg/mL in water at 25 °C and 0.2 mg/mL in alcohol. Acyclovir sodium has a solubility of greater than 100 mg/mL in water at 25 °C.

pK$_a$
Acyclovir has pK$_a$ values of 2.27 and 9.25.

Method 1
Dubhashi and Vavia developed an HPLC method for the quantitation of acyclovir in pharmaceutical dosage forms. The instrument consisted of a Jasco PU-980 intelligent pump with a 20-µL fixed-loop injector and a Jasco UV-975 intelligent detector. The stationary phase was a Jasco Finepak SiL C$_{18}$ column (250 × 4.6 mm, 5-µm particle size). The mobile phase was 5% (vol/vol) acetonitrile in 0.01 M monobasic potassium phosphate aqueous buffer solution (pH 4.8). The flow rate was 1.0 mL/min. UV detection was performed at 255 nm and 0.16 AUFS.

A sample of gel (about 100 mg) was accurately weighed, transferred into a 10-mL volumetric flask, and extracted in distilled water by ultrasonication for 30 minutes; the flask was then filled to the 10-mL mark with water. The aqueous layer was collected, diluted with water to a nominal acyclovir concentration of about 100 µg/mL, filtered, and assayed. For ointment samples, extraction of acyclovir was performed the same way as for the gel sample. The resulting mixture was centrifuged at 5000 rpm for 30 minutes. The lower, clear aqueous layer was collected and assayed. The injection volume was 20 µL. Under these conditions, the retention time for acyclovir was about 11 minutes.

The analytical method was evaluated to be stability indicating by accelerated decomposition of acyclovir. An acyclovir solution (2.5 mg/2.5 mL) was mixed with 10 mL of 1 N sodium hydroxide solution or 1 N sulfuric acid, diluted with 10 mL of water, heated to boiling for 10 minutes using a hot plate, cooled to room temperature, neutralized using 1 N sulfuric acid or 1 N sodium hydroxide solution, and assayed. The peak of the intact drug was resolved chromatographically from peaks of the degradation products.

A linear relationship between the acyclovir concentration and the peak area was obtained from 50 to 200 µg/mL. The correlation coefficient was greater than 0.9985. The limit of detection was 75 ng and the limit of quantification was 250 ng.

Reference

Dubhashi SS, Vavia PR. Stability indicating reverse phase HPLC method for acyclovir. *Indian Drugs*. 2000; 37: 464–8.

Method 2

Zhang et al. investigated the stability of acyclovir sodium 1, 7, and 10 mg/mL in 5% dextrose injection and 0.9% sodium chloride injection. The liquid chromatograph was a Waters model LC Module 1 Plus equipped with a multisolvent-delivery pump, a UV detector, and an autosampler. The stationary phase was a Waters µBondapak C_{18} column (300 × 3.9 mm, 10-µm particle size). The mobile phase was 5 mM sodium acetate in water adjusted to pH 3.0 with acetic acid. The flow rate was 1.45 mL/min. UV detection was performed at 254 nm and 0.5 AUFS.

Samples were appropriately diluted with HPLC-grade water. The injection volume was 20 µL. Under these conditions, the retention time for acyclovir was 8.6 minutes.

The analytical method was determined to be stability indicating by accelerated decomposition of acyclovir. Boiling the acyclovir sodium in 0.1 N hydrochloric acid for 5 minutes led to a 50% loss in peak area for intact acyclovir and the formation of a new peak at about 4.2 minutes. The degradation product peak did not interfere with the intact acyclovir peak.

Calibration curves for acyclovir were constructed from a linear plot of peak area versus concentration from 0.025 to 0.150 mg/mL; the correlation coefficient was greater than 0.9999. The intraday and interday coefficients of variation were 0.8 and 0.9%, respectively.

Reference

Zhang Y, Trissel LA, Martinez JF, et al. Stability of acyclovir sodium 1, 7, and 10 mg/mL in 5% dextrose injection and 0.9% sodium chloride injection. *Am J Health Syst Pharm*. 1998; 55: 574–7.

Method 3

Gupta et al. studied the chemical stability of acyclovir sodium in 5% dextrose and 0.9% sodium chloride injections. The chromatograph was a Waters ALC 202 liquid chromatograph coupled with a Rheodyne 7125 injector, a Schoeffel SF 770 multiple-wavelength detector, and an Omniscribe recorder. The stationary phase was a Waters µBondapak C_{18} column (300 × 3.9 mm). The mobile phase contained 3% acetonitrile (vol/vol) in 0.1 M monobasic potassium phosphate aqueous buffer solution. The isocratic flow rate was 2.0 mL/min. UV detection was performed at 252 nm and 0.04 AUFS. A 0.2% stock solution of salicylic acid was used as an internal standard.

Each sample was diluted with water. The injection volume was 30 μL. Under these conditions, retention times of acyclovir and the internal standard were 4.5 and 7 minutes, respectively (estimated from the published chromatogram).

The analytical method was determined to be stability indicating. A solution of acyclovir was mixed with either 1 N sulfuric acid or one pellet of sodium hydroxide, and the mixture was heated to boiling for 30 minutes. Degradation product peaks did not interfere with the intact acyclovir peak.

Reference
Gupta DV, Pramar Y, Bethea C. Stability of acyclovir sodium in dextrose and sodium chloride injections. *J Clin Pharm Ther*. 1989; 14: 451–6.

Adenosine

Chemical Names
9-β-D-Ribofuranosyl-9*H*-purin-6-amine
6-Amino-9-β-D-ribofuranosyl-9*H*-purine

Other Name
Adenocard

Form	Molecular Formula	MW	CAS
Adenosine	$C_{10}H_{13}N_5O_4$	267.2	58-61-7

Appearance
Adenosine occurs as a white crystalline powder.

Solubility
Adenosine is soluble in water and practically insoluble in alcohol.

Method 1
Ketkar et al. studied the stability of undiluted and diluted adenosine at three temperatures in syringes and bags. The HPLC system was a Shimadzu model LC-6A liquid chromato-

graph. The stationary phase was a Waters μBondapak C_{18} column (300 × 3.9 mm). The mobile phase was 8% acetonitrile in 15 mM monobasic potassium phosphate solution. The flow rate was 1 mL/min. UV detection was performed at 259 nm.

The adenosine samples were diluted with double-distilled water to an approximate concentration of 30 μg/mL. The injection volume was 10 μL. Under these conditions, the retention time for adenosine was 6.8 minutes.

The stability-indicating capability of the assay was demonstrated by accelerated decomposition of adenosine. Samples of adenosine were heated to 90 °C or treated with about 1 M hydrochloric acid, about 0.5 M sodium hydroxide, or about 4 M hydrogen peroxide. The degradation product peaks did not interfere with the intact adenosine peak.

A standard curve for adenosine was constructed from 3 to 40 μg/mL; the correlation coefficient was greater than 0.999.

Reference
Ketkar VA, Kolling WM, Nardviriyakul N, et al. Stability of undiluted and diluted adenosine at three temperatures in syringes and bags. *Am J Health Syst Pharm.* 1998; 55: 466–70.

Method 2
Lau et al. evaluated the stability of adenosine in cardioplegic solutions containing high and low concentrations of potassium in order to determine reasonable expiry dates for adenosine in these solutions. The HPLC system consisted of a Spectra-Physics model P100 isocratic solvent-delivery pump, a Waters model 715 WISP autoinjector, a Hewlett-Packard model 1050 variable-wavelength UV detector, and a Spectra-Physics model 4240 chromatographic integrator. The stationary phase was a Beckman Ultrasphere ODS C_{18} reversed-phase column (250 × 4.2 mm, 5-μm particle size). The mobile phase consisted of acetonitrile and 0.01 M monobasic potassium phosphate buffer (5:95, vol/vol) and was delivered isocratically at 1.0 mL/min. UV detection was performed at 220 nm. The injection volume was 1 μL. Under these conditions, the retention time for adenosine was between 270 and 360 seconds.

The analytical method was determined to be stability indicating by accelerated decomposition of adenosine. At pH 0.8, 38.4% of adenosine degraded within 2 hours, and it completely decomposed after 10 hours. The major degradation product was adenine. The peaks of the degradation products did not interfere with the quantification of adenosine.

Standard curves were constructed from 0.04 to 0.2 mg/mL. The intraday and interday coefficients of variation were 1 and 4.8%, respectively.

Reference
Lau DWC, Walker SE, Fremes SE, et al. Adenosine stability in cardioplegic solutions. *Can J Hosp Pharm.* 1995; 48: 167–71.

Albuterol

Chemical Names

α^1-[[(1,1-Dimethylethyl)amino]methyl]-4-hydroxy-1,3-benzenedimethanol

α^1-[(*tert*-Butylamino)methyl]-4-hydroxy-*m*-xylene-α,α'-diol

Other Names

Proventil, Salbutamol, Ventolin

Form	Molecular Formula	MW	CAS
Albuterol	$C_{13}H_{21}NO_3$	239.3	18559-94-9
Albuterol sulfate	$(C_{13}H_{21}NO_3)_2 \cdot H_2SO_4$	576.7	51022-70-9

Appearance

Albuterol occurs as a white to almost white and almost tasteless crystalline powder. Albuterol sulfate is a white crystalline powder.

Solubility

Albuterol is sparingly soluble in water and soluble in alcohol. Albuterol sulfate is soluble in water and slightly soluble in alcohol.

pK$_a$

Albuterol has pK$_a$ values of 9.3 and 10.3.

Method

Jacobson and Peterson developed an HPLC assay for the simultaneous determination of ipratropium bromide, fenoterol, albuterol, and terbutaline in a nebulizer solution. The liquid chromatograph consisted of a Varian model 9010 solvent-delivery system, a Rheodyne model 7161 injector with a 10-µL external loop, a Varian model 9050 variable-wavelength UV-visible detector, and a Varian GC Star workstation. The stationary phase was a Waters Nova-Pak C_{18} Radial-Pak cartridge (100 × 8 mm, 4-µm particle size) inside a Waters RCM 8 × 10 compression module. Mobile phase A consisted of tetrahydrofuran and distilled water (40:60, vol/vol) containing 0.0025 M Waters Pic B-8 Reagent Low UV. Mobile phase B was distilled water and mobile phase C consisted of methanol and distilled water (50:50, vol/vol). The flow rate was 2.0 mL/min. A mixture of 50% mobile phase A and 50% mobile phase B was delivered up to 7.7 minutes; then the composition of the mixture was changed linearly to 60% mobile phase A, 15% mobile phase B, and 25% mobile phase C at 13.0 minutes. The run time was 13.0 minutes, with a 5.0-minute equilibration time. Mepivacaine hydrochloride 1% was used as an internal standard. The injection volume was 20 µL. Retention times for albuterol, terbutaline, ipratropium, mepivacaine, and fenoterol were 3.2, 4.3, 5.9, 8.2, and 12.7 minutes, respectively.

The assay was determined to be stability indicating by accelerated decomposition of albuterol using heat, hydrogen peroxide, acid, and base. In all cases, degradation product peaks did not interfere with the intact albuterol peak.

Reference

Jacobson GA, Peterson GM. High-performance liquid chromatographic assay for the simultaneous determination of ipratropium bromide, fenoterol, salbutamol, and terbutaline in nebulizer solution. *J Pharm Biomed Anal*. 1994; 12: 825–32.

Allantoin

Chemical Names

Glyoxyldiureide
2,5-Dioxoimidazolidin-4-ylurea

Other Names

Alasulf, Benegyn

Form	Molecular Formula	MW	CAS
Allantoin	$C_4H_6N_4O_3$	158.1	97-59-6

Appearance

Allantoin is a white crystalline powder.

Solubility

Allantoin is slightly soluble in water and very slightly soluble in alcohol.

Method

Trivedi described a liquid chromatographic procedure for the determination of allantoin in a cosmetic lotion. A Perkin-Elmer chromatograph consisting of a series 3B pump, a model LC-75 spectrophotometric detector, a model LC-420B autosampler, and a Sigma 15 chromatography datastation was used. The stationary phase was a DuPont Zorbax amino packing column (250 × 4.5 mm, 10-µm particle size). Solvent A was 10% of 0.1% (vol/vol) triethylamine in water, adjusted to pH 4.0 with phosphoric acid. Solvent B was acetonitrile. The mobile phase was 10% (vol/vol) solvent A in solvent B. The flow rate was 2 mL/min. UV detection was performed at 240 nm.

Samples containing about 50 mg of allantoin were dispersed by shaking in 30 mL of water in a 100-mL volumetric flask for 10 minutes. Fifty milliliters of acetonitrile was added and mixed for 15 minutes. The diluted samples were brought to volume with additional acetonitrile. Twenty-five milliliters of this dilution was centrifuged for 20 minutes at 25 °C. The supernatant was filtered and the filtrate was collected and assayed. The injection volume was 100 μL. Under these conditions, the retention time for allantoin was about 6.4 minutes.

The stability-indicating nature of the method was demonstrated by spiking the standard solution with the degradation products (urea, allantoic acid, and glyoxylic acid). The degradation products did not coelute with allantoin.

A standard curve for allantoin was generated from 0.20 to 1.0 mg/mL. The correlation coefficient was 0.997.

Reference
Trivedi RJ. Stability-indicating liquid chromatographic procedure for determination of allantoin in cosmetic lotion. *J Assoc Off Anal Chem*. 1988; 71: 290–4.

Allopurinol

Chemical Names
1,5-Dihydro-4*H*-pyrazolo[3,4-*d*]pyrimidin-4-one
1*H*-Pyrazolo[3,4-*d*]pyrimidin-4-ol

Other Name
Zyloprim

Form	Molecular Formula	MW	CAS
Allopurinol	$C_5H_4N_4O$	136.1	315-30-0

Appearance
Allopurinol is a fluffy white to off-white powder.

Solubility
Allopurinol has the following solubilities at 25 °C: 0.48 mg/mL in water, 0.30 mg/mL in ethanol, 0.60 mg/mL in chloroform, and 4.6 mg/mL in dimethyl sulfoxide.

pK_a
Allopurinol has a pK_a of 9.4.

Method
Allen and Erickson evaluated the chemical stability of allopurinol in extemporaneously compounded oral liquids. An automated Hewlett-Packard series 1050 chromatograph

consisted of a multisolvent mixing and pumping system, an autoinjector, a diode-array detector, and a computer with Chem Station software. The stationary phase was a Bakerbond C_{18} analytical column (250 × 4.6 mm). The mobile phase was 0.05 M mono-basic ammonium phosphate. The flow rate was 1.5 mL/min. UV detection was performed at 254 nm. Samples were diluted 1:100. The injection volume was 20 μL. Allopurinol eluted in 1.3 minutes.

The stability-indicating nature of the assay was confirmed by accelerated decom-position of allopurinol. Samples of allopurinol in water and in the commercial vehicles were subjected to stress treatments of heat, base, acid, oxidizing agent, and light. The chromatograms that were then obtained showed that the degradation product peaks did not interfere with the intact allopurinol peak.

A standard curve was constructed from 1 to 50 μg/mL. The intraday and interday coefficients of variation were 0.9 and 1.1%, respectively.

Reference
Allen LV, Erickson MA. Stability of acetazolamide, allopurinol, azathioprine, clonazepam, and flucytosine in extemporaneously compounded oral liquids. *Am J Health Syst Pharm.* 1996; 53: 1944–9.

Alprazolam

Chemical Names
8-Chloro-1-methyl-6-phenyl-4*H*-[1,2,4]triazolo[4,3-*a*][1,4]benzodiazepine
8-Chloro-1-methyl-6-phenyl-4*H*-*s*-triazolo[4,3-*a*][1,4]benzodiazepine

Other Name
Xanax

Form	Molecular Formula	MW	CAS
Alprazolam	$C_{17}H_{13}ClN_4$	308.8	28981-97-7

Appearance
Alprazolam is a white to off-white crystalline powder.

Solubility
Alprazolam is insoluble in water, soluble in alcohol, sparingly soluble in acetone, freely soluble in chloroform, and slightly soluble in ethyl acetate.

Method

Allen and Erickson evaluated the stability of alprazolam in extemporaneously compounded oral liquids. A Hewlett-Packard series 1050 automated high-performance liquid chromatograph included a multisolvent mixing and pumping system, an autoinjector, a diode-array detector, and a computer with Chem Station software. The stationary phase was a Bakerbond C_{18} column (250 × 4.6 mm, 5-μm particle size). The mobile phase consisted of methanol, acetonitrile, and 0.04 M sodium acetate in water (pH 2.4) (45:8:47, vol/vol/vol). The flow rate was 0.6 mL/min. UV detection was performed at 230 nm.

Samples were diluted 1:50 before injection. Under these conditions, the retention time for alprazolam was 9.9 minutes.

The analytical method was determined to be stability indicating. A composite chromatogram of alprazolam after accelerated decomposition showed that degradation product peaks did not interfere with the intact alprazolam peak.

A standard curve for alprazolam was constructed from 1 to 25 μg/mL. The intraday and interday coefficients of variation were 1.4 and 2.3%, respectively.

Reference

Allen LV Jr, Erickson MA III. Stability of alprazolam, chloroquine phosphate, cisapride, enalapril maleate, and hydralazine hydrochloride in extemporaneously compounded oral liquids. *Am J Health Syst Pharm.* 1998; 55: 1915–20.

Alprostadil

Chemical Names

(11α,3*E*,15*S*)-11,15-Dihydroxy-9-oxoprost-13-en-1-oic acid
3-Hydroxy-2-(3-hydroxy-1-octenyl)-5-oxocyclopentaneheptanoic acid

Other Name

Caverject

Form	Molecular Formula	MW	CAS
Alprostadil	$C_{20}H_{34}O_5$	354.5	745-65-3

Appearance

Alprostadil is a white to off-white crystalline powder.

Solubility

Alprostadil is soluble in water, freely soluble in alcohol, soluble in acetone, and very slightly soluble in chloroform and ether. It is slightly soluble in ethyl acetate.

Method

Shulman and Fyfe reported the shelf-life determination of alprostadil injections. The liquid chromatograph consisted of a Waters 501 solvent-delivery pump, a Waters 991 photodiode-array UV detector, and a Rheodyne 7125 syringe-loading injector with a 20-μL sample loop. The stationary phase was a Beckman Ultrasphere (IP) ODS C_{18} column (150 × 4.6 mm, 5-μm particle size). The mobile phase consisted of water and acetonitrile (55:45, vol/vol) in 60 mM potassium dihydrogen phosphate buffer (pH 3.0). The flow rate was 1.0 mL/min. UV detection was performed at 193 nm. Butylparaben 0.01% in methanol served as an internal standard.

The assay was determined to be stability indicating by accelerated decomposition of alprostadil. Alprostadil samples 20 μg/mL in 4% alcohol were sealed in glass ampules and stored at 47 °C for 7.9 days. Only 48% alprostadil remained. The degradation product peak did not interfere with the alprostadil and internal standard peaks.

Reference

Shulman NH, Fyfe RK. Shelf-life determination of prostaglandin E1 injections. *J Clin Pharm Ther*. 1995; 20: 41–4.

Amikacin

Chemical Name

(*S*)-*O*-3-Amino-3-deoxy-α-D-glucopyranosyl-(1→6)-*O*-[6-amino-6-deoxy-α-D-glucopyranosyl-(1→4)]-*N*[1]-(4-amino-2-hydroxy-1-oxobutyl)-2-deoxy-D-streptamine

Other Name

Amikin

Form	Molecular Formula	MW	CAS
Amikacin	$C_{22}H_{43}N_5O_{13}$	585.6	37517-28-5
Amikacin sulfate	$C_{22}H_{43}N_5O_{13}.2H_2SO_4$	781.8	39831-55-5

Appearance

Amikacin and amikacin sulfate occur as white crystalline powders.

Solubility

Amikacin is sparingly soluble in water. Amikacin sulfate is freely soluble in water.

Method

Allen et al. used HPLC to determine the stability of amikacin 5.0 mg/mL in the presence of cefpirome sulfate during simulated Y-site injection. One of two systems was utilized. One system included an Alcott model 728 Micromeritics autosampler, a Rheodyne 7010 injector with an Alcott 732 electrically actuated valve, a Waters model 501 solvent-delivery pump, a Waters model 441 UV detector, a Waters model 401 refractive index detector, and a Waters model 745 data module. The other system consisted of an Alcott model 728 Micromeritics autosampler, a Rheodyne 7010 injector with an Alcott 732 electrically actuated valve, a Shimadzu model LA-6A solvent-delivery pump, a Shimadzu model SPD-6A UV detector, and an Orion model 901 microprocessor ion analyzer. The stationary phase was a C_{18} analytical column. The mobile phase was 72% methanol and 28% of 0.27% monobasic potassium phosphate with pH adjusted to 6.5 with 0.4 N potassium hydroxide, which was delivered isocratically at 0.8 mL/min. UV detection was performed at 340 nm.

A 50-μL aliquot of an amikacin sample diluted to 1 mg/mL was derivatized with 2 mL of 2,4,6-trinitrobenzenesulfonic acid in water 10 mg/mL along with 3.2 mL of pyridine. The mixture was heated using a water bath at 75 °C for 45 minutes. The solution was cooled to room temperature and passed through a filter having a porosity of 0.5 μm or less. The injection volume was 20 μL. Under these conditions, the retention times for cefpirome and the amikacin derivative were 6.0 and 7.6 minutes, respectively.

The analytical method was demonstrated to be stability indicating by degrading amikacin with hydrochloric acid, sodium hydroxide solution, potassium chlorate, heat at 80 °C, or light (150 foot-candles from a tungsten filament source). In each of these cases, amikacin was suitably separated from its decomposition products on the chromatogram.

A standard curve was constructed from 1 to 100 μg/mL; the correlation coefficient was greater than 0.99. The intraday and interday coefficients of variation were 2.5 and 2.9%, respectively.

Reference

Allen LV, Stiles ML, Prince SJ, et al. Stability of cefpirome sulfate in the presence of commonly used intensive care drugs during simulated Y-site injection. *Am J Health Syst Pharm.* 1995; 52: 2427–33.

Aminophylline

Chemical Name

3,7-Dihydro-1,3-dimethyl-1*H*-purine-2,6-dione compounded with 1,2-ethanediamine (2:1)

Other Name
Aminophyllin

Form
Aminophylline

Molecular Formula
$(C_7H_8N_4O_2)_2 \cdot C_2H_4(NH_2)_2$

MW
420.4

CAS
317-34-0

Appearance
Aminophylline appears as white or slightly yellowish granules or powder.

Solubility
Aminophylline has an aqueous solubility of 1 g in about 5 mL of water. It is insoluble in alcohol and ether.

pK$_a$
At high pH, aminophylline has a pK_a of 8.79. At physiological pH, it has a pK_b of 13–14.

Method 1
Caufield and Stewart reported an HPLC assay for the simultaneous analysis of meropenem and aminophylline in an intravenous fluid mixture. The instrument was a Hewlett-Packard model 1090 system, including a pump, an autosampler, and a Gilson model 117 variable-wavelength UV detector. A Waters model 996 photodiode-array detector was used for the peak purity analysis. The stationary phase was a YMC ODS-AQ column (150 × 2.0 mm, 3-µm particle size). The mobile phase was aqueous acetic acid solution (pH 3) and acetonitrile (89:11, vol/vol). The flow rate was 0.2 mL/min. UV detection was performed at 270 nm.

Samples were diluted with mobile phase. The sample injection volume was 5 µL. Under these conditions, retention times of meropenem and aminophylline were about 4.4 and 5.9 minutes, respectively.

The method was confirmed to be stability indicating by accelerated degradation of aminophylline. Solutions of aminophylline were treated with 6 N hydrochloric acid at 95 °C for 5 days, 1 N sodium hydroxide solution at 60 °C for 3 days, or 3% hydrogen peroxide for 3 days, or the solutions were heated at 95 °C for 10 days. None of the degradation product peaks interfered with the intact drug peak.

A standard curve for aminophylline was constructed from 13 to 52 µg/mL. The correlation coefficient was 0.9996. Intraday and interday coefficients of variation were 0.69 and 0.38%, respectively. The limit of detection was 92 ng/mL.

Reference
Caufield WV, Stewart JT. HPLC separations of meropenem and selected pharmaceuticals using a polar endcapped octadecylsilane narrow bore column. *Chromatographia*. 2000; 51: 308–14.

Method 2

Parrish et al. studied the short-term compatibility and stability of aminophylline or theophylline and ceftriaxone sodium in an admixture containing various concentrations of both drugs. The chromatographic system consisted of a Rabbit-HP solvent-delivery system, a Micromeritics model 725 autosampler, an Isco model V-4 variable-wavelength UV detector, a Linseis model L6522 potentiometric recorder, an Apple IIe computer, and a Dynamax C_{18} analytical column (250 × 4.6 mm, 10-μm particle size) with matching guard column. The mobile phase was a mixture of 40% acetonitrile, 55.2% 0.0065 M tetraheptylammonium bromide, 4.4% 0.1 M phosphate buffer (pH 7.0), and 0.4% 0.1 M citrate buffer (pH 5.0). The flow rate was 1.0 mL/min. The UV detection wavelength was 271 nm. The sample injection volume was 20 μL. Under these conditions, the retention time for theophylline was 3.0 minutes.

The method was determined to be stability indicating by accelerated decomposition. Solutions of theophylline 2 mg/mL were adjusted to pH 2 with hydrochloric acid and also to pH 12 with sodium hydroxide solution and were then heated at 100 °C for 15 minutes. Another theophylline solution was spiked with a major degradation product (1,3-dimethyluric acid) and two degradation products of ceftriaxone (Ro 11-8390 and Ro 15-2254). Retention times for 1,3-dimethyluric acid, Ro 11-8390, ceftriaxone, and Ro 15-2254 were 3.9, 7.0, 14.9, and 22.4 minutes, respectively.

A standard curve was generated by analyzing standards of theophylline from 50 to 150% of the expected drug concentration in the admixture. The correlation coefficient was greater than 0.99. The coefficient of variation was 0.8%.

Reference

Parrish MA, Bailey LC, Medwick T. Stability of ceftriaxone sodium and aminophylline or theophylline in intravenous admixtures. *Am J Hosp Pharm*. 1994; 51: 92–4.

Method 3

Stewart et al. assessed the stability of aminophylline or theophylline with cefuroxime sodium in intravenous solutions. The HPLC system included a Beckman model 110B solvent-delivery pump, a Rheodyne model 7125 manual injector equipped with a 20-μL loop, a Beckman model 160 fixed-wavelength UV detector, a Hewlett-Packard model 3392A integrator, and a Beckman Ultrasphere octadecylsilane column (150 × 4.6 mm, 5-μm particle size). The mobile phase was a mixture of 10 parts of 0.1 M acetate buffer (pH 3.4) and 1 part of acetonitrile. The flow rate was 2 mL/min. UV detection was performed at 254 nm. 5-Methylresorcinol 1.5 mg/mL in water was used as an internal standard. Samples were diluted with water and internal standard solutions.

The method was determined to be stability indicating. Aqueous aminophylline and cefuroxime solutions were degraded at 80 °C for 4 hours, ambient temperature for 6 hours, and exposure to fluorescent light for 6 hours. No interference with the intact theophylline peak from degradation products was observed.

Calibration curves were generated by least-squares regression of the drug to internal standard ratios versus the concentration of the drug from 1 to 16 μg/mL for aminophylline and from 0.8 to 13 μg/mL for theophylline. The correlation coefficient was greater than 0.9998.

Reference

Stewart JT, Warren FW, Johnson SM. Stability of cefuroxime sodium and aminophylline or theophylline. *Am J Hosp Pharm*. 1994; 51: 809–11.

Method 4

Johnson et al. utilized an HPLC method to evaluate the stability of aminophylline in intravenous admixtures. The chromatographic system consisted of a Waters model 501 solvent-delivery system, a Waters model U6K variable-volume injector, a Waters model 484 UV detector, and a Hewlett-Packard model 3390 integrator-recorder. The stationary phase was a Waters μBondapak C_{18} chromatographic column. The mobile phase was a mixture of 7% acetonitrile and 93% 0.01 M sodium acetate buffer adjusted to pH 4 with acetic acid. The flow rate was 2 mL/min. β-Hydroxyethyl-theophylline 2.0 mg/mL was used as an internal standard. Samples were diluted with the mobile phase. The injection volume was 15 μL.

Johnson et al. verified that the method was stability indicating.

A standard curve was constructed by linear regression of theophylline concentration against the ratio of the peak height of theophylline to that of the internal standard from 60 to 100 μg/mL. The correlation coefficient was greater than 0.999. The intraday and interday coefficients of variation for aminophylline were less than 1%.

A similar method was used by Pleasants et al.

References

Johnson CJ, Jacobson PA, Pillen HA, et al. Stability and compatibility of fluconazole and aminophylline in intravenous admixtures. *Am J Hosp Pharm*. 1993; 50: 703–6.

Pleasants RA, Vaughan LM, Williams DM, et al. Compatibility of ceftazidime and aminophylline admixtures for different methods of intravenous infusion. *Ann Pharmacother*. 1992; 26: 1221–6.

Method 5

Using an accurate, reproducible, and stability-indicating HPLC method, Nahata et al. determined the stability of aminophylline diluted in bacteriostatic water for injection and stored in plastic syringes at room temperature and under refrigeration. The system included a Varian 2010 series pump, an autosampler, a variable-wavelength UV detector, and an Anspec D 200 integrator. The stationary phase was a Beckman Ultrasphere ODS column (250×4.6 mm, 5-μm particle size). The mobile phase consisted of 75% 0.01 M sodium acetate, pH adjusted to 4 with acetic acid, and 25% methanol. The isocratic flow rate was 1.0 mL/min. UV detection was performed at 280 nm. Caffeine was used as the internal standard.

Each sample was diluted 1:200 with the internal standard solution. The injection volume was 10 μL. Retention times for caffeine and aminophylline were approximately 5.4 and 9.5 minutes, respectively.

The HPLC method was determined to be stability indicating by subjecting aminophylline to base and acid hydrolysis. One milliliter of aminophylline 1 mg/mL was mixed with 1 mL of 1.0 M sodium hydroxide or 1.0 M sulfuric acid and heated to 80 °C for 30 minutes. The peaks from the degradation products did not interfere with the intact drug peak.

A standard curve was constructed for aminophylline from 1.25 to 10.0 mg/mL. The correlation coefficient of the standard curve was greater than 0.999.

Reference

Nahata MC, Morosco RS, Hipple TF. Stability of aminophylline in bacteriostatic water for injection stored in plastic syringes at two temperatures. *Am J Hosp Pharm*. 1992; 49: 2962–3.

Method 6

Schaaf et al. investigated the stability of aminophylline in 5% dextrose injection using an HPLC method. The chromatograph consisted of a Shimadzu model LC-6A pump, an LDC/Milton Roy Spectrometer III variable-wavelength UV detector, and an ES Industries reversed-phase C_{18} column (250 × 4.6 mm, 5-μm particle size). The mobile phase was a mixture of 38 mM aqueous monobasic potassium phosphate, methanol, and acetonitrile (58:27:15, vol/vol/vol) and was delivered isocratically at 1 mL/min. UV detection was performed at 215 nm. *o*-Chlorobenzyl alcohol was used as the internal standard.

Each sample was diluted with deionized distilled water and then with internal standard solution. The injection volume was 20 μL. Under these conditions, aminophylline had a retention time of 3.7 minutes.

The stability-indicating nature of this assay was confirmed by accelerated degradation. Solutions of aminophylline in 1 N sodium hydroxide or 1 N hydrochloric acid were refluxed, and the samples were chromatographed. Degradation product peaks did not interfere with the aminophylline peak.

Reference

Schaaf LJ, Robinson DH, Vogel GJ, et al. Stability of esmolol hydrochloride in the presence of aminophylline, bretylium tosylate, heparin sodium, and procainamide hydrochloride. *Am J Hosp Pharm*. 1990; 47: 1567–71.

Amiodarone

Chemical Names

(2-Butyl-3-benzofuranyl)[4-[2-(diethylamino)ethoxy]-3,5-diiodophenyl]methanone
2-Butyl-3-benzofuranyl 4-[2-(diethylamino)ethoxy]-3,5-diiodophenyl ketone

Other Name

Cordarone

Form	Molecular Formula	MW	CAS
Amiodarone	$C_{25}H_{29}I_2NO_3$	645.3	1951-25-3
Amiodarone hydrochloride	$C_{25}H_{29}I_2NO_3$.HCl	681.8	19774-82-4

Appearance

Amiodarone hydrochloride is a white to cream crystalline powder.

Solubility

Amiodarone hydrochloride has solubilities of approximately 0.72 mg/mL in water and 12.8 mg/mL in alcohol at 25 °C.

pK$_a$

Amiodarone has a pK$_a$ of approximately 6.6.

Method 1

Khan et al. reported an assay method for the determination of amiodarone hydrochloride and its related substances. A Shimadzu 2010A system was equipped with a quaternary gradient pump and a variable-wavelength UV detector or a photodiode-array detector. The stationary phase was a Waters Symmetry C$_8$ column (250 × 4.6 mm, 5-µm particle size). The mobile phase consisted of 750 mL of methanol, 240 mL of water, and 10 mL of glacial acetic acid; the pH was adjusted to 5.8 with diluted ammonium hydroxide solution. The flow rate was 1.0 mL/min. UV detection was carried out at 254 nm. The run time was 25 minutes.

Amiodarone hydrochloride bulk material was dissolved in methanol and diluted in mobile phase. The injection volume was 20 µL. Under these conditions, the retention time for amiodarone was about 17 minutes.

Amiodarone samples were exposed to various stress conditions such as 0.1 N hydrochloric acid, 0.1 N sodium hydroxide solution, 10% hydrogen peroxide, heat (105 °C), humidity (75% relative humidity), and light (1.2 million lux hours). A mixture of related substances was also assayed. No degradation products or impurities interfered with the analysis of amiodarone hydrochloride, confirming the stability-indicating nature of the analytical method.

A standard curve for amiodarone hydrochloride was obtained from 50 to 150 µg/mL with a correlation coefficient of 1.00. The coefficient of variation of the assay was 0.07%. The limits of detection and quantitation were 0.13 and 0.25 µg/mL, respectively.

Reference

Khan MA, Kumar S, Jayachandran J, et al. Validation of a stability indicating LC method for amiodarone HCl and related substances. *Chromatographia*. 2005; 61: 599–607.

Method 2

Christopherson et al. developed an HPLC method for the analysis of amiodarone hydrochloride injection. The Waters 2690 Alliance system was equipped with a model 486 variable-wavelength detector, a model 2487 dual-wavelength absorbance detector, or a model 996 photodiode-array detector. The stationary phase was a Phenomenex Luna C$_8$ column (75 × 4.6 mm, 3-µm particle size) with a Security Guard C$_8$ guard column. The column temperature was maintained at 40 °C. Mobile phase A consisted of 15 mM monobasic potassium phosphate and 30 mM triethylamine buffer (adjusted to pH 2.50 ± 0.05). Mobile phase B consisted of acetonitrile and methanol (1:1, vol/vol). The mobile phases were delivered at 1 mL/min gradiently as follows:

Time, minutes	Mobile Phase A, %	Mobile Phase B, %
0	50	50
21	25	75
26	25	75
27	15	85
31	15	85
32	50	50
40	50	50

UV detection was performed at 240 nm. The overall run time was 40 minutes.

Samples were diluted to a final concentration of 100 µg/mL with a diluent [mobile phase A + mobile phase B (1:1, vol/vol)]. Under these conditions, the retention time of amiodarone was 10 minutes.

Drug samples were intentionally degraded with acid, base, heat, light, and peroxide. Spectral purity analysis demonstrated that the analytical method was stability indicating.

A standard curve for amiodarone was constructed from 50 to 150 µg/mL. The correlation coefficient was 1.000. The limit of detection and the limit of quantitation were 0.025 and 0.050 µg/mL, respectively.

Reference
Christopherson MJ, Yoder KJ, Miller RB. Validation of a stability-indicating HPLC method for the determination of amiodarone HCl and its related substances in amiodarone HCl injection. *J Liq Chromatogr Rel Technol.* 2004; 27: 95–111.

Method 3
Nahata used HPLC to study the stability of an oral suspension of amiodarone hydrochloride that had been stored in plastic and glass prescription bottles under refrigeration and at room temperature. A Hewlett-Packard model 1050 system included a pump, an autosampler, and a variable-wavelength UV detector. The stationary phase was a Mac-Mod Analytical Zorbax CN column (150 × 4.6 mm, 5-µm particle size). The mobile phase was 1% methanol in 50 mM monobasic ammonium phosphate; it was delivered isocratically at 1.5 mL/min. UV detection was performed at 230 nm.

Samples were mixed with an equal volume of mobile phase before injection. The sample injection volume was 10 µL. The retention time for amiodarone was 8.8 minutes.

The method was determined to be stability indicating by accelerated degradation of amiodarone. An amiodarone solution 1 mg/mL was mixed with an equal volume of 2.0 M hydrochloric acid, 2.0 M sodium hydroxide, or 0.3% hydrogen peroxide at 60 °C. Degradation products did not interfere with the quantification of amiodarone.

A calibration curve for amiodarone was constructed from 0.05 to 2.5 mg/mL; the correlation coefficient was greater than 0.999. The intraday and interday coefficients of variation were less than 2%.

Reference
Nahata MC. Stability of amiodarone in an oral suspension stored under refrigeration and at room temperature. *Ann Pharmacother.* 1997; 31: 851–2.

Method 4

Campbell et al. determined the stability of amiodarone hydrochloride in 5% dextrose injection and 0.9% sodium chloride injection in both polyvinyl chloride and polyolefin containers. The high-performance liquid chromatograph consisted of a Beckman model 110A pump, a Beckman model 160 fixed-wavelength (254 nm) UV detector, and a Hewlett-Packard model 3390-A integrator. The stationary phase was a Beckman Ultrasphere-Si normal phase analytical column (250 × 4.6 mm, 5-μm particle size). The mobile phase consisted of methanol and ethyl ether (80:20, vol/vol) to which from 10 to 15 μL of triethylamine per liter of mobile phase was added. The flow rate was 2.2 mL/min. UV detection was performed at 254 nm and 0.25 AUFS. Triflupromazine was used as an internal standard.

Each sample was diluted 1:50 with methanol. The injection volume was 20 μL. Under these conditions, the retention times for amiodarone and the internal standard were 2.85 and 3.65 minutes, respectively.

The HPLC analytical method was determined to be stability indicating. A test solution was simultaneously heated at 75 °C and exposed to 254-nm UV radiation for up to 43 hours. A second test solution was acidified to pH 1.0 with hydrochloric acid. These test solutions were assayed by HPLC, and no interfering peaks were observed. The concentrations of amiodarone were calculated based on the peak height ratios of drug to internal standard.

Reference

Campbell S, Nolan PE, Bliss M, et al. Stability of amiodarone hydrochloride in admixtures with other injectable drugs. *Am J Hosp Pharm*. 1986; 43: 917–21.

Method 5

Weir and Ueda developed a rapid liquid chromatographic assay for the determination of amiodarone and its *N*-deethyl metabolite in plasma, urine, and bile. The HPLC system consisted of a Laboratory Data Control Constametric III pump, a Rheodyne injector fitted with a 200-μL loop, a Houston chart recorder, and a SpectroMonitor III variable-wavelength UV detector. The stationary phase was a Waters C_{18} reversed-phase column (300 × 3.9 mm, 10-μm particle size). The mobile phase was a mixture of methanol, water, and 58% ammonium hydroxide (94:4:2, vol/vol/vol). The flow rate was 1.5 mL/min. UV detection was performed at 244 nm. 2-Ethyl-3-(3,5-dibromo-4-γ-dipropylaminopropoxy-benzoyl)benzothiophene was used as the internal standard.

For plasma, 100 μL of plasma was mixed with 200 μL of the internal standard solution 1 μg/mL, vortexed for 30 seconds, and centrifuged at 9500 × *g* for 4 minutes. Two hundred microliters of the supernatant was injected into the column.

For urine and bile, 100 μL of urine or bile was mixed with 20 μL of concentrated hydrochloric acid, and then 600 μL of 2,2-dimethoxypropane was added. The resulting solution was cooled to –10 °C for 4 hours. Following centrifugation at 9500 × *g* for 4 minutes, the supernatant was evaporated to dryness under nitrogen. The residue was dissolved with 300 μL of the internal standard solution (10 μg/mL). The injection volume was 200 μL.

The analytical method was determined to be stability indicating. Retention times for desethylamiodarone (the *N*-deethyl metabolite of amiodarone), amiodarone, and the internal standard were 4.6, 5.5, and 6.8 minutes, respectively. The degradation product peak did not interfere with the intact amiodarone and internal standard peaks.

Forty-four drugs were tested for their potential to interfere with this assay (retention times are given in parentheses in minutes): acetaminophen (2.2), acetazolamide (1.5), amitriptyline (3.4), ampicillin (1.4), aspirin (1.7), caffeine (2.4), chlorcyclizine (3.1), chloroquine (4.6), chlorpromazine (3.9), cyclizine (2.9), cyproheptadine (3.3), desipramine (5.6), diphenhydramine (2.8), haloperidol (1.9), heparin (2.3), hydralazine (2.7), hydrocortisone (2.1), hydroxyzine (2.6), imipramine (3.7), lidocaine (2.5), lorazepam (2.9), loxapine (2.1), meclizine (3.4), mesoridazine (3.4), methapyrilene (2.7), nortriptyline (4.7), pentobarbital (1.7), perphenazine (2.8), phenobarbital (1.7), phenoxybenzamine (2.2), phenytoin (1.5), procainamide (2.5), procaine (2.7), propranolol (2.7), propylthiouracil (1.4), pyrilamine (2.8), quinidine (3.1), quinine (3.1), theophylline (2.4), thioridazine (4.2), thiothixene (4.9), tranylcypromine (3.3), trifluoperazine (3.4), and valproic acid (2.8). Only the desipramine (5.6) peak interfered with the amiodarone peak.

Standard curves for amiodarone were constructed from 0.1 to 10.0 μg/mL for plasma and from 1 to 100 μg/mL for urine and bile. The correlation coefficients were 0.991, 0.994, and 0.991, respectively.

References
Weir SJ, Ueda CT. Rapid liquid chromatographic assay for the determination of amiodarone and its *N*-deethyl metabolite in plasma, urine, and bile. *J Pharm Sci*. 1985; 74: 460–5.

Weir SJ, Myers VA, Bengtson KD, et al. Sorption of amiodarone to polyvinyl chloride infusion bags and administration sets. *Am J Hosp Pharm*. 1985; 42: 2679–83.

Amitriptyline

Chemical Name
3-(10,11-Dihydro-5*H*-dibenzo[*a,d*]cyclohepten-5-ylidene)propyldimethylamine

Other Names
Elavil, Endep

Form	Molecular Formula	MW	CAS
Amitriptyline	$C_{20}H_{23}N$	277.4	50-48-6
Amitriptyline embonate	$(C_{20}H_{23}N)_2.C_{23}H_{16}O_6$	943.2	17086-03-2
Amitriptyline hydrochloride	$C_{20}H_{23}N.HCl$	313.9	549-18-8

Appearance

Amitriptyline embonate is a pale yellow to brownish-yellow, odorless or almost odorless powder. Amitriptyline hydrochloride occurs as odorless or almost odorless, colorless crystals or white or almost white powder.

Solubility

Amitriptyline embonate is practically insoluble in water, slightly soluble in alcohol, and freely soluble in chloroform. Amitriptyline hydrochloride is freely soluble in water, alcohol, chloroform, and methylene chloride. It is practically insoluble in ether.

pK_a

Amitriptyline hydrochloride has a pK_a of 9.4.

Method

Walker reported the HPLC determination of amitriptyline hydrochloride in tablets and injectables. The system was composed of a Tracor model 950 solvent pump, a model 970A variable-wavelength detector, a model 26325 recorder, and a Rheodyne model 7120 loop injector. The stationary phase was a Waters µBondapak CN stainless steel column (300 × 3.9 mm, 10-µm particle size). The mobile phase was a mixture of methanol and 0.005 M ammonium acetate (90:10, vol/vol). The flow rate was 1.33 mL/min. UV detection was performed at 239 nm and 16 AUFS. Trifluoperazine hydrochloride 0.5 mg/mL in methanol was used as an internal standard.

For tablets, a sample equivalent to 10 mg of amitriptyline hydrochloride was mixed with 25 mL of internal standard for 15 minutes and then filtered. The filtrate was further diluted with methanol to an amitriptyline hydrochloride concentration of about 0.04 mg/mL. For injectables, a sample equivalent to 10 mg of amitriptyline hydrochloride was mixed with 25 mL of internal standard and then diluted with methanol to about 0.04 mg/mL of amitriptyline hydrochloride. The injection volume was 20 µL. Under these conditions, the retention time for amitriptyline hydrochloride was about 7 minutes.

The stability-indicating nature of the method was demonstrated by assaying a mixture of amitriptyline hydrochloride, dibenzosuberone (its major degradation product), and the internal standard. Amitriptyline hydrochloride was resolved satisfactorily from dibenzosuberone and trifluoperazine hydrochloride.

Reference

Walker ST. Rapid high pressure liquid chromatographic determination of amitriptyline hydrochloride in tablets and injectables: Collaborative study. *J Assoc Off Anal Chem.* 1983; 66: 1196–202.

Amlodipine Besylate

Chemical Name

3-Ethyl 5-methyl 2-(2-aminoethoxymethyl)-4-(2-chlorophenyl)-1,4-dihydro-6-methyl-pyridine-3,5-dicarboxylate monobenzenesulfonate

Other Name
Norvasc

Form	Molecular Formula	MW	CAS
Amlodipine besylate	$C_{20}H_{25}ClN_2O_5 \cdot C_6H_6O_3S$	567.1	111470-99-6

Appearance
Amlodipine besylate occurs as a white to pale yellow crystalline powder.

Solubility
Amlodipine besylate is slightly soluble in water and sparingly soluble in alcohol.

Method 1
Kamat and Chaturvedi developed a method for the analysis of amlodipine in tablets. A Waters Alliance system included a quaternary gradient pump and a model 996 photo-diode-array detector. The stationary phase was an ODS column (150 × 4.6 mm, 10-μm particle size). The column temperature was maintained at 40 °C. The mobile phase consisted of a buffer, methanol, and acetonitrile (50:35:15 vol/vol/vol), where the buffer was 0.05 M phosphoric acid; the pH was adjusted to 3.0 with triethylamine. The flow rate was 1.5 mL/min. The UV detector was set at 361 nm. The injection volume was 20 μL.

Amlodipine samples were exposed to strong acid, base, oxidizing agent, heat, humidity, and sunlight. Degradation products were well separated from the intact drug on the chromatogram.

A standard curve for amlodipine was constructed from 25 to 175 μg/mL. The coefficient of variation of the assay was 0.59%.

Reference
Kamat K, Chaturvedi SC. Stability indicating assay method for amlodipine tablets. *Indian J Pharm Sci.* 2005; 67:236–9.

Method 2
Naidu et al. presented a method for the simultaneous determination of amlodipine and benazepril hydrochloride in tablets. A Thermo Separation liquid chromatographic system was equipped with a model P1000 pump, a model AS100 autosampler, and a model 6000LP photodiode-array detector. The stationary phase was a Zorbax SB C_{18} column (250 × 4.6 mm, 5-μm particle size). The column temperature was maintained at 28 °C. The

mobile phase consisted of 0.05 M monobasic potassium phosphate buffer and acetonitrile (65:35, vol/vol); the pH was adjusted to 7.0 with 1.0 N potassium hydroxide solution. The flow rate was 1.0 mL/min. UV detection was performed at 240 nm.

A portion of the powder from 10 tablets, equivalent to one tablet content, was weighed, transferred into a 50-mL volumetric flask, mixed with 10 mL of methanol, shaken for 10 minutes, ultrasonicated for 5 minutes, diluted to the mark with mobile phase, centrifuged at 10,000 rpm, filtered, and assayed. The injection volume was 25 μL. Under these conditions, the retention times of benazepril and amlodipine were about 8.3 and 21.0 minutes, respectively.

This method was demonstrated to be stability indicating by accelerated degradation studies. Samples were exposed to 0.5 N hydrochloric acid for 48 hours, 0.1 N sodium hydroxide solution for 2 hours, or 3% hydrogen peroxide for 48 hours. Sample solutions were also irradiated with UV radiation at 254 and 366 nm. Peaks of degradation products were well separated from the peaks of the intact drugs.

A calibration curve for amlodipine was constructed from 6.0 to 14.0 μg/mL. The correlation coefficient was 0.9999. The interday coefficient of variation was less than 0.91%.

Reference
Naidu KR, Kale UN, Shingare MS. Stability indicating RP-HPLC method for simultaneous determination of amlodipine and benazepril hydrochloride from their combination drug product. *J Pharm Biomed Anal.* 2005; 39: 147–55.

Method 3
Nahata et al. used HPLC to investigate the stability of amlodipine besylate in two liquid dosage forms. The Hewlett-Packard model 1050 liquid chromatograph included a pump, an autosampler, a variable-wavelength detector, and a model 3396 integrator. The stationary phase was a Mac-Mod Analytical Zorbax CN column (150 × 3.0 mm). The mobile phase consisted of 40 mM ammonium acetate, methanol, and acetonitrile (35:15:50, vol/vol/vol). The flow rate was 0.4 mL/min. UV detection was performed at 240 nm. Desipramine hydrochloride 20.0 μg/mL in mobile phase was used as an internal standard.

Each sample (100 μL) was mixed with 5.0 mL of internal standard solution and centrifuged. The supernatant was collected and assayed. The injection volume was 10 μL. Under these conditions, retention times for amlodipine and desipramine were about 6.4 and 8.8 minutes, respectively.

The stability-indicating nature of the assay was established by accelerated degradation of amlodipine. Samples of amlodipine 1.0 mg/mL were treated with one of the following: 2.0 M hydrochloric acid, 2.0 M sodium hydroxide solution, or 0.03% hydrogen peroxide at 60 °C. The quantification of amlodipine was not compromised by its degradation products.

A standard curve for amlodipine was generated from 0.10 to 1.50 mg/mL. The correlation coefficient was greater than 0.999. The intraday and interday coefficients of variation were less than 1.9 and 2.6%, respectively.

Reference
Nahata MC, Morosco RS, Hipple TF. Stability of amlodipine besylate in two liquid dosage forms. *J Am Pharm Assoc.* 1999; 39: 375–7.

Amoxicillin

Chemical Name
[2S-[2α,5α,6β(S*)]]-6-[[Amino(4-hydroxyphenyl)acetyl]amino]-3,3-dimethyl-7-oxo-4-thia-1-azabicyclo[3.2.0]heptane-2-carboxylic acid

Other Name
Amoxycillin

Form	Molecular Formula	MW	CAS
Amoxicillin	$C_{16}H_{19}N_3O_5S$	365.4	26787-78-0
Amoxicillin sodium	$C_{16}H_{18}N_3NaO_5S$	387.4	34642-77-8
Amoxicillin trihydrate	$C_{16}H_{19}N_3O_5S.3H_2O$	419.5	61336-70-7

Appearance
Amoxicillin is a white practically odorless crystalline powder. Amoxicillin sodium is a white or almost white, very hygroscopic powder. Amoxicillin trihydrate is a white or almost white crystalline powder.

Solubility
Amoxicillin has solubilities of 4.0 mg/mL in water, 7.5 mg/mL in methanol, and 3.4 mg/mL in absolute alcohol. It is insoluble in hexane, benzene, ethyl acetate, and acetonitrile. Amoxicillin sodium is very soluble in water, sparingly soluble in dehydrated alcohol, very slightly soluble in acetone, and practically insoluble in chloroform and ether. Amoxicillin trihydrate is slightly soluble in water, methanol, and alcohol. It is practically insoluble in carbon tetrachloride and chloroform.

pK$_a$
Amoxicillin has three dissociable protons with pK$_a$ values of 2.63, 7.55, and 9.64 at 23 °C.

Method 1
Khedr and Darwish developed an HPLC method for amoxicillin. The apparatus consisted of a Spectra-Physics model SP8810 pump, a TSK model UV-8 II spectrophotometer detector, a Shimadzu model C-R3A integrator, a DuPont column oven, and a Rheodyne model 7125 injector with a 50-μL loop. The stationary phase was a Beckman Ultrasphere ODS column (250 × 4.6 mm, 5-μm particle size). The mobile phase was a mixture of 4 mL of methanol, 96.5 mL of 0.01 monobasic potassium phosphate, and 0.5 mL of 4% triethylamine that was adjusted to pH 3.6 with 0.1 M phosphoric acid. The flow rate was 0.9 mL/min. UV detection was performed at 242 nm.

The drug substance was dissolved in water. The injection volume was 50 µL. Under these conditions, the retention time of amoxicillin was 10.2 minutes.

The stability-indicating nature of the method was shown by accelerated degradation. Amoxicillin solutions were stored at 60 °C for 3 days, heated in 0.1 N hydrochloric acid at 100 °C for 5 minutes, heated in 0.1 N sodium carbonate at 100 °C for up to 60 minutes, or exposed to daylight for 60 days. The intact amoxicillin peak was well separated from its degradation products.

A linear relationship between the peak area and the concentration was obtained with the correlation coefficient of 0.9997. The limit of detection and limit of quantitation were 1.0 and 2.0 µg/mL, respectively.

Reference
Khedr A, Darwish I. A validated stability-indicating high performance liquid chromatographic assay of amoxicillin. *Bull Pharm Sci*. 2000; 23: 11–21.

Method 2
Shakoor et al. developed an HPLC analytical method for assaying amoxicillin in oral pharmaceutical preparations. The HPLC system consisted of a Waters model 6000A isocratic chromatography pump, a Jasco 875 variable-wavelength UV detector, a BBC Goerz Servogor recorder, and a Rheodyne 7125 injection valve with a 20-µL fixed-volume loop. The stationary phase was a laboratory-made Hypersil ODS column (100 × 2 mm, 5-µm particle size). The mobile phase consisted of 3% methanol in 50 mM aqueous phosphate buffer adjusted to pH 7.0 with phosphoric acid. The flow rate was 0.4 mL/min. The detector was set at 230 nm.

Sample solutions were filtered through a 0.2-µm filter and diluted with water. The retention time for amoxicillin was about 6.4 minutes (estimated from the published chromatogram).

The analytical method was determined to be stability indicating. A solution of amoxicillin (0.1% mass/vol) was allowed to decompose at room temperature and exposed to sunlight for 14 days. HPLC assay results indicated no interference from degradation product peaks with the intact drug peak. Also, an amoxicillin sample was spiked with the main starting materials (4-hydroxyphenyl glycine and 6-aminopenicillanic acid), and the chromatogram showed that there was adequate resolution among all of these compounds.

A standard curve was constructed by preparing aqueous amoxicillin solutions from 0.021 to 0.169 mg/mL. The correlation coefficient of the calibration curve was 0.998.

A similar analytical method was used by Tu et al.

References
Shakoor O, Taylor RB, Moody RR. Analysis of amoxycillin in capsules and oral suspensions by high-performance liquid chromatography. *Analyst*. 1995; 120: 2191–4.

Tu Y-H, Stiles ML, Allen LV, et al. Stability of amoxicillin trihydrate-potassium clavulanate in original containers and unit dose oral syringes. *Am J Hosp Pharm*. 1988; 45: 1092–9.

Method 3
Concannon et al. used HPLC to evaluate the stability of aqueous solutions of amoxicillin sodium in the frozen and liquid states. The chromatographic system included a Pye Unicam L3 UV detector and a Hewlett-Packard 3380A integrator. The stationary phase was a

Waters μBondapak C_{18} column. The mobile phase was 6% methanol in phosphate buffer (pH 5.5). UV detection was performed at 269 nm. Phenoxyacetic acid 4 mg/mL was used as an internal standard.

Each sample was diluted 1:5 with distilled water. The injection volume was 20 μL. The stability-indicating nature of the assay was stated to have been confirmed.

A standard curve for amoxicillin was constructed from 0.2 to 2 mg/mL; the correlation coefficient was 0.999. Coefficients of variation ranged from 0.5 to 2%.

Reference
Concannon J, Lovitt H, Ramage M, et al. Stability of aqueous solutions of amoxicillin sodium in the frozen and liquid states. *Am J Hosp Pharm*. 1986; 43: 3027–30.

Amphotericin B

Chemical Name
[1*R*-(1*R**,3*S**,5*R**,6*R**,9*R**,11*R**,15*S**,16*R**,17*R**,18*S**,19*E*,21*E*,23*E*,25*E*,27*E*,29*E*,31*E*,-33*R**,35*S**,36*R**,37*S**)]-33-[(3-Amino-3,6-dideoxy-β-D-mannopyranosyl)oxy]-1,3,5,6,9,11,17,37-octahydroxy-15,16,18-trimethyl-13-oxo-14,39-dioxabicyclo[33.3.1]-nonatriaconta-19,21,23,25,27,29,31-heptaene-36-carboxylic acid

Other Name
Fungizone

Form	Molecular Formula	MW	CAS
Amphotericin B	$C_{47}H_{73}NO_{17}$	924.1	1397-89-3

Appearance
Amphotericin B is a yellow to orange odorless or practically odorless powder.

Solubility
Amphotericin B is insoluble in water at pH 6–7 but is soluble in water to about 0.1 mg/mL at pH 2 or 11. It is insoluble in ethanol. Solubility in dimethyl sulfoxide is 30–40 mg/mL.

Method 1
Wilkinson et al. reported an HPLC assay for amphotericin B in a hydrophilic colloidal paste base. The apparatus consisted of a Waters model 501 pump, a Waters model 484

variable-wavelength detector, a Rheodyne model 7125 injector, and a Hewlett-Packard model 3396A integrator. The stationary phase was a Waters µBondapak C_{18} column (250 × 4.6 mm, 10-µm particle size) with an Alltech C_{18} guard column. The mobile phase was a mixture of methanol, acetonitrile, and 0.0025 M disodium ethylenediamine-tetraacetic acid in water (50:35:20, vol/vol/vol). The flow rate was 1.6 mL/min. UV detection was performed at 405 nm.

An amphotericin B sample of approximately 1 g of antibiotic formulation was mixed with 20 mL of dimethylformamide and 15 mL of cyclohexane, shaken until dissolved, and centrifuged. The lower dimethylformamide layer was collected and diluted with the mobile phase. The injection volume was 20 µL. Under these conditions, the retention time for amphotericin B was about 4.7 minutes (estimated from the published chromatogram).

The method was determined to be stability indicating because the paste base, other antibiotics in the formulation, and its degradation product did not interfere with the analysis of amphotericin B.

A standard curve for amphotericin B was generated from 0.2 to 3.0% (wt/wt). The correlation coefficient was 0.9995.

Reference

Wilkinson JM, McDonald C, Parkin JE, et al. A high-performance liquid-chromatographic assay for amphotericin B in a hydrophilic colloidal paste base. *J Pharm Biomed Anal.* 1998; 17: 751–5.

Method 2

Lopez et al. studied the stability of amphotericin B in an extemporaneously prepared intravenous fat emulsion. The HPLC system included a Kontron model 325 solvent-delivery pump, a Kontron model 465 autosampler, a Kontron model 432 variable-wavelength UV detector, and an Acer 1120SX personal computer with Kontron PC integrator software v3.00. The stationary phase was a Perkin-Elmer C_{18} column (30 × 4.6 mm, 3-µm particle size). The mobile phase consisted of 2.5 mM disodium ethylenedia-minetetraacetic acid and acetonitrile (70:30, vol/vol). The flow rate was 1.0 mL/min. UV detection was performed at 405 nm and 0.05 AUFS.

Samples were extracted with an equal volume of polysorbate 80 (Tween 80) in sodium chloride and then with an equal volume of a mixture of chloroform and methanol (1:1, vol/vol), centrifuged at 3000 rpm for 20 minutes, dissolved in a mixture of dimethyl sulfoxide and methanol (1:1, vol/vol), and diluted with methanol to an amphotericin B concentration of approximately 12 µg/mL. The injection volume was 20 µL. Under these conditions, the retention time for amphotericin B was 1.5–2 minutes.

The assay was determined to be stability indicating by accelerated degradation testing. Solutions of amphotericin B 5 µg/mL were exposed to 1 N sodium hydroxide (pH ≥12) and to 5 N hydrochloric acid (pH ≤2) for 20 hours at 85 °C. In both cases, none of the new degradation product peaks interfered with the amphotericin B peak.

A standard curve for amphotericin B was constructed from 1 to 20 µg/mL. The correlation coefficient of the standard curve was 0.99. The intrarun and interrun variabil-ities were 1.4 and 2.0%, respectively.

Reference

Lopez RM, Ayestaran A, Pou L, et al. Stability of amphotericin B in an extemporaneously prepared i.v. fat emulsion. *Am J Health Syst Pharm.* 1996; 53: 2724–7.

Method 3

Allen et al. used a stability-indicating HPLC method to evaluate the chemical stability of amphotericin B 0.1 mg/mL in the presence of cefpirome sulfate during simulated Y-site injection. HPLC analysis was conducted using one of two systems. One unit included an Alcott model 728 Micromeritics autosampler, a Rheodyne 7010 injector with an Alcott 732 electrically actuated valve, a Waters model 501 solvent-delivery pump, a Waters model 441 UV detector, a Waters model 401 refractive index detector, and a Waters model 745 data module. The other unit consisted of an Alcott model 728 Micromeritics autosampler, a Rheodyne 7010 injector with an Alcott 732 electrically actuated valve, a Shimadzu model LA-6A solvent-delivery pump, a Shimadzu model SPD-6A UV detector, and an Orion model 901 microprocessor ion analyzer. The stationary phase was a C_{18} analytical column. The mobile phase was a mixture of methanol and dimethylformamide (1:4, vol/vol) and was delivered isocratically at 1.5 mL/min. UV detection was performed at 280 nm.

Samples were diluted before injection. The injection volume was 20 μL. The retention times for amphotericin B and cefpirome were 2.2 and 3.3 minutes, respectively.

This method was determined to be stability indicating by accelerated degradation of amphotericin B with hydrochloric acid, sodium hydroxide, potassium chlorate, heat at 80 °C, or light (150 foot-candles from a tungsten filament source). In each case, amphotericin B was suitably separated from its degradation products.

A calibration curve was constructed from 1 to 10 μg/mL; the correlation coefficient was greater than 0.99. The intraday and interday coefficients of variation were 4.0 and 3.5%, respectively.

Reference

Allen LV, Stiles ML, Prince SJ, et al. Stability of cefpirome sulfate in the presence of commonly used intensive care drugs during simulated Y-site injection. *Am J Health Syst Pharm*. 1995; 52: 2427–33.

Method 4

Wiest et al. used an HPLC assay to investigate the stability of amphotericin B in four concentrations of dextrose injection. The chromatographic system consisted of a Waters model 510 solvent-delivery system, a Waters model 712 WISP autosampler, a Waters model 481 variable-wavelength UV detector, and a Waters model 745 data module. The stationary phase was a Waters μBondapak C_{18} reversed-phase analytical column (300 × 3.9 mm). The mobile phase was a mixture of 40% acetonitrile in 0.01 M ethylenediaminetetraacetic acid. The flow rate was 1.3 mL/min. UV detection was performed at 382 nm. 1-Amino-4-nitronaphthalene in methanol 1 μg/mL was added to each sample as the internal standard. Retention times were 5.1 and 10.6 minutes for amphotericin B and the internal standard, respectively.

The HPLC assay was determined to be stability indicating by subjecting amphotericin B reference standard to extremes of pH and heat. The drug samples were adjusted to pH 0.08 with 2 N hydrochloric acid and to pH 12.2 with 1 N potassium hydroxide. After being stored at room temperature for 2 hours, these two samples were neutralized with potassium hydroxide or acetic acid, respectively. Another sample solution was heated to 85 °C for 1 hour. HPLC analysis showed that none of the degradation products interfered with the intact amphotericin B.

A standard curve for amphotericin B was generated by linear least-squares regression of the ratio of the peak area of the internal standard to the peak area of the drug from 40 to 120 µg/mL. The correlation coefficient for the standard curve was greater than 0.995. The intraday and interday coefficients of variation for the reference standards were 3.46 and 3.71%, respectively.

Reference

Wiest DB, Maish WA, Garner SS, et al. Stability of amphotericin B in four concentrations of dextrose injection. *Am J Hosp Pharm.* 1991; 48: 2430–3.

Method 5

Kintzel and Kennedy studied the stability of amphotericin B in 5% dextrose injection at concentrations used for administration through a central venous line. The HPLC system included a Beckman model 110A single pump, a Kratos model GM 770R variable-wavelength UV detector, a Rheodyne model 7125 manual injector fitted with a 20-µL injection loop, a Perkin-Elmer LCI-100 computer integrator, and a Beckman ODS C_{18} reversed-phase column (250 × 4.6 mm, 5-µm particle size). The mobile phase was a mixture of 37% acetonitrile, 18% methanol, and 45% aqueous solution containing 0.050 M sodium acetate and 0.003 M disodium ethylenediaminetetraacetic acid. The flow rate was 1.2 mL/min. UV detection was performed at 405 nm. Under these conditions, the retention times for amphotericin A, amphotericin X, and amphotericin B were 4.5, 5.4, and 7.8 minutes, respectively.

Amphotericin B solutions were decomposed to demonstrate the stability-indicating capability of the assay by adding 50 µL of 0.1 N hydrochloric acid or 0.1 N sodium hydroxide to 5-mL samples of amphotericin B solution. The solutions were allowed to stand for 2 hours at room temperature before being neutralized with 0.1 N sodium hydroxide or acetic acid, respectively. Another 5 mL of amphotericin B solution was heated at 90 °C for 1 hour. For these three samples, the degradation product peaks were chromatographically resolved from the amphotericin B peak.

A calibration curve was constructed for amphotericin B from 15.7 to 250.0 µg/mL. The correlation coefficient was greater than 0.999. Intraday and interday coefficients of variation were 1.1 and 3.1%, respectively. The coefficients of variation were 4.6% at 15.7 µg/mL and 0.3% at 250.0 µg/mL. The lower limit of detection was 625 ng/mL, and the upper limit was 400 µg/mL.

Similar HPLC methods were used by Miltrano et al. and Owen et al.

References

Kintzel PE, Kennedy PE. Stability of amphotericin B in 5% dextrose injection at concentrations used for administration through a central venous line. *Am J Hosp Pharm.* 1991; 48: 283–5.

Miltrano FP, Outman WR, Baptista RJ, et al. Chemical and visual stability of amphotericin B in 5% dextrose injection stored at 4 °C for 35 days. *Am J Hosp Pharm.* 1991; 48: 2635–7.

Owen D, Fleming RA, Restino MS, et al. Stability of amphotericin B 0.05 and 0.5 mg/mL in 20% fat emulsion. *Am J Health Syst Pharm.* 1997; 54: 683–6.

Method 6

Raymond and Davis used an HPLC method to investigate the physical compatibility and chemical stability of amphotericin B in combination with magnesium sulfate in 5% dextrose injection. The chromatograph consisted of two Isco model 2300 pumps, an Isco V4 variable-wavelength UV detector, and a Zenith Z-150 personal computer running ChromAdapt-PC software. The stationary phase was an Altex Ultrasphere ODS column (150 × 4.6 mm, 5-μm particle size) with a Rainin Perisorb RP18 precolumn (30–40-μm pellicular packing). The mobile phase consisted of 65% 0.01 M phosphate buffer at pH 6.4 and 35% acetonitrile. The flow rate was 1.5 mL/min. UV detection was performed at 405 nm and 0.05 AUFS. Each sample was diluted with purified water. The injection volume was 100 μL.

The assay was determined to be stability indicating by accelerated decomposition of amphotericin B. A solution of amphotericin B in 5% dextrose injection was stored at 60 °C for 120 hours. Degradation product peaks did not interfere with the intact amphotericin B peak.

Standard curves for amphotericin B were constructed from 8.4 to 13.8 μg/mL; the correlation coefficients were greater than 0.9945. The interday coefficient of variation was less than 2.5%.

Reference

Raymond GG, Davis RL. Physical compatibility and chemical stability of amphotericin B in combination with magnesium sulfate in 5% dextrose injection. *Ann Pharmacother.* 1991; 25: 123–6.

Ampicillin

Chemical Name

[2S-[2α,5α,6β(S*)]]-6-[(Aminophenylacetyl)amino]-3,3-dimethyl-7-oxo-4-thia-1-azabicyclo[3.2.0]heptane-2-carboxylic acid

Other Name

Omnipen

Form	Molecular Formula	MW	CAS
Ampicillin	$C_{16}H_{19}N_3O_4S$	349.4	69-53-4
Ampicillin sodium	$C_{16}H_{18}N_3NaO_4S$	371.4	69-52-3

Appearance

Ampicillin is a white and practically odorless crystalline powder. Ampicillin sodium is a white to off-white, odorless or practically odorless crystalline hygroscopic powder.

Solubility

Ampicillin is sparingly soluble in water at room temperature. Ampicillin sodium is very soluble in water.

Method 1

Abdel-Moety et al. developed an HPLC assay for the determination of ampicillin and dicloxacillin in bulk forms, in admixtures, and in capsules. The Shimadzu LC-10AD liquid chromatograph consisted of a model SPD-10A tunable UV detector, a model CTO-10A column oven controller, a model DGU-3A mechanical degasser, a model C-R4A data unit, and a Rheodyne 20-µL injector. The stationary phase was a Merck reversed-phase LiChrosorb RP18 column (250 × 4 mm, 10-µm particle size). The mobile phase consisted of acetonitrile and 1% aqueous acetic acid (75:25, vol/vol) and was delivered at 1.5 mL/min. UV detection was performed at 240 nm. The injection volume was 20 µL. Under these conditions, retention times for dicloxacillin and ampicillin were 1.15 and 2.76 minutes, respectively.

The assay was reported to be stability indicating since it simultaneously determined dicloxacillin and ampicillin in admixtures in the presence of their degradation products.

A standard curve for ampicillin was constructed from 50 to 300 µg/mL. The correlation coefficient was 0.9998.

Reference

Abdel-Moety EM, Al-Rashood KA, Al-Deeb OA, et al. Stability-indicating HPLC method for determination of ampicillin and dicloxacillin in bulk forms, admixtures and in capsules. *Sci Pharm.* 1995; 63: 7–15.

Method 2

Belliveau et al. reported the stability of ampicillin sodium 20 mg/mL with aztreonam 10 mg/mL and sulbactam sodium 10 mg/mL in 0.9% sodium chloride injection. The HPLC system consisted of a Waters model 510 pump, a Waters model 710A WISP autosampler, an LDC analytical model SM4000 variable-wavelength UV detector, and a Hewlett-Packard model HP 3396 series II integrator. The stationary phase was a Waters µBondapak C_{18} column (300 × 3.9 mm). The mobile phase consisted of 0.005 M tetra-butylammonium hydrogen sulfate, 7% acetonitrile, and 14% methanol in a 0.010 M phosphate buffer (pH 2.6–2.7). The flow rate was 1 mL/min. UV detection was performed at 225 nm. Cimetidine was used as the internal standard.

Samples were diluted 1:8 with water. The injection volume was 20 µL. Under these conditions, retention times for cimetidine, ampicillin, sulbactam, and aztreonam were 3.27, 4.35, 5.79, and 7.05 minutes, respectively.

The stability-indicating capability of the assay was demonstrated by analyzing the ampicillin solution 4 mg/mL after storage in a water bath (70 °C) for 8 days. The degradation peaks increased in height and area; however, these peaks did not interfere with the intact ampicillin peak.

Calibration curves for ampicillin were constructed from 1 to 4 mg/mL. The correlation coefficient was greater than 0.9977. Intraday and interday coefficients of variation were less than 7%.

Reference

Belliveau PP, Nightingale CH, Quintiliani R. Stability of aztreonam and ampicillin sodium-sulbactam sodium in 0.9% sodium chloride injection. *Am J Hosp Pharm.* 1994; 51: 901–4.

Method 3

Allwood and Brown used an HPLC method to investigate the stability of ampicillin infusions in unbuffered and buffered saline. The stationary phase was a Spherisorb ODS column (100 × 4.6 mm, 5-µm particle size). The mobile phase was methanol and 0.067 M potassium dihydrogen phosphate. The flow rate was 1.5 mL/min. UV detection was performed at 225 nm.

Each sample was diluted 1:20 with cold water (5 °C) and maintained in an iced water bath until analysis.

The method was determined to be stability indicating by comparing chromatograms of heat-stressed (at 60 °C for 2 hours) solutions of ampicillin with untreated solutions. No interference of degradation product peaks with the intact ampicillin peak occurred.

Linearity was obtained with ampicillin concentrations from 62.5 to 250 µg/mL. The correlation coefficient was 0.999; the coefficient of variation of repeated injections ($n = 5$) was 0.54%.

Reference

Allwood MC, Brown PW. Stability of ampicillin infusions in unbuffered and buffered saline. *Int J Pharm.* 1993; 97: 219–22.

Method 4

James and Riley used HPLC to study the stability of aztreonam with ampicillin in intravenous admixtures. The liquid chromatograph consisted of a ConstaMetric III G pump, a SpectroMonitor D detector, a Negretti model 190 injection valve fitted with a 20-µL loop, and a series 500 Fisher Recordall dual-pen recorder. The stationary phase was a Shandon ODS Hypersil column (150 × 4.6 mm) with a presaturator Permaphase ODS column. The mobile phase consisted of 22.5% methanol in an aqueous solution of 5 mM tetrabutylammonium hydrogen sulfate and 5 mM ammonium sulfate (pH 2.6). The flow rate was 1.5 mL/min and the UV detector was set at 238 nm. Under these conditions, the retention time of ampicillin was 10.5 minutes.

The stability-indicating capability of the assay was demonstrated by accelerated decomposition of ampicillin. Ampicillin solutions were heated to 70 °C at pH 2.0 and 10.0 until the ampicillin peak decreased to about 5% of its initial height. Degradation product peaks did not interfere with the ampicillin peak.

Calibration curves were constructed for ampicillin from 20 to 200 µg/mL. The intraday and interday reproducibilities of these peak areas were 2.00 and 3.53%, respectively.

Reference

James MJ, Riley CM. Stability of intravenous admixtures of aztreonam and ampicillin. *Am J Hosp Pharm.* 1985; 42: 1095–100.

Method 5

Using a stability-indicating HPLC method, Gupta and Stewart studied the chemical stability of ampicillin sodium when mixed with metronidazole injection for intravenous infusion. A Waters ALC 202 high-performance liquid chromatograph was equipped with a Schoeffel SF 770 multiple-wavelength detector, an Omniscribe recorder, and a Waters μBondapak C_{18} column. The mobile phase was 7% acetonitrile and 0.1 M monobasic potassium phosphate in water with the pH adjusted to 4.1 with 1% phosphoric acid in water. The flow rate was 2.0 mL/min. The detector was set at 240 nm and 0.1 AUFS. About 16 μg was injected.

The method was stated to be stability indicating.

References

Gupta VD, Stewart KR. Chemical stability of hydrocortisone sodium succinate and several antibiotics when mixed with metronidazole injection for intravenous infusion. *J Parenter Sci Technol.* 1985; 39: 145–8.

Gupta VD, Shah KA, Dela Torre M. Stability of ampicillin sodium and penicillin G potassium solutions using high pressure liquid chromatography. *Can J Pharm Sci.* 1981; 16: 61.

Amyl Nitrite

Chemical Name
Isopentyl nitrite

and

Other Names
Emergent-Ez, Isoamyl Nitrite

Form	Molecular Formula	MW	CAS
Amyl nitrite	$C_5H_{11}NO_2$	117.1	110-46-3

Appearance
Amyl nitrite is a clear yellow volatile flammable liquid with a fragrant odor.

Solubility
Amyl nitrite is practically insoluble in water but is miscible with alcohol and ether.

Method

Chen et al. developed an HPLC method for isoamyl nitrite. The Shimadzu liquid chromatograph consisted of a model LC-6A solvent-delivery system, a model SIL-6A autoinjector with a 20-µL loop, a model SPD-6AV variable-wavelength UV detector, and a model C-R3A integrator. A Hewlett-Packard 1040M photodiode-array detector was used for peak purity analysis. The stationary phase was a Nucleosil C_{18} column (300 × 4.6 mm, 5-µm particle size). The mobile phase consisted of tetrabutylammonium hydroxide, methanol, and 0.05 M phosphate buffer (pH 6.3) (0.05:30:70, vol/vol/vol). The flow rate was 1.0 mL/min. UV detection was performed at 356 nm and 0.02 AUFS.

Each sample was diluted with absolute alcohol. Under these conditions, the retention time of isoamyl nitrite was 12.3 minutes.

The method was reported to be stability indicating. Isoamyl nitrite samples were degraded in a clear glass bottle exposed to daylight for 6 months, in 1% acetic acid at 60 °C, and in 0.1 M glycine buffer (pH 11.25) at 60 °C. No degradation products interfered with the analysis of isoamyl nitrite.

A standard curve for isoamyl nitrite was generated from 0.47 to 3.78 mg/mL. The correlation coefficient was 0.998. The coefficient of variation for the analysis of isoamyl nitrite was 1.97%.

Reference

Chen G-L, Hu M-K, Chan L-Y, et al. A stability-indicating method for isoamyl nitrite by high-performance liquid chromatography/diode-array detector. *Chin Pharm J.* 1989; 41: 331–7.

Anakinra

Chemical Name

N^2-L-Methionylinterleukin 1 receptor antagonist (human isoform x reduced)

Other Names

Antril, Kineret

Form	Molecular Formula	MW	CAS
Anakinra	$C_{759}H_{1186}N_{208}O_{232}S_{10}$	17,258 (Daltons)	143090-92-0

Method

Nahata et al. used an HPLC method to study the stability of anakinra and clindamycin phosphate in 0.9% sodium chloride injection. The chromatographic apparatus was composed of a Hewlett-Packard 1050 series pump, an autosampler, a variable-wavelength detector, and a model 3396A integrator. The stationary phase was a Bio-Rad Bio-Gel SP-5-PW column (75 × 7.5 mm). Buffer A was 20 mM 2-(morpholino)ethanesulfonic acid monohydrate buffer (pH 5.5), and buffer B was a mixture of 20 mM 2-(morpholino)-ethanesulfonic acid monohydrate buffer (pH 5.5) and 1.0 M sodium chloride buffer (pH 5.5). The mobile phase was gradiently delivered at a flow rate of 0.5 mL/min from 1 to 60% buffer B over 60 minutes.

Samples were diluted in citrate-buffered saline with ethylenediaminetetraacetic acid (EDTA) (10 mM citrate buffer, 140 mM sodium chloride, and 0.5 mM ethylenediaminetetraacetic acid at pH 6.5). UV detection was performed at 280 nm. The injection volume was 100 μL.

The stability-indicating nature of this method was evaluated by the intentional decomposition of anakinra through lowering the pH of an anakinra solution to 5.3. The decomposition products did not interfere with the analysis of anakinra.

A linear relationship between anakinra concentration and detector response was obtained with a correlation coefficient greater than 0.999.

References
Nahata MC, Morosco RS, Sabados BK, et al. Stability and compatibility of anakinra and clindamycin phosphate injections in 0.9% sodium chloride injection. *J Appl Ther Res*. 1998; 2: 87–9.

Nahata MC, Morosco RS, Sabados BK, et al. Stability and compatibility of anakinra with ceftriaxone sodium injection in 0.9% sodium chloride or 5% dextrose injection. *J Clin Pharm Ther*. 1997; 22: 167–9.

Ancitabine Hydrochloride

Chemical Name
2,2′-Anhydro-(1β-D-arabinofuranosyl)cytosine hydrochloride

Other Name
Cyclocytidine Hydrochloride

Form	Molecular Formula	MW	CAS
Ancitabine hydrochloride	$C_9H_{11}N_3O_4.HCl$	261.7	10212-25-6

Method
Tuncel et al. developed an isocratic reversed-phase HPLC method for the study of cyclocytidine hydrolysis. The liquid chromatograph consisted of an Altex model 332 pump, a model 210 20-μL loop injection port, and a model 110A 254-nm UV detector. The stationary phase was an Altex Spherisorb ODS C_{18} column (46 × 4.2 mm, 5-μm particle size). The mobile phase was prepared by mixing 1000 mL of 0.005 M aqueous 1-heptanesulfonic acid with 30 mL of methanol and adjusting the pH to 2.9 with glacial acetic acid. The flow rate was 2 mL/min. UV detection was performed at 254 nm. Thymidine 6.44×10^{-5} M was used as an internal standard. Under these conditions, retention times for arabinosyluracil, thymidine, cyclocytidine, and arabinosylcytosine were about 0.7, 1.5, 3.0, and 5.8 minutes, respectively.

The method was reported to be stability indicating since it separated the parent drug from its degradation products.

Reference

Tuncel M, Notari RE, Malspeis L. A rapid stability-indicating HPLC assay for the arabinosylcytosine prodrug, cyclocytidine. *J Liq Chromatogr*. 1981; 4: 887–96.

Antazoline

Chemical Names

4,5-Dihydro-*N*-phenyl-*N*-(phenylmethyl)-1*H*-imidazole-2-methanamine
2-(*N*-Benzylanilinomethyl)-2-imidazoline

Other Name

Antistine

Form	Molecular Formula	MW	CAS
Antazoline	$C_{17}H_{19}N_3$	265.4	91-75-8
Antazoline hydrochloride	$C_{17}H_{19}N_3.HCl$	301.8	2508-72-7
Antazoline mesylate	$C_{17}H_{19}N_3.CH_3SO_3H$	361.5	3131-32-6
Antazoline phosphate	$C_{17}H_{19}N_3.H_3PO_4$	363.4	154-68-7
Antazoline sulfate	$(C_{17}H_{19}N_3)_2.H_2SO_4.2H_2O$	664.8	24359-81-7

Appearance

Antazoline phosphate is a white to off-white crystalline powder.

Solubility

Antazoline phosphate is soluble in water, sparingly soluble in methanol, and practically insoluble in ether.

Method

Ruckmick et al. identified the primary degradation product of antazoline phosphate and naphazoline hydrochloride in a commercial ophthalmic formulation. The liquid chromatograph consisted of a Waters 600 E low-pressure mixing gradient solvent-delivery controller and pump, a Waters 715 ultra-WISP autosampler, and a Waters 490 multichannel variable-wavelength UV detector connected to a PE Nelson series 900 A/D converter interface to a VAX 6210 computer. The stationary phase was a Beckman Ultrasphere C_{18} column (250 × 4.6 mm, 5-µm particle size). The mobile phase consisted of 57% methanol in water containing 22 mM heptanesulfonic acid, 0.1% dibutylamine, and 1% acetic acid. The flow rate was 1 mL/min. UV detection was performed at 280 nm.

Samples were diluted 1:10 with water. The injection volume was 20 μL. Retention times for naphazoline and antazoline were about 8.5 and 20 minutes, respectively (estimated from the published chromatogram).

The method was determined to be stability indicating by spiking the sample with the known degradation product, N-[(N-benzylanilino)acetyl]ethylenediamine. No interference with the intact drug peak occurred.

Standard curves were constructed from 100 to 1000 μg/mL. Linearity was good, with the correlation coefficient greater than 0.9999.

Reference

Ruckmick SC, Marsh DF, Duong ST. Synthesis and identification of the primary degradation product in a commercial ophthalmic formulation using NMR, MS, and a stability-indicating HPLC method for antazoline and naphazoline. *J Pharm Sci.* 1995; 84: 502–7.

Antipyrine

Chemical Names

1,2-Dihydro-1,5-dimethyl-2-phenyl-3*H*-pyrazol-3-one

2,3-Dimethyl-1-phenyl-3-pyrazolin-5-one

Other Name

Phenazone

Form

Form	Molecular Formula	MW	CAS
Antipyrine	$C_{11}H_{12}N_2O$	188.2	60-80-0
Antipyrine salicylate	$C_{11}H_{12}N_2O.C_7H_6O_3$	326.4	520-07-0

Appearance

Antipyrine occurs as colorless odorless crystals or as a white crystalline powder.

Solubility

One gram of antipyrine dissolves in less than 1 mL of water, 1.3 mL of alcohol, 1 mL of chloroform, and 43 mL of ether.

Method

Mutch and Hutson used HPLC to determine the stability of antipyrine and caffeine in an intravenous solution under conditions of stress. The stationary phase was a Waters

Nova-Pak RCSS phenyl column (4-μm particle size) with a C_{18} guard column. The mobile phase consisted of 0.01 M potassium phosphate, methanol, and tetrahydrofuran (79:20:1, vol/vol/vol) at pH 3.0. The flow rate was 2 mL/min. Absorbance was measured at 214 nm by a Beckman model 163 UV detector.

Samples were diluted 1:200 with deionized distilled water. The sample injection volume was 20 μL. Retention times for caffeine and antipyrine were 10.4 and 12.6 minutes, respectively.

The stability-indicating capability of this method was demonstrated by accelerated decomposition. Samples were subjected to heating in water baths at 40 or 60 °C or in a steam autoclave (121 °C). Samples also were exposed to fluorescent light. Decomposition products did not interfere with the peak of the intact drug.

The intraday and interday coefficients of variation were 0.92 and 1.49%, respectively.

Reference
Mutch RS, Hutson PR. Stability of antipyrine plus caffeine in intravenous solution. *Am J Hosp Pharm.* 1991; 48: 1267–70.

Apaziquone

Chemical Name
5-(1-Aziridinyl)-3-(hydroxymethyl)-2-[(1*E*)-3-hydroxy-1-propenyl]-1-methyl-1*H*-indole-4,7-dione

Other Name
EO9

Form	Molecular Formula	MW	CAS
Apaziquone	$C_{15}H_{16}N_2O_4$	288.3	114560-48-4

Method
De Vries et al. reported an HPLC study on the chemical stability of the novel indoloquinone antitumor agent apaziquone. The Waters liquid chromatograph included a model M6000 solvent-delivery system, a model U6K septumless injector, and a model 440 dual-wavelength UV detector. The stationary phase was a laboratory-made LiChrosorb RP8 column (125 × 4 mm, 5-μm particle size) with a LiChrosorb RP8 precolumn (20 × 4 mm, 10-μm particle size). The mobile phase consisted of methanol and

water (30:70, wt/wt) containing 1% (vol/wt) 0.5 M sodium phosphate buffer (pH 7.0). The flow rate was 1.0 mL/min. UV detection was performed at 254 nm and 0.05 AUFS.

Samples were analyzed with dilution. The injection volume was 25 μL. Under these conditions, the retention time of apaziquone was about 6 minutes (estimated from the published chromatogram).

The stability-indicating nature of the method was demonstrated by the degradation of apaziquone in acidic and alkaline media. Apaziquone was partly degraded at pH greater than 10 or pH less than 7. One degradation product was obtained at the retention time of 4.4 minutes.

Calibration curves of apaziquone in water were constructed with correlation coefficients greater than 0.999.

Reference
De Vries JD, Winkelhorst J, Underberg WJM, et al. A systematic study on the chemical stability of the novel indoloquinone antitumour agent E09. *Int J Pharm*. 1993; 100: 181–8.

Aplidine

Chemical Name
1-(1,2-Dioxopropyl)-L-prolyl-*N*-methyl-D-leucyl-L-threonyl-(3*S*,4*R*,5*S*)-4-amino-3-hydroxy-5-methylheptanoyl-(2*S*,4*S*)-4-hydroxy-2,5-dimethyl-3-oxohexanoyl-L-leucyl-L-prolyl-*N,O*-dimethyl-L-tyrosine (8→3)-lactone

Form	Molecular Formula	MW	CAS
Aplidine	$C_{57}H_{87}N_7O_{15}$	1110.3	137219-37-5

Solubility
Aplidine is very poorly soluble (<0.1 mg/mL) in water but very soluble (>10 mg/mL) in absolute alcohol, dimethylacetamide, and dimethyl sulfoxide.

Method 1

Waterval et al. investigated the degradation kinetics of aplidine using reversed-phase HPLC. The liquid chromatographic system consisted of a Waters model M6000A pump, a Perkin-Elmer ISS-100 autosampler, and an Applied Biosystems model 785A UV detector. The stationary phase was a Waters Symmetry reversed-phase C_{18} column (100 × 4.6 mm, 3.5-µm particle size) coupled with a SecurityGuard C_{18} precolumn (4 × 3 mm). The mobile phase consisted of acetonitrile–water (65:35, vol/vol). The flow rate was 1 mL/min. UV detection was performed at 208 nm. The sample injection volume was 20 µL. Under these conditions, retention times for the *cis* and *trans* isomers of aplidine were about 4.1 and 5.2 minutes, respectively (estimated from the published chromatogram).

The method was reported to be stability indicating.

Reference

Waterval JCM, Bloks JC, Sparidans RW, et al. Degradation kinetics of aplidine, a new marine antitumoural cyclic peptide, in aqueous solution. *J Chromatogr B*. 2001; 754: 161–8.

Method 2

Nuijen et al. developed an HPLC method to characterize the aplidine raw material. The instrument consisted of a Thermo Separation model SP8800 pump, a model Spectra 200 UV-visible detector, a model SP8880 autosampler, and a Jones model 7971 column heater. The stationary phase was a Waters Zorbax-SB C_{18} analytical column (150 × 4.6 mm, 3.5-µm particle size). Mobile phase A was acetonitrile containing 0.04% trifluoroacetic acid (TFA), and mobile phase B was water containing 0.04% TFA. The flow rate was 0.6 mL/min. Mobile phase A was linearly increased from 35 to 70% in 15 minutes. The total run time was 30 minutes. UV detection was performed at 225 nm. The column temperature was set at 80 °C.

Samples were diluted with a mixture of acetonitrile–water (1:1, vol/vol). The injection volume was 20 µL. Under these conditions, the retention time for aplidine was 21 minutes.

The assay was reported to be stability indicating. Impurities of the aplidine bulk drug eluted at approximately 14 and 20 minutes.

A standard curve for aplidine was constructed from 5 to 300 µg/mL. Least-squares regression analysis was performed based on peak areas versus concentration.

Reference

Nuijen B, Bouma M, Henrar REC, et al. Pharmaceutical development of a parenteral lyophilized formulation of the novel antitumor agent aplidine. *PDA J Pharm Sci Technol.* 2000; 54: 193–208.

Apomorphine Hydrochloride

Chemical Name

(6a*R*)-5,6,6a,7-Tetrahydro-6-methyl-4*H*-dibenzo[*de,g*]quinoline-10,11-diol hydrochloride hemihydrate

•HCl • 1/2 H$_2$O

Other Names
Apokyn, Apomine, Ixense, Uprima

Form	Molecular Formula	MW	CAS
Apomorphine hydrochloride	$C_{17}H_{17}NO_2 \cdot HCl \cdot \frac{1}{2}H_2O$	312.8	41372-20-7

Appearance
Apomorphine hydrochloride occurs as odorless and minute white or grayish-white glistening crystals or as a white powder.

Solubility
Apomorphine hydrochloride is sparingly soluble in water and alcohol.

Method 1
Ingram et al. developed an assay of apomorphine for use in clinical pharmacokinetic studies in Parkinson's disease. A ThermoElectron ConstaMetric 3200 isocratic solvent-delivery system was coupled with a Jasco AS-950 autosampler and an FP-920 spectrofluorometer (excitation and emission wavelengths of 270 and 450 nm, respectively). The peak purity was evaluated by a ThermoElectron VU6000P photodiode-array detector (wavelength range of 198–798 nm). The stationary phase was a Phenomenex Columbus C_{18} analytical column (150 × 4.6 mm, 5-μm particle size) with a C_{18} precolumn (4 × 3 mm). The mobile phase consisted of a buffer and methanol (60:40, vol/vol), where the buffer contained 0.25 M monobasic sodium phosphate, 0.25% (wt/vol) heptanesulfonic acid (adjusted to pH 3.30 with phosphoric acid), and 0.003% (wt/vol) ethylenediaminetetraacetic acid (EDTA). The flow rate was 1 mL/min. *R*-Propylnorapomorphine hydrochloride was used as an internal standard.

Standards and quality control samples were prepared by spiking apomorphine in blank plasma. Internal standard was added to the plasma to give a final concentration of 100 ng/mL and the sample was mixed gently. Apomorphine was extracted using a Varian Bond-Elut C_{18} 100-mg solid phase extraction column (1 mL) under vacuum. The column was conditioned with methanol (2 mL) followed by water (2 mL); loaded with a plasma sample (0.25–2.5 mL); and washed with water (2 mL) followed sequentially by 10% (vol/vol) methanol in water (1 mL), 20% (vol/vol) methanol in water (1 mL), 50% (vol/vol) methanol in water (1 mL), and 2% (wt/vol) sodium metabisulfite in 0.10% (wt/vol) EDTA and 0.15% (wt/vol) ascorbic acid in water (diluent A, 200 μL). Analytes were eluted with an eluting solution (800 μL) into an autosampler vial containing 2% (wt/vol) sodium metabisulfate in diluent A (200 μL). The eluting solution consisted of 0.25 M monobasic sodium phosphate buffer (adjusted to pH 3.30 with phosphoric acid) and methanol (60:40, vol/vol). The sample was then analyzed by HPLC. The retention times of apomorphine and the internal standard were about 7.3 and 11.2 minutes, respectively.

The stability-indicating nature of the method was established by accelerated decomposition of apomorphine. The drug in water (1 mL) and the internal standard in water (1 mL) were incubated separately in 0.1 M hydrochloric acid (1 mL), in 0.1 M sodium hydroxide solution (1 mL), hydrogen peroxide (1 mL), and distilled water (1 mL) at 25 and/or 60 °C for up to 30 minutes. Both the intact drug and the internal standard were well separated from degradation products. Furthermore, no endogenous compounds in plasma interfered with the analysis of apomorphine.

A standard curve for apomorphine was constructed from 0.50 to 70 ng/mL. The correlation coefficient was 0.9975. Intraday and interday coefficients of variation were less than 8.2 and 5.8%, respectively. The coefficients of variation of the analysis were less than 0.36% at 1 ng/mL and 0.52% at 20 ng/mL, respectively. The limit of detection and the limit of quantitation were 0.03 and 0.5 ng/mL, respectively.

Reference

Ingram WM, Priston MJ, Sewell GJ. Improved assay for R(–)-apomorphine with application to clinical pharmacokinetics studies in Parkinson's disease. *J Chromatogr B*. 2006; 831: 1–7.

Method 2

Kuehl et al. described a method for the determination of apomorphine (apomine) in a topical cream formulation. A Waters 2690 separation module was coupled with a Waters 996 photodiode-array detector. The stationary phase was an Alltech Altima C_{18} column (150 × 2.1 mm, 5-μm particle size). The mobile phase was delivered in a gradient mode as follows:

Time, minutes	Flow Rate, mL/min	Acetonitrile, %	Water, %	Isopropanol, %
0	0.6	65	35	0
9	0.6	65	35	0
10	0.35	0	0	100
19	0.35	0	0	100
23	0.6	65	35	0
29	0.6	65	35	0

Typical column back pressure throughout the gradient mode ranged from 2750 to 3250 psi. UV detection was carried out at 210 nm.

Apomorphine stock solution was prepared in a mixture of acetonitrile and water (65:35, vol/vol). For the cream solution, the sample was diluted with tetrahydrofuran and sonicated for 10 minutes. The injection volume was 5 μL. Under these conditions, the retention time of apomorphine was about 8.2 minutes.

With the use of a photodiode-array detector, the specificity of this assay was verified by mass spectrometry and library UV spectra matching.

A standard curve was obtained from 22.6 to 452.0 μg/mL, with a correlation coefficient of ≥ 0.9995. Intraday and interday coefficients of variation were less than 0.53 and 0.12%, respectively. The limit of detection and the limit of quantitation were 1.56 and 4.74 μg/mL, respectively.

Reference

Kuehl PJ, Angersbach BS, Stratton SP, et al. Development of an HPLC method for the analysis of Apomine in a topical cream formulation. *J Pharm Biomed Anal*. 2006; 40: 975–80.

Artemether

Chemical Name
(3R,5aS,6R,8aS,9R,10S,12R,12aR)-Decahydro-10-methoxy-3,6,9-trimethyl-3,12-epoxy-12H-pyrano[4,3-j]-1,2-benzodioxepin

Other Name
Artemetherum

Form
Artemether

Molecular Formula
$C_{16}H_{26}O_5$

MW
298.4

CAS
71963-77-4

Method
Atemnkeng et al. developed a method for the determination of artemether, methylparaben, and propylparaben in a pharmaceutical dosage form. The liquid chromatographic system consisted of a Merck-Hitachi model L-6000 pump, a Perkin-Elmer model LC 90 UV spectrophotometric detector, and a Merck-Hitachi model D-2500 Chromato-Integrator. The stationary phase was a Macherey-Nagel Nucleosil 120-5 C_{18} column (125 × 4 mm, 5-μm particle size). The mobile phase was a mixture of acetonitrile, 0.05 M monobasic potassium phosphate buffer (pH 5.0), and water (48:32:10, vol/vol/vol). The flow rate was 1.0 mL/min. UV detection was performed at 215 nm.

An artemether standard solution was prepared in methanol. For drug suspensions, the sample was mixed with methanol and water (80:20, vol/vol), ultrasonicated, and centrifuged at 3000 rpm (1512 × g) for 15 minutes. The supernatant was collected and diluted by a factor of 25 with methanol. The injection volume was 20 μL. Under these conditions, the retention time of artemether was about 9 minutes.

The method was reported to be stability indicating.

A linear calibration curve for artemether was constructed from 1 to 10 mg/mL, with a correlation coefficient of 1.00. Intraday and interday coefficients of variation were 0.23 and 3%, respectively.

Reference
Atemnkeng MA, Marchand E, Plaizier-Vercammen J. Assay of artemether, methylparaben and propylparaben in a formulated paediatric antimalarial dry suspension. *J Pharm Biomed Anal.* 2007; 43: 727–32.

Ascorbic Acid (Vitamin C)

Chemical Name

2,3-Didehydro-L-*threo*-hexono-1,4-lactone

Other Name

Cenolate

Form	Molecular Formula	MW	CAS
Ascorbic acid	$C_6H_8O_6$	176.1	50-81-7
Calcium ascorbate	$(C_6H_7O_6)_2Ca.2H_2O$	426.3	5743-27-1
Sodium ascorbate	$C_6H_7NaO_6$	198.1	134-03-2

Appearance

Ascorbic acid occurs as odorless or almost odorless colorless crystals or as a white or slightly yellow crystalline powder. Calcium ascorbate is a white or slightly yellow crystalline powder. Sodium ascorbate occurs as odorless or practically odorless, white or very faintly yellow crystals or crystalline powder.

Solubility

Ascorbic acid is freely soluble in water and sparingly soluble in alcohol. Calcium ascorbate is very soluble in water. Sodium ascorbate is freely soluble in water, slightly soluble in alcohol, and practically insoluble in chloroform and ether.

pK_a

Ascorbic acid has pK_a values of 4.2 and 11.6.

Method 1

Kearney et al. used HPLC to study the stability of ascorbic acid in total parenteral nutrition (TPN) mixtures. The chromatograph consisted of a Kontron pump, a model 332 UV detector, and a model 360 autosampler. The stationary phase was a Phenomenex Techsphere ODS column (250 × 4.6 mm, 5-µm particle size). The mobile phase was a mixture of methanol and 0.15 M phosphate buffer (pH 7.8) (1:1, vol/vol) with 0.05 M cetrimide as an ion-pair reagent. The mobile phase was delivered isocratically at 1 mL/min. UV detection was performed at 278 nm. The injection volume was 20 µL. Under these conditions, the retention time for ascorbic acid was about 3.4 minutes.

The method was shown to be stability indicating by spiking the ascorbic acid solution with dehydroascorbic acid. Dehydroascorbic acid was not detected under these conditions.

A calibration curve was generated from 20 to 100 µg/mL ascorbic acid in a TPN mixture. The correlation coefficient was 0.996. The intraday and interday coefficients of variation were 2.4 and 3.3%, respectively. The limit of detection was approximately 1 µg/mL.

Reference

Kearney MCJ, Allwood MC, Martin H, et al. The influence of amino acid source on the stability of ascorbic acid in TPN mixture. *Nutrition.* 1998; 14: 173–8.

Method 2

Van der Horst et al. reported the determination by HPLC of ascorbic acid in a TPN solution. The liquid chromatographic system consisted of a Waters model M6000A pump, a WISP autosampler, a model 481 variable-wavelength detector, a Perkin-Elmer model 204-A fluorescence spectrophotometer, and a Shimadzu Chromatopac C-R3A integrator. The stationary phase was a LiChrosorb RP8 analytical column (250 × 4.6 mm, 10-µm particle size) with a RCSS Guard-Pak C_{18} precolumn (4 × 6 mm). The mobile phase was a mixture of 0.1% triethylamine in 0.067 M monobasic potassium phosphate buffer (pH 7.8) and methanol (70:30, vol/vol). The flow rate was 2 mL/min. Fluorescence detection was performed at an excitation wavelength of 355 nm and an emission wavelength of 415 nm.

An aliquot of 100 µL was diluted with 1 mL of metaphosphoric acid, mixed with 25 µL of 0.01 M iodine solution, and allowed to stand for 1 minute to oxidize the ascorbic acid. Then the solution was allowed to react with 25 µL of 0.01 M thiosulfate and 150 µL of 4.5 M sodium acetate, mixed with 250 µL of 1 mg/mL *o*-phenylenediamine solution, and allowed to stand for 15 minutes in the dark at 20 °C. Twenty microliters of this solution was injected onto the column. Under these conditions, the retention time for the ascorbic acid derivative was about 2.0 minutes (estimated from the published chromatogram).

The method was demonstrated to be stability indicating by accelerated degradation of ascorbic acid. The acid solutions were treated with 1 M sodium hydroxide solution or 1 M hydrochloric acid and heated to boiling for about 30 minutes. Degradation products were separated from the peak of the ascorbic acid derivative.

A standard curve for ascorbic acid was generated from 1.1 to 9.4 mg/L. The correlation coefficient was 0.9994. The limit of detection was 0.1 mg/L.

Reference

Van der Horst A, Martens HJM, de Goede PNFC. Analysis of water-soluble vitamins in total parenteral nutrition solution by high pressure liquid chromatography. *Pharm Weekbl [Sci].* 1989; 11: 169–74.

Asoxime Chloride

Chemical Name

1-[[[4-(aminocarbonyl)pyridino]methoxy]methyl]-2-[(hydroxylimino)]pyridinium dichloride

Other Name
HI-6

Form | Molecular Formula | MW | CAS

Form	Molecular Formula	MW	CAS
Asoxime chloride	$C_{14}H_{16}N_4O_3 \cdot 2Cl$	359.2	34433-31-3

Method

Wang et al. investigated by HPLC the stability of asoxime chloride (HI-6) in solutions. A liquid chromatographic system consisted of a Shimadzu model LC-6A single-piston pump, a model SPD-M6A photodiode-array detector, and a model C-R6A recorder. The stationary phase was a Merck LiChroCART C_{18} column (150 × 3.9 mm, 5-μm particle size). The mobile phase consisted of acetonitrile and 0.01 M 1-heptanesulfonic acid sodium salt aqueous solution (20:80, vol/vol), adjusted to pH 3.2 with acetic acid. The flow rate was 1.0 mL/min. UV detection was performed at 305 nm. Under these conditions, the retention time for asoxime chloride was 14.5 minutes.

The method was demonstrated to be stability indicating by accelerated degradation of asoxime chloride. The sample was adjusted to pH 2.54 with hydrochloric acid and maintained at 70 °C for 100 hours. The degradation products were well separated from asoxime chloride on the chromatogram.

A standard curve for asoxime chloride was generated from 2.2 to 5.6 μg/mL. The correlation coefficient was 0.999. The intraday and interday coefficients of variation were 2.51 and 0.14%, respectively.

Reference

Wang D-P, Lee J-D, Lin R-A. Stability of HI-6 in solution. *Drug Dev Ind Pharm*. 1995; 21: 509–16.

Aspartame

Chemical Name
3-Amino-*N*-(α-methoxycarbonylphenethyl)succinamic acid

Other Name
Aspartin

Form
Form	Molecular Formula	MW	CAS
Aspartame	$C_{14}H_{18}N_2O_5$	294.3	22839-47-0

Appearance
Aspartame occurs as a white odorless slightly hygroscopic crystalline powder with an intensely sweet taste.

Solubility
Aspartame is sparingly soluble in water and practically insoluble in dichloromethane and hexane.

Method
Prankerd et al. investigated the degradation of aspartame in acidic aqueous media and its stabilization by complexation with cyclodextrins or modified cyclodextrins. The apparatus consisted of a Kratos Spectroflow 400 solvent pump, a Rheodyne model 7125 20-µL loop injector, a Kratos Spectroflow 757 variable-wavelength detector, and a Hewlett-Packard model 3392A integrator. The stationary phase was a Keystone reversed-phase C_8 column (150 × 4.6 mm, 5-µm particle size). The mobile phase consisted of acetonitrile and 0.18 M monobasic sodium phosphate buffer (adjusted to pH 2.1 with phosphoric acid) (12.5:87.5, vol/vol), containing 0.020 M sodium heptanesulfonate and 0.114 M potassium chloride. The flow rate was 1.2 mL/min. UV detection was performed at 210 nm. Under these conditions, the retention time for aspartame was 19.2 minutes.

The method was reported to be stability indicating. Aspartame was well resolved from its potential degradation products. Retention times in minutes were 4.2 for 3-methylenecarboxyl-6-benzyl-2,5-diketopiperazine, 5.1 for L-phenylalanine, 6.6 for L-phenylalanyl-α-L-aspartic acid, 7.9 for α-L-aspartyl-L-phenylalanine, 12.5 for L-phenylalanine methyl ester, 15 for β-L-aspartyl-L-phenylalanine methyl ester, and 19.2 for aspartame.

Reference
Prankerd RJ, Stone HW, Sloan KB, et al. Degradation of aspartame in acidic aqueous media and its stabilization by complexation with cyclodextrins or modified cyclodextrins. *Int J Pharm.* 1992; 88: 189–99.

Aspirin

Chemical Name
Acetylsalicylic acid

Other Name
ASA

Form	Molecular Formula	MW	CAS
Aspirin	$C_9H_8O_4$	180.2	50-78-2

Appearance
Aspirin occurs as white crystals or as a white crystalline powder.

Solubility
Aspirin is slightly soluble in water and freely soluble in alcohol.

pK$_a$
Aspirin has a pK$_a$ of 3.5.

Method 1
Montgomery et al. described the development and validation of an HPLC method for simultaneous determination of aspirin and warfarin in tablets. The chromatographic system consisted of a Hewlett-Packard 1050 series pump, autosampler, column oven, and detector. The stationary phase was a Zorbax C_8 column (250 × 4.0 mm, 5-µm particle size). The column temperature was maintained at 40 °C. Mobile phase A was water that was adjusted to pH 2.6 with formic acid. Mobile phase B was methanol, and mobile phase C was acetonitrile. The flow rate was 1.0 mL/min. Mobile phases (A:B:C) were initially delivered at a ratio of 68:17:15 (vol/vol/vol) for 11 minutes, increased linearly to 56:17:27 (vol/vol/vol) at 15 minutes, held at this ratio for 38 minutes, then returned in 0.5 minute to the initial ratio, and equilibrated for 6 minutes. The sample solvent mixture was made of acetonitrile, chloroform, formic acid, citric acid, and butylparaben (49.95:49.95:0.1:0.1:0.05, vol/vol/vol/vol/vol). UV detection was performed at 280 nm. The injection volume was 20 µL. Butylparaben was used as an internal standard. Under these conditions, retention times for aspirin and warfarin were about 13 and 42 minutes, respectively (estimated from the published chromatogram).

The assay was demonstrated to be stability indicating. A sample of aspirin and warfarin was spiked with known potential degradation products and the formulation ingredients. These did not interfere with the analysis of aspirin and warfarin.

A calibration curve for aspirin was generated from 0.77 to 4.65 mg/mL. The correlation coefficient was 0.9996. The limit of detection for aspirin was 0.004%.

Reference
Montgomery ER, Taylor S, Segretario J, et al. Development and validation of a reversed-phase liquid chromatographic method for analysis of aspirin and warfarin in a combination tablet formulation. *J Pharm Biomed Anal.* 1996; 15: 73–82.

Method 2
Abuirjeie et al. developed a simultaneous analytical measurement of acetaminophen, acetyl-salicylic acid, caffeine, and *d*-propoxyphene. The liquid chromatograph consisted of a Beckman model 114M single-piston pump, a model 165 variable-wavelength UV detector, a 20-μL loop injector, and an SP4270 integrator. The stationary phase was an Alltech Synchropack RP C_8 column (250 × 4.1 mm, 6.5-μm particle size). The mobile phase consisted of 0.01 M sodium acetate solution and methanol (85:15, vol/vol) adjusted to pH 4.1. The flow rate was 1.5 mL/min. UV detection was performed at 254 nm and 0.1 AUFS. Phenacetin was used as an internal standard.

Samples were mixed with internal standard and then methanol. The injection volume was 20 μL. Under these conditions, retention times of acetaminophen, aspirin, *d*-propoxyphene, caffeine, and phenacetin were about 2.3, 4.1, 4.5, 6.4, and 11.3 minutes, respectively (estimated from the published chromatogram).

The method was reported to be stability indicating since it determined aspirin in the presence of salicylic acid, a potential major degradation product.

A standard curve for aspirin was constructed from 37.5 to 125 μg/mL. The correlation coefficient was 0.9903. The limit of detection was 9.6 μg/mL. The coefficient of variation for the analysis of aspirin was 1.02–2.33%.

Reference
Abuirjeie MA, Abdel-Hamid ME, Ibrahim EA. Simultaneous high-performance liquid chromatographic assay of acetaminophen, acetylsalicylic acid, caffeine, and *d*-propoxyphene hydrochloride. *Anal Lett.* 1989; 22: 365–75.

Atenolol

Chemical Names
4-[2-Hydroxy-3-[(1-methylethyl)amino]propoxy]benzeneacetamide
2-[*p*-[2-Hydroxy-3-(isopropylamino)propoxy]phenyl]acetamide

Other Name
Tenormin

Form	Molecular Formula	MW	CAS
Atenolol	$C_{14}H_{22}N_2O_3$	266.3	29122-68-7

Appearance
Atenolol is a white crystalline powder.

Solubility
Atenolol has a solubility of 26.5 mg/mL in water at 37 °C. It is freely soluble in methanol, soluble in acetic acid and dimethyl sulfoxide, sparingly soluble in 96% ethanol, slightly soluble in isopropanol, and very slightly soluble in acetone and dioxane. It is practically insoluble in acetonitrile, ethyl acetate, and chloroform.

pK_a
Atenolol has a pK_a of 9.6.

Method 1
Garner et al. studied the stability of atenolol in an extemporaneously compounded oral liquid. The liquid chromatograph consisted of a Waters model 510 solvent-delivery system, a Waters model 712 WISP autosampler, a Waters model 481 variable-wavelength UV detector, and a Waters model 745 data module. The stationary phase was a Waters Nova-Pak C_{18} Radial-Pak cartridge (100 × 5 mm) with a Nova-Pak C_{18} Guard-Pak. The mobile phase consisted of 10 mM potassium phosphate buffer adjusted to pH 3.2 with 0.2 M phosphoric acid and acetonitrile (4:1, vol/vol) containing 3 mM 1-octanesulfonic acid. The flow rate was 1.0 mL/min. UV detection was performed at 226 nm and 0.02 AUFS.

The extraction buffer was a saturated solution of sodium tetraborate adjusted to pH 9 with hydrochloric acid. The extraction solvent was a solution of dichloromethane and isopropanol (3:1, vol/vol). Sotalol 2.5 μg/mL was used as an internal standard. The injection volume was 30 μL. Retention times for atenolol and the internal standard were 3.70 and 5.56 minutes, respectively.

The stability-indicating capability of the method was demonstrated by accelerated degradation of atenolol. A solution of atenolol was adjusted to pH 0.67 with hydrochloric acid and stored for 2.5 hours at room temperature. A second solution of atenolol was alkalinized to pH 12.3 with potassium hydroxide solution and stored for 2.5 hours at room temperature. A third solution of atenolol was frozen at −10 °C for 14 hours. A fourth solution of atenolol was heated in an oven at 100 °C for 3 hours. In all of these cases, degradation product peaks did not interfere with the intact atenolol peak.

Standard curves for atenolol were constructed from 0.5 to 2.5 μg/mL; correlation coefficients were greater than 0.99. The intraday and interday coefficients of variation for the assay were 2.15 and 4.45%, respectively.

Reference
Garner SS, Wiest DB, Reynolds ER. Stability of atenolol in an extemporaneously compounded oral liquid. *Am J Hosp Pharm*. 1994; 51: 508–11.

Method 2

Sasa et al. reported the determination by reversed-phase HPLC of atenolol in combination with hydrochlorothiazide and chlorthalidone in tablet formulations. The chromatographic apparatus included a Varian model 2010 pump, a Rheodyne model 7125 10-μL loop injector, a Varian model 2050 spectrophotometric detector, and a Varian model 4290 integrator. The stationary phase was a Supelco Supelcosil LC-8-DB reversed-phase column (250 × 4.6 mm, 5-μm particle size). The mobile phase was a mixture of 1.0 mM ammonium acetate and 2.0 mM sodium octanesulfonate in acetonitrile and water (25:75, vol/vol) and was adjusted to pH 3.5 with glacial acetic acid. The flow rate was 1.5 mL/min. UV detection was performed at 254 nm and 0.2 AUFS. Methyl p-hydroxy-benzoate 10 μg/mL in methanol was used as an internal standard.

Twenty tablets were ground. A portion of this powder equivalent to one tablet was weighed, transferred into a 25-mL volumetric flask, sonicated with 20 mL of the internal standard for 3 minutes, diluted to 1.00 mg/mL of atenolol with internal standard, and filtered through a 0.45-μm membrane filter. The injection volume was 10 μL. Under these conditions, the retention times of hydrochlorothiazide, atenolol, chlorthalidone, and the internal standard were 3.5, 5.2, 5.9, and 8.4 minutes, respectively.

The method was verified to be stability indicating by studying a synthetic mixture of the drugs and their degradation products. All compounds were separated. No interference from any of the excipients, the other drugs, or the degradation products was observed.

A standard curve for atenolol was generated from 5 to 15 μg/mL. The correlation coefficient was greater than 0.999. The limit of detection was 200 pg.

Reference

Sasa SI, Jalal IM, Khalil HS. Determination of atenolol combinations with hydrochlorothiazide and chlorthalidone in tablet formulations by reverse phase HPLC. *J Liq Chromatogr*. 1988; 11: 1673–96.

Atorvastatin Calcium

Chemical Name

2-(4-Fluorophenyl)-β,δ-dihydroxy-5-(1-methylethyl)-3-phenyl-4-[(phenylamino)-carbo-nyl]-1H-pyrrole-1-heptanoic acid calcium

Other Name

Lipitor

Form	Molecular Formula	MW	CAS
Atorvastatin calcium	$C_{66}H_{68}CaF_2N_4O_{10}\cdot 3H_2O$	1209.4	134523-03-8

Appearance

Atorvastatin calcium occurs as a white to off-white crystalline powder.

Solubility

Atorvastatin calcium is very slightly soluble in distilled water, pH 7.4 phosphate buffer, and acetonitrile; slightly soluble in ethanol; and freely soluble in methanol.

Method

Mohammadi et al. reported a reversed-phase HPLC method for the simultaneous analysis of atorvastatin and amlodipine in tablets. The liquid chromatographic system consisted of a Knauer Smartline model 1000 pump, a Knauer injection system, a Knauer solvent degasser, and a Waters model 486 tunable absorbance detector. The stationary phase was a Perfectsil Target ODS-3 column (250 × 4.6 mm, 5-μm particle size). The mobile phase consisted of acetonitrile and 0.025 M monobasic ammonium phosphate buffer (pH 4.5) (55:45, vol/vol). The isocratic flow rate was 1 mL/min. The eluent was monitored at 237 nm.

Atorvastatin stock solution was prepared in methanol. For tablets, a portion of powder from 20 tablets, equivalent to the weight of one tablet, was accurately weighed into each of nine 100-mL volumetric flasks, mixed with 20 mL of methanol, sonicated for 20 minutes, made up to volume with water, filtered through a 0.45-μm nylon filter, and further diluted by a factor of 10 with mobile phase. The injection volume was 20 μL. Under these conditions, retention times of amlodipine and atorvastatin were about 4.2 and 9.5 minutes, respectively.

The method was demonstrated to be stability indicating by accelerated degradation studies. An atorvastatin solution was prepared in methanol and 0.1 M hydrochloric acid (20:80, vol/vol) and incubated for 5 minutes. An atorvastatin solution was prepared in methanol and 0.1 M sodium hydroxide solution and incubated for 5 minutes. Another solution was prepared in methanol and water (20:80, vol/vol) and incubated at 90 °C for 20 minutes. Atorvastatin tablets were exposed to dry heat at 100 °C in a convection oven for 8 hours. Atorvastatin solutions were also exposed to UV light at 320–400 nm at 25 °C for 40 hours. In all cases, the peak of the intact drug was well resolved from peaks of degradation products.

Linear calibration curves for atorvastatin were constructed from 2 to 30 μg/mL. The correlation coefficients were greater than 0.999. Intraday and interday coefficients of variation were less than 2.4 and 3.7%, respectively. The limit of detection and the limit of quantitation were less than 0.65 and 2 μg/mL, respectively.

Reference

Mohammadi A, Rezanour N, Ansari Dogaheh M, et al. A stability-indicating high performance liquid chromatographic (HPLC) assay for the simultaneous determination of atorvastatin and amlodipine in commercial tablets. *J Chromatogr B*. 2007; 846: 215–21.

Atracurium Besylate

Chemical Name

2,2′-(3,11-Dioxo-4,10-dioxatridecamethylene)bis(1,2,3,4-tetrahydro-6,7-dimethoxy-2-
   methyl-1-veratrylisoquinolinium) di(benzenesulfonate)

Other Name
Tracrium

Form

Form	Molecular Formula	MW	CAS
Atracurium besylate	$C_{53}H_{72}N_2O_{12}.2C_6H_5O_3S$	1243.5	64228-81-5

Appearance
Atracurium besylate is a white to pale yellow powder.

Solubility
Atracurium besylate has solubilities of 50, 200, and 35 mg/mL in water, alcohol, and 0.9% sodium chloride, respectively, at 25 °C.

Method
Pramar et al. investigated by HPLC the chemical stability and adsorption of atracurium besylate injections in disposable plastic syringes. A Waters model 590 liquid chromatograph was equipped with a Waters model 710 WISP autosampler, a Hitachi model 100-40 multiple-wavelength detector, and a Kipp & Zonen model BD41 recorder. The stationary phase was a Waters Partisil 5 column (300 × 3.9 mm). The mobile phase consisted of acetonitrile, water, and phosphoric acid (specific gravity 1.75) (900:90:10, vol/vol/vol). The flow rate was 0.6 mL/min. UV detection was performed at 280 nm and 0.05 AUFS. Each sample was diluted with mobile phase. The injection volume was 15 μL.

The method was stated to be stability indicating.

A standard curve for atracurium besylate was constructed and the correlation coefficient was 0.997.

A similar method was used by Russell and Meyer-Witting.

References
Pramar YV, Loucas VA, Word D. Chemical stability and adsorption of atracurium besylate injections in disposable plastic syringes. *J Clin Pharm Ther*. 1996; 21: 173–5.

Russell WJ, Meyer-Witting M. The stability of atracurium in clinical practice. *Anaesth Intens Care*. 1990; 18: 550–2.

Atrasentan

Chemical Name

4-(1,3-Benzodioxol-5-yl)-1-[2-(dibutylamino)-2-oxoethyl]-2-(4-methoxyphenyl)-3-pyrro-lidinecarboxylic acid monohydrochloride

Other Name

ABT-627

Form	Molecular Formula	MW	CAS
Atrasentan hydrochloride	$C_{29}H_{38}N_2O_6$.HCl	547.1	195733-43-8

Method

Morley et al. described the determination by HPLC analysis of the endothelin receptor antagonist atrasentan and related substances. The instrument consisted of a Thermo-Separation model P-2000 binary pump, a model 100 variable-wavelength detector, a Shimadzu model SIL-10A autosampler, and a Shimadzu model C-R4AX datastation. The stationary phase was a Phenomenex Inersil ODS-2 column (150 × 4.6 mm, 5-μm particle size). The mobile phase was 0.05 M potassium phosphate buffer (adjusted to pH 3.0 with phosphoric acid) and acetonitrile. It was gradiently delivered at a flow rate of 1 mL/min as follows:

Time, minutes	Buffer, %	Acetonitrile, %
0	53	47
15	53	47
25	20	80
50	20	80
55	53	47
70	53	47

UV detection was performed at 234 nm and 0.10 AUFS. *p*-Diethoxybenzene 1.2 mg/mL in a diluent was used as an internal standard.

The diluent was a mixture of 0.01 M phosphate buffer and acetonitrile (1:1, vol/vol). For capsules, the composite was extracted by shaking with the diluent for

15 minutes, and this solution was then diluted to approximately 1.2 mg/mL of the drug. The injection volume was 20 µL. Under these conditions, retention times of atrasentan and the internal standard were about 8.3 and 14.0 minutes, respectively.

The stability-indicating nature of the method was assessed by spiking the solution with known related substances. Atrasentan was separated from other compounds.

A standard curve was constructed from 24 to 118 µg/mL. The correlation coefficient was greater than 0.9999.

Reference

Morley JA, Elrod L Jr., Schmit JL, et al. Determination of the endothelin receptor antagonist ABT-627 and related substances by high performance liquid chromatography. *J Pharm Biomed Anal*. 1999; 19: 777–84.

Atropine

Chemical Name

endo-(±)-α-(Hydroxymethyl)benzeneacetic acid 8-methyl-8-azabicyclo[3.2.1]oct-3-yl ester

Form	Molecular Formula	MW	CAS
Atropine	$C_{17}H_{23}NO_3$	289.4	51-55-8
Atropine sulfate	$(C_{17}H_{23}NO_3)_2.H_2SO_4.H_2O$	694.8	5908-99-6

Appearance

Atropine appears as usually needlelike white crystals or as a white crystalline powder. Atropine sulfate occurs as colorless crystals or as a white crystalline powder.

Solubility

Atropine has solubilities of ~ 21.7 mg/mL in water and 0.5 g/mL in alcohol at 25 °C. Atropine sulfate has solubilities of ~ 2 g/mL in water and 0.2 g/mL in alcohol at 25 °C.

pK$_a$

Atropine has a pK$_a$ of 9.8.

Method 1

Wu et al. reported an HPLC method for the determination of atropine in morphine–atropine injections. The Shimadzu liquid chromatograph consisted of a model

LC-6A solvent-delivery pump, a model SIL-6B autoinjector, a model C-R4A recorder, a model SCL-6B system controller, and a model SPD-M6A photodiode-array detector. The stationary phase was a Nucleosil C_{18} column (300 × 4.0 mm, 10-μm particle size). The mobile phase consisted of methanol and an aqueous solution (30:70, vol/vol), composed of 1% ammonium acetate, 0.8% triethylamine, and 0.017% sodium 1-heptanesulfonate in water. The flow rate was 1.0 mL/min. UV detection was performed at 258 nm. The injection volume was 20 μL. Under these conditions, the retention time of atropine was 11.1 minutes.

The stability-indicating nature of the method was demonstrated by accelerated degradation of atropine. Atropine sulfate was treated with 0.1% hydrogen peroxide at 80 °C for 6 hours, 0.01 N sodium hydroxide at 30 °C for 4 hours, or 0.1 N hydrochloric acid at 80 °C for 25 hours and was exposed to light (15-Watt UV lamp and two 15-Watt fluorescent lamps) for 162 days. The atropine peak was well separated from its degradation product peaks.

A linear relationship between peak area and concentration of atropine was obtained with a correlation coefficient of 0.9999. Intraday and interday coefficients of variation were all less than 2%.

Reference
Wu S-C, Chen F-S, Li J-H. Stability-indicating HPLC method for the determination of morphine and atropine in morphine–atropine injection. *Chin Pharm J*. 1995; 47: 457–68.

Method 2
Wilson and Forde determined the stability of atropine sulfate using an HPLC method. The liquid chromatograph consisted of a Varian model 5000 pump and autosampler, a Waters model 441 UV detector, a Fisher Scientific model 5000 strip-chart recorder, and a Hewlett-Packard model 3354 laboratory automation system. The stationary phase was a Whatman PXS ODS-3 stainless steel column (250 × 4.6 mm, 10-μm particle size). The mobile phase was a mixture of 250 mL of acetonitrile and 750 mL of an aqueous solution. This aqueous solution contained 1.08 g of sodium octanesulfonate in 1000 mL of water with the pH adjusted to 3.5 with acetic acid. The flow rate was 2.0 mL/min. UV detection was performed at 229 nm and 0.01 AUFS.

Samples were diluted to about 0.2 mg/mL with water. The injection volume was 20 μL. Retention times for milrinone, atropine excipient, and atropine were about 3, 3.5, and 5 minutes, respectively.

This method was stated to be stability indicating.

A standard curve was constructed from 0 to 0.331 mg/mL of atropine sulfate.

Reference
Wilson TD, Forde MD. Stability of milrinone and epinephrine, atropine sulfate, lidocaine hydrochloride, or morphine sulfate injection. *Am J Hosp Pharm*. 1990; 47: 2504–7.

Azacitidine

Chemical Name
4-Amino-1-β-D-ribofuranosyl-1,3,5-triazine-2(1*H*)-one

Other Name
Vidaza

Form	Molecular Formula	MW	CAS
Azacitidine	$C_8H_{12}N_4O_5$	244.2	320-67-2

Method 1
Using HPLC, Hartigh et al. studied the stability of azacitidine 0.5 and 2.0 mg/mL in lactated Ringer's injection frozen in polypropylene syringes. A Spectra-Physics liquid chromatograph model 3500B was equipped with a Rheodyne model 7125 syringe-loading sample injector with a fixed-volume loop injector valve, a Pye Unicam LC-UV variable-wavelength detector, and a Spectra-Physics 4270 integrator. The stationary phase was a Chrompack Chromspher C_{18} column (100 × 3 mm, 5-μm particle size). The mobile phase consisted of a 0.01 M phosphate buffer at pH 6.5. The flow rate was 0.4 mL/min. UV detection was performed at 240 nm.

The stability-indicating capability of the method was confirmed by chromatographing an aqueous solution of azacitidine that was stored at room temperature for 24 hours. The peaks of the decomposition products, *N*-formylribosylguanylurea and ribosylguanylurea, did not interfere with the peak of azacitidine. Retention times for ribosylguanylurea, *N*-formylribosylguanylurea, and azacitidine were 110, 140, and 220 seconds, respectively.

Reference
Hartigh JD, Brandenburg HCR, Vermeij P. Stability of azacitidine in lactated Ringer's injection frozen in polypropylene syringes. *Am J Hosp Pharm.* 1989; 46: 2500–5.

Method 2
Cheung et al. used HPLC to evaluate the stability of azacitidine in several infusion fluids. The chromatograph included a Spectra-Physics model 3500B constant-flow pump, a Valco model CV-6-UHPa-N6O rotary-valve injector with a 10-μL injection loop, a Spectra-Physics model SP8200 UV detector, an Omniscribe strip-chart recorder, and a Waters C_{18} Radial-Pak cartridge (115 × 8 mm, 5-μm particle size). The mobile phase was a 0.01 M phosphate buffer (pH 6) at a flow rate of 2 mL/min. UV detection was performed at 254 nm. Uridine was used as an internal standard.

Samples were diluted with internal standard solution. Retention volumes were 8.0, 14, and 19 mL for the hydrolysis product, azacitidine, and uridine, respectively.

The HPLC method was determined to be stability indicating using partially and totally decomposed solutions of azacitidine. The azacitidine peak was well separated from all other degradation product peaks.

A calibration curve for azacitidine was constructed by plotting the peak height ratios of azacitidine to the internal standard from 50 to 200 µg/mL. The correlation coefficient was greater than 0.998.

Reference

Cheung Y-W, Vishnuvajjala BR, Morris NL, et al. Stability of azacitidine in infusion fluids. *Am J Hosp Pharm.* 1984; 41: 1156–9.

Azathioprine

Chemical Names

6-[(1-Methyl-4-nitro-1*H*-imidazol-5-yl)thio]-1*H*-purine

6-(1-Methyl-4-nitro-5-imidazolyl)mercaptopurine

Other Name

Imuran

Form	Molecular Formula	MW	CAS
Azathioprine	$C_9H_7N_7O_2S$	277.3	446-86-6
Azathioprine sodium	$C_9H_6N_7NaO_2S$	299.3	55774-33-9

Appearance

Azathioprine is a pale yellow odorless powder. Azathioprine sodium is a bright yellow amorphous mass or cake.

Solubility

Azathioprine is slightly soluble in water, ethanol, and chloroform. Azathioprine sodium is soluble in water.

pK$_a$

The pK$_a$ has been cited as 8.2 and 7.87.

Method 1

Allen and Erickson studied the stability of azathioprine 50 mg/mL in extemporaneously compounded oral liquids. The chromatographic system consisted of a Hewlett-Packard series 1050 automated liquid chromatograph including a multisolvent mixing and pumping system, an autosampler, a diode-array detector, and a computer with Chem Station software. The stationary phase was a Bakerbond C_{18} analytical column (250 × 4.6 mm). The mobile phase consisted of 700 mL of water, 300 mL of methanol, and 1.1 g of sodium 1-heptanesulfonate with pH adjusted to 3.5 with 1 N hydrochloric acid. The flow rate was 2.0 mL/min. UV detection was performed at 254 nm.

Samples were diluted 1:100. The injection volume was 20 µL. The retention time for azathioprine was 3.7 minutes.

The stability-indicating ability of the method was demonstrated by accelerated degradation of azathioprine. Samples of azathioprine were subjected to heat, base, acid, an oxidizing agent, or light. The resulting chromatogram showed no interference with the intact azathioprine peak from the decomposition product peaks.

The standard curve was constructed from 100 to 600 µg/mL. The intraday and interday coefficients of variation were 1.3 and 1.4%, respectively.

Reference

Allen LV, Erickson MA. Stability of acetazolamide, allopurinol, azathioprine, clonazepam, and flucytosine in extemporaneously compounded oral liquids. *Am J Health Syst Pharm.* 1996; 53: 1944–9.

Method 2

Fell et al. reported a simple and precise reversed-phase HPLC assay for the rapid analysis of azathioprine and 6-mercaptopurine. The chromatograph included an Applied Chromatography Systems constant flow pump and a Cecil Instruments CE-212 variable-wavelength UV monitor. The stationary phase was a Shandon-Southern Instruments Hypersil ODS stainless steel column (100 × 5 mm, 5-µm particle size). The mobile phase was a mixture of methanol, 25 mM monobasic potassium phosphate, and glacial acetic acid (20:79.5:0.5, vol/vol/vol), adjusted to pH 4.5. The flow rate was 1.5 mL/min. Theophylline was used as an internal standard. UV detection was performed at 240 nm and 0.02 AUFS.

Samples of azathioprine injection were dissolved in 0.02 M sodium hydroxide solution. Samples of azathioprine tablets were crushed, mixed with 0.02 M sodium hydroxide solution, and filtered. Solutions of azathioprine in 0.02 M sodium hydroxide solution were added to theophylline 150 µg/mL and diluted with 0.02 M sodium hydroxide solution to about 2 mg/mL of azathioprine before injection. The injection volume was 1.5 µL. Under these conditions, the retention times for 6-mercaptopurine, theophylline, and azathioprine were about 1.7, 3.5, and 4.2 minutes, respectively (estimated from the published chromatogram).

The assay was stability indicating since azathioprine was well resolved from all its known potential impurities and degradation products.

The limit of detection for azathioprine was 0.5 ng.

Reference

Fell AF, Plag SM, Neil JM. Stability-indicating assay for azathioprine and 6-mercaptopurine by reversed-phase high-performance liquid chromatography. *J Chromatogr.* 1979; 186: 691–704.

Azintamide

Chemical Name
2-[(6-Chloro-3-pyridazinyl)thio]-*N*,*N*-diethylacetamide

Other Names
Colerin, Oragalin, Ora-Gallin

Form	Molecular Formula	MW	CAS
Azintamide	$C_{10}H_{14}ClN_3OS$	259.8	1830-32-6

Method
Al-Deeb et al. established a sensitive HPLC method for the determination of azintamide in the presence of its degradation products. The Waters chromatograph included a model U6K injector, model 486 tunable absorbance detector, a model 746 data module, and a model 600E system controller. The stationary phase was a Merck LiChrospher 100RP18 column (150 × 4.6 mm, 5-µm particle size). The mobile phase was a mixture of methanol and water (49:51, vol/vol) and was isocratically delivered at 1 mL/min. UV detection was performed at 254 nm.

A powder of ground tablets equivalent to 10 mg of azintamide was weighed, transferred into a 100-mL volumetric flask, mixed with 75 mL of the mobile phase, extracted for 5 minutes, brought to volume, and filtered. The filtrate was diluted with the mobile phase to about 4 µg/mL of azintamide. Under these conditions, the retention time for azintamide was about 3.7 minutes.

The method was evaluated to be stability indicating by accelerated degradation of azintamide. The azintamide sample was incubated in 1 N hydrochloric acid at 40 °C for 30 minutes and a yellow coloration appeared. The intact azintamide had a retention time of 3.7 minutes and was well resolved from its degradation products with retention times of 1.1 and 1.9 minutes.

Reference
Al-Deeb OA, El-Subbagh HI, Abdel-Moety EM, et al. Stability-indicating quantitation of azintamide in dosage formulations. *Sci Pharm*. 1998; 66: 47–57.

Azithromycin

Chemical Name
13-[(2,6-Dideoxy-3-*C*-methyl-3-*O*-methyl-α-L-*ribo*-hexopyranosyl)oxy]-2-ethyl-3,4,10-trihydroxy-3,5,6,8,10,12,14-heptamethyl-11-[[3,4,6-trideoxy-3-(dimethylamino)-β-D-*xylo*-hexopyranosyl]oxy]-1-oxa-6-azacyclopentadecan-15-one dihydrate

• 2H$_2$O

Other Names
Robitrex, Trozocina, Zithromax

Form	Molecular Formula	MW	CAS
Azithromycin	C$_{38}$H$_{72}$N$_2$O$_{12}$	749.0	83905-01-5
Azithromycin dihydrate	C$_{38}$H$_{72}$N$_2$O$_{12}$.2H$_2$O	785.0	117772-70-0

Appearance
Azithromycin dihydrate is a white or almost white powder.

Solubility
Azithromycin dihydrate has a solubility of 39 mg/mL in water (pH 7.4) at 37 °C.

Method
Zubata et al. reported a method for the quantitation of azithromycin. The instrument included a Konik model KNK-500G dual-piston reciprocating pump, a model KNK-029-757 UV-visible detector, a model SP 4600 integrator, and a Rheodyne model 7125 injector. The stationary phase was a Merck LiChrospher 100 RPS select B column (125 × 4.6 mm, 5-μm particle size). The mobile phase was a mixture of phosphate buffer, acetonitrile, and methanol (60:20:20, vol/vol/vol) adjusted to pH 8.0 with phosphoric acid. The phosphate buffer was prepared by dissolving 2.88 g of dibasic ammonium phosphate in 500 mL of water, adding 45.6 mL of a 10% solution of tetrabutyl-ammonium phosphate in water, and diluting with water to 1000 mL. The flow rate was 1.0 mL/min. UV detection was performed at 215 nm.

A portion of ground tablets (20) was accurately weighed, suspended in mobile phase, sonicated for 2 minutes, and filtered through a 0.45-μm membrane filter. The injection volume was 20 μL. Under these conditions, the retention time for azithromycin was about 5 minutes.

The method was evaluated to be stability indicating by accelerated degradation of the drug. Azithromycin was partially degraded following alkaline and acid hydrolysis, reduction, oxidation, and photolysis. Neither formulation ingredients nor degradation products interfered with the determination of azithromycin.

A calibration curve for azithromycin was constructed from 0.6 to 3.0 mg/mL. The correlation coefficient was 0.9994.

Reference

Zubata P, Ceresole R, Rosasco MA, et al. A new HPLC method for azithromycin quantitation. *J Pharm Biomed Anal.* 2002; 27: 833–6.

Azlocillin

Chemical Name

Sodium (6R)-6-[D-2-(2-oxoimidazolidine-1-carboxamido)-2-phenylacetamido]penicillanate

Other Name

Securopen

Form	Molecular Formula	MW	CAS
Azlocillin sodium	$C_{20}H_{22}N_5NaO_6S$	483.5	37091-65-9

Appearance

Azlocillin sodium occurs as a white to pale yellow hygroscopic powder.

Solubility

Azlocillin sodium is soluble in water and very soluble in methanol. It is practically insoluble in acetone, chloroform, and ether.

Method

Using HPLC, Grover et al. investigated the degradation rate of azlocillin at 35 °C in borate buffer (pH 9.2). A Shimadzu LC-10A liquid chromatograph included model LC-10AS pumps, a model SPD-10A dual-wavelength detector, a model C-R7A data processor, and a Rheodyne model 7125 injector with a 20-μL loop. The stationary phase was a Phenomenex C_{18} column (300 × 3.9 mm, 5-μm particle size). The mobile phase was a mixture of 86% water containing 20 mM monobasic potassium phosphate and 10 mM tetramethylammonium chloride (pH 5.0) and 14% acetonitrile. The flow rate was 1.0 mL/min. UV detection was performed at 220 and 240 nm.

Samples were incubated at 35 °C and then assayed. Under these conditions, the retention time for azlocillin was 14.8 minutes. The method was described as stability indicating.

A standard curve for azlocillin was generated from 5 to 150 μg/mL. The correlation coefficient was greater than 0.99. The coefficient of variation was less than 1%.

Reference

Grover M, Gulati M, Singh B, et al. Correlation of penicillin structure with rate constants for basic hydrolysis. *Pharm Pharmacol Commun.* 2000; 6: 355–63.

Aztreonam

Chemical Name
[2S-[2α,3α(Z)]]-2-[[[1-(2-Amino-4-thiazolyl)-2-[(2-methyl-4-oxo-1-sulfo-3-azetidinyl)-amino]-2-oxoethylidene]amino]oxy]-2-methylpropanoic acid

Other Names
Azactam, Primbactam, Urobactam

Form	Molecular Formula	MW	CAS
Aztreonam	$C_{13}H_{17}N_5O_8S_2$	435.4	78110-38-0

Appearance
Aztreonam is a white crystalline powder.

Solubility
Aztreonam has solubilities of 10 mg/mL in water and 0.2 mg/mL in alcohol at 25 °C. It is soluble in dimethylformamide and dimethyl sulfoxide. It is practically insoluble in toluene, chloroform, and ethyl acetate.

pK$_a$
Aztreonam has pK$_a$ values of –0.7, 2.75, and 3.91 at 25 °C.

Method 1
Trissel et al. used HPLC to evaluate the compatibility and stability of vancomycin hydrochloride and aztreonam in 5% dextrose and 0.9% sodium chloride injections. The chromatographic system consisted of a Waters model 600E multisolvent-delivery pump, a Waters 490E programmable multiple-wavelength UV detector, a Waters model 712 WISP autosampler, and a Separations Group Vydac reversed-phase C$_8$ analytical column (250 × 4.6 mm, 5-μm particle size). The system was controlled and integrated by an NEC PowerMate SX/16 personal computer running Waters Maxima 820 chromatography manager. The mobile phase consisted of 85% 0.05 M monobasic potassium phosphate buffer with the pH adjusted to 3.0 with phosphoric acid and 15% methanol. The flow rate was 1.0 mL/min. UV detection was performed at 270 nm.

Samples were diluted with the mobile phase. The injection volume was 15 μL. Under these conditions, retention times for vancomycin and aztreonam were 6.6 and 7.8 minutes, respectively.

The HPLC assay was determined to be stability indicating by accelerated degradation of aztreonam. Adding 0.5 N sodium hydroxide to a sample, followed by boiling for 2 hours, led to a reduction in the peak for the intact aztreonam and the formation of a new peak at 4.5 minutes. Addition of 1 N hydrochloric acid to the aztreonam solution, followed by boiling for 2 hours, led to a reduction of the intact aztreonam peak and new peaks at about 4 and 6 minutes.

A standard curve was constructed from a linear plot of peak area versus aztreonam concentration from 0.025 to 0.150 mg/mL. The correlation coefficient was greater than 0.9998. The intraday and interday coefficients of variation were 3.90 and 3.11%, respectively.

Reference
Trissel LA, Xu QA, Martinez JF. Compatibility and stability of aztreonam and vancomycin hydrochloride. *Am J Health Syst Pharm.* 1995; 52: 2560–4.

Method 2
Belliveau et al. studied the stability of aztreonam with ampicillin sodium and sulbactam sodium in 0.9% sodium chloride injection. The HPLC system consisted of a Waters model 510 pump, a Waters model 710A WISP autosampler, an LDC analytical model SM 4000 variable-wavelength detector, and a Hewlett-Packard model 3396 series II integrator. The stationary phase was a Waters μBondapak C_{18} column (300 × 3.9 mm). The mobile phase was a mixture of 0.005 M tetrabutylammonium hydrogen sulfate, 7% acetonitrile, and 14% methanol in a 0.010 M phosphate buffer (pH 2.6–2.7). The flow rate was 1 mL/min. UV detection was performed at 225 nm.

Samples were diluted 1:8 with HPLC-grade water. The injection volume was 20 μL. Cimetidine was used as the internal standard. Under these conditions, the retention times for cimetidine, ampicillin, sulbactam, and aztreonam were 3.27, 4.35, 5.79, and 7.05 minutes, respectively.

The HPLC method was determined to be stability indicating. Solutions of aztreonam 2 mg/mL were stored in a water bath (70 °C) for 8 days and then were analyzed by the HPLC method. The degradation peaks increased in height and area over 8 days, but they did not interfere with the intact drug peak.

Standard curves were constructed for aztreonam from 0.5 to 2 mg/mL. The correlation coefficient was greater than 0.9977. Intraday and interday coefficients of variation were less than 7%.

Reference
Belliveau PP, Nightingale CH, Quintiliani R. Stability of aztreonam and ampicillin sodium–sulbactam sodium in 0.9% sodium chloride injection. *Am J Hosp Pharm.* 1994; 51: 901–4.

Method 3
Using an HPLC method, Marble et al. investigated the stability of aztreonam with clindamycin phosphate in 5% dextrose injection and 0.9% sodium chloride injection. The system consisted of an Altex model 110A solvent-metering pump, a model 210 injector (or a Waters WISP 710B autoinjector), a Waters Lambda-Max model 481 LC spectro-photometer, an Altex model 153 analytical UV detector, a Linear Instruments model 555 recorder, or a Hewlett-Packard model 3392A integrator. The stationary phase was an

Alltech C_8 reversed-phase analytical column (250 × 4.6 mm, 10-μm particle size). The mobile phase was 23% acetonitrile in a 0.1 M acetic acid aqueous buffer (710 mL of 0.1 M acetic acid and 60 mL of sodium acetate per liter) to which 1.79 g of tetrabutylammonium hydrogen sulfate was added. Its pH was adjusted to 3.5 with 5 N sodium hydroxide. The mobile phase was delivered isocratically at 2 mL/min. UV detection was performed at 254 nm and 0.64 AUFS. Cefoperazone sodium 2 mg/mL was used as the internal standard.

Samples were diluted 1:10 with internal standard solution. The injection volume was 10 μL. Under these conditions, retention times for aztreonam and the internal standard were 8 and 13.5 minutes, respectively.

The stability-indicating capability of the assay was verified by an accelerated degradation study of aztreonam. Acidification (pH 2.0) or heating of aztreonam solutions resulted in degradation products that did not interfere with the determination of the intact drug.

Intrarun and interrun coefficients of variation were 2.0 and 2.6%, respectively.

A similar method was used by Bosso et al.

References

Marble DA, Bosso JA, Townsend RJ. Stability of clindamycin phosphate with aztreonam, ceftazidime sodium, ceftriaxone sodium, or piperacillin sodium in two intravenous solutions. *Am J Hosp Pharm.* 1986; 43: 1732–6.

Bosso JA, Prince RA, Fox JL. Compatibility of ondansetron hydrochloride with fluconazole, ceftazidime, aztreonam, and cefazolin sodium under simulated Y-site conditions. *Am J Hosp Pharm.* 1994; 51: 389–91.

Method 4

James and Riley studied by HPLC the stability of aztreonam with ampicillin in intravenous admixtures. The liquid chromatograph consisted of a ConstaMetric III G pump, a SpectroMonitor D detector, a Negretti model 190 injection valve fitted with a 20-μL loop, and a Fisher series 500 Recordall dual-pen recorder. The stationary phase was a Shandon ODS Hypersil column (150 × 4.6 mm) with a presaturator Permaphase ODS column. The mobile phase consisted of 22.5% methanol in an aqueous solution of 5 mM tetrabutylammonium hydrogen sulfate and 5 mM ammonium sulfate (pH 2.6). The flow rate was 1.5 mL/min and the UV detector was set at 238 nm.

Samples were centrifuged at 2500 rpm for 5 minutes and then diluted 1:100 with 5 mM phosphate buffer (pH 6.0). Under these conditions, the retention times for ampicillin and aztreonam were 4.8 and 10.5 minutes, respectively.

The method was confirmed to be stability indicating. The products generated from the reaction of aztreonam with dextrose were well separated from aztreonam.

Calibration curves were constructed for aztreonam from 20 to 200 μg/mL. The intraday and interday reproducibilities of peak areas were 1.20 and 3.32%, respectively.

Similar methods were used by the other researchers cited here.

References

James MJ, Riley CM. Stability of intravenous admixtures of aztreonam and ampicillin. *Am J Hosp Pharm.* 1985; 42: 1095–100.

McLaughlin JP, Simpson C. The stability of reconstituted aztreonam. *Pract Res.* 1990; 328–34.

Riley CM, James MJ. Stability of intravenous admixtures containing aztreonam and cefazolin. *Am J Hosp Pharm.* 1986; 43: 925–7.

Riley CM, Lipford LC. Interaction of aztreonam with nafcillin in intravenous admixtures. *Am J Hosp Pharm.* 1986; 43: 2221–4.

Bell RG, Lipford LC, Massanari MJ, et al. Stability of intravenous admixtures of aztreonam and cefoxitin, gentamicin, metronidazole, or tobramycin. *Am J Hosp Pharm.* 1986; 43: 1444–53.

Method 5

James and Riley evaluated the stability of intravenous admixtures of aztreonam and clindamycin phosphate using HPLC. The liquid chromatograph included a ConstaMetric III G pump, a SpectroMonitor D detector, a Negretti model 190 injection valve fitted with a 20-µL loop, and a Fisher series 5000 Recordall dual-pen recorder. The stationary phase was a Varian Micropak Si-5 silica column. The mobile phase was 0.1% phosphoric acid and 3% acetonitrile in water. The flow rate was 1.0 mL/min. UV detection was performed at 205 nm.

Samples were centrifuged at 2500 rpm for 5 minutes, and then the supernatants were diluted 1:100 in phosphate buffer (pH 5.5). Under these conditions, the retention times for aztreonam, arginine, benzyl alcohol, and clindamycin were 2.90, 3.45, 3.65, and 14.60 minutes, respectively.

This analytical method was determined to be stability indicating by spiking the sample solution with authentic samples of the *E*-isomer and the open-ring form of aztreonam. The impurity and the degradation product peaks did not interfere with the intact aztreonam peak.

Calibration curves were constructed for aztreonam from 0 to 200 µg/mL. The correlation coefficient was greater than 0.999.

Reference

James MJ, Riley CM. Stability of intravenous admixtures of aztreonam and clindamycin phosphate. *Am J Hosp Pharm.* 1985; 42: 1984–6.

Bacitracin

$$CH_3$$

(chemical structure)

H_3C

H_2N

S

N

O

His — D-Asp — Asn   Leu

D-Phe   ε   D-Glu

Ile — D-Orn — Lys — α — Ile

Form	Molecular Formula	MW	CAS
Bacitracin A	$C_{66}H_{103}N_{17}O_{16}S$	1422.7	1405-87-4

Appearance

Bacitracin occurs as a white to pale buff hygroscopic powder.

Solubility

Bacitracin is freely soluble in water and soluble in alcohol. It is practically insoluble in ether, chloroform, and acetone.

Method

Pavli and Kmetec reported a stability-indicating HPLC method for stability testing of bacitracin. The Hewlett-Packard model 1050 chromatograph consisted of a binary gradient pump, a diode-array detector, an autosampler, and a column heater. The stationary phase was a Kromasil C_8 column (150×4.6 mm, 5-μm particle size). Mobile phase A consisted of 51 parts of 0.05 M monobasic potassium phosphate buffer (pH 6.0) and 49 parts of a mixture of methanol and acetonitrile (1:1, vol/vol). Mobile phase B consisted of 40 parts of 0.05 M monobasic potassium phosphate buffer (pH 6.0) and 60 parts of a mixture of methanol and acetonitrile (1:1, vol/vol). The flow rate was 1.4 mL/min. The mobile phase was 100% mobile phase A from 0 to 7.5 minutes, linearly changed to 100% mobile phase B from 7.5 to 12 minutes, held at 100% mobile phase B from 12 to 18 minutes, and then equilibrated again with 100% mobile phase A from 18.1 to 22 minutes. UV detection was performed at 254 nm. The injection volume was 10 μL. The column temperature was 30 ℃. Under these conditions, retention times for bacitracin B_1, B_2, and A were about 6.6, 7.9, and 10.4, minutes, respectively.

The assay was demonstrated to be stability indicating based on peak purity analysis with photodiode-array detection. There was no interference among bacitracin A, B_1, B_2, impurities, decomposition products (bacitracin in water at 50 ℃ for 3 days), and ballast materials in the entire UV range.

A standard curve was constructed from 80 to 155% of the stated potency of bacitracin. The correlation coefficient was 0.999.

A similar method was reported by the same authors.

References

Pavli V, Kmetec V. Optimization of HPLC method for stability testing of bacitracin. *J Pharm Biomed Anal*. 2001; 24: 977–82.

Pavli V, Kmetec V. Fast separation of bacitracin on monolithic silica columns. *J Pharm Biomed Anal*. 2004; 36: 257–64.

Baclofen

Chemical Names

β-(Aminomethyl)-4-chlorobenzenepropanoic acid
β-(Aminomethyl)-*p*-chlorohydrocinnamic acid

Other Name

Lioresal

Form	Molecular Formula	MW	CAS
Baclofen	$C_{10}H_{12}ClNO_2$	213.7	1134-47-0

Appearance

Baclofen occurs as white to off-white crystals.

Solubility

Baclofen is slightly soluble in water and very slightly soluble in methanol.

pK$_a$

Baclofen has pK$_a$ values of 5.4 and 9.5.

Method

Using a stability-indicating HPLC assay, Johnson and Hart studied the stability of an extemporaneously compounded baclofen oral liquid 5 mg/mL. The chromatograph consisted of a Waters model 501 pump, a Waters model U6K variable-volume injector, a Waters model 484 UV detector, and a Hewlett-Packard model 3394 recording integrator. The stationary phase was an Applied Biosystems Spheri-5 ODS column (250 × 4.6 mm, 5-μm particle size). The mobile phase was a mixture of 0.05 M monobasic sodium phosphate buffer and acetonitrile (80:20, vol/vol) adjusted to pH 3.5 with 85% phosphoric acid. The flow rate was 1.0 mL/min. UV detection was performed at 220 nm and 0.1 AUFS.

Samples were diluted with deionized distilled water to about 5 µg/mL and filtered through 0.22-µm Millex-GV filters. The injection volume was 15 µL. The retention time for baclofen was about 5.5 minutes (estimated from the published chromatogram).

The HPLC assay was determined to be stability indicating by accelerated decomposition of baclofen. The induced degradation of baclofen by heating an aqueous solution of baclofen 1 mg/mL to 100 °C for 60 minutes after adjustment to pH 2 with 1 N sulfuric acid or to pH 12 with 1 N sodium hydroxide did not produce any degradation product peaks that interfered with the peak of the intact baclofen.

A standard curve was constructed on each day of sample analysis by linear regression of the peak heights of baclofen against its concentration from 2 to 8 µg/mL. The correlation coefficient was greater than 0.997. The intraday and interday coefficients of variation were 1.9 and 3.0%, respectively.

A similar method was used by Allen and Erickson.

References

Johnson CE, Hart SM. Stability of an extemporaneously compounded baclofen oral liquid. *Am J Hosp Pharm.* 1993; 50: 2353–5.

Allen LV, Erickson MA. Stability of baclofen, captopril, diltiazem hydrochloride, dipyridamole, and flecainide acetate in extemporaneously compounded oral liquids. *Am J Health Syst Pharm.* 1996; 53: 2179–84.

Beclomethasone Dipropionate

Chemical Name
9α-Chloro-11β,17α,21-trihydroxy-16β-methylpregna-1,4-diene-3,20-dione 17,21-dipropionate

Other Name
Beconase

Form	Molecular Formula	MW	CAS
Beclomethasone dipropionate	$C_{28}H_{37}ClO_7$	521.1	5534-09-8

Appearance
Beclomethasone dipropionate is a white to creamy white powder.

Solubility
Beclomethasone dipropionate has solubilities of less than 5 µg/mL in water and 22 mg/mL in alcohol at 25 °C.

Method
Cornarakis-Lentzos and Cowin used HPLC analysis to evaluate the stability of diluted betamethasone valerate and beclomethasone dipropionate creams and ointments. The Waters chromatograph consisted of a model 6000 pump, a model U6K injector, a

model 440 UV detector equipped with a Taper cell of 12.5-µL capacity and 10-mm pathlength, and a model 730 data module. The stationary phase was a µBondapak stainless steel column (300 × 3.9 mm, 10-µm particle size). The mobile phase consisted of methanol, acetonitrile, and water (49:25:26, vol/vol/vol) and was delivered isocratically at 2.0 mL/min. UV detection was performed at 254 nm and 0.1 AUFS. Betamethasone 17-valerate was used as an internal standard.

Samples of the creams and ointments equivalent to 0.2–0.3 mg of beclomethasone dipropionate were dispersed in 50 mL of hexane (hot solvent for ointments), extracted with 10 mL of ethanol (85% for the cream and 65% for the ointment) into a 50-mL volumetric flask via a cotton wool plug five times, and further diluted with the internal standard and then ethanol (65%). The sample injection volume was 250 µL. Under these conditions, retention times were about 2.30 minutes for betamethasone, 3.22 minutes for beclomethasone monoproprionate, 3.97 minutes for betamethasone 17-valerate, 4.48 minutes for betamethasone 21-valerate, and 5.19 minutes for beclomethasone dipropionate.

The method was reported to be stability indicating because of the resolution of the intact drug and its degradation products.

A standard curve for beclomethasone dipropionate in ethanol (65%) was constructed from 0.0003 to 0.005% (wt/vol). The correlation coefficient was 0.9999.

Reference

Cornarakis-Lentzos M, Cowin PR. Dilutions of corticosteroid creams and ointments–a stability study. *J Pharm Biomed Anal*. 1987; 5: 707–16.

Benazepril

Chemical Name

[S-(R*,R*)]-3-[[1-(Ethoxycarbonyl)-3-phenylpropyl]amino]-2,3,4,5-tetrahydro-2-oxo-1H-1-benzazepine-1-acetic acid

Other Names

Lotensin, Lotrel

Form	Molecular Formula	MW	CAS
Benazepril	$C_{24}H_{28}N_2O_5$	424.5	86541-75-5
Benazepril hydrochloride	$C_{24}H_{28}N_2O_5$.HCl	461.0	86541-74-4

Appearance

Benazepril hydrochloride occurs as a white to off-white crystalline powder.

Solubility

Benazepril hydrochloride is soluble (>100 mg/mL) in water, ethanol, and methanol.

Method 1

Naidu et al. presented a method for the simultaneous determination of amlodipine and benazepril hydrochloride in tablets. A Thermo Separation liquid chromatographic system was equipped with a model P1000 pump, a model AS100 autosampler, and a model 6000LP photodiode-array detector. The stationary phase was a Zorbax SB C_{18} column (250 × 4.6 mm, 5-μm particle size). The column temperature was maintained at 28 °C. The mobile phase consisted of 0.05 M monobasic potassium phosphate buffer and acetonitrile (65:35, vol/vol); the pH was adjusted to 7.0 with 1.0 N potassium hydroxide solution. The flow rate was 1.0 mL/min. UV detection was performed at 240 nm.

A portion of the powder from 10 tablets, equivalent to one tablet content, was weighed, transferred into a 50-mL volumetric flask, mixed with 10 mL of methanol, shaken for 10 minutes, ultrasonicated for 5 minutes, diluted to the mark with mobile phase, centrifuged at 10,000 rpm, filtered, and assayed. The injection volume was 25 μL. Under these conditions, the retention times of benazepril and amlodipine were about 8.3 and 21.0 minutes, respectively.

This method was demonstrated to be stability indicating by accelerated degradation studies. Samples were exposed to 0.5 N hydrochloric acid for 48 hours, 0.1 N sodium hydroxide solution for 2 hours, or 3% hydrogen peroxide for 48 hours. Sample solutions were also irradiated with UV radiation at 254 and 366 nm. Peaks of degradation products were well separated from the peaks of the intact drugs.

A calibration curve for benazepril was constructed from 12.0 to 28.0 μg/mL. The correlation coefficient was 0.9999. The interday coefficient of variation was less than 0.96%.

Reference

Naidu KR, Kale UN, Shingare MS. Stability indicating RP-HPLC method for simultaneous determination of amlodipine and benazepril hydrochloride from their combination drug product. *J Pharm Biomed Anal.* 2005; 39: 147–55.

Method 2

Radhakrishna et al. described an HPLC procedure for the separation and quantification of benazepril and benazepril-related compounds. The Waters LC Module 1 Plus chromatograph consisted of a pump and a variable-wavelength UV-visible detector and was integrated by Waters Millennium 2010 Chromatography Manager software. The stationary phase was a Hichrom RPB C_{18} column (250 × 4.6 mm, 5-μm particle size). Solvent A was a mixture of 500 mL of 0.025 M monobasic sodium phosphate buffer (adjusted to pH 2.8 with phosphoric acid), 300 mL of acetonitrile, and 200 mL of methanol. Solvent B was acetonitrile. The mobile phase was delivered at 100% A for the first 15 minutes, linearly decreased to 60% A from 15 to 30 minutes, and held at 60% A and 40% B from 30 to 45 minutes. The flow rate was 1.2 mL/min. UV detection was performed at 210 nm. 5-Methyl-2-nitrophenol was used as an internal standard.

Twenty benazepril hydrochloride tablets were extracted into methanol and centrifuged. The supernatant was then diluted appropriately with mobile phase before injection. The injection volume was 20 μL. Under these conditions, retention times for benazepril and the internal standard were about 8.0 and 11.2 minutes, respectively (estimated from the published chromatogram).

The assay was reported to be stability indicating since benazepril was well resolved from all possible known impurities and excipients.

A standard curve for benazepril was generated from 50 to 800 μg/mL. The correlation coefficient was greater than 0.999. The interday and intraday coefficients of variation for the analysis of benazepril hydrochloride were less than 1.6%.

Reference

Radhakrishna T, Sreenivas Rao D, Vyas K, et al. A validated method for the determination and purity evaluation of benazepril hydrochloride in bulk and in pharmaceutical dosage forms by liquid chromatography. *J Pharm Biomed Anal.* 2000; 22: 641–50.

Bendroflumethiazide

Chemical Name

3,4-Dihydro-3-(phenylmethyl)-6-(trifluoromethyl)-2*H*-1,2,4-benzothiadiazine-7-sulfon-amide 1,1-dioxide

Other Names

Bendrofluazide, Naturetin

Form	Molecular Formula	MW	CAS
Bendroflumethiazide	$C_{15}H_{14}F_3N_3O_4S_2$	421.4	73-48-3

Appearance

Bendroflumethiazide is a white- to cream-colored finely divided crystalline powder.

Solubility

Bendroflumethiazide is insoluble in water, chloroform, benzene, and ether; it is soluble in acetone and alcohol.

pK$_a$

Bendroflumethiazide has a pK$_a$ of 8.5.

Method

Using a stability-indicating HPLC method, Barnes and Nash investigated the stability of bendroflumethiazide in a low-dose extemporaneously prepared capsule 1.25 mg. The liquid chromatograph included a Cecil CE1100 pump, a Severn analytical SA6500 variable-wavelength UV detector, a Talbot ASI-4 autosampler equipped with a Rheodyne 7010 injection valve, and a Shimadzu C-R3A integrator. The HPLC analytical column was a Spherisorb hexyl column (100 × 4.5 mm, 5-μm particle size) with a guard column (10 × 4.6 mm) of the same material. The mobile phase consisted of methanol and water (50:50, vol/vol) run isocratically at 1.5 mL/min. UV detection was performed at 270 nm. The injection volume was 10 μL.

The assay was determined to be stability indicating by acid and alkali degradation of a capsule extract.

A standard curve was constructed by plotting the peak areas of bendroflumethiazide against bendroflumethiazide concentration from 0.06 to 0.18 mg/mL. The correlation coefficients were 0.9998. The intraday coefficient of variation was 2.6%.

References

Barnes AR, Nash S. Stability of bendrofluazide in a low-dose extemporaneously prepared capsule. *J Clin Pharm Ther.* 1994; 19: 89–93.

Barnes AR, Nash S. HPLC determination of bendrofluazide in capsules. *Int J Pharm.* 1993; 94: 231–4.

Benoxinate Hydrochloride

Chemical Name
4-Amino-3-butoxybenzoic acid 2-(diethylamino)ethyl ester monohydrochloride

Other Names
Dorsacaine, Oxybuprocaine

Form	Molecular Formula	MW	CAS
Benoxinate	$C_{17}H_{28}N_2O_3$	308.4	99-43-4
Benoxinate hydrochloride	$C_{17}H_{28}N_2O_3 \cdot HCl$	344.9	5987-82-6

Appearance
Benoxinate hydrochloride occurs as white or slightly off-white or colorless crystals or as a crystalline powder.

Solubility
Benoxinate hydrochloride is very soluble in water and freely soluble in alcohol and chloroform.

Method

Chorny et al. reported a simple HPLC method for the determination of benoxinate hydrochloride and its degradation products in ophthalmic solutions. The liquid chromatograph was a Hewlett-Packard model 1090L with a photodiode-array detector. The stationary phase was a Phenomenex Spherisorb phenyl column (250 × 3.2 mm, 10-μm particle size). The mobile phase was a mixture of acetonitrile and a buffer (35:65, vol/vol); the buffer was a mixture of 50 mM monobasic sodium phosphate, 2.5 mM sodium hydrogen sulfate, and 5 mM 1-heptanesulfonic acid sodium salt, which was adjusted to pH 3.5 with 1 N sodium hydroxide solution. The flow rate was 0.8 mL/min. UV detection was performed at 308 nm.

Samples were diluted by a factor of 100 with water and then by a factor of 5 with mobile phase. The injection volume was 10 μL. Under these conditions, the retention times for the degradation product, 4-amino-3-butoxybenzoic acid, and benoxinate were about 4.5 and 5.5 minutes, respectively.

Benoxinate was intentionally degraded by adding 2 mL of 0.1 N sodium hydroxide solution or 0.3 N hydrochloric acid to 5 mL of a 0.4% benoxinate ophthalmic solution and incubating these solutions at 85 °C for 20 hours. Degradation products were well separated from the intact drug on the chromatogram.

A five-point calibration curve for benoxinate was constructed from 4.8 to 11.2 μg/mL. The correlation coefficient was 0.9999. The interday coefficient of variation for the analysis of benoxinate was 0.4%.

Reference

Chorny M, Levy D, Schumacher I, et al. Development and validation of a stability-indicating high performance liquid chromatographic assay for benoxinate. *J Pharm Biomed Anal.* 2003; 32: 189–96.

Benzalkonium Chloride

Chemical Name

A mixture of alkyldimethylbenzylammonium chlorides

R = C_8H_{17} to $C_{18}H_{37}$

Other Name

Zephiran Chloride

Form	Molecular Formula	CAS
Benzalkonium chloride	$C_9H_{13}NRCl$ (R = C_8H_{17} to $C_{18}H_{37}$)	8001-54-5

Appearance

Benzalkonium chloride occurs as a white or yellowish-white powder, thick gel, or gelatinous pieces with a mild aromatic odor.

Solubility

Benzalkonium chloride is very soluble in water and alcohol.

Method 1

Parhizkari et al. reported a sensitive, accurate, and reproducible reversed-phase HPLC method for the determination of benzalkonium chloride in 0.5% tramadol ophthalmic solution. The chromatographic apparatus consisted of a Waters model 600E system controller and pump, a WISP 712D autosampler, and a model 486 variable-wavelength UV detector. The stationary phase was a Keystone CPS Hypersil-1 cyano stainless steel column (150 × 4.6 mm, 3-μm particle size). The mobile phase was a mixture of 50 mM sodium propionate (adjusted to pH 5.3 with sulfuric acid) and acetonitrile (35:65, vol/vol). The flow rate was 1.3 mL/min. UV detection was performed at 214 nm.

Samples were diluted 1:5 with the mobile phase. The injection volume was 100 μL. Under these conditions, the retention times of the C_{12} and C_{14} homologues of benzalkonium chloride were 10.0 and 11.7 minutes, respectively. The run time was 15 minutes.

The stability-indicating capability of the method was evaluated by accelerated degradation of benzalkonium chloride. One aliquot (5 mL) of benzalkonium chloride was exposed to UV light for 13 hours, and other aliquots were adjusted to either pH 2 with hydrochloric acid or pH 12 with sodium hydroxide solution and stored at 80 °C for up to 2 months. The observed degradation products did not interfere with the determination of benzalkonium chloride.

A standard curve for benzalkonium chloride was constructed from 50 to 150% of its label claim in 0.5% tramadol solution. The correlation coefficient was 1.000.

Reference

Parhizkari G, Delker G, Miller RB, et al. A stability-indicating HPLC method for the determination of benzalkonium chloride in 0.5% tramadol ophthalmic solution. *Chromatographia*. 1995; 40: 155–8.

Method 2

Miller et al. described the determination of benzalkonium chloride in Vasocidin ophthalmic solution. The instrument consisted of a Waters model 600E system controller and pump, a WISP 712D autosampler, and a model 486 variable-wavelength UV detector. The stationary phase was a Waters μBondapak phenyl stainless steel column (300 × 3.9 mm, 10-μm particle size). The mobile phase was a mixture of acetonitrile and buffer (65:35, vol/vol), where the buffer consisted of 50 mM monobasic potassium phosphate monohydrate and 57 mM sodium hexanesulfonate, adjusted to pH 6.3 with 1 N sodium hydroxide solution. The flow rate was 1.8 mL/min. UV detection was performed at 215 nm. The sample injection volume was 100 μL. Under these conditions, retention times for the C_{12}, C_{14}, and C_{16} homologues of benzalkonium chloride were 7.4, 9.5, and 12.4 minutes, respectively. The run time was 16 minutes.

The method was demonstrated to be stability indicating by accelerated degradation of benzalkonium chloride. Vasocidin ophthalmic solution was treated with heat, acid (pH 2 at 120 °C for 15 hours), base (pH 12 at 120 °C for 15 hours), and UV radiation

(200–400 nm, 40 mWatt/cm^2 for 15 hours). No interfering peaks from the degradation of benzalkonium chloride were observed in any of the treated solutions.

A calibration curve for benzalkonium chloride was constructed from 50 to 150% of its labeled concentration in Vasocidin. The correlation coefficient was greater than 0.998.

References

Miller RB, Chen C, Sherwood CH. High-performance liquid chromatographic determination of benzalkonium chloride in Vasocidin ophthalmic solution. *J Liq Chromatogr*. 1993; 16: 3801–11.

Parhizkari G, Miller RB, Chen C. A stability-indicating HPLC method for the determination of benzalkonium chloride in phenylephrine HCl 10% ophthalmic solution. *J Liq Chromatogr*. 1995; 18: 553–63.

Benzoic Acid

Chemical Name
Benzoic acid

Other Name
Drazoic

Form	Molecular Formula	MW	CAS
Benzoic acid	$C_6H_5.CO_2H$	122.1	65-85-0
Sodium benzoate	$C_6H_5.CO_2Na$	144.1	532-32-1

Appearance
Benzoic acid occurs as colorless or white crystals or as white scales or needles or as a crystalline powder. Sodium benzoate occurs as a white, odorless or practically odorless, granular or crystalline, slightly hygroscopic powder or flakes.

Solubility
Benzoic acid is slightly soluble in water and freely soluble in alcohol, chloroform, and ether. Sodium benzoate is freely soluble in water and sparingly soluble in alcohol.

Method 1
Hewala described an HPLC method for paracetamol, guaifenesin, sodium benzoate, and oxomemazine in the presence of degradation products. A Beckman Gold system consisting of a model 125 programmable pump, a model 166 programmable UV detector, and a

Rheodyne 20-μL loop injector was used. The stationary phase was a stainless steel C_{18} ODS column (250 × 4.6 mm, 5-μm particle size) with a guard column (50 × 4.6 mm) of the same packing material. Mobile phase A was a mixture of methanol and water (18:82, vol/vol) adjusted to pH 3.9 with phosphoric acid, and mobile phase B was a mixture of methanol and water (80:20, vol/vol) adjusted to pH 3.9 with phosphoric acid. Mobile phase A was delivered from 0 to 12 minutes and mobile phase B from 12 to 22 minutes. The flow rate was 1.5 mL/min. UV detection was performed at 235 nm and 0.05 AUFS. Metronidazole was used as an internal standard.

A portion of the cough syrup (10 mL) was diluted to 100 mL with methanol, mixed with internal standard solution, and further diluted with methanol to 20 μg/mL of metronidazole in the final solution. Under these conditions, retention times for par-acetamol, metronidazole, sodium benzoate, guaifenesin, and oxomemazine were about 2.6, 5.5, 8.6, 10.7, and 16.8 minutes, respectively (estimated from the published chromatogram).

The method was evaluated to be stability indicating by assaying a mixture of active compounds and their possible degradation products. Retention times for two possible degradants, 4-aminophenol and guaicol, were about 3.7 and 7.0 minutes, respectively (estimated from the published chromatogram).

A standard curve for sodium benzoate was constructed from 5.0 to 30.0 μg/mL. The correlation coefficient was 0.9998. Intraday and interday coefficients of variation were 2.01 and 2.13%, respectively.

Reference
Hewala II. Stability-indicating HPLC assay for paracetamol, guaiphenesin, sodium benzoate and oxomemazine in cough syrup. *Anal Lett.* 1994; 27: 71–93.

Method 2
Tan et al. reported a reversed-phase HPLC method for the determination of benzoic acid and benzyl alcohol in dog plasma following solid-phase extraction. An Altex model 330 liquid chromatograph was equipped with a model 110A pump, a model 210 sample injection valve with a 20-μL loop, a Beckman model 153 fixed-wavelength UV detector, and a Varian model 4270 electronic integrator. The stationary phase was a Beckman Ultrasphere ODS column (250 × 4.6 mm, 5-μm particle size) with a Whatman Co:Pell ODS guard column (40 × 4.6 mm, 25–37-μm particle size). The mobile phase was a mixture of acetonitrile, water, and glacial acetic acid (120:380:2.5, vol/vol/vol) and was isocratically delivered at 2 mL/min. UV detection was performed at 254 nm and 0.02–0.08 AUFS. Benzocaine was used as an internal standard.

A plasma sample (0.5 mL) was diluted with 0.5 mL of 0.9% sodium chloride injection and 1.0 mL of internal standard solution, transferred onto a dry EM Science Extrelut QE solid-phase extraction cartridge, and eluted twice with 4 mL of ethyl acetate. The eluate was collected in a calibrated 2-mL tube, evaporated to 0.5 mL using a stream of nitrogen gas, and brought to volume with acetonitrile before injection. The injection volume was 20 μL. Under these conditions, retention times for benzyl alcohol and benzoic acid were 3.4 and 4.8 minutes, respectively.

The method was evaluated to be stability indicating by assaying a solution containing benzyl alcohol, benzaldehyde, benzoic acid, hippuric acid, and benzocaine. Retention times for hippuric acid, benzyl alcohol, benzoic acid, benzaldehyde, and benzocaine were 2.2, 3.4, 4.8, 6.6, and 9.5 minutes, respectively.

A calibration curve for benzoic acid was constructed from 0.5 to 4 µg/mL. The correlation coefficient was 0.998.

Reference

Tan HIS, Manning MA, Hahn M, et al. Determination of benzyl alcohol and its metabolite in plasma by reversed-phase high-performance liquid chromatography. *J Chromatogr*. 1991; 568: 145–55.

Method 3

Schieffer et al. reported a simultaneous determination of phenylephrine hydrochloride, phenylpropanolamine hydrochloride, guaifenesin, and sodium benzoate in dosage forms. A Waters model ALC 204 system equipped with a Waters model 6000A reciprocating pump, a model 440 absorbance detector, a Rheodyne model 7125 injector, and a Hewlett-Packard model 3352B integrator was used. The stationary phase was a Whatman Partisil-10 C_8 column. The mobile phase consisted of 50 mL of methanol, 170 mL of acetonitrile, 755 mL of water, and 25 mL of pentanesulfonic acid sodium salt in glacial acetic acid. The flow rate was 2.0 mL/min. UV detection was performed at 254 nm.

For liquid forms, samples were diluted with water. For solid forms, samples were leached with water or mobile phase and then diluted. Under these conditions, retention times for phenylephrine hydrochloride, phenylpropanolamine hydrochloride, guaifenesin, and benzoic acid were about 2.8, 4.4, 7.4, and 14.4 minutes, respectively (estimated from the published chromatogram).

The stability-indicating nature of the method was demonstrated by spiking the sample solution with possible impurities and degradation products. Phenylephrine hydrochloride, phenylpropanolamine hydrochloride, guaifenesin, and benzoic acid were separated from α-aminopropiophenone, 2-(2-methoxyphenoxy)-1,3-propanediol, *m*-hydroxybenzaldehyde, guaicol, and benzaldehyde.

A standard curve for benzoic acid was constructed from 0 to 0.12 µg/mL. The correlation coefficient was 0.9999.

References

Schieffer GW, Smith WO, Lubey GS, et al. Determination of the structure of a synthetic impurity in guaifenesin: Modification of a high-performance liquid chromatographic method for phenylephrine hydrochloride, phenylpropanolamine hydrochloride, guaifenesin, and sodium benzoate in dosage forms. *J Pharm Sci*. 1984; 73: 1856–8.

Schieffer GW, Hughes DE. Simultaneous stability-indicating determination of phenylephrine hydrochloride, phenylpropanolamine hydrochloride, and guaifenesin in dosage forms by reversed-phase paired-ion high-performance liquid chromatography. *J Pharm Sci*. 1983; 72: 55–9.

Method 4

Heidemann presented an HPLC method for the determination of theophylline, guaifenesin, and benzoic acid in liquid and solid pharmaceutical dosage forms. A DuPont model 830 liquid chromatograph equipped with a DuPont model 837 variable-wavelength detector, a Rheodyne model 7105 injector, and a Spectra-Physics autolab system 1V integrator was used. The stationary phase was a Whatman Partisil-10-ODS reversed-phase column (250 × 4.6 mm, 10-µm particle size). The mobile phase was a mixture of 0.001 M sodium citrate–citric acid buffer (adjusted to pH 4.15) and acetonitrile (9:1, vol/vol). The flow rate was 2 mL/min. UV detection was performed at 230 nm. Methylparaben was used as an internal standard.

For liquid dosage forms, samples were diluted with water and then internal standard. For solid dosage forms, tablets were ground, wetted with alcohol, diluted with water, stirred for 1 hour, and filtered. The filtrate was diluted with water and the internal standard. The injection volume was 20 µL. Under these conditions, retention times for benzoic acid, theophylline sodium glycinate, guaifenesin, and methylparaben were about 6.8, 9.4, 11.6, and 16.0 minutes, respectively.

The method was reported to be stability indicating.

Reference

Heidemann DR. Rapid, stability-indicating, high-pressure liquid chromatographic determination of theophylline, guaifenesin, and benzoic acid in liquid and solid pharmaceutical dosage forms. *J Pharm Sci*. 1979; 68: 530–2.

Benzoyl Peroxide

Chemical Name
Dibenzoyl peroxide

Other Name
Debroxide

Form	Molecular Formula	MW	CAS
Benzoyl peroxide	$C_{14}H_{10}O_4$	242.2	94-36-0

Appearance
Benzoyl peroxide is a white amorphous or granular powder.

Solubility
Benzoyl peroxide is sparingly soluble in water and alcohol; it is soluble in benzene, chloroform, and ether.

Method
Gaddipati et al. described a stability-indicating HPLC procedure for the quantitative determination of benzoyl peroxide in pharmaceutical dosage forms. The liquid chromatograph consisted of a Waters model 6000 solvent-delivery system, a Waters model 710A

automatic sampler system, a Waters model 450 variable-wavelength UV detector, and a Hewlett-Packard model 3385A electronic integrator. The analytical column was a Waters μBondapak C_{18} reversed-phase column (300 × 3.9 mm). The mobile phase consisted of methanol and water (75:25, vol/vol) filtered through a 0.45-μm membrane filter. The flow rate was 1.0 mL/min. UV detection was performed at 238 nm. Acenaphthylene in acetonitrile at 2.0 mg/mL was used as an internal standard.

Samples were diluted with acetonitrile. The injection volume for the standard and sample preparations was 20 μL. The retention time for benzoyl peroxide was 9.46 minutes.

The stability-indicating capacity of the assay was demonstrated. Analyses of the synthetic placebos for benzoyl peroxide lotion and wash formulations showed no interference from the degradation product and impurity peaks with the intact benzoyl peroxide peak. Degradation products and impurities follow, with their retention times in minutes given in parentheses: homophthalic acid (2.20), benzoic acid (2.76), 4-biphenylcarboxylic acid (3.38), phenol (3.65), benzaldehyde (4.12), benzene (6.00), phenyl benzoate (8.46), acenaphthylene (11.96), biphenyl (13.87), *o*-terphenyl (25.82), and *p*-terphenyl (55.94).

Quantitation was based on the benzoyl peroxide–internal standard peak ratio. Calibration curves were established for concentrations from 13.5 to 76.4 μg of benzoyl peroxide. The correlation coefficient was 0.9999 and the coefficient of variation was 0.74%.

Reference
Gaddipati N, Volpe F, Anthony G. Quantitative determination of benzoyl peroxide by high-performance liquid chromatography and comparison to the iodometric method. *J Pharm Sci.* 1983; 72: 1398–400.

Benzyl Alcohol

Chemical Name
Benzyl alcohol

Other Name
Alcohol Benzylicus

Form	Molecular Formula	MW	CAS
Benzyl alcohol	C_7H_8O	108.1	100-51-6

Appearance
Benzyl alcohol is a clear colorless oily refractive liquid.

Solubility
Benzyl alcohol is soluble in water and miscible with alcohol, chloroform, and ether.

Method 1
Tan et al. reported a reversed-phase HPLC method for the determination of benzyl alcohol and benzoic acid in dog plasma following solid-phase extraction. An Altex model 330 liquid chromatograph was equipped with a model 110A pump, a model 210 sample injection valve with a 20-μL loop, a Beckman model 153 fixed-wavelength UV detector, and a Varian model 4270 electronic integrator. The stationary phase was a Beckman Ultrasphere ODS column (250 × 4.6 mm, 5-μm particle size) with a Whatman Co:Pell ODS guard column (40 × 4.6 mm, 25–37-μm particle size). The mobile phase was a mixture of acetonitrile, water, and glacial acetic acid (120:380:2.5, vol/vol/vol) and was isocratically delivered at 2 mL/min. UV detection was performed at 254 nm and 0.02–0.08 AUFS. Benzocaine was used as an internal standard.

A plasma sample (0.5 mL) was diluted with 0.5 mL of 0.9% sodium chloride injection and 1.0 mL of internal standard solution, transferred onto a dry EM Science Extrelut QE solid-phase extraction cartridge, and eluted twice with 4 mL of ethyl acetate. The eluate was collected in a calibrated 2-mL tube, evaporated to 0.5 mL using a stream of nitrogen gas, and brought to volume with acetonitrile before injection. The injection volume was 20 μL. Under these conditions, retention times for benzyl alcohol and benzoic acid were 3.4 and 4.8 minutes, respectively.

The method was demonstrated to be stability indicating by assaying a solution containing benzyl alcohol, benzaldehyde, benzoic acid, hippuric acid, and benzocaine. Retention times were 2.2 minutes for hippuric acid, 3.4 minutes for benzyl alcohol, 4.8 minutes for benzoic acid, 6.6 minutes for benzaldehyde, and 9.5 minutes for benzocaine.

A calibration curve for benzyl alcohol was constructed from 4 to 32 μg/mL. The correlation coefficient was 0.999.

Reference
Tan HIS, Manning MA, Hahn M, et al. Determination of benzyl alcohol and its metabolite in plasma by reversed-phase high-performance liquid chromatography. *J Chromatogr.* 1991; 568: 145–55.

Method 2
Menon and Norris reported a simultaneous determination by HPLC of hydroxyzine hydrochloride and benzyl alcohol in injection solutions. A liquid chromatographic system consisting of a Waters model 6000A dual-piston reciprocating pump, a model U6K universal injector, and a Perkin-Elmer model LC-55 or Schoeffel model SF 770 Spectroflow variable-wavelength UV detector was used. The stationary phase was a Waters μBondapak C_{18} column (300 × 4 mm, 10-μm particle size). The mobile phase was a mixture of water, acetonitrile, and methanol (60:25:15, vol/vol/vol, pH 2.6), containing 0.06% (vol/vol) sulfuric acid, 0.5% (wt/vol) sodium sulfate, and 0.02% (wt/vol) heptane-sulfonic acid sodium salt. The flow rate was 2 mL/min. UV detection was performed at 257 nm and 0.2 AUFS. *p*-Nitroacetophenone 0.2 mg/mL and isobutyrophenone 2.5 mg/mL in methanol were used as an internal standard.

Samples (5 mL) were mixed with 5 mL of the internal standard and diluted to 50 mL with methanol. The injection volume was 10 μL. Under these conditions, retention times for benzyl alcohol, *p*-nitroacetophenone, hydroxyzine hydrochloride, and isobutyrophenone were about 3.1, 5.6, 9.3, and 12.5 minutes, respectively (estimated from the published chromatogram).

The stability-indicating ability of the method was demonstrated by assaying a synthetic mixture of benzyl alcohol, hydroxyzine hydrochloride, and their potential degradation products. Benzyl alcohol and hydroxyzine hydrochloride were separated from their degradation products. Retention times in minutes were about 3.1 for benzyl alcohol, 3.3 for benzoic acid, 4.0 for benzaldehyde, 5.6 for *p*-nitroacetophenone, 6.7 for *p*-chlorobenzoic acid, 7.6 for *p*-chlorobenzaldehyde, 9.3 for hydroxyzine hydrochloride, 12.5 for isobutyrophenone, and 46.0 for *p*-chlorobenzophenone (estimated from the published chromatogram).

A standard curve for benzyl alcohol was constructed from 0.54 to 1.8 mg/mL. The correlation coefficient was 0.9999.

Reference
Menon CN, Norris BJ. Simultaneous determination of hydroxyzine hydrochloride and benzyl alcohol in injection solutions by high-performance liquid chromatography. *J Pharm Sci.* 1981; 70: 697–8.

Bepridil Hydrochloride

Chemical Names
N-Benzyl-*N*-(3-isobutoxy-2-pyrrolidin-1-ylpropyl)aniline hydrochloride monohydrate
β-[(2-Methylpropoxy)methyl]-*N*-phenyl-*N*-(phenylmethyl)-1-pyrrolidineethanamine monohydrochloride monohydrate

•HCl •H$_2$O

Other Names
Cordium, Vascor

Form	Molecular Formula	MW	CAS
Bepridil hydrochloride	C$_{24}$H$_{34}$N$_2$O.HCl.H$_2$O	421.0	74764-40-2

Method
Renzi et al. presented a stability-indicating assay for bepridil hydrochloride drug substance and drug products. The DuPont model 850 liquid chromatograph included an autosampler, a 20-μL loop injector, a fixed-wavelength 254-nm detector, and a Spectra-

Physics model 4270 integrator. The stationary phase was a Waters μBondapak C_{18} column (300 × 4.6 mm, 10-μm particle size). The column temperature was set at 35 °C. The mobile phase consisted of 405 mL of aqueous solution containing 1.1 g of sodium 1-heptanesulfonate adjusted to pH 2.37 with glacial acetic acid and 580 mL of acetonitrile. The flow rate was 1.3 mL/min.

Drug product samples were diluted with acetonitrile and filtered through 0.22-μm filter paper. Drug substance samples were diluted with the mobile phase. The injection volume was 20 μL. Under these conditions, the retention time for bepridil hydrochloride was about 6.8 minutes.

The HPLC method was shown to be stability indicating by spiking the drug solution with its impurities and degradation products. Retention times for benzoic acid, benzaldehyde, debenzylated bepridil hydrochloride, and N-benzylaniline were 2.6, 3.2, 4.2, and 5.0 minutes, respectively.

A standard curve for bepridil hydrochloride was constructed from 0 to 400% of the normal amount of bepridil hydrochloride. The correlation coefficient was 0.9999. The limit of detection for the analysis was 0.01 μg.

Reference
Renzi NL, Fronheiser ME, Duong HT, et al. Stability-indicating high-performance liquid chromatography assay for bepridil hydrochloride drug substance and drug products. *J Chromatogr.* 1989; 462: 398–405.

Betamethasone

Chemical Name
(11β,16β)-9-Fluoro-11,17,21-trihydroxy-16-methylpregna-1,4-diene-3,20-dione

Other Names
Celestone, Flubenisolone

Form	Molecular Formula	MW	CAS
Betamethasone	$C_{22}H_{29}FO_5$	392.5	378-44-9
Betamethasone benzoate	$C_{29}H_{33}FO_6$	496.6	22298-29-9
Betamethasone 17-valerate	$C_{27}H_{37}FO_6$	476.6	2152-44-5

Appearance

Betamethasone occurs as a white to practically white crystalline powder. Betamethasone benzoate appears as a white to practically white powder. Betamethasone 17-valerate occurs as a white to almost-white odorless crystalline powder.

Solubility

Betamethasone is insoluble in water and sparingly soluble in alcohol. Betamethasone benzoate is insoluble in water and soluble in alcohol. Betamethasone 17-valerate is practically insoluble in water and soluble in alcohol and chloroform. It is freely soluble in acetone.

Method 1

Smith and Haigh developed an HPLC assay for betamethasone 17-valerate in purified isopropyl myristate receptor phase. The chromatographic system consisted of a Waters model 6000A solvent-delivery system, a Rheodyne model 7125 injector equipped with a Brownlee LiChrosorb RP18 loop column (30 × 4.6 mm), a Kratos model SF 769 variable-wavelength UV detector, and a Hewlett-Packard model 100A integrator. The stationary phase was an HPLC Technology Techsil analytical column (250 × 4.6 mm, 10-μm particle size). The mobile phase was a mixture of 45 parts of water and 55 parts of acetonitrile and was isocratically delivered at 1.5 mL/min. UV detection was performed at 239 nm. Norethisterone 3.5 μg/mL in mobile phase was used as an internal standard.

An aliquot of 40 μL was sampled from the receptor chamber of the diffusion cell, mixed with 40 μL of internal standard solution, vortexed for 30 seconds, and centrifuged for 60 seconds. The lower, mobile phase layer was collected and 10 μL of this solution was injected onto the column. Prior to the next injection, lipophilic components trapped on the loop column were flushed to waste using 1 mL of methanol followed by 5 mL of mobile phase. Under these conditions, retention times for norethisterone and betamethasone 17-valerate were 4.2 and 5.6 minutes, respectively.

The stability-indicating capability of the method was demonstrated by assaying a solution containing betamethasone 17-valerate, its degradation product, and preservatives. Retention times for methyl hydroxybenzoate, chlorocresol, norethisterone, betamethasone 17-valerate, and betamethasone 21-valerate were 2.9, 3.6, 4.2, 5.6, and 6.5 minutes, respectively.

Calibration curves for betamethasone 17-valerate were generated from 2.5 to 40.0 μg/mL. The correlation coefficient was 0.9992.

Reference

Smith EW, Haigh JM. *In vitro* diffusion cell design and validation. I. A stability-indicating high-performance liquid chromatographic assay for betamethasone 17-valerate in purified isopropyl myristate receptor phase. *Pharm Res*. 1989; 6: 431–5.

Method 2

Maron et al. used a stability-indicating HPLC method to assay betamethasone 17-benzoate in lipophilic bases. The HPLC system consisted of a Waters model 6000A pump, a Waters WISP autosampler, a Waters model 440 UV detector, and a Hewlett-Packard 3390 integrator. The stationary phase was a Waters reversed-phase μBondapak C$_{18}$ column (300 × 3.9 mm, 10-μm particle size) with a Waters Guard-Pak precolumn module equipped with a disposable μBondapak C$_{18}$ precolumn insert. The mobile phase was a mixture of methanol, acetonitrile, and water (7:2:4, vol/vol/vol). The flow rate was 1.5 mL/min. UV detection was performed at 254 nm.

Betamethasone 17-benzoate was extracted from the lipophilic vehicles with a mixture of methanol and acetonitrile (7:2, vol/vol). 17-Methyltestosterone was used as the internal standard. The injection volume was 20 μL.

The method was determined to be stability indicating by spiking the sample with potential degradation products of the drug, betamethasone and betamethasone 21-benzoate. All peaks were well separated. The retention times for excipients, betamethasone, internal standard, betamethasone 21-benzoate, and betamethasone 17-benzoate were 2.5, 3.5, 6.3, 8.0, and 9.2 minutes, respectively.

Calibration curves from 25 to 62.5 μg/mL were constructed based on the peak height ratio of the drug to that of the internal standard. The correlation coefficient was 0.9999, and the coefficient of variation was 0.3%.

Reference
Maron N, Cristi EA, Ramos AA. Determination of betamethasone 17-benzoate in lipophilic vehicles by reversed-phase high-performance liquid chromatography. *J Pharm Sci.* 1988; 77: 638–9.

Method 3
Using HPLC analysis, Cornarakis-Lentzos and Cowin evaluated the stability of diluted betamethasone valerate and beclomethasone dipropionate creams and ointments. The Waters chromatograph consisted of a model 6000 pump, a model U6K injector, a model 440 UV detector equipped with a Taper cell of 12.5-μL capacity and 10-mm pathlength, and a model 730 data module. The stationary phase was a μBondapak stainless steel column (300 × 3.9 mm, 10-μm particle size). The mobile phase consisted of methanol, acetonitrile, and water (49:25:26, vol/vol/vol) and was delivered isocratically at 2.0 mL/min. Beclomethasone dipropionate was used as an internal standard. UV detection was performed at 254 nm and 0.1 AUFS.

Samples of the creams and ointments equivalent to 1–1.2 mg of betamethasone 17-valerate were dispersed in 50 mL of hexane (hot solvent for ointments), extracted with 10 mL of ethanol (85% for the cream and 65% for the ointment) into a 50-mL volumetric flask via a cotton wool plug five times, and further diluted with the internal standard and ethanol (65%). The sample injection volume was 200 μL. Under these conditions, retention times for betamethasone, beclomethasone monopropionate, betamethasone 17-valerate, betamethasone 21-valerate, and beclomethasone dipropionate were about 2.30, 3.22, 3.97, 4.48, and 5.19 minutes, respectively.

The method was reported to be stability indicating because of the resolution of the intact drug and its degradation products.

A standard curve for betamethasone 17-valerate in ethanol (65%) was constructed from 0.0003 to 0.005% (wt/vol). The correlation coefficient was 0.9999.

Reference
Cornarakis-Lentzos M, Cowin PR. Dilutions of corticosteroid creams and ointments—a stability study. *J Pharm Biomed Anal.* 1987; 5: 707–16.

Method 4

Mehta and Calvert investigated the stability of betamethasone 17-valerate by HPLC. The system included an Altex pump, a Varian UV detector, and a Smiths Servoscribe recorder. The stationary phase was a Hypersil ODS column (5-μm particle size). The mobile phase consisted of acetonitrile and water (50:50, vol/vol). The flow rate was 2 mL/min. UV detection was performed at 240 nm. Hydrocortisone acetate (0.005%) in the mobile phase was used as an internal standard.

An ointment sample (5 g) was partitioned between equal volumes of hexane and the internal standard solution. The aqueous phase was collected and analyzed. The injection volume was 100 μL. The run time was 15 minutes.

The analytical method was reported to be stability indicating since it separated betamethasone 17-valerate from betamethasone, betamethasone 21-valerate, and hydro-cortisone acetate.

A calibration curve for betamethasone 17-valerate was constructed from 0 to 0.03% (wt/vol).

Reference

Mehta AC, Calvert RT. Betamethasone 17-valerate—an investigation into its stability in Betnovate after dilution with emulsifying ointment: Quantitation of degradation products. *Br J Pharm Prac*. 1982; 4: 10–3.

Betaxolol Hydrochloride

Chemical Name

1-[4-[2-(Cyclopropylmethoxy)ethyl]phenoxy]-3-(isopropylamino)propan-2-ol hydrochloride

Other Names

Betoptic, Kerlone

Form	Molecular Formula	MW	CAS
Betaxolol hydrochloride	$C_{18}H_{29}NO_3$.HCl	343.9	63659-19-8

Appearance

Betaxolol hydrochloride occurs as a white or almost white crystalline powder.

Solubility

Betaxolol hydrochloride is very soluble to freely soluble in water and freely soluble in alcohol, chloroform, and methanol. It is practically insoluble in ether.

Method

Mahalaxmi et al. reported an HPLC method for the quantification of betaxolol in its dosage forms. A Perkin-Elmer series 4 liquid chromatograph comprising a model LC-4 pump, a model LC-85B variable-wavelength spectrophotometric detector, and a model LCI-100 laboratory computing integrator was used. The stationary phase was a Perkin-Elmer C_{18} column (150 × 3.9 mm, 5-µm particle size). The mobile phase consisted of methanol, acetonitrile, and 0.1% diethylamine (adjusted to pH 3.0 with phosphoric acid). The flow rate was 1.0 mL/min. UV detection was performed at 222 nm and 1.0 AUFS. Trimethoprim 1.0 µg/mL in mobile phase was used as an internal standard.

For tablets, five tablets were ground and mixed with 35 mL of methanol. This mixture was stirred for 5 minutes, diluted to 50.0 mL with mobile phase, and filtered. The filtrate (1.0 mL) was mixed with 1.0 mL of internal standard solution and diluted with mobile phase to approximately 10.0 µg/mL of betaxolol. For gels, a gel sample equivalent to 10.0 mg of betaxolol was mixed with 30 mL of mobile phase in a 50.0-mL volumetric flask, vortexed for 10 minutes, brought to volume with mobile phase, vortexed for another 10 minutes, filtered, and diluted with additional mobile phase. The injection volume was 100 µL. Under these conditions, retention times for betaxolol and trimethoprim were about 3.7 and 5.8 minutes, respectively (estimated from the published chromatogram).

The stability-indicating nature of the assay was demonstrated by accelerated degradation of betaxolol. A betaxolol solution was treated with sulfuric acid or 2.5 N hydrochloric acid and heated to boiling. The degradation products did not interfere with the analysis of betaxolol.

A calibration curve for betaxolol was generated from 0.1 to 1.0 µg/mL. The correlation coefficient was 0.9999.

Reference

Mahalaxmi D, Samarth MM, Shiravadekar HS, et al. Stability-indicating HPLC method for betaxolol HCl and its pharmaceutical dosage forms. *Drug Dev Ind Pharm.* 1996; 22: 1037–9.

Bethanechol Chloride

Chemical Name

2-[(Aminocarbonyl)oxy]-*N,N,N*-trimethyl-1-propanaminium chloride

Other Names

Duvoid, Urecholine

Form	Molecular Formula	MW	CAS
Bethanechol chloride	$C_7H_{17}ClN_2O_2$	196.7	590-63-6

Appearance
Bethanechol chloride occurs as colorless or white crystals or as a white crystalline powder.

Solubility
Bethanechol chloride is freely soluble in water and alcohol.

Method
Allen and Erickson investigated the stability of bethanechol chloride in extemporaneously compounded oral liquids. A Hewlett-Packard series 1050 automated high-performance liquid chromatograph included a multisolvent mixing and pumping system, an autoinjector, a diode-array detector, and a computer with Chem Station software. The stationary phase was a Bakerbond phenylethyl column (250 × 4.6 mm, 5-µm particle size). The mobile phase consisted of acetonitrile and water (33:67, vol/vol). It was delivered isocratically at 0.7 mL/min. UV detection was performed at 200 nm.

Samples were diluted 1:10. Under these conditions, the retention time for bethanechol chloride was 3.0 minutes.

This method was determined to be stability indicating. A composite chromatogram of bethanechol chloride after accelerated degradation showed that degradation product peaks did not interfere with the intact bethanechol chloride peak.

A standard curve for bethanechol chloride was constructed from 100 to 500 µg/mL. The intraday and interday coefficients of variation were 3.1 and 4.2%, respectively.

Reference
Allen LV Jr, Erickson MA III. Stability of bethanechol chloride, pyrazinamide, quinidine sulfate, rifampin, and tetracycline hydrochloride in extemporaneously compounded oral liquids. *Am J Health Syst Pharm.* 1998; 55:1804–9.

Bromazepam

Chemical Name
7-Bromo-1,3-dihydro-5-(2-pyridyl)-1,4-benzodiazepin-2-one

Other Names
Lectopam, Lexotan

Form	Molecular Formula	MW	CAS
Bromazepam	$C_{14}H_{10}BrN_3O$	316.2	1812-30-2

Appearance
Bromazepam is a white or yellowish crystalline powder.

Solubility
Bromazepam is practically insoluble in water and sparingly soluble in alcohol and in methylene chloride.

Method
Ellaithy et al. reported an HPLC method for the determination of bromazepam. The instrument was a Perkin-Elmer liquid chromatograph equipped with a Perkin-Elmer model LC 410 pump and a model LC 235 diode-array detector. The stationary phase was a Phenomenex Prodigy ODS column (150 × 4.6 mm, 5-µm particle size). The mobile phase was 40% (vol/vol) acetonitrile in water. The flow rate was 2 mL/min.

A powder prepared from ground tablets equivalent to 10 mg of bromazepam was accurately weighed, transferred into a flask, and shaken with 30 mL of acetonitrile for 20 minutes. The solution was filtered into a 100-mL volumetric flask and the residue was washed twice with acetonitrile. The final volume was brought to 100 mL with acetonitrile. This solution was further diluted by a factor of 10 with acetonitrile before injection. The sample injection volume was 20 µL. Under these conditions, the retention times for bromazepam and its degradation product, 2-amino-5-bromobenzoylpyridine, were about 2.0 and 6.6 minutes, respectively.

The stability-indicating nature of the method was demonstrated by the analysis of different samples of bromazepam in the presence of its degradation product. The degradation product did not interfere with the analysis of bromazepam.

A calibration curve for bromazepam was generated from 5 to 25 µg/mL. The correlation coefficient was 0.999.

Reference
Ellaithy MM, Abdelkawy M, Tolba RM. Stability indicating methods for the determination of bromazepam. *Bull Fac Pharm Cairo Univ.* 2001; 39: 271–82.

Bryostatin 1

Chemical Name
(1S,3S,5Z,7R,8E,11S,13E,15S,17R,21R,23R,25S)-25-(Acetyloxy)-1,11,21-trihydroxy-17-[(1R)-1-hydroxyethyl]-5,13-bis(2-methoxy-2-oxoethylidene)-10,10,26,26-tetramethyl-19-oxo-18,27,28,29-tetraoxatetracyclo[21.3.1.13,7.111,15]nonacos-8-en-12-yl ester 2,4-octadienoic acid

Form	Molecular Formula	MW	CAS
Bryostatin 1	$C_{47}H_{68}O_{17}$	905.0	83314-01-6

Solubility

Bryostatin 1 has solubilities of 0.77 mg/L in water, 0.72 mg/L in 0.9% sodium chloride, and greater than 4000 mg/L in ethanol.

Method 1

Cheung et al. used an HPLC assay to evaluate the compatibility and stability of bryostatin 1 in infusion devices. The Hewlett-Packard 1050 chromatograph included a quaternary pump, an autosampler, and a fixed-wavelength UV detector equipped with an HP-1040A photodiode-array detector. The stationary phase was a Phenomenex Spherisorb ODS C_{18} column (250 × 4.6 mm, 5-μm particle size). The mobile phase consisted of 75% acetonitrile and 25% water. The flow rate was 1.0 mL/min. Pyrene solution 0.16 mg in acetonitrile–0.9% sodium chloride (1:25, vol/vol) was used as an internal standard. UV detection was performed at 260 nm. Under these conditions, retention times for bryostatin 1 and pyrene were about 11.3 and 14.2 minutes, respectively (estimated from the published chromatogram).

The stability-indicating nature of the assay was shown by accelerated degradation of drug solutions. The first bulk bryostatin 1 sample was heated for 72 hours at 70 °C and then dissolved in methanol. The second sample was exposed to UV light for 72 hours and then dissolved in methanol. The third sample was dissolved in methanol, diluted with 15% 1 N hydrochloric acid, and stored at room temperature for 24 hours. The last sample was dissolved in methanol, diluted with 15% 1 N sodium hydroxide, and then heated at 70 °C for 72 hours. In all cases, the degradation product peaks were well resolved from the peak of the intact bryostatin 1.

A calibration curve for bryostatin 1 was constructed from 0.1 to 10 μg/mL. The correlation coefficient was greater than 0.999.

Reference

Cheung AP, Hallock YF, Vishnuvajjala BR, et al. Compatibility and stability of bryostatin 1 in infusion devices. *Invest New Drugs*. 1999; 16: 227–36.

Method 2

Baer et al. developed an HPLC assay for bryostatin 1. The chromatographic system consisted of three Waters 510 pumps, a 490 UV detector, and a 710B WISP autosampler controlled by an 840 datastation. The stationary phase was a Waters μBondapak phenyl column (8 mm ID, 10-μm particle size). The mobile phase consisted of methanol and distilled water (85:15, vol/vol). The flow rate was 1.5 mL/min. UV detection was performed at 266 nm and 0.2 AUFS. Cholest-4-en-3-one was used as an internal standard. The injection volume was 10 μL. Under these conditions, retention times for bryostatin 1 and the internal standard were about 4.7 and 8.0 minutes, respectively (estimated from the published chromatogram).

The method was verified to be stability indicating by the intentional degradation of bryostatin 1. The drug was heated at 70 °C in solutions containing 0.1 M hydrochloric acid, 0.1 M sodium hydroxide, 0.1 M hydrogen peroxide, or water for 4 hours. All of the degradation product peaks were resolved from the peak of the intact bryostatin 1.

Calibration curves were generated from 25 to 75 µg/mL. The correlation coefficient was greater than 0.998. The intraday and interday coefficients of variation for the analysis were 1.1 and 3.1%, respectively. The limit of detection was 5 ng.

Reference

Baer JC, Slack JA, Pettit GR. Stability-indicating high-performance liquid chromatography assay for the anticancer drug bryostatin 1. *J Chromatogr.* 1989; 467: 332–5.

Budesonide

Chemical Name

16α,17α-Butylidenedioxy-11β,21-dihydroxypregna-1,4-diene-3,20-dione

and epimer at C*

Other Name

Rhinocort

Form	Molecular Formula	MW	CAS
Budesonide	$C_{25}H_{34}O_6$	430.5	51333-22-3

Method

Hou et al. developed an HPLC method for budesonide. The HPLC instrument consisted of a Shimadzu LC-10AD VP pump, a SIL-10AD VP autoinjector, an Applied Biosystems 783A UV detector, and a Shimadzu C-R5A Chromatopac integrator. The stationary phase was a Hypersil C_{18} column (150 × 4.6 mm, 5-µm particle size). The mobile phase consisted of ethanol, acetonitrile, and 25.6 mM phosphate buffer (pH 3.4) (2:30:68, vol/vol/vol). The flow rate was 1.5 mL/min. UV detection was performed at 240 nm. The injection volume was 20 µL. Under these conditions, retention times for the *R*-epimer and *S*-epimer of budesonide were 18 and 20 minutes, respectively.

This method was evaluated to be stability indicating by accelerated decomposition of budesonide. Solution samples of budesonide were stored at 40 °C for 3 months with continuous oxygenation. All seven major decomposition products were well resolved from the epimers of budesonide.

A calibration curve for budesonide was constructed from 2.5 to 25.0 µg/mL. The correlation coefficient was 1.00. The intraday and interday coefficients of variation for the budesonide analysis were 1.1 and 1.6%, respectively, at 10 µg/mL of budesonide. The limit of detection was 0.30 µg/mL.

Reference
Hou S, Hindle M, Byron PR. A stability-indicating HPLC assay method for budesonide. *J Pharm Biomed Anal.* 2001; 24: 371–80.

Bumetanide

Chemical Names
3-(Aminosulfonyl)-5-(butylamino)-4-phenoxybenzoic acid
3-(Butylamino)-4-phenoxy-5-sulfamoylbenzoic acid

Other Name
Bumex

Form	Molecular Formula	MW	CAS
Bumetanide	$C_{17}H_{20}N_2O_5S$	364.4	28395-03-1

Appearance
Bumetanide occurs as a practically white crystalline powder with a slightly bitter taste.

Solubility
Bumetanide has solubilities of 0.1 mg/mL in water and 30.6 mg/mL in alcohol at 25 °C.

pK$_a$
Bumetanide has pK$_a$ values of 0.3, 4, and 10.

Method
Cornish et al. studied the stability of bumetanide in 5% dextrose injection. The chromatographic system consisted of a Waters model 501 constant-flow solvent-delivery system, a Waters model 712 WISP variable-volume injector, a Waters model 991 PAD UV detector, and a recorder with Millipore Millenium 2010 Chromatography Manager. The stationary phase was a Waters µBondapak C_{18} column (300 × 3.9 mm, 10-µm particle size). The mobile phase consisted of methanol, water, and acetic acid (70:30:1, vol/vol/vol). The flow rate was 1.0 mL/min. UV detection was performed at 232 nm. The sample injection volume was 10 µL. The retention time for bumetanide was 5.5 minutes.

The stability-indicating capability of the method was demonstrated by accelerated decomposition of bumetanide. Bumetanide 1.0 mg/mL was adjusted to pH 12 with 1 N sodium hydroxide or to pH 2 with 1 N sulfuric acid, heated at 100 °C for 1 hour, and then stored at room temperature for 3 days. Bumetanide was degraded by acid; the degradation product peaks did not interfere with the intact bumetanide peak. No degradation of bumetanide occurred after treatment with base.

Standard curves for bumetanide were constructed from 0.015 to 0.025 mg/mL and from 0.15 to 0.3 mg/mL. Correlation coefficients were 0.998. The intraday and interday coefficients of variation were less than 2.6%.

Reference

Cornish LA, Montgomery PA, Johnson CE. Stability of bumetanide in 5% dextrose injection. *Am J Health Syst Pharm.* 1997; 54: 422–3.

Bupivacaine

Chemical Names

1-Butyl-*N*-(2,6-dimethylphenyl)-2-piperidinecarboxamide
dl-1-Butyl-2′,6′-pipecoloxylidide

Other Names

Marcaine, Sensorcaine

Form	Molecular Formula	MW	CAS
Bupivacaine	$C_{18}H_{28}N_2O$	288.4	2180-92-9
Bupivacaine hydrochloride	$C_{18}H_{29}ClN_2O$	324.9	18010-40-7

Appearance

Bupivacaine hydrochloride is a white odorless crystalline powder.

Solubility

Bupivacaine hydrochloride is freely soluble in water and alcohol.

pK_a

Bupivacaine hydrochloride has a pK_a of 8.1.

Method 1

Grassby and Roberts used an HPLC method to investigate the stability of epidural opiate solutions in 0.9% sodium chloride infusion bags. The chromatographic system consisted of a Kratos SF 400 pump, a Philips PU4700 20-μL loop injector, and a Uvikon 740LC UV detector. The stationary phase was a Chrompak Spherisorb ODS-2 column (250 × 4.6 mm, 5-μm particle size). The mobile phase consisted of 0.01 M heptanesulfonic acid (55%) and acetonitrile (45%) containing 7.7 mM dimethyloctylamine and adjusted to pH 3.5 with phosphoric acid. The flow rate was 1 mL/min. UV detection was performed at 254 nm.

Samples were analyzed without dilution. The retention times for diamorphine and bupivacaine hydrochloride were approximately 3.5 and 5.5 minutes, respectively.

The analytical method was determined to be stability indicating by accelerated degradation of bupivacaine hydrochloride. A bupivacaine sample was heated on a water bath for 2 hours. The degradation product did not interfere with the intact bupivacaine peak.

A standard curve for bupivacaine was constructed from 0.875 to 1.375 mg/mL. The correlation coefficient was 0.996, and the coefficient of variation was 2.52%.

Reference
Grassby PF, Roberts DE. Stability of epidural opiate solution in 0.9 percent sodium chloride infusion bags. *Int J Pharm Pract*. 1995; 3: 174–7.

Method 2
Jones and Davis evaluated the stability of bupivacaine hydrochloride injection 1.25 mg/mL in 0.9% sodium chloride injection in polypropylene syringes. The chromatograph was composed of a Beckman model 110B pump, a Beckman model 160 UV detector, an Alcott model 738 HPLC autosampler, and a Gateway 2000 386SX computer running Maxima 820 chromatography workstation software. The stationary phase was a Beckman Ultrasphere XL-ODS C_{18} column (75 × 4.6 mm, 3-μm particle size). The mobile phase was 40% acetonitrile in 0.1 M monobasic potassium phosphate buffer (pH 5). The flow rate was 1.5 mL/min. UV detection was performed at 254 nm.

The HPLC assay was determined to be stability indicating. Under the conditions of the assay, the decomposition products did not coelute with the intact bupivacaine.

The standard curve for bupivacaine was constructed from 0.31 to 1.25 mg/mL. The intraday and interday coefficients of variation were 1.2 and 2.2%, respectively.

A similar method was used by Johnson et al.

References
Jones JW, Davis AT. Stability of bupivacaine hydrochloride in polypropylene syringes. *Am J Hosp Pharm*. 1993; 50: 2364–5.

Johnson CE, Christen C, Perez MM, et al. Compatibility of bupivacaine hydrochloride and morphine sulfate. *Am J Health Syst Pharm*. 1997; 54: 61–4.

Method 3
Using an HPLC method, Tu et al. studied the stability of bupivacaine hydrochloride and fentanyl citrate in portable pump reservoirs. The liquid chromatograph consisted of a Waters model 6000A pump, a Beckman model 210A loop injector, and a Beckman model 163 variable-wavelength UV detector. The stationary phase was a Waters Radial-Pak phenyl column (100 × 3.9 mm, 4-μm particle size). The mobile phase consisted of 65% methanol and 35% 0.005 M monobasic potassium phosphate buffer at pH 4.8,

including 0.0014 M tetrabutylammonium hydroxide. The flow rate was 3.0 mL/min. UV detection was performed at 210 nm.

Samples were analyzed without dilution. The injection volume was 20 μL. Under these conditions, retention times for bupivacaine hydrochloride and fentanyl citrate were 7.5 and 11.6 minutes, respectively.

The stability-indicating capability of the method was demonstrated by accelerated decomposition of bupivacaine. A solution of bupivacaine hydrochloride (2 mg/mL in 0.3 N hydrochloric acid) was stored at 140 °C for 44 hours. The degradation products eluted separately and their peaks did not interfere with the intact bupivacaine peak.

Standard curves for bupivacaine hydrochloride were constructed each day from 200 to 2200 μg/mL; the correlation coefficients were 0.9999. The intraday and interday coefficients of variation were 1.87 and 0.99%, respectively.

Similar methods were used by the other researchers cited here.

References

Tu Y-H, Stiles ML, Allen LV. Stability of fentanyl citrate and bupivacaine hydrochloride in portable pump reservoirs. *Am J Hosp Pharm.* 1990; 47: 2037–40.

Christen C, Johnson CE, Walters JR. Stability of bupivacaine hydrochloride and hydromorphone hydrochloride during simulated epidural coadministration. *Am J Health Syst Pharm.* 1996; 53: 170–3.

Allen LV, Stiles ML, Wang DP, et al. Stability of bupivacaine hydrochloride, epinephrine hydrochloride, and fentanyl citrate in portable infusion-pump reservoirs. *Am J Hosp Pharm.* 1993; 50: 714–5.

Buspirone Hydrochloride

Chemical Name
8-[4-[4-(2-Pyrimidinyl)-1-piperazinyl]butyl]-8-azaspiro[4.5]decane-7,9-dione monohydrochloride

Other Names
Ansial, Bespar, Buspar

Form	Molecular Formula	MW	CAS
Buspirone	$C_{21}H_{31}N_5O_2$	385.5	36505-84-7
Buspirone hydrochloride	$C_{21}H_{31}N_5O_2 \cdot HCl$	422.0	33386-08-2

Appearance

Buspirone hydrochloride occurs as a white crystalline powder.

Solubility

Buspirone hydrochloride is very soluble in water, sparingly soluble in alcohol and acetonitrile, and freely soluble in methyl alcohol.

Method

Khedr and Sakr described an assay of buspirone hydrochloride, its impurities, and its degradation products. The liquid chromatographic system consisted of an Beckman 126 pump, a Beckman 507e autosampler with a 20-μL sample loop, a Rheodyne model 7010-122 injector, and a Beckman 168 diode-array detector. The stationary phase was a Beckman Ultrasphere ODS column (250 × 4.6 mm, 5-μm particle size) with an Upchurch C_{18} precolumn (20 × 2 mm, 5-μm particle size). The column temperature was maintained at 40 °C using a Beckman 235 block heater. Mobile phase A was prepared by dissolving 1.36 g of monobasic potassium phosphate in 1 liter of water; it was adjusted to pH 6.9 with 10% sodium hydroxide solution. Mobile phase B was a mixture of methanol and acetonitrile (17:13, vol/vol). The mobile phase was delivered at 65% A for 5 minutes and then linearly decreased to 46% A in another 5.5 minutes. The flow rate was 1.4 mL/min. UV detection was performed at 244 nm.

The sample injection volume was 20 μL. Under these conditions, the retention times of buspirone acid hydrochloride, 1,4-bis[4-(2-pyrimidinyl)piperazine-1-yl]-butane, 8-(4-chlorobutyl)-8-azaspiro[4,5]decane-7,9-dione, buspirone hydrochloride, and 1,4-bis(8-azaspiro[4,5]decane-7,9-dione-8-yl)-butane were 4.5, 18.1, 21.7, 23.6, and 30.2 minutes, respectively.

The method was demonstrated to be stability indicating. Buspirone hydrochloride was (1) stored at 60 °C for 14 days, (2) prepared in 5 N hydrochloric acid and heated at 100 °C for 1 hour, (3) prepared in 0.1 N sodium hydroxide solution and 0.1 N sodium carbonate basic media and heated for up to 15 minutes, (4) prepared in 1% hydrogen peroxide and heated at 100 °C for 30 minutes, or (5) exposed to the sunlight for 30 days. In all cases, buspirone hydrochloride was well resolved chromatographically from its impurities and degradation products.

A calibration curve for buspirone hydrochloride was constructed from 5.06 to 202.40 ng/μL, using a least-squares regression of the peak area versus concentration. The correlation coefficient was 0.9999. The limit of detection was 0.5 ng/μL, and the limit of quantitation was 2.53 ng/μL.

Reference

Khedr A, Sakr A. Stability-indicating high-performance liquid chromatographic assay of buspirone HCl. *J Chromatogr Sci.* 1999; 37: 462–8.

Busulfan

Chemical Names

1,4-Butanediol dimethanesulfonate

1,4-Bis(methanesulfonoxy)butane

Other Names
Busulphan, Myleran

Form	Molecular Formula	MW	CAS
Busulfan	$C_6H_{14}O_6S_2$	246.3	55-98-1

Appearance
Busulfan is a white crystalline powder.

Solubility
Busulfan has solubilities of 2.4 g/100 mL in acetone at 25 °C and 0.1 g/100 mL in alcohol. It is practically insoluble in water but will dissolve slowly as hydrolysis occurs.

Method
Xu et al. evaluated the stability of busulfan 0.5 and 0.1 mg/mL in 5% dextrose injection and 0.9% sodium chloride injection. The liquid chromatograph consisted of a Waters 600E multisolvent-delivery pump, a Waters 490E programmable multiwavelength UV detector, and a Waters 712 WISP autosampler. The system was controlled and integrated by an NEC PowerMate SX/16 personal computer. The stationary phase was a Customsil C_{18} reversed-phase HPLC analytical column (250 × 4.6 mm, 5-µm particle size). The mobile phase consisted of acetonitrile, tetrahydrofuran, and water (55:20:25, vol/vol/vol) and was delivered isocratically at 1 mL/min. UV detection was performed at 254 nm and 0.5 AUFS.

Busulfan was derivatized with sodium diethyldithiocarbamate, extracted with ethyl acetate, and then centrifuged at 3000 rpm for 10 minutes. Two milliliters of the ethyl acetate layer was evaporated to dryness and reconstituted with 2 mL of mobile phase. The injection volume was 20 µL.

The HPLC analytical method was determined to be stability indicating by accelerated decomposition of busulfan. After heating a busulfan solution at 95 °C for 10 minutes, a 90% reduction in peak area for the intact busulfan derivative was observed. No new peaks interfered with the busulfan derivative peak.

The busulfan standard curve was constructed from a linear plot of peak area versus concentration of the busulfan reference standard from 25 to 150 µg/mL. The correlation coefficient of the working curves was greater than 0.9998. For a nominal 0.1-mg/mL busulfan solution, the mean ± *SD* precision of assay, determined from 10 replicate injections, was 99.3 ± 1.2 µg/mL. The intraday and interday coefficients of variation were 1.7 and 2.9%, respectively.

A similar method was reported by Chow et al.

References
Xu QA, Zhang YP, Trissel LA, et al. Stability of busulfan injection admixtures in 5% dextrose injection and 0.9% sodium chloride injection. *J Oncol Pharm Pract*. 1996; 2: 101–5.

Chow DS, Bhagwatwar HP, Phadungpojna S, et al. Stability-indicating high-performance liquid chromatographic assay of busulfan in aqueous and plasma samples. *J Chromatogr B*. 1997; 704: 277–88.

Butabarbital Sodium

Chemical Name
Sodium 5-*sec*-butyl-5-ethylbarbiturate

Other Names
Butisol Sodium, Secbutobarbitone Sodium

Form	Molecular Formula	MW	CAS
Butabarbital	$C_{10}H_{16}N_2O_3$	212.2	125-40-6
Butabarbital sodium	$C_{10}H_{15}N_2NaO_3$	234.2	143-81-7

Appearance
Butabarbital is a white odorless crystalline powder. Butabarbital sodium is a white powder.

Solubility
Butabarbital is very slightly soluble in water and soluble in alcohol, chloroform, ether, and aqueous solutions of alkali hydroxides and carbonates. Butabarbital sodium is freely soluble in water and alcohol.

pK_a
Butabarbital sodium has a pK_a of 8.0.

Method
Scott reported an HPLC method coupled with postcolumn ionization for the analysis of butabarbital sodium elixir. The system consisted of a Waters model 6000A solvent-delivery system, a model 720 system controller, a model U6K injector, a model 440 absorbance detector with a 254-nm filter, and a model 730 data module. Another Waters model 6000A solvent pump was installed in the line between the column and the detector. The stationary phase was a Waters μBondapak C_{18} column (300 × 3.9 mm, 10-μm particle size). The mobile phase consisted of methanol–distilled water (35:65, vol/vol). The flow rate was 1.5 mL/min. The ionization solvent was prepared by mixing 250 mL of 0.2 M boric acid, 250 mL of 0.2 M potassium chloride, and 220 mL of 0.2 M sodium hydroxide and diluting to 1 L with distilled water. This buffer (pH 10.0) was delivered into the mobile phase at a flow rate of 0.1 mL/min. UV detection was performed at 254 nm and 0.1 AUFS. The injection volume was about 7.5 μL. Phenobarbital in methanol 3 mg/mL was used as an internal standard. Under these conditions, retention times for phenobarbital and butabarbital were about 9.0 and 12.8 minutes, respectively.

The assay was shown to be stability indicating by thermal degradation of butabarbital sodium. The degradation products capuride and valnoctamide did not interfere with the intact drug on the chromatogram.

A standard curve for butabarbital sodium was constructed from 0.35 to 1.05 mg/mL. The correlation coefficient was 0.9999.

Reference
Scott EP. Application of postcolumn ionization in the high-performance liquid chromatographic analysis of butabarbital sodium elixir. *J Pharm Sci*. 1983; 72: 1089–91.

Butorphanol

Chemical Names
17-(Cyclobutylmethyl)morphinan-3,14-diol
(−)-N-Cyclobutylmethyl-3,14-dihydroxymorphinan

Other Name
Stadol

Form	Molecular Formula	MW	CAS
Butorphanol	$C_{21}H_{29}NO_2$	327.5	42408-82-2
Butorphanol tartrate	$C_{21}H_{29}NO_2.C_4H_6O_6$	477.6	58786-99-5

Appearance
Butorphanol tartrate is a white powder.

Solubility
Butorphanol tartrate is sparingly soluble in water and insoluble in alcohol.

pK$_a$
Butorphanol tartrate has a pK$_a$ of 8.6.

Method
Willey et al. developed an HPLC analytical method for the quantitative determination of butorphanol using fluorescence detection. The chromatograph consisted of a Waters model 510 solvent pump, a Waters model 712 WISP autosampler, a Kratos 980 fluorescence detector, a Linear Instruments model 585 flatbed recorder, and a Hewlett-Packard model 3357 Laboratory Automation System computer. The stationary phase was a Jones Chromatography C$_8$ column (250 × 4.6 mm, 5-μm particle size). The mobile phase was a mixture of 20 mL of a 1.0 M tetramethylammonium hydroxide–1.0 M ammonium acetate solution (pH 6.0 adjusted with acetic acid), 1380 mL of deionized water, and 400 mL of acetonitrile. The mobile phase was filtered through a Millipore 0.22-μm Durapore filter. The flow rate was 1.0 mL/min. The excitation and emission wavelengths of the fluorescence detector were set at 200 and 325 nm, respectively.

Four milliliters of sample solution was evaporated to dryness, and the residue was reconstituted with 150 μL of acetonitrile, methanol, and water (20:10:70, vol/vol/vol) containing 10 mM ammonium acetate and 10 mM tetramethylammonium hydroxide (pH 5.0 with acetic acid). The injection volume was 75 μL. Under these conditions, the retention time for butorphanol was 23.4 minutes.

The HPLC method was evaluated to be stability indicating by examining the separation of butorphanol, hydroxybutorphanol, and norbutorphanol. No interference with butorphanol was observed. The retention times for hydroxybutorphanol, norbutorphanol, and butorphanol were 5.5, 9.0, and 23.4 minutes, respectively.

Calibration curves were constructed from 1 to 100 ng/mL. The correlation coefficient was 0.995. The intraday and interday variations were less than 8 and 10%, respectively.

Reference

Willey TA, Duncan GF, Tay LK, et al. High-performance liquid chromatographic method for the quantitative determination of butorphanol, hydroxybutorphanol, and norbutorphanol in human urine using fluorescence detection. *J Chromatogr B*. 1994; 652: 171–8.

Butylparaben

Chemical Name

Butyl hydroxybenzoate

Form	Molecular Formula	MW	CAS
Butylparaben	$C_{11}H_{14}O_3$	194.2	94-26-8

Appearance

Butylparaben occurs as small colorless crystals or as a white or almost white crystalline powder.

Solubility

Butylparaben is very slightly soluble in water. It is freely soluble in alcohol, acetone, ether, and methanol.

Method

Radus and Gyr described an HPLC method for the determination of methylparaben and butylparaben in pharmaceutical formulations. The liquid chromatograph consisted of a Waters model 6000A pump or a Milton-Roy minipump, a Rheodyne model 7126 injector, a Waters model 440 or LDC model 1203 UV detector, and an Upjohn autosampler. The stationary phase was a Waters μBondapak C_{18} column. The mobile phase was a mixture of water, acetonitrile, and glacial acetic acid (58:40:2, vol/vol/vol) and was delivered isocratically at 2 mL/min. Calusterone in water–acetonitrile was used as an internal standard.

An ointment sample (1.0 g) was mixed with 10 mL of dimethylformamide and 20 mL of internal standard, shaken at 50 °C, and centrifuged. The injection volume was 10 μL. Under these conditions, retention times for methylparaben, ethylparaben, propylparaben, and butylparaben were about 3.7, 4.6, 6.3, and 9.3 minutes, respectively (estimated from the published chromatogram).

The method was reported to be stability indicating. The primary degradation product, *p*-hydroxybenzoic acid, eluted with the solvent front.

A standard curve for butylparaben was constructed from 0.9 to 2.7 mg/g.

Similar methods were used by these authors to analyze butylparaben in creams, lotions, sterile solutions, and suspensions.

Reference
Radus TP, Gyr G. Determination of antimicrobial preservatives in pharmaceutical formulations using reverse-phase liquid chromatography. *J Pharm Sci*. 1983; 72: 221–4.

Caffeine

Chemical Names
3,7-Dihydro-1,3,7-trimethyl-1*H*-purine-2,6-dione
1,3,7-Trimethylxanthine

Other Name
Cafedrine

Form	Molecular Formula	MW	CAS
Caffeine	$C_8H_{10}N_4O_2$	194.2	58-08-2
Caffeine citrate	$C_8H_{10}N_4O_2 \cdot C_6H_8O_7$	386.3	69-22-7

Appearance
Caffeine and caffeine citrate occur as white powders.

Solubility
Caffeine is sparingly soluble in water and alcohol. Caffeine citrate is freely soluble in water and soluble in alcohol.

Method 1
Mutch and Hutson used HPLC to evaluate the stability of caffeine with antipyrine in an intravenous solution. A Waters Nova-Pak RCSS phenyl 4-μm column was used with a C_{18} guard column. The mobile phase consisted of 0.01 M potassium phosphate buffer, methanol, and tetrahydrofuran (79:20:1, vol/vol/vol) at pH 3.0. The flow rate was 2 mL/min. UV detection was performed at 214 nm by a Beckman model 163 UV detector.

Samples were diluted 1:200 with deionized distilled water; injection volume was 20 μL. Retention times for caffeine and antipyrine were 10.4 and 12.6 minutes, respectively.

The stability-indicating nature of this assay was demonstrated by accelerated decomposition. Samples were heated with a water bath (40 or 60 °C) or a steam autoclave (121 °C). Samples were also exposed to fluorescent light. Degradation product peaks did not interfere with the intact caffeine peak.

The intraday and interday coefficients of variation for the test solutions of caffeine were 1.06 and 1.01%, respectively.

Reference

Mutch RS, Hutson PR. Stability of antipyrine plus caffeine in an intravenous solution. *Am J Hosp Pharm*. 1991; 48: 1267–70.

Method 2

Using a specific stability-indicating HPLC method, Nahata et al. investigated the stability of caffeine citrate injection 10 mg/mL in intravenous admixtures and parenteral nutrition solutions. The stationary phase was an Altex Ultrasphere ODS 5-µm analytical column (150 × 4.6 mm). The mobile phase consisted of 5 mM tetrabutylammonium hydrogen sulfate and 10 mM sodium acetate in 13.5% methanol. A Varian model 2050 UV detector was set at 274 nm.

The method was stability indicating; the degradation peaks of caffeine citrate exposed to pH 12.8 did not interfere with the peak of caffeine.

Standard curves for caffeine citrate were constructed. The coefficient of variation was less than 4%.

References

Nahata MC, Zingarelli J, Durrell DE. Stability of caffeine citrate injection in intravenous admixtures and parenteral nutrition solutions. *J Clin Pharm Ther*. 1989; 14: 53–5.

Nahata MC, Zingarelli JR, Durrell DE. Stability of caffeine injection in intravenous admixtures and parenteral nutrition solutions. *Ann Pharmacother*. 1989; 23: 466–7.

Calcitriol

Chemical Names

(1α,3β,5Z,7E)-9,10-Secocholesta-5,7,10(19)-triene-1,3,25-triol
1α,25-Dihydroxycholecalciferol

Other Name
Calcijex

Form	Molecular Formula	MW	CAS
Calcitriol	$C_{27}H_{44}O_3$	416.6	32222-06-3

Appearance
Calcitriol occurs as colorless crystals.

Solubility
Calcitriol is insoluble in water and slightly soluble in methanol, ethanol, ethyl acetate, and tetrahydrofuran.

Method
Pecosky et al. used HPLC to study the stability of calcitriol in infusion solutions. The chromatograph consisted of a Shimadzu model LC-6A liquid chromatographic pump, a model SIL-9A autoinjector, a model SPD-6A variable-wavelength UV detector, and a model CR501 integrator. The stationary phase was a Waters Resolve reversed-phase nonpolar column (150 × 3.9 mm, 5-µm particle size). The mobile phase was 69% methanol (vol/vol) in water. The flow rate was 1.4 mL/min. UV detection was performed at 264 nm. The sample injection volume was 150 µL. Calcitriol retention times were between 15 and 20 minutes.

The stability-indicating nature of this HPLC method was demonstrated by assaying a sample of calcitriol injection that was exposed to air for 2 days. The chromatogram showed no interference from the degradation product peaks with the intact calcitriol peak.

Standard curves were constructed from a linear plot of peak area versus calcitriol concentration from 0.4 to 0.8 µg/mL. The correlation coefficient was 0.9990; the coefficient of variation of the assay was 3.0%.

Reference
Pecosky DA, Parasrampuria J, Li LC, et al. Stability and sorption of calcitriol in plastic tuberculin syringes. *Am J Hosp Pharm.* 1992; 49: 1463–6.

Calcium Cyanamide

Chemical Name
Calcium cyanamide

$$N \equiv CN = Ca$$

Other Names
Abstem, Calcium Carbimide, Colme, Temposil

Form	Molecular Formula	MW	CAS
Calcium cyanamide	$CCaN_2$	80.1	156-62-7

Solubility
Calcium cyanamide is essentially insoluble in water but undergoes partial hydrolysis to the soluble calcium hydrogen cyanamide.

Method
Chen et al. developed a stability-indicating HPLC method for the analysis of calcium cyanamide. The HPLC system consisted of a SSI 222B pump, a Waters WISP 712 autosampler, a Kratos 783 variable-wavelength UV detector, and a Linear Instruments model 1200 chart recorder. The stationary phase was a Waters µBondapak C_{18} analytical column (150 × 3.9 mm, 10-µm particle size). The acetonitrile concentration and the pH of the mobile phase were modified to obtain optimum separation. Calcium cyanamide in samples was converted into dansyl cyanamide with dansyl chloride. An increase in the acetonitrile concentration or in pH decreased the retention time of dansyl cyanamide. The mobile phase consisted of 0.01 M sodium phosphate (pH 6.3) and acetonitrile (75:25, vol/vol). The flow rate was 1 mL/min. The UV detector was set at 254 nm and 0.05 AUFS. The sample injection volume was 10 µL. The retention time for dansyl cyanamide was about 5.3 minutes (estimated from the published chromatogram).

The HPLC method was determined to be stability indicating by accelerated degradation of calcium cyanamide samples by UV light, heat, base, acid, and hydrogen peroxide. Chromatograms showed that calcium cyanamide was stable after acid, base, and UV treatment but lost 20 and 100% after heat and hydrogen peroxide treatments, respectively.

Calibration curves were constructed from peak area versus calcium cyanamide concentration from 0.1 to 0.5 mg/mL. The correlation coefficient was 0.9999. The intraday assay variations for the bulk material and tablets were 0.20 and 0.67%, respectively. The interday coefficient of variation was 1.13%.

Reference
Chen S, Ocampo AP, Kucera PJ. Liquid chromatographic method for the determination of calcium cyanamide using precolumn derivatization. *J Chromatogr*. 1991; 558: 141–6.

Calcium Pantothenate

Chemical Name
N-(2,4-Dihydroxy-3,3-dimethyl-1-oxobutyl)-β-alanine calcium salt

Other Name
Calpanate

Form	Molecular Formula	MW	CAS
Calcium pantothenate	$C_{18}H_{32}CaN_2O_{10}$	476.5	6381-63-1

Appearance
Calcium pantothenate is a slightly hygroscopic odorless white powder.

Solubility
Calcium pantothenate has an aqueous solubility of 1 g in 2.8 mL. It is soluble in glycerol and slightly soluble in alcohol and acetone.

Method
Timmons et al. developed a reversed-phase HPLC method for the assay of calcium pantothenate in commercial multivitamin tablet formulations and raw materials. The liquid chromatographic system consisted of a Beckman model 110A reciprocating piston pump, a Rheodyne model 7126 fixed-loop injector fitted with a 20-μL loop, a Kratos model 773 variable-wavelength UV detector, an Omniscribe D-5000 strip-chart recorder, a DuPont model 834 autosampler, and a Hewlett-Packard model 3357 digital computer with chromatographic data integration software. The stationary phase was an Alltech Hypersil ODS reversed-phase C_{18} analytical column (150 × 4.6 mm, 5-μm particle size). The mobile phase was a mixture of 3% acetonitrile and 97% 0.25 M monobasic sodium phosphate aqueous buffer which had been adjusted to pH 2.5 with phosphoric acid. The flow rate was 2.0 mL/min. UV detection was performed at 205 nm and 0.1 AUFS.

Samples were dissolved in water and filtered through a 0.45-μm filter. The retention time for pantothenate was about 4.6 minutes (estimated from the published chromatogram).

The stability-indicating capacity of the analytical method was confirmed by assaying samples degraded under accelerated conditions (1 month at 80 °C). Related substances were also examined. None of the peaks interfered with the intact pantothenate peak.

The coefficient of variation was 0.4%. The intraday coefficient of variation ranged from 0.3 to 2.0%.

Reference
Timmons JA, Meyer JC, Steibble OJ, et al. Reversed-phase liquid chromatographic assay for calcium pantothenate in multivitamin preparations and raw materials. *J Assoc Off Anal Chem.* 1987; 70: 510–3.

Captopril

Chemical Names
(*S*)-1-(3-Mercapto-2-methyl-1-oxopropyl)-L-proline
(2*S*)-1-(3-Mercapto-2-methylpropionyl)-L-proline

Other Name
Capoten

Form	Molecular Formula	MW	CAS
Captopril	$C_9H_{15}NO_3S$	217.3	62571-86-2

Appearance
Captopril is a white to off-white crystalline powder.

Solubility
Captopril is freely soluble in water, alcohol, chloroform, and methylene chloride.

pK_a
Captopril has pK_a values of 3.7 and 9.8.

Method 1
Allen and Erickson studied the stability of captopril 0.75 mg/mL in extemporaneously compounded oral liquids by using HPLC. A Hewlett-Packard series 1050 automated high-performance liquid chromatograph with Chem Station software consisted of a multisolvent mixing and pumping system, an autoinjector, a diode-array detector, and a computer. The stationary phase was a Bakerbond C_{18} analytical column (250 × 4.6 mm, 5-μm particle size). The mobile phase was methanol and water (11:9, vol/vol) with 0.5 mL of phosphoric acid. The flow rate was 1.1 mL/min. UV detection was performed at 220 nm.

Samples were diluted 1:100. Under these conditions, the retention time for captopril was 5.1 minutes.

The HPLC assay was determined to be stability indicating. A composite chromatogram of captopril after degradation by heat, acid, base, oxidizing agent, and light showed that the peak of the intact captopril was well separated from the peaks of the degradation products.

Standard curves were constructed for captopril from 1 to 10 μg/mL. The intraday and interday coefficients of variation were 0.9 and 1.4%, respectively.

Reference

Allen LV, Erickson MA. Stability of baclofen, captopril, diltiazem hydrochloride, dipyridamole, and flecainide acetate in extemporaneously compounded oral liquids. *Am J Health Syst Pharm.* 1996; 53: 2179–84.

Method 2

Nahata et al. used HPLC to investigate the stability of captopril 1 mg/mL in three liquid dosage forms: syrup, distilled water, and sodium ascorbate in distilled water. The chromatographic system consisted of a Hewlett-Packard 1050 series pump, an auto-sampler, a variable-wavelength UV detector, and a 3396A integrator. The stationary phase was a Waters μBondapak C_{18} analytical column (300 × 3.9 mm, 10-μm particle size). The mobile phase consisted of 0.1% phosphoric acid and methanol (45:55, vol/vol) which had been filtered through a 0.45-μm nylon 66 filter. The flow rate was 1.0 mL/min. The UV detector was set at 260 nm. Hydrochlorothiazide was used as the internal standard.

Samples were diluted with the mobile phase. The injection volume was 10 μL. Retention times for the internal standard and captopril were 5.6 and 7.1 minutes, respectively.

The stability-indicating ability of this assay was determined by accelerated degradation of captopril. Captopril sample solution was mixed with 0.03% hydrogen peroxide and incubated at 22 °C for 1 hour. The chromatogram showed that the intact captopril peak at 7.1 minutes was well separated from its degradation product peaks at 9.4 and 14.7 minutes.

Calibration curves were constructed from a linear plot of the captopril concentration versus the ratio of the peak height for captopril to that of the internal standard from 0.25 to 2.00 mg/mL. The correlation coefficient was greater than 0.999. The coefficient of variation was less than 3%.

Reference

Nahata MC, Morosco RS, Hipple TF. Stability of captopril in three liquid dosage forms. *Am J Hosp Pharm.* 1994; 51: 95–6.

Method 3

Taketomo et al. evaluated the stability of captopril powder using HPLC. The chromato-graphic system included a Perkin-Elmer model 410 LC pump, a model ISS-100 autosampler, a model LC90 UV detector, and a Nelson analytical series 3000 chromato-graphy data reduction system. The stationary phase was a Partisil ODS-2 analytical column (250 × 4.6 mm, 10-μm particle size). The mobile phase was a mixture of methanol, deionized water, and phosphoric acid (550:450:0.5, vol/vol/vol). The flow rate was 1.0 mL/min. UV detection was performed at 212 nm. Hydrochlorothiazide was used as an internal standard.

Samples were centrifuged before analysis. The injection volume was 4 μL. Under these conditions, retention times for the internal standard, captopril, and the degradation product (captopril disulfide) were 2.9, 4.9, and 8.4 minutes, respectively.

The HPLC method was reported to be stability indicating and capable of separating captopril from pharmaceutical excipients, degradation products, and impurities.

The standard curve for captopril was constructed by linear least-squares regression analysis of the peak area versus the ratio of the concentration of the internal standard to the concentration of captopril from 100 to 500 μg/mL. The intraday coefficient of variation was 3.5%.

Similar stability-indicating HPLC methods were reported by other researchers.

References

Taketomo CK, Chu SA, Cheng MH, et al. Stability of captopril in powder papers under three storage conditions. *Am J Hosp Pharm*. 1990; 47: 1799–801.

Anaizi NH, Swenson C. Instability of aqueous captopril solutions. *Am J Hosp Pharm*. 1993; 50: 486–8.

Sam WJ, Ho PC. Stability of captopril in invert sugar solution. *J Clin Pharm Ther*. 1998; 23: 451–6.

Carbamazepine

Chemical Names

5*H*-Dibenz[*b,f*]azepine-5-carboxamide
5-Carbamoyl-5*H*-dibenz[*b,f*]azepine

Other Name

Tegretol

Form	Molecular Formula	MW	CAS
Carbamazepine	$C_{15}H_{12}N_2O$	236.3	298-46-4

Appearance

Carbamazepine is a white to off-white powder.

Solubility

Carbamazepine is practically insoluble in water but soluble in alcohol, acetone, and propylene glycol.

Method 1

Handa et al. reported an HPLC method for the determination of carbamazepine. A Jasco liquid chromatograph was equipped with a model 880-PU pump, a model 875 variable-wavelength UV detector, and a Rheodyne model 7125 injector with a 20-µL fixed loop. The stationary phase was a Finepak C_8 analytical column (250 × 4.6 mm) with a Corasil C_{18} guard column (30 mm, 35–45-µm particle size). The mobile phase was methanol and water (70:30, vol/vol). The flow rate was 1 mL/min. UV detection was performed at 254 nm. Sodium phenytoin was used as an internal standard. Under these conditions, retention times for sodium phenytoin and carbamazepine were about 3.7 and 5.4 minutes, respectively.

The method was demonstrated to be stability indicating by accelerated decomposition of carbamazepine. A carbamazepine sample was refluxed with 1 N hydrochloric acid for 2 hours. Carbamazepine was well separated from its degradation products.

A linear relationship between the peak area and carbamazepine concentration from 0.04 to 20 µg/mL was obtained. The correlation coefficient was 0.999.

Reference
Handa AK, Shedbalkar VP, Bhalla HL. Stability indicating HPLC method for carbamazepine. *Indian Drugs*. 1996; 33: 559–62.

Method 2
Lowe et al. used HPLC to study the stability of carbamazepine in a commercially available suspension repackaged in various containers. The liquid chromatograph consisted of a Varian model 5020 pump, a Gilson model 116 variable-wavelength UV detector, and a Shimadzu model C-R6A integrator. The stationary phase was an Alltech Spherisorb ODS-2 reversed-phase column (250 × 4.6 mm). The mobile phase was a mixture of 65% methanol in distilled water. The flow rate was 1.0 mL/min. UV detection was performed at 212 nm and 0.2 AUFS. Carbamazepine 10,11-epoxide was used as an internal standard.

Samples were diluted with methanol. The injection volume was 50 µL.

The stability-indicating nature of the HPLC assay was demonstrated by accelerated degradation. A chromatogram of carbamazepine treated with concentrated hydrochloric acid and autoclaving showed that breakdown products of carbamazepine did not interfere with the intact carbamazepine peak.

A calibration curve was constructed for carbamazepine from 5 to 30 µg/mL.

A similar HPLC method was described by Clark-Schmidt et al.

References
Lowe DR, Fuller SH, Pesko LJ, et al. Stability of carbamazepine suspension after repackaging into four types of single-dose containers. *Am J Hosp Pharm*. 1989; 46: 982–4.

Clark-Schmidt AL, Garnett WR, Lowe DR, et al. Loss of carbamazepine suspension through nasogastric feeding tubes. *Am J Hosp Pharm*. 1990; 47: 2034–7.

Carbenicillin

Chemical Name
[2S-(2α,5α,6β)]-6-[(Carboxyphenylacetyl)amino]-3,3-dimethyl-7-oxo-4-thia-1-azabicyclo[3.2.0]heptane-2-carboxylic acid

Other Name
Geocillin

Form

Form	Molecular Formula	MW	CAS
Carbenicillin	$C_{17}H_{18}N_2O_6S$	378.4	4697-36-3
Carbenicillin disodium	$C_{17}H_{16}N_2Na_2O_6S$	422.4	4800-94-6
Carbenicillin indanyl sodium	$C_{26}H_{25}N_2NaO_6S$	516.6	26605-69-6

Appearance
Carbenicillin disodium is a white to off-white hygroscopic powder. Carbenicillin indanyl sodium is a white to off-white powder.

Solubility
Carbenicillin disodium is freely soluble in water. It is soluble in alcohol and methanol but is practically insoluble in chloroform and ether. Carbenicillin indanyl sodium is soluble in water and alcohol.

Method 1
Grover et al. determined by HPLC the degradation rate of carbenicillin at 35 °C in borate buffer (pH 9.2). The Shimadzu LC-10A liquid chromatograph included model LC-10AS pumps, a model SPD-10A dual-wavelength detector, a model C-R7A data processor, and a Rheodyne model 7125 injector with a 20-μL loop. The stationary phase was a Phenomenex C_{18} column (300 × 3.9 mm, 5-μm particle size). The mobile phase was a mixture of 92% water containing 20 mM monobasic potassium phosphate and 10 mM tetramethylammonium chloride (pH 5.0) and 8% methanol. The flow rate was 1.2 mL/min. UV detection was performed at 220 and 240 nm.

Samples were incubated at 35 °C and assayed. Under these conditions, the retention time for carbenicillin was 29.3 minutes. The method was reported to be stability indicating.

A standard curve for carbenicillin was generated from 5 to 150 μg/mL. The correlation coefficient was greater than 0.99. The coefficient of variation for the analysis of carbenicillin was less than 1%.

Reference
Grover M, Gulati M, Singh B, et al. Correlation of penicillin structure with rate constants for basic hydrolysis. *Pharm Pharmacol Commun*. 2000; 6: 355–63.

Method 2
Using a stability-indicating HPLC method, Gupta and Stewart studied the stability of carbenicillin disodium in 0.9% sodium chloride injection and 5% dextrose injection. The high-

performance liquid chromatograph consisted of a Waters model ALC 202 system with a U6K universal injector, a Schoeffel Spectroflow monitor SF 770 multiple-wavelength detector, a Houston Omniscribe recorder, and a Spectra-Physics autolab minigrator-integrator. The stationary phase was a Waters µBondapak phenyl column (300 × 4 mm). The mobile phase was 0.01 M ammonium acetate in water. The flow rate was 1.6 mL/min. UV detection was performed at 245 nm and 0.04 AUFS. Samples were diluted with water. The injection volume was 20 µL.

The analytical method was stated to be stability indicating.

The coefficient of variation was 1.69%.

References
Gupta VD, Stewart KR. Quantitation of carbenicillin disodium, cefazolin sodium, cephalothin sodium, nafcillin sodium, and ticarcillin disodium by high-pressure liquid chromatography. *J Pharm Sci.* 1980; 69: 1264–7.

Gupta VD, Stewart KR. Chemical stabilities of hydrocortisone sodium succinate and several antibiotics when mixed with metronidazole injection for intravenous infusion. *J Parenter Sci Technol.* 1985; 39: 145–9.

Carbenicillin Phenyl Sodium

Chemical Name
Sodium (6*R*)-6-(2-phenoxycarbonyl-2-phenylacetamido)penicillanate

Other Names
Carfecillin, Urocarf

Form	Molecular Formula	MW	CAS
Carbenicillin phenyl sodium	$C_{23}H_{21}N_2NaO_6S$	476.5	21649-57-0

Method
Grover et al. studied by HPLC the degradation rate of carbenicillin phenyl sodium (carfecillin) at 35 °C in borate buffer (pH 9.2). The Shimadzu LC-10A liquid chromatograph included model LC-10AS pumps, a model SPD-10A dual-wavelength detector, a model C-R7A data processor, and a Rheodyne model 7125 injector with a 20-µL loop. The stationary phase was a Phenomenex C_{18} column (300 × 3.9 mm, 5-µm particle size).

The mobile phase was a mixture of 71% water containing 20 mM monobasic potassium phosphate and 10 mM tetramethylammonium chloride (pH 5.0) and 29% acetonitrile. The flow rate was 1.0 mL/min. UV detection was performed at 220 and 240 nm.

Samples were incubated at 35 ℃ and assayed. Under these conditions, the retention time for carbenicillin phenyl sodium was 9.8 minutes. The method was reported to be stability indicating.

A standard curve for carbenicillin phenyl sodium was generated from 5 to 150 µg/mL. The correlation coefficient was greater than 0.99. The coefficient of variation was less than 1%.

Reference

Grover M, Gulati M, Singh B, et al. Correlation of penicillin structure with rate constants for basic hydrolysis. *Pharm Pharmacol Commun*. 2000; 6: 355–63.

Carbidopa

Chemical Name

(+)-2-(3,4-Dihydroxybenzyl)-2-hydrazinopropionic acid monohydrate

Other Name

Lodosyn

Form	Molecular Formula	MW	CAS
Carbidopa	$C_{10}H_{14}N_2O_4.H_2O$	244.2	38821-49-7

Appearance

Carbidopa is a white to creamy white odorless or almost odorless powder.

Solubility

Carbidopa is slightly soluble in water and methanol. It is practically insoluble in alcohol. It is freely soluble in 3 M hydrochloric acid.

Method

Kafil and Dhingra developed a method for the separation of levodopa, carbidopa, and the related impurities. The liquid chromatograph was equipped with a Waters model 510 pump, a model 710B WISP autosampler, and an Environmental Sciences model 5100A Coulochem dual-electrode coulimetric detector. The stationary phase was a Waters µBondapak C_{18} reversed-phase column (250 × 4.6 mm, 10-µm particle size) with a guard column containing a 0.2-µm filter. The mobile phase was 0.05 M ammonium acetate with 0–2% methanol and adjusted to pH 4.1 with 0.6 M acetic acid. The flow rate was 0.9 mL/min. The electrochemical detector was set to applied potentials of +0.3 V for the first (screen) electrode and +0.6 V for the second (sample) electrode.

Samples were prepared by dissolving the compound in the mobile phase, after which this solution was filtered. Under these conditions, retention times for levodopa and carbidopa were 6.35 and 11.41 minutes, respectively.

The stability-indicating capability of the assay was demonstrated by assaying a mixture of levodopa, carbidopa, and their potential degradation products. Retention times in minutes were 5.49 for 6-hydroxydopa, 6.35 for levodopa, 8.59 for methyldopa, 11.41 for carbidopa, 14.31 for 3-methoxytyrosine, and 22.68 for 3-*o*-methylcarbidopa. The limit of detection was 2 ng/mL. Intraday and interday coefficients of variation were 2.0 and 1.6%, respectively.

Reference
Kafil JB, Dhingra BS. Stability-indicating method for the determination of levodopa, levodopa–carbidopa and related impurities. *J Chromatogr A*. 1994; 667: 175–81.

Carbocysteine

Chemical Name
S-Carboxymethyl-L-cysteine

Other Names
Carbocisteine, Mucofan

Form	Molecular Formula	MW	CAS
Carbocysteine	$C_5H_9NO_4S$	179.2	2387-59-9

Appearance
Carbocysteine occurs as a white crystalline powder.

Solubility
Carbocysteine is practically insoluble in water, alcohol, and ether. It dissolves in dilute mineral acids and in dilute solutions of alkali hydroxides.

Method
Melucci et al. reported the determination by HPLC of carbocysteine in syrup formulations. The system consisted of a Waters model 590 pump, a model 710B autoinjector, a model 757 variable-wavelength UV detector, and a Hewlett-Packard model HP-3356 integrator. The stationary phase was an aminopropyl silica column (150 × 4 mm, 5-μm particle size).

The mobile phase consisted of 750 mL of 0.01 M monobasic potassium phosphate buffer and 250 mL of acetonitrile. The flow rate was 1.0 mL/min. UV detection was performed at 220 nm and 0.1 AUFS.

Samples were diluted 1:10 with 0.1 M sodium hydroxide and then further diluted to an appropriate concentration with water. The injection volume was 50 µL. Under these conditions, the retention time for carbocysteine was about 6.4 minutes (estimated from the published chromatogram).

The HPLC method was evaluated to be stability indicating by accelerated decomposition of carbocysteine. Samples of the cough syrup formulation were exposed to heat (80 °C for 4 days in a sealed glass ampule) or to light (4000 foot-candles for 26 days in a centrifuge tube). In all cases, decomposition products, impurities, and other components were resolved from the intact carbocysteine on its chromatogram.

Reference

Melucci CK, Lyman GW, Bond AD, et al. Determination of *S*-carboxymethylcysteine in syrup formulations by high-performance liquid chromatography. *J Chromatogr*. 1987; 391: 321–4.

Carboplatin

Chemical Name

(*SP*-4-2)-Diammine[1,1-cyclobutanedicarboxylato(2-)-*O,O′*]platinum

Other Name

Paraplatin

Form	Molecular Formula	MW	CAS
Carboplatin	$C_6H_{12}N_2O_4Pt$	371.3	41575-94-4

Appearance

Carboplatin occurs as a white to off-white crystalline powder.

Solubility

Carboplatin is soluble in water.

Method 1

Rochard et al. determined the stability of carboplatin 1 mg/mL using HPLC. The chromatographic system included a Merck L6000 pump, a Waters WISP 710B sampler injector, a Waters 990 photodiode-array UV-visible detector, and an APC model IV NEC personal computer with Waters 990 software and plotter. The stationary phase was a Merck reversed-phase Nucleosil C_{18} column (150 × 4.2 mm, 5-μm particle size). The mobile phase consisted of 0.01 M phosphate buffer (pH 7.0) containing 0.00055 M hexadecyltrimethylammonium bromide and was delivered isocratically. UV detection was performed at 216 nm.

Samples were diluted in water for injection. Carboplatin and cisplatin eluted at approximately 3 and 8 minutes, respectively.

The stability-indicating nature of the assay was demonstrated using known degradation products and accelerated degradation of carboplatin samples. Samples of carboplatin solution were subjected to 1 N hydrochloric acid, 1 N sodium hydroxide, and hydrogen peroxide heated at 70 °C for 2 hours. Carboplatin eluted without interference from its degradation products. Retention times of transplatin and diaquodiammineplatinum were 1.3 and 1.5 minutes, respectively.

Good linearity between peak area and concentration from 5 to 60 mg/L was obtained. The correlation coefficients were at least 0.990. The intraday and interday coefficients of variation were 0.87 and 2.06%, respectively.

References

Rochard E, Barthes D, Courtois P. Stability and compatibility study of carboplatin with three portable infusion pump reservoirs. *Int J Pharm.* 1994; 101: 257–62.

Rochard E, Boutelet H, Griesemann E, et al. Simultaneous high performance liquid chromatographic analysis of carboplatin and cisplatin in infusion fluids. *J Liq Chromatogr.* 1993; 16: 1505–16.

Method 2

Prat et al. used HPLC to investigate the stability of carboplatin in 5% dextrose injection in glass containers, polyethylene bags, and polypropylene bags. The system consisted of a Hewlett-Packard 1090 isocratic pump, a diode-array UV detector, a Hewlett-Packard 3396 integrator, and a Hewlett-Packard ThinkJet printer. The stationary phase was a Tecknocroma Spherisorb phenyl analytical column (250 × 4.6 mm, 5-μm particle size). The mobile phase was a mixture of water and methanol (98:2, vol/vol) and was delivered isocratically at 0.8 mL/min. The detector was set at 210 nm. The retention time for carboplatin was 3.1 minutes.

The HPLC method was determined to be stability indicating by accelerated carboplatin degradation. Samples of carboplatin 3.2 mg/mL in 5% dextrose injection were heated at 60 °C for 1 month, mixed with 0.1 M hydrochloric acid, or mixed with 0.05 M sodium hydroxide. In all cases, the degradation product absorption peaks were well resolved from the intact drug peak.

Calibration curves were generated from 0.4 to 4 mg/mL of carboplatin in 5% dextrose injection. The correlation coefficient was greater than 0.999.

Reference

Prat J, Pujol M, Girona V, et al. Stability of carboplatin in 5% glucose solution in glass, polyethylene and polypropylene containers. *J Pharm Biomed Anal.* 1994; 12: 81–4.

Method 3

Using HPLC, Sewell et al. evaluated the stability of carboplatin in continuous infusion preparations. The system consisted of a Beckman model 110A pump, an LDC-Milton Roy UV monitor fixed-wavelength detector fitted with a zinc lamp, an Altex model 210 injection valve with a 20-µL loop, and a Shimadzu C-R3A integrator/printer-plotter. The stationary phase was a Zorbax ODS analytical column (150 × 4.9 mm). The mobile phase was 0.02 M phosphate buffer (pH 6.5) delivered isocratically at 1 mL/min. UV detection was performed at 214 nm.

Samples were diluted before analysis. The retention time for carboplatin was about 2.6 minutes (estimated from the published chromatogram).

The HPLC method was determined to be stability indicating by accelerated carboplatin degradation. An ampule containing 5 mL of carboplatin solution 0.1 mg/mL was incubated at 70 °C for 24 hours. The degradation product peaks did not interfere with the intact drug peak.

Standard curves were constructed from 0.0025 to 0.025 mg/mL. The correlation coefficient was 0.999. The coefficient of variation was 1.11%.

A similar method was used by Northcott et al.

References

Sewell GJ, Riley CM, Rowland CG. The stability of carboplatin in ambulatory continuous infusion regimes. *J Clin Pharm Ther*. 1987; 12: 427–32.

Northcott M, Allsopp MA, Powell H, et al. The stability of carboplatin, diamorphine, 5-fluorouracil and mitozantrone infusions in an ambulatory pump under storage and prolonged 'in-use' conditions. *J Clin Pharm Ther*. 1991; 16: 123–9.

Method 4

Cheung et al. reported the stability of carboplatin in commonly used infusion vehicles. A Spectra-Physics model 3500B chromatograph was equipped with a reciprocating piston pump with flow feedback control, a Valco rotary injector valve with a 10-µL sample loop, a Spectra-Physics model 770 variable-wavelength UV detector, and an Omniscribe strip-chart recorder. The stationary phase was a Spectra-Physics Spheri-5 PR-8 cartridge (100 × 4.6 mm, 5-µm particle size). The mobile phase was water delivered isocratically at 1 mL/min. UV detection was performed at 210 nm. Uridine 180 µg/mL was used as the internal standard. Under these conditions, the retention times for carboplatin and uridine were 3.6 and 5.6 minutes, respectively.

The stability-indicating nature of the assay was verified using decomposed and partially decomposed solutions of carboplatin. Decomposition was accelerated by heat and by the addition of base or acid. In all cases, the peak for the intact carboplatin was adequately resolved from the degradation product peaks.

The standard curve was constructed from 7.5 to 75 µg/mL. The correlation coefficient was greater than 0.999 and the relative standard deviation based on peak height or peak area ratios was 0.4%.

Similar methods were reported by other researchers, and a similar method was used by Mayron and Gennaro.

References

Cheung Y-W, Cradock JC, Vishnuvajjala BR, et al. Stability of cisplatin, iproplatin, carboplatin, and tetraplatin in commonly used intravenous solutions. *Am J Hosp Pharm.* 1987; 44: 124–30.

Trisssel LA, Zhang Y. Physical and chemical stability of palonosetron HCl with cisplatin, carboplatin, and oxaliplatin during simulated Y-site administration. *J Oncol Pharm Pract.* 2004; 10: 191–5.

Mayron D, Gennaro AR. Stability and compatibility of granisetron hydrochloride in i.v. solutions and oral liquids and during simulated Y-site injection with selected drugs. *Am J Health Syst Pharm.* 1996; 53: 294–304.

Carprofen

Chemical Names

6-Chloro-α-methyl-9*H*-carbazole-2-acetic acid
2-(6-Chloro-2-carbazolyl)propanoic acid

Form	Molecular Formula	MW	CAS
Carprofen	$C_{15}H_{12}ClNO_2$	273.7	53716-49-7

Method

Wu et al. developed an HPLC method for monitoring the degradation of carprofen under various conditions. The Shimadzu LC-10AT chromatograph included an SPD-10A UV-visible detector, a Jasco 851-AS autosampler, and a C-R6A integrator. The stationary phase was a Cosmosil 5C18-AR column (250 × 4.6 mm, 5-µm particle size). The mobile phase consisted of acetonitrile, water, and acetic acid (50:49:1, vol/vol/vol). The flow rate was 1 mL/min. UV detection was performed at 260 and 270 nm. Indomethacin was used as an internal standard. The injection volume was 20 µL. Under these conditions, retention times for carprofen and indomethacin were about 8.9 and 10.4 minutes, respectively.

The assay was demonstrated to be stability indicating by accelerated decomposition of carprofen. A 500-µg/mL solution of carprofen in ethanol was diluted to 100 µg/mL with 0.2 N hydrochloric acid or 0.2 N sodium hydroxide solution and incubated for 3 days at 60 °C. Another 500-µg/mL solution of carprofen in ethanol was diluted to 100 µg/mL with distilled water and exposed to sunlight for 10 minutes. In all cases, the intact carprofen was separated from its impurities and decomposition products on its chromatogram.

A calibration curve for carprofen was constructed from 2.5 to 80 µg/mL in ethanol containing 40 µg/mL indomethacin. The correlation coefficient was greater than 0.9999. The intraday and interday coefficients of variation for the analysis at 50 µg/mL were 0.9 and 1.1%, respectively.

Reference
Wu A-B, Chen C-Y, Chu S-D, et al. Stability-indicating high-performance liquid chromatographic assay method and photostability of carprofen. *J Chromatogr Sci.* 2001; 39: 7–11.

Carzelesin

Chemical Name
N-[2-[[(1*S*)-1-(Chloromethyl)-1,6-dihydro-8-methyl-5-[[(phenylamino)carbonyl]oxy]-
    benzo[1,2-*b*:4,3-*b'*]dipyrrol-3(2*H*)-yl]carbonyl]-1*H*-indol-5-yl]-6-(diethylamino)-2-
    benzofurancarboxamide

Other Name
U-80244

Form	Molecular Formula	MW	CAS
Carzelesin	$C_{41}H_{37}ClN_6O_5$	729.2	119813-10-4

Method
Vries et al. reported the systematic study on the chemical stability of the prodrug antitumor agent carzelesin. A liquid chromatograph included a Waters model 510 pump, a Perkin-Elmer model ISS-100 autosampler, and a Thermo Separation Products model SP4270 integrator. The stationary phase was a Waters µBondapak C_{18} analytical column (300 × 3.9 mm, 10-µm particle size). The mobile phase was a mixture of 2 mM phosphate buffer (pH 6.5) and acetonitrile (25:65, vol/vol). The flow rate was 1.0 mL/min. UV detection was performed at 254 nm.

Samples were prepared in a mixture of 0.02 M buffer solution and acetonitrile (1:1, vol/vol). The injection volume was 20 µL. Under these conditions, the retention time for carzelesin was about 10 minutes (estimated from the published chromatogram).

The stability-indicating nature of the method was demonstrated by accelerated degradation of carzelesin. The carzelesin was partially degraded in 0.1 M perchloric acid buffer solution (pH 1.5) or 0.03 M phosphate buffer (pH 7.2). The carzelesin peak was well separated from the degradation product peaks.

A standard curve for carzelesin was constructed with a correlation coefficient greater than 0.990.

Reference
Vries JDJ, Doppenberg WG, Henrar REC, et al. Systematic study on the chemical stability of the prodrug antitumor agent carzelesin (U-80244). *J Pharm Sci.* 1996; 85: 1227–33.

Cefaclor

Chemical Name
(7*R*)-3-Chloro-7-(α-D-phenylglycylamino)-3-cephem-4-carboxylic acid monohydrate

Other Name
Ceclor

Form	Molecular Formula	MW	CAS
Cefaclor	$C_{15}H_{14}ClN_3O_4S.H_2O$	385.8	70356-03-5

Appearance
Cefaclor is a white to off-white or slightly yellow crystalline powder.

Solubility
Cefaclor is slightly soluble in water and practically insoluble in methanol.

Method
Mathew et al. developed an HPLC method for the quantitation of cefaclor in pharmaceutical products. A Waters model ALC 202 liquid chromatograph was equipped with a Rheodyne model 7125 universal injector, a Schoeffel model SF 770 multiple-wavelength detector, and an Omniscribe recorder. The stationary phase was a Waters C_{18} column (300 × 3.9 mm). The mobile phase consisted of acetonitrile and 0.2% glacial acetic acid in water (pH 2.85) (12:88, vol/vol). The flow rate was 1.5 mL/min. UV detection was performed at 262 nm and 0.1 AUFS. Cefazolin sodium 2 mg/mL was used as an internal standard.

For capsules, a powder sample of 50.0 mg of cefaclor was accurately weighed, mixed with 40 mL of water, stirred for 5 minutes, filled to 50 mL with water, and filtered. The filtrate was diluted with water. For suspensions, the sample was diluted with water. The injection volume was 20 μL.

The stability-indicating capacity of the method was demonstrated by intentional degradation of cefaclor. The cefaclor samples were mixed with 1 N sulfuric acid or sodium hydroxide solution and heated to boiling for about 5 minutes. The degradation products did not interfere with the analysis of cefaclor.

A calibration curve for cefaclor was generated from 24 to 72 μg/mL. The correlation coefficient was 0.9997.

Reference
Mathew M, Gupta VD, Bethea C. Quantitation of cefaclor in pharmaceutical dosage forms using high performance liquid chromatography. *Drug Dev Ind Pharm*. 1993; 19: 1723–9.

Cefadroxil

Chemical Name
[6R-[6α,7β(R*)]]-7-[[Amino-(4-hydroxyphenyl)acetyl]amino]-3-methyl-8-oxo-5-thia-1-azabicyclo[4.2.0]oct-2-ene-2-carboxylic acid monohydrate

Other Name
Duricef

Form	Molecular Formula	MW	CAS
Cefadroxil	$C_{16}H_{17}N_3O_5S.H_2O$	381.4	66592-87-8

Appearance
Cefadroxil is a white to yellowish-white crystalline powder.

Solubility
Cefadroxil is soluble in water and slightly soluble in alcohol.

pK$_a$
Cefadroxil has pK$_a$ values of 2.64, 7.30, and 9.69 at 35 °C.

Method
Nahata and Jackson evaluated the stability of cefadroxil in a reconstituted suspension under refrigeration and at room temperature. Cefadroxil concentrations at each time point were measured by HPLC. The stationary phase was an Ultrasphere C_8 column. The mobile phase was a mixture of 15% methanol and 85% monobasic sodium phosphate in deionized water adjusted to pH 2.6 with phosphoric acid. The flow rate was 2 mL/min. UV detection was performed at 230 nm.

Samples were diluted 1:10,000 with deionized water. Cefadroxil and cefaclor (as the internal standard) eluted at 4.2 and 10.7 minutes, respectively.

The analytical method was determined to be stability indicating by accelerated degradation of cefadroxil. A cefadroxil sample was mixed with 1 M hydrochloric acid and warmed to 45 °C. The two degradation product peaks did not interfere with the intact cefadroxil peak.

The coefficient of variation of the assay was less than 4.5%.

Reference
Nahata MC, Jackson DS. Stability of cefadroxil in reconstituted suspension under refrigeration and at room temperature. *Am J Hosp Pharm.* 1991; 48: 992–3.

Cefamandole

Chemical Name

[6R-[6α,7β(R*)]]-7-[(Hydroxyphenylacetyl)amino]-3-[[(1-methyl-1H-tetrazol-5-yl)-thio]methyl]-8-oxo-5-thia-1-azabicyclo[4.2.0]oct-2-ene-2-carboxylic acid

Other Names

Cephamandole, Mandol

Form	Molecular Formula	MW	CAS
Cefamandole	$C_{18}H_{18}N_6O_5S_2$	462.5	34444-01-4
Cefamandole nafate	$C_{19}H_{17}N_6NaO_6S_2$	512.5	42540-40-9

Appearance

Cefamandole nafate is a white to off-white crystalline powder.

Solubility

Cefamandole nafate is soluble in water and methanol; it is practically insoluble in ether, chloroform, benzene, and cyclohexane.

Method 1

Bosso and Townsend determined the stability and compatibility of cefamandole nafate with clindamycin phosphate in 5% dextrose injection and 0.9% sodium chloride injection. The liquid chromatograph consisted of an Altex model 110A solvent-metering pump, an Altex model 210 injector (or a Waters model 710B autoinjector), a Waters Lambda-Max model 481LC spectrophotometer, an Altex model 153 analytical UV detector, and a Linear Instruments model 555 recorder. The stationary phase was an Alltech Versapak C_{18} reversed-phase column (300 × 4.1 mm, 10-µm particle size). The mobile phase consisted of 28% methanol in 0.01 M sodium acetate buffer. The pH of the buffer was adjusted to 5.2 with acetic acid. The mobile phase was delivered isocratically at 1 mL/min. UV detection was performed at 254 nm and 1.0 AUFS. Cefazolin sodium 10 mg/mL was used as the internal standard.

Samples were diluted 1:2 with the internal standard solution. The injection volume was 4 µL. Retention times for cefazolin and cefamandole were 4 and 7 minutes, respectively.

This method was reported to be stability indicating because it separated the intact cefamandole nafate from its own breakdown products generated by acidification and heating.

Intrarun and interrun coefficients of variation were 3.9%.

Reference
Bosso JA, Townsend RJ. Stability of clindamycin phosphate and ceftizoxime sodium, cefoxitin sodium, cefamandole nafate, or cefazolin sodium in two intravenous solutions. *Am J Hosp Pharm.* 1985; 42: 2211–4.

Method 2
Gupta et al. conducted a chemical stability study of cefamandole nafate using a stability-indicating HPLC method. A Waters ALC 202 chromatographic system was equipped with a U6K universal injector, a Schoeffel SF 770 multiple-wavelength detector, and a Houston Omniscribe recorder. The stationary phase was a Waters µBondapak phenyl column (300 × 3.9 mm). The mobile phase was a mixture of 40% methanol and 60% 0.02 M aqueous solution of ammonium acetate and was delivered at 2.0 mL/min. UV detection was performed at 254 nm and 0.4 AUFS.

Samples were diluted to 800 µg/mL of cefamandole nafate with water. Retention times for cefamandole and cefamandole nafate were 3.3 and 6.0 minutes, respectively.

The analytical method was stated to be stability indicating because it separated cefamandole nafate from cefamandole.

References
Gupta VD, Stewart KR, Torre MD. Chemical stabilities of cefamandole nafate and metronidazole when mixed together for intravenous infusion. *J Clin Hosp Pharm.* 1985; 10: 379–83.

Gupta VD, Stewart KR. Stability of cefamandole nafate and cefoxitin sodium solutions. *Am J Hosp Pharm.* 1981; 38: 875–9.

Cefazolin

Chemical Name
(6R-trans)-3-[[(5-Methyl-1,3,4-thiadiazol-2-yl)thio]methyl]-8-oxo-7-[[(1H-tetrazol-1-yl)-acetyl]amino]-5-thia-1-azabicyclo[4.2.0]oct-2-ene-2-carboxylic acid

Other Names
Ancef, Cephazolin, Kefzol

Form	Molecular Formula	MW	CAS
Cefazolin	$C_{14}H_{14}N_8O_4S_3$	454.5	25953-19-9
Cefazolin sodium	$C_{14}H_{13}N_8NaO_4S_3$	476.5	27164-46-1

Appearance

Cefazolin and cefazolin sodium occur as white to off-white crystalline powders.

Solubility

Cefazolin is slightly soluble in water and alcohol. Cefazolin sodium is freely soluble in water and slightly soluble in methanol and ethanol. It is practically insoluble in benzene, acetone, and chloroform.

Method 1

Xu et al. used HPLC to evaluate the chemical stability of linezolid injection at 2 mg/mL admixed with three common cephalosporin antibiotics and stored for 7 days at 4 and 23 °C. The Waters liquid chromatograph consisted of a model 600E multi-solvent-delivery pump, a model 996 photodiode-array detector, and a WISP model 712 autosampler. The stationary phase was a YMC-Pack ODS-AM column (150 × 4.6 mm, 5-μm particle size). The mobile phase consisted of 820 mL of water, 180 mL of acetonitrile, 1 mL of trifluoroacetic acid, and 1 mL of triethylamine. The flow rate was 1.5 mL/min. UV detection was performed at 254 nm and 0.5 AUFS.

Samples were diluted 1:100 with water. The injection volume was 10 μL. Retention times for cefazolin and linezolid were 6.1 and 10.6 minutes, respectively.

The method was shown to be stability indicating by accelerated degradation. Sample solutions were treated with 1 N sodium hydroxide, 1 N hydrochloric acid, or 3% hydrogen peroxide or subjected to heating. The degradation product peaks did not interfere with the intact drug peak.

A standard curve for cefazolin was constructed from 25 to 150 μg/mL. The correlation coefficient was greater than 0.9999. Intraday and interday coefficients of variation were 0.3 and 0.1%, respectively.

Reference

Xu QA, Trissel LA, Williams KY. Compatibility and stability of linezolid injection admixed with three cephalosporin antibiotics. *J Am Pharm Assoc*. 2000; 40: 509–14.

Method 2

Using one of two HPLC systems, Allen et al. determined the chemical stability of cefazolin sodium with cefpirome sulfate during simulated Y-site injection. One unit included an Alcott model 728 Micromeritics autosampler, a Rheodyne 7010 injector with an Alcott 732 electrically actuated valve, a Waters model 501 solvent-delivery pump, a Waters model 441 UV detector, a Waters model 401 refractive index detector, and a Waters model 745 data module. The other unit consisted of an Alcott model 728 Micromeritics autosampler, a Rheodyne 7010 injector with an Alcott 732 electrically actuated valve, a Shimadzu model LA-6A solvent-delivery pump, a Shimadzu model SPD-6A UV detector, and an Orion model 901 microprocessor ion analyzer. Cefazolin was separated on a C_{18} analytical column. The mobile phase was a mixture of acetonitrile (900 mL) and water (100 mL) containing 0.9 g of anhydrous disodium phosphate and 1.298 g of citric acid monohydrate. The flow rate was 2.0 mL/min. UV detection was performed at 254 nm.

Samples were diluted before injection. The injection volume was 50 μL. The retention times for cefazolin and cefpirome were 3.1 and 5.3 minutes, respectively.

The stability-indicating ability of the assay was confirmed by accelerated degradation of cefazolin sodium with hydrochloric acid, sodium hydroxide, potassium chlorate, heat at 80 °C, and light (150 foot-candles from a tungsten filament source). In each case, the intact cefazolin was well separated from its degradation products.

The standard curve was constructed from 10 to 100 μg/mL; the correlation coefficient was greater than 0.99. The intraday and interday coefficients of variation were 1.4 and 2.2%, respectively.

A similar method was used by How et al.

References

Allen LV, Stiles ML, Prince SJ, et al. Stability of cefpirome sulfate in the presence of commonly used intensive care drugs during simulated Y-site injection. *Am J Health Syst Pharm.* 1995; 52: 2427–33.

How TH, Loo WY, Yow KL, et al. Stability of cefazolin sodium eye drops. *J Clin Pharm Ther.* 1998; 23: 41–7.

Method 3

Rivers et al. studied the stability of cefazolin sodium at 8 °C in an intravenous admixture. A fully automated, computer-controlled Waters liquid chromatographic system consisted of a model 510 pump, a model 712 refrigerated autosampler, and a model 481 variable-wavelength UV detector. The stationary phase was a Waters μBondapak C_{18} analytical column (150 × 3.9 mm, 4-μm particle size). The mobile phase was a mixture of acetonitrile and 0.02 M potassium dihydrogen phosphate with 0.005 M triethylamine adjusted to pH 4.8 with sodium hydroxide (7:93, vol/vol). The flow rate was 1.75 mL/min. UV detection was performed at 270 nm. The injection volume was 20 μL. The run time was 12.5 minutes and the retention times for metronidazole and cefazolin were 4.46 and 9.25 minutes, respectively.

The HPLC analytical method was determined to be stability indicating by accelerated decomposition of the sample with acid and base. One test solution was mixed with 0.1 M sodium hydroxide and the other with 0.05 M sulfuric acid. Both solutions were refluxed for 1 hour. The chromatograms of each solution indicated no interference of the degradation products with the drug peak.

Reference

Rivers TE, McBride HA, Trang JM. Stability of cefazolin sodium and metronidazole at 8 °C for use as an IV admixture. *J Parenter Sci Technol.* 1993; 47: 135–7.

Method 4

Nahata and Ahalt used HPLC to investigate the stability of cefazolin in two commonly used peritoneal dialysis solutions. The stationary phase was an Ultrasphere ODS C_8 analytical column. The mobile phase consisted of 18% acetonitrile and 82% 0.01 M sodium phosphate buffer (pH 5.3). The flow rate was 2 mL/min. UV detection was performed at 230 nm. The retention times for cefazolin and 7-hydroxycoumarin (the internal standard) were 5.6 and 6.7 minutes, respectively.

The stability-indicating ability of the method was confirmed by accelerated decomposition of cefazolin samples. Cefazolin sample solution was mixed with 1 M sodium hydroxide. The chromatogram showed a reduction of the intact cefazolin peak and the appearance of several new degradation product peaks that did not interfere with the cefazolin peak.

The intraday and interday coefficients of variation were less than 4%.

A similar method was used by Walker et al.

References
Nahata NC, Ahalt PA. Stability of cefazolin sodium in peritoneal dialysis solutions. *Am J Hosp Pharm.* 1991; 48: 291–2.

Walker SE, DeAngelis C, Iazzetta J. Stability and compatibility of combinations of hydromorphone and a second drug. *Can J Hosp Pharm.* 1991; 44: 289–94.

Method 5

Using HPLC, Stiles et al. evaluated the stability of cefazolin in portable pump reservoirs. The chromatographic system consisted of a Waters model 501 pump, a model U6K injector, and a model 441 UV detector. The stationary phase was a Waters Nova-Pak C_{18} analytical column (150 × 3.9 mm, 5-µm particle size). The mobile phase consisted of 20% methanol and 80% 0.005 M phosphate buffer at pH 7.5 and was delivered isocratically at 1.0 mL/min. UV detection was performed at 254 nm.

Samples were diluted with distilled water. Cefazolin eluted in 2.7 minutes.

The assay was determined to be stability indicating by accelerated decomposition of the cefazolin sodium after storage at 45 °C for 72 hours; the degradation products and excipients in the formulation were well separated from the intact cefazolin peak.

Standard curves for cefazolin were constructed from 2.0 to 8.0 mg/mL. The correlation coefficient was greater than 0.99. The coefficient of variation was less than 5%.

A similar method was used by Wang et al.

References
Stiles ML, Tu Y-H, Allen LV. Stability of cefazolin sodium, cefoxitin sodium, ceftazidime, and penicillin G sodium in portable pump reservoirs. *Am J Hosp Pharm.* 1989; 46: 1408–12.

Wang D-P, Chang L-C, Wong C-Y, et al. Stability of cefazolin sodium–famotidine admixture. *Am J Hosp Pharm.* 1994; 51: 2205, 2209.

Method 6

Zbrozek et al. reported the stability of cefazolin sodium when mixed with clindamycin phosphate and gentamicin sulfate. The chromatographic system included an Altex model 110A solvent-metering pump, a Waters WISP 710B autoinjector, a Waters Lambda-Max model 481 LC spectrophotometer, and a Linear Instruments model 555 recorder. The stationary phase was an Altex Ultrasphere ODS column (250 × 4.6 mm, 5-µm particle size). The mobile phase consisted of 25% acetonitrile in a 0.01 M sodium acetate buffer adjusted to pH 6.5 with 0.1 M sodium hydroxide and was delivered at 1 mL/min. UV detection was performed at 254 nm and 0.64 AUFS. Cefamandole sodium 0.5 g/mL was used as the internal standard.

Samples were diluted 1:20 with internal standard solution. The injection volume was 10 µL. Under these conditions, the internal standard and cefazolin eluted in 4 and 8.25 minutes, respectively.

The HPLC assay was determined to be stability indicating by accelerated degradation of cefazolin. Samples of cefazolin solution were subjected to 1 N hydrochloric acid (pH 2) or heat (80 °C). No breakdown product peaks were observed on the chromatogram after either acidification or heating.

Standard curves were constructed for cefazolin from 1.25 to 20 mg/mL. The correlation coefficient of the five-point standard curve was 0.998. The intrarun and interrun coefficients of variation were 4.8 and 3.9%, respectively.

A similar method was used by Bosso et al.

References

Zbrozek AS, Marble DA, Bosso JA. Compatibility and stability of cefazolin sodium, clindamycin phosphate, and gentamicin sulfate in two intravenous solutions. *Drug Intell Clin Pharm.* 1988; 22: 873–5.

Bosso JA, Prince RA, Fox JL. Compatibility of ondansetron hydrochloride with fluconazole, ceftazidime, aztreonam, and cefazolin sodium under simulated Y-site conditions. *Am J Hosp Pharm.* 1994; 51: 389–91.

Method 7

Using an HPLC method, Ahmed and Day investigated the stability of cefazolin sodium in various artificial tear solutions and aqueous vehicles. The stationary phase was a Waters μBondapak ODS column (300 × 3.9 mm). The mobile phase was 30% methanol in 1% acetic acid and was adjusted to pH 3.5 with sodium hydroxide. The flow rate was 1 mL/min. UV detection was performed at 254 nm and 0.5 AUFS.

Samples were diluted 1:10 with distilled water. The retention time for cefazolin was 7 minutes.

The assay was demonstrated to be stability indicating by accelerated degradation of cefazolin sodium. Cefazolin sodium 3.33-mg/mL solutions were refluxed in 0.1 N sodium hydroxide and 0.1 N hydrochloric acid for 8 hours. Two degradation product peaks were observed at 3 and 5 minutes and did not interfere with the intact cefazolin peak.

A standard curve for cefazolin sodium was constructed from 0.01 to 0.1 mg/mL; the correlation coefficient was 0.999. The coefficient of variation of the assay was 1.26%.

Similar methods were used by the other researchers cited here.

References

Ahmed I, Day P. Stability of cefazolin sodium in various artificial tear solutions and aqueous vehicles. *Am J Hosp Pharm.* 1987; 44: 2287–90.

Gupta VD, Stewart KR. Quantitation of carbenicillin disodium, cefazolin sodium, cephalothin sodium, nafcillin sodium, and ticarcillin disodium by high-pressure liquid chromatography. *J Pharm Sci.* 1980; 69: 1264–7.

Bosso JA, Townsend RJ. Stability of clindamycin phosphate and ceftizoxime sodium, cefoxitin sodium, cefamandole nafate, or cefazolin sodium in two intravenous solutions. *Am J Hosp Pharm.* 1985; 42: 2211–4.

Cefdinir

Chemical Name
[6R-[6α,7β(Z)]]-7-[[(2-Amino-4-thiazolyl)(hydroxyimino)acetyl]amino]-3-ethenyl-8-oxo-5-thia-1-azabicyclo[4.2.0]oct-2-ene-2-carboxylic acid

Other Names
Cefzon, Omnicef, Sefdin

Form	Molecular Formula	MW	CAS
Cefdinir	$C_{14}H_{13}N_5O_5S_2$	395.4	91832-40-5

pK_a
Cefdinir has the following values: pK_{a1} of 1.9 for carboxylate, pK_{a2} of 3.3 for the amino group in the 2-aminothiazole ring, and pK_{a3} of 9.9 for hydroxyimino.

Method
Okamoto et al. described the development and validation of the HPLC method used for the determination of cefdinir and its related substances. The liquid chromatograph consisted of several autoinjectors, including a Waters WISP 710B, a Kyowaseimitsu KST-KMH, and a Shimadzu SIL-6A with an SCL-6A system controller; Waters 6000A and Shimadzu LC-6A pumps; Waters 440 and Shimadzu SPD 2A and 6A detectors; and Shimadzu C-R1B, C-R4AX, and C-R5A automatic data processors. The stationary phase was a Tosoh TSKgel ODS-80 Tm column (75 × 4.6 mm, 5-μm particle size). The mobile phase consisted of a citrate (33 mM)–phosphate buffer solution (pH 2.0), methanol, and dioxane (36:4:1, vol/vol/vol) delivered isocratically. The flow rate was adjusted to achieve the retention time of about 4 minutes for cefdinir. UV detection was performed at 254 nm. The injection volume was 5 μL.

The stability-indicating nature of the HPLC method was demonstrated by chromatographing cefdinir and m-hydroxybenzoic acid (as the internal standard) spiked with substances related to cefdinir. Cefdinir and its internal standard were well separated from related substances.

Calibration curves for cefdinir were constructed from 0.1 to 0.3 mg/mL with the internal standard at 0.7 mg/mL. The correlation coefficient was 0.9999. The coefficient of variation of the assay was 0.24%.

Reference
Okamoto Y, Itoh K, Namiki Y, et al. Method development for the determination of cefdinir and its related substances by high-performance liquid chromatography. *J Pharm Biomed Anal.* 1996; 14: 739–48.

Cefepime Hydrochloride

Chemical Name

[6R-[6α,7β(Z)]]-1-[[7-[[(2-Amino-4-thiazolyl)(methoxyimino)acetyl]amino]-2-carboxy-8-oxo-5-thia-1-azabicyclo[4.2.0]oct-2-en-3-yl]methyl]-1-methylpyrrolidinium chloride monohydrochloride monohydrate

Other Name

Maxipime

Form | Molecular Formula | MW | CAS

Form	Molecular Formula	MW	CAS
Cefepime hydrochloride	$C_{19}H_{25}ClN_6O_5S_2 \cdot HCl \cdot H_2O$	571.5	123171-59-5

Appearance

Cefepime hydrochloride is a white to pale yellow powder.

Solubility

Cefepime hydrochloride is highly soluble in water.

pK$_a$

Cefepime hydrochloride has pK$_a$ values of 1.5–1.6 and 3.1–3.2.

Method

Ling and Gupta used HPLC to investigate the stability of cefepime hydrochloride in 0.9% sodium chloride injection after storage in polypropylene syringes. A Waters ALC 202 chromatograph was equipped with a Rheodyne model 7125 universal injector, a Schoeffel SF 770 multiple-wavelength UV detector, and a Houston Omniscribe recorder. The stationary phase was a Phenomenex Ultrasphere column (150 × 4.6 mm, 5-μm particle size). The mobile phase consisted of 6.5% (vol/vol) acetonitrile in 0.01 M ammonium acetate buffer in water (pH 6.5) at 1.3 mL/min. UV detection was performed at 290 nm and 0.2 AUFS.

Sample solutions were diluted 1:200 with water before injection. Under these conditions, the retention time for cefepime was about 3.6 minutes (estimated from the published chromatogram). L-Arginine in the powder for the injection was not detected at 290 nm.

The method was shown to be stability indicating by intentional degradation of cefepime hydrochloride. A solution of cefepime (0.1 mg/mL) was boiled for 10 minutes. The degradation product did not interfere with the intact cefepime.

Peak heights of cefepime were related to concentration from 60 to 20 μg/mL.

A similar method was reported by other researchers.

References

Ling J, Gupta VD. Stability of cefepime hydrochloride after reconstitution in 0.9% sodium chloride injection and storage in polypropylene syringes for pediatric use. *Int J Pharm Compound.* 2001; 5: 151–2.

Trissel LA, Xu QA. Stability of cefepime hydrochloride in Autodose infusion system bags. *Ann Pharmacother.* 2003; 37: 804–7.

Cefmetazole

Chemical Name

(6*R*-*cis*)-7-[[[(Cyanomethyl)thio]acetyl]amino]-7-methoxy-3-[[(1-methyl-1*H*-tetrazol-5-yl)thio]methyl]-8-oxo-5-thia-1-azabicyclo[4.2.0]oct-2-ene-2-carboxylic acid

Other Names

Cefadel, Cefmetazon, Cemetol, Zefazone

Form	Molecular Formula	MW	CAS
Cefmetazole	$C_{15}H_{17}N_7O_5S_3$	471.5	56796-20-4
Cefmetazole sodium	$C_{15}H_{16}N_7NaO_5S_3$	493.5	56796-39-5

Solubility

Cefmetazole sodium is very soluble in water and methanol. It is soluble in acetone but practically insoluble in chloroform.

Method 1

Using HPLC, Lee et al. studied the stability of cefmetazole sodium with famotidine in 5% dextrose injection at 4 and 25 °C. The chromatograph consisted of a Shimadzu model LC-6A liquid chromatograph, a Rheodyne model 7125 single-piston 20-µL loop injector, a Shimadzu model SPD-6AV dual UV-visible detector, and a Waters model 745 data module. The stationary phase was a Waters Nova-Pak C_{18} analytical column. The mobile phase consisted of acetonitrile, 0.1% acetic acid, and 0.01 M dibasic ammonium phosphate solution at pH 7.8 (10:23:74, vol/vol/vol) and was delivered isocratically at 1.0 mL/min. The UV detector was set at 300 nm. Cefmetazole and famotidine had retention times of 8.1 and 13.9 minutes, respectively.

The method was demonstrated to be stability indicating by analyzing a cefmetazole sodium–famotidine admixture stored at 55 °C for 15 hours. Degradation products eluted separately and did not interfere with peaks of the intact cefmetazole and famotidine.

Calibration curves were generated from 0.1 to 10 mg/mL; the correlation coefficients were greater than 0.999. The intraday and interday coefficients of variation for the assay ($n = 6$) were 0.56 and 0.88%, respectively.

Reference
Lee DKT, Wong C-Y, Wang D-P, et al. Stability of cefmetazole sodium and famotidine. *Am J Health Syst Pharm.* 1996; 53: 432, 442.

Method 2
Inagaki et al. reported the stability of cefmetazole sodium with ranitidine hydrochloride during simulated Y-site administration. The liquid chromatograph consisted of a Hitachi model 6200 intelligent pump, an L-4200 UV-visible detector, a Hitachi model AS-2000 autosampler, and a D-2500 chromato-integrator. The stationary phase was an Alltech Adsorbosphere C_{18} analytical column (250 × 4.6 mm, 5-μm particle size) with a guard column of the same packing material. The mobile phase consisted of methanol and 50 mM ammonium phosphate (pH 4.5) containing 5 mM tetrabutylammonium hydroxide (30:70, vol/vol) and was delivered isocratically at 1.0 mL/min. UV absorption was monitored at 254 nm and 0.128 AUFS.

Samples were diluted 1:50 with 0.9% sodium chloride injection. The injection volume was 50 μL. Under these conditions, the retention time of cefmetazole was 9.8 minutes.

The assay was determined to be stability indicating by accelerated decomposition of the sample solutions. The cefmetazole solution was acidified to pH 2.0 with hydrochloric acid, alkalinized to pH 13 with sodium hydroxide, exposed to 1% hydrogen peroxide at room temperature, and heated for 1 hour at 80 °C. The resultant solutions were assayed. The method could differentiate not only between degradation products of cefmetazole and the intact cefmetazole but also between degradation products of cefmetazole and the intact ranitidine hydrochloride.

Standard curves were constructed from the plot of the peak area versus the concentration from 2.5 to 20 mg/mL. The correlation coefficients were greater than 0.999.

Reference
Inagaki K, Gill MA, Okamoto MP, et al. Chemical compatibility of cefmetazole sodium with ranitidine hydrochloride during simulated Y-site administration. *J Parenter Sci Technol.* 1993; 47: 35–9.

Cefonicid

Chemical Name
[6*R*-[6α,7β(*R**)]]-7-[(Hydroxyphenylacetyl)amino]-8-oxo-3-[[[1-(sulfomethyl)-1*H*-tetrazol-5-yl]thio]methyl]-5-thia-1-azabicyclo[4.2.0]oct-2-ene-2-carboxylic acid

Other Name
Monocid

Form

Form	Molecular Formula	MW	CAS
Cefonicid	$C_{18}H_{18}N_6O_8S_3$	542.6	61270-58-4
Cefonicid sodium	$C_{18}H_{16}N_6Na_2O_8S_3$	586.5	61270-78-8

Appearance
Cefonicid sodium is a white to off-white lyophilized powder.

Solubility
Cefonicid sodium has solubilities of 750 and 170 mg/mL in water and methanol at 25 °C, respectively.

pK$_a$
Cefonicid has pK$_a$ values of 0–1 and 2.7.

Method
Using HPLC, Wong et al. determined the chemical stability of cefonicid sodium in different infusion solutions. The liquid chromatographic system included a Micromeritics model 725 automatic injector/motorized valve with a fixed-size 10-µL sample loop, a Kratos Spectroflow 773 variable-wavelength UV detector, and an LDC model 308 computing integrator. Cefonicid was chromatographed using a µBondapak C$_{18}$ column (300 × 3.9 mm, 10-µm particle size). The mobile phase was a mixture of methanol, 0.1 M ammonium dihydrogen phosphate, and water (5:4:31, vol/vol/vol) and was delivered isocratically. UV detection was performed at 254 nm. The retention time for cefonicid was 10 minutes.

The stability-indicating capability was demonstrated using cefonicid solution spiked with its degradation product, tetrazole. No interference of tetrazole with the intact cefonicid peak occurred. Retention times for tetrazole and cefonicid were 2 and 10 minutes, respectively.

Reference
Wong WW, Maderich AB, Poli GP, et al. Stability of cefonicid sodium in infusion fluids. *Am J Hosp Pharm*. 1985; 42: 1980–3.

Cefoperazone

Chemical Name

[6R-[6α,7β(R*)]]-7-[[[[(4-Ethyl-2,3-dioxo-1-piperazinyl)carbonyl]amino](4-hydroxy-phenyl)acetyl]amino]-3-[[(1-methyl-1H-tetrazol-5-yl)thio]methyl]-8-oxo-5-thia-1-azabicyclo[4.2.0]oct-2-ene-2-carboxylic acid

Other Name

Cefobid

Form

Form	Molecular Formula	MW	CAS
Cefoperazone	$C_{25}H_{27}N_9O_8S_2$	645.7	62893-19-0
Cefoperazone sodium	$C_{25}H_{26}N_9NaO_8S_2$	667.7	62893-20-3

Appearance

Cefoperazone sodium is a white crystalline powder.

Solubility

Cefoperazone sodium is freely soluble in water and poorly soluble in alcohol.

pK$_a$

Cefoperazone has a pK$_a$ of 2.55.

Method

Gupta et al. used HPLC to investigate the chemical stability of cefoperazone sodium and ceftazidime in 5% dextrose and 0.9% sodium chloride injections. The chromatograph consisted of a Waters ALC 202 pump system, a Schoeffel SF 770 multiple-wavelength detector, and a Houston Omniscribe recorder. The stationary phase was a Waters μBondapak phenyl semipolar column (300 × 3.9 mm). The mobile phase consisted of 18% acetonitrile in a 0.01 M aqueous solution of ammonium acetate and was delivered isocratically at 2.0 mL/min. UV detection was performed at 240 nm and 0.04 AUFS. Oxacillin 0.16% in water was used as the internal standard.

Samples were diluted 1:50 with water. The injection volume was 20 μL. Retention times for cefoperazone and oxacillin were 4.8 and 8.0 minutes, respectively (estimated from the published chromatogram).

The method was demonstrated to be stability indicating. Cefoperazone stock solutions were mixed with 1 N sodium hydroxide or 1 N sulfuric acid and heated on a hot plate for 5 minutes. These solutions were then cooled to room temperature, neutralized with sulfuric acid or sodium hydroxide, respectively, and assayed. Degradation product peaks did not interfere with the intact cefoperazone peak.

Reference

Gupta VD, Bethea C, Torre MD. Chemical stabilities of cefoperazone sodium and ceftazidime in 5% dextrose and 0.9% sodium chloride injections. *J Clin Pharm Ther*. 1988; 13: 199–205.

Cefotaxime

Chemical Name

[6*R*-[6α,7β(Z)]]-3-[(Acetyloxy)methyl]-7-[[(2-amino-4-thiazolyl)(methoxyimino)-acetyl]amino]-8-oxo-5-thia-1-azabicyclo[4.2.0]oct-2-ene-2-carboxylic acid

Other Name

Claforan

Form	Molecular Formula	MW	CAS
Cefotaxime	$C_{16}H_{17}N_5O_7S_2$	455.5	63527-52-6
Cefotaxime sodium	$C_{16}H_{16}N_5NaO_7S_2$	477.5	64485-93-4

Appearance

Cefotaxime sodium is an off-white to pale yellow crystalline powder.

Solubility

Cefotaxime sodium is freely soluble in water and slightly soluble in alcohol.

pK$_a$

Cefotaxime has a pK$_a$ of 3.4.

Method 1

Gupta used HPLC to investigate the stability of cefotaxime sodium in 0.9% sodium chloride injection when packaged in polypropylene syringes and stored at 5 and 25 °C. The liquid chromatographic system consisted of a Waters model M45 pump, a model 484 multiple-wavelength detector, a Rheodyne model 7125 injector, and an Omniscribe recorder. The stationary phase was a Supelco C$_8$ column (150 × 4.6 mm, 5-μm particle size). The mobile phase consisted of 9.5% (vol/vol) acetonitrile in water containing 0.02 M ammonium acetate buffer (pH 6.9). The flow rate was 1.3 mL/min. UV detection was performed at 290 nm and 0.7 AUFS.

Samples were diluted with water. The injection volume was 80 μL. Under these conditions, the retention time for cefotaxime was about 4.7 minutes (estimated from the published chromatogram).

The method was demonstrated to be stability indicating by thermal degradation of cefotaxime. A cefotaxime sodium solution was boiled on a hot plate for 40 seconds. The chromatogram showed that the intact cefotaxime was well resolved from its degradation product peaks.

Reference

Gupta VD. Stability of cefotaxime sodium after reconstitution in 0.9% sodium chloride injection and storage in polypropylene syringes for pediatric use. *Int J Pharm Compound*. 2002; 6: 234–6.

Method 2

Using an HPLC method, Belliveau et al. evaluated the stability of cefotaxime sodium and metronidazole in 0.9% sodium chloride injection or in ready-to-use metronidazole bags. The HPLC system consisted of a Waters model 510 pump, a Waters model 710A WISP autosampler, a Waters model M440 UV detector, and a Hewlett-Packard model 3396 series II integrator. The stationary phase was a Waters Resolve C_{18} column (150×3.9 mm, 5-μm particle size). The mobile phase consisted of an acetate buffer and acetonitrile (86:16, vol/vol). The buffer contained 2.46 g of anhydrous sodium acetate, 8 mL of acetic acid, and 0.2 g of tetrabutylammonium hydrogen sulfate per 1000 mL with the pH adjusted to 3.0. The flow rate was 1.2 mL/min. UV detection was performed at 254 nm. Cefoxitin was used as the internal standard.

Samples were diluted 1:1 with internal standard and then further diluted with water. The injection volume was 20 μL. Under these conditions, retention times for metronidazole, cefotaxime, and cefoxitin were 1.8, 2.3, and 3.0 minutes, respectively.

The assay was determined to be stability indicating. A solution of cefotaxime, cefoxitin, and metronidazole was heated at 80 °C for 26 hours. This heating caused complete decomposition of cefotaxime; its decomposition products eluted before 1.8 minutes.

Standard curves for cefotaxime were constructed from 4 to 12 μg/mL and were linear with a correlation coefficient of 0.9971. The interday coefficient of variation was 1.45%.

Reference

Belliveau PP, Nightingale CH, Quintiliani R. Stability of cefotaxime sodium and metronidazole in 0.9% sodium chloride injection or in ready-to-use metronidazole bags. *Am J Health Syst Pharm*. 1995; 52: 1561–6.

Method 3

Rivers et al. examined the stability of cefotaxime sodium and metronidazole in an intravenous admixture. An HPLC system consisted of a Waters 510 pump, a Waters model 712 refrigerated autosampler, and a Waters model 481 variable-wavelength UV detector. The analytical column was a Waters μBondapak column (150×3.9 mm, 4-μm particle size). The mobile phase was 7% acetonitrile and 93% 0.02 M potassium dihydrogen phosphate with 0.005 M triethylamine adjusted to pH 4.8 with sodium hydroxide. The flow rate was 1.5 mL/min. UV detection was performed at 270 nm. The injection volume was 20 μL. The run time was 13.5 minutes. Cefotaxime retention time was 8.44 minutes.

The analytical method was determined to be stability indicating by intentional degradation of cefotaxime. Calibrator solutions containing both cefotaxime sodium and metronidazole were mixed with 0.1 M sodium hydroxide or 0.05 M sulfuric acid and heated under reflux conditions for 1 hour. Subsequent chromatograms showed there was no interference with the intact cefotaxime peak by degradation product peaks.

Reference
Rivers TE, McBride HA, Trang JM. Stability of cefotaxime sodium and metronidazole in an I.V. admixture at 8 °C. *Am J Hosp Pharm*. 1991; 48: 2638–40.

Method 4
Paap and Nahata used an HPLC method to study the stability of cefotaxime in aqueous solutions at various pH values, temperatures, and ionic strengths as well as in commonly used intravenous admixtures. The HPLC system consisted of a Beckman model 110A pump, an Isco variable-wavelength UV detector, an Anspec D-2000 chromato-integrator, a Waters model 712 WISP autosampler, and a Waters μBondapak 10-μm C_{18} analytical column. The mobile phase consisted of 20% acetonitrile and 80% 0.01 M acetate buffer containing 0.007 M tetrabutylammonium hydrogen sulfate. The pH of the mobile phase was adjusted to 5.2 with sodium hydroxide. The flow rate was 2.0 mL/min. UV detection was performed at 254 nm. Under these conditions, the retention times for cefotaxime and the internal standard (cefoxitin) were about 4.5 and 7.5 minutes, respectively.

The analytical method was determined to be stability indicating. The accelerated degradation with acid and with heat produced a major metabolite, desacetylcefotaxime, and unknown degradation product peaks that did not interfere with the intact cefotaxime peak.

Standard curves were determined daily from 500 to 1500 mg/L. Intraday and interday variations ($n = 6$) were 2.8 and 5.5%, respectively.

Reference
Paap CM, Nahata MC. Stability of cefotaxime in two peritoneal dialysis solutions. *Am J Hosp Pharm*. 1990; 47: 147–50.

Method 5
Using a stability-indicating HPLC method, Gupta investigated the stability of cefotaxime sodium at various pH values and in intravenous admixtures. A Waters model ALC 202 liquid chromatograph was equipped with a U6K universal injector, a Spectroflow SF 770 multiple-wavelength detector, an Omniscribe recorder, and a Spectra-Physics Autolab minigrator-integrator. The stationary phase was a Waters μBondapak phenyl column (300 × 4 mm). The mobile phase contained 0.02 M ammonium acetate and 18% (vol/vol) of methanol in water. The flow rate was 2.5 mL/min. UV detection was performed at 254 nm and 0.1 AUFS.

Samples were diluted with water. The injection volume was 20 μL. The retention time for cefotaxime was about 4.3 minutes (estimated from the published chromatogram).

The HPLC method was stability indicating. Degradation products in a solution of pH 1.5 did not interfere with the intact cefotaxime peak.

Reference
Gupta VD. Stability of cefotaxime sodium as determined by high-performance liquid chromatography. *J Pharm Sci*. 1984; 73: 565–7.

Cefotetan

Chemical Name

[6R-(6α,7α)]-7-[[[4-(2-Amino-1-carboxy-2-oxoethylidene)-1,3-dithietan-2-yl]carbonyl]-
amino]-7-methoxy-3-[[(1-methyl-1H-tetrazol-5-yl)thio]methyl]-8-oxo-5-thia-1-azabi-
cyclo[4.2.0]oct-2-ene-2-carboxylic acid

Other Name

Cefotan

Form	Molecular Formula	MW	CAS
Cefotetan	$C_{17}H_{17}N_7O_8S_4$	575.6	69712-56-7
Cefotetan disodium	$C_{17}H_{15}N_7Na_2O_8S_4$	619.6	74356-00-6

Appearance

Cefotetan disodium is a white to pale yellow lyophilized powder.

Solubility

Cefotetan disodium is very soluble in water (769 mg/mL) and slightly soluble in alcohol
(1.4 mg/mL) at 20 °C.

pK$_a$

Cefotetan has pK$_a$ values of 2.58 and 3.19.

Method

Gupta et al. used HPLC to study the chemical stability of cefotetan disodium in 5% dextrose
and 0.9% sodium chloride injections. A Waters ALC 202 HPLC system was used with a
Rheodyne model 7125 injector, a Schoeffel SF 770 multiple-wavelength detector, and a
Houston Omniscribe recorder. The stationary phase was a Waters μBondapak C$_{18}$ column (300
× 3.9 mm). The mobile phase consisted of 6% acetonitrile in 0.015 M monobasic potassium
phosphate aqueous buffer solution. The flow rate was 2.5 mL/min. UV detection was performed at 238 nm and 0.1 AUFS. Benzoic acid in methanol was used as the internal standard.

Samples were diluted with water. The injection volume was 20 μL. Retention
times for cefotetan and the internal standard were about 4.0 and 7.3 minutes, respectively
(estimated from the published chromatogram).

The assay was determined to be stability indicating. Stock solutions of cefotetan
disodium were mixed with 1 N sulfuric acid or 1 N sodium hydroxide and heated to
boiling for 15 minutes. These solutions were cooled to room temperature and neutralized
using 1 N sodium hydroxide or 1 N sulfuric acid, respectively, and assayed. The degradation product peaks did not interfere with the intact cefotetan peak.

Reference
Gupta VD, Pramar Y, Odom C, et al. Chemical stability of cefotetan disodium in 5% dextrose and 0.9% sodium chloride injections. *J Clin Pharm Ther.* 1990; 15: 109–14.

Cefoxitin

Chemical Name
(6*R-cis*)-3-[[(Aminocarbonyl)oxy]methyl]-7-methoxy-8-oxo-7-[(2-thienylacetyl)amino]-5-thia-1-azabicyclo[4.2.0]oct-2-ene-2-carboxylic acid

Other Name
Mefoxin

Form	Molecular Formula	MW	CAS
Cefoxitin	$C_{16}H_{17}N_3O_7S_2$	427.5	35607-66-0
Cefoxitin sodium	$C_{16}H_{16}N_3NaO_7S_2$	449.4	33564-30-6

Appearance
Cefoxitin sodium is a somewhat hygroscopic white to off-white powder or granules.

Solubility
Cefoxitin sodium is very soluble in water and soluble in methanol; it is sparingly soluble in ethanol and acetone.

Method 1
Using HPLC, Stiles et al. determined the stability of cefoxitin sodium in portable pump reservoirs. The system was composed of a Waters model 501 pump, a model U6K universal injector, and a model 441 UV detector. The stationary phase was a Waters Nova-Pak C_{18} column (150 × 3.9 mm, 5-µm particle size). The mobile phase consisted of 20% methanol and 80% 0.005 M phosphate buffer at pH 7.5. The flow rate was 1.0 mL/min. UV detection was performed at 254 nm.

Samples were diluted with distilled water. Cefoxitin sodium eluted in 2.0 minutes.

The stability-indicating ability of the method was demonstrated by accelerated decomposition of cefoxitin sodium samples. Cefoxitin solution was heated at 45 °C for 72 hours. No interference with the intact cefoxitin peak by the degradation products occurred.

Standard curves for cefoxitin were constructed from 1.0 to 4.5 mg/mL. The correlation coefficient was greater than 0.99. The coefficient of variation was less than 5%.

Reference
Stiles ML, Tu Y-H, Allen LV. Stability of cefazolin sodium, cefoxitin sodium, ceftazidime, and penicillin G sodium in portable pump reservoirs. *Am J Hosp Pharm.* 1989; 46: 1408–12.

Method 2
Bell et al. used HPLC to study the stability of intravenous admixtures of cefoxitin with other drugs. The chromatograph consisted of a Beckman model 110B pump, an LDC/Milton Roy SpectroMonitor D variable-wavelength detector, a Rheodyne model 7125 injector fitted with a 20-μL loop, and a Fisher model 5000 strip-chart recorder. A Zorbax ODS analytical column (150 × 4.6 mm) was used as the stationary phase. The mobile phase consisted of 18% methanol in 0.05 M monobasic potassium phosphate adjusted to pH 3.0 with phosphoric acid. The flow rate was 2 mL/min. UV detection was performed at 238 nm.

Samples were diluted 1:100 with 5 mM phosphate buffer (pH 6.0). Under these conditions, retention times for aztreonam and cefoxitin were 2.2 and 6.8 minutes, respectively.

The stability-indicating capability of the method was demonstrated by accelerated decomposition of cefoxitin. Storing an aqueous solution of cefoxitin 200 μg/mL (pH 5.0) at room temperature for several weeks resulted in a 50% decrease in the peak height of cefoxitin and the appearance of three new peaks attributed to degradation products of cefoxitin. These three peaks had retention times of 1.3, 1.7, and 2.9 minutes and were well separated from the intact cefoxitin peak.

Reference
Bell RG, Lipford LC, Massanari MJ, et al. Stability of intravenous admixtures of aztreonam and cefoxitin, gentamicin, metronidazole, or tobramycin. *Am J Hosp Pharm.* 1986; 43: 1444–53.

Method 3
Bosso and Townsend determined the stability and compatibility of cefoxitin sodium with clindamycin phosphate in 5% dextrose injection and 0.9% sodium chloride injection. The liquid chromatographic system included an Altex model 110A solvent-metering pump, an Altex model 210 injector (or a Waters model 710B autoinjector), a Waters Lambda-Max model 481LC spectrophotometer, an Altex model 153 analytical UV detector, and a Linear Instruments model 555 recorder. The stationary phase was an Alltech Versapak C_{18} reversed-phase column (300 × 4.1 mm, 10-μm particle size). The mobile phase was 14% acetonitrile in a 0.02 M monopotassium phosphate buffer. The pH of the buffer was adjusted to 2.7 with phosphoric acid. The mobile phase was delivered isocratically at 1.5 mL/min. UV detection was performed at 254 nm and 0.64 AUFS. Cefotaxime sodium 20 mg/mL was used as the internal standard.

A 200-μL aliquot of sample solution was diluted with 100 μL of internal standard solution. The injection volume was 2 μL. Under these conditions, retention times for cefoxitin and the internal standard were 5 and 11 minutes, respectively.

This method was stated to be stability indicating because it separated breakdown products generated by acidification or heating from the intact cefoxitin peak.

Intrarun and interrun coefficients of variation were 4.4 and 3.1%, respectively.

Reference

Bosso JA, Townsend RJ. Stability of clindamycin phosphate and ceftizoxime sodium, cefoxitin sodium, cefamandole nafate, or cefazolin sodium in two intravenous solutions. *Am J Hosp Pharm*. 1985; 42: 2211–4.

Method 4

Gupta and Stewart used a stability-indicating HPLC method to investigate the chemical stability of cefoxitin sodium when mixed with metronidazole injection for intravenous infusion. A Waters ALC 202 high-performance liquid chromatograph was equipped with a Schoeffel SF 770 multiple-wavelength detector, an Omniscribe recorder, and a Waters μBondapak phenyl column. The mobile phase consisted of 20% methanol and 80% 0.02 M ammonium acetate buffer. The flow rate was 2.0 mL/min. UV detection was performed at 254 nm and 0.2 AUFS. A solution containing approximately 24 μg of cefoxitin was injected. The retention time for cefoxitin was about 3.0 minutes (estimated from the published chromatogram).

The method was stated to be stability indicating since a chromatogram of a 3-day-old solution of cefoxitin stored at 25 °C showed that degradation product peaks did not interfere with the cefoxitin peak.

Reference

Gupta VD, Stewart KR. Chemical stability of hydrocortisone sodium succinate and several antibiotics when mixed with metronidazole injection for intravenous infusion. *J Parenter Sci Technol*. 1985; 39: 145–8.

Cefpirome

Chemical Name

[6R-[6α,7β(Z)]]-1-[[7-[[(2-Amino-4-thiazolyl)(methoxyimino)acetyl]amino]-2-carboxy-8-oxo-5-thia-1-azabicyclo[4.2.0]oct-2-en-3-yl]methyl]-6,7-dihydro-5H-cyclo-penta[b]pyrindinium inner salt

Other Names

Cefrom, Metran

Form	Molecular Formula	MW	CAS
Cefpirome	$C_{22}H_{22}N_6O_5S_2$	514.6	84957-29-9
Cefpirome sulfate	$C_{22}H_{22}N_6O_5S_2.H_2SO_4$	612.7	98753-19-6

Solubility

Cefpirome sulfate is highly soluble in water at pH 5–7.

Method 1

Using HPLC, Khalfi et al. evaluated the compatibility of cefpirome and cephalothin with PVC bags during simulated infusion and storage. A Hewlett-Packard 1090 liquid chromatograph included a variable-volume injector, an automatic sampling system, an HP 79994 UV diode-array detector, and an HP 9000 model 300 integrator. The stationary phase was a Kromasil C_{18} BDS column (150 × 4.6 mm, 5-μm particle size). The mobile phase consisted of acetonitrile and 0.1% triethylamine buffer (adjusted to pH 3.5 with acetic acid) (20:80, vol/vol). The flow rate was 0.8 mL/min. UV detection was performed at 270 nm. Cefuroxime was used as an internal standard. Samples were diluted with mobile phase. The injection volume was 20 μL. Under these conditions, retention times for cefpirome and cefuroxime were about 2.4 and 3.4 minutes, respectively (estimated from the published chromatogram).

The stability-indicating nature of the method was determined by accelerated degradation of cefpirome. Samples of cefpirome were treated with dilute solutions of hydrochloric acid, sodium hydroxide, or hydrogen peroxide. The intact drug was satisfactorily separated from its degradation products on the chromatogram.

Calibration curves for cefpirome were constructed from 1.25 to 7.5 μg/mL.

Reference

Khalfi F, Dine T, Gressier B, et al. Compatibility of cefpirome and cephalothin with PVC bags during simulated infusion and storage. *Pharmazie*. 1998; 53: 112–6.

Method 2

Allen et al. used HPLC to evaluate the stability of cefpirome sulfate during simulated Y-site injection with drugs commonly used in the intensive care unit. The analysis was conducted utilizing one of two chromatographic systems. One unit included an Alcott model 728 Micromeritics autosampler, a Rheodyne 7010 injector with an Alcott 732 electrically actuated valve, a Waters model 501 solvent-delivery pump, a Waters model 441 UV detector, a Waters model 401 refractive index detector, and a Waters model 745 data module. The other unit consisted of an Alcott model 728 Micromeritics auto-sampler, a Rheodyne 7010 injector with an Alcott 732 electrically actuated valve, a Shimadzu model LA-6A solvent-delivery pump, a Shimadzu model SPD-6A UV detector, and an Orion model 901 microprocessor ion analyzer. The stationary phase was a C_{18} analytical column. The mobile phase consisted of acetonitrile and 0.2% ammonium acetate buffer with pH adjusted to 4.65 with acetic acid (7:1, vol/vol). The flow rate was 1.5 mL/min. UV detection was performed at 254 nm.

Samples were diluted before injection. The injection volume was 20 μL. The retention time for cefpirome was 3.9 minutes.

The assay was determined to be stability indicating by accelerated degradation of cefpirome with hydrochloric acid, sodium hydroxide, potassium chlorate, heating at 80 °C, or light exposure (150 foot-candles from a tungsten filament source). In each case, cefpirome was suitably separated from its degradation products.

Standard curves were constructed from 10 to 100 μg/mL; the correlation coefficient was greater than 0.99. The intraday and interday coefficients of variation of the assay were 1.0 and 2.1%, respectively.

Reference
Allen LV, Stiles ML, Prince SJ, et al. Stability of cefpirome sulfate in the presence of commonly used intensive care drugs during simulated Y-site injection. *Am J Health Syst Pharm.* 1995; 52: 2427–33.

Cefsulodin

Chemical Name
[6*R*-[6α,7β(*R**)]]-4-(Aminocarbonyl)-1-[[2-carboxy-8-oxo-7-[(phenylsulfoacetyl)amino]-5-thia-1-azabicyclo[4.2.0]oct-2-en-3-yl]methyl]pyridinium hydroxide inner salt

Other Names
Cefomonil, Monaspor, Pyocefal, Takesulin

Form	Molecular Formula	MW	CAS
Cefsulodin	$C_{22}H_{20}N_4O_8S_2$	532.6	62587-73-9
Cefsulodin sodium	$C_{22}H_{19}N_4NaO_8S_2$	554.5	52152-93-9

Appearance
Cefsulodin occurs as colorless needles.

Method
Gupta and Stewart evaluated the stability of cefsulodin in aqueous solutions using an HPLC method. The liquid chromatograph consisted of a Waters ALC 202 equipped with a U6K universal injector, a Spectroflow Monitor SF 770 multiple-wavelength detector, and an Omniscribe recorder. The stationary phase was a Waters semipolar μBondapak phenyl column (300 × 4 mm). The mobile phase contained 3.5% methanol and 1.5% acetonitrile (vol/vol) in 0.02 M ammonium acetate in water; the pH was adjusted to about

4.1 with acetic acid. The flow rate was 2.0 mL/min. UV detection was performed at 280 nm and 0.4 AUFS.

Samples were diluted with water. The injection volume was 20 μL. The retention time for cefsulodin was about 3.0 minutes (estimated from the published chromatogram).

The method was reported to be stability indicating. Cefsulodin solution was stored at pH 7.5 for 6 days; degradation products from its partial decomposition did not interfere with the intact cefsulodin peak.

Reference
Gupta VD, Stewart KR. Stability of cefsulodin in aqueous buffered solutions and some intravenous admixtures. *J Clin Hosp Pharm.* 1984; 9: 21–7.

Ceftazidime

Chemical Name
[6*R*-[6α,7β(Z)]]-1-[[7-[[(2-Amino-4-thiazolyl)[(1-carboxy-1-methylethoxy)imino]acetyl]-amino]-2-carboxy-8-oxo-5-thia-1-azabicyclo[4.2.0]oct-2-en-3-yl]methyl]pyridinium] inner salt

Other Names
Ceptaz, Fortaz, Tazicef, Tazidime

Form	Molecular Formula	MW	CAS
Ceftazidime	$C_{22}H_{22}N_6O_7S_2$	546.6	72558-82-8

Appearance
Ceftazidime is a white to off-white powder.

Solubility
Ceftazidime has solubilities of 5 mg/mL in water and less than 1 mg/mL in alcohol.

pK$_a$
Ceftazidime has pK$_a$ values of 1.9, 2.7, and 4.1.

Method 1

Xu et al. used HPLC to evaluate the chemical stability of linezolid injection at 2 mg/mL admixed with three common cephalosporin antibiotics and stored for 7 days at 4 and 23 °C. The Waters liquid chromatograph consisted of a model 600E multisolvent-delivery pump, a model 996 photodiode-array detector, and a WISP model 712 autosampler. The stationary phase was a YMC-Pack ODS-AM column (150 × 4.6 mm, 5-μm particle size). The mobile phase consisted of 30% water and 70% of a mixture of 820 mL of water, 180 mL of acetonitrile, 1 mL of trifluoroacetic acid, and 1 mL of triethylamine. The flow rate was 1.0 mL/min. UV detection was performed at 312 nm and 0.5 AUFS.

Samples were diluted 1:100 with water. The injection volume was 10 μL. Under these conditions, the retention time for ceftazidime was 5.7 minutes.

The method was shown to be stability indicating by accelerated degradation. Sample solutions were treated with 1 N sodium hydroxide, 1 N hydrochloric acid, or 3% hydrogen peroxide, or they were subjected to heating. There was no interference of the degradation product peaks with the intact drug peak.

A standard curve for ceftazidime was constructed from 50 to 300 μg/mL. The correlation coefficient was greater than 0.9999. Intraday and interday coefficients of variation were 0.7 and 0.2%, respectively.

Reference

Xu QA, Trissel LA, Williams KY. Compatibility and stability of linezolid injection admixed with three cephalosporin antibiotics. *J Am Pharm Assoc.* 2000; 40: 509–14.

Method 2

Rivers and Webster studied the stability of ceftazidime with metronidazole in an admixture. A fully automated computer-controlled HPLC system consisted of a Waters model 510 solvent-delivery pump, a Waters 712 WISP refrigerated autosampler, a Waters model 490E programmable multiple-wavelength detector, and a Waters Nova-Pak analytical column (150 × 3.9 mm). The mobile phase contained 0.02 M potassium dihydrogen phosphate and acetonitrile (93:7, vol/vol) with 0.01 M triethylamine and was adjusted to pH 4.8 with hydrochloric acid. The flow rate was 1.5 mL/min. UV detection was performed at 270 nm.

Each sample was diluted 1:50 with distilled water. The injection volume was 20 μL. Under these conditions, the retention times for ceftazidime and metronidazole were 1.7 and 3.1 minutes, respectively.

The method was demonstrated to be stability indicating by accelerated decomposition of ceftazidime. One solution was diluted 1:5 with 0.1 M sodium hydroxide, and the other one was diluted 1:5 with 0.05 M sulfuric acid. They were then heated under reflux conditions for 1 hour. Degradation product peaks did not interfere with the intact drug peak. In acidic conditions, reflux resulted in peaks at 0.9, 1.1, 1.3, 1.6, 2.8, and 4.1 minutes. In alkaline solution, reflux resulted in peaks at 0.8, 1.1, 1.9, 2.1, 2.3, 2.9, and 4.5 minutes.

A standard curve for ceftazidime was constructed from 0.1 to 0.4 mg/mL. The correlation coefficient was 0.996.

Similar methods were used by the other researchers cited here.

References

Rivers TE, Webster AA. Stability of ceftizoxime sodium, ceftriaxone sodium, and ceftazidime with metronidazole in ready-to-use metronidazole bags. *Am J Health Syst Pharm.* 1995; 52: 2568–70.

Mayron D, Gennaro AR. Stability and compatibility of granisetron hydrochloride in i.v. solutions and oral liquids and during simulated Y-site injection with selected drugs. *Am J Health Syst Pharm.* 1996; 53: 294–304.

Walker SE, DeAngelis C, Iazzetta J. Stability and compatibility of combinations of hydromorphone and a second drug. *Can J Hosp Pharm.* 1991; 44: 289–94.

Walker SE, Dranitsaris G. Ceftazidime stability in normal saline, in dextrose, and in water. *Can J Hosp Pharm.* 1988; 41: 65–71.

Method 3

Mason et al. evaluated the stability of ceftazidime and tobramycin sulfate together in peritoneal dialysis solution. The HPLC system consisted of a Waters model 501 constant-flow solvent-delivery system, a Waters model U6K variable-volume injector, a Waters model 484 tunable UV detector, and a Hewlett-Packard model 3390 integrator recorder. The stationary phase was a Waters µBondapak C_{18} column (300 × 4 mm, 10-µm particle size). The mobile phase consisted of 7% acetonitrile, 20 mL of acetic acid, and distilled deionized water to produce a final volume of 2000 mL. The mobile phase was adjusted to pH 4.0 with 6 N sodium hydroxide. UV detection was performed at 254 nm and 0.02 AUFS. Hydrochlorothiazide 8 µg/mL was used as the internal standard. The injection volume was 15 µL. Retention times for ceftazidime and the internal standard were about 8.4 and 11.0 minutes, respectively (estimated from the published chromatogram).

The stability-indicating capability of the assay was demonstrated by analysis of samples of ceftazidime after accelerated degradation. One ceftazidime solution was adjusted to pH 2 with sulfuric acid and heated to 100 °C for 30 minutes. The other ceftazidime solution was adjusted to pH 12 with 1 N sodium hydroxide and heated to 100 °C for 30 minutes. These solutions were then cooled, neutralized, and assayed. No peaks interfered with the peak of the intact ceftazidime.

A standard curve was constructed and was linear ($r > 0.999$) over the working concentration range of the study. The intraday coefficient of variation was less than 0.8%.

This method, slightly modified, was used by the other researchers cited here.

References

Mason NA, Johnson CE, O'Brien MA. Stability of ceftazidime and tobramycin sulfate in peritoneal dialysis solution. *Am J Hosp Pharm.* 1992; 49: 1139–42.

Manduru M, Fariello A, White RL, et al. Stability of ceftazidime sodium and teicoplanin sodium in a peritoneal dialysis solution. *Am J Health Syst Pharm.* 1996; 53: 2731–4.

Barnes AR. Determination of ceftazidime and pyridine by HPLC: Application to a viscous eye drop formulation. *J Liq Chromatogr.* 1995; 18: 3117–28.

Zhou M, Notari RE. Influence of pH, temperature, and buffers on the kinetics of ceftazidime degradation in aqueous solutions. *J Pharm Sci.* 1995; 84: 534–8.

Bednar DA, Klutman NE, Henry DW, et al. Stability of ceftazidime (with arginine) in an elastomeric infusion device. *Am J Health Syst Pharm.* 1995; 52: 1912–4.

Bosso JA, Prince RA, Fox JL. Compatibility of ondansetron hydrochloride with fluconazole, ceftazidime, aztreonam, and cefazolin sodium under simulated Y-site conditions. *Am J Hosp Pharm.* 1994; 51: 389–91.

Stewart JT, Warren FW, Johnson SM, et al. Stability of ceftazidime in plastic syringes and glass vials under various storage conditions. *Am J Hosp Pharm.* 1992; 49: 2765–8.

Pleasants RA, Vaughan LM, Williams DM, et al. Compatibility of ceftazidime and aminophylline admixtures for different methods of intravenous infusion. *Ann Pharmacother.* 1992; 26: 1221–6.

Stiles ML, Allen LV Jr, Fox JL. Stability of ceftazidime (with arginine) and of cefuroxime sodium in infusion-pump reservoirs. *Am J Hosp Pharm.* 1992; 49: 2761–4.

Wade CS, Lampasona V, Mullins RE, et al. Stability of ceftazidime and amino acids in parenteral nutrient solutions. *Am J Hosp Pharm.* 1991; 48: 1515–9.

Method 4

Nahata et al. used an HPLC method to study the stability of ceftazidime (with arginine) stored in plastic syringes at three temperatures. The stationary phase was a Beckman Ultrasphere ODS column (250 × 4.6 mm, 5-μm particle size). The mobile phase was 89% 0.01 M ammonium acetate buffer and 11% methanol. UV detection was performed at 254 nm.

This method was determined to be stability indicating by accelerated base and acid hydrolysis. Ceftazidime 100 mg/mL (with arginine) was diluted 1:2 with 1.0 M sodium hydroxide or with 1.0 M sulfuric acid and was incubated at 80 °C for 1 hour. The degradation products did not interfere with the intact ceftazidime peak.

A calibration curve was constructed from 50 to 150% of the working concentration; the correlation coefficient was 0.999. The coefficient of variation was less than 1%.

Reference

Nahata MC, Morosco RS, Fox JL. Stability of ceftazidime (with arginine) stored in plastic syringes at three temperatures. *Am J Hosp Pharm.* 1992; 49: 2954–6.

Method 5

Using HPLC, Stiles et al. assessed the stability of ceftazidime in portable pump reservoirs. The liquid chromatographic system included a Waters model 501 pump, a Waters U6K universal injector, a Waters model 441 UV detector, and a Waters 745 data module. The stationary phase was a Waters Nova-Pak C_{18} analytical column (150 × 3.9 mm, 5-μm particle size). The mobile phase consisted of 10% methanol and 90% 0.005 M phosphate buffer at pH 7.5 and was delivered isocratically at 0.6 mL/min. UV detection was performed at 254 nm.

Samples were diluted with distilled water. The retention time for ceftazidime was 2.3 minutes.

The method was shown to be stability indicating by intentionally degrading the sample solution at 45 °C for 72 hours. The decomposition products and excipients eluted separately and were detected without apparent interference with the intact ceftazidime peak.

Standard curves for ceftazidime were constructed from 1.0 to 4.0 mg/mL; the correlation coefficient was greater than 0.99. The coefficient of variation of the assay was less than 5%.

A similar method was used by Manduru et al.

References

Stiles ML, Tu Y-H, Allen LV. Stability of cefazolin sodium, cefoxitin sodium, ceftazidime, and penicillin G sodium in portable pump reservoirs. *Am J Hosp Pharm.* 1989; 46: 1408–12.

Manduru M, Fariello A, White RL, et al. Stability of ceftazidime sodium and teicoplanin sodium in a peritoneal dialysis solution. *Am J Health Syst Pharm.* 1996; 53: 2731–4.

Method 6

Gupta et al. used HPLC to study the chemical stability of ceftazidime in 5% dextrose and 0.9% sodium chloride injections. The system consisted of a Waters ALC 202 pump, Schoeffel SF 770 multiple-wavelength detector, and a Houston Omniscribe recorder. The stationary phase was a Waters μBondapak phenyl semipolar column (300 × 3.9 mm). The mobile phase consisted of 6% acetonitrile in a 0.01 M aqueous solution of ammonium acetate and was delivered isocratically at 2.0 mL/min. The detection wavelength was 240 nm and the sensitivity was 0.04 AUFS. Cephalexin 0.15% in water was used as the internal standard.

Samples were diluted 1:80 with water, and the injection volume was 20 μL. Retention times for ceftazidime and the internal standard were about 4.2 and 9.8 minutes, respectively (estimated from the published chromatogram).

The stability-indicating nature of the assay was shown by accelerated decomposition of ceftazidime. Ceftazidime solutions were mixed with 1 N sodium hydroxide or 1 N sulfuric acid and heated on a hot plate for 5 minutes. These solutions were then cooled to room temperature, neutralized with sulfuric acid or sodium hydroxide, respectively, and were assayed by HPLC. Degradation products of ceftazidime did not interfere with the intact ceftazidime peak.

A similar method was used by Vaughan and Poon.

References

Gupta VD, Bethea C, Torre MD. Chemical stabilities of cefoperazone sodium and ceftazidime in 5% dextrose and 0.9% sodium chloride injections. *J Clin Pharm Ther.* 1988; 13: 199–205.

Vaughan LM, Poon CY. Stability of ceftazidime and vancomycin alone and in combination in heparinized and nonheparinized peritoneal dialysis solution. *Ann Pharmacother.* 1994; 28: 572–6.

Method 7

Using an HPLC method, Marble et al. investigated the stability of ceftazidime with clindamycin phosphate in 5% dextrose injection and 0.9% sodium chloride injection. The system consisted of an Altex model 110A solvent-metering pump, an Altex model 210 injector (or a Waters WISP 710B autoinjector), a Waters Lambda-Max model 481 LC spectrophotometer, an Altex model 153 analytical UV detector, and a Linear Instruments model 555 recorder or a Hewlett-Packard model 3392A integrator. The stationary phase was an Alltech Partisil SCX column (250 × 4.1 mm, 10-μm particle size). The mobile phase consisted of 4% acetonitrile and 5% methanol in an aqueous phase consisting of 97.5% 0.1 M acetic acid and 2.5% 0.1 M sodium acetate. The pH was adjusted to 3.7 with 5 N sodium hydroxide. The mobile phase was delivered isocratically at 1.5 mL/min. UV detection was performed at 254 nm and 0.64 AUFS. Cephaloridine sodium 0.5 mg/mL in water was used as the internal standard.

Samples were diluted 1:20 with internal standard solution. The injection volume was 10 µL. Under these conditions, retention times for the internal standard and ceftazidime were 4.5 and 9 minutes, respectively.

The stability-indicating capability of the assay was verified by accelerated degradation of ceftazidime. Acidification (pH 2.0) or heating of ceftazidime solutions resulted in degradation products that did not interfere with the determination of the intact drug.

Intrarun and interrun coefficients of variation were 0.4 and 3.6%, respectively.

Reference
Marble DA, Bosso JA, Townsend RJ. Stability of clindamycin phosphate with aztreonam, ceftazidime sodium, ceftriaxone sodium, or piperacillin sodium in two intravenous solutions. *Am J Hosp Pharm.* 1986; 43: 1732–6.

Ceftizoxime

Chemical Name
[6R-[6α,7β(Z)]]-7-[[(2-Amino-4-thiazolyl)(methoxyimino)acetyl]amino]-8-oxo-5-thia-1-azabicyclo[4.2.0]oct-2-ene-2-carboxylic acid

Other Name
Cefizox

Form	Molecular Formula	MW	CAS
Ceftizoxime	$C_{13}H_{13}N_5O_5S_2$	383.4	68401-81-0
Ceftizoxime sodium	$C_{13}H_{12}N_5NaO_5S_2$	405.4	68401-82-1

Appearance
Ceftizoxime sodium is a white to pale yellow crystalline powder.

Solubility
Ceftizoxime sodium has a solubility of 500 mg/mL in water at 25 °C.

pK_a
Ceftizoxime has pK_a values of 2.1 and 2.7.

Method 1
Nahata et al. investigated the stability and compatibility of ceftizoxime sodium and metronidazole in containers stored at 4 and 25 °C. The liquid chromatograph included a Varian model 2010 solvent-delivery pump, a Shimadzu model SIL-10A autosampler with a sample cooler, a Varian model 2050 variable-wavelength detector, a Hewlett-

Packard 3396A integrator, and a Mac-Mod Zorbax Reliance SB C_{18} analytical column (160 × 4 mm, 5-μm particle size) with a guard column (12.5 × 4 mm). The mobile phase consisted of 82% 0.05 M acetate buffer and 18% methanol with 4 mL of 40% (wt/wt) tetrabutylammonium hydroxide per liter. The final pH of the mobile phase was adjusted to 5.1 with acetic acid. The flow rate was 1.0 mL/min. UV detection was performed at 240 nm. The temperature of vials in the autosampler was maintained at 10 °C, while that of the column was at ambient temperature. Cefaclor 1 mg/mL was used as the internal standard.

Samples were diluted 1:10 with water. The injection volume was 10 μL. Under these conditions, retention times for metronidazole, cefaclor, and ceftizoxime were 3.5, 7.8, and 9.8 minutes, respectively.

The stability-indicating capability of the assay was determined by accelerated hydrolysis of ceftizoxime with an acid and a base. Solutions of the drug admixture were mixed with 1.0 M hydrochloric acid or 1.0 M sodium hydroxide solution for 1 hour at 60 °C. The degradation product peaks appeared before the intact ceftizoxime peak.

Standard curves were generated by linear regression analysis of drug concentration versus peak height ratio for the drug and the internal standard from 0.10 to 1.50 mg/mL. The correlation coefficient was greater than 0.999. The intraday and interday coefficients of variation were less than 0.62%.

Reference

Nahata MC, Edmonds JJ, Morosco RS. Stability of metronidazole and ceftizoxime sodium in ready-to-use metronidazole bags stored at 4 and 25 °C. *Am J Health Syst Pharm.* 1996; 53: 1046–8.

Method 2

Rivers and Webster studied the stability of ceftizoxime sodium with metronidazole in an admixture. A fully automated computer-controlled HPLC system consisted of a Waters model 510 solvent-delivery pump, a Waters 712 WISP refrigerated autosampler, a Waters model 490E programmable multiple-wavelength detector, and a Waters Nova-Pak analytical column (150 × 3.9 mm). The mobile phase contained 0.02 M potassium dihydrogen phosphate and acetonitrile (93:7, vol/vol) with 0.01 M triethylamine and was adjusted to pH 4.8 with hydrochloric acid. The flow rate was 1.5 mL/min. UV detection was performed at 270 nm.

Each sample was diluted 1:50 with distilled water. The injection volume was 20 μL. Under these conditions, the retention times for ceftizoxime and metronidazole were 2.4 and 3.1 minutes, respectively.

The assay was demonstrated to be stability indicating by accelerated decomposition of ceftizoxime and metronidazole. One solution was diluted 1:5 with 0.1 M sodium hydroxide, and the other one was diluted 1:5 with 0.05 M sulfuric acid. They were then heated under reflux conditions for 1 hour. In each case, degradation product peaks were stated not to interfere with the intact drug peak. In acidic conditions, reflux resulted in peaks at 0.7, 0.9, 1.1, 1.3, 1.6, 2.1, 2.9, and 4.4 minutes. In alkaline conditions, reflux resulted in peaks at 0.2, 1.0, 1.1, 1.5, 1.6, 2.2, and 2.8 minutes.

A standard curve was constructed for ceftizoxime from 0.1 to 0.4 mg/mL. The correlation coefficient was 0.998.

Reference

Rivers TE, Webster AA. Stability of ceftizoxime sodium, ceftriaxone sodium, and ceftazidime with metronidazole in ready-to-use metronidazole bags. *Am J Health Syst Pharm.* 1995; 52: 2568–70.

Method 3

Bosso and Townsend determined the stability and compatibility of ceftizoxime sodium with clindamycin phosphate in 5% dextrose injection and 0.9% sodium chloride injection. The liquid chromatograph consisted of an Altex model 110A solvent-metering pump, an Altex model 210 injector (or a Waters model 710B auto-injector), a Waters Lambda-Max model 481LC spectrophotometer, an Altex model 153 analytical UV detector, and a Linear Instruments model 555 recorder. The stationary phase was an Alltech Versapak C_{18} reversed-phase column (300 × 4.1 mm, 10-µm particle size). The mobile phase consisted of 13% acetonitrile in a 0.02 M monopotassium phosphate buffer. The pH of the buffer was adjusted to 2.6 with phosphoric acid. The mobile phase was delivered isocratically at 1.2 mL/min. UV detection was performed at 254 nm and 0.64 AUFS. Cefotaxime sodium 20 mg/mL was used as the internal standard.

Samples were diluted 1:1 with internal standard solution. The injection volume was 2.5 µL. Under these conditions, retention times for ceftizoxime and the internal standard were 6 and 10 minutes, respectively.

This method was stability indicating because it separated degradation product peaks generated by acidification and heating from the intact drug peak.

Intrarun and interrun coefficients of variation were 1.8 and 4.4%, respectively.

Reference

Bosso JA, Townsend RJ. Stability of clindamycin phosphate and ceftizoxime sodium, cefoxitin sodium, cefamandole nafate, or cefazolin sodium in two intravenous solutions. *Am J Hosp Pharm.* 1985; 42: 2211–4.

Ceftriaxone

Chemical Name

[6*R*-[6α,7β(*Z*)]]-7-[[(2-Amino-4-thiazolyl)(methoxyimino)acetyl]amino]-8-oxo-3-[[(1,2,5,6-tetrahydro-2-methyl-5,6-dioxo-1,2,4-triazin-3-yl)thio]methyl]-5-thia-1-azabicyclo[4.2.0]oct-2-ene-2-carboxylic acid

Other Name
Rocephin

Form

Form	Molecular Formula	MW	CAS
Ceftriaxone	$C_{18}H_{18}N_8O_7S_3$	554.6	73384-59-5
Ceftriaxone sodium	$C_{18}H_{16}N_8Na_2O_7S_3.3\frac{1}{2}H_2O$	661.6	104376-79-6

Appearance
Ceftriaxone sodium is an almost white to yellowish and slightly hygroscopic crystalline powder.

Solubility
Ceftriaxone sodium has a solubility of approximately 40 g/100 mL in water at 25 °C.

pK$_a$
Ceftriaxone sodium has pK$_a$ values of 3, 3.2, and 4.1.

Method 1
Xu et al. used HPLC to study the chemical stability of linezolid injection at 2 mg/mL admixed with three common cephalosporin antibiotics and stored for 7 days at 4 and 23 °C. The Waters liquid chromatograph consisted of a model 600E multisolvent-delivery pump, a model 996 photodiode-array detector, and a model 712 WISP autosampler. The stationary phase was a YMC-Pack ODS-AM column (150 × 4.6 mm, 5-μm particle size). The mobile phase consisted of 15% water and 85% of a mixture consisting of 820 mL of water, 180 mL of acetonitrile, 1 mL of trifluoroacetic acid, and 1 mL of triethylamine. The flow rate was 1.5 mL/min. UV detection was performed at 254 nm and 0.5 AUFS.

Samples were diluted 1:100 with water. The injection volume was 10 μL. Under these conditions, retention times for ceftriaxone and linezolid were 5.4 and 18.0 minutes, respectively.

The method was shown to be stability indicating by accelerated degradation. Sample solutions were treated with 1 N sodium hydroxide, 1 N hydrochloric acid, or 3% hydrogen peroxide or were subjected to heating. The degradation product peaks did not interfere with the intact drug peak.

A standard curve for ceftriaxone was constructed from 25 to 150 μg/mL. The correlation coefficient was greater than 0.9999. Intraday and interday coefficients of variation were 0.7 and 0.2%, respectively.

Reference
Xu QA, Trissel LA, Williams KY. Compatibility and stability of linezolid injection admixed with three cephalosporin antibiotics. *J Am Pharm Assoc*. 2000; 40: 509–14.

Method 2
Plumridge et al. determined the stability of ceftriaxone in sterile water for injection stored in polypropylene syringes. The liquid chromatograph included a GBC LC1610 autosampler, a GBC LC1210 multiple-wavelength detector, an integrator using GBC DP900 software, and a Waters Nova-Pak C$_{18}$ analytical column (100 × 4.6 mm, 4-μm particle size). The mobile phase consisted of 40% acetonitrile and 60% 0.0075 M phosphate buffer (pH 7.0) with 2 g of the counterion tetrabutylammonium phosphate added per liter. The flow rate was 1.5 mL/min. UV detection was performed at 242 nm. Ceftazidime served as an internal standard.

Samples were diluted 1:500 with water. The injection volume was 10 μL. The retention times for ceftazidime and ceftriaxone were 1.6 and 2.3 minutes, respectively.

The stability-indicating nature of this method was demonstrated by exposing a sample of ceftriaxone to a high temperature (90 °C for 1 hour) and assaying the resulting solution. The drug breakdown products were well separated from the intact ceftriaxone peak.

Standard curves were generated for ceftriaxone from 0.010 to 0.400 mg/mL; the correlation coefficient was 0.99. The interday coefficient of variation of the method was 0.9%.

References

Plumridge RJ, Rieck AM, Annus TP, et al. Stability of ceftriaxone sodium in polypropylene syringes at −20, 4, and 20 °C. *Am J Health Syst Pharm.* 1996; 53: 2320–3.

Plumridge RJ, Rieck AM, Annus TP, et al. Stability of ceftriaxone sodium reconstituted with lidocaine hydrochloride and stored in polypropylene syringes. *Am J Health Syst Pharm.* 1996; 53: 2323–5.

Method 3

Rivers and Webster evaluated the stability of ceftriaxone sodium with metronidazole in an admixture. A fully automated computer-controlled HPLC system consisted of a Waters model 510 solvent-delivery pump, a Waters 712 WISP refrigerated autosampler, a Waters model 490E programmable multiple-wavelength detector, and a Waters Nova-Pak analytical column (150 × 3.9 mm). The mobile phase contained 0.02 M potassium dihydrogen phosphate and acetonitrile (93:7, vol/vol) with 0.01 M triethylamine and was adjusted to pH 4.8 with hydrochloric acid. The flow rate was 1.5 mL/min. UV detection was performed at 270 nm.

Each sample was diluted 1:50 with distilled water. The injection volume was 20 μL. Under these conditions, the retention times for ceftriaxone and metronidazole were 2.1 and 3.1 minutes, respectively.

The method was demonstrated to be stability indicating by accelerated decomposition of ceftriaxone and metronidazole. One sample was diluted 1:5 with 0.1 M sodium hydroxide, and the other sample was diluted 1:5 with 0.05 M sulfuric acid. They were then heated under reflux conditions for 1 hour. In each case, degradation product peaks were stated not to interfere with the intact drug peak. In acidic conditions, retention times for degradation products were 0.9, 1.1, 1.6, and 2.9 minutes. In alkaline conditions, retention times for degradation products were 1.1, 1.5, 2.2, 2.5, 2.8, and 3.8 minutes.

A five-point standard curve was constructed for ceftriaxone from 0.1 to 0.4 mg/mL. The correlation coefficient was 1.000.

Reference

Rivers TE, Webster AA. Stability of ceftizoxime sodium, ceftriaxone sodium, and ceftazidime with metronidazole in ready-to-use metronidazole bags. *Am J Health Syst Pharm.* 1995; 52: 2568–70.

Method 4

Parrish et al. evaluated the short-term compatibility and stability of ceftriaxone sodium and aminophylline or theophylline in an admixture containing various concentrations of both drugs. The chromatographic system consisted of a Rabbit-HP solvent-delivery system, a Micromeritics model 725 autosampler, an Isco model V-4 variable-wavelength UV detector, a Linseis model L6522 potentiometric recorder, an Apple IIe computer, and a Dynamax C_{18} analytical column (250 × 4.6 mm, 10-μm particle size) with matching guard column. The mobile phase was a mixture of 40% acetonitrile, 55.2% 0.0065 M tetraheptylammonium bromide, 4.4% 0.1 M phosphate buffer (pH 7.0), and 0.4% 0.1 M citrate buffer (pH 5.0). The flow rate was 1.0 mL/min. UV detection was performed at 271 nm. The sample injection volume was 20 μL. Under these conditions, the retention time for ceftriaxone was 14.9 minutes.

This method was stability indicating since it simultaneously quantitated ceftriaxone, its two known degradation products (Ro 11-8390 and Ro 15-2254), theophylline, and 1,3-dimethyluric acid. Retention times for aminophylline, 1,3-dimethyluric acid, Ro 11-8390, ceftriaxone, and Ro 15-2254 were 3.0, 3.9, 7.0, 14.9, and 22.4 minutes, respectively.

Standard curves were generated by analyzing standards of ceftriaxone at concentrations ranging from 50 to 150% of the expected drug concentration in the admixture. The correlation coefficient was greater than 0.99; the coefficient of variation was 1.9%.

A similar method was used by Bailey and Orosz.

References
Parrish MA, Bailey LC, Medwick T. Stability of ceftriaxone sodium and aminophylline or theophylline in intravenous admixtures. *Am J Hosp Pharm*. 1994; 51: 92–4.

Bailey LC, Orosz ST Jr. Stability of ceftriaxone sodium and metronidazole hydrochloride. *Am J Health Syst Pharm*. 1997; 54: 424–7.

Method 5
Walker and Dranitsaris investigated the stability of reconstituted ceftriaxone in dextrose and saline solutions. The HPLC system consisted of a Spectra-Physics model SP8770 isocratic solvent-delivery pump, a Schoeffel model SF 770 UV detector, and a Spectra-Physics SP4270 integrator. The stationary phase was a Brownlee Labs Spheri-10 reversed-phase C_{18} column (10-μm particle size). The mobile phase consisted of acetonitrile and 0.05 M phosphoric acid (30:70, vol/vol). The mobile phase also contained 0.005 M sodium lauryl sulfate. The flow rate was 2.0 mL/min. UV detection was performed at 254 nm.

Samples were diluted 1:30 with water. The injection volume was 15 μL.

The method was demonstrated to be stability indicating by accelerated decomposition of ceftriaxone. A solution of ceftriaxone in 5% dextrose injection was incubated in a water bath at 60 °C for 25 hours. The ceftriaxone concentration was reduced to less than 15% of the initial concentration. Degradation product peaks did not interfere with the intact ceftriaxone peak.

Standard curves for ceftriaxone were constructed from 0.667 to 1.667 mg/mL.

Reference
Walker SE, Dranitsaris G. Stability of reconstituted ceftriaxone in dextrose and saline solutions. *Can J Hosp Pharm*. 1987; 40: 161–6.

Method 6
Using an HPLC method, Marble et al. studied the stability of ceftriaxone sodium with clindamycin phosphate in 5% dextrose injection and 0.9% sodium chloride injection.

The system consisted of an Altex model 110A solvent-metering pump, an Altex model 210 injector (or a Waters WISP 710B autoinjector), a Waters Lambda-Max model 481 LC spectrophotometer, an Altex model 153 analytical UV detector, and a Linear Instruments model 555 recorder or a Hewlett-Packard model 3392A integrator. The stationary phase was an Altex Ultrasphere ODS analytical column (250 × 4.6 mm, 5-μm particle size). The mobile phase consisted of 50% acetonitrile and 50% 0.02 M potassium phosphate buffer to which 1.82 g (0.005 M) of hexadecyltrimethylammonium bromide was added. The pH was adjusted to 7.2 with phosphoric acid. The mobile phase was delivered isocratically at 1.5 mL/min. UV detection was performed at 254 nm and 0.64 AUFS. p-Nitrobenzene-sulfonamide 0.125 mg/mL in methanol was used as the internal standard.

Samples were diluted 1:20 with internal standard solution. The injection volume was 10 μL. Under these conditions, retention times for the internal standard and ceftriaxone were 2 and 4 minutes, respectively.

The stability-indicating capability of the assay was verified by accelerated degradation of ceftriaxone. Acidification (pH 2.0) or heating of ceftriaxone solutions resulted in degradation products that did not interfere with the determination of the intact drug.

Intrarun and interrun coefficients of variation were 2.9 and 4.3%, respectively.

Similar methods were used by Bailey et al. and Nahata.

References

Marble DA, Bosso JA, Townsend RJ. Stability of clindamycin phosphate with aztreonam, ceftazidime sodium, ceftriaxone sodium, or piperacillin sodium in two intravenous solutions. *Am J Hosp Pharm.* 1986; 43: 1732–6.

Bailey LC, Cappel KM, Orosz ST. Stability of ceftriaxone sodium in injectable solutions stored frozen in syringes. *Am J Hosp Pharm.* 1994; 51: 2159–61.

Bailey LC, Tang KT, Medwick T. Stability of ceftriaxone sodium in infusion-pump syringes. *Am J Hosp Pharm.* 1993; 50: 2092–9.

Nahata MC. Stability of ceftriaxone sodium in peritoneal dialysis solutions. *Ann Pharmacother.* 1991; 25: 741–2.

Cefuroxime

Chemical Name

[6R-[6α,7β(Z)]]-3-[[(Aminocarbonyl)oxy]methyl]-7-[[2-furanyl-(methoxyimino)-acetyl]amino]-8-oxo-5-thia-1-azabicyclo[4.2.0]oct-2-ene-2-carboxylic acid

Other Names
Kefurox, Zinacef

Form	Molecular Formula	MW	CAS
Cefuroxime	$C_{16}H_{16}N_4O_8S$	424.4	55268-75-2
Cefuroxime sodium	$C_{16}H_{15}N_4NaO_8S$	446.4	56238-63-2

Appearance
Cefuroxime sodium is a white to off-white powder.

Solubility
Cefuroxime sodium has an aqueous solubility of 500 mg/2.5 mL of water. In ethanol its solubility is about 1 mg/mL.

pK$_a$
Cefuroxime has a pK$_a$ of 2.45.

Method 1
Using HPLC, Wang and Notari reported cefuroxime hydrolysis kinetics and stability predictions in aqueous solution. The liquid chromatograph consisted of a Beckman 110B solvent-delivery module, a Waters M-441 280-nm fixed-wavelength detector, and a Hewlett-Packard 3394 A integrator. The stationary phase was a Keystone Spherisorb C$_6$ column (150 × 4.6 mm, 5-μm particle size). The mobile phase consisted of 9% methanol, 0.1% triethylamine as an ion-pairing agent, and 90.9% 0.1 M aqueous acetate buffer (pH 3.4). The flow rate was 1 mL/min. UV detection was performed at 280 nm. The sample injection volume was 20 μL. Under these conditions, the retention time for cefuroxime was about 13 minutes.

The method was shown to be stability indicating. Cefuroxime was partially hydrolyzed at 40.1 °C in 0.194 M acetate buffer at pH 4.53. Degradation product peaks did not interfere with the intact drug peak.

Linear calibration plots were constructed for cefuroxime from 0.8% 10^{-5} to 9% 10^{-5} M. The coefficient of variation was 0.893%.

Reference
Wang D, Notari RE. Cefuroxime hydrolysis kinetics and stability predictions in aqueous solution. *J Pharm Sci.* 1994; 83: 577–81.

Method 2

Stewart et al. evaluated the stability of cefuroxime sodium with theophylline in intravenous solutions. The HPLC system included a Beckman model 110B solvent-delivery pump, a Rheodyne model 7125 manual injector equipped with a 20-µL loop, a Beckman model 160 fixed-wavelength UV detector, a Hewlett-Packard model 3392A integrator, and a Beckman Ultrasphere octadecylsilane column (150 × 4.6 mm, 5-µm particle size). The mobile phase was a mixture of 0.1 M acetate buffer (pH 3.4) and acetonitrile (10:1, vol/vol). The flow rate was 2 mL/min. UV detection was performed at 254 nm. 5-Methylresorcinol 1.5 mg/mL in water served as an internal standard.

Samples were diluted before injection.

The method was determined to be stability indicating. Aqueous cefuroxime solutions were degraded at 80 °C for 4 hours, at ambient temperature for 6 hours, and by exposure to fluorescent light for 6 hours. Degradation product peaks did not interfere with the intact cefuroxime peak.

Calibration curves for cefuroxime were generated by least-squares regression of the drug to internal standard ratios versus the concentration of the drug from 7.5 to 120 µg/mL. The correlation coefficient was greater than 0.9998.

A similar method was used by Stiles et al.

References

Stewart JT, Warren FW, Johnson SM. Stability of cefuroxime sodium and aminophylline or theophylline. *Am J Hosp Pharm*. 1994; 51: 809–11.

Stiles ML, Allen LV, Fox JL. Stability of ceftazidime (with arginine) and of cefuroxime sodium in infusion-pump reservoirs. *Am J Hosp Pharm*. 1992; 49: 2761–4.

Method 3

Using HPLC, Barnes studied the chemical stability of cefuroxime sodium and metronidazole in an admixture for intravenous infusion. The chromatographic system consisted of a Cecil CE 1100 pump, a Pye LC3 variable-wavelength detector with an 8-µL cell, a Rheodyne 7125 or 7010 injection valve, and a Shimadzu C-R3A integrator. The stationary phase was a reversed-phase Spherisorb hexyl column (100 × 4.5 mm, 5-µm particle size). The mobile phase was a mixture of acetonitrile and 0.1 M sodium acetate buffer at pH 3.4 (7:93, vol/vol). The flow rate was 2 mL/min. UV detection was performed at 273 nm and 0.16 AUFS. Orcinol monohydrate 1.5 mg/mL in water was used as an internal standard.

Samples were diluted with water. The injection volume was 20 µL. Retention times for metronidazole, cefuroxime, and the internal standard were about 2.5, 3.0, and 4.0 minutes, respectively (estimated from the published chromatogram).

The stability-indicating capability of the method was evaluated by chromatographing a cefuroxime sample that had been stored for 18 days at 25 °C. Degradation product peaks did not interfere with the intact cefuroxime peak.

Reference

Barnes AR. Chemical stabilities of cefuroxime sodium and metronidazole in admixture for intravenous infusion. *J Clin Pharm Ther*. 1990; 15: 187–96.

Celecoxib

Chemical Name
4-[5-(4-Methylphenyl)-3-(trifluoromethyl)-1*H*-pyrazol-1-yl]benzenesulfonamide

Other Names
Artilog, Celebrex, Coxel, Solexa

Form	Molecular Formula	MW	CAS
Celecoxib	$C_{17}H_{14}F_3N_3O_2S$	381.4	169590-42-5

Appearance
Celecoxib is an odorless, white to off-white or pale yellow crystalline powder.

Solubility
Celecoxib has aqueous solubilities of about 5 μg/mL at 5–40 °C and pH < 9 and 0.8 mg/mL at 40 °C and pH 12. It has a solubility of 111 mg/mL in alcohol at room temperature.

Method 1
Jadhav and Shingare developed a reversed-phase method for the determination of celecoxib and its impurity. The system consisted of a Shimadzu model LC-10AT VP with dual-piston reciprocating pumps, a model SCL-10A autosampler, and a model SPD-10M VP PDA detector. The stationary phase was a YMC ODS-A C_{18} column (250 × 4.6 mm, 5-μm particle size). The mobile phase consisted of 40% 0.067 M monobasic potassium phosphate buffer and 60% acetonitrile. The flow rate was 1 mL/min. UV detection was performed at 254 nm.

Standard and sample solutions of celecoxib were prepared in mobile phase. The injection volume was 20 μL. Under these conditions, the retention times for celecoxib and its known impurity were 10.9 and 12.6 minutes, respectively.

The stability-indicating property of the method was verified by forced degradation studies. Celecoxib was treated with 1 N hydrochloric acid, 1 N sodium hydroxide solution, or 30% hydrogen peroxide solution. The known impurity and its degradation products did not interfere with the determination of celecoxib.

A standard curve was constructed for celecoxib from 1 to 150 μg/mL with a correlation coefficient of 0.9999. The coefficient of variation for the analysis of celecoxib was 0.16%.

Reference

Jadhav AS, Shingare MS. A new stability-indicating RP-HPLC method to determine assay and known impurity of celecoxib API. *Drug Dev Ind Pharm.* 2005; 31: 779–83.

Method 2

Dhabu and Akamanchi reported the development and validation of an HPLC method for celecoxib in capsules. The apparatus consisted of a Jasco model PU 1580 dual-piston pump, a model UV 1575 detector, and a Rheodyne injector. The stationary phase was a HiQSil RP C_{18} column (250 × 4.6 mm, 5-μm particle size). The mobile phase consisted of methanol and water (85:15, vol/vol) and was delivered isocratically at 0.8 mL/min. UV detection was performed at 251 nm.

A portion of powder from 10 capsules equivalent to 100 mg of celecoxib was weighed, transferred into a 100-mL volumetric flask, mixed with 70 mL of methanol at room temperature for 1 hour, diluted to 100 mL with methanol, and filtered. The injection volume was 20 μL. The run time was 10 minutes. Under these conditions, the retention time for celecoxib was about 5 minutes.

Intentional degradation of celecoxib was carried out by refluxing celecoxib with 1 N hydrochloric acid, 1 N sodium hydroxide solution, or 30% hydrogen peroxide for about 8 hours. The retention times of degradation products were about 3.1 and 6.1 minutes.

A standard curve for celecoxib was constructed from 2 to 50 μg/mL with a correlation coefficient of 0.9997. The coefficient of variation of the assay was 0.7%.

Reference

Dhabu PM, Akamanchi KG. A stability-indicating HPLC method to determine celecoxib in capsule formulations. *Drug Dev Ind Pharm.* 2002; 28: 815–21.

Cephaeline

Chemical Name

7′,10,11-Trimethoxyemetan-6′-ol

Other Name

Cepheline

Form	Molecular Formula	MW	CAS
Cephaeline	$C_{28}H_{38}N_2O_4$	466.6	483-17-0

Solubility

Cephaeline is practically insoluble in water. It is freely soluble in ethanol, methanol, acetic acid, dilute hydrochloric acid, and dilute sulfuric acid.

Method

Elvidge et al. reported an assay for the analysis of emetine and cephaeline at very low levels in pharmaceutical formulations. The system consisted of a Waters model 6000A pump, a WISP autosampler, a Perkin-Elmer LS4 fluorescence spectrophotometer, and an

LCI 100 integrator. The stationary phase was a Waters µBondapak C_{18} column (150 × 3.9 mm). The column temperature was kept at 35 °C. The mobile phase was prepared by dissolving 1.0 g of 1-heptanesulfonic acid sodium salt in water, mixing with 400 mL of methanol and 1 mL of phosphoric acid, and diluting to 1 L with water. The flow rate was 2 mL/min. The fluorescence detector was set at an excitation wavelength of 276 nm and an emission wavelength of 304 nm. The internal standard was prepared by dissolving 0.1 g of ethyl 4-hydroxybenzoate in 50 mL of acetonitrile and diluting to 500 mL with the mobile phase.

Liquid samples were diluted in the mobile phase, whereas pastilles were dissolved in a mixture of internal standard solution and mobile phase. The injection volume was 10 µL. Under these conditions, relative retention times for internal standard, cephaeline, and emetine were 1.0, 1.8, and 3.0, respectively.

The stability-indicating capability of the method was demonstrated by accelerated degradation of the drugs. Samples containing emetine and cephaeline were heated at 70 °C in water, 0.1 M hydrochloric acid, or 0.1 M sodium hydroxide. No significant amount of any degradation product was detected by HPLC.

A standard curve for cephaeline was constructed from 0.01 to 0.036 µg per injection.

Reference

Elvidge DA, Johnson GW, Harrison JR. Selective, stability-indicating assay of the major ipecacuanha alkaloids, emetine and cephaeline, in pharmaceutical preparations by high-performance liquid chromatography using spectrofluorometric detection. *J Chromatogr.* 1989; 463: 107–18.

Cephalexin

Chemical Name

[6R-[6α,7β(R*)]]-7-[(Aminophenylacetyl)amino]-3-methyl-8-oxo-5-thia-1-azabicyclo-[4.2.0]oct-2-ene-2-carboxylic acid

Other Names

Keflex, Keftab

Form	Molecular Formula	MW	CAS
Cephalexin	$C_{16}H_{17}N_3O_4S$	347.4	15686-71-2
Cephalexin hydrochloride	$C_{16}H_{17}N_3O_4S.HCl.H_2O$	401.9	105879-42-3

Appearance
Cephalexin and cephalexin hydrochloride occur as white to off-white crystalline powders.

Solubility
Cephalexin is slightly soluble in water and practically insoluble in alcohol. Cephalexin hydrochloride has solubilities of greater than 10 mg/mL in water and alcohol.

pK$_a$
Cephalexin has pK$_a$ values of 5.2 and 7.3.

Method
Strom and Miller determined the in vitro compatibility, chemical stability, and interaction of cephalexin, cimetidine, diazepam, and propranolol with three selected enteral nutrient products. A Micromeritics liquid chromatograph included a solvent-delivery system, a variable-wavelength UV-visible detector, and a variable-volume injector. Chromatograms were recorded and peak heights calculated with a Micromeritics microprocessor. The stationary phase was a C$_{18}$ analytical column. The mobile phase consisted of 85% 0.05 M potassium phosphate and 15% acetonitrile and was delivered isocratically at 1.0 mL/min. UV detection was performed at 240 nm. Theophylline 100 µg/mL was used as the internal standard.

Each sample was diluted with methanol and centrifuged for 2 minutes. The supernatant was filtered through a Gelman Alpha-450 0.45-µm filter. The injection volume was 25 µL.

The analytical method was demonstrated to be stability indicating by accelerated decomposition of cephalexin. Cephalexin sample solutions were prepared in 1 N hydrochloric acid or 1 N sodium hydroxide and were heated at 80 °C for 24 hours. Intact cephalexin was well separated from its degradation products and from other components in the enteral nutrient formulas.

Reference
Strom JG, Miller SW. Stability of drugs with enteral nutrient formulas. *Ann Pharmacother.* 1990; 24: 130–4.

Cephalothin

Chemical Name
(6R-*trans*)-3-[(Acetyloxy)methyl]-8-oxo-7-[(2-thienylacetyl)amino]-5-thia-1-azabicyclo[4.2.0]oct-2-ene-2-carboxylic acid

Other Name
Keflin

Form

Form	Molecular Formula	MW	CAS
Cephalothin	$C_{16}H_{16}N_2O_6S_2$	396.4	153-61-7
Cephalothin sodium	$C_{16}H_{15}N_2NaO_6S_2$	418.4	58-71-9

Appearance
Cephalothin sodium is a white to off-white crystalline powder.

Solubility
Cephalothin sodium is freely soluble in water and very slightly soluble in alcohol.

pK$_a$
The pK$_a$ of the conjugated acid is 2.22 at 35 °C.

Method 1
Khalfi et al. used HPLC to evaluate the compatibility of cephalothin and cefpirome with PVC bags during simulated infusion and storage. A Hewlett-Packard 1090 liquid chromatograph included a variable-volume injector, an automatic sampling system, an HP 79994 UV diode-array detector, and an HP 9000 model 300 integrator. The stationary phase was a Hypersil C_{18} ODS column (150 × 4.6 mm, 5-μm particle size). The mobile phase consisted of acetonitrile and 0.2% triethylamine buffer (adjusted to pH 4 with acetic acid) (18:82, vol/vol). The flow rate was 1.5 mL/min. UV detection was performed at 240 nm. Cefuroxime was used as an internal standard.

Samples were diluted with mobile phase. The injection volume was 20 μL. Under these conditions, retention times for cefuroxime and cephalothin were about 2.1 and 5.2 minutes, respectively (estimated from the published chromatogram).

The stability-indicating nature of the method was assessed by accelerated degradation of cephalothin. Samples of cephalothin were treated with dilute solutions of hydrochloric acid, sodium hydroxide, or hydrogen peroxide. The intact drug was satisfactorily separated from its degradation products on the chromatograms.

Calibration curves for cephalothin were constructed from 5 to 20 μg/mL.

Reference
Khalfi F, Dine T, Gressier B, et al. Compatibility of cefpirome and cephalothin with PVC bags during simulated infusion and storage. *Pharmazie*. 1998; 53: 112–6.

Method 2

Walker et al. investigated the chemical stability of cephalothin in infusion solutions. The chromatographic system consisted of a Spectra-Physics model 4200 isocratic solvent-delivery pump, a Waters WISP 715 autoinjector, an Applied Biosystems model 759A variable-wavelength UV detector, a Spectra-Physics model 4240 chromatographic integrator, and a Beckman Ultrasphere ODS C_{18} column (250 × 4.2 mm, 5-μm particle size). The mobile phase was a mixture of 20% acetonitrile and 80% 0.05 M monobasic potassium phosphate. The flow rate was 2.0 mL/min. UV detection was performed at 254 nm. The injection volume was 15 μL. The retention time for cephalothin was between 270 and 360 seconds.

The stability-indicating ability of the method was demonstrated by dissolving cephalothin in water and heating at 61 °C for 6.5 hours. Intact cephalothin and its degradation products were well separated.

Standard curves were generated from 50 to 200 mg/L. Intraday and interday coefficients of variation were 1 and 2.3%, respectively.

Reference

Walker SE, Paton TW, Oreopoulos DG. Cephalothin stability in normal saline, five percent dextrose in water, and Dianeal solution. *Can J Hosp Pharm*. 1992; 45: 237–42.

Method 3

Gupta and Stewart described a stability-indicating method for quantitation of cephalothin sodium. The liquid chromatograph consisted of a Waters model ALC 202 system, a U6K universal injector, a Schoeffel Spectroflow monitor SF 770 multiple-wavelength detector, a Houston Omniscribe recorder, and a Spectra-Physics autolab minigrator-integrator. The stationary phase was a Waters μBondapak phenyl column (300 × 4 mm). The mobile phase was 30% methanol and 70% 0.01 M ammonium acetate in water. The flow rate was 2.2 mL/min. UV detection was at 254 nm and 0.04 AUFS.

Samples were diluted with water; the injection volume was 20 μL.

The analytical method was stated to be stability indicating.

The coefficient of variation was 1.3%.

References

Gupta VD, Stewart KR. Quantitation of carbenicillin disodium, cefazolin sodium, cephalothin sodium, nafcillin sodium, and ticarcillin disodium by high-pressure liquid chromatography. *J Pharm Sci*. 1980; 69: 1264–7.

Gupta VD, Stewart KR. Chemical stability of hydrocortisone sodium succinate and several antibiotics when mixed with metronidazole injection for intravenous infusion. *J Parenter Sci Technol*. 1985; 39: 145–8.

Cephapirin Sodium

Chemical Name

Sodium (7*R*)-3-acetoxymethyl-7-[2-(4-pyridylthio)acetamido]-3-cephem-4-carboxylate

Other Names
Cefadyl, Cefapirin Sodium

Form	Molecular Formula	MW	CAS
Cephapirin sodium	$C_{17}H_{16}N_3NaO_6S_2$	445.5	24356-60-3

Appearance
Cephapirin sodium occurs as a white to off-white crystalline powder.

Solubility
Cephapirin sodium is very soluble in water and practically insoluble in most organic solvents.

Method
MacNeil et al. described an HPLC determination of cephapirin, desacetyl cephapirin, and cephapirin lactone in sodium cephapirin bulk and injectable formulations. The chromatograph was composed of a Waters model 6000A pump, a model 440 fixed-wave-length UV detector, a model 710B autosampler, and a model 730 data module. The stationary phase was a Waters μBondapak C_{18} analytical column (300 × 3.9 mm, 10-μm particle size). The mobile phase was an aqueous solution containing 2.25% dimethylfor-mamide, 0.09% acetic acid, and 0.0045% potassium hydroxide. The flow rate was 2.0 mL/min. UV detection was performed at 254 nm. Acetanilide 125 μg/mL in mobile phase was used as an internal standard.

Samples were diluted to about 200 μg/mL of cephapirin sodium before injection. The injection volume was 20 μL. Under these conditions, retention times for desacetyl cephapirin, cephapirin lactone, and cephapirin were about 4.0, 10.4, and 13.3 minutes, respectively. Acetanilide eluted at about 18.2 minutes (estimated from the published chromatogram).

The stability-indicating nature of the assay was confirmed by accelerated degradation of cephapirin. Cephapirin samples were treated with 1 M acetic acid and heated at 80 °C for 45 minutes, treated with 1 M sodium bicarbonate and heated at 80 °C for 90 minutes, heated at 80 °C for 11 days, or exposed to light at 254 nm for 24 days. The known impurities and degradation products did not interfere with the analysis of cephapirin.

A calibration curve for cephapirin was constructed from 100 to 300 μg/mL. The correlation coefficient was greater than 0.9999.

Reference
MacNeil L, Rice JJ, Muhammad N, et al. Stability-indicating liquid chromatographic determination of cephapirin, desacetyl cephapirin and cephapirin lactone in sodium cephapirin bulk and injectable formulations. *J Chromatogr*. 1986; 361: 285–90.

Cetirizine Hydrochloride

Chemical Name

[2-[4-[(4-Chlorophenyl)phenylmethyl]-1-piperazinyl]ethoxy]acetic acid dihydrochloride

• 2HCl

Other Name

Zyrtec

Form	Molecular Formula	MW	CAS
Cetirizine hydrochloride	$C_{21}H_{25}ClN_2O_3.2HCl$	461.8	83881-52-1

Appearance

Cetirizine hydrochloride occurs as a white or almost white powder.

Solubility

Cetirizine hydrochloride is freely soluble in water and practically insoluble in acetone and dichloromethane.

Method

Jaber et al. described a method for the simultaneous analysis of cetirizine hydrochloride and its related impurities in the presence of formulation excipients either in solution or solid formulation. A Thermo Separation system consisted of a model P1000 solvent module and a model UV1000 programmable detector module. The stationary phase was a Hypersil BDS C_{18} column (250 × 4.6 mm, 5-μm particle size). The mobile phase consisted of 0.05 M monobasic potassium phosphate buffer, acetonitrile, methanol, and tetrahydrofuran (60:25:10:5, vol/vol/vol/vol, pH 5.5). The flow rate was 1 mL/min. UV detection was performed at 230 nm.

Cetirizine hydrochloride standard solution was prepared in mobile phase. The injection volume was 20 μL. Under these conditions, the retention time of cetirizine was about 10 minutes (estimated from the published chromatogram).

Cetirizine solutions were prepared separately in 0.1 M hydrochloric acid, 0.1 M sodium hydroxide solution, and 1% hydrogen peroxide and incubated at 80 °C for 10 hours. Cetirizine solutions and powder were also exposed to a low intensity UV lamp and daylight for 65 hours and 2 days, respectively. Degradation products did not interfere with the determination of the intact drug, confirming the stability-indicating capability of the method.

The linear standard curve was obtained from 200 to 800 μg/mL with a correlation coefficient of 0.998. The coefficient of variation in accuracy was better than 4.0%. The limit of detection and the limit of quantitation were 0.10 and 0.34 μg/mL, respectively.

Reference
Jaber AMY, Al Sherife HA, Al Omari MM, et al. Determination of cetirizine dihydrochloride, related impurities and preservatives in oral solution and tablet dosage forms using HPLC. *J Pharm Biomed Anal.* 2004; 36: 341–50.

Chlorambucil

Chemical Name
4-[Bis(2-chloroethyl)amino]benzenebutanoic acid

Other Name
Leukeran

Form	Molecular Formula	MW	CAS
Chlorambucil	$C_{14}H_{19}Cl_2NO_2$	304.2	305-03-3

Appearance
Chlorambucil is an off-white slightly granular powder.

Solubility
Chlorambucil is practically insoluble in water. It is soluble in ether and soluble at 20 °C in 1.5 parts alcohol, 2.5 parts chloroform, and 2 parts acetone.

pK_a
Chlorambucil has apparent pK_a values of 1.3 and 5.8.

Method
Chatterji et al. developed an HPLC method to assay intact chlorambucil in the presence of its hydrolytic decomposition products. A Spectra-Physics model 3500B liquid chromatograph was equipped with a fixed-volume loop injector, a Spectra-Physics model 225 fixed-wavelength detector, and a DuPont Zorbax reversed-phase C_8 column (150 × 4.6 mm, 6-μm particle size). The mobile phase consisted of methanol, acetonitrile, and 0.01 M acetate buffer at pH 4.5 (65:5:30, vol/vol/vol) and was delivered isocratically at 1.6 mL/min. UV detection was performed at 254 nm and 0.32 AUFS. The injection volume was 10 μL. The retention time for chlorambucil was about 10.6 minutes.

This assay was determined to be stability indicating. Chlorambucil degraded about 50% in water alone and in 0.05 M acetate buffer (pH 3.0). Major intermediate degradation products did not interfere with the intact drug peak.

The peak height of chlorambucil was linearly related to its concentration from 10 to 200 µg/mL.

Reference
Chatterji DC, Yeager RL, Gallelli JF. Kinetics of chlorambucil hydrolysis using high-pressure liquid chromatography. *J Pharm Sci.* 1982; 71: 50–4.

Chloramphenicol

Chemical Name
2,2-Dichloro-*N*-[(α*R*, β*R*)-β-hydroxy-α-hydroxymethyl-4-nitrophenethyl]acetamide

Other Name
Chloromycetin

Form	Molecular Formula	MW	CAS
Chloramphenicol	$C_{11}H_{12}Cl_2N_2O_5$	323.1	56-75-7
Chloramphenicol palmitate	$C_{27}H_{42}Cl_2N_2O_6$	561.5	530-43-8
Chloramphenicol sodium succinate	$C_{15}H_{15}Cl_2N_2NaO_8$	445.2	982-57-0

Appearance
Chloramphenicol occurs as a white to grayish-white or yellowish-white fine crystalline powder or fine crystals, needles, or elongated plates. Chloramphenicol palmitate is a fine white or almost white and unctuous crystalline powder. Chloramphenicol sodium succinate occurs as a white or yellowish-white hygroscopic powder.

Solubility
Chloramphenicol has a solubility of approximately 2.5 mg/mL in water at 25 °C. It is freely soluble in alcohol, propylene glycol, acetone, and ethyl acetate. It is slightly soluble in ether. Chloramphenicol palmitate is practically insoluble in water and sparingly soluble in alcohol. It is freely soluble in acetone and chloroform. It is also soluble in ether and very slightly soluble in hexane. Chloramphenicol sodium succinate is freely soluble in water and alcohol. It is practically insoluble in chloroform and ether.

pK$_a$
Chloramphenicol has a pK$_a$ of 5.5.

Method 1

Rao et al. reported an assay method for ophthalmic solutions containing chloramphenicol and dexamethasone. A liquid chromatographic system included a Waters model 600E system controller, a model 715 autosampler, and a model 991 photodiode-array detector. The stationary phase was a Waters μBondapack C_{18} column (300 × 3.9 mm, 5-μm particle size). The mobile phase consisted of 5% glacial acetic acid aqueous solution, acetonitrile, and triethylamine (700:300:2, vol/vol/vol); the pH was adjusted to 5.0 with 10 M sodium hydroxide solution. The flow rate was 1.5 mL/min. UV detection was carried out at 254 nm and 0.5 AUFS.

Sample solutions were diluted with mobile phase. The injection volume was 20 μL. Under these conditions, retention times for chloramphenicol and dexamethasone were about 4.1 and 6.6 minutes, respectively.

The method was demonstrated to be stability indicating by assaying a synthetic mixture of chloramphenicol, its degradation product, and dexamethasone. The degradation product did not interfere with the determination of chloramphenicol.

Calibration curves for chloramphenicol were constructed from 10 to 100 μg/mL and 50 to 500 μg/mL. Correlation coefficients were greater than 0.998. The coefficient of variation of the assay was less than 0.95%. The limit of detection and the limit of quantitation were 10 and 20 ng, respectively.

Reference

Rao RM, Al-Ashban RM, Shah AH. Stability-indicating liquid chromatographic assay method for ophthalmic solutions containing a combination of dexamethasone and chloramphenicol. *J Saudi Chem Soc.* 2004; 8: 223–32.

Method 2

Al-Saidan et al. evaluated an analytical method for the stability study of chloramphenicol succinate ester in 0.01 M arginine solution (pH 10). The liquid chromatograph consisted of a Shimadzu model LC-10 AD single pump and a model SPD-M 10A diode-array detector connected to an Epson LQ 1170 printer. The stationary phase was a Schimpack GLC-ODS column (150 × 6 mm, 5-μm particle size). The mobile phase consisted of acetonitrile and deionized water (50:50, vol/vol), adjusted to pH 4.5 with glacial acetic acid. The flow rate was 1.5 mL/min. UV detection was performed at 270 nm.

Samples were appropriately diluted with mobile phase. The injection volume was 20 μL. Under these conditions, retention times for chloramphenicol and chloramphenicol succinate were 2.88 and 3.28 minutes, respectively.

The assay was evaluated to be stability indicating by intentional degradation of chloramphenicol. A chloramphenicol sample was treated with 0.1 N sodium hydroxide solution and heated at 40 °C in a water bath for 30 minutes. The degradation product peak (1.53 minutes) was separated from the peak of the intact drug (2.88 minutes).

Reference

Al-Saidan SM, Abdel-Hamid ME, Abdel-Aziz A. Determination of chloramphenicol in pharmaceutical formulations using spectrophotometric full spectrum quantitation (FSQ) and high performance liquid chromatography (HPLC). *Alex J Pharm Sci.* 1996; 10: 173–8.

Method 3

Borst and Kubala reported an HPLC determination of chloramphenicol in ophthalmic solutions. The system consisted of a Waters model 6000A pump, a model U6K injector, and a model 441 variable-wavelength UV detector. The stationary phase was a Waters C_{18} reversed-phase Radial-Pak column (100 mm, 10-µm particle size). The mobile phase consisted of Pic 7 buffer in methanol and water (40:60, vol/vol). The flow rate was 0.9 mL/min. UV detection was performed at 280 nm. Methylparaben 0.8 mg/mL was used as an internal standard. The injection volume was 10 µL. Under these conditions, retention times for 2-amino-(4-nitrophenyl)-1,3-propanediol (degradation product), chloramphenicol, and methylparaben were about 4.2, 6.5, and 7.5 minutes, respectively.

The assay was demonstrated to be stability indicating by spiking the chloramphenicol solution with its degradation product. The degradation product, preservatives, and additives did not interfere with the analysis.

A standard curve for chloramphenicol was constructed from 0.08 to 0.13 mg/mL. The correlation coefficient was 0.997.

Reference

Borst SI, Kubala T. High-pressure liquid chromatographic determination of chloramphenicol in ophthalmic solutions. *Pharm Acta Helv.* 1990; 65: 62–4.

Chlordiazepoxide

Chemical Name

7-Chloro-2-(methylamino)-5-phenyl-3*H*-1,4-benzodiazepine 4-oxide

Other Names

Cebrum, Equibral, Libritabs, Librium, Reposans, Risolid, Silibrin, Tropium

Form	Molecular Formula	MW	CAS
Chlordiazepoxide	$C_{16}H_{14}ClN_3O$	299.8	58-25-3
Chlordiazepoxide hydrochloride	$C_{16}H_{14}ClN_3O.HCl$	336.2	438-41-5

Appearance

Chlordiazepoxide occurs as an almost white or light yellow and practically odorless crystalline powder; it is sensitive to light. Chlordiazepoxide hydrochloride occurs as an odorless white or slightly yellow crystalline powder. It is also sensitive to light.

Solubility

Chlordiazepoxide is practically insoluble in water. It is soluble in 1 in 50 of alcohol, 1 in 6250 of chloroform, and 1 in 130 of ether. Chlordiazepoxide hydrochloride is soluble in water. It is sparingly soluble in alcohol and practically insoluble in chloroform, hexane, and ether.

pK$_a$

Chlordiazepoxide has a pK$_a$ of 4.8.

Method

Caufield and Stewart reported a method for the simultaneous determination of zidovudine and chlordiazepoxide. A Hewlett-Packard series 1090 system included a pump, an autosampler with a 25-μL loop, and a Gilson model 117 variable-wavelength UV detector or a Waters model 996 photodiode-array detector. The stationary phase was a Supelco Discovery RP-Amide C$_{16}$ column (250 x 4.6 mm, 5-μm particle size). The mobile phase consisted of 25 mM monobasic sodium phosphate monohydrate (pH adjusted to 3.0 with 0.1 M phosphoric acid) and acetonitrile (80:20, vol/vol). The flow rate was 1.0 mL/min. Solutions of zidovudine and chlordiazepoxide were prepared with an aqueous-acetonitrile diluent matching the mobile phase compositions. UV detection was performed at 265 nm. Under these conditions, retention times of zidovudine and chlordiazepoxide were 6.2 and 14.7 minutes, respectively.

To demonstrate that the method was stability indicating, solutions of zidovudine were subjected to acid hydrolysis (6 M hydrochloric acid), base hydrolysis (6 M sodium hydroxide solution), oxidation (0.3% hydrogen peroxide), heat (90 ℃), and radiation (254 nm). Solutions of chlordiazepoxide were also subjected to acid hydrolysis (1 M hydrochloric acid), base hydrolysis (1 M sodium hydroxide solution), oxidation (0.3% hydrogen peroxide), heat (60 ℃), and radiation (254 nm). In all cases, zidovudine and chlordiazepoxide were separated from their degradation products on their chromatograms.

A standard curve of chlordiazepoxide was constructed from 250 to 1000 μg/mL. The correlation coefficient was greater than 0.9999. Intraday and interday coefficients of variation were 0.08 and 0.25%, respectively.

Reference

Caufield WV, Stewart JT. HPLC separations of zidovudine and selected pharmaceuticals using a hexadecylsilane amide column. *Chromatographia.* 2001; 54: 561–8.

Chlorhexidine

Chemical Names

N,N''-Bis(4-chlorophenyl)-3,12-diimino-2,4,11,13-tetraazatetradecanediimidamide
1,1'-Hexamethylenebis[5-(*p*-chlorophenyl)biguanide]

Form	Molecular Formula	MW	CAS
Chlorhexidine	$C_{22}H_{30}Cl_2N_{10}$	505.5	55-56-1
Chlorhexidine acetate	$C_{22}H_{30}Cl_2N_{10}.2C_2H_4O_2$	625.6	56-95-1
Chlorhexidine gluconate	$C_{22}H_{30}Cl_2N_{10}.2C_6H_{12}O_7$	897.8	18472-51-0
Chlorhexidine hydrochloride	$C_{22}H_{30}Cl_2N_{10}.2HCl$	578.4	3697-42-5

Appearance

Chlorhexidine acetate is a white or almost white microcrystalline powder. Chlorhexidine hydrochloride is a white or almost white crystalline powder.

Solubility

Chlorhexidine is soluble in water at 20 °C. Chlorhexidine acetate is sparingly soluble in water, soluble in alcohol, and slightly soluble in glycerol and propylene glycol. Chlorhexidine gluconate is miscible with water and soluble in alcohol and acetone. Chlorhexidine hydrochloride is sparingly soluble in water and propylene glycol and very slightly soluble in alcohol.

pK_a

Chlorhexidine has pK_a values of 2.2 and 10.3.

Method 1

Ha and Cheung developed an HPLC assay for chlorhexidine. The liquid chromatograph consisted of a Varian 9010 pump, a Varian 9050 UV-visible detector, a Varian 9065 photodiode-array detector, a Rheodyne 7125 manual injector, and a Varian LC Star work-station. The stationary phase was an Alltech Hamilton PRP-1 stainless steel column (250 × 4.1 mm, 10-µm particle size). The mobile phase was acetonitrile in 0.02 M ammonium acetate or phosphate buffer (pH 5.0). The flow rate was 1.0 mL/min. The linear gradient was from 10% acetonitrile to 65% acetonitrile in buffer over 40 minutes. UV detection was performed at 235 nm. The injection volume was 100 µL. The retention time for chlorhexidine was 20.9 minutes.

The method was determined to be stability indicating by accelerated decomposition of chlorhexidine. Chlorhexidine in water was heated at 80 °C for 5 days. Chlorhexidine in 0.01 N hydrochloric acid and 0.01 N sodium hydroxide was heated at 80 °C for 2 days. Degradation product peaks did not interfere with the intact chlorhexidine peak.

The lowest limit of quantitation for chlorhexidine was less than 0.3 ppm when 50 µL of sample solution was injected.

Reference

Ha Y, Cheung AP. New stability-indicating high performance liquid chromatography assay and proposed hydrolytic pathways of chlorhexidine. *J Pharm Biomed Anal.* 1996; 14: 1327–34.

Method 2

Gadde et al. reported an HPLC analysis of chlorhexidine phosphanilate. The Waters liquid chromatograph consisted of a model 6000A pump, a model 710 WISP auto-sampler, and a model 440 absorbance detector. A Hewlett-Packard model 1040A photodiode-array detector was used for the peak purity analysis. The stationary phase was a Waters µBondapak C_{18} column (300 × 3.9 mm). The mobile phase consisted of methanol, tetrahydrofuran, and 0.1 M aqueous sodium sulfate (40:10:50, vol/vol/vol), adjusted to pH 2.2 with sulfuric acid. The flow rate was 1.0 mL/min. UV detection was performed at 254 nm. Ethyl benzoate 0.33 mg/mL in methanol was used as an internal standard.

Chlorhexidine samples were prepared in dimethyl sulfoxide and diluted in methanol. The injection volume was 10 μL. Under these conditions, the retention times of phosphanilic acid, chlorhexidine, and ethyl benzoate were about 1.3, 4.4, and 6.6 minutes, respectively (estimated from the published chromatogram).

The stability-indicating nature of the assay was demonstrated by accelerated degradation. Chlorhexidine phosphanilate samples were heated at 60 °C or exposed to 1000-foot candles light for 8 days. No degradation products interfered with the analysis of chlorhexidine.

A standard curve for chlorhexidine phosphanilate was obtained from 0.013 to 0.053 mg/mL. The correlation coefficient was greater than 0.999.

Reference

Gadde RR, McNiff EF, Peer MM. High-performance liquid chromatographic analysis of chlorhexidine phosphanilate, a new antimicrobial agent. *J Pharm Biomed Anal*. 1991; 9: 1031–6.

Method 3

Hu et al. reported an assay for the simultaneous determination of thimerosal and chlorhexidine gluconate in solutions. The liquid chromatographic system consisted of a Jasco model 880-PU pump, a model 870 variable-wavelength UV detector, an SIC Autosampler 23 automatic sampler, and an SIC Chromatocorder 12 integrator. The stationary phase was a Nucleosil C_{18} column (7-μm particle size) with a Nucleosil C_{18} precolumn (7-μm particle size). The mobile phase consisted of a mixture of 0.1 M monobasic potassium phosphate buffer (which had been adjusted to pH 3.5 with phosphoric acid) and methanol (40:60, vol/vol). The flow rate was 1.0 mL/min. UV detection was performed at 254 nm and 0.04 AUFS. Methylparaben 3 mg/mL was used as an internal standard.

A sample (9.5 mL) was mixed with 0.5 mL of the internal standard solution in a 10-mL volumetric flask. The injection volume was 20 μL. Under these conditions, retention times for methylparaben, thimerosal, and chlorhexidine gluconate were 7.8, 11.4, and 14.9 minutes, respectively.

The stability-indicating capability of the assay was shown by assaying a synthetic mixture of thimerosal, chlorhexidine gluconate, internal standard, and their degradation products. Thimerosal and chlorhexidine gluconate were separated from other compounds.

A standard curve for chlorhexidine was constructed from 2.72 to 436 μg/mL. The correlation coefficient was 0.999.

Reference

Hu OY-P, Wang S-Y, Fang Y-J, et al. Simultaneous determination of thimerosal and chlorhexidine in solutions for soft contact lenses and its applications in stability studies. *J Chromatogr*. 1990; 523: 321–6.

Chlorobutanol

Chemical Name
1,1,1-Trichloro-2-methylpropan-2-ol

$$
\begin{array}{c}
CH_3 \\
| \\
H_3C - C - CCl_3 \\
| \\
HO
\end{array}
$$

Other Name
Chloretone

Form | Molecular Formula | MW | CAS

Form	Molecular Formula	MW	CAS
Chlorobutanol	$C_4H_7Cl_3O$	177.5	57-15-8

Appearance
Chlorobutanol occurs as colorless or white crystals or as a white crystalline powder.

Solubility
Chlorobutanol is slightly soluble in water, very soluble in alcohol and ether, and freely soluble in chloroform.

Method
Dunn et al. described an analysis of chlorobutanol in ophthalmic ointments and aqueous solutions by reversed-phase HPLC. The HPLC system consisted of an Altex model 110A pump, a Schoeffel model SF 770 variable-wavelength UV detector, and a Waters model WISP 710B autosampler; the system was connected to a Houston Omniscribe model A 5111-1 recorder. The stationary phase was a Bio-Rad ODS-10 column (250 × 4.0 mm, 10-μm particle size) coupled with a Waters Bondapak C_{18}/Corasil precolumn (100 × 2.0 mm, 47-μm particle size). The mobile phase was a mixture of methanol and water (1:1, vol/vol). The mobile phase flow rate was 1.8 mL/min. UV detection was performed at 210 nm.

A sample of ointment containing about 20 mg of chlorobutanol was dissolved in 50 mL of hexane and extracted with 20 mL of methanol and water (75:25, vol/vol). The lower (methanolic) layer was collected. The sample was extracted two more times with 15 mL of methanol and water (75:25, vol/vol). The collected methanolic solution was appropriately diluted with methanol and water (75:25, vol/vol) before injection. Ophthalmic aqueous solutions of chlorobutanol were diluted with methanol to 0.5 mg/mL chlorobutanol. The injection volume was 100 μL. The retention time was about 6 minutes.

The HPLC method was evaluated to be stability indicating by intentional degradation of chlorobutanol. Several drops of concentrated hydrochloric acid, concentrated ammonium hydroxide, or 30% hydrogen peroxide or nothing was added to separate solutions of chlorobutanol. Then these solutions were heated at 110 °C overnight. Degradation products did not interfere with the intact chlorobutanol in the analysis.

A standard curve for chlorobutanol was generated from 0.28 to 0.48 mg/mL. The correlation coefficient was 0.999.

Reference
Dunn DL, Jones WJ, Dorsey ED. Analysis of chlorobutanol in ophthalmic ointments and aqueous solutions by reverse-phase high-performance liquid chromatography. *J Pharm Sci.* 1983; 72: 277–80.

Chloroquine

Chemical Names
N^4-(7-Chloro-4-quinolinyl)-N',N'-diethyl-1,4-pentanediamine
7-Chloro-4-(4-diethylamino-1-methylbutylamino)quinoline

Other Name
Aralen

Form	Molecular Formula	MW	CAS
Chloroquine	$C_{18}H_{26}ClN_3$	319.9	54-05-7
Chloroquine hydrochloride	$C_{18}H_{26}ClN_3.2HCl$	392.8	3545-67-3
Chloroquine phosphate	$C_{18}H_{26}ClN_3.2H_3PO_4$	515.9	50-63-5
Chloroquine sulfate	$C_{18}H_{26}ClN_3.H_2SO_4.H_2O$	436.0	132-73-0

Appearance
Chloroquine occurs as a white or slightly yellow odorless crystalline powder. Chloroquine phosphate occurs as a white or almost white hygroscopic crystalline powder. Chloroquine sulfate is a white or almost white crystalline powder.

Solubility
Chloroquine is very slightly soluble in water and soluble in chloroform, ether, and dilute acids. Chloroquine phosphate is freely soluble in water and very slightly soluble in alcohol, chloroform, and ether. Chloroquine sulfate is freely soluble in water and methanol, very slightly soluble in alcohol, and practically insoluble in chloroform and ether.

Method
Allen and Erickson evaluated the stability of chloroquine phosphate in extemporaneously compounded oral liquids. A Hewlett-Packard series 1050 automated high-performance liquid chromatograph included a multisolvent mixing and pumping system, an auto-injector, a diode-array detector, and a computer with Chem Station software. The

stationary phase was a Bakerbond C_{18} column (250 × 4.6 mm, 5-μm particle size). The mobile phase consisted of 0.02 M 1-heptanesulfonic acid (pH 3.4) and acetonitrile (66:34, vol/vol). The mobile phase was delivered isocratically at 1.5 mL/min. UV detection was performed at 340 nm.

Samples were diluted 1:100 before injection. Under these conditions, the retention time for chloroquine was 9.4 minutes.

The analytical method was determined to be stability indicating. A composite chromatogram of chloroquine after accelerated decomposition showed that degradation product peaks did not interfere with the intact chloroquine peak.

A standard curve for chloroquine was constructed from 10 to 150 μg/mL. The intraday and interday coefficients of variation were 0.8 and 1.9%, respectively.

Reference
Allen LV Jr, Erickson MA III. Stability of alprazolam, chloroquine phosphate, cisapride, enalapril maleate, and hydralazine hydrochloride in extemporaneously compounded oral liquids. *Am J Health Syst Pharm.* 1998; 55: 1915–20.

Chlorpheniramine

Chemical Names
γ-(4-Chlorophenyl)-*N,N*-dimethyl-2-pyridinepropanamine
2-[*p*-Chloro-α-(2-dimethylaminoethyl)benzyl]pyridine

Other Name
Chlor-Trimeton

Form	Molecular Formula	MW	CAS
Chlorpheniramine	$C_{16}H_{19}ClN_2$	274.8	132-22-9
Chlorpheniramine maleate	$C_{16}H_{19}ClN_2.C_4H_4O_4$	390.9	113-92-8

Appearance
Chlorpheniramine is a white odorless crystalline powder.

Solubility
Chlorpheniramine maleate has solubilities of 330 mg/mL in ethanol, 240 mg/mL in chloroform, 160 mg/mL in water, and 130 mg/mL in methanol at 25 °C. It is slightly soluble in benzene and ether.

pK$_a$
Chlorpheniramine has a pK$_a$ of approximately 9.2.

Method 1
Heidemann et al. described an HPLC analysis of phenylpropanolamine hydrochloride and chlorpheniramine maleate in cough-cold products. The chromatographic apparatus consisted of a Waters model 510 dual-piston pump, a model 481 variable-wavelength detector, and a model 710B WISP autosampler. The stationary phase was a DuPont Zorbax 300-SCX column (150 × 4.6 mm, 7–8-μm particle size). The mobile phase consisted of acetonitrile and ethylenediamine sulfate buffer (pH 4.52) (35:65, vol/vol). Ethylenediamine sulfate buffer was prepared by dissolving 225 mg of the salt in 1 L of water and adjusting the pH to 4.52 with 0.2 N sulfuric acid. The flow rate was 2 mL/min. UV detection was performed at 216 nm.

For immediate-release tablets, 10 tablets were placed in a 1000-mL volumetric flask containing 250 mL of 0.1 N hydrochloric acid in 50% alcohol, the flask was shaken for 2 hours, and the mixture was diluted to volume with water and filtered through glass-fiber paper. The injection volume was 10 μL. For sustained-release tablets, 10 tablets were powdered, transferred into a 1000-mL volumetric flask containing 600 mL of 0.01 N hydrochloric acid, boiled vigorously with magnetic stirring for 10–13 minutes, cooled to room temperature, diluted to volume with 0.01 N hydrochloric acid, filtered through glass-fiber paper, and diluted 1:10 with water. The injection volume was 20 μL. Under these conditions, retention times of phenylpropanolamine hydrochloride and chlorpheniramine maleate were about 2.6 and 9.5 minutes, respectively (estimated from the published chromatogram).

The method was shown to be stability indicating by the analysis of degraded samples of the different formulations. Degradation products did not interfere with the determination of the two drugs.

A standard curve for chlorpheniramine maleate was constructed from 80 to 120% of the labeled concentrations. The correlation coefficient was 1.000.

Reference
Heidemann DR, Groon KS, Smith JM. HPLC determination of cough-cold products. *LC-GC.* 1987; 5: 422–6.

Method 2
Yacobi et al. described a stability-indicating HPLC method for the simultaneous determination of chlorpheniramine and pseudoephedrine in pharmaceutical dosage forms. The HPLC system consisted of a Waters model ALC/GPC 204 liquid chromatograph equipped with a UV detector, a Houston Omniscribe recorder, and a Waters μBondapak C$_{18}$ nonpolar column (300 × 4 mm). The mobile phase was a mixture of acetonitrile, methanol, and sodium nitrate (35:40:25, vol/vol/vol) and 1-heptanesulfonic acid (0.001 M) with a pH of 5. The flow rate was 2 mL/min. UV detection was performed at 254 nm. Chlorpromazine was used as the internal standard. The injection volume was 5 μL. Retention times for chlorpheniramine and chlorpromazine were 4.6 and 7.3 minutes, respectively.

To determine the method to be stability indicating, solutions containing chlorpheniramine maleate 40 μg/mL and pseudoephedrine hydrochloride 600 μg/mL in water, 3 N sodium hydroxide, and buffers of various pH values (2.5, 4.5, 6.5, and 8) were heated either at 90 °C for 1 hour or at 100 °C for 24 hours. Degradation product peaks did not interfere with the intact chlorpheniramine peak.

A standard curve for chlorpheniramine was constructed from 20 to 240 µg/mL.

Reference
Yacobi A, Look ZM, Lai C-M. Simultaneous determination of pseudoephedrine and chlorpheniramine in pharmaceutical dosage forms. *J Pharm Sci.* 1978; 67: 1668–70.

Chlorpromazine

Chemical Name
3-(2-Chlorophenothiazin-10-yl)propyldimethylamine

Other Name
Thorazine

Form	Molecular Formula	MW	CAS
Chlorpromazine	$C_{17}H_{19}ClN_2S$	318.9	50-53-3
Chlorpromazine hydrochloride	$C_{17}H_{19}ClN_2S.HCl$	355.3	69-09-0

Appearance
Chlorpromazine occurs as a white crystalline solid. Chlorpromazine hydrochloride occurs as a white or slightly creamy white crystalline powder.

Solubility
Chlorpromazine is practically insoluble in water and freely soluble in alcohol. Chlorpromazine hydrochloride has approximate solubilities of 1 g/mL in water and 667 mg/mL in alcohol at 25 °C.

Method 1
Chagonda and Millership developed an HPLC assay for the determination of chlorpromazine and its degradation products in pharmaceutical formulations. The instrument comprised an Altex model 110A pump, a Rheodyne 7125 20-µL injector, a Perkin-Elmer LS-5 luminescence spectrometer, and a Perkin-Elmer 56 recorder. The stationary phase was a Merck LiChrosorb NH_2 column (250 × 4 mm, 5-µm particle size). The mobile phase was a mixture of 0.0001 M hydrochloric acid aqueous solution and acetonitrile (17:83, vol/vol) (pH 4.0). Hydrochloric acid aqueous solution contained 0.01% sodium metabisulfite, 0.01% D-isoascorbic acid, and 0.06% sodium chloride. The flow rate was 1 mL/min.

Fluorescence detection was performed at 450 nm with the excitation wavelength of 280 nm. Quinine hydrochloride 25 μg/mL in the mobile phase was used as an internal standard.

For tablets, a sample of powdered tablets was weighed and dissolved in the internal standard solution to yield approximately 150–200 μg/mL of chlorpromazine hydrochloride. For injectables, an aliquot was diluted with internal standard. The injection volume was 50 μL. Under these conditions, retention times for chlorpromazine and quinine were about 2.4 and 3.2 minutes, respectively (estimated from the published chromatogram).

The analytical method was reported to be stability indicating.

A calibration curve for chlorpromazine was obtained from 50 to 250 μg/mL. The correlation coefficient was more than 0.999. The limit of quantitation was 5.27 μg/mL.

A similar method was described by the same authors in an earlier paper.

References

Chagonda LFS, Millership JS. High-performance liquid chromatographic determination of chlorpromazine and its degradation products in pharmaceutical dosage forms: A stability-indicating assay. *Analyst*. 1988; 113: 233–7.

Chagonda LS, Millership JS. Stability indicating high pressure liquid chromatography method for chlorpromazine. *Dev Drugs Mod Med*. 1986; 244–50.

Chlorpropamide

Chemical Name
4-Chloro-*N*-[(propylamino)carbonyl]benzenesulfonamide

Other Name
Diabinese

Form	Molecular Formula	MW	CAS
Chlorpropamide	$C_{10}H_{13}ClN_2O_3S$	276.7	94-20-2

Appearance
Chlorpropamide is a white odorless or almost odorless crystalline powder.

Solubility
Chlorpropamide has a solubility of 2.2 mg/mL in water at pH 6. It is practically insoluble in water at pH 7.3. It is soluble in alcohol.

Method 1

Gupta described the quantitation of chlorpropamide in tablets by HPLC. A Waters model ALC 202 liquid chromatograph was equipped with a model U6K universal injector, a Schoeffel model SF 770 UV detector, a Spectra-Physics Autolab minigrator, and an Omniscribe recorder. The stationary phase was a Waters μBondapak phenyl column (300 × 4 mm). The mobile phase consisted of methanol, water, and glacial acetic acid (45:54.5:0.5, vol/vol/vol), containing 0.02 M ammonium acetate. The flow rate was 2.0 mL/min. UV detection was performed at 232 nm and 0.1 AUFS. Tolbutamide 1.0 mg/mL in methanol was used as an internal standard.

Ten tablets were ground. A portion of this powder equivalent to 50 mg of chlorpropamide was weighed, mixed with 40 mL of methanol, stirred for 4–5 minutes, brought to 50 mL with methanol, filtered, and finally diluted 20-fold with internal standard and water. The injection volume was 20 μL. Under these conditions, the retention times of chlorpropamide and tolbutamide were about 5.3 and 6.9 minutes, respectively (estimated from the published chromatogram).

The method was stated to be stability indicating since the known degradation product of chlorpropamide, p-chlorobenzenesulfonamide, did not influence the analysis of the drug.

A standard curve for chlorpropamide was constructed from 0.4 to 1.2 μg.

Reference

Gupta VD. Quantitation of chlorpropamide and tolbutamide in tablets by stability-indicating reverse phase high-performance liquid chromatography. *Anal Lett*. 1984; 17: 2119–28.

Method 2

Robertson et al. developed a stability-indicating HPLC method for chlorpropamide, tolbutamide, and their respective sulfonamide degradates. The liquid chromatograph consisted of a Varian model 4100 liquid chromatograph with a fixed-wavelength UV detector, a septumless injector port, and a Spectra-Physics computing integrator. The stationary phase was a Brinkmann LiChrosorb Si-60 column (250 × 3.2 mm, 10-μm particle size). The mobile phase was a mixture of 4% absolute ethanol, 8% tetrahydrofuran, and 0.06% acetic acid in n-hexane. The flow rate was 1 mL/min. UV detection was performed at 254 nm and 0.02 AUFS. Micronized prednisone in ethyl acetate 0.70 mg/mL was used as an internal standard.

Samples were extracted with 4.0 mL of 10% (vol/vol) aqueous hydrochloric acid and 25.0 mL of the internal standard solution. The injection volume was 5 μL. Retention times for chlorpropamide and the internal standard were about 3.2 and 16 minutes, respectively (estimated from the published chromatogram).

This method was confirmed to be stability indicating by accelerated degradation of chlorpropamide samples. Chlorpropamide solution was exposed to 60 °C and 70% relative humidity for 6 weeks. Chlorpropamide partially degraded but the degradation product, p-chlorobenzenesulfonamide, did not interfere with the intact chlorpropamide.

Calibration curves were constructed from 0.3 to 3.0 mg/mL. The correlation coefficient was 0.9998. The coefficient of variation was 1.42%.

Reference

Robertson DL, Butterfield AG, Kolasinski H, et al. Stability-indicating high-performance liquid chromatographic determination of chlorpropamide, tolbutamide, and their respective sulfonamide degradates. *J Pharm Sci*. 1979; 68: 577–80.

Chlorprothixene

Chemical Name
(Z)-3-(2-Chlorothioxanthen-9-ylidene)-N,N-dimethylpropylamine

Other Name
Taractan

Form	Molecular Formula	MW	CAS
Chlorprothixene	$C_{18}H_{18}ClNS$	315.9	113-59-7
Chlorprothixene hydrochloride	$C_{18}H_{18}ClNS.HCl$	352.3	
Chlorprothixene mesylate	$C_{19}H_{22}ClNO_3S_2.H_2O$	430.0	

Appearance
Chlorprothixene is a yellow crystalline powder. Chlorprothixene hydrochloride is a white or almost white crystalline powder.

Solubility
Chlorprothixene is very slightly soluble in water but soluble in alcohol. Chlorprothixene hydrochloride is soluble in water and alcohol. It is slightly soluble in dichloromethane.

Method
Kopelent-Frank and Mittlbock developed an HPLC assay for the simultaneous determination of dixyrazine and chlorprothixene in intravenous admixtures. The liquid chromatograph consisted of a Shimadzu model LC-10AS pump, a model SPD-M10A diode-array detector, a model CTO-10AC column oven, and a Rheodyne 20-µL injector. The stationary phase was a Superspher 100 RP18 column (125 × 4.6 mm, 4-µm particle size). Solvent A was methanol and solvent B was acetate buffer (pH 4.6). The mobile phase was a mixture of solvents A and B and was gradiently delivered as follows:

Time, minutes	Flow Rate, mL/min	Solvent A, %
0–25	0.9	55
25–27	0.9 → 1.2	55 → 75
27–33	1.2	75
33–35	1.2	75 → 55
35–38	1.2 → 0.9	55
38–38.5	0.9	55

UV detection was performed at 252 nm. Under these conditions, retention times for dixyrazine and chlorprothixene were about 24.3 and 26.3 minutes, respectively.

The stability-indicating nature of the assay was demonstrated by accelerated degradation of the drug. The sample was irradiated for 24 minutes. The degradation products did not interfere with the analysis of chlorprothixene.

A standard curve for chlorprothixene was generated from 40 to 100 µg/mL. The correlation coefficient was 0.9997. Intraday and interday coefficients of variation were 1.23 and 2.03%, respectively.

Reference

Kopelent-Frank H, Mittlbock M. Stability-indicating high-performance liquid chromatographic assay for the simultaneous determination of dixyrazine and chlorprothixene in intravenous admixtures. *J Chromatogr A*. 1996; 729: 201–6.

Chlorthalidone

Chemical Name

2-Chloro-5-(2,3-dihydro-1-hydroxy-3-oxo-1*H*-isoindol-1-yl)benzenesulfonamide

Other Name

Hygroton

Form	Molecular Formula	MW	CAS
Chlorthalidone	$C_{14}H_{11}ClN_2O_4S$	338.8	77-36-1

Appearance

Chlorthalidone is a white to yellowish-white crystalline powder.

Solubility

Chlorthalidone is practically insoluble in water and slightly soluble in alcohol.

pK$_a$

Chlorthalidone has a pK$_a$ of 9.4.

Method 1

Using reversed-phase HPLC, Sasa et al. determined atenolol in combination with hydrochlorothiazide and chlorthalidone in tablet formulations. The chromatographic apparatus included a Varian model 2010 pump, a Rheodyne model 7125 10-µL loop injector, a Varian model 2050 spectrophotometric detector, and a Varian model 4290 integrator. The stationary phase was a Supelco Supelcosil LC-8-DB reversed-phase column (250 × 4.6 mm, 5-μm particle size). The mobile phase was a mixture of 1.0 mM ammonium acetate and 2.0 mM sodium octanesulfonate in acetonitrile–water (25:75, vol/vol). The mobile phase was adjusted to pH 3.5 with glacial acetic acid. The flow rate was 1.5 mL/min. UV detection was performed at 254 nm and 0.2 AUFS. Methyl p-hydroxybenzoate 10 μg/mL in methanol was used as the internal standard.

Twenty tablets were ground. A portion of this powder equivalent to one tablet was weighed, transferred into a 25-mL volumetric flask, sonicated with 20 mL of the internal standard for 3 minutes, diluted to 1.00 mg/mL of atenolol with internal standard, and filtered through a 0.45-μm membrane filter. The injection volume was 10 μL. Under these conditions, the retention times of hydrochlorothiazide, atenolol, chlorthalidone, and the internal standard were 3.5, 5.2, 5.9, and 8.4 minutes, respectively.

The method was verified to be stability indicating by studying a synthetic mixture of the drugs and their degradation products. All compounds were separated. No interference from excipients, other drugs, or the degradation products was observed.

A standard curve for chlorthalidone was generated from 1.25 to 3.75 μg/mL. The correlation coefficient was greater than 0.999. The limit of detection was 50 pg.

Reference

Sasa SI, Jalal IM, Khalil HS. Determination of atenolol combinations with hydro-chlorothiazide and chlorthalidone in tablet formulations by reverse phase HPLC. *J Liq Chromatogr*. 1988; 11: 1673–96.

Method 2

Fogel et al. developed an assay for chlorthalidone in tablet formulations. A Hewlett-Packard 1084B liquid chromatograph was used. The stationary phase was a Waters μBondapak C_{18} column (300 × 3.9 mm, 10-μm particle size) with a Waters μBondapak C_{18}/Corasil guard column. The mobile phase consisted of methanol, water, and acetic acid (35:65:1, vol/vol/vol). The flow rate was 2.0 mL/min. UV detection was performed at 235 nm and 0.2 AUFS.

Twenty tablets were weighed and ground. A portion of powder equivalent to about 25 mg of chlorthalidone was weighed, transferred to a 50-mL conical centrifuge tube, mixed with 25 mL of methanol and water (80:20, vol/vol), vigorously shaken for 30 minutes, and centrifuged at 2000 rpm for 5 minutes. An aliquot of 2 mL of the super-natant was diluted to 25 mL with methanol–water (35:65, vol/vol), and filtered through a 0.5-μm membrane filter. The injection volume was 25 μL. Under these conditions, the retention time of chlorthalidone was about 5 minutes (estimated from the published chromatogram).

The assay was reported to be stability indicating since it separated the chlor-thalidone peak from its degradation product peaks.

A standard curve for chlorthalidone was constructed from 1.0 to 3.0 μg. The correlation coefficient was 0.9999.

Reference

Fogel J, Sisco J, Hess F. Validation of liquid chromatographic method for assay of chlorthalidone in tablet formulations. *J Assoc Off Anal Chem*. 1985; 68: 96–8.

Method 3

Bauer et al. described a stability-indicating assay for a chlorthalidone formulation. The instrumentation consisted of a Waters model 6000A pump, a Spectra-Physics model 4100 computing integrator, a DuPont variable-wavelength UV detector, and a Rheodyne model 7120 injector with a 20-μL loop. The stationary phase was a Waters C_{18} μBondapak column. The mobile phase was a mixture of acetonitrile and 2% acetic acid (30:70, vol/vol) and was delivered isocratically at 1.5 mL/min. UV detection was performed at 280 nm. *p*-Nitroaniline 5 mg/mL in methanol was used as the internal standard.

Samples were extracted with methanol. Retention times for chlorthalidone and the internal standard were 3.9 and 5.9 minutes, respectively.

The method was reported to be stability indicating, with the chlorthalidone well separated from its degradation product peaks.

Reference

Bauer J, Quick J, Krogh S, et al. Stability-indicating assay for chlorthalidone formulation: Evaluation of the USP analysis and a high-performance liquid chromatographic analysis. *J Pharm Sci.* 1983; 72: 924–8.

Cidofovir

Chemical Name

(*S*)-[[2-(4-Amino-2-oxo-1(2*H*)-pyrimidinyl)-1-(hydroxymethyl)ethoxy]methyl]phosphonic acid

Other Name

Vistide

Form	Molecular Formula	MW	CAS
Cidofovir	$C_8H_{14}N_3O_6P$	279.2	113852-37-2
Cidofovir dihydrate	$C_8H_{14}N_3O_6P.2H_2O$	315.2	149394-66-1

Appearance

Cidofovir is a fluffy white powder.

Method

Yuan et al. reported the stability and compatibility of cidofovir in intravenous infusion solutions. The liquid chromatograph comprised a Spectra-Physics model AS3000 autosampler, a model UV1000 UV detector, a model P4000 pump, and a Beckman PeakPro data acquisition system. The stationary phase was an Alltech Hypersil ODS C_{18} column (150 × 4.6 mm, 5-μm particle size) with a Direct-Connect refillable guard column (10 × 4.6 mm). The mobile phase was 3.5 mM dibasic sodium phosphate heptahydrate with 5.0 mM tetrabutylammonium dihydrogen phosphate adjusted to pH 6.0 with concentrated phosphoric acid. The flow rate was 2.0 mL/min. UV detection was performed at 280 nm.

Samples were diluted 1:50 with deionized water. The injection volume was 20 μL. The retention time for cidofovir was about 7.0 minutes (estimated from the published chromatogram).

The method was stability indicating. Cidofovir was separated from the degradation product, 1-[(S)-3-hydroxy-2-(phosphonomethoxy)propyl]uracil. The retention time for the degradation product was about 13.4 minutes (estimated from the published chromatogram).

Linearity of the assay was validated from 0.002 to 0.40 mg/mL; the correlation coefficient was 0.99998.

References

Yuan L-C, Samuels GJ, Visor GC. Stability of cidofovir in 0.9% sodium chloride injection and in 5% dextrose injection. *Am J Health Syst Pharm*. 1996; 53: 1939–43.

Oliyai R, Lee WA, Visor GC, et al. Enhanced chemical stability of the intracellular prodrug, 1-[((S)-2-hydroxy-2-oxo-1,4,2-dioxaphosphorinan-5-yl)methyl]cytosine, relative to its parent compound, cidofovir. *Int J Pharm*. 1999; 179: 257–65.

Cilastatin

Chemical Name

[R-[R*,S*-(Z)]]-7-[(2-Amino-2-carboxyethyl)thio]-2-[[(2,2-dimethylcyclopropyl)-carbonyl]amino]-2-heptenoic acid

Other Name

Component of Primaxin

Form	Molecular Formula	MW	CAS
Cilastatin	$C_{16}H_{26}N_2O_5S$	358.5	82009-34-5
Cilastatin sodium	$C_{16}H_{25}N_2NaO_5S$	380.4	81129-83-1

Appearance

Cilastatin sodium is an off-white to yellowish-white hygroscopic amorphous solid.

Solubility

Cilastatin sodium is very soluble in water and methanol.

Method 1

Using an HPLC method, Trissel and Xu investigated the stability of cilastatin in imipenem–cilastatin mixtures in AutoDose infusion system bags. A Waters LC Module-1 consisting of a multisolvent delivery system, an autosampler, and a UV light detector was used for drug analysis. The stationary phase was a Phenomenex Kromasil C_{18} analytical column (250 × 4.6 mm, 5-µm particle size) coupled with a guard column of the same material. The mobile phase consisted of 88% 0.040 M monobasic potassium phosphate aqueous solution and 12% acetonitrile and was delivered isocratically at 1.2 mL/min. UV detection was performed at 216 nm and 0.5 AUFS.

Samples were diluted appropriately with mobile phase to a nominal cilastatin concentration of 0.1 mg/mL for analysis. Injection volume was 15 µL. Under these conditions, retention times for imipenem and cilastatin were about 1.9 and 6.4 minutes, respectively.

The method was shown to be stability indicating by accelerated degradation. Cilastatin solution was mixed with 1 N sodium hydroxide solution, 1 N hydrochloric acid, or 3% hydrogen peroxide and then subjected to heating. Loss of cilastatin was observed, and there was no interference with the peak of the intact cilastatin by the degradation product peaks or the imipenem peaks

A calibration curve for cilastatin was constructed from 25 to 150 µg/mL; the correlation coefficient was greater than 0.9999. Intraday and interday coefficients of variation were 0.2 and 1%, respectively.

Reference

Trissel LA, Xu QA. Stability of imipenem and cilastatin sodium in AutoDose infusion system bags. *Hosp Pharm*. 2003; 38: 130–4.

Method 2

Hall et al. reported the compatibility of imipenem and cilastatin with filgrastim during simulated Y-site administration. The liquid chromatograph consisted of a Waters model 510 pump, a model Ultra 715 WISP autosampler, a model 481 UV detector, and a Shimadzu model C-R3A integrator. The stationary phase was a Waters µBondapak C_{18} column. The mobile phase was 8% acetonitrile in 0.02 M dibasic potassium phosphate (adjusted to pH 7.0 with 10% phosphoric acid). The flow rate was 2 mL/min. UV detection was performed at 254 nm. Under these conditions, the retention times for imipenem and cilastatin were 2.0 and 6.7 minutes, respectively.

The stability-indicating properties of the method were tested by intentional degradation of drugs. Samples of imipenem and cilastatin were acidified to pH 2, alkalinized to pH 8, or heated at 30 °C for 30 minutes. In all cases, degradation products did not influence the determination of cilastatin.

A standard curve for cilastatin was constructed from 1.0 to 3.0 mg/mL. Intrarun and interrun coefficients of variation were less than 3.1%.

Reference

Hall PD, Yui D, Lyons S, et al. Compatibility of filgrastim with selected antimicrobial drugs during simulated Y-site administration. *Am J Health Syst Pharm.* 1997; 54: 185–9.

Method 3

Using two HPLC systems, Allen et al. determined the stability of cilastatin in the latex reservoir of an elastomeric infusion device. One system consisted of a Waters model 501 pump, a Waters model 441 UV detector, a Waters model 745 data module, and a Rheodyne 7010 injector. The other system was composed of a Shimadzu model LC-6A pump, a Shimadzu model SPD-6A UV detector, a Shimadzu model CR-601 data module, a Shimadzu model SIL-9A autosampler, and a Rheodyne 7010 injector. The stationary phase was a Baker C_{18} column. Mobile phase A contained 70% acetonitrile in 0.1% phosphoric acid and mobile phase B was 0.1% phosphoric acid. The gradient was run from 15% of mobile phase A to 100% in 30 minutes. The flow rate was 2 mL/min. UV detection was performed at 210 nm.

Samples were diluted 1:50 with mobile phase. Under these conditions, the retention time for cilastatin was 10.5 minutes.

The stability-indicating nature of the assay was demonstrated by accelerated degradation of cilastatin with heat (to boiling) and acid (1 N sulfuric acid to pH 1.5–2.0). The degradation product peaks did not interfere with the intact drug peak. Retention times for degradation products were 1.0, 4.0, and 7.0 minutes.

A standard curve was constructed from 25 to 125 μg/mL; the correlation coefficient was greater than 0.99. Intraday and interday coefficients of variation for the assay were 1.3 and 1.2%, respectively.

Reference

Allen LV Jr, Stiles ML, Prince SJ, et al. Stability of 14 drugs in the latex reservoir of an elastomeric infusion device. *Am J Health Syst Pharm.* 1996; 53: 2740–3.

Method 4

Zaccardelli et al. used an HPLC method to study the stability of imipenem and cilastatin sodium in total parenteral nutrient solutions. The liquid chromatograph consisted of a Rheodyne model 125 injector, a Waters model 6000A solvent-delivery system, a Waters model Lambda-Max 481 UV detector, and a Hewlett-Packard model 3390A integrator. The stationary phase was a Hewlett-Packard RP8 column (200 × 4.6 mm). The mobile phase consisted of 0.0004 M 3-(N-morpholino)propanesulfonic acid buffer with sodium hexane sulfate (2 g), acetonitrile (4 mL), and methanol (5 mL) per liter. The mobile phase was adjusted to pH 7.00 with sodium hydroxide. The flow rate was 1.8 mL/min. UV detection was performed at 250 nm and 0.1 AUFS.

Samples were diluted 1:100 with mobile phase. The injection volume was 30 μL. Under these conditions, retention times for imipenem and cilastatin were 5.5 and 8.8 minutes, respectively.

The method was determined to be stability indicating by accelerated degradation of imipenem–cilastatin sodium. A standard solution was adjusted to pH 1.8 with phosphoric acid; another was evaporated on a hot plate and allowed to dry at excessive heat before reconstitution to the initial volume with water; and a third solution was heated on a boiling water bath for 15 minutes. Degradation product peaks did not interfere with the intact drug peaks.

A standard curve for cilastatin was constructed each day from 0 to 86.4 µg/mL; the correlation coefficient was 0.9995.

Reference

Zaccardelli DS, Krcmarik CS, Wolk R, et al. Stability of imipenem and cilastatin sodium in total parenteral nutrient solution. *J Parenter Enter Nutr*. 1990; 14: 306–9.

Cimetidine

Chemical Name

N-Cyano-*N'*-methyl-*N''*-[2-[[(5-methyl-1*H*-imidazol-4-yl)methyl]thio]ethyl]guanidine

Other Name

Tagamet

Form	Molecular Formula	MW	CAS
Cimetidine	$C_{10}H_{16}N_6S$	252.3	51481-61-9
Cimetidine hydrochloride	$C_{10}H_{16}N_6S \cdot HCl$	288.8	70059-30-2

Appearance

Cimetidine is a white to off-white crystalline powder. Cimetidine hydrochloride is a white crystalline powder.

Solubility

Cimetidine is sparingly soluble in water and soluble in alcohol. Cimetidine hydrochloride is very soluble in water and soluble in alcohol.

pK$_a$

Cimetidine has a pK$_a$ of 6.8. Cimetidine hydrochloride has a pK$_a$ of 7.11.

Method 1

Mayron and Gennaro used an HPLC method to evaluate the stability of cimetidine with granisetron in intravenous fluids and oral liquids. The liquid chromatographic system consisted of a piston pump with a pulse dampener, a rotary injection port with a 20-µL loop, a variable-wavelength spectrometric detector, and an integrator. The stationary phase was a C_{18} reversed-phase column (300 × 4.6 mm, 5-µm particle size). The mobile phase consisted of acetonitrile and 0.1 M monobasic sodium phosphate dihydrate (20:80, vol/vol) adjusted to pH 4.2 with phosphoric acid. The flow rate was 1.50 mL/min. UV detection was performed at 228 nm.

Samples were diluted with mobile phase. The retention times for cimetidine and granisetron were 2.71 and 6.20 minutes, respectively.

The stability-indicating capability of the analytical method was demonstrated by an accelerated degradation study. Solutions of cimetidine were adjusted to pH 2 and 11, boiled for 1 hour, then readjusted to pH 5, diluted with the mobile phase, and analyzed. Degradation product peaks did not interfere with the intact cimetidine peak.

Reference

Mayron D, Gennaro AR. Stability and compatibility of granisetron hydrochloride in i.v. solutions and oral liquids and during simulated Y-site injection with selected drugs. *Am J Health Syst Pharm*. 1996; 53: 294–304.

Method 2

Nahata et al. determined the stability of cimetidine hydrochloride injection diluted from glass vials at room temperature and under refrigeration. A Hewlett-Packard 1050 series liquid chromatograph consisted of a solvent-delivery pump, an autosampler, a variable-wavelength detector, and an integrator. The stationary phase was a Waters µBondapak C_{18} column (300 × 3.9 mm, 10-µm particle size). The mobile phase consisted of 67% 0.05 M sodium acetate and 33% acetonitrile and was delivered isocratically at 1.0 mL/min. UV detection was performed at 248 nm. Hydrocortisone 300 µg/mL was used as the internal standard.

Each sample was diluted 1:10 with mobile phase. The injection volume was 10 µL. Under these conditions, the retention times for hydrocortisone and cimetidine were 5.6 and 7.1 minutes, respectively.

To determine the stability-indicating nature of the assay, 1 mL of cimetidine 1 mg/mL was mixed with 1 mL of 1.0 M sodium hydroxide or 1.0 M sulfuric acid and heated to 60 °C for 30 minutes. The samples were then assayed. Degradation product peaks did not interfere with the cimetidine peak.

The standard curve was constructed by linear regression analysis of cimetidine concentrations against the peak height ratios of cimetidine and hydrocortisone from 1.5 to 22.5 mg/mL. The correlation coefficient was greater than 0.999; the coefficient of variation was less than 2.1%.

Reference

Nahata MC, Morosco RS, Hipple TF. Stability of cimetidine hydrochloride and of clindamycin phosphate in water for injection stored in glass vials at two temperatures. *Am J Hosp Pharm*. 1993; 50: 2559–63.

Method 3

Strom and Miller studied the in vitro compatibility and chemical stability of cephalexin, cimetidine, diazepam, and propranolol with three selected enteral nutrient products. A Micromeritics liquid chromatograph included a solvent-delivery system, a variable-wavelength UV-visible detector, and a variable-volume injector. Chromatograms were recorded and peak heights calculated with a Micromeritics microprocessor. The stationary phase was a cyano column. The mobile phase was a mixture of 0.025 M potassium phosphate, methanol, and triethylamine (74.9:24.9:0.2, vol/vol/vol) and was delivered isocratically at 0.9 mL/min. UV detection was performed at 228 nm. Caffeine 75 µg/mL was used as the internal standard.

Each sample was diluted with methanol and centrifuged for 2 minutes. The supernatant was filtered through a Gelman Alpha-450 0.45-µm filter. The injection volume was 25 µL.

The analytical method was determined to be stability indicating. Cimetidine sample solutions were prepared in 1 N hydrochloric acid and 1 N sodium hydroxide and were heated at 80 °C for 24 hours. Intact cimetidine was well separated from its degradation products and from other components in the enteral nutrient formulas.

The intraday coefficient of variation was less than 2.8%.

References

Strom JG, Miller SW. Stability of drugs with enteral nutrient formulas. *Ann Pharmacother.* 1990; 24: 130–4.

Strom JG Jr, Miller SW. Stability and compatibility of methylprednisolone sodium succinate and cimetidine hydrochloride in 5% dextrose injection. *Am J Hosp Pharm.* 1991; 48: 1237–41.

Method 4

Using an HPLC method, Baptista et al. conducted a stability study of cimetidine in a total nutrient admixture. The Waters HPLC system included a model 6000A solvent-delivery system, a WISP 710B automated sample processor, a model 441 UV detector fitted with a cadmium lamp and filters (229 nm), and a model 730 data module. The stationary phase was a Z-module with a 10-µm Radial-Pak Bondapak C_{18} column. The mobile phase was 5% acetonitrile in water, to which one 10-mL bottle of dibutyl-amine (PIC D4, Waters Associates, Inc.) was added. The flow rate was 1.0 mL/min. UV detection was performed at 229 nm and 0.2 AUFS.

Samples were extracted with a mixture of chloroform and methanol (1:1, vol/vol). The injection volume was 25 µL. The retention times for metiamide (the internal standard) and cimetidine were 6.4 and 7.0 minutes, respectively.

The method was reported to be stability indicating.

A similar method was used by Mihaly et al.

References

Baptista RJ, Palombo JD, Tahan SR, et al. Stability of cimetidine hydrochloride in a total nutrient admixture. *Am J Hosp Pharm.* 1985; 42: 2208–10.

Mihaly GW, Cockbain S, Jones DB, et al. High-pressure liquid chromatographic determination of cimetidine in plasma and urine. *J Pharm Sci.* 1982; 71: 590–2.

Ciprocinonide

Chemical Name

(6α,11β,16α)-21-[(cyclopropylcarbonyl)oxy]-6,9-difluoro-11-hydroxy-16,17-[(1-methylethylidene)-bis(oxy)]pregna-1,4-diene-3,20-dione

Other Name

Component of Trisyn

Form	Molecular Formula	MW	CAS
Ciprocinonide	$C_{28}H_{34}F_2O_7$	520.6	58524-83-7

Method

Shek et al. reported a reversed-phase HPLC method for a stability study of ciprocinonide cyclopropyl carboxylate in a cream. The instrument was a Spectra-Physics model 3500 system equipped with an oven, a Valco loop injector, and a SpectraMonitor III UV detector or a Spectra-Physics model 8300 UV detector. The stationary phase was a Spectra-Physics Spherisorb ODS stainless steel column (250 × 4.6 mm) with a Whatman Co:Pell ODS guard column (70 × 2.1 mm). The mobile phase was a mixture of tetrahydrofuran, acetonitrile, and water (1:3:6, vol/vol/vol) and was delivered isocratically at 1.2 mL/min. UV detection was performed at 254 nm and 0.02 AUFS. The column temperature was 45 °C. Flucloronide (190 µg/mL) in acetonitrile was used as an internal standard.

Two grams of cream was transferred to a mixture of 75 mL of isooctane and 25 mL of acetonitrile containing 1 mL of the internal standard, and the mixture was vigorously shaken for 1 minute. The lower acetonitrile layer was collected. This extraction was repeated three or four times. The combined acetonitrile solution was evaporated to dryness using a steam bath and a stream of nitrogen; the residue was redissolved in 10 mL of acetonitrile, diluted with 10 mL of water, and centrifuged at 3000 rpm for 10 minutes. The clear solution was collected and assayed. The injection volume was 50 µL. Under these conditions, retention times for the major degradation product (fluocinolone acetonide), the internal standard, and ciprocinonide were about 7.2, 13.3, and 30.0 minutes, respectively (estimated from the published chromatogram).

The stability-indicating nature of the method was demonstrated by assaying a mixture of the drug and its possible degradation products. The chromatogram showed the excellent separation of the intact drug from all of the degradation products.

Reference

Shek E, Bragonje J, Benjamin EJ, et al. A stability indicating high-performance liquid chromatography determination of a triple corticoid integrated system in a cream. *Int J Pharm.* 1982; 11: 257–69.

Ciprofloxacin

Chemical Name

1-Cyclopropyl-6-fluoro-1,4-dihydro-4-oxo-7-(1-piperazinyl)-3-quinolinecarboxylic acid

Other Name

Cipro

Form	Molecular Formula	MW	CAS
Ciprofloxacin	$C_{17}H_{18}FN_3O_3$	331.4	85721-33-1
Ciprofloxacin hydrochloride	$C_{17}H_{18}FN_3O_3 \cdot HCl \cdot H_2O$	385.8	86393-32-0
Ciprofloxacin lactate	$C_{17}H_{18}FN_3O_3 \cdot C_3H_6O_3$	421.4	97867-33-9

Appearance

Ciprofloxacin hydrochloride is a faintly yellowish to yellow crystalline powder.

Solubility

Ciprofloxacin hydrochloride has a solubility of approximately 36 mg/mL in water at 25 °C.

pK$_a$

Ciprofloxacin has pK$_a$ values of 6 and 8.8.

Method 1

Trissel et al. investigated by HPLC the stability of ciprofloxacin in AutoDose infusion system bags. A Waters Alliance model 2690 chromatograph included a solvent-delivery pump, an autosampler, and a model 996 photodiode-array detector. The stationary phase was a Vydac C$_{18}$ column (250 × 4.6 mm, 5-μm particle size). The mobile phase consisted of 0.01 M monobasic sodium phosphate buffer and acetonitrile (82:18, vol/vol), adjusted to pH 3.9 with phosphoric acid. The flow rate was 1.2 mL/min. UV detection was performed at 254 nm and 0.5 AUFS.

Samples were diluted 1:40 with 0.9% sodium chloride injection. The injection volume was 15 μL. Under these conditions, the retention time for ciprofloxacin was 7.2 minutes.

The stability-indicating nature of the method was demonstrated by accelerated degradation of the drug. Ciprofloxacin samples were treated with 1 N hydrochloric acid, 1 N sodium hydroxide solution, or 3% hydrogen peroxide or were heated to boiling for 4 hours. The degradation product peaks did not interfere with the peak of the intact drug.

A standard curve for ciprofloxacin was generated from 25 to 150 µg/mL. The correlation coefficient was greater than 0.9999. Intraday and interday coefficients of variation were both 0.3%.

References

Trissel LA, Xu QA, Zhang Y, et al. Stability of ciprofloxacin and vancomycin hydrochloride in AutoDose Infusion System bags. *Hosp Pharm.* 2001; 36: 1170–3.

Zhang Y, Xu QA, Trissel LA, et al. Compatibility and stability of linezolid injection admixed with three quinolone antibiotics. *Ann Pharmacother.* 2000; 34: 996–1001.

Method 2

Thoppil and Amin described an HPLC method for the determination of ciprofloxacin as bulk drug and in pharmaceutical formulations. The chromatographic system consisted of a Jasco PV980 pump, a U975 UV-visible intelligent detector, and a Rheodyne 7125 20-µL loop injector. The stationary phase was a LiChrospher 100 RP18 column (250 × 4 mm, 10-µm particle size). The mobile phase was a mixture of water, acetonitrile, and triethylamine (80:20:0.6, vol/vol/vol), adjusted to pH 3.0 with phosphoric acid. The flow rate was 1.5 mL/min.

The stability-indicating nature of the method was established by intentional degradation. Ciprofloxacin samples were treated with 2 M hydrochloric acid or 2 M sodium hydroxide solution and refluxed for 1 hour. The degradation product peaks were resolved from the intact drug on the chromatogram.

Standard curves for ciprofloxacin were generated from 20 to 200 µg/mL and from 200 to 1200 µg/mL. The correlation coefficients were 0.995 and 0.999, respectively.

Reference

Thoppil SO, Amin PD. Stability indicating reversed-phase liquid chromatographic determination of ciprofloxacin as bulk drug and in pharmaceutical formulations. *J Pharm Biomed Anal.* 2000; 22: 699–703.

Method 3

Argekar and Powar reported the simultaneous determination by HPLC of ciprofloxacin and tinidazole in tablet dosage forms. The instrument included a TOSHO CCPE pump, a universal injector, a UV detector, and an ORACLE-2 integrator. The stationary phase was an Inertsil C_{18} column (250 × 4.0 mm, 5-µm particle size). The mobile phase was a mixture of 0.1% triethanolamine in water and acetonitrile (78:22, vol/vol), adjusted to pH 2.6 with 1% phosphoric acid. The flow rate was 1.0 mL/min. UV detection was performed at 310 nm.

A sample of ground tablets equivalent to 50 mg of ciprofloxacin and 600 mg of tinidazole was accurately weighed, sonicated for 30 minutes in a 100-mL volumetric flask containing distilled water, filled to the mark with water, and then filtered through a Whatman No. 42 filter. The filtrate was diluted 1:10 with mobile phase. The injection volume was 20 μL. Under these conditions, retention times for ciprofloxacin and tinidazole were 4.2 and 8.2 minutes, respectively.

The assay was demonstrated to be stability indicating. Samples were mixed with 0.1 N hydrochloric acid, 0.1 N sodium hydroxide solution, or 30% hydrogen peroxide or were exposed to sunlight for a week, or were stored at 45 °C for a month. The degradation products did not interfere with the analysis of ciprofloxacin and tinidazole.

A calibration curve for ciprofloxacin was constructed from 10 to 500 μg/mL. The limits of detection and quantitation were 1 and 3 μg/mL, respectively.

Reference

Argekar AP, Powar SG. Simultaneous determination of ciprofloxacin and tinidazole in pharmaceutical preparations by RP-HPLC. *Indian Drugs.* 1999; 36: 399–402.

Method 4

Lacroix et al. developed an HPLC method for the assay of ciprofloxacin hydrochloride and the determination of related compounds. A Varian Vista 5560 HPLC system was equipped with a model UV200 programmable variable-wavelength detector, a model 8085 auto-sampler, a model 402 data processor, and a Hewlett-Packard 7470A plotter. The stationary phase was a Keystone Inertsil ODS2 column (150 × 4.6 mm, 5-μm particle size). The mobile phase was a mixture of tetrahydrofuran, acetonitrile, and 1-hexanesulfonic acid sodium (0.005 M, pH 3.0 with 0.1 M phosphoric acid) (10:5:85, vol/vol/vol). The flow rate was 1 mL/min. UV detection was performed at 254 nm. The injection volume was 10 μL. The retention time for ciprofloxacin was about 14.7 minutes (estimated from the published chromatogram).

The method was demonstrated to be stability indicating by spiking a solution of ciprofloxacin with the related compounds. Ciprofloxacin was well separated from the related compounds on the chromatogram. The retention times in minutes for related compounds estimated from the published chromatogram were as follows: 1-cyclopropyl-6-fluoro-1,4-dihydro-4-oxo-7-(1-piperazinyl)-3-quinoline (3.4), 7-chloro-1-cyclopropyl-1,4-dihydro-4-oxo-6-(1-piperazinyl)-3-quinolinecarboxylic acid (7.3), 1-cyclopropyl-1,4-dihydro-4-oxo-7-(1-piperazinyl)fluoro-1,4-dihydro-4-oxo-3-quinolinecarboxylic acid (8.0), 7-[(2-aminoethyl)amino]-1-cyclopropyl-6-fluoro-1,4-dihydro-4-oxo-3-quinolinecar-boxylic acid (12.5), and 7-chloro-1-cyclopropyl-6-fluoro-1,4-dihydro-4-oxo-3-quino-linecarboxylic acid methyl ester (29.0).

The standard curve was constructed from 50 to 150% of the assay concentration (from 1 to 3 mg/mL) with a correlation coefficient of 0.9996. The coefficient of variation was 2.4%.

Reference

Lacroix PM, Curran NM, Sears CR. High-pressure liquid chromatographic methods for ciprofloxacin hydrochloride and related compounds in raw materials. *J Pharm Biomed Anal.* 1996; 14: 641–54.

Method 5

Kane et al. studied the stability of ciprofloxacin in peritoneal dialysis solutions containing 1.5 and 4.25% dextrose at refrigerated, room, and body temperatures for 2 days, 7 days, and 2 weeks, respectively. The HPLC system was equipped with a Waters model 717 autosampler, a Perkin-Elmer model LS-1 fluorescence detector, and a Varian model 9176 recorder. The stationary phase was a Waters Nova-Pak C_{18} column with a Whatman guard cartridge system. The mobile phase consisted of acetonitrile and 0.02 M phosphoric acid adjusted to pH 3.0 with tetrabutylammonium hydroxide. The flow rate was 1 mL/min. The fluorescence detector was set at an excitation wavelength of 254 nm and an emission wavelength of 425 nm.

The analytical method was determined to be stability indicating by spiking the ciprofloxacin solution with the known degradation product, 7-(2-aminoethylamino)-1-cyclopropyl-6-fluoro-1,4-dihydro-4-oxo-3-quinolinecarboxylic acid. Retention times for the known degradation product and ciprofloxacin were 6.35 and 9.2 minutes, respectively.

A standard curve for ciprofloxacin was constructed from 5 to 100 μg/mL; the correlation coefficient was 0.995. The assay coefficient of variation was 1.94%.

Reference
Kane MP, Bailie GR, Moon DG, et al. Stability of ciprofloxacin injection in peritoneal dialysis solutions. *Am J Hosp Pharm.* 1994; 51: 373–7.

Method 6
Jim evaluated the physical and chemical compatibility of ciprofloxacin with cyclosporine, digoxin, furosemide, gentamicin, heparin, metoclopramide, metronidazole, midodrine, netilmicin, prednisolone, ranitidine, sodium bicarbonate, sodium chloride, and teicoplanin during simulated Y-site administration. A Waters liquid chromatographic system included a model 501 solvent-delivery pump, a model 680 automated gradient controller, a model 470 scanning fluorescence detector, a model U6K universal injector, and a dual-pen recorder. The stationary phase was a μBondapak C_{18} cartridge (10-μm particle size). The mobile phase consisted of acetonitrile and 0.1 M sodium dihydrogen phosphate adjusted to pH 3.3 with phosphoric acid (20:80, vol/vol). The flow rate was 2.5 mL/min. Ciprofloxacin was monitored by a fluorescence detector set at an excitation wavelength of 277 nm and an emission wavelength of 453 nm.

Each sample was diluted 1:10 with deionized water. The injection volume was 10 μL. The retention time for ciprofloxacin was about 4.8 minutes (estimated from the published chromatogram).

The method was determined to be stability indicating. Solutions of ciprofloxacin 100 μg/mL were mixed with 1 N hydrochloric acid or 1 N sodium hydroxide and were heated at 80 °C for 6 hours. Degradation product peaks did not interfere with the intact ciprofloxacin peak.

The intraday and interday coefficients of variation were less than 2%.

A similar method was used by Nahata et al.

References
Jim LK. Physical and chemical compatibility of intravenous ciprofloxacin with other drugs. *Ann Pharmacother.* 1993; 27: 704–7.

Jim LK, El-Sayed N, Al-Khamis KI. A simple high-performance chromatographic assay for ciprofloxacin in human serum. *J Clin Pharm Ther.* 1992; 17: 111–5.

Nahata MC, Morosco RS, Hipple TF. Development of stable oral suspensions of ciprofloxacin. *J Appl Ther Res.* 2000; 3: 61–5.

Cisapride

Chemical Name
cis-4-Amino-5-chloro-*N*-[1-[3-(4-fluorophenoxy)propyl]-3-methoxy-4-piperidinyl]-2-methoxybenzamide

Other Name
Propulsid

Form
Cisapride

Molecular Formula
$C_{23}H_{29}ClFN_3O_4$

MW
466.0

CAS
81098-60-4

Appearance
Cisapride is a white to slightly beige odorless powder.

Solubility
Cisapride is practically insoluble in water.

Method
Nahata et al. determined the stability of cisapride in a liquid dosage form stored in plastic bottles at room temperature and under refrigeration. A Hewlett-Packard 1050 series chromatograph included a pump, an autosampler, a variable-wavelength detector, and a Hewlett-Packard 3396A integrator. The stationary phase was a Zorbax Rx-C$_8$ column (160 × 4.0 mm, 5-μm particle size). The mobile phase was a mixture of 35% deionized water with 0.02% triethylamine and 65% acetonitrile. The flow rate was 1.0 mL/min. UV detection was performed at 276 nm.

Samples were diluted with methanol. The injection volume was 10 μL. Under these conditions, the retention times for cisapride and the internal standard, *cis*-4-amino-5-chloro-*N*-[1-[5-(4-fluorophenoxy)pentyl]-3-methoxy-4-piperidinyl]-2-methoxybenzamide monohydrate, were 2.7 and 3.8 minutes, respectively.

This method was demonstrated to be stability indicating. One-milliliter aliquots of cisapride standard 1 mg/mL were mixed with 1 mL of 1.0 M hydrochloric acid or 1.0 M sodium hydroxide and heated to 60 °C for 90 minutes. One milliliter of the cisapride 1-mg/mL standard was also mixed with 1 mL of 0.3% hydrogen peroxide. The resulting solutions were assayed by HPLC. The degradation product peaks did not interfere with the intact cisapride peak.

The standard curve was generated by linear regression analysis of the cisapride concentration versus peak area ratios of cisapride and the internal standard from 0.10

to 1.50 mg/mL. The correlation coefficient was 0.999, and the coefficient of variation was less than 3%.

Reference
Nahata MC, Morosco RS, Hipple TF. Stability of cisapride in a liquid dosage form at two temperatures. *Ann Pharmacother.* 1995; 29: 125–6.

Cisatracurium

Chemical Name
[1*R*-[1α,2α(1′*R**,2′*R**)]]-2,2′-[1,5-pentanediylbis[oxy(3-oxo-3,1-propanediyl)]]bis[1-[(3,4-dimethoxyphenyl)methyl]-1,2,3,4-tetrahydro-6,7-dimethoxy-2-methylisoquinolinium] dibenzenesulfonate

Other Name
Nimbex

Form	Molecular Formula	MW	CAS
Cisatracurium besylate	$C_{65}H_{82}N_2O_{18}S_2$	1243.5	96946-42-8

Method 1
Xu et al. studied the stability of cisatracurium (as the besylate) 10 mg/mL in original vials and 2 mg/mL in repackaged plastic syringes, as well as 0.1, 2, and 5 mg/mL in 5% dextrose injection and 0.9% sodium chloride injection stored at 4 and 23 °C both in the dark and exposed to fluorescent light. A Waters LC-Module 1 chromatograph consisted of a multisolvent-delivery pump, a UV detector, and an autosampler. The system was controlled and integrated by a Digital Venturis 575 personal computer with Waters Millennium 2010 chromatography manager. The stationary phase was a Phenomenex Kromasil C_{18} column (250 × 4.6 mm, 5-μm particle size). The mobile phase consisted of 600 mL of water, 200 mL of acetonitrile, and 200 mL of methanol with 10 mL of formic acid and 20 g of ammonium formate and was delivered isocratically at 1.5 mL/min. UV detection was performed at 280 nm and 0.5 AUFS.

Samples were properly diluted with a diluent containing 600 mL of water, 200 mL of acetonitrile, 200 mL of methanol, and 0.4 mL of formic acid. The injection volume was 10 μL. Under these conditions, the retention time for cisatracurium was about 22 minutes.

The analytical method was demonstrated to be stability indicating by accelerated degradation of cisatracurium besylate. Exposure of cisatracurium besylate solutions to 1 N hydrochloric acid, 1 N sodium hydroxide, or 3% hydrogen peroxide and then boiling for 2 hours resulted in a reduction in the intact cisatracurium peak and the formation of two new peaks at retention times of 4 and 34 minutes. The peaks for degradation products did not interfere with the intact cisatracurium peak.

Calibration curves were constructed from a linear plot of cisatracurium peak areas versus concentration of freshly prepared cisatracurium besylate reference standard from 0.025 to 0.150 mg/mL. The correlation coefficient of the standard curve was greater than 0.9997. The intraday and interday coefficients of variation were 2.2 and 1.7%, respectively.

Reference

Xu QA, Zhang Y-P, Trissel LA, et al. Stability of cisatracurium besylate in vials, syringes, and infusion admixtures. *Am J Health Syst Pharm.* 1998; 55: 1037–41.

Method 2

Zhang et al. presented an HPLC method for the determination of cisatracurium besylate and propofol in a mixture. The apparatus consisted of a Beckman model 110B pump, an Alcott model 738 autosampler, a Waters Lambda-Max model 481 LC spectrophotometer, and a Hewlett-Packard model 3394A integrator. The stationary phase was an Alltech Spherisorb ODS-2 column (250 × 4.6 mm, 5-μm particle size). The mobile phase was a mixture of acetonitrile and 0.3 M ammonium formate (adjusted to pH 5.2 with formic acid) (50:50, vol/vol). The flow rate was 1.0 mL/min. UV detection was performed at 280 nm.

Cisatracurium besylate was dissolved in acetonitrile. The injection volume was 20 μL. Under these conditions, retention times for cisatracurium besylate and propofol were about 8.8 and 13.1 minutes, respectively.

The method was evaluated to be stability indicating. Mixtures of cisatracurium besylate and propofol were treated with 0.01 N hydrochloric acid, 0.01 N sodium hydroxide solution, or 3–30% hydrogen peroxide. The degradation product peaks did not interfere with the analysis of cisatracurium besylate and propofol.

A standard curve for cisatracurium besylate was generated from 8 to 128 μg/mL. The correlation coefficient was 0.9999.

Reference

Zhang H, Wang P, Bartlett MG, et al. HPLC determination of cisatracurium besylate and propofol mixtures with LC-MS identification of degradation products. *J Pharm Biomed Anal.* 1998; 16: 1241–9.

Cisplatin

Chemical Names

(*SP*-4-2)-Diamminedichloroplatinum

cis-Diamminedichloroplatinum

Other Name

Platinol

Form	Molecular Formula	MW	CAS
Cisplatin	$Cl_2H_6N_2Pt$	300.1	15663-27-1

Appearance

Cisplatin is a white powder.

Solubility

Cisplatin has a solubility of 0.253 g/100 g in water at 25 °C. It is insoluble in most common organic solvents but soluble in dimethylformamide.

Method 1

Using an HPLC method, Mayron and Gennaro studied the stability of cisplatin with granisetron hydrochloride during simulated Y-site administration. The liquid chromatographic system consisted of a piston pump with a pulse dampener, a rotary injection port with a 20-µL loop, a variable-wavelength spectrometric detector, and an integrator. The stationary phase was an amino reversed-phase column (250 × 4.6 mm, 5-µm particle size). The mobile phase was a mixture of acetonitrile and water (90:10, vol/vol). The flow rate was 2.50 mL/min. UV detection was performed at 300 nm.

Samples were diluted with the mobile phase. The retention times for cisplatin and granisetron were 3.16 and 7.51 minutes, respectively.

The stability-indicating capability of this analytical method was demonstrated by an accelerated degradation study. Solutions of cisplatin and granisetron hydrochloride were adjusted to pH 2 and 11, boiled for 1 hour, readjusted to pH 5, diluted with the mobile phase, and analyzed. Degradation product peaks did not interfere with the intact cisplatin or granisetron peak.

Reference

Mayron D, Gennaro AR. Stability and compatibility of granisetron hydrochloride in i.v. solutions and oral liquids and during simulated Y-site injection with selected drugs. *Am J Health Syst Pharm.* 1996; 53: 294–304.

Method 2

Zieske et al. used HPLC to characterize cisplatin degradation as affected by pH and light. The liquid chromatographic systems consisted of Waters model 501, Perkin-Elmer series 10, and Applied Biosystems Spectroflow 400 pumps; Waters WISP 710B and Micromeritics model 728 autosamplers; Applied Biosystems variable-wavelength Spectroflow 757 detectors; and Hewlett-Packard 1000/3357 laboratory automation systems. The stationary phase was a Hamilton reversed-phase column with polystyrene–divinylbenzene packing (150 × 4.1 mm, 10-µm particle size). The mobile phase was 0.5% hexadecyltrimethylammonium chloride. The flow rate was 0.5 mL/min. UV detection was performed at 280 nm and 0.02 AUFS. The injection volume was 5 µL. Under these conditions, the retention time for cisplatin was 4.8 minutes.

The analytical method was determined to be stability indicating by spiking solutions with known degradation products. Degradation product peaks did not interfere with the cisplatin peak.

A standard curve for cisplatin was constructed from 0.5 to 2.0 mg/mL. The coefficient of variation was less than 1.0%.

Reference

Zieske PA, Koberda M, Hines JL, et al. Characterization of cisplatin degradation as affected by pH and light. *Am J Hosp Pharm.* 1991; 48: 1500–6.

Method 3

Stewart and Hampton studied the stability of cisplatin and etoposide in intravenous admixtures. The HPLC system consisted of an LDC Constametric III pump, an LDC SpectroMonitor III variable-wavelength detector, a Rheodyne 7126 injector, and a Shimadzu C-R3A computing integrator. The stationary phase was a Hamilton PRP-1 reversed-phase analytical column (250 × 4.1 mm, 10-µm particle size). The mobile phase was a mixture of 2.8–4.5% tetrabutylammonium hydroxide in water, adjusted to pH 7.0 with phosphoric acid. The isocratic flow rate was 1.0 mL/min. UV detection was performed at 313 nm and 0.01–0.005 AUFS.

Each sample was diluted before injection. Under these conditions, the cisplatin retention time ranged from 6.5 to 7.5 minutes.

The method was reported to be stability indicating. Etoposide eluted with the solvent front.

Intraday and interday coefficients of variation were 2.6 and 3.4%, respectively.

A similar method was reported by other researchers.

References

Stewart CF, Hampton EM. Stability of cisplatin and etoposide in intravenous admixtures. *Am J Hosp Pharm*. 1989; 46: 1400–4.

Fleming RA, Stewart CF. Stability-indicating high-performance liquid chromatographic method for the simultaneous determination of cisplatin and 5-fluorouracil in 0.9% sodium chloride for injection. *J Chromatogr*. 1990; 528: 517–25.

Trissel LA, Zhang Y. Physical and chemical stability of palonosetron HCl with cisplatin, carboplatin, and oxaliplatin during simulated Y-site administration. *J Oncol Pharm Pract*. 2004; 10: 191–5.

Method 4

Cheung et al. investigated the stability of cisplatin, iproplatin, carboplatin, and tetraplatin in commonly used intravenous solutions. The HPLC system consisted of a Spectra-Physics model SP8100 pump, a Valco autoinjector with a 10-µL sample loop, a Spectra-Physics model SP8440 variable-wavelength UV detector, and a Spectra-Physics model SP4270 integrator. The stationary phase was an Alltech LiChrosorb amino cartridge (250 × 4.6 mm). The mobile phase consisted of 90% acetonitrile and 10% water (vol/vol). The flow rate was 2 mL/min. UV detection was performed at 210 nm. Adenine hydrochloride 70 µg/mL was used as an internal standard. Under these conditions, retention times for cisplatin and the internal standard were 3.8 and 6.2 minutes, respectively.

The stability-indicating capability of the method was demonstrated using decomposed and partially decomposed solutions of cisplatin. Decomposition of cisplatin was accelerated by heat and the addition of acid or base. In all cases, degradation product peaks did not interfere with the intact cisplatin peak.

A standard curve for cisplatin was constructed from 24 to 240 µg/mL. The correlation coefficient was greater than 0.999. The coefficient of variation was 0.5%.

A similar method was used by Henry et al.

References
Cheung Y-W, Cradock JC, Vishnuvajjala BR, et al. Stability of cisplatin, iproplatin, carboplatin, and tetraplatin in commonly used intravenous solutions. *Am J Hosp Pharm.* 1987; 44: 124–30.

Henry DW, Marshall JL, Nazzaro D, et al. Stability of cisplatin and ondansetron hydrochloride in admixtures for continuous infusion. *Am J Health Syst Pharm.* 1995; 52: 2570–3.

Clanfenur

Chemical Name
N-[[(4-Chlorophenyl)amino]carbonyl]-2-(dimethylamino)-6-fluorobenzamide

Other Name
DU 113901

Form	Molecular Formula	MW	CAS
Clanfenur	$C_{16}H_{15}ClFN_3O_2$	335.8	51213-99-1

Method
Sluiter et al. studied by HPLC analysis the degradation of the investigational anticancer drug clanfenur. The liquid chromatograph included a Waters model M6000 pump, a model U6K injector with a 20-μL loop, a model 440 detector, and a Spectra-Physics model SP4290 integrator. A Hewlett-Packard 1040A photodiode-array detector was also used in the peak purity analysis. The stationary phase was a Merck LiChrospher 100 RP8 analytical column (125 × 4 mm, 5-μm particle size). The mobile phase consisted of 10 mM citric acid buffer (pH 2.8) and methanol (300:700, vol/vol), containing 10 mM heptanesulfonic acid sodium salt. The flow rate was 0.5 mL/min. UV detection was performed at 254 nm. Under these conditions, the retention time for clanfenur was about 9.7 minutes (estimated from the published chromatogram).

The method was reported to be stability indicating since the clanfenur peak was well resolved from its degradation product peaks.

A standard curve for clanfenur was constructed from 1.0×10^{-7} to 5.0×10^{-4} M. The intraday and interday coefficients of variation were 1.3 and 1.6%, respectively. The limit of detection was 0.8 ng.

Reference
Sluiter C, Kettenes-van den Bosch JJ, Hop E, et al. Degradation study of the investigational anticancer drug clanfenur. *Int J Pharm.* 1999; 185: 227–35.

Clarithromycin

Chemical Name

(2R,3S,4S,5R,6R,8R,10R,11R,12S,13R)-3-(2,6-Dideoxy-3-C,30-dimethyl-α-L-*ribo*-hex-opyranosyloxy)-11,12-dihydroxy-6-methoxy-2,4,6,8,10,12-hexamethyl-9-oxo-5-(3,4,6-trideoxy-3-dimethylamino-β-D-*xylo*-hexopyranosyloxy)pentadecan-13-olide

Other Names

Biaxin, Klacid, Klaricid, Zeclar

Form	Molecular Formula	MW	CAS
Clarithromycin	$C_{38}H_{69}NO_{13}$	748.0	81103-11-9

Method 1

Abuga et al. developed an HPLC method for the separation of clarithromycin and related substances in bulk samples. The liquid chromatograph consisted of a Spectra-Physics P4000 pump, a Spectra-Physics AS3000 autoinjector with a 50-μL loop, a Thermo Quest/Linear UVIS detector, and a Hewlett-Packard 3396 series II integrator. The stationary phase was a Waters Xterra RP18 column (250 × 4.6 mm, 5-μm particle size). The mobile phase consisted of 0.2 M dibasic potassium phosphate adjusted to pH 6.8, acetonitrile, and water (3.5:40:56.5, vol/vol/vol). The flow rate was 1 mL/min. UV detection was performed at 205 nm.

Samples were dissolved in a mixture of acetonitrile and water (40:60, vol/vol). Under these conditions, the retention time of clarithromycin was about 24.1 minutes.

The method was evaluated to be stability indicating by accelerated degradation of clarithromycin. Clarithromycin solutions were adjusted to pH 1.4 using 0.1 M phosphoric acid or to pH 10.0 using 0.1 M dibasic potassium phosphate buffer and then incubated at 37 °C, or they were treated with hydrogen peroxide. Degradation products were well separated from the intact drug on the chromatogram.

A standard curve for clarithromycin was constructed from 0.25 to 1.50 mg/mL. The correlation coefficient was 0.9995. The limit of detection and limit of quantitation were 0.05 and 0.15 mg/mL, respectively.

Reference

Abuga KO, Chepkwony HK, Roets E, et al. A stability-indicating HPLC method for the separation of clarithromycin and related substances in bulk samples. *J Sep Sci*. 2001; 24: 849–55.

Method 2

Morgan et al. reported an HPLC method for the determination of clarithromycin in various dosage forms. A liquid chromatograph consisted of a Spectra-Physics model SP 8700 pump, an IBM model LC/9505 autosampler, a Kratos Spectroflow 783 detector, and a Spectra-Physics model 4270 integrator. The stationary phase was an Alltech Nucleosil C_{18} column (150 × 4.6 mm, 5-μm particle size). The mobile phase consisted of 0.067 M monobasic potassium phosphate and methanol (40:60, vol/vol), adjusted to pH 3.5 with phosphoric acid. The flow rate was 1 mL/min. UV detection was performed at 210 nm and 0.04 AUFS. Biphenyl 2.6 μg/mL was used as an internal standard.

For bulk and injectable drug, samples were mixed with internal standard and diluted in mobile phase. For tablets and capsules, samples were extracted in methanol by shaking for 30 minutes, mixed with internal standard, and diluted with mobile phase. For oral suspensions, samples were dispersed in 0.067 M monobasic potassium phosphate buffer, and extracted in methanol. All samples were filtered through a 0.4–0.5-μm polycarbonate membrane before assay. The injection volume was 50 μL. Under these conditions, retention times for clarithromycin and biphenyl were 7.5 and 19.5 minutes, respectively.

The method was proven to be stability indicating. Clarithromycin was degraded with 0.1 M hydrochloric acid, 0.1 M sodium hydroxide in methanol, hydrogen peroxide, or high-intensity UV radiation. Degradation products did not influence the analysis of clarithromycin.

A standard curve for clarithromycin was constructed from 98.6 to 986 μg/mL. The correlation coefficient was 0.9997.

Reference

Morgan DK, Brown DM, Rotsch TD, et al. A reversed-phase high-performance liquid chromatographic method for the determination and identification of clarithromycin as the drug substance and in various dosage forms. *J Pharm Biomed Anal*. 1991; 9: 261–9.

Clavulanic Acid

Chemical Name

3-(2-Hydroxyethylidene)-7-oxo-4-oxa-1-azabicyclo[3.2.0]heptane-2-carboxylic acid

Other Name
Component of Augmentin

Form	Molecular Formula	MW	CAS
Clavulanic acid	$C_8H_9NO_5$	199.2	58001-44-8
Clavulanate potassium	$C_8H_8KNO_5$	237.3	61177-45-5

Appearance
Clavulanate potassium is an off-white crystalline powder.

Solubility
Clavulanate potassium is very soluble in water and slightly soluble in alcohol at room temperature.

pK_a
Clavulanic acid has a pK_a of 2.7.

Method 1
Mayron and Gennaro used an HPLC method to study the stability of clavulanate (as a component of Augmentin) with granisetron in intravenous fluids and oral liquids. The liquid chromatographic system consisted of a piston pump with a pulse dampener, a rotary injection port with a 20-µL loop, a variable-wavelength spectrometric detector, and an integrator. The stationary phase was a cyano reversed-phase column (250 × 4.6 mm, 5-µm particle size). The mobile phase was a mixture of acetonitrile and 0.02 M sodium acetate trihydrate (26:74, vol/vol) adjusted to pH 4.0 with acetic acid. The flow rate was 1.50 mL/min. UV detection was performed at 300 nm.

Samples were diluted with the mobile phase. The retention times for clavulanate and granisetron were 2.71 and 7.22 minutes, respectively.

The stability-indicating capability of the analytical method was demonstrated by an accelerated degradation study. Solutions of clavulanate were adjusted to pH 2 and 11, boiled for 1 hour, readjusted to pH 5, diluted with the mobile phase, and analyzed. Degradation product peaks did not interfere with the intact clavulanate peak.

Reference
Mayron D, Gennaro AR. Stability and compatibility of granisetron hydrochloride in i.v. solutions and oral liquids and during simulated Y-site injection with selected drugs. *Am J Health Syst Pharm.* 1996; 53: 294–304.

Method 2
Tu et al. studied the stability of reconstituted amoxicillin trihydrate–potassium clavulanate suspension in the original containers and in unit-dose oral syringes. The concentrations of potassium clavulanate were determined using reversed-phase HPLC. The chromatograph consisted of a Waters model 6000A HPLC system, a Waters model U6K injector, a Waters model 440 dual UV detector, and an Omniscribe strip-chart recorder. The stationary phase was a Waters Nova-Pak C_{18} analytical column (150 × 3.9 mm, 5-µm particle size). The mobile phase was 3% methanol in 0.05 M potassium dihydrogen phosphate buffer at pH 4.0. The flow rate was 0.8 mL/min. UV detection was performed at 254 nm.

Each of the samples was diluted and filtered. Under these conditions, retention times for potassium clavulanate and amoxicillin trihydrate were 2.5 and 4.5 minutes, respectively.

This assay was determined to be stability indicating. The degradation compounds and other inactive ingredients in the formulation did not interfere with the drug on the chromatogram.

Standard curves were constructed daily and the correlation coefficients for these curves were greater than 0.99. The intraday and interday coefficients of variation of clavulanate potassium were within 2%.

Reference

Tu Y-H, Stiles ML, Allen LV, et al. Stability of amoxicillin trihydrate-potassium clavulanate in original containers and unit dose oral syringes. *Am J Hosp Pharm.* 1988; 45: 1092–9.

Clindamycin

Chemical Name

(2S-*trans*)-Methyl 7-chloro-6,7,8-trideoxy-6-[[(1-methyl-4-propyl-2-pyrrolidinyl)-car-bonyl]amino]-1-thio-L-*threo*-α-D-*galacto*-octopyranoside

Other Name

Cleocin

Form	Molecular Formula	MW	CAS
Clindamycin	$C_{18}H_{33}ClN_2O_5S$	425.0	18323-44-9
Clindamycin hydrochloride	$C_{18}H_{33}ClN_2O_5S.HCl$	461.4	21462-39-5
Clindamycin phosphate	$C_{18}H_{34}ClN_2O_8PS$	505.0	24729-96-2

Appearance

Clindamycin phosphate is a white to off-white hygroscopic crystalline powder.

Solubility

Clindamycin phosphate has a solubility of about 400 mg/mL in water at 25 °C.

pK_a
Clindamycin has a pK_a of 7.45.

Method 1
Allen et al. investigated the stability of clindamycin phosphate in the presence of cefpirome sulfate during simulated Y-site injection. The liquid chromatograph consisted of an Alcott model 728 Micromeritics autosampler, a Rheodyne 7010 injector with an Alcott 732 electrically actuated valve, a Waters model 501 solvent-delivery pump, a Waters model 441 UV detector, a Waters model 401 refractive index detector, and a Waters model 745 data module. The stationary phase was a C_{18} analytical column. The mobile phase contained 2 g of *dl*-10-camphorsulfonic acid, 1 g of ammonium acetate, 1 mL of acetic acid, and sufficient water to make 200 mL plus sufficient methanol to make 500 mL with pH adjusted to 6.0. The flow rate was 1.0 mL/min.

Samples were diluted before injection. The injection volume was 25 µL. Under these conditions, the retention times for cefpirome and clindamycin were 3.9 and 12.1 minutes, respectively.

The method was determined to be stability indicating by accelerated degradation of clindamycin with sodium hydroxide, hydrochloric acid, potassium chlorate, heat at 80 °C, or light (150 foot-candles from a tungsten filament source). In each case, clindamycin was suitably resolved from its breakdown products on the chromatogram.

A calibration curve for clindamycin was constructed from 10 to 200 µg/mL, and the correlation coefficient was greater than 0.99. The intraday and interday coefficients of variation of the method were 2.4 and 2.2%, respectively.

Reference
Allen LV, Stiles ML, Prince SJ, et al. Stability of cefpirome sulfate in the presence of commonly used intensive care drugs during simulated Y-site injection. *Am J Health Syst Pharm*. 1995; 52: 2427–33.

Method 2
Nahata et al. determined the stability of clindamycin phosphate injection at room temperature and under refrigeration. A Hewlett-Packard 1050 series liquid chromatograph consisted of a solvent-delivery pump, an autosampler, a variable-wavelength UV detector, and an integrator. The stationary phase was a Waters µBondapak C_{18} column (300 × 3.9 mm, 10-µm particle size). The mobile phase consisted of 45% 0.01 M monobasic potassium phosphate, 0.0005 M 1-heptanesulfonic acid, and 55% methanol. The flow rate was 1.2 mL/min. UV detection was performed at 214 nm. Propylparaben 50 µg/mL was used as the internal standard.

Each sample was diluted 1:10 with mobile phase. The injection volume was 10 µL. Under these conditions, the retention times for propylparaben and clindamycin were 6.2 and 8.1 minutes, respectively.

To determine the stability-indicating nature of the assay, 1 mL of clindamycin phosphate 1 mg/mL was mixed with 1 mL of 1.0 M sodium hydroxide or 1.0 M sulfuric acid and heated to 60 °C for 30 minutes. The samples were then assayed. The degradation products did not interfere with the clindamycin peak.

A standard curve was established by linear regression analysis of clindamycin concentrations against the peak height ratios of clindamycin and propylparaben from

1.0 to 22.5 mg/mL. The correlation coefficient was greater than 0.999; the coefficient of variation was less than 2.1%.

Reference
Nahata MC, Morosco RS, Hipple TF. Stability of cimetidine hydrochloride and of clindamycin phosphate in water for injection stored in glass vials at two temperatures. *Am J Hosp Pharm.* 1993; 50: 2559–63.

Method 3
Gupta et al. evaluated the physical and chemical stabilities of clindamycin phosphate in dextrose and saline solutions. A Waters ALC 202 liquid chromatograph was equipped with an Omniscribe recorder, a Schoeffel SF 770 variable-wavelength detector, and a Waters μBondapak C_{18} reversed-phase column (300 × 3.9 mm). The mobile phase was 23% acetonitrile in 0.01 M aqueous potassium dihydrogen phosphate buffer containing 0.005 M heptanesulfonic acid sodium salt. The flow rate was 2.5 mL/min. UV detection was performed at 214 nm and 0.1 AUFS. Nafcillin sodium 0.6 mg/mL was used as an internal standard.

Samples were diluted with water. The injection volume was 30 μL. The retention times for benzyl alcohol, clindamycin phosphate, and nafcillin were about 3.2, 4.5, and 7.7 minutes, respectively (estimated from the published chromatogram).

The stability-indicating capability of the assay was demonstrated by accelerated decomposition of clindamycin. Decomposition of clindamycin phosphate was induced by mixing with 1 N sulfuric acid or 1 N sodium hydroxide and boiling for 10 minutes. The solutions were cooled and neutralized with 1 N sodium hydroxide or 1 N sulfuric acid, respectively, and assayed. Degradation product peaks were well separated from the clindamycin peak.

Interday variation was less than 1%.

Reference
Gupta VD, Parasrampuria J, Bethea C, et al. Stability of clindamycin phosphate in dextrose and saline solutions. *Can J Hosp Pharm.* 1989; 42: 109–12.

Method 4
Zbrozek et al. studied the compatibility and stability of clindamycin phosphate when mixed with cefazolin sodium and gentamicin sulfate in 5% dextrose injection or 0.9% sodium chloride injection. The chromatograph consisted of an Altex model 110A solvent-metering pump, a Waters WISP 710B autoinjector, a Waters Lambda-Max model 481 LC spectrophotometer, and a Linear Instruments model 555 recorder. The stationary phase was an Alltech Versapak C_{18} reversed-phase column (300 × 4.1 mm, 10-μm particle size). The mobile phase consisted of 50% methanol in a 0.01 M phosphate buffer aqueous solution with 0.005 M pentanesulfonic acid at pH 6.3. The flow rate was 1.2 mL/min. UV detection was performed at 214 nm and 0.02 AUFS. Propylparaben 20 μg/mL was used as an internal standard.

Samples were diluted 1:20 with internal standard solution. The injection volume was 10 μL. Retention times for clindamycin and propylparaben were 9 and 15 minutes, respectively.

This assay was stated to be stability indicating because it separated three degradation products from the intact clindamycin peak.

Standard curves for clindamycin were constructed from 1.25 to 20 mg/mL. The correlation coefficient was 0.999. Intrarun and interrun coefficients of variation were 3.1 and 2.1%, respectively.

Similar methods were used by the other researchers cited here.

References

Zbrozek AS, Marble DA, Bosso JA. Compatibility and stability of cefazolin sodium, clindamycin phosphate, and gentamicin sulfate in two intravenous solutions. *Drug Intell Clin Pharm.* 1988; 22: 873–5.

Sarkar MA, Rogers E, Reinhard M, et al. Stabilities of clindamycin phosphate, ranitidine hydrochloride, and piperacillin sodium in polyolefin containers. *Am J Hosp Pharm.* 1991; 48: 2184–6.

Marble DA, Bosso JA, Townsend RJ. Stability of clindamycin phosphate with aztreonam, ceftazidime sodium, ceftriaxone sodium, or piperacillin sodium in two intravenous solutions. *Am J Hosp Pharm.* 1986; 43: 1732–5.

Bosso JA, Townsend RJ. Stability of clindamycin phosphate and ceftizoxime sodium, cefoxitin sodium, cefamandole nafate, or cefazolin sodium in two intravenous solutions. *Am J Hosp Pharm.* 1985; 42: 2211–4.

Foley PT, Bosso JA, Bair JN, et al. Compatibility of clindamycin phosphate with either cefotaxime sodium or netilmicin sulfate in small volume parenteral solutions. *Am J Hosp Pharm.* 1985; 42: 839–43.

Method 5

James and Riley reported the stability of intravenous admixtures of clindamycin phosphate and aztreonam by analyzing the drug concentration using HPLC. The chromatographic system consisted of a ConstaMetric III G pump, an LDC SpectroMonitor D UV detector, a Negretti model 190 injection valve fitted with a 20-μL loop, and a Fisher series 5000 Recordall dual-pen recorder. The stationary phase was a Varian Micropak Si-5 silica column. The mobile phase was 0.1% phosphoric acid and 3% acetonitrile in water. The flow rate was 1.0 mL/min. UV detection was performed at 205 nm.

Samples were centrifuged at 2500 rpm for 5 minutes and then diluted 1:100 in pH 5.5 phosphate buffer. Under these conditions, the retention times for aztreonam, arginine, benzyl alcohol, and clindamycin phosphate were 2.90, 3.45, 3.65, and 14.6 minutes, respectively.

The stability-indicating nature of this analytical method was confirmed by spiking clindamycin phosphate sample solution with authentic degradation product. The degradation product peak did not interfere with the intact clindamycin phosphate peak.

The peak areas were related linearly to the concentrations of clindamycin phosphate from 0 to 200 μg/mL. The correlation coefficient was greater than 0.999. The coefficient of variation was less than 2%.

Reference

James MJ, Riley CM. Stability of intravenous admixtures of aztreonam and clindamycin phosphate. *Am J Hosp Pharm.* 1985; 42: 1984–6.

Clomesone

Chemical Name
2-Chloroethyl(methylsulfonyl)methanesulfonate

Other Name
NSC-338947

Form	Molecular Formula	MW	CAS
Clomesone	$C_4H_9ClO_5S_2$	236.7	88343-72-0

Method
Using an HPLC method, Kennedy et al. investigated the degradation of clomesone in different conditions. The stationary phase was a PRP-1 column (150 × 4.6 mm, 10-μm particle size). The mobile phase consisted of 22% acetonitrile and 78% 0.1 M phosphate buffer (pH 11). The flow rate was 1 mL/min. UV detection was performed at 214 nm. Under these conditions, the retention time for clomesone was 5 minutes.

The method was stated to be stability indicating.

Reference
Kennedy PE, Riley CM, Stella VJ. Degradation of the antineoplastic drug, clomesone (2-chloroethyl(methylsulfonyl)methanesulfonate, NSC-338947). A kinetic and mechanistic study. *Int J Pharm.* 1988; 48: 179–88.

Clonazepam

Chemical Names
5-(*o*-Chlorophenyl)-1,3-dihydro-7-nitro-2*H*-1,4-benzodiazepin-2-one
7-Nitro-5-(2-chlorophenyl)-3*H*-1,4-benzodiazepin-2(1*H*)-one

Other Name
Klonopin

Form	Molecular Formula	MW	CAS
Clonazepam	$C_{15}H_{10}ClN_3O_3$	315.7	1622-61-3

Appearance

Clonazepam is an off-white to light yellow crystalline powder.

Solubility

Clonazepam has the following solubilities at 25 °C: < 0.1 mg/mL in water, 31 mg/mL in acetone, 15 mg/mL in chloroform, 8.6 mg/mL in methanol, and 0.7 mg/mL in ether.

pK_a

Clonazepam has pK_a values of 1.5 and 10.5.

Method 1

Using smallbore HPLC, Spell and Stewart reported the analysis of clonazepam in a tablet dosage form. The liquid chromatographic system consisted of a Micromeritics model 760 pump, an Alcott model 708 autosampler equipped with a 10-µL loop, a Varian model 2550 variable-wavelength UV detector, and a Hewlett-Packard model 3392A integrator. A Waters model 996 photodiode-array detector was used for the peak purity analysis. The stationary phase was a YMC Slimbore C_{18} column (150 × 3.0 mm, 3.0-µm particle size). The mobile phase consisted of water, methanol, and acetonitrile (40:30:30, vol/vol/vol). The flow rate was 400 µL/min. UV detection was performed at 254 nm. 1,2-Dichlorobenzene was used as an internal standard. Under these conditions, retention times of clonazepam and 1,2-dichlorobenzene were 4.0 and 12.5 minutes, respectively.

The method was demonstrated to be stability indicating by accelerated degradation of clonazepam. Clonazepam standard samples were treated with 0.1 N hydrochloric acid at 25 and 80 °C, 0.1 N sodium hydroxide at 25 and 80 °C, or 15% hydrogen peroxide at 25 and 80 °C, or they were heated in water at 80 °C for up to 120 minutes. No degradation products or preservatives interfered with the analysis of clonazepam.

A standard curve for clonazepam was generated from 50 to 200 µg/mL. The correlation coefficients were greater than 0.9990.

Reference

Spell JC, Stewart JT. Analysis of clonazepam in a tablet dosage form using smallbore HPLC. *J Pharm Biomed Anal.* 1998; 18: 453–60.

Method 2

Allen and Erickson described a study of the stability of clonazepam in extemporaneously compounded oral liquids. A Hewlett-Packard series 1050 automated high-performance liquid chromatograph included a multisolvent mixing and pumping system, an autosampler, a diode-array detector, and a computer with Chem Station software. A Bakerbond C_{18} analytical column (250 × 4.6 mm) served as the stationary phase. The mobile phase was a mixture of water, methanol, and acetonitrile (4:3:3, vol/vol/vol) and was delivered isocratically at 1.0 mL/min. UV detection was performed at 254 nm.

Samples were diluted 1:4. The injection volume was 20 µL. Under these conditions, clonazepam eluted in 6.9 minutes.

The stability-indicating capacity of the assay was demonstrated by accelerated degradation of clonazepam. Samples of clonazepam were subjected to heat, base, acid, oxidizing agent, and light. The degradation products did not interfere with the intact clonazepam peak.

The standard curve was constructed from 1 to 50 µg/mL. The intraday and interday coefficients of variation were 1.8 and 2.0%, respectively.

Reference

Allen LV, Erickson MA. Stability of acetazolamide, allopurinol, azathioprine, clonazepam, and flucytosine in extemporaneously compounded oral liquids. *Am J Health Syst Pharm.* 1996; 53: 1944–9.

Clonidine

Chemical Names

2,6-Dichloro-*N*-2-imidazolidinylidenebenzenamine

2-[(2,6-Dichlorophenyl)amino]-2-imidazoline

Other Name

Catapres

Form	Molecular Formula	MW	CAS
Clonidine	$C_9H_9Cl_2N_3$	230.1	4205-90-7
Clonidine hydrochloride	$C_9H_9Cl_2N_3.HCl$	266.6	4205-91-8

Appearance

Clonidine hydrochloride is a white or almost white crystalline powder.

Solubility

Clonidine hydrochloride is soluble in water (1 in 13), soluble in dehydrated alcohol, and slightly soluble in chloroform.

Method 1

Trissel et al. described the HPLC analysis of clonidine hydrochloride injections designed for epidural and intrathecal administration. The liquid chromatograph consisted of a Waters 600E multisolvent-delivery pump, a Waters 490E programmable multiple-wavelength UV detector, and a Waters 712 WISP autosampler. The system was controlled and integrated by an NEC PowerMate SX/16 personal computer. The stationary phase was an Alltech Hypersil BDS C_8 analytical column (150 × 4.6 mm, 5-µm particle size). The mobile phase consisted of 500 mL of water, 500 mL of methanol, 1.1 g sodium 1-octanesulfonate, and 1 mL of phosphoric acid; pH was adjusted to 3.0. The flow rate was 1.0 mL/min, and UV detection was performed at 220 nm and 0.5 AUFS. Injection volume was 15 µL. The retention time for clonidine was about 5.8 minutes.

The HPLC method was determined to be stability indicating by accelerating the decomposition of clonidine hydrochloride using heat and extremely acidic pH as well as oxidation. A solution of clonidine hydrochloride was mixed with 1.0 N hydrochloric acid and boiled for 2 hours. Another solution of clonidine hydrochloride was mixed with 3% hydrogen peroxide for 1 hour at room temperature. In each case, degradation product peaks did not interfere with the intact clonidine peak. The potential decomposition products and contaminants 2-imidazolidone, 2-butanone, and 2,6-dichloroaniline eluted at about 1.8, 2.5, and 11.7 minutes, respectively.

Calibration curves were constructed from linear plots of peak area versus concentration of the clonidine hydrochloride reference standard (from 0.025 to 0.25 mg/mL). The correlation coefficient of the standard curves was greater than 0.9999; the coefficient of variation was 0.6%.

Reference

Trissel LA, Xu QA, Hassenbusch SJ. Development of clonidine hydrochloride injections for epidural and intrathecal administration. *Int J Pharm Compound*. 1997; 1: 274–7.

Method 2

Using an HPLC method, Levinson and Johnson determined the stability of an extemporaneously compounded clonidine hydrochloride oral liquid. The liquid chromatographic system included a Waters model 501 constant-flow solvent-delivery pump, a Waters U6K variable-volume injector, a Waters model 484 tunable UV detector, and a Hewlett-Packard model 3394 integrator-recorder. The stationary phase was a DuPont Zorbax trimethylsilyl column (250×4.6 mm, 5-μm particle size). The mobile phase was a mixture of 65% methanol and 35% phosphate buffer solution (0.0022 M monobasic potassium phosphate and 0.016 M dibasic sodium phosphate, pH 7.9) and was delivered isocratically at 1.0 mL/min. UV detection was performed at 254 nm. Guanabenz acetate was used as the internal standard.

Each sample was diluted with deionized water. The injection volume was 15 μL. The retention times for clonidine and the internal standard were about 5.7 and 6.7 minutes, respectively (estimated from the published chromatogram).

The method was demonstrated to be stability indicating by accelerated decomposition of clonidine. Aqueous solutions of clonidine hydrochloride 200 μg/mL were adjusted to pH 12 with 1 N sodium hydroxide or to pH 2 with 1 N sulfuric acid and were then heated to 100 °C for 30 minutes. These solutions were cooled, adjusted to neutral, and assayed. Degradation product peaks did not interfere with the intact clonidine hydrochloride peak.

Each day of sample analysis a standard curve was constructed from 3 to 7 μg/mL by linear regression of the peak height ratio of clonidine hydrochloride to internal standard against the clonidine hydrochloride concentration. The correlation coefficient was greater than 0.999. The intraday and interday coefficients of variation for the assay were 2.3 and 3.6%, respectively.

A similar method was used by Walters and Stonys.

References

Levinson ML, Johnson CE. Stability of an extemporaneously compounded clonidine hydrochloride oral liquid. *Am J Hosp Pharm*. 1992; 49: 122–5.

Walters SM, Stonys DB. Determination of chlorthalidone and clonidine hydrochloride in tablets by HPLC. *J Chromatogr Sci.* 1983; 21: 43–5.

Clopidogrel Bisulfate

Chemical Name

α-(2-Chlorophenyl)-6,7-dihydrothieno[3,2-c]pyridine-5(4*H*)-acetic acid methyl ester sulfate

Other Name

Plavix

Form	Molecular Formula	MW	CAS
Clopidogrel	$C_{16}H_{16}ClNO_2S$	321.8	113665-84-2
Clopidogrel bisulfate	$C_{16}H_{16}ClNO_2S.H_2SO_4$	419.9	120202-66-6

Appearance

Clopidogrel bisulfate occurs as a white to off-white powder.

Solubility

Clopidogrel bisulfate is freely soluble in water and methanol. It is practically insoluble in ether.

Method

Mitakos and Panderi reported a rapid reversed-phase HPLC method for the quality control of clopidogrel in pharmaceutical preparations. The instrument included a GBC model LC1126 pump, a GBC model LC1210 UV-visible detector, and a Rheodyne model 7725i injector with a 5-μL loop. The stationary phase was a Shandon reversed-phase BDS C_8 column (250 × 2.1 mm, 5-μm particle size). The mobile phase consisted of 10 mM dibasic sodium phosphate (pH 3.0 with phosphoric acid) and acetonitrile (35:65, vol/vol). The flow rate was 0.30 mL/min. UV detection was performed at 235 nm. Naproxen (100 ng/mL) was used as an internal standard.

A portion of ground tablets (20) equivalent to 75.0 mg of clopidogrel was accurately weighed, dissolved in 40 mL of acetonitrile in a 50-mL volumetric flask by sonicating for 20 minutes, diluted to volume with acetonitrile, and centrifuged at 4000 rpm (2890 *g*) for 5 minutes. The clear liquid was diluted by a factor of 100 with water and then by a factor of 10 with mobile phase containing internal standard (100 ng/mL).

The injection volume was 5 μL. Under these conditions, retention times for naproxen and clopidogrel were about 3.1 and 6.3 minutes, respectively.

The method was demonstrated to be stability indicating by accelerated degradation of clopidogrel by heat, acid (1 N hydrochloric acid), base (1 N sodium hydroxide), or oxidation (5% hydrogen peroxide). The degradation product peak was separated from that of the intact drug.

Calibration curves for clopidogrel were constructed from 1.00 to 3.00 μg/mL. The correlation coefficients were greater than 0.9995. The limit of detection and the limit of quantitation were 0.12 and 0.39 μg/mL, respectively. The intraday and interday coefficients of variation were less than 0.99 and 1.96%, respectively.

Reference

Mitakos A, Panderi I. A validated LC method for the determination of clopidogrel in pharmaceutical preparations. *J Pharm Biomed Anal.* 2002; 28: 431–8.

Clorazepic Acid

Chemical Name

7-Chloro-2,3-dihydro-2,2-dihydroxy-5-phenyl-1*H*-1,4-benzodiazepine-3-carboxylic acid

Other Names

Clorazecaps, Tranxene

Form	Molecular Formula	MW	CAS
Clorazepic acid	$C_{16}H_{13}ClN_2O_4$	332.7	20432-69-3
Clorazepate dipotassium	$C_{16}H_{11}ClK_2N_2O_4$	408.9	57109-90-7
Clorazepate monopotassium	$C_{16}H_{10}ClKN_2O_3$	352.8	5991-71-9

Appearance

Clorazepate dipotassium occurs as a white or light yellow crystalline powder.

Solubility

Clorazepate dipotassium is very soluble in water and very slightly soluble in alcohol.

Method

Elrod et al. reported the HPLC analysis of clorazepate dipotassium and monopotassium in solid dosage forms. The chromatographic apparatus consisted of a Waters model M6000A pump, a model 450 variable-wavelength detector, a Rheodyne model 7120 injector, and a Hewlett-Packard 3385A integrator. The stationary phase was a Waters μBondapak C_{18} column (300 × 4 mm). The mobile phase was prepared by diluting 300 mL of acetonitrile to 1000 mL with 0.005 M tetra-*n*-butylammonium hydroxide solution (adjusted to pH 7.5 with phosphoric acid). The flow rate was 1.8 mL/min. UV detection was performed at 230 nm and 0.2 AUFS. 2,6-Dimethylaniline hydrochloride 0.3 mg/mL in 0.04% (wt/vol) sodium hydroxide solution was used as an internal standard.

Sodium hydroxide (0.04% wt/vol) was used as a sample solvent. For capsules, the powder contents from four capsules were weighed, mixed with 200 mL of the sample solvent, sonicated for 5 minutes, diluted to 250 mL with the solvent, filtered through 0.4–0.5-μm polycarbonate membrane, mixed with the internal standard, and diluted appropriately with the sample solvent. For tablets, ground powder samples containing 75 mg of clorazepate dipotassium were mixed with 200 mL of the sample solvent, homogenized, filtered, and diluted. The injection volume was 20 μL. Under these conditions, retention times for clorazepate and internal standard were about 6.2 and 9.8 minutes, respectively (estimated from the published chromatogram).

The stability-indicating capability of the method was shown by intentional degradation of clorazepate. Clorazepate was degraded at elevated temperature. The degradation product did not interfere with the analysis of clorazepate.

A standard curve for clorazepate was generated from 0 to 0.030 mg/mL with a correlation coefficient of 0.999.

Reference

Elrod L Jr., Shada DM, Taylor VE. High-performance liquid chromatographic analysis of clorazepate dipotassium and monopotassium in solid dosage forms. *J Pharm Sci.* 1981; 70: 793–5.

Clotrimazole

Chemical Names
1-[(2-Chlorophenyl)diphenylmethyl]-1*H*-imidazole
1-(*o*-Chloro-α,α-diphenylbenzyl)imidazole

Other Name
Lotrimin

Form	Molecular Formula	MW	CAS
Clotrimazole	$C_{22}H_{17}ClN_2$	344.8	23593-75-1

Appearance
Clotrimazole is a white to pale yellow crystalline powder.

Solubility
Clotrimazole is soluble in alcohol (1 in 10), chloroform (1 in 10), and ether (1 in 100). It is freely soluble in acetone and methanol but is practically insoluble in water.

Method

Hoogerheide et al. described a stability-indicating HPLC method for the determination of clotrimazole and its major degradation product in pharmaceutical dosage forms and drug substances. The liquid chromatograph consisted of a Waters M6000A pump, a Waters model 710A autosampler, a Waters model 440 fixed-wavelength detector, a Linear model 485 recorder, and a PDP 11/34 minicomputer utilizing Peak-11 software. The stationary phase was a Waters μBondapak C_{18} column (300 × 3.9 mm) with an MPLC RP18 guard column (30 × 4.6 mm). The mobile phase was a mixture of 0.025 M potassium hydrogen phosphate aqueous buffer and HPLC-grade methanol (1:3, vol/vol). The flow rate was 1.0 mL/min. UV detection was performed at 254 nm and 0.2 AUFS. Testosterone propionate 0.7 mg/mL in mobile phase was used as the internal standard.

Samples were diluted with mobile phase. The injection volume was 20 μL. Under these conditions, retention times for clotrimazole and testosterone propionate were about 9 and 14 minutes, respectively.

The assay was demonstrated to be stability indicating. Clotrimazole tablets, cream, and solution were stored from 10 to 14 days at 75 or 95 °C. Under these conditions, the major degradation product, (o-chlorophenyl)diphenylmethanol, along with several minor decomposition products, was formed. However, all degradation product peaks were well separated from clotrimazole and the internal standard peaks.

The standard curve was constructed from 0.1 to 2.0 mg/mL. The correlation coefficient was greater than 0.9999. Interday relative standard deviation was 2.4%.

Reference

Hoogerheide JG, Strusiak SH, Taddei CR, et al. High performance liquid chromatographic determination of clotrimazole in pharmaceutical formulations. *J Assoc Off Anal Chem.* 1981; 64: 864–9.

Cloxacillin

Chemical Names

[2S-(2α,5α,6β)]-6-[[[3-(2-Chlorophenyl)-5-methyl-4-isoxazolyl]carbonyl]amino]-3,3-dimethyl-7-oxo-4-thia-1-azabicyclo[3.2.0]heptane-2-carboxylic acid
[3-(o-Chlorophenyl)-5-methyl-4-isoxazolyl]penicillin

Other Name

Cloxapen

Form	Molecular Formula	MW	CAS
Cloxacillin	$C_{19}H_{18}ClN_3O_5S$	435.9	61-72-3
Cloxacillin sodium	$C_{19}H_{17}ClN_3NaO_5S.H_2O$	475.9	7081-44-9

Appearance

Cloxacillin sodium is a white odorless crystalline powder.

Solubility

Cloxacillin is soluble in water, methanol, ethanol, pyridine, and ethylene glycol.

pK_a

Cloxacillin sodium has a pK_a of 2.7.

Method 1

Grover et al. described an HPLC method for analysis of cloxacillin in the presence of its degradation products. A Shimadzu model LC-10A liquid chromatograph included model LC-10AS pumps, a model SPD-10A dual-wavelength UV-visible detector, a model C-R7A data processor, and a Rheodyne model 7125 injector with a 20-μL loop. The stationary phase was a Phenomenex Resolve C_{18} column (300 × 3.9 mm, 5-μm particle size). The mobile phase was 18% acetonitrile in 20 mM monobasic potassium phosphate and 10 mM tetramethylammonium chloride in water (pH 5.0). The flow rate was 1 mL/min. UV detection was performed at 220 and 240 nm. Under these conditions, the retention time for cloxacillin was 31.2 minutes.

The stability-indicating nature of the assay was demonstrated by degrading the drug. Cloxacillin was degraded at pH 2.0. The intact cloxacillin peak was separated satisfactorily from its degradation product peaks.

A standard curve for cloxacillin was constructed from 5 to 150 μg/mL. The correlation coefficient was greater than 0.99.

References

Grover M, Gulati M, Singh S. Stability-indicating analysis of isoxazolyl penicillins using dual wavelength high-performance liquid chromatography. *J Chromatogr B.* 1998; 708: 153–9.

Grover M, Gulati M, Singh B, et al. Correlation of penicillin structure with rate constants for basic hydrolysis. *Pharm Pharmacol Commun.* 2000; 6: 355–63.

Method 2

Walker et al. evaluated the chemical stability of cloxacillin using an HPLC method. The chromatographic system consisted of a Spectra-Physics SP4200 ternary gradient solvent-delivery pump, a Schoeffel SF 770 variable-wavelength UV detector, and a Beckman Ultrasphere ODS C_{18} reversed-phase column (250 × 4.2 mm, 5-μm particle size). The mobile phase consisted of 35% acetonitrile and 65% 0.05 M phosphoric acid in water that also contained 1 mg/mL of heptanesulfonic acid as a counterion. The flow rate was 2.0 mL/min. UV detection was performed at 230 nm.

The method was demonstrated to be stability indicating by accelerated decomposition of cloxacillin. A solution of cloxacillin 100 mg/mL in water was incubated at 90 °C for 170 minutes and then assayed. The degradation products did not interfere with the determination of the intact cloxacillin.

The intraday coefficient of variation was 0.34%.

Reference

Walker SE, DeAngelis C, Iazzetta J. Stability and compatibility of combinations of hydromorphone and a second drug. *Can J Hosp Pharm.* 1991; 44: 289–95.

Clozapine

Chemical Name

8-Chloro-11-(4-methyl-1-piperazinyl)-5*H*-dibenzo[*b*,*e*][1,4]diazepine

Other Names

Clozaril, Leponex

Form	Molecular Formula	MW	CAS
Clozapine	$C_{18}H_{19}ClN_4$	326.8	5786-21-0

Appearance

Clozapine is a yellow crystalline powder.

Solubility

Clozapine is practically insoluble in water; soluble in alcohol; and freely soluble in acetone, chloroform, ethyl acetate, and dichloromethane.

pK$_a$

Clozapine has a pK_{a1} of 3.70 and a pK_{a2} of 7.60.

Method

Hasan et al. developed an HPLC method for the determination of clozapine in raw materials and pharmaceutical formulations. A Shimadzu class LC-10 liquid chromatograph was equipped with a Shimadzu SPD-10A photodiode-array detector. The stationary phase was a Zorbax C_{18} column (5-μm, particle size). The mobile phase was a mixture of acetonitrile and water (40:60, vol/vol). The flow rate was 1 mL/min. UV detection was performed at 230 nm.

A portion of 10 powdered tablets equivalent to 100 mg of clozapine was accurately weighed, stirred in 70 mL of methanol for 30 minutes, filtered through a filter paper, and diluted to the 100-mL mark with methanol. The injection volume was 20 µL. Under these conditions, the retention time of clozapine was about 4.4 minutes.

The stability-indicating ability of the method was confirmed by assaying a mixture of clozapine and its degradation product. The retention time for the degradation product was about 2.6 minutes.

A calibration curve for clozapine was constructed from 5 to 100 µg/mL. The correlation coefficient was 0.9996. The limit of detection and the limit of quantitation were 1.65 and 4.97 µg/mL, respectively.

Reference

Hasan NY, Eldawy MA, Elzeany BE, et al. Stability-indicating methods for the determination of clozapine. *J Pharm Biomed Anal.* 2002; 30: 35–47.

Cocaine

Chemical Names

3-(Benzoyloxy)-8-methyl-8-azabicyclo[3.2.1]octane-2-carboxylic acid methyl ester

(1*R*,2*R*,3*S*,5*S*)-2-Methoxycarbonyltropan-3-yl benzoate

Form	Molecular Formula	MW	CAS
Cocaine	$C_{17}H_{21}NO_4$	303.4	50-36-2
Cocaine hydrochloride	$C_{17}H_{21}NO_4$.HCl	339.8	53-21-4

Appearance

Cocaine occurs as colorless or white crystals or as a white crystalline powder. Cocaine hydrochloride occurs as odorless hygroscopic colorless crystals or as a white crystalline powder.

Solubility

Cocaine is practically insoluble in water but freely soluble in alcohol and ether. Cocaine hydrochloride is very soluble in water and freely soluble in alcohol.

Method

Gupta studied by HPLC the stability of cocaine hydrochloride solutions at various pH values. A Waters ALC 202 liquid chromatograph was equipped with a Schoeffel SF

770 multiple-wavelength detector and an Omniscribe recorder. The stationary phase was a Waters μBondapak CN semipolar column (300 × 4 mm). The mobile phase was 25% (vol/vol) methanol in 0.02 M aqueous ammonium acetate buffer (pH ~7). The flow rate was 2.0 mL/min. UV detection was performed at 275 nm and 0.04 AUFS. Ethyl *p*-aminobenzoate 5.0 μg/mL in water was used as an internal standard.

Samples were diluted before assay. The injection volume was 20 μL. Under these conditions, retention times for benzoylecgonine (the degradation product), the internal standard, and cocaine were about 1.6, 3.5, and 6.1 minutes, respectively (estimated from the published chromatogram).

The method was reported to be stability indicating since the intact cocaine was separated from its excipients and degradation products.

Reference

Gupta VD. Stability of cocaine hydrochloride solutions at various pH values as determined by high-pressure liquid chromatography. *Int J Pharm*. 1982; 10: 249–57.

Codeine

Chemical Name
(5α,6α)-7,8-Didehydro-4,5-epoxy-3-methoxy-17-methylmorphinan-6-ol

Other Name
Methylmorphine

Form	Molecular Formula	MW	CAS
Codeine	$C_{18}H_{21}NO_3$	299.4	76-57-3
Codeine hydrochloride	$C_{18}H_{21}NO_3 \cdot HCl \cdot 2H_2O$	371.9	1422-07-7
Codeine phosphate	$C_{18}H_{21}NO_3 \cdot H_3PO_4 \cdot \frac{1}{2}H_2O$	406.4	41444-62-6
Codeine sulfate	$(C_{18}H_{21}NO_3)_2 \cdot H_2SO_4 \cdot 3H_2O$	750.9	6854-40-6

Appearance
Codeine occurs as colorless or white crystals or as a white crystalline powder. Codeine hydrochloride occurs as small colorless crystals or as a white crystalline powder. Codeine phosphate occurs as fine white needle-shaped crystals or as a white crystalline powder. Codeine sulfate occurs as white crystals, usually needlelike, or as a white crystalline powder.

Solubility

Codeine is slightly soluble in water and freely soluble in alcohol. Codeine hydrochloride is soluble in water and slightly soluble in alcohol. Codeine phosphate is freely soluble in water and slightly soluble in alcohol. Codeine sulfate is soluble in water and very slightly soluble in alcohol.

Method

Sisco et al. developed an HPLC method for the simultaneous determination of acetaminophen, codeine phosphate, and sodium benzoate in an elixir formulation. The chromatographic system consisted of a DuPont model 850 high-pressure liquid chromatograph equipped with a DuPont automatic sampler with a 20-μL loop, a DuPont column oven, a DuPont 4100 integrator, and a Waters model 440 UV detector. A Hewlett-Packard model 1040A diode-array detector was used for specificity studies of stressed and unstressed samples. The stationary phase was a Waters μBondapak C_{18} column (300 × 3.9 mm). The mobile phase consisted of a buffer and methanol (80:20, vol/vol). The buffer contained 2.4 g of 1-butanesulfonic acid sodium salt (0.015 M), 2.04 g of monobasic potassium phosphate (0.015 M), and 2 mL of triethylamine per liter of water. The pH of this solution was adjusted to 4.8 ± 0.1 with dilute phosphoric acid. The flow rate was 2.0 mL/min. UV detection was performed at 214 nm.

Samples were diluted 1:40 with distilled water. The injection volume was 20 μL. Under these conditions, retention times for acetaminophen, sodium benzoate, and codeine phosphate were 2.6, 3.8, and 4.7 minutes, respectively.

The HPLC method was determined to be stability indicating by accelerated decomposition of codeine phosphate. An elixir sample was placed in a clear flint-glass bottle, capped, and stored in a 60 °C oven for 2 weeks. A standard solution of acetaminophen and codeine phosphate was also spiked with potential degradation products. Retention times for p-aminophenol, codeine N-oxide, and codeinone were 1.8, 6.4, and 8.8 minutes, respectively. Degradation product peaks did not interfere with peaks of acetaminophen, sodium benzoate, or codeine phosphate.

A standard curve for codeine was constructed from 0.03 to 0.09 mg/mL (estimated from the published figure). The correlation coefficient was 0.9998.

Reference

Sisco WR, Rittenhouse CT, Everhart LA, et al. Simultaneous high-performance liquid chromatographic stability indicating analysis of acetaminophen, codeine phosphate, and sodium benzoate in elixirs. *J Chromatogr*. 1986; 354: 355–66.

Colchicine

Chemical Name

(*S*)-*N*-(5,6,7,9-Tetrahydro-1,2,3,10-tetramethoxy-9-oxobenzo[*a*]heptalen-7-yl)acetamide

Other Names

Colchiquim, Colgout, Goutnil

Form	**Molecular Formula**	**MW**	**CAS**
Colchicine	$C_{22}H_{25}NO_6$	399.4	64-86-8

Appearance

Colchicine appears as pale yellow scales or powder. It darkens on exposure to light.

Solubility

Colchicine is soluble 1 in 22 to 25 of water, 1 in 220 of ether, and 1 in 100 of benzene. It is freely soluble in alcohol and chloroform.

pK_a

Colchicine has a pK_a of 12.35 at 20 °C.

Method

Korner and Kohn developed an HPLC method for assaying Colchicum dry extract. The stationary phase was a Merck Chromolith RP-18e column (100 × 4.6 mm). The phosphate buffer (0.05 M) was prepared by dissolving 13.6092 g of monobasic potassium phosphate and 1.7534 g of ethylenediaminetetraacetic acid (EDTA) (3 mM) in 2000 mL of water; the buffer was adjusted to pH 6.0 with 1 M sodium hydroxide solution. The mobile phase consisted of methanol and phosphate buffer (27:73, vol/vol). The flow rate was 3.0 mL/min. UV detection was performed at 245 nm.

The standard solution of colchicine was prepared in methanol and water (1:1, vol/vol) and the sample solution of a Colchicum dry extract was prepared by dissolving 300.0 mg of the extract in 20.0 mL of methanol and water (1:1, vol/vol) with the addition of 2.92 g of EDTA. Under these conditions, retention times of colchicine and its degradation product, colchiceine, were about 11.2 and 14.0 minutes, respectively.

The acidic hydrolysis of colchicine was carried out by refluxing colchicine in hydrochloric acid. The peak of colchicine was well separated from the peak of its degradation product.

A standard curve was constructed for colchicine from 0.15 to 7.63 µg/mL by linear regression of the peak area against concentration. The correlation coefficient was 0.9999. The coefficient of variation of the analysis was 0.83%.

Reference

Korner A, Kohn S. Development and optimization of a stability indicating method on a monolithic reversed-phase column for Colchicum dry extract. *J Chromatogr A.* 2005; 1089: 148–57.

Cromolyn Sodium

Chemical Name

Disodium 4,4'-dioxo-5,5'-(2-hydroxytrimethylenedioxy)di(4*H*-chromene-2-carboxylate)

Other Names

Crolom, Opticrom

Form	Molecular Formula	MW	CAS
Cromolyn sodium	$C_{23}H_{14}Na_2O_{11}$	512.3	15826-37-6

Appearance

Cromolyn sodium occurs as a white or almost white odorless hygroscopic crystalline powder.

Solubility

Cromolyn sodium is soluble in water and insoluble in alcohol.

pK_a

Cromolyn sodium has pK_{a1} and pK_{a2} values of 2.0.

Method 1

Barnes et al. developed an HPLC method for cromolyn sodium and its known impurities. The apparatus included a Hewlett-Packard model 1050 solvent-delivery system, a variable-wavelength detector, and a variable-volume injector. The stationary phase was a Zorbax SB-C_8 column (150 × 4.6 mm, 5-μm particle size). The column temperature was 40 °C. The mobile phase was prepared by adding 20 g of myristyltrimethylammonium bromide and 5 g of monobasic potassium phosphate to 900 mL of water, adjusting to pH 6.5 with sodium hydroxide or hydrochloric acid, and mixing with 1100 mL of methanol. The flow rate was 2.0 mL/min. UV detection was performed at 326 nm. The total run time was 30 minutes.

Cromolyn sodium drug substance was weighed and diluted with a diluent of acetonitrile and water (30:70, vol/vol) to about 0.5 mg/mL of cromolyn sodium. The injection volume was 20 μL.

The stability-indicating nature of the method was verified by accelerated degradation. Cromolyn sodium was treated with 0.1 N sodium hydroxide at 50 °C for 15 hours, treated with 30% hydrogen peroxide at 50 °C for 15 hours, stored at 50 °C for 9 days, or exposed to light at 765 Watts/m² for 20 minutes. No interference from degradation products was observed.

A standard curve for cromolyn sodium was constructed from 0.375 to 6.250 mg/mL. The correlation coefficient was 1.0000. The limit of quantitation was 0.25 μg/mL.

Reference

Barnes M, Mansfield R, Thatcher S. The selection of an ion pairing reagent for developing and validating a stability-indicating HPLC method for cromolyn sodium and its known impurities. *J Liq Chromatogr Rel Technol.* 2002; 25: 1721–45.

Method 2

Mansfield et al. reported the development and validation of an HPLC method for the determination of cromolyn sodium and its related substances in cromolyn sodium drug substance and cromolyn sodium inhalation solution 1.0%. The liquid chromatograph consisted of a Hewlett-Packard model 1100 pump, a variable-wavelength detector, and a variable-volume injector. The stationary phase was a Waters Nova-Pak C_8 column (150 × 3.9 mm). The mobile phase consisted of 550 mL of tetrabutylammonium dihydrogen phosphate (TBA) buffer and 450 mL of methanol. The TBA buffer was prepared by adding 40 mL of 1.0 M TBA to 1 L of water. The flow rate was 1.0 mL/min. UV detection was performed at 326 nm.

The drug product was accurately diluted to 1 mg/mL and then to 0.2 mg/mL with water. The injection volume was 10 μL. Under these conditions, the retention time of cromolyn sodium was 6.2 minutes.

The method was reported to be stability indicating. Cromolyn sodium solution was subjected to basic (at pH 12 for 60 hours), acidic (at pH 2), oxidative (with 30% hydrogen peroxide for 68 hours), thermal (at 60 °C for 63 hours), and UV light (at 250 Watts/m² for 16 hours) environments. Retention times of diethyl-4,4'-dioxo-5,5'-(2-hydroxytrimethylenedioxy) di(chromene-2-carboxylate) ester and 1,3-bis(2-acetyl-3-hydroxyphenoxy)-2-propanol were 15.2 and 47.5 minutes, respectively.

A standard curve for cromolyn sodium was constructed from 91.3 to 274.0 μg/mL. The correlation coefficient was 1.000. The limit of quantitation was 0.1 μg/mL.

Reference

Mansfield R, Huang J, Thatcher S, et al. Development and validation of a stability-indicating HPLC method for the determination of cromolyn sodium and its related substances in cromolyn sodium drug substance and cromolyn sodium inhalation solution, 1.0%. *J Liq Chromatogr Rel Technol.* 1999; 22: 2187–209.

Cyclizine

Chemical Name

1-Benzhydryl-4-methylpiperazine

Other Name

Marezine

Form	Molecular Formula	MW	CAS
Cyclizine	$C_{18}H_{22}N_2$	266.4	82-92-8
Cyclizine hydrochloride	$C_{18}H_{22}N_2 \cdot HCl$	302.8	303-25-3
Cyclizine lactate	$C_{18}H_{22}N_2 \cdot C_3H_6O_3$	356.5	5897-19-8

Appearance

Cyclizine is a white or creamy white almost odorless crystalline powder. Cyclizine hydrochloride occurs as a white odorless or almost odorless crystalline powder or as small colorless crystals.

Solubility

Cyclizine is practically insoluble in water, freely soluble in alcohol and ether, and very soluble in chloroform. Cyclizine hydrochloride is slightly soluble in water and alcohol, sparingly soluble in chloroform, and practically insoluble in ether.

Method

Jalal et al. described an HPLC method for the determination of ergotamine and cyclizine in commercial products. The system consisted of a DuPont model 8800 pump, a Rheodyne 7125 injector, a DuPont variable-wavelength detector, and a Spectra-Physics model SP4100 integrator. The stationary phase was a Merck reversed-phase C_{18} polymeric column (250 × 4.6 mm, 5-µm particle size). The mobile phase was a mixture of acetonitrile, water, and triethylamine (35:64:1, vol/vol/vol) containing 0.01 M ammonium acetate and adjusted to pH 3.7 with glacial acetic acid. The flow rate was 1.5 mL/min. UV detection was performed at 254 nm and 0.08 AUFS. Ethylparaben 5.6 µg/mL in methanol was used as an internal standard.

A powder sample was weighed, sonicated for 3 minutes with 20 mL of internal standard solution in a 25-mL volumetric flask, and filled to volume with internal standard. It was diluted again with internal standard solution to approximately 0.032 mg/mL of ergotamine and 0.800 mg/mL of cyclizine. The injection volume was 10 µL. Under these conditions, retention times for ergotamine, ethylparaben, and cyclizine were about 3.5, 4.1, and 4.9 minutes, respectively.

The method was stability indicating since no interference due to excipients was detected.

A calibration curve for cyclizine was constructed from 0.40 to 1.20 mg/mL. The limit of detection was 35.0 ng.

Reference

Jalal IM, Sasa SI, Yasin TA. Determination of ergotamine tartrate and cyclizine hydrochloride in pharmaceutical tablets by reverse phase HPLC. *Anal Lett.* 1988; 21: 1561–77.

Cyclophosphamide

Chemical Names

N,N-Bis(2-chloroethyl)tetrahydro-2*H*-1,3,2-oxazaphosphorin-2-amine 2-oxide
2-[Bis(2-chloroethyl)amino]tetrahydro-2*H*-1,3,2-oxazaphosphorine 2-oxide

Other Names

Cytoxan, Neosar

Form	Molecular Formula	MW	CAS
Cyclophosphamide	$C_7H_{15}Cl_2N_2O_2P$	261.1	50-18-0

Appearance
Cyclophosphamide is a white or almost white crystalline powder.

Solubility
Cyclophosphamide has a solubility of 40 g/L in water. It is slightly soluble in alcohol, benzene, ethylene glycol, carbon tetrachloride, and dioxane and sparingly soluble in ether and acetone.

Method 1
Using an HPLC method, Mayron and Gennaro studied the stability of cyclophosphamide with granisetron hydrochloride during simulated Y-site administration. The liquid chromatographic system consisted of a piston pump with a pulse dampener, a rotary injection port with a 20-µL loop, a variable-wavelength spectrometric detector, and an integrator. The stationary phase was a C_{18} reversed-phase column (300 × 4.6 mm, 5-µm particle size). The mobile phase was a mixture of acetonitrile and 0.1 M monobasic sodium phosphate dihydrate (20:80, vol/vol) adjusted to pH 4.2 with phosphoric acid. The flow rate was 1.75 mL/min. UV detection was performed at 198 nm.

Samples were diluted with the mobile phase. The retention times for cyclophosphamide and granisetron were 2.98 and 4.98 minutes, respectively.

The stability-indicating capability of the analytical method was demonstrated by an accelerated degradation study. Solutions of cyclophosphamide and granisetron hydrochloride were adjusted to pH 2 and 11, boiled for 1 hour, readjusted to pH 5, diluted with the mobile phase, and analyzed. Degradation product peaks did not interfere with the intact cyclophosphamide or granisetron peaks.

Reference
Mayron D, Gennaro AR. Stability and compatibility of granisetron hydrochloride in i.v. solutions and oral liquids and during simulated Y-site injection with selected drugs. *Am J Health Syst Pharm.* 1996; 53: 294–304.

Method 2
Fleming et al. investigated the stability of cyclophosphamide and ondansetron hydrochloride in injectable solutions. The high-performance liquid chromatograph consisted of an LDC 4100 multisolvent pump, an LDC 4100 autosampler, and an LDC 5000 photodiode-array detector and was controlled by an IBM-compatible 386/25-MHz computer with LDC analytical LC Talk chromatographic software. The stationary phase was a Beckman Ultrasphere reversed-phase column (250 × 4.6 mm, 5-µm particle size). The mobile phase consisted of 40% acetonitrile and 60% deionized water. The flow rate was 1.5 mL/min. UV detection was performed at 200 nm. The injection volume was 20 µL. Under these conditions, retention times for cyclophosphamide and ondansetron were 2.75 and 3.3 minutes, respectively.

The stability-indicating capability of the assay was demonstrated by accelerated decomposition of cyclophosphamide. Solutions of cyclophosphamide were adjusted to pH 2.0 with phosphoric acid and to pH 12 with sodium hydroxide and were then heated at 100 °C for 10 minutes. Degradation product peaks did not interfere with the intact cyclophosphamide peak.

Standard curves for cyclophosphamide were constructed daily from 0.05 to 5 mg/mL. The intraday and interday coefficients of variation of the assay were less than 5%.

Reference
Fleming RA, Olsen DJ, Savage PD, et al. Stability of ondansetron hydrochloride and cyclophosphamide in injectable solutions. *Am J Health Syst Pharm.* 1995; 52: 514–6.

Method 3
Kirk et al. examined the stability of cyclophosphamide injection in various containers when stored under refrigeration or frozen, with thawing by microwave irradiation. The chromatograph utilized a Pye Unicam LC-XPS pump equipped with a Rheodyne 7125 20-μL fixed-volume injection valve, a Pye Unicam LC-UV variable-wavelength UV detector, an LDL 308 integrator, and a chart recorder. The stationary phase was a Hypersil octadecylsilane column (250 × 4.3 mm, 5-μm particle size). The mobile phase was 27% acetonitrile in 100 mM phosphate buffer adjusted to pH 7.0. The flow rate was 1.5 mL/min. UV detection was performed at 200 nm and 0.04 AUFS. Hydrocortisone 400 μg/mL was used as an internal standard.

Samples were diluted 1:20 with distilled water. The injection volume was 20 μL. The retention time for cyclophosphamide was about 7.7 minutes (estimated from the published chromatogram).

The method was demonstrated to be stability indicating by accelerated decomposition of cyclophosphamide. Solutions of cyclophosphamide were adjusted to pH 2 or 14 or heated at 100 °C for 48 hours. The resulting solutions were assayed. The degradation product peaks did not interfere with the intact drug peak.

A calibration curve was constructed for cyclophosphamide from 0.28 to 1.4 mg/mL.

Reference
Kirk B, Melia CD, Wilson JV, et al. Chemical stability of cyclophosphamide injection. *Br J Parenter Ther.* 1984; 5: 90–7.

Method 4
Kensler et al. reported the use of an HPLC method for the analysis of cyclophosphamide. The liquid chromatograph consisted of a Constametric II G reciprocating pump, a Valco 25-μL loop injector, a Varian Vari-Chrom detector, and a Varian model 9176 strip-chart recorder. The stationary phase was a Waters μBondapak C_{18} reversed-phase column (300 × 3.9 mm). The mobile phase was a mixture of 30% acetonitrile and 70% water and was delivered isocratically at 1.5 mL/min. UV detection was performed at 200 nm and 0.64 AUFS.

Samples were diluted with water. The injection volume was 25 μL. The retention time for cyclophosphamide was 5.4 minutes.

The method was determined to be stability indicating by accelerated decomposition of cyclophosphamide. A sample of cyclophosphamide was stored at 50 °C for 11 hours in pH 7.0 buffer and assayed. The degradation product peak was well separated from the intact cyclophosphamide peak.

A standard curve was generated by a least-squares analysis of integrator counts versus concentration from 1.038 to 4.436 mg/mL. The correlation coefficient was 0.9993. The coefficient of variation was 1.5%.

Reference
Kensler TT, Behme RJ, Brooke D. High-performance liquid chromatographic analysis of cyclophosphamide. *J Pharm Sci.* 1979; 68: 172–4.

Method 5

Wantland and Hersh developed a stability-indicating HPLC assay for cyclophosphamide in raw materials and parenteral dosage forms. The chromatograph included a Hewlett-Packard model 1084A variable-volume injector, a Schoeffel SF 770 spectroflow variable-wavelength detector, and an Altex Lichrosorb C_{18} reversed-phase column (250 × 3.2 mm, 10-μm particle size). The mobile phase contained 45% methanol in water and was delivered isocratically at 1.0 mL/min. UV detection was performed at 200 nm.

Samples were diluted with water. The injection volume was 200 μL. The retention time for cyclophosphamide was 5.1 minutes.

The stability-indicating capability of the assay was demonstrated by accelerated decomposition of cyclophosphamide. A solution of cyclophosphamide 1% in water was heated to 83 °C for 2 hours and assayed. The degradation product peaks did not interfere with the intact drug peak.

Reference

Wantland LR, Hersh SD. High-performance liquid chromatographic assay of cyclophosphamide in raw materials and parenteral dosage forms. *J Pharm Sci.* 1979; 68: 1144–6.

Cyclosporine

Chemical Name

Cyclo[[4-(*E*)-but-2-enyl-*N*,4-dimethyl-L-threonyl]-L-homoalanyl-(*N*-methylglycyl)-(*N*-methyl-L-leucyl)-L-valyl-(*N*-methyl-L-leucyl)-L-alanyl-D-alanyl-(*N*-methyl-L-leucyl)-(*N*-methyl-L-leucyl)-(*N*-methyl-L-valyl)]

Other Names

Ciclosporin, Cyclosporin A, Sandimmune

Form	Molecular Formula	MW	CAS
Cyclosporine	$C_{62}H_{111}N_{11}O_{12}$	1202.6	59865-13-3

Appearance

Cyclosporine is a white or essentially white fine crystalline powder.

Solubility

Cyclosporine is soluble in methanol, ethanol, acetone, ether, and chloroform and slightly soluble in water and saturated hydrocarbons.

Method 1

Kumar et al. reported an HPLC method for cyclosporine in oral solution. The chromatographic apparatus included a Waters model 510 pump, a Rheodyne 7725i injector, and a Waters model 996 photodiode-array detector. The stationary phase was a Nucleosil RP-2 column. Mobile phase A was 0.8 mL of phosphoric acid in 1000 mL of water, and mobile phase B was 0.8 mL of phosphoric acid in 1000 mL of acetonitrile. The mobile phase was delivered linearly from 35% B to 60% B in 55 minutes.

Samples were diluted with isopropanol. The injection volume was 20 μL. UV detection was performed at 210 nm. Under these conditions, the retention time for cyclosporine was about 23.5 minutes (estimated from the published chromatogram).

The assay was reported to be stability indicating. Cyclosporine samples were treated with 1 N hydrochloric acid at 80 °C, 1 N sodium hydroxide solution at 80 °C, or 15% hydrogen peroxide at 80 °C or were exposed to light (1000 foot-candles) for 72 hours. There was no interference from known degradation products, impurities, and placebo. The peak purity of cyclosporine was also confirmed by a photodiode-array detector.

A standard curve for cyclosporine was constructed from 100 to 2000 ppm. The correlation coefficient was greater than 0.9999. The limit of detection and limit of quantitation were 1 and 3 ppm, respectively.

Reference

Kumar M, Singhal SK, Singh A. Development and validation of a stability-indicating HPLC assay method for cyclosporine in cyclosporine oral solution. *J Pharm Biomed Anal.* 2001; 25: 9–14.

Method 2

Using an HPLC method, Jacobson et al. evaluated the compatibility of cyclosporine with fat emulsion. The chromatograph included a Waters model 501 pump, a Waters WISP autosampler, a Waters model 481 UV detector, and a Hewlett-Packard model 3390A electronic integrator. The stationary phase was a Waters μBondapak C_{18} column (150 × 3.9 mm, 10-μm particle size), heated to 55 °C. The mobile phase was 70% acetonitrile in 20 mM ammonium phosphate. The flow rate was 1 mL/min. UV detection was performed at 214 nm.

Twenty microliters of the sample was mixed with 2.98 mL of methanol. The injection volume was 100 μL. Under these conditions, the retention time for cyclosporine was 3 minutes.

The analytical method was determined to be stability indicating by accelerated degradation of cyclosporine. Solutions of cyclosporine 2 mg/mL were adjusted to pH 13 with 4 N sodium hydroxide or to pH 1 with 85% nitric acid and allowed to stand for 30 minutes. The solutions were diluted and assayed. The decomposition product peaks did not interfere with the intact cyclosporine peak.

The standard curve for cyclosporine was linear ($r = 0.997$) from 0.125 to 2 mg/mL. The intraday and interday coefficients of variation were less than 5%.

A similar stability-indicating method was reported by Shea et al.

References

Jacobson PA, Maksym CJ, Landvay A, et al. Compatibility of cyclosporine with fat emulsion. *Am J Hosp Pharm*. 1993; 50: 687–90.

Shea BF, Ptachcinski RJ, O'Neill S, et al. Stability of cyclosporine in 5% dextrose injection. *Am J Hosp Pharm*. 1989; 46: 2053–5.

Method 3

Using an HPLC method, Ptachcinski et al. reported the stability of cyclosporine in 5% dextrose injection or 0.9% sodium chloride injection. The chromatographic system included a Waters model 710B WISP autosampler, a Waters model 441 UV detector, and a Supelco C_{18} column (150 mm, 5-μm particle size). The mobile phase was 68% acetonitrile in water. The flow rate was 1.5 mL/min. UV absorption was monitored at 214 nm. The injection volume was 5 μL. The column temperature was 70 °C. The retention time for cyclosporine was 7.4 minutes.

The analytical method was determined to be stability indicating by accelerated degradation of cyclosporine. A cyclosporine sample was intentionally degraded with 12 N hydrochloric acid at 24 °C for 2 hours and assayed. Degradation product peaks did not interfere with the intact cyclosporine peak.

The standard curve for cyclosporine was constructed from 0.5 to 5 mg/mL. Its correlation coefficient was 0.999. The assay had a mean coefficient of variation of 1.8%.

References

Ptachcinski RJ, Logue LW, Burckart GJ, et al. Stability and availability of cyclosporine in 5% dextrose injection or 0.9% sodium chloride injection. *Am J Hosp Pharm*. 1986; 43: 94–7.

Ptachcinski RJ, Walker S, Burckart GJ, et al. Stability and availability of cyclosporine stored in plastic syringes. *Am J Hosp Pharm*. 1986; 43: 692–4.

Cyproterone Acetate

Chemical Name

6-Chloro-1β,2β-dihydro-17α-hydroxy-3′*H*-cyclopropa[1,2]pregna-1,4,6-triene-3,20-
   dione acetate

Other Name

Cyprostat

Form	Molecular Formula	MW	CAS
Cyproterone acetate	$C_{24}H_{29}ClO_4$	416.9	427-51-0

Appearance

Cyproterone acetate occurs as a white or almost white crystalline powder.

Solubility

Cyproterone acetate is practically insoluble in water, sparingly soluble in dehydrated alcohol, and freely soluble in acetone.

Method

Segall et al. described an HPLC assay for the determination of cyproterone acetate in tablet formulations. The liquid chromatograph consisted of a Spectra-Physics model ISO pump, a Hewlett-Packard model 1050 UV-visible detector, a Rheodyne model 7125 injector, and a Hewlett-Packard series 3395 integrator. The stationary phase was a Merck LiChrosorb 100 RP18 column (250 × 4 mm, 10-μm particle size) with a Merck LiChrosorb RP18 guard column (4 × 4 mm, 10-μm particle size). The mobile phase consisted of 60% acetonitrile and 40% water. The flow rate was 1.2 mL/min. UV detection was performed at 254 nm and 1 AUFS. The total run time was 12 minutes. Progesterone or hydrocortisone 0.5 mg/mL was used as an internal standard.

Twenty tablets were weighed and ground. A portion of powder equivalent to one tablet was transferred into a 100-mL volumetric flask, mixed with 70 mL of mobile phase, sonicated for 5 minutes, diluted to volume with mobile phase, and filtered through a Whatman No. 42 filter paper. The injection volume was 20 μL. Under these conditions, the retention times of hydrocortisone, cyproterone, and progesterone were 4, 12, and 17 minutes, respectively.

The assay was evaluated for stability-indicating ability by accelerated degradation. Cyproterone acetate was refluxed in 1 N hydrochloric acid for 30 minutes, 1 N sodium hydroxide for 30 minutes, hydrogen peroxide for 30 minutes, or 1 N hydrochloric acid and zinc for 30 minutes, or it was heated at 110 °C for 24 hours, or it was exposed to daylight for 24 hours. Degradation products did not interfere with the analysis of cyproterone acetate.

A standard curve for cyproterone acetate was obtained from 50 to 650 μg/mL. The correlation coefficient was 0.9999.

Reference

Segall A, Vitale M, Perez V, et al. A stability-indicating HPLC method to determine cyproterone acetate in tablet formulations. *Drug Dev Ind Pharm*. 2000; 26: 867–72.

Cytarabine

Chemical Names

4-Amino-1-β-D-arabinofuranosyl-2(1*H*)-pyrimidinone
1-β-D-Arabinofuranosylcytosine

Other Name
Cytosar-U

Form	**Molecular Formula**	**MW**	**CAS**
Cytarabine	$C_9H_{13}N_3O_5$	243.2	147-94-4

Appearance
Cytarabine is an odorless white to off-white crystalline powder.

Solubility
Cytarabine has a solubility of 100 mg/mL in water and is very slightly soluble in alcohol.

pK$_a$
Cytarabine has a pK$_a$ of 4.35.

Method 1
Zhang et al. used HPLC to study the chemical stability of cytarabine 3 mg/mL in Elliott's B solution, an artificial cerebrospinal fluid. The chromatograph consisted of a multisolvent-delivery pump, a multiple-wavelength UV detector, and an autosampler in one unit (Waters Module 1 Plus). The system was controlled and integrated by a Digital workstation running Waters Millenium 2010 Chromatography Manager. The stationary phase was an Alltech reversed-phase C_{18} analytical column (250 × 4.6 mm, 5-μm particle size). The mobile phase was a mixture of methanol (50 mL) and aqueous phosphate buffer (950 mL) composed of 1.34 g of dibasic sodium phosphate and 0.69 g of monobasic sodium phosphate. The flow rate was 1 mL/min. UV detection was performed at 254 nm and 0.500 AUFS.

Samples were diluted 1:30 with mobile phase. The injection volume was 20 μL. Under these conditions, the retention time for cytarabine was 5.8 minutes.

The method was determined to be stability indicating by accelerated decomposition of cytarabine. Exposure of the cytarabine solution to 3% hydrogen peroxide for 24 hours at room temperature produced a 30% reduction in the intact drug peak and the appearance of a new peak at 2.9 minutes. The degradation product peak did not interfere with the intact cytarabine peak.

Calibration curves were constructed from a linear plot of peak area versus concentration for cytarabine from 25 to 150 μg/mL. The correlation coefficient was greater than 0.9999. The intraday and interday coefficients of variation were 0.3 and 0.6%, respectively.

Reference
Zhang Y, Xu QA, Trissel LA, et al. Physical and chemical stability of methotrexate sodium, cytarabine, and hydrocortisone sodium succinate in Elliott's B solution. *Hosp Pharm.* 1996; 31: 965–70.

Method 2
Mayron and Gennaro studied the stability of cytarabine with granisetron hydrochloride during Y-site administration. The liquid chromatographic system consisted of a piston pump with a pulse dampener, a rotary injection port with a 20-μL loop, a variable-wavelength spectrometric detector, and an integrator. The stationary phase was a C_{18} reversed-phase column (300 × 4.6 mm, 5-μm particle size). The mobile phase was a mixture of acetonitrile and 0.1 M monobasic

sodium phosphate dihydrate (20:80, vol/vol) adjusted to pH 4.2 with phosphoric acid. The flow rate was 2.50 mL/min. UV detection was performed at 300 nm.

Samples were diluted with the mobile phase. The retention times for cytarabine and granisetron were 1.45 and 5.16 minutes, respectively.

The stability-indicating capability of the analytical method was demonstrated by an accelerated degradation study. Solutions of cytarabine and granisetron hydrochloride were adjusted to pH 2 and 11, boiled for 1 hour, readjusted to pH 5, diluted with the mobile phase, and analyzed. Degradation product peaks did not interfere with the intact cytarabine or granisetron peaks.

Reference

Mayron D, Gennaro AR. Stability and compatibility of granisetron hydrochloride in i.v. solutions and oral liquids and during simulated Y-site injection with selected drugs. *Am J Health Syst Pharm.* 1996; 53: 294–304.

Method 3

Rochard et al. evaluated the stability of cytarabine in ethylene vinyl acetate portable infusion-pump reservoirs. The liquid chromatograph was equipped with a Waters model 510 reciprocating piston pump, a Waters model 710A WISP autosampler, a Waters model 484 variable-wavelength UV detector, and a Waters model 810 computer integrator. The stationary phase was a Nucleosil C_{18} column (150 × 4.6 mm, 5-μm particle size). The mobile phase was a mixture of 0.05 M ammonium dihydrogen phosphate and aceto-nitrile (96:4, vol/vol) and was delivered isocratically at 1.0 mL/min. UV detection was performed at 270 nm.

The stability-indicating capability of the method was demonstrated by accelerated degradation of cytarabine with 1 N sodium hydroxide and with 1 N hydrochloric acid. The intact cytarabine peak was well separated from the solvent front and the degradation product peaks. Retention times for cytarabine and arabinosyluracil were 4.0 and 6.1 minutes, respectively.

Standard curves for cytarabine were generated each day for calibration from 4 to 14 μg/mL. Their correlation coefficients were greater than 0.990. Intraday and interday coefficients of variation for the assay were 1.10 and 2.40%, respectively.

Reference

Rochard EB, Barthes DMC, Courtois PY. Stability of fluorouracil, cytarabine, or doxorubicin hydrochloride in ethylene vinylacetate portable infusion-pump reservoirs. *Am J Hosp Pharm.* 1992; 49: 619–23.

Method 4

Cheung et al. determined the stability of cytarabine in Elliott's B solution, 0.9% sodium chloride injection, 5% dextrose injection, and lactated Ringer's injection. The liquid chromatograph was equipped with a Spectra-Physics model 3500B reciprocating piston pump with flow feedback control, a Valco rotary injector valve fitted with a 10-μL loop, a Spectra-Physics model 770 variable-wavelength UV detector, and an Omniscribe strip-chart recorder. The stationary phase was an Alltech C_{18} stainless steel column (250 × 4.6 mm, 10-μm particle size). The mobile phase consisted of 1% acetic acid, 70% distilled water, and 29% acetonitrile and was delivered at 2 mL/min. UV detection was performed at 240 nm and 0.1 AUFS. Propylparaben was used as an internal standard.

Each sample was diluted with the internal standard solution. The injection volume was 10 μL. Under these conditions, retention volumes for cytarabine and propyl-paraben were 9.6 and 40.4 mL, respectively.

The method was demonstrated to be stability indicating using drug solutions that were degraded by heating on a steam bath for 6 hours. Decomposition product peaks did not interfere with the intact cytarabine peak.

The standard curve for cytarabine was linear ($r > 0.999$) from 108 to 430 μg/mL. Precision determined on two consecutive days was about 2%.

Reference

Cheung Y-W, Vishnuvajjala BR, Flora KP. Stability of cytarabine, methotrexate sodium, and hydrocortisone sodium succinate admixture. *Am J Hosp Pharm.* 1984; 41: 1802–6.

Dacarbazine

Chemical Names

5-(3,3-Dimethyl-1-triazenyl)-1*H*-imidazole-4-carboxamide
5(or 4)-(Dimethyltriazeno)imidazole-4(or 5)-carboxamide

Other Name

DTIC-Dome

Form	Molecular Formula	MW	CAS
Dacarbazine	$C_6H_{10}N_6O$	182.2	4342-03-4

Appearance

Dacarbazine is a colorless to ivory-colored crystalline solid.

Solubility

Dacarbazine is slightly soluble in water and alcohol.

pK$_a$

Protonated dacarbazine has a pK$_a$ of 4.42.

Method

Mayron and Gennaro used an HPLC method to assess the stability of dacarbazine with granisetron hydrochloride during simulated Y-site administration. The liquid chromatograph consisted of a piston pump with a pulse dampener, a rotary injection port with a 20-μL loop, a variable-wavelength spectrometric detector, and an integrator. The stationary phase was a C_{18} reversed-phase column (300 × 4.6 mm, 5-μm particle size). The mobile phase was a mixture of acetonitrile and 0.1 M monobasic sodium phosphate dihydrate (20:80, vol/vol) adjusted to pH 4.2 with phosphoric acid. The flow rate was 1.5 mL/min. UV detection was performed at 300 nm.

Samples were diluted with mobile phase. The retention times for dacarbazine and granisetron were 2.32 and 5.10 minutes, respectively.

The stability-indicating capability of the analytical method was demonstrated by an accelerated degradation study. Solutions of dacarbazine and granisetron hydrochloride were adjusted to pH 2 and 11, boiled for 1 hour, readjusted to pH 5, diluted with the mobile phase, and analyzed. The degradation product peaks did not interfere with the intact dacarbazine and granisetron peaks.

Reference
Mayron D, Gennaro AR. Stability and compatibility of granisetron hydrochloride in i.v. solutions and oral liquids and during simulated Y-site injection with selected drugs. *Am J Health Syst Pharm.* 1996; 53: 294–304.

Danazol

Chemical Name
17α-Pregna-2,4-dien-20-yno[2,3-*d*]isoxazol-17-ol

Other Name
Danocrine

Form	Molecular Formula	MW	CAS
Danazol	$C_{22}H_{27}NO_2$	337.5	17230-88-5

Appearance
Danazol is a white or pale yellow crystalline powder.

Solubility
Danazol is practically insoluble in water, sparingly soluble in alcohol, and freely soluble in chloroform.

Method

Kariem et al. used HPLC for a photodegradation kinetic study of danazol. A Waters liquid chromatograph included a model 600E system controller, a model U6K injector, a model 486 tunable absorbance detector, and a model 746 data module. The stationary phase was a Merck LiChrosphere 100 RP18 column (125 × 4.0 mm, 5-μm particle size). The mobile phase consisted of acetonitrile and water (55:45, vol/vol) and was delivered at a flow rate of 1.3 mL/min. UV detection was performed at 287 nm. Imperatorin 300 μg/mL in methanol was used as an internal standard.

The powder from capsule contents equivalent to 25 mg of danazol was shaken in 80 mL of methanol for 10 minutes, the flask was filled to 100 mL with methanol, and the solution was filtered. The filtrate was appropriately mixed with the internal standard and diluted with mobile phase. The injection volume was 20 μL. Under these conditions, retention times for imperatorin and danazol were 4.60 and 6.46 minutes, respectively.

The stability-indicating nature of the method was shown by the photo-degradation study of danazol. Danazol was irradiated with a UV source at 254 nm for 24 hours. The intact danazol peak was well resolved from its degradation product peaks.

A standard curve for danazol was constructed from 0.5 to 200 μg/mL. The correlation coefficient was 0.9998. Intraday and interday coefficients of variation were 0.67 and 3.5%, respectively.

A similar method was described.

References

Gad Kariem EA, Abounassif MA, Hagga ME, et al. Photodegradation kinetic study and stability-indicating assay of danazol using high-performance liquid chromatography. *J Pharm Biomed Anal*. 2000; 23: 413–20.

Gad Kariem EA, El-Obeid HA, Abounassif MA, et al. Effects of alkali and simulated gastric and intestinal fluids on danazol stability. *J Pharm Biomed Anal*. 2003; 31: 743–51.

Dantrolene

Chemical Names

1-[[[5-(4-Nitrophenyl)-2-furanyl]methylene]amino]-2,4-imidazolidinedione

1-[[5-(*p*-Nitrophenyl)furfurylidene]amino]hydantoin

Other Name

Dantrium

Form	Molecular Formula	MW	CAS
Dantrolene	$C_{14}H_{10}N_4O_5$	314.3	7261-97-4
Dantrolene sodium	$C_{14}H_9N_4NaO_5 \cdot 3\frac{1}{2}H_2O$	399.3	24868-20-0

Appearance

Dantrolene sodium is an orange powder.

Solubility

Dantrolene sodium is slightly soluble in water and more soluble in alkaline solutions.

pK$_a$

Dantrolene has a pK$_a$ of 7.5.

Method

Fawcett et al. studied the chemical stability of a dantrolene oral suspension prepared from capsules. The HPLC system included a Jasco model 880-PU pump, a model 851-AS refrigerated automatic sampler fitted with a 50-μL fixed-loop injector, a model 875 variable-wavelength UV detector, and a computerized chromatography data analysis system. The stationary phase was an Alltech Applied Science Labs Nucleosil CN stainless steel column (150 × 4.6 mm, 5-μm particle size) maintained at 30 °C with an Eppendorf model CH-30 column heater. The mobile phase consisted of 25% acetonitrile and 75% phosphate buffer (pH 6.8). Phosphate buffer was 50% of a dibasic sodium phosphate solution (11.88 g/L) and 50% of a monobasic potassium phosphate solution (9.08 g/L). The mobile phase was delivered isocratically at 1.5 mL/min. UV detection was performed at 375 nm from 0 to 2.5 minutes and at 315 nm from 2.5 to 3.7 minutes and then returned to 375 nm. Methyl hydroxybenzoate 0.15% was used as an internal standard.

The sample suspension was diluted 1:10 with dimethylformamide and was further diluted 1:50 with mobile phase. Under these conditions, the retention times for the internal standard and dantrolene were 2.7 and 4.2 minutes, respectively.

The stability-indicating capability of the assay was investigated by heating a suspension of dantrolene sodium powder in syrup BP at 60 °C. Degradation product peaks did not interfere with the intact dantrolene peak.

A calibration curve was generated from 0 to 10 μg/mL and was linear ($r = 0.995$). The intraday and interday coefficients of variation were 3.6 and 4.6%, respectively.

Reference

Fawcett JP, Stark G, Tucker IG, et al. Stability of dantrolene oral suspension prepared from capsules. *J Clin Pharm Ther*. 1994; 19: 349–53.

Dapsone

Chemical Name

Bis(4-aminophenyl)sulfone

Other Names
Avlosulfon, Sulfona

Form	Molecular Formula	MW	CAS
Dapsone	$C_{12}H_{12}N_2O_2S$	248.3	80-08-0

Appearance
Dapsone occurs as a white or creamy white crystalline powder.

Solubility
Dapsone is very slightly soluble in water but freely soluble in alcohol.

Method
Nahata et al. used HPLC to evaluate the stability of dapsone in two oral liquid dosage forms. The Hewlett-Packard 1050 liquid chromatograph consisted of a pump, an autosampler, a variable-wavelength detector, and a model 3366 integrator. The stationary phase was a Mac-Mod Analytical Products Zorbax SB C_{18} column (150 × 3.0 mm) that was maintained at 40 °C. The mobile phase consisted of 50 mM ammonium phosphate and acetonitrile (88:12, vol/vol) (pH 4.6). The flow rate was 0.7 mL/min. The UV detector was set at 295 nm. Diazoxide in mobile phase was used as an internal standard. The injection volume was 10 μL. Under these conditions, retention times for dapsone and diazoxide were about 8.9 and 12.9 minutes, respectively.

The method was established to be stability indicating by accelerated degradation. Dapsone samples were treated with 2.0 M hydrochloric acid, 2.0 M sodium hydroxide solution, or 0.3% hydrogen peroxide or were heated at 80 °C. The analysis of dapsone was not influenced by degradation products.

A calibration curve for dapsone was generated from 6.25 to 62.5 μg/mL. The correlation coefficient was greater than 0.999. Intraday and interday coefficients of variation were less than 2.3 and 3.1%, respectively.

Reference
Nahata MC, Morosco RS, Trowbridge JM. Stability of dapsone in two oral liquid dosage forms. *Ann Pharmacother.* 2000; 34: 848–50.

Daptomycin

Chemical Name
N-Decanoyl-L-tryptophyl-L-asparaginyl-L-aspartyl-L-threonylglycyl-L-ornithyl-L-aspartyl-D-alanyl-L-aspartylglycyl-D-seryl-*threo*-3-methyl-L-glutamyl-3-anthraniloyl-L-alanine ε_1-lactone

Other Name
Cubicin

Form	Molecular Formula	MW	CAS
Daptomycin	$C_{72}H_{101}N_{17}O_{26}$	1620.7	103060-53-3

Appearance
Daptomycin occurs as a yellow to light brown material.

Method
Using an HPLC method, Lai and Brodeur evaluated the chemical stability of daptomycin in 0.9% sodium chloride injection admixed with commonly administered intravenous medications. An Agilent 1100 series system consisted of a quaternary pump, a vacuum degasser, a thermostat-controlled column compartment, a thermostat-controlled autosampler, and a diode-array detector. The stationary phase was an IB-SIL 5 C_8 column (250 × 4.6 mm) with an IB-SIL 5 C_8 precolumn (30 × 4.6 mm). The mobile phase was 38.5% acetonitrile in 0.45% ammonium dihydrogen phosphate buffer (pH 3.25). The flow rate was 1.5 mL/min. UV detection was performed at 214 nm. The injection volume was 20 μL. Under these conditions, the retention time for daptomycin was 15 minutes.

The method was validated as a stability-indicating assay.

Reference
Lai J-J, Brodeur SK. Physical and chemical compatibility of daptomycin with nine medications. *Ann Pharmacother.* 2004; 38: 1612–6.

Daunorubicin

Chemical Name
(8*S*-*cis*)-8-Acetyl-10-[(3-amino-2,3,6-trideoxy-α-L-*lyxo*-hexopyranosyl)oxy]-7,8,9,10-tetrahydro-6,8,11-trihydroxy-1-methoxy-5,12-naphthacenedione

Other Names
Cerubidine, Daunomycin, Rubidomycin

Form

Form	Molecular Formula	MW	CAS
Daunorubicin	$C_{27}H_{29}NO_{10}$	527.5	20830-81-3
Daunorubicin hydrochloride	$C_{27}H_{29}NO_{10}.HCl$	564.0	23541-50-6

Appearance
Daunorubicin hydrochloride is an orange-red hygroscopic crystalline powder.

Solubility
Daunorubicin hydrochloride is freely soluble in water and methanol, slightly soluble in alcohol, and practically insoluble in acetone.

pK$_a$
Daunorubicin hydrochloride has a pK$_a$ of 10.3.

Method 1
Wood et al. used HPLC to determine the shelf life of daunorubicin hydrochloride when reconstituted with water for injection, 5% dextrose injection, and 0.9% sodium chloride injection. The system included an Altex 100A pump, a Rheodyne model 7125 sample injector equipped with a 10-μL injection loop, a Pye Unicam variable-wavelength detector, and a JJ Instruments CR452 chart recorder. The stationary phase was a Shandon ODS Hypersil reversed-phase stainless steel column (100 × 4.6 mm, 5-μm particle size). The mobile phase was a mixture of acetonitrile and water (55:45, vol/vol). Ten drops of diethylamine were added to each liter of mobile phase. It was then adjusted to pH 2.5 with 10% phosphoric acid. The flow rate was 1.5 mL/min. UV detection was performed at 290 nm and 0.16 AUFS.

   The analytical method was determined to be stability indicating by comparison of daunorubicin hydrochloride standard solution with partially degraded and fully degraded solutions. Degradation was accelerated by subjecting the solutions to extremes of temperature and pH. Degradation products did not interfere with the quantification of daunorubicin.

References

Wood MJ, Irwin WJ, Scott DK. Stability of doxorubicin, daunorubicin, and epirubicin in plastic syringes and minibags. *J Clin Pharm Ther.* 1990; 15: 279–89.

Wood MJ, Irwin WJ, Scott DK. Photodegradation of doxorubicin, daunorubicin, and epirubicin measured by high-performance liquid chromatography. *J Clin Pharm Ther.* 1990; 15: 291–300.

Method 2

Bekers et al. used HPLC analysis to investigate the effect of cyclodextrins on daunorubicin stability in acidic aqueous media. The liquid chromatograph consisted of a Waters model M-6000 pump, a model U6K injector, a Spectroflow 757 absorbance detector, and a Spectra-Physics model 4270 integrator. The stationary phase was a Merck LiChrosorb RP8 column (300 × 3.9 mm, 10-µm particle size). The mobile phase consisted of 0.01 M sodium chloride (adjusted to pH 2.25 with perchloric acid) and acetonitrile (45:55, vol/vol). The flow rate was 1.5 mL/min. UV detection was performed at 480 nm. The sample injection volume was 20 µL.

    The method was reported to be stability indicating.

    Standard curves were generated with a correlation coefficient greater than 0.999.

Reference

Bekers O, Beijnen JH, Groot Bramel EH, et al. Effect of cyclodextrins on anthracycline stability in acidic aqueous media. *Pharm Weekbl [Sci].* 1988; 10: 207–12.

Deferoxamine

Chemical Names

N'-[5-[[4-[[5-(Acetylhydroxyamino)pentyl]amino]-1,4-dioxobutyl]hydroxyamino]pentyl]-N-(5-aminopentyl)-N-hydroxybutanediamide

N-[5-[3-[(5-Aminopentyl)hydroxycarbamoyl]propionamido]pentyl]-3-[[5-(N-hydroxy-acetamido)pentyl]carbamoyl]propionohydroxamic acid

Other Names

Desferal, Desferrioxamine

Form	Molecular Formula	MW	CAS
Deferoxamine	$C_{25}H_{48}N_6O_8$	560.7	70-51-9
Deferoxamine mesylate	$C_{25}H_{48}N_6O_8.CH_4SO_3$	656.8	138-14-7

Appearance
Deferoxamine mesylate is a white to off-white powder.

Solubility
Deferoxamine mesylate is freely soluble in water and soluble in alcohol.

Method
Using an HPLC assay, Stiles et al. determined the stability of deferoxamine mesylate in polypropylene infusion-pump syringes. The liquid chromatograph included a Waters model 501 pump, a Waters model 441 UV detector, a Waters model 745 data module, an Alcott model 728 autosampler, and a Rheodyne 7010 injector. The stationary phase was a Bakerbond C_{18} analytical column. The mobile phase was 10% acetonitrile in 0.5 M sodium phosphate buffer (pH adjusted to 6.6) containing 4 mM EDTA (edetate) and 1 M ammonium acetate. The flow rate was 1.0 mL/min. UV detection was performed at 240 nm.

Samples were diluted 1:1000 before injection. Under these conditions, the retention time for deferoxamine mesylate was 3.2 minutes.

The stability-indicating nature of the assay was investigated by accelerated decomposition of deferoxamine mesylate with sulfuric acid, 1 N sodium hydroxide, or heat at 80 °C for 24 hours. Degradation product peaks did not interfere with the intact deferoxamine mesylate peak.

The standard curve was constructed for deferoxamine mesylate from 100 to 500 µg/mL. Intraday and interday coefficients of variation of the assay were 0.4 and 0.5%, respectively.

Reference
Stiles ML, Allen LV, Prince SJ. Stability of deferoxamine mesylate, floxuridine, fluorouracil, hydromorphone hydrochloride, lorazepam, and midazolam hydrochloride in polypropylene infusion-pump syringes. *Am J Health Syst Pharm.* 1996; 53: 1583–8.

Desoxycorticosterone Acetate

Chemical Names
21-Acetyloxypregn-4-ene-3,20-dione
21-Hydroxypregn-4-ene-3,20-dione 21-acetate

Other Names
Deoxycorticosterone Acetate, Deoxycortone Acetate, Desoxycortone Acetate

Form	Molecular Formula	MW	CAS
Desoxycorticosterone acetate	$C_{23}H_{32}O_4$	372.5	56-47-3

Appearance
Desoxycorticosterone acetate occurs as colorless crystals or as a white or creamy white odorless crystalline powder.

Solubility

Desoxycorticosterone acetate is slightly soluble in alcohol, methanol, acetone, ether, and propylene glycol (10 mg/mL). It is practically insoluble in water.

Method

Smith developed an HPLC method for determining desoxycorticosterone acetate in oil injections. The liquid chromatograph consisted of a Spectra-Physics model 3500B with a Valco 7000 psi injection valve with a 10-µL injection loop, a Spectra-Physics model 230 UV detector, and a Spectra-Physics model 4000 data integration system. The analytical column was a Spherisorb ODS-10 column (250 × 3 mm). The mobile phase was 60% methanol in water and was delivered isocratically at 1.2 mL/min. UV detection was performed at 254 nm and 0.01 AUFS. Progesterone was used as an internal standard.

Desoxycorticosterone acetate was separated from oil using reversed-phase partition chromatography and then diluted with methanol. The injection volume was 10 µL. Under these conditions, the retention time for desoxycorticosterone acetate was 13.8 minutes.

The stability-indicating capability of the method was demonstrated by accelerated decomposition of desoxycorticosterone. The chromatograms of aged samples showed the presence of desoxycorticosterone as well as a decrease in the desoxycorticosterone acetate content.

Reference

Smith E. Liquid chromatographic determination of desoxycorticosterone acetate in oil injections. *J Assoc Off Anal Chem.* 1979; 62: 812–7.

Dexamethasone

Chemical Name

(11β,16α)-9-Fluoro-11,17,21-trihydroxy-16-methylpregna-1,4-diene-3,20-dione

Other Name

Decadron

Form	Molecular Formula	MW	CAS
Dexamethasone	$C_{22}H_{29}FO_5$	392.5	50-02-2
Dexamethasone acetate	$C_{24}H_{31}FO_6$	434.5	1177-87-3
Dexamethasone sodium phosphate	$C_{22}H_{28}FNa_2O_8P$	516.4	2392-39-4

Appearance

Dexamethasone is a white to practically white odorless crystalline powder. Dexamethasone acetate is a clear white to off-white odorless powder. Dexamethasone sodium phosphate occurs as a white or slightly yellow crystalline powder.

Solubility

Dexamethasone is practically insoluble in water and sparingly soluble in alcohol. Dexamethasone acetate is practically insoluble in water. Dexamethasone sodium phosphate is freely soluble in water and soluble in alcohol.

Method 1

Mayron and Gennaro used HPLC to assess the stability of dexamethasone sodium phosphate with granisetron hydrochloride during simulated Y-site administration. The liquid chromatograph consisted of a piston pump with a pulse dampener, a rotary injection port with a 20-μL loop, a variable-wavelength spectrometric detector, and an integrator. The stationary phase was a cyano reversed-phase column (250 × 4.6 mm, 10-μm particle size). The mobile phase was a mixture of acetonitrile and 0.1 M monobasic sodium phosphate dihydrate (20:80, vol/vol) which had been adjusted to pH 4.2 with phosphoric acid. The flow rate was 2.0 mL/min. UV detection was performed at 228 nm.

Samples were diluted with mobile phase. The retention times for dexamethasone and granisetron were 2.36 and 5.60 minutes, respectively.

The stability-indicating capability of the analytical method was demonstrated by an accelerated degradation study. Solutions of dexamethasone sodium phosphate and granisetron hydrochloride were adjusted to pH 2 and 11, boiled for 1 hour, readjusted to pH 5, diluted with the mobile phase, and analyzed. The degradation product peaks did not interfere with the intact dexamethasone or granisetron peaks.

Reference

Mayron D, Gennaro AR. Stability and compatibility of granisetron hydrochloride in i.v. solutions and oral liquids and during simulated Y-site injection with selected drugs. *Am J Health Syst Pharm.* 1996; 53: 294–304.

Method 2

Allen et al. used HPLC to study the stability of dexamethasone sodium phosphate in the presence of cefpirome sulfate during simulated Y-site injection. The method utilized one of two systems. One unit included an Alcott model 728 Micromeritics autosampler, a Rheodyne 7010 injector with an Alcott 732 electrically actuated valve, a Waters model 501 solvent-delivery pump, a Waters model 441 UV detector, a Waters model 401 refractive index detector, and a Waters model 745 data module. The other unit consisted of an Alcott model 728 Micromeritics autosampler, a Rheodyne 7010 injector with an Alcott 732 electrically actuated valve, a Shimadzu model LA-6A solvent-delivery pump, a Shimadzu model SPD-6A UV detector, and an Orion model 901 microprocessor ion analyzer. A C_8 analytical column was used as the stationary phase. The mobile phase was 70% water and 30% acetonitrile and was delivered isocratically at 2.0 mL/min. The UV absorption was monitored at 254 nm.

Samples were diluted before injection. The injection volume was 30 μL. Under these conditions, the retention times for dexamethasone and cefpirome were 2.7 and 3.6 minutes, respectively.

The stability-indicating nature of the method was shown by accelerated decomposition of standard solutions of dexamethasone sodium phosphate with hydrochloric acid, sodium hydroxide, potassium chlorate, heat at 80 °C, or light (150 foot-candles from a tungsten filament source). In each case, dexamethasone was well separated from its breakdown products on the chromatogram.

The calibration curve was generated from 5 to 50 µg/mL and its correlation coefficient was greater than 0.99. The intraday and interday coefficients of variation of the assay were 1.5%.

Reference

Allen LV, Stiles ML, Prince SJ, et al. Stability of cefpirome sulfate in the presence of commonly used intensive care drugs during simulated Y-site injection. *Am J Health Syst Pharm.* 1995; 52: 2427–33.

Method 3

Lugo and Nahata evaluated the stability of dexamethasone sodium phosphate injection diluted from 4 to 1 mg/mL with bacteriostatic 0.9% sodium chloride injection. The liquid chromatograph consisted of a Varian 2010 solvent-delivery system, a Varian 2050 variable-wavelength UV detector, and a Hitachi Anspec D-2000 chromatointegrator. The stationary phase was a Beckman Ultrasphere reversed-phase C_{18} column (250 × 4.6 mm, 5-µm particle size). The mobile phase was a mixture of 30% acetonitrile and 70% 0.05 M monobasic potassium phosphate buffer and was delivered isocratically at 1 mL/min. UV detection was performed at 230 nm. Hydrocortisone was used as an internal standard.

An aliquot of 50 µL of sample was mixed with 50 µL of internal standard solution, followed by 650 µL of mobile phase. The injection volume was 10 µL. Under these conditions, the retention times of dexamethasone sodium phosphate and hydrocortisone were 4.2 and 8.0 minutes, respectively.

The stability-indicating capability of the assay was determined by accelerated degradation of dexamethasone sodium phosphate. Solutions of dexamethasone sodium phosphate were mixed with equal volumes of 2 M hydrochloric acid or 2 N sodium hydroxide and incubated at 80 °C. Degradation product peaks did not interfere with the intact dexamethasone peak.

The calibration curve was constructed for dexamethasone from 0.6 to 1.4 mg/mL by least-squares regression analysis of the peak height ratio of dexamethasone and hydrocortisone at each dexamethasone concentration. The correlation coefficient was 0.9993, and the interday coefficient of variation was less than 1.3%.

Similar analytical methods were used by the other researchers cited here.

References

Lugo RA, Nahata MC. Stability of diluted dexamethasone sodium phosphate injection at two temperatures. *Ann Pharmacother.* 1994; 28: 1018–9.

Evrard B, Ceccato A, Gaspard O, et al. Stability of ondansetron hydrochloride and dexamethasone sodium phosphate in 0.9% sodium chloride injection and in 5% dextrose injection. *Am J Health Syst Pharm.* 1997; 54: 1065–8.

McGuire TR, Narducci WA, Fox JL. Compatibility and stability of ondansetron hydrochloride, dexamethasone, and lorazepam in injectable solutions. *Am J Hosp Pharm.* 1993; 50: 1410–4.

Method 4

Walker et al. studied the stability of dexamethasone at various concentrations at room temperature. The HPLC method used a chromatographic system consisting of a Spectra-Physics SP8700 ternary gradient solvent-delivery pump, a Schoeffel SF 770 variable-wavelength UV detector, and a Beckman Ultrasphere reversed-phase C_{18} column (250 × 4.2 mm, 5-μm particle size). The mobile phase consisted of acetonitrile and 0.05 M phosphoric acid (pH 2.2) and contained heptanesulfonic acid 1 mg/mL as a counterion. Acetonitrile began at 13% from 0.0 to 2.5 minutes and was increased linearly to 47% over 2.5 to 16 minutes, remaining at 47% until 20 minutes (estimated from the published chromatogram). UV detection was performed at 230 nm. The sample injection volume was 20 μL.

The stability-indicating nature of the method was demonstrated by adding acid or base to aqueous solutions of dexamethasone in 30-mL multidose vials and incubating them in a water bath at 90 °C. The intact drug peak was well separated from degradation product peaks and other constituents of the formulation on its liquid chromatogram.

Reference

Walker SE, DeAngelis C, Iazzetta J, et al. Compatibility of dexamethasone sodium phosphate with hydromorphone hydrochloride or diphenhydramine hydrochloride. *Am J Hosp Pharm.* 1991; 48: 2161–6.

Dextromethorphan

Chemical Name

(9*S*,13*S*,14*S*)-6,18-Dideoxy-7,8-dihydro-3-*o*-methylmorphine

Form	Molecular Formula	MW	CAS
Dextromethorphan	$C_{18}H_{25}NO$	271.4	125-71-3
Dextromethorphan hydrobromide	$C_{18}H_{25}NO \cdot HBr \cdot H_2O$	370.3	6700-34-1

Appearance

Dextromethorphan occurs as a practically white to slightly yellow odorless crystalline powder. Dextromethorphan hydrobromide occurs as a white or almost white crystalline powder.

Solubility

Dextromethorphan is practically insoluble in water but freely soluble in chloroform. Dextromethorphan hydrobromide is sparingly soluble in water, freely soluble in alcohol and chloroform, and practically insoluble in ether.

Method 1

Fong and Eickhoff described an HPLC method for the determination of pseudoephedrine hydrochloride, doxylamine succinate, and dextromethorphan hydrobromide in a commercial cough-cold liquid formulation. The instrument consisted of two Beckman model 112 pumps, a model 165 variable-wavelength UV-visible detector, a Waters WISP 710B autosampler, and a Beckman model 450 data system. The stationary phase was an Analytichem Sepralyte CN column (150 × 4.6 mm, 5-μm particle size). The mobile phase consisted of acetonitrile, methanol, and an aqueous buffer (75:10:15, vol/vol/vol). The buffer solution contained 5 mM dibasic potassium phosphate and 59 mM triethylamine, adjusted to pH 5.3 with glacial acetic acid. The flow rate was 2.0 mL/min. UV detection was performed at 262 nm and 0.03 AUFS.

The sample was diluted 1:10 with water and centrifuged. The injection volume was 20 μL. Under these conditions, the retention times for pseudoephedrine hydrochloride, doxylamine succinate, and dextromethorphan hydrobromide were about 4.0, 4.9, and 5.7 minutes, respectively.

The method was reported to be stability indicating since it separated chemically similar compounds.

Standard curves were generated from 10 to 200% of the normal working concentration levels. Intraday and interday coefficients of variation were less than 1 and 2.9%, respectively. The limit of detection was 2 μg/mL.

Reference

Fong GW, Eickhoff WM. Liquid chromatographic determination of amines in complex cough-cold formulations. *Int J Pharm.* 1989; 53: 91–7.

Method 2

Heidemann et al. presented an HPLC analysis of phenylpropanolamine hydrochloride, pseudoephedrine hydrochloride, dextromethorphan hydrobromide, diphenhydramine hydrochloride, and chlorpheniramine maleate in cough-cold products. The apparatus consisted of a Waters model 510 dual-piston pump, a model 481 variable-wavelength detector, and a model WISP 710B autosampler. The stationary phase was a DuPont Zorbax 300-SCX column (150 × 4.6 mm, 7–8-μm particle size). The mobile phase consisted of acetonitrile and ethylenediamine sulfate buffer (pH 4.52) (35:65, vol/vol). Ethylenediamine sulfate buffer was prepared by dissolving 225 mg of the salt in 1 L of water and adjusting the pH to 4.52 with 0.2 N sulfuric acid. The flow rate was 2 mL/min. UV detection was performed at 216 nm.

Ten packets of drug product granules were placed in a 1000-mL volumetric flask, wetted and diluted to volume with water, and filtered through a glass-fiber paper. The injection volume was 50 μL. Under these conditions, retention times of pseudoephedrine hydrochloride, dextromethorphan hydrobromide, and chlorpheniramine maleate were about 3.0, 7.3, and 9.5 minutes, respectively (estimated from the published chromatogram).

The method was demonstrated to be stability indicating by the analysis of degraded samples of the different formulations. Degradation products did not interfere with the determination of drugs.

A standard curve for dextromethorphan hydrobromide was constructed from 80 to 120% of the labeled concentrations. The correlation coefficient was 1.000.

Reference

Heidemann DR, Groon KS, Smith JM. HPLC determination of cough-cold products. *LC-GC.* 1987; 5: 422–6.

Diacerein

Chemical Name

4,5-Bis(acetyloxy)-9,10-dihydro-9,10-dioxo-2-anthracenecarboxyic acid

Other Names

Artrodar, Benedar, Diacerhein, Diadar, Fisiodar

Form	Molecular Formula	MW	CAS
Diacerein	$C_{19}H_{12}O_8$	368.3	13739-02-1

Appearance

Diacerein occurs as a yellow crystalline material.

Method

Giannellini et al. developed an HPLC assay for the quality control of diacerein bulk drug substance. A Perkin-Elmer LC200 series chromatograph consisted of a quaternary pump, an autosampler, a diode-array detector, and a Peltier column oven. The stationary phase was a Phenomenex reversed-phase Luna $C_{18}(2)$ column (150 × 4.6 mm, 5-μm particle size). The column temperature was maintained at 40 °C. The mobile phase consisted of 0.1 M phosphoric acid and methanol (40:60, vol/vol). The flow rate was 1.0 mL/min. UV detection was performed at 254 nm.

A standard solution of diacerein was prepared by dissolving the drug in *N,N*-dimethylacetamide and then diluting to an appropriate concentration with methanol. The sample injection volume was 20 µL. Under these conditions, the retention time of diacerein was about 5.9 minutes (estimated from the published chromatogram).

The stability-indicating nature of the method was demonstrated by accelerated degradation studies of diacerein. The drug was treated with 1 N sodium hydroxide solution or 1 N hydrochloric acid at room temperature and 37 °C for 3 hours, heated to 105 °C, exposed to UV light for 24 hours, or treated with 5% hydrogen peroxide at room temperature for 3 hours. The peak of the intact diacerein was well resolved from peaks of its degradation products.

Three calibration curves were constructed from 162.4 to 243.6 µg/mL. The average correlation coefficient was 0.9991. Intraday and interday coefficients of variation were 2.65 and 1.22%, respectively.

Reference

Giannellini V, Salvatore F, Bartolucci G, et al. A validated HPLC stability-indicating method for the determination of diacerhein in bulk drug substance. *J Pharm Biomed Anal.* 2005; 39: 776–80.

Diacetolol

Chemical Name
N-[3-acetyl-4-[2-hydroxy-3-[(1-methylethyl)amino]propoxy]phenyl]acetamide

Other Names
Acetanilide, Acetylacebutolol

Form	Molecular Formula	MW	CAS
Diacetolol	$C_{16}H_{24}N_2O_4$	308.4	22568-64-5

Method
Schieffer reported an HPLC method with differential pulse polarographic detection for diacetolol in feed. A Waters model ALC/GPC 202 liquid chromatograph was equipped with a model U6K injector, a model 6000 reciprocating pump, a model 440 absorbance detector at 254 nm, and a Princeton Applied Research model 310 DME (dropping mercury electrode) polarographic detector controlled by a model 174A polarographic analyzer. The stationary phase was a Phase Separations microparticulate reversed-phase Spherisorb ODS column (10-µm particle size). The mobile phase consisted of methanol and 1% (wt/vol) ammonium acetate aqueous solution (35:65, vol/vol). The flow rate was 1.5 mL/min. The electrochemical detector was operated in the DME mode with a small drop size and a drop time of 1 sec. The analyzer was operated in the differential pulse mode with the initial potential of −1.275 V and a modulation amplitude of 50 mV at a

scan rate of 2 mV/sec. Potentials were reported compared to an Ag/AgCl reference electrode saturated in potassium chloride.

The feed sample was run through a glass column (2.0 cm ID) containing a small glass wool plug with methanol as the solvent. The eluent solution was evaporated to dryness at 50–60 °C with a stream of nitrogen, reconstituted with 0.01 M phosphoric acid, and filtered. The injection volume was 100 μL. Under these conditions, the retention time for diacetolol was about 6.3 minutes (estimated from the published chromatogram).

The stability-indicating capability of the method was demonstrated by accelerated degradation of diacetolol. Feed samples (0.005%) were stored in an oven at 65 °C for 2 weeks, in 0.1 M hydrochloric acid for 2 months, or in 0.1 M sodium hydroxide solution for 6 days at room temperature. The diacetolol peak was separated from its degradation product peak.

A standard curve for diacetolol was obtained from 1 to 9 μg/mL. The correlation coefficient was 0.9999. The limit of detection was 10 ng.

Reference

Schieffer GW. Reversed-phase high-performance liquid chromatography with differential pulse polarographic detection for assaying drugs in feed: Stability-indicating assay of diacetolol. *J Chromatogr.* 1980; 202: 405–12.

Diacetylmorphine

Chemical Name
(5α,6α)-7,8-Didehydro-4,5-epoxy-17-methylmorphinan-3,6-diol diacetate (ester)

Other Names
Diamorphine, Heroin

Form	Molecular Formula	MW	CAS
Diacetylmorphine	$C_{21}H_{23}NO_5$	369.4	561-27-3
Diacetylmorphine hydrochloride	$C_{21}H_{23}NO_5.HCl.H_2O$	423.9	1502-95-0

Appearance
Diacetylmorphine is an almost white crystalline powder.

Solubility
One gram of diacetylmorphine dissolves in 1.5 mL of chloroform, 31 mL of alcohol, 100 mL of ether, and 1700 mL of water. Diacetylmorphine hydrochloride is freely soluble in water and chloroform; it is soluble in alcohol.

Method 1
Using an HPLC method, Grassby and Roberts investigated the stability of epidural opiate solutions in 0.9% sodium chloride infusion bags. The chromatographic system consisted of a Kratos SF 400 pump, a Philips PU4700 20-µL loop injector, and a Uvikon 740LC UV detector. The stationary phase was a Chrompak Spherisorb ODS-2 column (250 × 4.6 mm, 5-µm particle size). The mobile phase consisted of 0.01 M heptanesulfonic acid (55%) and acetonitrile (45%) containing 7.7 mM dimethyloctylamine and adjusted to pH 3.5 with phosphoric acid. The flow rate was 1 mL/min. UV detection was performed at 254 nm.

Samples were analyzed after dilution. The retention times for diacetylmorphine and bupivacaine were approximately 3.5 and 5.5 minutes, respectively.

The analytical method was demonstrated to be stability indicating by accelerated degradation of diacetylmorphine. A diacetylmorphine sample was heated on a water bath for 2 hours. The degradation product did not interfere with the intact diacetylmorphine peak.

A standard curve for diacetylmorphine was constructed from 0.0875 to 0.1375 mg/mL. The correlation coefficient was 0.989. The coefficient of variation was 2.37%.

Reference
Grassby PF, Roberts DE. Stability of epidural opiate solution in 0.9 percent sodium chloride infusion bags. *Int J Pharm Pract.* 1995; 3: 174–7.

Method 2
Allwood et al. used an HPLC assay to determine the stability of injections containing diacetylmorphine (diamorphine) hydrochloride and midazolam in plastic syringes. The chromatograph consisted of a Kontron model 325 pump, a Kontron model 332 UV detector, a Rheodyne model 7125 fixed-volume loop injector, and a Techsphere ODS column (250 × 4.6 mm, 5-µm particle size). The mobile phase was acetonitrile and 0.1 M ammonium acetate (70:30, vol/vol), with the addition of 24 mmol/L of diethylamine. The mobile phase was delivered isocratically at 1 mL/min. UV detection was performed at 220 nm and 0.2 AUFS.

Samples were diluted with distilled water. The injection volume was 20 µL. The retention times for diacetylmorphine and midazolam were about 5.0 and 7.5 minutes, respectively (estimated from the published chromatogram).

To confirm the stability-indicating capability of the assay, solutions of diacetylmorphine (diamorphine) hydrochloride 25 µg/mL were heated at 50 °C for 2 hours in distilled water, 0.01 M hydrochloric acid, 0.01 M sodium hydroxide, or hydrogen peroxide. Chromatograms of the resulting solutions showed that degradation product peaks did not interfere with the intact diacetylmorphine hydrochloride peak.

The standard curve for diacetylmorphine hydrochloride was constructed from 10 to 40 µg/mL. It was linear and its correlation coefficient was 0.9989. The coefficient of variation was 2.27% ($n = 7$).

Reference
Allwood MC, Brown PW, Lee M. Stability of injections containing diamorphine and midazolam in plastic syringes. *Int J Pharm Pract.* 1994; 3: 57–9.

Method 3

Northcott et al. used HPLC to study the chemical stability of diacetylmorphine (diamorphine) hydrochloride infusion (5 mg/mL) in an ambulatory pump under storage and prolonged in-use conditions. The liquid chromatograph included a Constametric 3000 pump, a SpectroMonitor 3100 variable-wavelength UV-visible detector, a Rheodyne 7125 loop valve fitted with a 20-μL loop, and a CI-10B integrator/printer plotter. The stationary phase was a Kontron S5CN analytical column (250 × 4.9 mm). The mobile phase consisted of 15% acetonitrile and 85% 0.01 M phosphate buffer, adjusted to pH 3.5 with acetic acid. The flow rate was 2.0 mL/min. UV detection was performed at 274 nm.

Samples were diluted 1:50 with water. The injection volume was 20 μL.

The stability-indicating ability of the HPLC method was demonstrated by accelerated degradation. Four sets of diacetylmorphine hydrochloride samples were subjected to 1 M hydrochloric acid, 1 M sodium hydroxide, hydrogen peroxide, or heat. Each stress treatment reduced the intact diacetylmorphine peak height. In all cases, the intact diacetylmorphine peak clearly resolved from the decomposition product peaks.

Calibration curves were constructed from linear plots of the diacetylmorphine hydrochloride peak height versus its concentration from 0.02 to 0.5 mg/mL. The correlation coefficient was 0.999 and the coefficient of variation of the diacetylmorphine hydrochloride peak height was 0.88% ($n = 8$).

A similar method was used by Poochikian et al.

References

Northcott M, Allsopp MA, Powell H, et al. The stability of carboplatin, diacetylmorphine, 5-fluorouracil and mitozantrone infusions in an ambulatory pump under storage and prolonged 'in use' conditions. *J Clin Pharm Ther.* 1991; 16: 123–9.

Poochikian GK, Cradock JC, Davignon JP. Heroin: Stability and formulation approaches. *Int J Pharm.* 1983; 13: 219–26.

Method 4

Kleinberg et al. evaluated the stability of diacetylmorphine hydrochloride in infusion devices and containers for intravenous administration. The HPLC system consisted of a Waters model M6000A pump, a Waters model 710B WISP autosampler, a Waters model 480 UV detector, and a Nelson model 4430 chromatography data system. The stationary phase was a Waters μBondapak C_{18} column (250 × 3.9 mm, 5-μm particle size) preceded by an Upchurch Scientific guard column (45 × 10 mm). The mobile phase consisted of 40% acetonitrile and 60% 0.06 M ammonium acetate aqueous buffer. The flow rate was 1.5 mL/min. UV detection was performed at 280 nm. Butyrophenone in methanol was used as the internal standard.

Samples were diluted with internal standard solution. The injection volume was 10 μL. Retention times for diacetylmorphine and the internal standard were 8.5 and 12.5 minutes, respectively.

The analytical method was determined to be stability indicating by accelerated degradation of diacetylmorphine by heat, acid, base, or oxidation (0.0073 mM hydrogen peroxide). In all cases, degradation product peaks were separated from the intact diacetylmorphine peak. The stability-indicating capability of the method was further demonstrated by spiking the diacetylmorphine hydrochloride solution with authentic standards of known decomposition products (6-acetylmorphine and morphine). Retention times for morphine,

6-acetylmorphine, and diacetylmorphine hydrochloride were 3.0, 4.8, and 8.5 minutes, respectively.

Standard curves were constructed for calibration and were linear ($r > 0.999$) from 0.16 to 2.99 mg/mL.

Reference

Kleinberg KL, Duafala ME, Nacov C, et al. Stability of heroin hydrochloride in infusion devices and containers for intravenous administration. *Am J Hosp Pharm*. 1990; 47: 377–81.

Diaminopyridine

Chemical Name
3,4-Pyridinediamine

Form	Molecular Formula	MW	CAS
3,4-Diaminopyridine	$C_5H_7N_3$	109.1	54-96-6

Method

Trissel et al. investigated by HPLC analysis the stability of 3,4-diaminopyridine oral capsules. The Waters LC Module 1 Plus liquid chromatograph consisted of a multi-solvent-delivery system, an autosampler, and a UV light detector. A Waters Alliance 2960 liquid chromatograph equipped with a Waters model 996 photodiode-array detector was also used to confirm the drug peak purity. The stationary phase was a Phenomenex Prodigy C_{18} analytical column (250 × 4.6 mm, 5-μm particle size). The mobile phase was a mixture of 350 mL of acetonitrile and 1650 mL of aqueous buffer consisting of 50 mM monobasic sodium phosphate and 4 mM 1-octanesulfonic acid. The flow rate was 0.8 mL/min. UV detection was performed at 290 nm and 0.5 AUFS.

The powder content of each capsule was transferred into a glass test tube, mixed with 10 mL of water, vortexed for 30 seconds, filtered through a 0.45-μm filter, and diluted 20-fold with water. The injection volume was 10 μL. Under these conditions, the retention time of 3,4-diaminopyridine was 6.5 minutes.

The stability-indicating nature of the method was demonstrated by accelerated degradation. The sample solutions were mixed with 0.1 N sodium hydroxide solution, 0.1 N hydrochloric acid, or 3% hydrogen peroxide and heated. No degradation product peaks interfered with the peak of the intact 3,4-diaminopyridine. Photodiode-array purity analysis of the peak confirmed the peak purity of 3,4-diaminopyridine throughout the study.

A standard curve for 3,4-diaminopyridine was constructed from 10 to 40 μg/mL. The correlation coefficient was greater than 0.9998. Intraday and interday coefficients of variation were 1.4 and 1.2%, respectively.

Reference

Trissel LA, Zhang Y, Xu QA. Stability of 4-aminopyridine and 3,4-diaminopyridine oral capsules. *Int J Pharm Compound.* 2002; 6: 155–7.

Diazepam

Chemical Name
7-Chloro-1,3-dihydro-1-methyl-5-phenyl-2*H*-1,4-benzodiazepin-2-one

Other Name
Valium

Form
Diazepam

Molecular Formula
$C_{16}H_{13}ClN_2O$

MW
284.7

CAS
439-14-5

Appearance
Diazepam is an off-white to yellow practically odorless crystalline powder.

Solubility
Diazepam is slightly soluble in water but soluble in chloroform, dimethylformamide, benzene, acetone, and alcohol.

pK$_a$
Diazepam has a pK$_a$ of 3.4.

Method 1
Mannucci et al. described a method for the simultaneous determination of otilonium bromide, diazepam, and related compounds in tablets. A Hewlett-Packard 1090L

system was equipped with an autosampler and a model 1040M diode-array detector. The stationary phase was a Merck octyl derivative silica column (250 × 4.0 mm, 5-μm particle size) with a precolumn (4 × 4 mm) of the same packing. The column temperature was maintained at 50 °C. The mobile phase consisted of 0.5 M sodium acetate trihydrate buffer containing 5 mM 1-heptanesulfonic acid monohydrate sodium salt and methanol (30:70, vol/vol), adjusted to pH 6.0 with glacial acetic acid. The flow rate was 1 mL/min. n-Butyl p-hydroxybenzoate 1 mg/mL in acetonitrile was used as an internal standard. Diazepam, n-butyl p-hydroxybenzoate, and otilonium bromide were monitored at 230, 254, and 290 nm, respectively.

A tablet was accurately weighed and placed in a 50-mL volumetric flask, dissolved with 5 mL of water with stirring, mixed with 5 mL of internal standard, diluted to volume with acetonitrile, and centrifuged at 3500 rpm (2472 × g) for 5 minutes. The injection volume of the supernatant was 10 μL. The total run time was 15 minutes. Under these conditions, retention times for internal standard, diazepam, and otilonium bromide were 3.98, 4.72, and 10.48 minutes, respectively.

The method was verified to be stability indicating by assaying related compounds in solution. The intact drug peaks were well separated from the peaks of their related compounds. Furthermore, the excipients in the formulation did not interfere with the analysis of drugs.

A standard curve for diazepam was obtained from 0.010 to 0.080 mg/mL. The correlation coefficient was 1.0.

Reference

Mannucci C, Bertini J, Cocchini A, et al. High-performance liquid chromatographic method for assay of otilonium bromide, diazepam, and related compounds in finished pharmaceutical forms. *J Pharm Sci.* 1993; 82: 367–70.

Method 2

Walker et al. evaluated the chemical stability of diazepam using an HPLC method. The chromatographic system included a Spectra-Physics SP4200 ternary gradient solvent-delivery pump, a Schoeffel SF 770 variable-wavelength UV detector, and a Beckman Ultrasphere ODS C_{18} reversed-phase column (250 × 4.2 mm, 5-μm particle size). The mobile phase consisted of acetonitrile and 0.05 M phosphoric acid, which also contained 1 mg/mL heptanesulfonic acid as a counterion. It was gradiently delivered from 14 to 39% acetonitrile in 16 minutes at 2 mL/min. The UV detection wavelength was set at 230 nm.

The method was determined to be stability indicating by accelerated decomposition of diazepam. A solution of diazepam 1 mg/mL in water was incubated at 90 °C for 22 hours and then assayed. Degradation product peaks did not interfere with the intact diazepam peak.

The intraday coefficient of variation for the assay was 0.99%.

Reference

Walker SE, DeAngelis C, Iazzetta J. Stability and compatibility of combinations of hydromorphone and a second drug. *Can J Hosp Pharm.* 1991; 44: 289–95.

Method 3

Strom and Miller determined the compatibility, chemical stability, and interactions of cephalexin, cimetidine, diazepam, and propranolol with three selected enteral nutrient products. A Micromeritics liquid chromatograph included a solvent-delivery system, a variable-wavelength UV-visible detector, and a variable-volume injector. Chromatograms were recorded and peak heights calculated with a Micromeritics microprocessor. The stationary phase was a cyano analytical column. The mobile phase was 70% 0.020 M sodium phosphate and 30% acetonitrile and was delivered isocratically at 1.0 mL/min. UV detection was performed at 230 nm. Cimetidine 65 µg/mL was used as the internal standard.

Each sample was diluted with methanol and centrifuged for 2 minutes. The supernatants were filtered through Gelman Alpha-450 0.45-µm filters. The injection volume was 25 µL.

The analytical method was demonstrated to be stability indicating. Diazepam sample solutions were prepared in 1 N hydrochloric acid or 1 N sodium hydroxide and were heated at 80 °C for 24 hours. The intact diazepam was well separated from its degradation products and from other components in the enteral nutrient formulas.

The intraday coefficient of variation was less than 2.8%.

References
Strom JG, Miller SW. Stability of drugs with enteral nutrient formulas. *Ann Pharmacother.* 1990; 24: 130–4.

Strom JG, Kalu AU. Formulation and stability of diazepam suspension compounded from tablets. *Am J Hosp Pharm.* 1986; 43: 1489–91.

Diaziridinyl Benzoquinone

Chemical Name
2,5-Bis(1-aziridinyl)-3,6-dimethyl-2,5-cyclohexadiene-1,4-dione

Form	Molecular Formula	MW	CAS
Diaziridinyl benzoquinone	$C_{12}H_{14}N_2O_2$	218.3	18735-47-2

Method
Cheung et al. investigated by HPLC analysis the stability of diaziridinyl benzoquinone. The liquid chromatograph included a Waters model 600 pump, a Thermoseparation model AS3000 autosampler, and a Hewlett-Packard model 1050 photodiode-array detector. The stationary phase was a Phenomenex Inertsil C_8 stainless steel column (250 × 4.6 mm, 5-µm particle size). The mobile phase was a mixture of 20 mM ammonium acetate (pH 4.0) and methanol (70:30, vol/vol) and was delivered at a flow rate of 1 mL/min. UV detection was performed at 330 nm. *p*-Nitroaniline 0.13 mg/mL in water was used as an internal standard. The injection volume was 20 µL. Under these conditions, the retention times for diaziridinyl benzoquinone and *p*-nitroaniline were about 11 and 17 minutes, respectively (estimated from the published chromatogram).

The method was shown to be stability indicating by degradation study of diaziridinyl benzoquinone. Diaziridinyl benzoquinone instantaneously decomposed in 0.1 N hydrochloric acid and in 0.1 N sodium hydroxide solution. The degradation products did not interfere with the separation of diaziridinyl benzoquinone.

A standard curve for diaziridinyl benzoquinone was constructed from 0.0583 to 0.2473 mg/mL. The correlation coefficient was 0.9998. Intraday and interday coefficients of variation were both 0.8%.

Reference
Cheung AP, Struble E, Nguyen N, et al. Stability-indicating HPLC assay and solution stability of a new diaziridinyl benzoquinone. *J Pharm Biomed Anal*. 2001; 24: 957–66.

Diclofenac

Chemical Name
[2-(2,6-Dichloroanilino)phenyl]acetic acid

Other Names
Cataflam, Voltaren

Form	Molecular Formula	MW	CAS
Diclofenac	$C_{14}H_{11}Cl_2NO_2$	296.1	15307-86-5
Diclofenac diethylamine	$C_{18}H_{22}Cl_2N_2O_2$	369.3	78213-16-8
Diclofenac potassium	$C_{14}H_{10}Cl_2KNO_2$	334.2	15307-81-0
Diclofenac sodium	$C_{14}H_{10}Cl_2NNaO_2$	318.1	15307-79-6

Appearance
Diclofenac potassium and diclofenac sodium occur as faintly yellowish white to light beige, practically odorless and slightly hygroscopic crystalline powders.

Solubility
Diclofenac sodium is sparingly soluble in water, soluble in alcohol, slightly soluble in acetone, practically insoluble in ether, and freely soluble in methyl alcohol.

pK$_a$
Diclofenac potassium and diclofenac sodium have a pK$_a$ value of approximately 4 at 25 °C.

Method 1
Kubala et al. described an HPLC method for the analysis of diclofenac sodium in raw materials and pharmaceutical solid dosage forms. The chromatographic instrument included a Waters model 6000A pump, a model U6K injector, a model 441 variable-wavelength detector, and a Spectra-Physics model SP4290 electronic integrator. The stationary phase was a Beckman Ultrasphere IP ODS C$_{18}$ column (150 × 4.6 mm, 5-μm

particle size). The mobile phase was a mixture of acetonitrile, methanol, and Waters Pic B-6 (25:25:50, vol/vol/vol). The flow rate was 2.0 mL/min for 28 minutes and then was stepped up to 3.0 mL/min for the remainder of the run. Diflorasone diacetate (0.587 mg/mL) in a diluent of methanol and acetonitrile (1:1, vol/vol) was used as an internal standard. UV detection was performed at 229 nm.

For tablets, a sample of ground tablets equivalent to 50 mg of diclofenac sodium was accurately weighed and added to the diluent, sonicated, diluted to a final volume of 100 mL with the diluent, and filtered. Three milliliters of the filtrate was mixed with 3 mL of the internal standard and diluted to a final volume of 25 mL with the diluent. For the raw material, 50 mg of the material was accurately weighed, dissolved in 100 mL of the diluent, and diluted to 160 µg/mL of diclofenac sodium with the diluent. The sample injection volume was 10 µL. Under these conditions, the retention time for diclofenac sodium was about 18 minutes.

The analytical method was evaluated to be stability indicating. Tablets and raw material samples were stored in an oven at 40 °C and 50% humidity over 28 days, and at 90 °C and 55% humidity over 20 days. The degradation products did not interfere with the determination of diclofenac sodium.

A calibration curve for diclofenac sodium was obtained from 0.070 to 0.140 mg/mL.

Reference
Kubala T, Gambhir B, Borst SI. A specific stability indicating HPLC method to determine diclofenac sodium in raw materials and pharmaceutical solid dosage forms. *Drug Dev Ind Pharm.* 1993; 19: 749–57.

Method 2
Wang and Yeh reported a simple assay for the determination of diclofenac sodium in pharmaceutical dosage forms. A Waters model 440 system was equipped with a model 745 data module and a 254-nm UV detector. The stationary phase was a Waters µBondapak C_{18} column (150 × 4.6 mm). The mobile phase consisted of methanol and 0.1% acetic acid (80:130, vol/vol). The flow rate was 2.0 mL/min. UV detection was performed at 254 nm. Methyltestosterone 1 mg/mL in water was used as an internal standard.

For tablets, an amount of tablet powder from 20 tablets equivalent to 17.5 mg of diclofenac sodium was accurately weighed, extracted with 20 mL of mobile phase, mixed with 1 mL of internal standard, diluted to 25 mL with mobile phase, and centrifuged. The supernatant was assayed. For injection, an aliquot equivalent to 17.5 mg of diclofenac sodium was mixed with 1 mL of internal standard, diluted to 25 mL with mobile phase, and assayed. The injection volume was 20 µL. Under these conditions, retention times of methyltestosterone and diclofenac sodium were 5.25 and 10.25 minutes, respectively.

A stock solution of diclofenac sodium was incubated at 90 °C for 5 days. Degradation products did not interfere with the determination of the intact drug.

A calibration curve for diclofenac sodium was constructed from 0.1 to 0.9 mg/mL. Its correlation coefficient was 0.9996.

Reference
Wang D-P, Yeh M-K. High-performance liquid chromatographic assay of diclofenac sodium in pharmaceutical dosage forms. *Chin Pharm J.* 1990; 42: 485–91.

Dicloxacillin

Chemical Name
(6R)-6-[3-(2,6-Dichlorophenyl)-5-methylisoxazole-4-carboxamido]penicillanic acid

Other Names
Diclocil, Dycill, Dynapen

Form	Molecular Formula	MW	CAS
Dicloxacillin	$C_{19}H_{17}Cl_2N_3O_5S$	470.3	3116-76-5
Dicloxacillin sodium	$C_{19}H_{16}Cl_2N_3NaO_5S.H_2O$	510.3	13412-64-1

Appearance
Dicloxacillin sodium occurs as a white or almost white hygroscopic crystalline powder.

Solubility
Dicloxacillin sodium is freely soluble in water and soluble in alcohol and methanol.

pK$_a$
Dicloxacillin sodium has a pK$_a$ of 2.7–2.8.

Method 1
Grover et al. described an HPLC method for the analysis of dicloxacillin in the presence of its degradation products. The Shimadzu model LC-10A liquid chromatograph included model LC-10AS pumps, a model SPD-10A dual-wavelength UV-visible detector, a model C-R7A data processor, and a Rheodyne model 7125 injector with a 20-µL loop. The stationary phase was a Phenomenex Resolve C_{18} column (300 × 3.9 mm, 5-µm particle size). The mobile phase was 21% acetonitrile in 20 mM monobasic potassium phosphate and 10 mM tetramethylammonium chloride in water (pH 5.0). The flow rate was 1 mL/min. UV detection was performed at 220 and 240 nm. Under these conditions, the retention time for dicloxacillin was 31.8 minutes.

The stability-indicating nature of the assay was demonstrated by degrading the drug at pH 2.0. The intact dicloxacillin peak was separated from its degradation product peaks.

A standard curve for dicloxacillin was obtained from 5 to 150 µg/mL. The correlation coefficient was greater than 0.99.

Reference
Grover M, Gulati M, Singh S. Stability-indicating analysis of isoxazolyl penicillins using dual wavelength high-performance liquid chromatography. *J Chromatogr B*. 1998; 708: 153–9.

Method 2
Abdel-Moety et al. developed an HPLC assay for the determination of dicloxacillin and ampicillin in bulk forms, admixtures, and capsules. The Shimadzu LC-10AD liquid chromatograph consisted of a model SPD-10A tunable UV detector, a model CTO-10A column oven controller, a model DGU-3A mechanical degasser, a model C-R4A data

unit, and a Rheodyne 20-µL injector. The stationary phase was a Merck reversed-phase LiChrosorb RP18 column (250 × 4 mm, 10-µm particle size). The mobile phase consisted of acetonitrile and 1% aqueous acetic acid (75:25, vol/vol) and was delivered at 1.5 mL/min. UV detection was performed at 240 nm. The injection volume was 20 µL. Under these conditions, retention times for dicloxacillin and ampicillin were 1.15 and 2.76 minutes, respectively.

The assay was reported to be stability indicating since it simultaneously determined dicloxacillin and ampicillin in admixtures in the presence of their degradation products.

A standard curve for dicloxacillin was constructed from 50 to 300 µg/mL. The correlation coefficient was 0.9998.

Reference
Abdel-Moety EM, Al-Rashood KA, Al-Deeb OA, et al. Stability-indicating HPLC method for determination of ampicillin and dicloxacillin in bulk forms, admixtures and in capsules. *Sci Pharm.* 1995; 63: 7–15.

Didanosine

Chemical Name
2′,3′-Dideoxyinosine

Other Name
Videx

Form	Molecular Formula	MW	CAS
Didanosine	$C_{10}H_{12}N_4O_3$	236.2	69655-05-6

Appearance
Didanosine occurs as a white crystalline powder.

Solubility
Didanosine has a solubility of approximately 27.3 mg/L in water at pH 6 at 25 °C.

pKa
Didanosine has a pK$_a$ of 9.13.

Method

Bekers et al. evaluated by HPLC the chemical stability of didanosine in aqueous solution. The instrument consisted of a Waters model 510 pump, a Kratos model 757 Spectroflow variable-wavelength detector, a Spectra-Physics model 8880 automatic sample injector, and a Spectra-Physics model 4270 integrator. The stationary phase was a Hypersil ODS analytical column. The mobile phase consisted of 5 mM phosphate buffer (pH 6.0) and methanol (90:10, vol/vol). The flow rate was 1.0 mL/min. UV detection was performed at 254 nm.

Samples were assayed without dilution. The injection volume was 10 μL. Under these conditions, the retention time of didanosine was about 7 minutes (estimated from the published chromatogram).

The method was reported to be stability indicating.

References

Bekers O, Beijnen JH, Tank MJTK, et al. 2′,3′-Dideoxyinosine (ddI): Its chemical stability and cyclodextrin complexation in aqueous media. *J Pharm Biomed Anal*. 1993; 11: 489–93.

Beijnen JH, Meenhorst PL, Rosing H, et al. Analysis of 2′,3′-dideoxyinosine (ddI) in plasma by isocratic high performance liquid chromatography with ultraviolet detection. *J Drug Dev*. 1990; 3: 127–33.

Diethylamine Salicylate

Chemical Name

Diethylamine salicylate

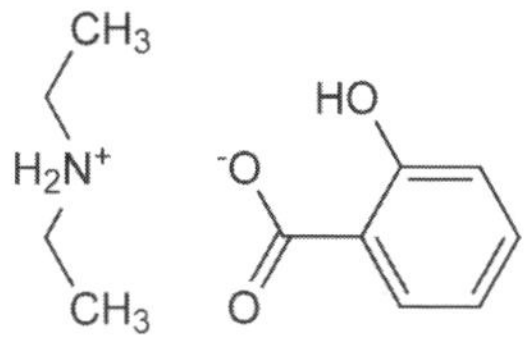

Other Name

Algesal

Form	Molecular Formula	MW	CAS
Diethylamine salicylate	$C_{11}H_{17}NO_3$	211.3	4419-92-5

Appearance

Diethylamine salicylate occurs as white or almost white, odorless or almost odorless crystals.

Solubility

Diethylamine salicylate is very soluble in water and freely soluble in alcohol and chloroform.

Method

Abounassif et al. described an HPLC analytical method for the simultaneous determination of diethylamine salicylate and methyl nicotinate in the presence of parabens. A Varian model 5000 liquid chromatograph was equipped with a model UV50 variable-wavelength detector, a Rheodyne model 7125 injector with a 20-µL loop, a Varian model 9176 recorder, and a model DSC 111L data system controller. The stationary phase was a Waters µBondapak C_{18} stainless steel column (300 × 4 mm, 10-µm particle size). Solvent A was acetonitrile and solvent B was 1% aqueous acetic acid. The mobile phase was delivered in 15% A and 85% B at 1.70 mL/min for 12 minutes and stepped to 75% A and 25% B at 2.5 mL/min for another 6 minutes. UV detection was performed at 254 nm. Acetaminophen (paracetamol) 8 µg/mL in water was used as an internal standard.

A sample of the ointment (150–250 mg) was accurately weighed and transferred into a 50-mL flask followed by 40 mL of mobile phase and 4 mL of the internal standard, shaken for 5 minutes, brought to volume using the mobile phase, and filtered. The injection volume was 20 µL. Under these conditions, retention times for acetaminophen, methyl nicotinate, and diethylamine salicylate were about 2.5, 4.2, and 8.6 minutes, respectively (estimated from the published chromatogram).

The analytical method was reported to be stability indicating since the intact drug peaks were resolved from their degradation product peaks.

A calibration curve for diethylamine salicylate was constructed from 100 to 500 µg/mL. The correlation coefficient was 0.9995.

Reference

Abounassif MA, Abdel-Moety EM, Gad-Kariem RA. HPLC-quantification of diethylamine salicylate and methyl nicotinate in ointments. *J Liq Chromatogr*. 1992; 15: 625–36.

Diethylpropion Hydrochloride

Chemical Name

2-Diethylaminopropiophenone hydrochloride

Other Names

Anorex, Tenuate, Tepanil, Tylinal

Form	Molecular Formula	MW	CAS
Diethylpropion	$C_{13}H_{19}NO$	205.3	90-84-6
Diethylpropion hydrochloride	$C_{13}H_{19}NO.HCl$	241.8	134-80-5

Appearance

Diethylpropion hydrochloride is a white to off-white fine crystalline powder.

Solubility

Diethylpropion hydrochloride is very soluble in water and freely soluble in alcohol and chloroform but is practically insoluble in ether.

Method

Using HPLC, Mey et al. studied the kinetics of racemization of (+)- and (−)-diethylpropi- on in various conditions. The instrument was a Hewlett-Packard 1090 chromatograph equipped with a UV detector. The stationary phase was a Daicel Chemical Chiralcel OD column (250 × 4.6 mm, 10-μm particle size). The column temperature was 5 °C. The mobile phase consisted of *n*-heptane, 2-propanol, and diethylamine (98:2:0.1, vol/vol/vol) and was delivered isocratically at 0.9 mL/min. UV detection was performed at 254 nm.

Aliquots of 1 mL of samples were extracted with *n*-heptane. The injection volume was 5 μL.

The HPLC method was reported to be stability indicating.

Reference

Mey B, Paulus H, Lamparter E, et al. Kinetics of racemization of (+)- and (−)-diethyl- propion: Studies in aqueous solution, with and without the addition of cyclodextrins, in organic solvents and in human plasma. *Chirality*. 1998; 10: 307–15.

Digoxin

Chemical Name

(3β,5β,12β)-3-[(*O*-2,6-Dideoxy-β-D-*ribo*-hexopyranosyl-(1→4)-*O*-2,6-dideoxy-β-D-*ribo*-hexopyranosyl-(1→4)-2,6-dideoxy-β-D-*ribo*-hexopyranosyl)oxy]-12,14-dihydroxy-card-20(22)-enolide

Other Name

Lanoxin

Form	Molecular Formula	MW	CAS
Digoxin	$C_{41}H_{64}O_{14}$	780.9	20830-75-5

Appearance

Digoxin occurs as clear to white crystals or as a white crystalline powder.

Solubility

Digoxin is soluble in diluted alcohol, pyridine, or a mixture of chloroform and alcohol, but it is almost insoluble in ether, acetone, ethyl acetate, chloroform, and water.

Method

Using an HPLC method, Riley and Junkin reported the stability of digoxin in intravenous admixtures. The chromatographic system consisted of a Kratos Spectroflow 400 solvent-delivery system, a Waters model 450 variable-wavelength UV detector, a Micromeritics model 728 autosampler, a Valco VICI six-port electronically actuated injection valve fitted with a 20-μL loop, a Fisher Scientific series 5000 Recordall strip-chart recorder, and a Shimadzu model C-R3A electronic integrator. The stationary phase was an Alltech Partisil ODS III analytical column (250 × 4.6 mm, 10-μm particle size). The mobile phase was a mixture of acetonitrile, water, and phosphoric acid (35:65:0.1, vol/vol/vol). The flow rate was 2 mL/min. UV detection was performed at 220 nm.

Samples were diluted 1:10 with the mobile phase.

The HPLC assay was confirmed to be stability indicating and free from interference by formulatory excipients and degradation products.

A standard curve was constructed for digoxin and was linear from 12.5 to 187.5 μg/mL. The coefficient of variation was less than 1%.

Reference

Riley CM, Junkin P. Stability of amrinone, digoxin, procainamide hydrochloride, propranolol hydrochloride, sodium bicarbonate, potassium chloride, or verapamil hydrochloride in intravenous admixtures. *Am J Hosp Pharm*. 1991; 48: 1245–52.

Diloxanide Furoate

Chemical Name

4-(*N*-Methyl-2,2-dichloroacetamido)phenyl 2-furoate

Other Names

Furamid, Furamide

Form	Molecular Formula	MW	CAS
Diloxanide	$C_9H_9Cl_2NO_2$	234.1	579-38-4
Diloxanide furoate	$C_{14}H_{11}Cl_2NO_4$	328.2	3736-81-0

Appearance

Diloxanide furoate occurs as a white or almost white and odorless or almost odorless crystalline powder.

Solubility

Diloxanide furoate is very slightly soluble in water and slightly soluble in alcohol. It is freely soluble in chloroform.

Method 1

Mishal and Sober reported an HPLC method for the simultaneous determination of diloxanide furoate and metronidazole benzoate in suspension form. The liquid chromatographic system consisted of a Shimadzu model SIL-10Advp pump and a model SPD-M10Avp photodiode-array detector. The stationary phase was a Supelco LC-18 DB column (150 × 4.6 mm, 5-μm particle size). The mobile phase consisted of 0.005 M monobasic potassium phosphate and acetonitrile (70:30, vol/vol); the pH was adjusted to 2.5 with phosphoric acid. The flow rate was 2.0 mL/min. UV detection was performed at 260 nm.

A suspension equivalent to 31.25 mg of diloxanide furoate and 40 mg of metronidazole was transferred into a 50-mL volumetric flask, mixed with 25 mL of acetonitrile–deionized water (1:1, vol/vol), sonicated for 10 minutes, and diluted to volume with the same solvent mixture. The injection volume was 20 μL. Under these conditions, retention times of methylparaben, metronidazole benzoate, propylparaben, and diloxanide furoate were about 3.0, 8.4, 10.2, and 23.0 minutes, respectively (estimated from the published chromatogram).

To confirm the stability-indicating nature of the analytical method, samples of placebo, active ingredients, and the prepared suspension were heated, treated with 1 N hydrochloric acid, hydrolyzed with 1 N sodium hydroxide solution, or oxidized with 10% hydrogen peroxide at 85 °C for 2 hours. All degradation products were well resolved from the intact drug on its chromatogram.

A calibration curve for diloxanide furoate was constructed from 0.158 to 0.947 mg/mL with a correlation coefficient of 0.9999.

Reference

Mishal A, Sober D. Stability indicating reversed-phase liquid chromatographic determination of metronidazole benzoate and diloxanide furoate as bulk drug and in suspension dosage form. *J Pharm Biomed Anal.* 2005; 39: 819–23.

Method 2

Hasan et al. established a stability-indicating HPLC method for the determination of diloxanide furoate in the presence of its degradation products. A Shimadzu class LC-10 liquid chromatographic system equipped with a Shimadzu SPD-10A photodiode-array detector was used. The stationary phase was a Zorbax C_{18} column (150 × 4.6 mm, 5-μm particle size). The mobile phase consisted of methanol and water (80:20, vol/vol). The flow rate was 1 mL/min. UV detection was performed at 258 nm.

A portion of 10 ground tablets, equivalent to 100 mg of diloxanide furoate, was accurately weighed, transferred into a 250-mL beaker with 70 mL of absolute ethanol and stirred for about 30 minutes, filtered through a filter paper into a 100-mL volumetric flask, and then filled to the mark with ethanol. The sample was appropriately diluted before assay. The injection volume was 20 μL. Under these conditions, the retention time for diloxanide was about 4 minutes.

In order to degrade diloxanide furoate, 0.3 g of the pure compound in 20 mL of 1 N sodium hydroxide solution was refluxed for 3 hours. The intact drug was well resolved from its degradation products on the chromatogram, confirming the stability-indicating capability of the analytical method.

A standard curve for diloxanide furoate in mobile phase was constructed from 2 to 50 µg/mL. The correlation coefficient was 0.9999. The limit of detection and the limit of quantitation of the assay were 0.65 and 1.97 µg/mL, respectively.

Reference

Hasan NY, Elkawy MA, Elzeany BE, et al. Stability-indicating methods for the determination of diloxanide furoate. *J Pharm Biomed Anal.* 2002; 28: 187–97.

Method 3

Gadkariem et al. reported the development of an HPLC assay for diloxanide furoate in the presence of its photodegradation products. A Waters liquid chromatograph consisted of a Waters model 600E system controller, a model 486 tunable absorbance detector, a model 746 data module, and a Rheodyne 7161 injector. The stationary phase was a LiChrospher 100 RP-18 column (150 × 4.6 mm, 5-µm particle size). The mobile phase consisted of water and acetonitrile (30:70, vol/vol) and was delivered isocratically at 1 mL/min. UV detection was performed at 258 nm. Ammidin 300 µg/mL in methanol was used as an internal standard.

A portion of powdered tablets equivalent to 20 mg of diloxanide furoate was transferred into a 100-mL volumetric flask, shaken for 20 minutes with about 80 mL of mobile phase filled to volume with mobile phase, then diluted by a factor of 10 with mobile phase and mixed with the internal standard solution. The injection volume was 20 µL. Under these conditions, the retention time of diloxanide furoate was about 3 minutes.

The method was demonstrated to be stability indicating by assaying a UV-degraded diloxanide furoate solution. The intact drug peak was separated from its degradation product peaks.

A standard curve for diloxanide furoate was constructed from 0.05 to 200 µg/mL. The correlation coefficient was 0.9998.

A similar method was reported in another study by the same researchers.

References

Gadkariem EA, Belal F, Abounassif MA. Photodegradation kinetic studies and stability-indicating assay of diloxanide furoate in dosage forms using high-performance liquid chromatography. *J Liq Chromatogr Rel Technol.* 2002; 25: 2947–64.

Gadkariem EA, El-Obeid H, Belal F, et al. Chemical and photochemical stability studies on diloxanide furoate in carbohydrates and polyols solutions. *J Food Drug Anal.* 2004; 12: 299–305.

Diltiazem

Chemical Name

(2S-cis)-3-(Acetyloxy)-5-[2-(dimethylamino)ethyl]-2,3-dihydro-2-(4-methoxyphenyl)-1,5-benzothiazepin-4(5H)-one

Other Names
Cardizem, Latiazem

Form

Form	Molecular Formula	MW	CAS
Diltiazem	$C_{22}H_{26}N_2O_4S$	414.5	42399-41-7
Diltiazem hydrochloride	$C_{22}H_{26}N_2O_4S \cdot HCl$	451.0	33286-22-5

Appearance
Diltiazem hydrochloride is a white to off-white crystalline powder.

Solubility
Diltiazem hydrochloride is freely soluble in water, methanol, and chloroform but slightly soluble in absolute ethanol. It is insoluble in benzene.

Method 1
Allen and Erickson used HPLC to study the stability of diltiazem hydrochloride 12 mg/mL in extemporaneously compounded oral liquids. A Hewlett-Packard series 1050 automated high-performance liquid chromatograph consisted of a multisolvent mixing and pumping system, an autoinjector, a diode-array detector, and a computer with Chem Station data-handling software. The HPLC analytical column was a Bakerbond C_{18} column (250 × 4.6 mm, 5-μm particle size). The mobile phase was acetonitrile, methanol, and buffer (1:1:2, vol/vol/vol); the buffer consisted of 1.16 g of d-10-camphor-sulfonic acid in 1000 mL of 0.1 M sodium acetate with the pH adjusted to 6.2 with sodium hydroxide. The mobile phase was delivered isocratically at 1.5 mL/min. UV detection was performed at 240 nm.

Samples were diluted 1:100. Under these conditions, diltiazem hydrochloride had a retention time of 9.6 minutes.

The stability-indicating capability of the assay was demonstrated by accelerated degradation of diltiazem hydrochloride. A composite chromatogram of diltiazem hydrochloride after intentional degradation by heat, acid, base, oxidizing agent, and light showed that the peak of the degradation product did not interfere with the intact diltiazem hydrochloride peak.

Standard curves for diltiazem hydrochloride were constructed from 10 to 150 μg/mL. The intraday and interday coefficients of variation were 1.3 and 1.9%, respectively.

Reference
Allen LV, Erickson MA. Stability of baclofen, captopril, diltiazem hydrochloride, dipyridamole, and flecainide acetate in extemporaneously compounded oral liquids. *Am J Health Syst Pharm*. 1996; 53: 2179–84.

Method 2

Shivram et al. described an HPLC method for the analysis of diltiazem hydrochloride in tablets. The Shimadzu model LC-6A system included a model SPD-6AV variable-wavelength detector. The stationary phase was a Rexchrome ODS column (250 × 4.6 mm). The mobile phase consisted of acetonitrile, methanol, and 0.05 M monobasic potassium phosphate (25:20:55, vol/vol/vol). The flow rate was 2 mL/min. UV detection was performed at 240 nm. Cyproheptadine hydrochloride 0.5 mg/mL in methanol was used as an internal standard.

Twenty tablets were weighed and ground. A portion of this powder equivalent to 50 mg of the drug was dissolved in 100 mL of methanol, filtered, diluted 1:1 with internal standard, and then diluted fivefold with water. The injection volume was 50 μL. Under these conditions, the retention times of diltiazem and cyproheptadine hydrochloride were about 5.5 and 7.1 minutes, respectively (estimated from the published chromatogram).

The method was evaluated to be stability indicating by accelerated degradation. Diltiazem solutions were mixed with 0.1 N sodium hydroxide, 0.1 N hydrochloric acid, or 3% hydrogen peroxide and heated in a boiling water bath for 30 minutes. Diltiazem was resolved from its degradation products.

Calibration curves for diltiazem were obtained with correlation coefficients of 0.9996. The limit of quantitation was 10 μg/mL.

Reference

Shivram K, Shah AC, Newalkar BL, et al. Stability indicating high performance liquid chromatographic method for the assay of diltiazem hydrochloride in tablets. *J Liq Chromatogr.* 1992; 15: 2417–22.

Method 3

Abdel-Hamid et al. presented the HPLC determination of diltiazem hydrochloride. The Varian model 2010 liquid chromatograph included a model 2050 variable-wavelength detector, a model 7125 injector with a 50-μL loop, and a Lloyd Instrument model JJ CR452 recorder. The stationary phase was a Micropak MCH-5 reversed-phase column (150 × 4.6 mm). The mobile phase consisted of acetonitrile and water (48:52, vol/vol) and was adjusted to pH 3.3 with phosphoric acid. The flow rate was 1 mL/min. UV detection was performed at 239 nm. Clonazepam was used as an internal standard.

Samples were diluted with acetonitrile. The injection volume was 40 μL. Under these conditions, the retention times of clonazepam and diltiazem were 3 and 9 minutes, respectively.

The method was reported to be stability indicating. A diltiazem solution was stored at pH 2 and 40 °C under UV light for 96 hours. Degradation products did not interfere with the peak of the intact diltiazem.

A standard curve for diltaizem was constructed from 1 to 10 μg/mL. The correlation coefficient was 0.9990. Intrarun and interrun coefficients of variation were less than 3.09 and 2.45%, respectively. The limit of detection was 0.2 μg/mL.

Reference

Abdel-Hamid ME, Suleiman MS, Najib NM, et al. Stability-indicating high performance liquid chromatographic determination of diltiazem hydrochloride. *Anal Lett.* 1988; 21: 2263–75.

Dimenhydrinate

Chemical Names

8-Chloro-3,7-dihydro-1,3-dimethyl-1*H*-purine-2,6-dione compounded with 2-(diphenyl methoxy)-*N*,*N*-dimethylethanamine (1:1)

2-(Benzhydryloxy)-*N*,*N*-dimethylethylamine 8-chlorotheophyllinate

Other Name

Dramamine

Form

Form	Molecular Formula	MW	CAS
Dimenhydrinate	$C_{24}H_{28}ClN_5O_3$	470.0	523-87-5

Appearance

Dimenhydrinate is a white odorless crystalline powder.

Solubility

Dimenhydrinate has a solubility of about 3 mg/mL in water. It is freely soluble in alcohol and chloroform and soluble in benzene. It is almost insoluble in ether.

Method 1

Using an HPLC method, Donnelly studied the chemical stability of dimenhydrinate stored at various conditions. The Shimadzu chromatograph included a model LC-10AS isocratic solvent-delivery pump, a model SPD-M6A photodiode-array detector, and a model SIL10A$_{XL}$ autoinjector. The stationary phase was a Phenomenex Luna C$_{18}$ column (250 × 4.0 mm, 5-μm particle size). The mobile phase consisted of 25 mM monobasic potassium phosphate buffer containing 2 mL/L of triethylamine and acetonitrile (65:35, vol/vol); the pH was adjusted to 2.2 with phosphoric acid. The flow rate was 1.5 mL/min. Methyl hydroxybenzoate 1.0 mg/mL was used as an internal standard. UV detection was performed at 230 nm.

Samples were mixed with internal standard and diluted with mobile phase before injection. The injection volume was 25 μL. Under these conditions, retention times of dimenhydrinate and the internal standard were about 7.5 and 12 minutes, respectively (estimated from the published chromatogram).

The stability-indicating nature of the method was demonstrated by forced degradation of dimenhydrinate. Solutions were adjusted to pH 1 with hydrochloric acid or to pH 12 with 1 N sodium hydroxide solution and heated to 80 °C. The third solution was exposed to fluorescent light for 165 days. The intact drug peak was separated from its degradation product peaks.

A standard curve for dimenhydrinate was constructed from 0.128 to 1.025 mg/mL. The correlation coefficient was 0.9998. Intraday and interday coefficients of variation for the analysis of dimenhydrinate were 0.12 and 2.45%, respectively.

Reference
Donnelly RF. Chemical stability of dimenhydrinate in minibags and polypropylene syringes. *Can J Hosp Pharm.* 2002; 55: 307–12.

Method 2
Walker et al. evaluated the stability of dimenhydrinate using an HPLC assay. The chromatograph consisted of a Schoeffel SF 770 UV detector, a Spectra-Physics SP4200 chromatographic integrator, and a Beckman Ultrasphere C_{18} reversed-phase column (250 × 4.2 mm, 5-μm particle size). The mobile phase consisted of acetonitrile and phosphate buffer containing 1 mg/mL of heptanesulfonic acid and was pumped gradiently at 2.0 mL/min from 13 to 40% acetonitrile in 14 minutes. Since dimenhydrinate is the 8-chlorotheophylline salt of diphenhydramine, the dimenhydrinate concentration was not determined. Instead, the concentrations of 8-chlorotheophylline and diphenhydramine were determined. UV detection was performed at 230 nm.

The analytical method was demonstrated to be stability indicating by accelerated degradation of dimenhydrinate using acid, base, and heat with inspection of the chromatograms for the appearance of additional peaks and for changes in retention times and peak shapes. Intraday and interday coefficients of variation for the assay were 0.40 and 1.97%, respectively.

Reference
Walker SE, Iazzetta J, De Angelis C, et al. Stability and compatibility of combinations of hydromorphone and dimenhydrinate, lorazepam or prochlorperazine. *Can J Hosp Pharm.* 1993; 46: 61–5.

Diphenhydramine

Chemical Names
2-Diphenylmethoxy-*N,N*-dimethylethanamine
2-(Benzhydryloxy)-*N,N*-dimethylethylamine

Other Name
Benadryl

Form	Molecular Formula	MW	CAS
Diphenhydramine	$C_{17}H_{21}NO$	255.4	58-73-1
Diphenhydramine citrate	$C_{17}H_{21}NO.C_6H_8O_7$	447.5	88637-37-0
Diphenhydramine hydrochloride	$C_{17}H_{21}NO.HCl$	291.8	147-24-0

Appearance

Diphenhydramine hydrochloride is a white or almost white, odorless or almost odorless, crystalline powder. It slowly darkens on exposure to light.

Solubility

One gram of diphenhydramine hydrochloride dissolves in 1 mL of water, 2 mL of alcohol, 2 mL of chloroform, and 50 mL of acetone. It is very slightly soluble in benzene and ether.

pK_a

Diphenhydramine has a pK_a of approximately 9.

Method 1

Mayron and Gennaro, using an HPLC method, assessed the stability of diphenhydramine with granisetron hydrochloride during simulated Y-site administration. The liquid chromatograph consisted of a piston pump with a pulse dampener, a rotary injection port with a 20-μL loop, a variable-wavelength spectrometric detector, and an integrator. The stationary phase was a cyano reversed-phase column (250 × 4.6 mm, 10-μm particle size). The mobile phase was a mixture of acetonitrile and 0.1 M monobasic sodium phosphate dihydrate (20:80, vol/vol) adjusted to pH 4.2 with phosphoric acid. The flow rate was 2.50 mL/min. UV detection was performed at 200 nm.

Samples were diluted with mobile phase. The retention times for granisetron and diphenhydramine were 4.00 and 6.97 minutes, respectively.

The stability-indicating capability of the analytical method was demonstrated by an accelerated degradation study. Solutions of diphenhydramine and granisetron hydrochloride were adjusted to pH 2 and 11, boiled for 1 hour, readjusted to pH 5, diluted with the mobile phase, and analyzed. The degradation product peaks did not interfere with the intact diphenhydramine or granisetron peaks.

Reference

Mayron D, Gennaro AR. Stability and compatibility of granisetron hydrochloride in i.v. solutions and oral liquids and during simulated Y-site injection with selected drugs. *Am J Health Syst Pharm.* 1996; 53: 294–304.

Method 2

Walker et al. studied the stability of diphenhydramine hydrochloride at various concentrations at room temperature. The HPLC method used a chromatographic system consisting of a Spectra-Physics SP8700 ternary gradient solvent-delivery pump, a Schoeffel SF 770 variable-wavelength UV detector, and a Beckman Ultrasphere reversed-phase C_{18} column (250 × 4.2 mm, 5-μm particle size). The mobile phase consisted of acetonitrile and 0.05 M phosphoric acid (pH 2.2) containing heptanesulfonic acid 1 mg/mL as a counterion. Acetonitrile began at 13% from 0.0 to 2.5 minutes and was increased linearly to 47% over 2.5 to 16 minutes, remaining at 47% until 20 minutes (estimated from the published chromatogram).

UV detection was performed at 230 nm. The sample injection volume was 20 μL. The retention times for creatinine, methylparaben, dexamethasone, and diphenhydramine were about 2.0, 9.7, 15.2, and 17.2 minutes, respectively (estimated from the published chromatogram).

The stability-indicating nature of the assay was demonstrated by adding acid or base to aqueous solutions of diphenhydramine hydrochloride in 30-mL multidose vials and incubating in a water bath at 90 °C. The intact drug peak was well separated from degradation product peaks and other constituents of the formulation on its liquid chromatogram.

Reference

Walker SE, DeAngelis C, Iazzetta J, et al. Compatibility of dexamethasone sodium phosphate with hydromorphone hydrochloride or diphenhydramine hydrochloride. *Am J Hosp Pharm.* 1991; 48: 2161–6.

Dipyridamole

Chemical Name

2,2′,2″,2‴-[(4,8-di-1-piperidinylpyrimido[5,4-*d*]pyrimidine-2,6-diyl)dinitrilo]tetrakis-ethanol

Other Name

Persantine

Form	Molecular Formula	MW	CAS
Dipyridamole	$C_{24}H_{40}N_8O_4$	504.6	58-32-2

Appearance

Dipyridamole is an intensely yellow crystalline powder.

Solubility

Dipyridamole is slightly soluble in water; it is very soluble in methanol, ethanol, and chloroform.

Method 1

Zhang et al. described an assay for the determination of dipyridamole in dipyridamole injection. A Hewlett-Packard model 1050 liquid chromatograph consisted of a solvent-delivery system, a variable-wavelength UV-visible detector, and a variable-volume injector. The stationary phase was a Waters μBondapak C_{18} column (300 × 3.9 mm, 10-μm particle size). The mobile phase was prepared by dissolving 2.38 g of sodium acetate in 350 mL of water (adjusted to pH 5.1 with 36% acetic acid) and mixing with 650 mL of methanol. The flow rate was 1.0 mL/min. UV detection was performed at 276 nm.

Each sample was diluted with mobile phase. The injection volume was 15 μL. Under these conditions, the retention time of dipyridamole was 23 minutes. The total run time was 30 minutes.

The stability-indicating ability of the assay was established by accelerated degradation. Dipyridamole samples were stored at 70 °C, adjusted to pH 2 with concentrated hydrochloric acid, adjusted to pH 12 with 50% sodium hydroxide solution, and subjected to 30% hydrogen peroxide or exposed to 500–700 foot-candles of radiation. No interfering degradation product peaks at the retention time of 23 minutes of dipyridamole were observed in any of the stressed samples.

A standard curve for dipyridamole was generated from 0.50 to 1.61 mg/mL. The correlation coefficient was 1.000. The coefficient of variation for the analysis of dipyridamole was less than 2%.

Reference

Zhang J, Miller RB, Russell S, et al. Validation of a stability-indicating HPLC method for the determination of dipyridamole in dipyridamole injection. *J Liq Chromatogr Rel Technol.* 1997; 20: 2109–21.

Method 2

Allen and Erickson reported an HPLC study of the chemical stability of dipyridamole 10 mg/mL in extemporaneously compounded oral liquids. A Hewlett-Packard series 1050 automated high-performance liquid chromatograph with Chem Station software included a multisolvent mixing and pumping system, an autoinjector, a diode-array detector, and a computer. The stationary phase was a Bakerbond C_{18} column (250 × 4.6 mm, 5-μm particle size). The mobile phase consisted of methanol and buffer (3:1, vol/vol) with the buffer composed of 250 mg of dibasic sodium phosphate in 250 mL of water and pH adjusted to 4.6 with dilute phosphoric acid. The flow rate was 1.3 mL/min. UV detection was performed at 288 nm.

Samples were diluted 1:100. Under these conditions, dipyridamole eluted in 7.3 minutes.

The analytical method was shown to be stability indicating by accelerated degradation of dipyridamole. A composite chromatogram of dipyridamole after intentional degradation by heat, acid, base, oxidizing agent, and light showed no interference with the intact dipyridamole from the degradation products.

Standard curves of dipyridamole were constructed from 10 to 150 μg/mL. The intraday and interday coefficients of variation of the assay were 2.3 and 2.5%, respectively.

Reference

Allen LV, Erickson MA. Stability of baclofen, captopril, diltiazem hydrochloride, dipyridamole, and flecainide acetate in extemporaneously compounded oral liquids. *Am J Health Syst Pharm.* 1996; 53: 2179–84.

Dipyrone

Chemical Name
Sodium *N*-(2,3-dimethyl-5-oxo-1-phenyl-3-pyrazolin-4-yl)-*N*-methylaminomethanesul-
fonate monohydrate

Other Names
Novaldin, Novalgine

Form	Molecular Formula	MW	CAS
Dipyrone	$C_{13}H_{16}N_3NaO_4S.H_2O$	351.4	5907-38-0

Appearance
Dipyrone is a white or almost white crystalline powder.

Solubility
Dipyrone is very soluble in water and soluble in alcohol.

Method
Eddine et al. reported an assay for dipyrone. The Spectra-Physics model SP8000 liquid chromatograph included a Schoeffel model SF 770 variable-wavelength detector and a 10-μL autosampler. The stationary phase was a Merck reversed-phase RP18 column (250 × 4.6 mm, 10-μm particle size) with an RP18 precolumn (50 mm, 10-μm particle size). The mobile phase was a mixture of methanol, water, and triethylamine (50:50:0.025, vol/vol/vol). The flow rate was 0.3 mL/min. UV detection was performed at 228 nm and 0.04 AUFS. Phenacetin 2 μg/mL in methanol was used as an internal standard. The injection volume was 10 μL. Under these conditions, the retention times for dipyrone and phenacetin were 3.26 and 11.83 minutes, respectively.

The method was stated to be stability indicating since it easily separated the intact dipyrone from its major degradation products. Retention times for dipyrone, 4-hydroxyantipyrine, 4-aminoantipyrine, and 4-methylaminoantipyrine were 3.26, 5.58, 7.66, and 8.83 minutes, respectively.

A standard curve for dipyrone was constructed from 1.5 to 6.0 μg/mL. The correlation coefficient was greater than 0.998. The limit of detection was 0.004 μg.

Reference
Eddine NH, Bressolle F, Mandrou B, et al. Stability indicating assay for dipyrone—Part II. Separation and quantitative determination of dipyrone and its degradation products by high-performance liquid chromatography. *Analyst*. 1982; 107: 67–70.

Disulfiram

Chemical Names
Tetraethylthioperoxydicarbonic diamide
Bis(diethylthiocarbamoyl) disulfide

Other Name
Antabuse

Form	Molecular Formula	MW	CAS
Disulfiram	$C_{10}H_{20}N_2S_4$	296.5	97-77-8

Appearance
Disulfiram is a white or almost white odorless crystalline powder.

Solubility
Disulfiram is practically insoluble in water (0.02 g/100 mL) but soluble in alcohol (3.82 g/100 mL), ether (7.14 g/100 mL), acetone, benzene, chloroform, and carbon disulfide.

Method
Phillips et al. used an HPLC method to evaluate the stability of an injectable disulfiram formulation sterilized by gamma irradiation. A Waters chromatograph consisted of a model 6000A solvent-delivery system, a model U6K injector, a model 440 UV detector, an Omniscribe recorder, and a Microporasil column (300 × 3.9 mm). The mobile phase was a mixture of n-heptane, tetrahydrofuran, and methanol (97.6:2.2:0.2, vol/vol/vol). The flow rate was 2.5 mL/min. UV detection was performed at 254 nm.

Samples were diluted before injection; the injection volume was 25 μL. The retention time was about 4.7 minutes (estimated from the published chromatogram).

To determine the stability-indicating ability of this method, a stock solution of disulfiram was spiked with diethyldithiocarbamate methyl ester, the degradation product. The degradation product peak did not interfere with the intact disulfiram peak.

Reference
Phillips M, Agarwal RP, Brodeur RJ, et al. Stability of an injectable disulfiram formulation sterilized by gamma irradiation. *Am J Hosp Pharm.* 1985; 42: 343–5.

Dithranol

Chemical Name
1,8-Dihydroxy-9(10*H*)-anthracenone

Other Names
Anthralin, Lasan, Miconal

Form	Molecular Formula	MW	CAS
Dithranol	$C_{14}H_{10}O_3$	226.2	1143-38-0

Appearance
Dithranol is a yellow to yellowish-brown odorless crystalline powder.

Solubility
Dithranol is practically insoluble in water and slightly soluble in alcohol.

Method 1
Cheah et al. reported a reversed-phase HPLC method for the determination of dithranol and its degradation products. An LDC Constametric model III liquid chromatograph was equipped with a Rheodyne 7125 100-µL loop injector, a Perkin-Elmer LC-75 spectrophotometer, and a Perkin-Elmer M-2 integrator. The stationary phase was a DuPont Zorbax ODS column (250 × 4.6 mm, 6-µm particle size) with a Newguard RP18 precolumn. The mobile phase was a mixture of acetonitrile, glacial acetic acid, and distilled water (68.5:1.5:30, vol/vol/vol). The flow rate was 2 mL/min. UV detection was performed at 354 nm. 1,8-Dihydroxy-3-methylanthraquinone 1 mg/mL in dichloroethane was used as an internal standard.

Dithranol samples were mixed with 4 mL of dichloroethane and 1 mL of internal standard, sonicated for 5 minutes (for an ointment) or homogenized for 20 seconds (for a cream), and filtered through a 1.0-µm Whatman GF/B glass microfiber filter. The injection volume was 10 µL. Under these conditions, retention times for dithranol and the internal standard were about 6.4 and 8.8 minutes, respectively (estimated from the published chromatogram).

The method was stability indicating since it resulted in the complete separation of dithranol from its degradation products, danthron and dianthrone. Retention times for danthron, dithranol, and dianthrone were about 5.6, 6.4, and 14 minutes, respectively (estimated from the published chromatogram).

A standard curve for dithranol was constructed from 0 to 100 µg/mL. The limit of detection was 4.8 ng.

Reference

Cheah ICL, Sitaram BR, Pappas A, et al. Normal-phase and reversed-phase liquid chromatographic techniques for the determination of dithranol and its degradation products. *J Chromatogr.* 1989; 467: 414–22.

Method 2

Cheah et al. also described a normal-phase HPLC method for the analysis of dithranol and its degradation products. An LDC Constametric model III liquid chromatograph was equipped with a Rheodyne 7125 100-μL loop injector, a Perkin-Elmer LC-75 spectrophotometer, and a Perkin-Elmer M-2 integrator. The stationary phase was a DuPont Sil silica column (250 × 4.6 mm, 6-μm particle size) with a Newguard silica precolumn. The mobile phase was 2,2,4-trimethylpentane, 1,2-dichloroethane, and glacial acetic acid (91:6:3, vol/vol/vol). The flow rate was 2 mL/min. UV detection was performed at 354 nm. 1,4-Naphthoquinone in dichloroethane was used as an internal standard.

Dithranol samples were mixed with 4 mL of dichloroethane and 1 mL of internal standard, sonicated for 5 minutes (for an ointment) or homogenized for 20 seconds (for a cream), and filtered through a 1.0-μm Whatman GF/B glass microfiber filter. The injection volume was 10 μL. Under these conditions, retention times for dithranol and the internal standard were about 3.2 and 6.2 minutes, respectively (estimated from the published chromatogram).

The method was stability indicating since it separated dithranol from its degradation products, danthron and dianthrone. Retention times for dithranol, danthron, and dianthrone were about 3.2, 4.2, and 8.8 minutes, respectively (estimated from the published chromatogram).

A standard curve for dithranol was constructed from 0 to 100 μg/mL. The limit of detection was 1.5 ng.

Reference

Cheah ICL, Sitaram BR, Pappas A, et al. Normal-phase and reversed-phase liquid chromatographic techniques for the determination of dithranol and its degradation products. *J Chromatogr.* 1989; 467: 414–22.

Dixyrazine

Chemical Name

2-[2-[4-[2-Methyl-3-(phenothiazin-10-yl)propyl]piperazin-1-yl]ethoxy]ethanol

Other Name
Esucos

Form	**Molecular Formula**	**MW**	**CAS**
Dixyrazine	$C_{24}H_{33}N_3O_2S$	427.6	2470-73-7

Method

Kopelent-Frank and Mittlbock described an HPLC assay for the simultaneous determination of dixyrazine and chlorprothixene in intravenous admixtures. The liquid chromatograph consisted of a Shimadzu model LC-10AS pump, a model SPD-M10A diode-array detector, a model CTO-10AC column oven, and a Rheodyne 20-μL injector. The stationary phase was a Superspher 100 RP18 column (125 × 4.6 mm, 4-μm particle size). Solvent A was methanol and solvent B was acetate buffer (pH 4.6). The mobile phase was a mixture of solvents A and B and was gradiently delivered as follows:

Time, minutes	Flow Rate, mL/min	Solvent A, %
0–25	0.9	55
25–27	0.9→1.2	55→75
27–33	1.2	75
33–35	1.2	75→55
35–38	1.2→0.9	55
38–38.5	0.9	55

UV detection was performed at 252 nm. Under these conditions, retention times for dixyrazine and chlorprothixene were about 24.3 and 26.3 minutes, respectively.

The stability-indicating nature of the assay was demonstrated by accelerated degradation of the drug. The sample was irradiated for 24 minutes. The degradation products did not interfere with the analysis of dixyrazine.

A standard curve for dixyrazine was generated from 20 to 50 μg/mL. The correlation coefficient was 0.9998. Intraday and interday coefficients of variation were 1.12 and 1.76%, respectively.

Reference
Kopelent-Frank H, Mittlbock M. Stability-indicating high-performance liquid chromatographic assay for the simultaneous determination of dixyrazine and chlorprothixene in intravenous admixtures. *J Chromatogr A.* 1996; 729: 201–6.

Dobutamine

Chemical Names
(±)-4-[2-[[3-(4-Hydroxyphenyl)-1-methylpropyl]amino]ethyl]-1,2-benzenediol

(±)-4-[2-[[3-(*p*-Hydroxyphenyl)-1-methylpropyl]amino]ethyl]pyrocatechol

Other Name
Dobutrex

Form

Form	Molecular Formula	MW	CAS
Dobutamine	$C_{18}H_{23}NO_3$	301.4	34368-04-2
Dobutamine hydrochloride	$C_{18}H_{23}NO_3$.HCl	337.8	49745-95-1

Appearance
Dobutamine hydrochloride is a white or almost white crystalline powder.

Solubility
Dobutamine hydrochloride is freely soluble in water and methanol, sparingly soluble in acetone and dichloromethane, and practically insoluble in ether and chloroform.

pK$_a$
Dobutamine has a pK$_a$ of 9.4.

Method 1
Caufield and Stewart reported a method for the simultaneous determination of zidovudine and dobutamine. A Hewlett-Packard series 1090 system included a pump, an autosampler with a 25-μL loop, and a Gilson model 117 variable-wavelength UV detector or a Waters model 996 photodiode-array detector. The stationary phase was a Supelco Discovery RP-Amide C_{16} column (250 × 4.6 mm, 5-μm particle size). The mobile phase consisted of 25 mM monobasic sodium phosphate monohydrate (pH adjusted to 3.0 with 0.1 M phosphoric acid) and acetonitrile (84:16, vol/vol). The flow rate was 1.0 mL/min. Solutions of zidovudine and dobutamine were prepared with an aqueous-acetonitrile diluent matching the mobile phase composition. UV detection was performed at 280 nm. Under these conditions, retention times of zidovudine and dobutamine were 7.5 and 10.8 minutes, respectively.

To demonstrate that the method was stability indicating, solutions of zidovudine were subjected to acid hydrolysis (6 M hydrochloric acid), base hydrolysis (6 M sodium hydroxide), oxidation (0.3% hydrogen peroxide), heat (90 °C), and radiation (254 nm). Solutions of dobutamine were also subjected to acid hydrolysis (6 M hydrochloric acid), base hydrolysis (0.001 M sodium hydroxide solution), oxidation (0.3% hydrogen peroxide), heat (60 °C), and radiation (254 nm). In all cases, zidovudine and dobutamine were separated from their degradation products on their chromatograms.

A standard curve of dobutamine was constructed from 50 to 200 μg/mL. The correlation coefficient was greater than 0.9999. Intraday and interday coefficients of variation were 0.43 and 0.52%, respectively.

Reference
Caufield WV, Stewart JT. HPLC separations of zidovudine and selected pharmaceuticals using a hexadecylsilane amide column. *Chromatographia.* 2001; 54: 561–8.

Method 2

Gora et al. used HPLC to evaluate the stability of dobutamine hydrochloride at various concentrations in dialysis solutions containing dextrose 1.5 or 4.25% under various storage conditions. The chromatographic system consisted of a Scientific Systems model 222B isocratic pump, a Scientific Systems model 02-0249 injector port with a 50-µL sample loop, a Varian model 2050 variable-wavelength UV detector, and a Beckman strip-chart recorder. The stationary phase was a Rainin Microsorb CN column (250 mm, 5-µm particle size). The mobile phase was a mixture of 240 µL of 9 M sulfuric acid, 280 mL of acetonitrile, and 520 mL of water. The flow rate was 0.9 mL/min. UV detection was performed at 290 nm and 0.005 AUFS.

Each sample was diluted 1:4 with sterile water. The injection volume was 50 µL. Under these conditions, the retention time for dobutamine hydrochloride was 12.8 minutes.

To demonstrate the stability-indicating nature of the assay, a solution of dobutamine hydrochloride 1.25 mg/mL was mixed with 5 N sodium hydroxide and boiled for 5 minutes. None of the degradation product peaks interfered with the intact dobutamine peak; the degradation product with the longest retention time appeared at 8.8 minutes.

Standard curves were constructed daily from 2.5 to 7.5 µg/mL. The correlation coefficient was 0.99. Coefficients of variation for triplicate injections ranged from 1.2 to 5.2%.

Reference
Gora ML, Seth S, Visconti JA, et al. Stability of dobutamine hydrochloride in peritoneal dialysis solutions. *Am J Hosp Pharm.* 1991; 48: 1234–7.

Method 3

Gupta and Stewart investigated the stability of dobutamine hydrochloride and verapamil hydrochloride when mixed together in 0.9% sodium chloride and 5% dextrose injections. A Waters model ALC 202 liquid chromatograph was equipped with a Waters model U6K universal injector, a Spectroflow SF 770 multiple-wavelength detector, an Omniscribe recorder, and a Waters µBondapak phenyl semipolar column (300 × 4.0 mm). The mobile phase was 20% (vol /vol) acetonitrile in water with 0.02 M monobasic potassium phosphate and 0.3% (vol/vol) acetic acid (pH 3.6 ± 0.1) and was delivered isocratically at 2.0 mL/min. UV detection was performed at 278 nm and 0.04 AUFS. The sample injection volume was 20 µL. Methapyrilene hydrochloride 400 µg/mL was used as an internal standard. The retention times for dobutamine hydrochloride and methapyrilene hydrochloride were 3.5 and 8.2 minutes, respectively (estimated from the published chromatogram).

To determine if the HPLC method was stability indicating, a solution of dobutamine hydrochloride 1.0 mg/mL was mixed with 1 N sodium hydroxide and boiled for about 30 minutes. The intact dobutamine peak was well separated from its degradation product peaks.

A standard curve for dobutamine was constructed from 1 to 3.5 µg/mL. The coefficient of variation was 1.25%.

A similar stability-indicating method was used by Sautou-Miranda et al.

References

Gupta VD, Stewart KR. Stability of dobutamine hydrochloride and verapamil hydrochloride in 0.9% sodium chloride and 5% dextrose injections. *Am J Hosp Pharm*. 1984; 41: 686–9.

Sautou-Miranda V, Gremeau I, Chamard I, et al. Stability of dopamine hydrochloride and of dobutamine hydrochloride in plastic syringes and administration sets. *Am J Health Syst Pharm*. 1996; 53: 186, 193.

Docetaxel

Chemical Name

tert-Butyl[(1*S*,2*S*)-2-[(2*S*,5*R*,7*S*,10*R*,13*S*)-4-acetoxy-2-benzoyloxy-1,7,10-trihydroxy-9-oxo-5,20-epoxytax-11-en-13-yloxy-carbonyl]-2-hydroxy-1-phenylethyl]carbamate

Other Name

Taxotere

Form	Molecular Formula	MW	CAS
Docetaxel	$C_{43}H_{53}NO_{14}$	807.9	114977-28-5

Appearance

Docetaxel occurs as a white to almost white powder.

Solubility

Docetaxel is highly lipophilic and practically insoluble in water.

Method 1

Rao et al. reported a method for docetaxel. An Agilent 1100 series liquid chromatograph included a variable-wavelength detector and a photodiode-array detector. The stationary phase was a Hichrom RPB column (250 × 4.6 mm, 5-μm particle size). Solvent A was water and solvent B was acetonitrile. The mobile phase was delivered at 1.0 mL/min in a gradient mode:

Time, minutes	Solvent A, %	Solvent B, %
0	65	35
15	35	65
25	25	75
30	5	95
35	0	100
39	0	100
40	65	35
45	65	35

UV detection was performed at 230 nm.

Samples were prepared in a mixture of water and acetonitrile (1:1, vol/vol). The injection volume was 10 μL. Under these conditions, retention time of docetaxel was 14.4 minutes.

To verify the stability-indicating nature of the analytical method, docetaxel bulk material was exposed to UV light at 254 nm, heat at 60 °C, 0.5 N hydrochloric acid, 0.005 N sodium hydroxide solution, and 3% hydrogen peroxide. Retention times for the docetaxel impurity (DCT-1), docetaxel, and its degradation product (7-epimer) were 12.1, 14.4, and 17.3 minutes, respectively. Peak purity test results obtained by a photodiode-array detector also confirmed that the intact docetaxel peak was homogeneous and pure in all cases.

A calibration curve for docetaxel was obtained from 250 to 750 μg/mL. The correlation coefficient was greater than 0.999.

Reference

Rao BM, Chakraborty A, Srinivasu MK, et al. A stability-indicating HPLC assay method for docetaxel. *J Pharm Biomed Anal.* 2006; 41: 676–81.

Method 2

Thiesen and Kramer determined by HPLC the stability of docetaxel over prolonged periods. The liquid chromatograph consisted of a Waters model 510 pump, a model 717 plus autosampler, and a model 996 photodiode-array detector. The stationary phase was a Waters Nova-Pak C_{18} column (150 × 3.9 mm, 4-μm particle size). The mobile phase consisted of water, methanol, and acetonitrile (42:32:26, vol/vol/vol). The flow rate was 1.2 mL/min. UV detection was performed at 232 nm.

Samples were diluted with 0.9% sodium chloride injection. The injection volume was 10 μL. The total run time was 12 minutes. Under these conditions, the retention time for docetaxel was about 7.4 minutes (estimated from the published chromatogram).

The method was established to be stability indicating by accelerated decomposition of docetaxel. Docetaxel solutions were adjusted to pH 11 with 2 N sodium hydroxide solution and kept for 2 hours or adjusted to pH 0.7 with 1 N hydrochloric acid and kept for 36 hours. The degradation product peaks did not influence the determination of docetaxel.

A calibration curve for docetaxel was obtained from 0.125 to 0.75 mg/mL. The correlation coefficient was 0.9999. Intraday and interday coefficients of variation were 0.5 and 2.5%, respectively.

Reference

Thiesen J, Kramer I. Physico-chemical stability of docetaxel premix solution and docetaxel infusion solutions in PVC bags and polyolefin containers. *Pharm World Sci.* 1999; 21: 137–41.

Dolasetron Mesylate

Chemical Name

(2α,6α,8α,9aβ)-Octahydro-3-oxo-2,6-methano-2*H*-quinolizin-8-yl-1*H*-indole-3-carboxylate acid monomethanesulfonate

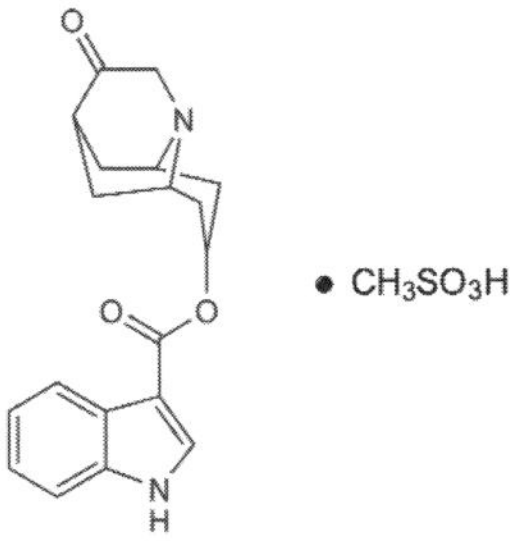

● CH₃SO₃H

Other Names

Anzemet, Zamanon

Form	Molecular Formula	MW	CAS
Dolasetron	$C_{19}H_{20}N_2O_3$	324.4	115956-12-2
Dolasetron mesylate	$C_{19}H_{20}N_2O_3 \cdot CH_4O_3S$	420.5	115956-13-3

Appearance

Dolasetron mesylate occurs as a white to off-white powder.

Solubility

Dolasetron mesylate is freely soluble in water and propylene glycol. It is slightly soluble in ethanol and normal saline.

Method

Using HPLC, Johnson et al. investigated the stability of dolasetron in two oral liquid vehicles. The instrument consisted of a Waters model 501 solvent-delivery system, a model U6K injector, a model 486 variable-wavelength UV detector, and a Hewlett-Packard model 3394 integrator. The stationary phase was a Spherisorb CN column (150 × 4.6 mm, 3-μm particle size), and the column temperature was maintained at 30 °C. The mobile phase consisted of 0.05 M ammonium acetate aqueous solution adjusted to pH 7.5 with dilute ammonium hydroxide and acetonitrile (76:24, vol/vol). The flow rate was 0.8 mL/min. UV detection was performed at 280 nm.

Samples were diluted to 10 μg/mL with a diluent of acetonitrile and water (24:76, vol/vol), shaken thoroughly for 15 seconds, and centrifuged at 1000 rpm for 2 minutes. Five microliters of clear solutions was injected. Under these conditions, the retention time of dolasetron was 6.9 minutes.

The assay was demonstrated to be stability indicating by intentional degradation of dolasetron. Dolastron samples in syrup and in sugar-free vehicle, adjusted to pH 12 and pH 2 in each vehicle, were exposed to sunlight for 90 days and then heated at 60 °C for 2 hours. Degradation of dolasetron occurred but did not interfere with the determination of the drug.

A standard curve of dolasetron was constructed from 8 to 12 μg/mL. The correlation coefficient was 0.999. The intraday and interday coefficients of variation for the assay were 1.3 and 2.7%, respectively.

Reference
Johnson CE, Wagner DS, Bussard WE. Stability of dolasetron in two oral liquid vehicles. *Am J Health Syst Pharm*. 2003; 60: 2242–4.

Domperidone

Chemical Name
5-Chloro-1-[1-[3-(2-oxobenzimidazolin-1-yl)propyl]-4-piperidyl]benzimidazolin-2-one

Other Name
Motilium

Form	Molecular Formula	MW	CAS
Domperidone	$C_{22}H_{24}ClN_5O_2$	425.9	57808-66-9
Domperidone maleate	$C_{22}H_{24}ClN_5O_2.C_4H_4O_4$	542.0	99497-03-7

Appearance
Domperidone occurs as a white or almost white powder. Domperidone maleate is a white or almost white powder.

Solubility
Domperidone is practically insoluble in water and slightly soluble in alcohol and methanol. It is soluble in dimethylformamide. Domperidone maleate is very slightly soluble in water and alcohol. It is sparingly soluble in dimethylformamide and slightly soluble in methanol.

Method
Ali et al. reported a method for the simultaneous determination of domperidone, methylparaben, and propylparaben. The Shimadzu system LC-2010A consisted of a four-liquid

gradient pump, an autosampler, a column oven, and a UV-visible detector. The stationary phase was an Optimapak OP C_8 analytical column (150 × 4.6 mm, 5-μm particle size). The mobile phase was a mixture of 5 g/L ammonium acetate buffer and methanol (40:60, vol/vol). The flow rate was 1.0 mL/min. UV detection was carried out at 280 nm.

Ten milliliters of an oral suspension was transferred into a 100-mL volumetric flask, treated with 50 mL of dimethylformamide and 1 mL of 0.1 N hydrochloric acid, mixed by vortexing for 30 seconds, sonicated for 5 minutes, shaken for 10 minutes, filled to the mark with methanol, and centrifuged at 3500 rpm for 5 minutes. A clear supernatant was collected. The injection volume was 10 μL. Under these conditions, retention times of methylparaben, propylparaben, and domperidone were about 3.3, 7.0, and 9.1 minutes, respectively (estimated from the published chromatogram).

There were no interferences due to the presence of excipients and degradation products.

A standard curve for domperidone was obtained from 50 to 150 μg/mL. The correlation coefficient was 0.9999. The intraday coefficient of variation was 0.13%.

Reference
Ali MS, Ghori M, Khatri AR. Stability indicating simultaneous determination of domperidone (DP), methylparaben (MP) and propylparaben by high performance liquid chromatography (HPLC). *J Pharm Biomed Anal.* 2006; 41: 358–65.

Donepezil Hydrochloride

Chemical Name
(±)-2-[(1-Benzyl-4-piperidyl)methyl]-5,6-dimethoxy-1-indanone hydrochloride

Other Name
Aricept

Form	Molecular Formula	MW	CAS
Donepezil	$C_{24}H_{29}NO_3$	379.5	120014-06-4
Donepezil hydrochloride	$C_{24}H_{29}NO_3 \cdot HCl$	416.0	120011-70-3

Appearance
Donepezil hydrochloride is a white crystalline powder.

Solubility
Donepezil hydrochloride is soluble in water and glacial acetic acid, freely soluble in chloroform, slightly soluble in alcohol and acetonitrile, and practically insoluble in ethyl acetate and *n*-hexane.

Method

Pappa et al. described an HPLC method for the determination of donepezil hydrochloride. The liquid chromatographic system consisted of a Spectra-Physics model ISO pump, a Rheodyne model 7125 injector, a Hewlett-Packard model 1050 UV-visible detector, and a Hewlett-Packard series 3395 integrator. The stationary phase was a Varian Microsorb-MV RP18 column (5-μm particle size). The mobile phase consisted of methanol and 0.02 M phosphate buffer (50:50, vol/vol), where the phosphate buffer was prepared by dissolving 13.8 g of monobasic sodium phosphate in 900 mL of water, mixing with 10 mL of triethylamine, and adjusting to pH 2.7 with phosphoric acid. The flow rate was 1.0 mL/min. UV detection was performed at 268 nm and 2 AUFS.

A portion of the powder of crushed tablets (30) equivalent to 25 mg of donepezil was weighed, transferred into a 25-mL volumetric flask, mixed with 20 mL of mobile phase, sonicated for 5 minutes, diluted to volume with mobile phase, and filtered through a Whatman No. 42 filter. The filtrate was diluted 1:25 (vol/vol) with mobile phase and filtered through a 0.45-μm membrane filter. The injection volume was 20 μL. Under these conditions, the retention time for donepezil was about 9 minutes.

The stability-indicating ability of the method was demonstrated by accelerated degradation of donepezil hydrochloride. Donepezil hydrochloride samples were treated with 1 N hydrochloric acid for 4 hours, 1 N sodium hydroxide solution for 4 hours, or hydrogen peroxide for 4 hours or were exposed to daylight for 24 hours. Neither degradation products nor formulation ingredients influenced the analysis of donepezil.

A standard curve for donepezil was constructed from 10 to 60 μg/mL. The correlation coefficient was 0.9995.

Reference

Pappa H, Farru R, Vilanova PO, et al. A new HPLC method to determine donepezil hydrochloride in tablets. *J Pharm Biomed Anal.* 2002; 27: 177–82.

Dopamine

Chemical Names

4-(2-Aminoethyl)-1,2-benzenediol
4-(2-Aminoethyl)pyrocatechol

Other Names

Dopastat, Intropin

Form	Molecular Formula	MW	CAS
Dopamine	$C_8H_{11}NO_2$	153.2	51-61-6
Dopamine hydrochloride	$C_8H_{11}NO_2.HCl$	189.6	62-31-7

Appearance

Dopamine hydrochloride is a white or almost white crystalline powder.

Solubility

Dopamine hydrochloride is freely soluble in water and is soluble in methanol, hot 95% ethanol, and aqueous solutions of alkali hydroxides. It is practically insoluble in ether, petroleum ether, chloroform, benzene, and toluene.

Method 1

Using an HPLC method, Braenden et al. evaluated the stability of dopamine hydrochloride in polypropylene syringes. The Shimadzu system consisted of a model LC10-AD pump, a model SIL-10A autoinjector, a model SPD-M10A diode-array detector, and a model DGU-3A degasser. The stationary phase was a Supelco Supelcosil C_{18} column (150 × 4.6 mm, 3-µm particle size). The mobile phase consisted of 0.005 M phosphate buffer (pH 2.4) containing 0.625 M heptanesulfonic acid sodium salt and methanol (80:20, vol/vol). The flow rate was 1.0 mL/min. UV detection was performed at 280 nm.

Dopamine was diluted to 10 µg/mL with 5% dextrose injection. The injection volume was 50 µL.

The analytical method was demonstrated to be stability indicating by accelerated degradation of dopamine. Dopamine solutions were adjusted to pH 2 or pH 11 and stored at 90 °C for 1 hour or at room temperature for 1.5 hours. No interfering degradation products were observed.

A calibration curve for dopamine was obtained from 5 to 17 µg/mL. The correlation coefficient was greater than 0.999. The limit of quantitation was 0.9 µg/mL. Intraday and interday coefficients of variation of the analysis were 0.7 and 1.8%, respectively.

Reference

Braenden JU, Stendal TL, Fagernaes CB. Stability of dopamine hydrochoride 0.5 mg/mL in polypropylene syringes. *J Clin Pharm Ther.* 2003; 28: 471–4.

Method 2

Caufield and Stewart reported an HPLC assay for the simultaneous analysis of meropenem and dopamine in an intravenous fluid mixture. The instrument was a Hewlett-Packard model 1090 system including a pump, an autosampler, and a Gilson model 117 variable-wavelength UV detector. A Waters model 996 photodiode-array detector was used for the peak purity analysis. The stationary phase was a Rainin Microsorb ODS column (150 × 2.1 mm, 5-µm particle size). The mobile phase consisted of aqueous acetic acid solution (pH 3) and acetonitrile (85:15, vol/vol). The flow rate was 0.2 mL/min. UV detection was performed at 280 nm.

Samples were diluted with a diluent of 0.9 mM ethylenediaminetetraacetic acid disodium salt aqueous solution and acetonitrile (85:15, vol/vol). The sample injection volume was 5 µL. Under these conditions, retention times of dopamine and meropenem were about 5.7 and 9.4 minutes, respectively.

The method was confirmed to be stability indicating by accelerated degradation of dopamine. Solutions of dopamine were treated with 0.1 N hydrochloric acid for

2 hours, 0.0001 N sodium hydroxide solution for 2 hours, or 3% hydrogen peroxide for 1 hour or were heated at 60 °C for 1.5 hours. None of the degradation product peaks interfered with the intact drug peak.

A standard curve for dopamine was constructed from 10 to 40 µg/mL. The correlation coefficient was 0.9996–1.0. Intraday and interday coefficients of variation were 1.04 and 3.00%, respectively. The limit of detection was 123 ng/mL.

Reference

Caufield WV, Stewart JT. HPLC separations of meropenem and selected pharmaceuticals using a polar endcapped octadecylsilane narrow bore column. *Chromatographia*. 2000; 51: 308–14.

Method 3

Peddicord et al. evaluated the stability of dopamine hydrochloride in 5% dextrose injection at ambient temperature and humidity. The chromatograph included a Shimadzu model LC-10AS solvent-delivery system, a Shimadzu model SIL-10A autoinjector, a Shimadzu model CTO-10A column oven, a Shimadzu model SPD-10A variable-wavelength UV spectrophotometer, and a Phenomenex C_{18} column (300 × 3.9 mm, 5-µm particle size). The mobile phase consisted of water, methanol, and acetic acid (72.3:25:2.7, vol/vol/vol) and 1.34 g of 1-heptanesulfonic acid. The flow rate was 1.6 mL/min. UV detection was performed at 254 nm. The oven temperature was 50 °C.

Each sample was diluted with 5% dextrose injection. The injection volume was 50 µL.

The method was determined to be stability indicating by accelerated decomposition of dopamine. Solutions of dopamine were adjusted to pH 2 with 1.0 N hydrochloric acid and pH 12 with 10 N sodium hydroxide and were raised to 100 °C. Degradation product peaks did not interfere with the intact dopamine peak.

A calibration curve was constructed for dopamine from 0 to 200.0 µg/mL. The correlation coefficient was greater than 0.999. The intraday and interday coefficients of variation were 2.1 and 8.7%, respectively.

A similar method was used by Williams et al.

References

Peddicord TE, Olsen KM, ZumBrunnen TL, et al. Stability of high-concentration dopamine hydrochloride, norepinephrine bitartrate, epinephrine hydrochloride, and nitroglycerin in 5% dextrose injection. *Am J Health Syst Pharm*. 1997; 54: 1417–9.

Williams DA, Fung EYY, Newton DW. Ion-pair high-performance liquid chromatography of terbutaline and catecholamines with aminophylline in intravenous solutions. *J Pharm Sci*. 1982; 71: 956–8.

Method 4

Sautou-Miranda et al. studied the stability of dopamine hydrochloride using an HPLC method. The liquid chromatograph consisted of a Merck-Hitachi model L6200 Intelligent pump, a Rheodyne model 7125 injector, a Merck-Hitachi LCM variable-wavelength UV-visible detector, and a Merck-Hitachi model D2500 chromato-integrator. The stationary phase was a Phase Separations LiChrospher reversed-phase endcapped

C_{18} column (150 × 4.6 mm, 5-µm particle size). The mobile phase consisted of 0.05 M potassium phosphate buffer, acetonitrile, and acetic acid (70:20:10, vol/vol/vol). UV detection was performed at 280 nm.

Samples were diluted with mobile phase; the injection volume was 20 µL.

The analytical method was determined to be stability indicating by accelerated degradation of dopamine. A solution of dopamine was mixed with 5 N sodium hydroxide and boiled for 5 minutes. Degradation product peaks did not interfere with the intact dopamine peak.

A standard curve for dopamine was constructed from 0.1 to 0.3 mg/mL. The correlation coefficient was greater than 0.999. The intraday and interday coefficients of variation were less than 2.0%.

Reference

Sautou-Miranda V, Gremeau I, Chamard I, et al. Stability of dopamine hydrochloride and of dobutamine hydrochloride in plastic syringes and administration sets. *Am J Health Syst Pharm.* 1996; 53: 186, 193.

Method 5

Allen et al. used HPLC to evaluate the chemical stability of dopamine hydrochloride in the presence of cefpirome sulfate during simulated Y-site injection. The analysis was conducted utilizing one of two systems. One unit included an Alcott model 728 Micromeritics autosampler, a Rheodyne 7010 injector with an Alcott 732 electrically actuated valve, a Waters model 501 solvent-delivery pump, a Waters model 441 UV detector, a Waters model 401 refractive index detector, and a Waters model 745 data module. The other unit consisted of an Alcott model 728 Micromeritics autosampler, a Rheodyne 7010 injector with an Alcott 732 electrically actuated valve, a Shimadzu model LA-6A solvent-delivery pump, a Shimadzu model SPD-6A UV detector, and an Orion model 901 microprocessor ion analyzer. The stationary phase was a C_8 analytical column. The mobile phase consisted of 260 mL of acetonitrile and 1740 mL of 1:1000 sodium 1-octanesulfonate in 1:100 acetic acid and was delivered isocratically at 1.5 mL/min. UV detection was performed at 280 nm.

Samples were diluted before injection. The injection volume was 20 µL. Under these conditions, the retention times for cefpirome and dopamine were 9.8 and 16.5 minutes, respectively.

The stability-indicating capability of the analytical method was demonstrated by accelerated decomposition of standard solutions of dopamine hydrochloride with sodium hydroxide, hydrochloric acid, potassium chlorate, heat at 80 °C, or light (150 foot-candles from a tungsten filament source). In each case, the intact dopamine peak was suitably separated from its breakdown products on the chromatogram.

The standard curve was generated from 100 to 500 µg/mL; the correlation coefficient was greater than 0.99. The intraday and interday coefficients of variation of the assay were 3.1%.

Reference

Allen LV, Stiles ML, Prince SJ, et al. Stability of cefpirome sulfate in the presence of commonly used intensive care drugs during simulated Y-site injection. *Am J Health Syst Pharm.* 1995; 52: 2427–33.

Method 6

Dandurand and Stennett assessed the stability of dopamine hydrochloride exposed to blue-light phototherapy. The chromatographic system consisted of a Waters model U-6000 pump with a model 440 fixed-wavelength UV detector and a WISP 710B autosampler. The stationary phase was a new Waters μBondapak C_{18} reversed-phase column (300×3.9 mm, 10-μm particle size). The mobile phase was 7% acetonitrile in aqueous buffer. The aqueous buffer contained 0.1 M citric acid and 0.2 M disodium phosphate adjusted to pH 4. The flow rate was 2.0 mL/min. UV detection was performed at 280 nm and 0.05 AUFS. β-Hydroxyethyltheophylline was used as an internal standard.

Each sample was diluted 1:2 with internal standard solution. The injection volume was 1 μL. Under these conditions, the retention times for dopamine and β-hydroxy-ethyltheophylline were about 4 and 11 minutes, respectively.

The method was determined to be stability indicating. Dopamine hydrochloride solution 0.5 mg/mL was mixed with sodium bicarbonate and was stored at room temperature for 8 months. Two new peaks were detected with retention times of 1.5 and 1.7 minutes when degraded drug solution alone was injected. The degradation product peaks did not interfere with dopamine or the internal standard peaks.

Standard curves were generated from 900 to 1050 μg/mL and had a correlation coefficient of 0.9940 and a coefficient of variation of 1.76%.

A similar stability-indicating method was used by Brown and Jenke.

References

Dandurand KR, Stennett DJ. Stability of dopamine hydrochloride exposed to blue-light phototherapy. *Am J Hosp Pharm*. 1985; 42: 595–7.

Brown DS, Jenke DR. Application of fast LC to pharmaceutical analysis: Dopamine HCl. *J Chromatogr Sci*. 1987; 25: 494–500.

Doretinel

Chemical Name

1,2,3,4-Tetrahydro-7-[(1*E*)-2-[4-(hydroxymethyl)phenyl]-1-methylethenyl]-1,1,4,4-tetramethyl-2-naphthalenol

Form	Molecular Formula	MW	CAS
Doretinel	$C_{24}H_{30}O_2$	350.5	104561-36-6

Method

Krailler et al. reported an HPLC method for the quantitation of doretinel in a topical gel. A Waters chromatographic system consisted of a model 590 pump, a model WISP 712 autosampler, and an ABI Applied Biosystems Spectroflow 783 variable-wavelength detector. The stationary phase was an Alltech Spherisorb ODS-II column (150×4.6 mm, 5-μm particle size). The mobile phase was prepared by mixing 500 mL of methanol with 100 mL of tetrahydrofuran in a 1000-mL volumetric flask and filling to the mark with 0.025 M monobasic potassium phosphate aqueous buffer. The flow rate was 1.5 mL/min. UV detection was performed at 300 nm.

A 1–2-g sample of doretinel gel was accurately weighed and mixed with 5 mL of water. The mixture was passed by force through a Sep-Pak C_{18} cartridge that had been preconditioned by rinsing with 2 mL of methanol followed by 10 mL of water. Approximately five or six rinses or 50 mL of water was required. Finally, doretinel was eluted from the cartridge with 1–5 mL of tetrahydrofuran followed by 5 mL of methanol. The collected sample was diluted to approximately 0.2 µg/mL of doretinel with an appropriate combination of solvents to mimic the mobile phase. The injection volume was 500 µL. Under these conditions, the retention time for doretinel was about 10 minutes.

The stability-indicating nature of the method was demonstrated by chromatographic examination of degraded doretinel samples. Doretinel samples were subjected to various conditions of stress such as acid hydrolysis (2 days at 70 °C in 0.3 N hydrochloric acid), base hydrolysis (2 days at 70 °C in 0.3 N sodium hydroxide solution), chemical oxidation (m-chloroperoxybenzoic acid), heat (5 days at 80 °C), or light (5 days at 1000 foot-candles). The intact doretinel peak was well resolved from all degradation product peaks.

A calibration curve of doretinel was generated from 0.00478 to 0.382 µg/mL. The correlation coefficient was greater than 0.999. The limit of detection was 0.00478 µg/mL.

Reference

Krailler RE, Adams PJ, Lane PA. Quantitation of doretinel in a topical gel using high performance liquid chromatography with solid phase extraction sample clean up. *J Liq Chromatogr.* 1991; 14: 2383–98.

Doxazosin Mesylate

Chemical Name

1-(4-Amino-6,7-dimethoxy-2-quinazolinyl)-4-[(2,3-dihydro-1,4-benzodioxin-2-yl)carbonyl]piperazine monomethanesulfonate

Other Names

Cardenalin, Cardura, Doxadura, Zoxan

Form	Molecular Formula	MW	CAS
Doxazosin	$C_{23}H_{25}N_5O_5$	451.5	74191-85-8
Doxazosin mesylate	$C_{23}H_{25}N_5O_5 \cdot CH_4O_3S$	547.6	77883-43-3

Solubility

Doxazosin mesylate is slightly soluble in water (0.8% at 25 °C). It is freely soluble in dimethylsulfoxide, soluble in dimethylformamide, and slightly soluble in methanol and ethanol.

Method

Bakshi et al. developed a stability-indicating HPLC method for the determination of doxazosin in the presence of its degradation products. The liquid chromatograph consisted of a Waters model 600E pump, a model 996 photodiode-array detector, and a model 717 autosampler. The stationary phase was a Waters Spherisorb C_{18} column (250 × 4.6 mm, 5-μm particle size). The mobile phase was acetonitrile, water, glacial acetic acid, and diethylamine (65:5:1:0.02, vol/vol/vol). The flow rate was 1 mL/min. UV detection was performed at 254 nm. The injection volume was 5 μL. Under these conditions, the retention time for doxazosin was about 15.3 minutes.

The doxazosin solutions were (1) heated at 80 °C for 10 days, (2) treated with 0.1 N hydrochloric acid and heated at 80 °C for 90 hours, (3) treated with 0.1 N sodium hydroxide solution and heated at 80 °C for 30 minutes, (4) treated with 3% hydrogen peroxide at room temperature for 6 hours, (5) treated with 30% hydrogen peroxide for 24 hours, (6) exposed to the light, or (7) subjected to dry heat at 50 °C for 3 months. The peak of the intact drug was well resolved from peaks of its degradation products. The absence of any coeluting peak was also confirmed by photodiode-array peak purity analysis.

A standard curve for doxazosin was constructed from 50 to 500 μg/mL. The correlation coefficient was greater than 0.9996. Intraday and interday coefficients of variation were less than 1.3 and 1.9%, respectively.

Reference

Bakshi M, Ojha T, Singh S. Validated specific HPLC methods for determination of prazosin, terazosin and doxazosin in the presence of degradation products formed under ICH-recommended stress conditions. *J Pharm Biomed Anal*. 2004; 34: 19–26.

Doxorubicin

Chemical Name

(8S-cis)-10-[(3-Amino-2,3,6-trideoxy-α-L-lyxo-hexopyranosyl)oxy]-7,8,9,10-tetrahydro-6,8,11-trihydroxy-8-(hydroxyacetyl)-1-methoxy-5,12-naphthacenedione

Other Names

Adriamycin, Rubex

Form	Molecular Formula	MW	CAS
Doxorubicin	$C_{27}H_{29}NO_{11}$	543.5	23214-92-8
Doxorubicin hydrochloride	$C_{27}H_{29}NO_{11}$.HCl	580.0	25316-40-9

Appearance

Doxorubicin hydrochloride is an orange-red hygroscopic crystalline powder.

Solubility

Doxorubicin hydrochloride is soluble in water and methanol but practically insoluble in acetone, benzene, chloroform, and ethyl ether.

Method 1

Mayron and Gennaro used an HPLC method to assess the stability of doxorubicin hydrochloride with granisetron hydrochloride during simulated Y-site administration. The liquid chromatograph consisted of a piston pump with a pulse dampener, a rotary injection port with a 20-μL loop, a variable-wavelength UV detector, and an integrator. The stationary phase was a C_{18} reversed-phase column (300 × 4.6 mm, 5-μm particle size). The mobile phase was a mixture of acetonitrile and 0.1 M monobasic sodium phosphate dihydrate (20:80, vol/vol) adjusted to pH 4.2 with phosphoric acid. The flow rate was 2.5 mL/min. UV detection was performed at 300 nm.

Samples were diluted with mobile phase. The retention times for granisetron and doxorubicin were 3.32 and 6.48 minutes, respectively.

The stability-indicating capability of the analytical method was demonstrated by an accelerated degradation study. Solutions of doxorubicin hydrochloride and granisetron hydrochloride were adjusted to pH 2 and 11, boiled for 1 hour, readjusted to pH 5, diluted with the mobile phase, and analyzed. The degradation product peaks did not interfere with the intact doxorubicin or granisetron peaks.

Reference

Mayron D, Gennaro AR. Stability and compatibility of granisetron hydrochloride in i.v. solutions and oral liquids and during simulated Y-site injection with selected drugs. *Am J Health Syst Pharm*. 1996; 53: 294–304.

Method 2

Using HPLC, Nyhammar et al. determined the stability of doxorubicin hydrochloride and vincristine sulfate in two portable infusion-pump reservoirs. The computer-controlled liquid chromatographic system consisted of an LDC/Milton ConstaMetric III pump, a Waters 717 autosampler, a Shimadzu SPD-6AV variable-wavelength UV-visible light detector, and a Hichrom Nucleosil 100-5CN column (150 × 4.6 mm, 5-μm particle size). The mobile phase was a mixture of acetonitrile, methanol, and 20 mM ammonium dihydrogen phosphate aqueous solution (pH 4.5) (20:20:60, vol/vol/vol) containing sodium heptanesulfonate at a final concentration of 10 mM as an ion-pairing agent. The flow rate was 1.0 mL/min. UV detection was performed at 297 nm.

Each sample was diluted 1:50 with 0.9% sodium chloride injection. The injection volume was 25 μL. Under these conditions, retention times for doxorubicin and vincristine were 4.1 and 11.8 minutes, respectively.

To demonstrate that the assay was stability indicating, solutions of doxorubicin hydrochloride were mixed with 0.1 M sodium hydroxide (final pH 13) or 0.1 M hydrochloric acid (final pH 1). Degradation product peaks did not interfere with the intact doxorubicin peak.

Calibration curves for doxorubicin hydrochloride were constructed and were linear for drug concentrations from 5 to 100 µg/mL. The correlation coefficient was 0.999. The intraday and interday coefficients of variation were 0.45 and 2.8%, respectively.

Reference
Nyhammar EK, Johansson SV, Seiving BE. Stability of doxorubicin hydrochloride and vincristine sulfate in two portable infusion-pump reservoirs. *Am J Health Syst Pharm.* 1996; 53: 1171–3.

Method 3
Rochard et al. evaluated the stability of doxorubicin hydrochloride in ethylene vinyl acetate portable infusion-pump reservoirs. The chromatograph included a Waters model 510 reciprocating piston pump, a Waters model 710A WISP autosampler, a Waters model 990 photodiode-array detector, and a Waters model 990 computing integrator. The stationary phase was a Nucleosil C_{18} column (150 × 4.6 mm, 5-µm particle size). The mobile phase consisted of 70% methanol and 30% 1-heptanesulfonic acid solution (0.9 g/L). The flow rate was 1.0 mL/min. The UV detection wavelength was 232 nm. The retention time was 4.9 minutes.

The stability-indicating nature of the assay was demonstrated by an accelerated degradation study. Doxorubicin hydrochloride was partially decomposed by the addition of 0.1 N hydrochloric acid. The doxorubicin hydrochloride peak was detected without interference from its degradation product.

Standard curves were constructed daily from 2 to 20 µg/mL. Their correlation coefficients were greater than 0.990. Intraday and interday coefficients of variation of the assay were 1.38 and 1.28%, respectively.

Reference
Rochard EB, Barthes DMC, Courtois PY. Stability of fluorouracil, cytarabine, or doxorubicin hydrochloride in ethylene vinylacetate portable infusion-pump reservoirs. *Am J Hosp Pharm.* 1992; 49: 619–23.

Method 4
Stiles and Allen determined the stability of doxorubicin hydrochloride in portable pump reservoirs. The high-performance liquid chromatographic system consisted of a Waters model 501 pump, a model U6K universal injector, a model 441 UV detector, and a model 745 data module. The stationary phase was a Waters Nova-Pak C_{18} column. The mobile phase consisted of 65% methanol and 35% 0.05 M monobasic potassium phosphate at pH 4.8 and included 0.0014 M tetrabutylammonium hydroxide. The flow rate was 1.0 mL/min. UV detection was performed at 254 nm and 0.02 AUFS.

Samples were diluted, and the injection volume was 15 µL.

The method was determined to be stability indicating by placing samples of doxorubicin hydrochloride solution in glass ampules and storing them at 90 °C. Degradation product peaks did not interfere with the intact doxorubicin peak.

Standard curves were constructed and had correlation coefficients of 0.99.

References

Stiles ML, Allen LV Jr. Stability of doxorubicin hydrochloride in portable pump reservoirs. *Am J Hosp Pharm.* 1991; 48: 1976–7.

Allen LV, Stiles ML, Prince SJ, et al. Stability of 14 drugs in the latex reservoir of an elastomeric infusion device. *Am J Health Syst Pharm.* 1996; 53: 2740–3.

Method 5

Walker et al. studied the stability of doxorubicin hydrochloride in syringes and glass vials. The liquid chromatograph included a Waters model EP600 isocratic solvent-delivery pump, a Waters model 712 WISP autosampler, a Kratos SF783 variable-wavelength detector, and a Spectra-Physics SP4270 integrator. The stationary phase was a Beckman Ultrasphere ODS reversed-phase C_{18} column (250 × 4.2 mm, 5-μm particle size). The mobile phase was acetonitrile–0.2% triethylammonium phosphate buffer adjusted to pH 2.5 with phosphoric acid (30:70, vol/vol). The flow rate was 1.5 mL/min. Column effluent was monitored at 230 nm.

Samples were injected without dilution. The injection volume was 1 μL. The retention time for doxorubicin was between 130 and 140 seconds.

The stability-indicating capability of the method was documented by accelerated degradation of doxorubicin hydrochloride with 1 N hydrochloric acid in a water bath at 90 °C for up to 300 minutes. The doxorubicin peak was separated from the peaks of its degradation products.

A standard curve for doxorubicin hydrochloride was constructed from 0.25 to 1.5 mg/mL.

Reference

Walker S, Lau D, DeAngelis C, et al. Doxorubicin stability in syringes and glass vials and evaluation of chemical contamination. *Can J Hosp Pharm.* 1991; 44: 71–88.

Method 6

Wood et al. investigated the shelf life of doxorubicin hydrochloride when reconstituted with water for injection, 5% dextrose injection, and 0.9% sodium chloride injection. The HPLC system included an Altex 100A pump, a Rheodyne model 7125 sample injector equipped with a 10-μL injection loop, a Pye Unicam variable-wavelength detector, and a JJ Instruments CR452 chart recorder. The stationary phase was a Shandon ODS Hypersil reversed-phase stainless steel column (100 × 4.6 mm, 5-μm particle size). The mobile phase was a mixture of acetonitrile and water (40:60, vol/vol). Ten drops of diethylamine were added to each liter of mobile phase, which was then adjusted to pH 2.5 with 10% phosphoric acid. The flow rate was 1.4 mL/min. UV detection was performed at 290 nm and 0.16 AUFS. The retention time for doxorubicin was about 3.7 minutes (estimated from the published chromatogram).

The analytical method was demonstrated to be stability indicating by comparison of doxorubicin hydrochloride standard solution with partially degraded and fully degraded solutions. Degradation was accelerated by subjecting the solutions of doxorubicin hydrochloride to extremes of temperature and pH. No degradation product peaks interfered with the quantification of doxorubicin.

References

Wood MJ, Irwin WJ, Scott DK. Stability of doxorubicin, daunorubicin, and epirubicin in plastic syringes and minibags. *J Clin Pharm Ther.* 1990; 15: 279–89.

Wood MJ, Irwin WJ, Scott DK. Photodegradation of doxorubicin, daunorubicin, and epirubicin measured by high-performance liquid chromatography. *J Clin Pharm Ther*. 1990; 15: 291–300.

Method 7

Bekers et al. evaluated by HPLC analysis the effect of cyclodextrins on doxorubicin stability in acidic aqueous media. The liquid chromatograph consisted of a Waters model M6000 pump, a model U6K injector, a Spectroflow 757 absorbance detector, and a Spectra-Physics model 4270 integrator. The stationary phase was a Merck LiChrosorb RP8 column (300 × 3.9 mm, 10-μm particle size). The mobile phase consisted of 0.01 M sodium chloride (adjusted to pH 2.25 with perchloric acid) and acetonitrile (45:55, vol/vol). The flow rate was 1.5 mL/min. UV detection was performed at 480 nm. The sample injection volume was 20 μL.

The method was reported to be stability indicating.

Standard curves with correlation coefficients of greater than 0.999 were generated.

Reference

Bekers O, Beijnen JH, Groot Bramel EH, et al. Effect of cyclodextrins on anthracycline stability in acidic aqueous media. *Pharm Weekbl [Sci]*. 1988; 10: 207–12.

Doxylamine Succinate

Chemical Name

N,N-Dimethyl-2-[α-methyl-α-(2-pyridyl)benzyloxy]ethylamine hydrogen succinate

Other Name

Unisom

Form	Molecular Formula	MW	CAS
Doxylamine succinate	$C_{17}H_{22}N_2O.C_4H_6O_4$	388.5	562-10-7

Appearance

Doxylamine succinate is a white or creamy white powder.

Solubility

Doxylamine succinate has solubilities of approximately 1 g/mL in water and 0.5 g/mL in alcohol at 25 °C.

pK$_a$
Doxylamine succinate has pK$_a$ values of 5.8 and 9.3.

Method
Fong and Eickhoff reported an HPLC method for the determination of pseudoephedrine hydrochloride, doxylamine succinate, and dextromethorphan hydrobromide in a commercial cough-cold liquid formulation. The instrument consisted of two Beckman model 112 pumps, a model 165 variable-wavelength UV-visible detector, a Waters WISP 710B autosampler, and a Beckman model 450 data system. The stationary phase was an Analytichem Sepralyte CN column (150 × 4.6 mm, 5-µm particle size). The mobile phase consisted of acetonitrile, methanol, and an aqueous buffer (75:10:15, vol/vol/vol). The buffer solution contained 5 mM dibasic potassium phosphate and 59 mM triethylamine, adjusted to pH 5.3 with glacial acetic acid. The flow rate was 2.0 mL/min. UV detection was performed at 262 nm and 0.03 AUFS.

The sample was diluted 1:10 with water and centrifuged. The injection volume was 20 µL. Under these conditions, the retention times for pseudoephedrine hydrochloride, doxylamine succinate, and dextromethorphan hydrobromide were about 4.0, 4.9, and 5.7 minutes, respectively.

The method was reported to be stability indicating since it separated chemically similar compounds.

Standard curves were constructed from 10 to 200% of the normal working concentration levels. Intraday and interday coefficients of variation were less than 1 and 1.2%, respectively. The limit of detection was 0.5 µg/mL.

Reference
Fong GW, Eickhoff WM. Liquid chromatographic determination of amines in complex cough-cold formulations. *Int J Pharm*. 1989; 53: 91–7.

Droperidol

Chemical Name
1-[1-[3-(4-Fluorobenzoyl)propyl]-1,2,3,6-tetrahydro-4-pyridyl]benzimidazolin-2-one

Other Names
Dridol, Droleptan, Inapsine

Form	Molecular Formula	MW	CAS
Droperidol	C$_{22}$H$_{22}$FN$_3$O$_2$	379.4	548-73-2

Appearance
Droperidol is a white to light tan amorphous or microcrystalline powder.

Solubility
Droperidol is practically insoluble in water. It is slightly soluble in alcohol.

Method 1
Trabelsi et al. developed an HPLC method to study the degradation of droperidol. The liquid chomatograph consisted of a Shimadzu model LC-10AT VP pump, a model SPD-10A VP variable UV-visible detector, a model C-R8A integrator, and a Rheodyne model 7725i injector. The stationary phase was a Supelco LC-ABZ$^+$ column (250 × 4.6 mm). The mobile phase consisted of methanol and 0.05 M monobasic sodium phosphate (40:60, vol/vol), adjusted to pH 4.5. The flow rate was 1.0 mL/min. UV detection was performed at 248 nm.

Sample injection volume was 20 µL. Under these conditions, the retention time of droperidol was about 15 minutes (estimated from the published chromatogram).

Droperidol samples were treated with 1 N sodium hydroxide or 1 N hydrochloric acid and refluxed for 5 hours or treated with hydrogen peroxide and heated at 80 °C for 2 hours. Degradation product peaks were well separated from the intact drug peak.

A standard curve for droperidol was constructed from 10 to 40 µg/mL. The correlation coefficient was 0.9994. The limit of detection and limit of quantitation were 0.1 and 0.3 µg/mL, respectively.

Reference
Trabelsi H, Guettat S, Bouzouita K, et al. LC determination and degradation study of droperidol. *J Pharm Biomed Anal.* 2002; 28: 453–62.

Method 2
Dolezalova studied the effect of temperature on the stability of droperidol injection. The HPLC system consisted of a Varian model 8500 pump, a Valco model UHP-7K injector with a 5-µL loop, a Varichrom model UV50 variable-wavelength detector or a Variscan model 635 UV detector, and a Spectra-Physics model SP4100 integrator. The stationary phase was an Ultrasphere ODS column (250 × 4.6 mm, 5-µm particle size). The mobile phase was a mixture of methanol and 0.02 M phosphate buffer (pH 6.8) (65:35, vol/vol). The flow rate was 1.5 mL/min. UV detection was performed at 230 nm and 2.0 AUFS.

Droperidol injection solutions were diluted 1:1 before injection. Under these conditions, the retention time for droperidol was about 14.2 minutes (estimated from the published chromatogram).

The stability-indicating nature of the method was shown by thermal degradation of droperidol. Droperidol injection solutions were stored at room temperature for 20 months, at 60 °C for 4 months, and at 70 °C for 4 months. The intact droperidol peak was well separated from its degradation product peaks. Retention times for *p*-hydroxybenzoic acid, 2-benzimidazolinone, methylparaben, 4′-fluoro-4-(4-oxopiperidino)-butyrophenone, propylparaben, and droperidol were about 1.6, 4.2, 4.7, 6.0, 6.8, and 14.2 minutes, respectively (estimated from the published chromatogram).

A standard curve for droperidol was generated from 0.38 to 1.31 mg/mL. The correlation coefficient was 0.9996.

Reference

Dolezalova M. Separation and determination of droperidol, methyl- and propylparaben and their degradation products by high-performance liquid chromatography. *J Chromatogr.* 1984; 286: 323–30.

Dyclonine Hydrochloride

Chemical Name

4′-Butoxy-3-piperidinopropiophenone hydrochloride

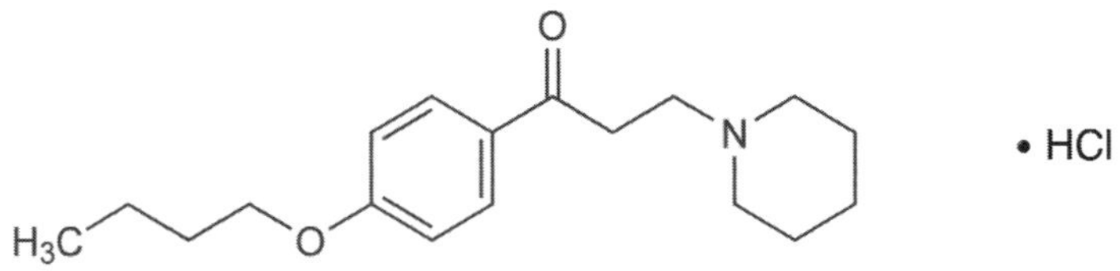

• HCl

Other Names

Dyclone, Sucrets

Form	Molecular Formula	MW	CAS
Dyclonine hydrochloride	$C_{18}H_{27}NO_2$.HCl	325.9	536-43-6

Appearance

Dyclonine hydrochloride occurs as white crystals or as a white crystalline powder.

Solubility

Dyclonine hydrochloride is soluble in water and alcohol.

Method 1

Bhagat et al. reported an HPLC determination of dyclonine hydrochloride in the presence of its degradation products. The instrument consisted of a Waters model 6000A pump, a model U6K loop injector, a Perkin-Elmer model LC-55 variable-wavelength UV detector, and a Hewlett-Packard model 3390A integrator-plotter. The stationary phase was a Waters µBondapak phenyl column (300 × 3.9 mm, 10-µm particle size). The mobile phase was a mixture of acetonitrile, tetrahydrofuran, and 0.2 M ammonium acetate buffer (pH 6.0) (46:9:45, vol/vol/vol) and was delivered isocratically at a flow rate of 2.0 mL/min. UV detection was performed at 282 nm and 0.64 AUFS.

Sample solutions were prepared by diluting 1.0 mL of the sample in a 25-mL volumetric flask with 18 mL of 30% tetrahydrofuran in acetonitrile; the solution was then adjusted to pH 6.0 with 1.0 M ammonium hydroxide solution and diluted to the final volume of 25 mL with 0.1 M ammonium acetate buffer (pH 6.0). The injection volume was 20 µL. Under these conditions, the retention time for dyclonine was about 5.8 minutes.

The method was demonstrated to be stability indicating by assaying dyclonine with its degradation products. The intact dyclonine peak was well resolved from its degradation product peaks.

A standard curve for dyclonine was generated from 80 to 180 µg/mL. The correlation coefficient was 0.9999.

Reference
Bhagat HR, Bhargava HN, Williams DA. High-performance liquid chromatographic determination of dyclonine hydrochloride in the presence of its degradation products. *J Pharm Biomed Anal.* 1989; 7: 441–6.

Method 2
Palermo described an HPLC determination of dyclonine hydrochloride and its degradation products in cough lozenges and liquids. The apparatus consisted of a Waters model 6000A pump, a Rheodyne model 7210 injector, a Waters model 440 UV detector, and a Houston Instruments Omniscribe recorder. The stationary phase was a Waters µBondapak C_{18} analytical column (300 × 3.9 mm, 10-µm particle size). The mobile phase was prepared by dissolving 7.7 g of ammonium acetate in 700 mL of distilled water, mixing with 50 mL of glacial acetic acid, and diluting to 2000 mL with methanol. The flow rate was 1.5 mL/min. UV detection was performed at 280 nm and 0.1 AUFS. 4-*n*-Butoxybenzoic acid was used as an internal standard.

Liquid samples were first run through a Florisil column (magnesium silicate, 100–200) to remove extraneous materials. Lozenge samples were dissolved in mobile phase. The sample injection volume was 20 µL.

The method was reported to be stability indicating.

Reference
Palermo PJ. High pressure liquid chromatographic determination of dyclonine hydrochloride and its degradation products in cough lozenges and liquids. *J Pharm Biomed Anal.* 1987; 5: 423–30.

Econazole Nitrate

Chemical Name
(±)-1-[2,4-Dichloro-β-(4-chlorobenzyloxy)phenethyl]imidazole nitrate

Other Name
Spectazole

Form	Molecular Formula	MW	CAS
Econazole nitrate	$C_{18}H_{15}Cl_3N_2O.HNO_3$	444.7	24169-02-6

Appearance
Econazole nitrate is a white or almost white and almost odorless crystalline powder.

Solubility
Econazole nitrate is very slightly soluble in water, soluble in methanol, slightly soluble in alcohol, and sparingly soluble in chloroform and dichloromethane.

Method

Christinat and Zulliger developed an HPLC method for the determination of econazole nitrate in cream and lotion formulations. A Hewlett-Packard model 1081B chromatograph was equipped with a Pye Unicam model PU 4020 UV detector. The stationary phase was a Waters μBondapak C_{18} column (300 × 4 mm). The mobile phase consisted of 200 mL of freshly prepared ammonium carbonate aqueous solution (1 g/1000 mL distilled water), 780 mL of methanol, and 20 mL of tetrahydrofuran. The flow rate was 2.0–2.3 mL/min. UV detection was performed at 220 nm. Miconazole nitrate (60 μg/mL in methanol) was used as an internal standard.

One gram of cream or lotion was weighed into a 100-mL volumetric flask, shaken with 2 mL of tetrahydrofuran, mixed with 20 mL of internal standard solution, brought to volume with a mixture of methanol and water (5:2, vol/vol), stored in a refrigerator for 45 minutes, and then filtered through a 0.22-μm Millipore Teflon filter before injection. The sample injection volume was 10 μL. Under these conditions, the retention times for econazole nitrate and miconazole nitrate were about 4.5 and 7.3 minutes, respectively (estimated from the published chromatogram).

The assay was stated to be stability indicating since the intact drug was resolved from its degradation products as well as from the inactive ingredients.

A standard curve for econazole nitrate was constructed from 0.5 to 3.0 μg per injection. The correlation coefficient was 1.000.

Reference

Christinat R, Zulliger HW. Stability indicating HPLC method for the determination of econazole nitrate in cream and lotion formulations. *Drug Res.* 1984; 34: 551–3.

Edetate Disodium (EDTA)

Chemical Name

Disodium dihydrogen ethylenediaminetetraacetate dihydrate

Other Names

Disodium EDTA, Disotate, Limclair

Form	Molecular Formula	MW	CAS
Edetate disodium	$C_{10}H_{14}N_2Na_2O_8.2H_2O$	372.2	6381-92-6
Edetate trisodium	$C_{10}H_{13}N_2Na_3O_8$	358.2	150-38-9

Appearance

Disodium edetate occurs as a white odorless crystalline powder.

Solubility

Edetate disodium is soluble in water, sparingly soluble in alcohol, and practically insoluble in ether.

Method 1

Kord et al. developed a novel HPLC method for the determination of EDTA in a cataract-inhibiting ophthalmic drug. The Perkin-Elmer liquid chromatograph consisted of a model 250 pump and a model 295 variable-wavelength detector. The stationary phase was an Alltech Anion/R column (250 × 4.1 mm, 10-μm particle size). The mobile phase consisted of 25% (vol/vol) acetonitrile in a solution of 2 mM cupric nitrate and 11 mM nitric acid in deionized water (pH 3.0). The flow rate was 1 mL/min. UV detection was performed at 250 nm. The injection volume was 20 μL. Under these conditions, the retention time for EDTA was about 8.2 minutes (estimated from the published chromatogram).

The method was reported to be stability indicating since no peaks from the ophthalmic formulation ingredients, drugs, or its degradation products interfered with the peak of EDTA.

A standard curve for EDTA was constructed from 52 to 78 μg/mL. The correlation coefficient was 0.9999.

Reference

Kord AS, Tumanova I, Matier WL. A novel HPLC method for determination of EDTA in a cataract inhibiting ophthalmic drug. *J Pharm Biomed Anal*. 1995; 13: 575–80.

Method 2

Bauer et al. described an HPLC assay for disodium EDTA in ophthalmic preparations. The liquid chromatographic system consisted of a Waters model 6000A pump, a DuPont variable-wavelength detector, and a Hewlett-Packard 3388 integrator. The stationary phase was a Waters μBondapak C_{18} column (300 × 4 mm). The mobile phase was prepared by adding 10 mL of 1 M tetra-*n*-butylammonium hydroxide solution to 910 mL of water, adjusting the solution to pH 7.5 with phosphoric acid, and mixing with 80 mL of methanol. UV detection was performed at 254 nm.

Cupric sulfate solution 0.2% was prepared by dissolving 3 g of cupric sulfate pentahydrate in 1 L of water. The internal standard solution was prepared by dissolving approximately 200 mg of nitrilotriacetic acid disodium salt in water and then diluting with water to 100 mL. A sample (5 mL) was transferred into a 25-mL volumetric flask, mixed with 5 mL of internal standard solution, and diluted to volume with 0.2% cupric sulfate solution. The injection volume was 30 μL. Under these conditions, retention times for EDTA and nitrilotriacetic acid were about 3.7 and 8.9 minutes, respectively.

The method was reported to be stability indicating.

A standard curve for EDTA was generated from 0.2 to 2.1 mg/mL. The correlation coefficient was 0.9999.

Reference

Bauer J, Heathcote D, Krogh S. High-performance liquid chromatographic, stability-indicating assay for disodium EDTA in ophthalmic preparations. *J Chromatogr*. 1986; 369: 422–5.

Efavirenz

Chemical Name

(S)-6-Chloro-4-(cyclopropylethynyl)-1,4-dihydro-4-(trifluoromethyl)-2H-3,1-benzox-azin-2-one

Other Name

Sustiva

Form	Molecular Formula	MW	CAS
Efavirenz	$C_{14}H_9ClF_3NO_2$	315.7	154598-52-4

Appearance

Efavirenz is a white to slightly pink crystalline powder.

Solubility

Efavirenz is practically insoluble in water.

pK$_a$

Efavirenz has a pK$_a$ of 10.2.

Method

Montgomery et al. developed an assay for the analysis of efavirenz and its related substances in the drug substance and in a capsule formulation. Several systems were used. One was a Hewlett-Packard 1050 series that included a pump, an autosampler, a column oven, and a detector. The stationary phase was a Zorbax SB-CN column (150 × 4.6 mm). Solvent A was a mixture of water containing 0.05% trifluoroacetic acid and methanol (90:10, vol/vol). Solvent B was a mixture of water containing 0.05% trifluoroacetic acid and methanol (10:90, vol/vol). The mobile phase was delivered linearly from 60% A to 50% A over 16 minutes, to 35% A over 7 minutes, to 30% A over 5 minutes, then to 20% A over 1 minute; it was then held for 2 minutes, returned to 60% A in 1 minute, and equilibrated for 8 minutes. The flow rate was 1.5 mL/min. Column temperature was maintained at 40 °C. UV detection was performed at 250 nm. The injection volume was 30 µL. The run time was 40 minutes. Under these conditions, the retention time for efavirenz was about 14 minutes (estimated from the published chromatogram).

The stability-indicating nature of the method was demonstrated by spiking an efavirenz sample with its known synthetic impurities and other related substances. Efavirenz was resolved from these compounds.

Standard curves for efavirenz were constructed from 0.02 to 2.0 μg/mL and from 120 to 360 μg/mL. Correlation coefficients were greater than 0.9999. The limit of detection was less than 0.01% and the limit of quantitation was less than 0.05%.

Reference
Montgomery ER, Edmanson AL, Cook SC, et al. Development and validation of a reverse-phase HPLC method for analysis of efavirenz and its related substances in the drug substance and in a capsule formulation. *J Pharm Biomed Anal*. 2001; 25: 267–84.

Emetine Hydrochloride

Chemical Name
(2*S*,3*S*,11b*S*)-3-Ethyl-1,3,4,6,7,11b-hexahydro-9,10-dimethoxy-2-[(1*R*)-1,2,3,4-tetra-hydro-6,7-dimethoxy-1-isoquinolylmethyl]-2*H*-benzo[*a*]quinolizine dihydrochloride heptahydrate

Other Names
Cophylac, Ipeca

Form	Molecular Formula	MW	CAS
Emetine hydrochloride	$C_{29}H_{40}N_2O_4.2HCl.7H_2O$	679.7	79300-08-6

Appearance
Emetine hydrochloride occurs as a white or slightly yellow odorless crystalline powder.

Solubility
Emetine hydrochloride is freely soluble in water and alcohol.

Method
Elvidge et al. reported an assay for the analysis of emetine and cephaeline at very low levels in pharmaceutical formulations. The system consisted of a Waters model 6000A pump, a WISP autosampler, a Perkin-Elmer LS4 fluorescence spectrophotometer, and an LCI 100 integrator. The stationary phase was a Waters μBondapak C_{18} column (150 × 3.9 mm). The column temperature was kept at 35 °C. The mobile phase was prepared by dissolving 1.0 g of 1-heptanesulfonic acid sodium salt in water, mixing with

400 mL of methanol and 1 mL of phosphoric acid, and diluting to 1 L with water. The flow rate was 2 mL/min. The fluorescence detector was set at an excitation wavelength of 276 nm and an emission wavelength of 304 nm. The internal standard was prepared by dissolving 0.1 g of ethyl 4-hydroxybenzoate in 50 mL of acetonitrile and diluting to 500 mL with mobile phase.

Liquid samples were diluted in mobile phase; pastilles were dissolved in a mixture of internal standard solution and mobile phase. The injection volume was 10 μL. Under these conditions, relative retention times for internal standard, cephaeline, and emetine were 1.0, 1.8, and 3.0, respectively.

The stability-indicating capability of the method was demonstrated by accelerated degradation of the drugs. Samples containing emetine and cephaeline were heated at 70 °C in water, 0.1 M hydrochloric acid, or 0.1 M sodium hydroxide. No significant amount of degradation products was detected by HPLC.

A standard curve for emetine was constructed from 0.007 to 0.025 μg per injection.

Reference

Elvidge DA, Johnson GW, Harrison JR. Selective, stability-indicating assay of the major ipecacuanha alkaloids, emetine and cephaeline, in pharmaceutical preparations by high-performance liquid chromatography using spectrofluorometric detection. *J Chromatogr.* 1989; 463: 107–18.

Enalapril

Chemical Names

(S)-1-[N-[1-(Ethoxycarbonyl)-3-phenylpropyl]-L-alanyl]-L-proline
1-[N-[(S)-1-Carboxy-3-phenylpropyl]-L-alanyl]-L-proline 1'-ethyl ester

Other Name

Vasotec

Form	Molecular Formula	MW	CAS
Enalapril	$C_{20}H_{28}N_2O_5$	376.5	75847-73-3
Enalaprilat	$C_{18}H_{24}N_2O_5.2H_2O$	384.4	76420-72-9
Enalapril maleate	$C_{20}H_{28}N_2O_5.C_4H_4O_4$	492.5	76095-16-4

Appearance

Enalaprilat is a white or nearly white hygroscopic crystalline powder. Enalapril maleate is a white to off-white crystalline powder.

Solubility

Enalaprilat is slightly soluble in water and sparingly soluble in methanol. Enalapril maleate has solubilities of 25 mg/mL in water and 80 mg/mL in alcohol at room temperature.

pK_a

Enalapril has a pK_{a1} of 3.0 and a pK_{a2} of 5.4 at 25 °C.

Method

Allen and Erickson evaluated the stability of enalapril maleate in extemporaneously compounded oral liquids. A Hewlett-Packard series 1050 automated high-performance liquid chromatograph included a multisolvent mixing and pumping system, an auto-injector, a diode-array detector, and a computer with Chem Station software. The stationary phase was a Bakerbond C_8 column (250 × 4.6 mm, 5-μm particle size). The mobile phase consisted of water, acetonitrile, and buffer solution (34:15:1, vol/vol/vol). The buffer solution contained 136 g of monobasic potassium phosphate in 800 mL of water, which was adjusted to pH 4.0 and diluted to 1000 mL with water. The mobile phase was delivered isocratically at 1.5 mL/min. UV detection was performed at 215 nm.

Samples were diluted 1:10 before injection. Under these conditions, the retention time for enalapril was 8.8 minutes.

The analytical method was determined to be stability indicating. A composite chromatogram of enalapril after accelerated decomposition showed that degradation product peaks did not interfere with the intact enalapril peak.

A standard curve for enalapril was constructed from 10 to 100 μg/mL. The intraday and interday coefficients of variation were 2.1 and 2.4%, respectively.

Reference

Allen LV Jr, Erickson MA III. Stability of alprazolam, chloroquine phosphate, cisapride, enalapril maleate, and hydralazine hydrochloride in extemporaneously compounded oral liquids. *Am J Health Syst Pharm.* 1998; 55: 1915–20.

Epicillin

Chemical Name

[2S-[2α,5α,6β(S*)]]-6-[(Amino-1,4-cyclohexadien-1-ylacetyl)amino]-3,3-dimethyl-7-oxo-4-thia-1-azabicyclo[3.2.0]heptane-2-carboxylic acid

Other Name

Dexacillin

Form	Molecular Formula	MW	CAS
Epicillin	$C_{16}H_{21}N_3O_4S$	351.4	26774-90-3

Method

Grover et al. determined by HPLC the degradation rate of epicillin at 35 °C in borate buffer (pH 9.2). The Shimadzu LC-10A liquid chromatograph included model LC-10AS pumps, a model

SPD-10A dual-wavelength detector, a model C-R7A data processor, and a Rheodyne model 7125 injector with a 20-µL loop. The stationary phase was a Phenomenex C_{18} column (300 × 3.9 mm, 5-µm particle size). The mobile phase was a mixture of 90% water containing 20 mM monobasic potassium phosphate and 10 mM tetramethylammonium chloride (pH 5.0) and 10% acetonitrile. The flow rate was 1.0 mL/min. UV detection was performed at 220 and 240 nm.

Samples were incubated at 35 °C and assayed. Under these conditions, the retention time for epicillin was 9.6 minutes.

The method was reported to be stability indicating.

A standard curve for epicillin was generated from 5 to 150 µg/mL. The correlation coefficient was greater than 0.99. The coefficient of variation for the analysis of epicillin was less than 1%.

Reference
Grover M, Gulati M, Singh B, et al. Correlation of penicillin structure with rate constants for basic hydrolysis. *Pharm Pharmacol Commun*. 2000; 6: 355–63.

Epinephrine

Chemical Names
(*R*)-4-[1-Hydroxy-2-(methylamino)ethyl]-1,2-benzenediol
(−)-3,4-Dihydroxy-α-[(methylamino)methyl]benzyl alcohol

Other Name
Adrenalin

Form	Molecular Formula	MW	CAS
Epinephrine	$C_9H_{13}NO_3$	183.2	51-43-4
Epinephrine bitartrate	$C_9H_{13}NO_3.C_4H_6O_6$	333.3	51-42-3
Epinephrine hydrochloride	$C_9H_{13}NO_3.HCl$	219.7	55-31-2

Appearance
Epinephrine is a white to nearly white microcrystalline powder or granules. Epinephrine hydrochloride occurs as a fine white powder.

Solubility
Epinephrine is only very slightly soluble in water and alcohol. Epinephrine hydrochloride is readily soluble in water and sparingly soluble in absolute alcohol.

Method 1

Peddicord et al. evaluated the stability of epinephrine hydrochloride in 5% dextrose injection at ambient temperature and humidity. The chromatograph consisted of a Shimadzu model LC-10AS solvent-delivery system, a Shimadzu model SIL-10A autoinjector, a Shimadzu model CTO-10A column oven, a Shimadzu model SPD-10A variable-wavelength UV spectrophotometer, and a Phenomenex C_{18} column (300 × 3.9 mm, 5-µm particle size). The mobile phase consisted of water, methanol, and acetic acid (72.3:25:2.7, vol/vol/vol) and 1.34 g of 1-heptanesulfonic acid. The flow rate was 1.6 mL/min. UV detection was performed at 280 nm. The oven temperature was 50 °C.

Each sample was diluted with 5% dextrose injection. The injection volume was 50 µL.

This method was determined to be stability indicating by accelerated decomposition of epinephrine. Solutions of epinephrine were adjusted to pH 2 with 1.0 N hydrochloric acid and to pH 12 with 10 N sodium hydroxide and then raised to 100 °C. No degradation product peaks interfered with the intact epinephrine peak.

A calibration curve for epinephrine was constructed from 0 to 100.0 µg/mL. The correlation coefficient was greater than 0.999. The intraday and interday coefficients of variation were 0.5 and 0.8%, respectively.

A similar method was used by Williams et al.

References

Peddicord TE, Olsen KM, ZumBrunnen TL, et al. Stability of high-concentration dopamine hydrochloride, norepinephrine bitartrate, epinephrine hydrochloride, and nitroglycerin in 5% dextrose injection. *Am J Health Syst Pharm.* 1997; 54: 1417–9.

Williams DA, Fung EYY, Newton DW. Ion-pair high-performance liquid chromatography of terbutaline and catecholamines with aminophylline in intravenous solutions. *J Pharm Sci.* 1982; 71: 956–8.

Method 2

Larson et al. determined the stability of epinephrine hydrochloride in an extemporaneously compounded topical anesthetic solution. The liquid chromatograph consisted of a Waters model 600E multisolvent-delivery system, a model 712 WISP autosampler, and a model 490E UV detector. The stationary phase was a Waters µBondapak C_{18} column (300 × 3.9 mm, 10-µm particle size) that was maintained at 35 °C. The mobile phase was a mixture of acetonitrile and an aqueous solution (200:800, vol/vol), where the aqueous solution was composed of acetic acid in water (50:930, vol/vol) adjusted to pH 3.0 with 1 N sodium hydroxide. The mobile phase was delivered isocratically at 1.5 mL/min. UV detection was performed at 254 nm. The injection volume was 15 µL. Under these conditions, retention times for epinephrine, lidocaine, and tetracaine were approximately 2.8, 6.3, and 9.5 minutes, respectively.

The assay was determined to be stability indicating. Epinephrine hydrochloride was degraded by hydrogen peroxide, resulting in a degradation product with a retention time of about 4.5 minutes.

Standard curves were constructed from 1.80 to 2.475 mg/mL. The correlation coefficient was 0.999. The intraday and interday coefficients of variation were less than 5%.

Reference

Larson TA, Uden DL, Shilling CG. Stability of epinephrine hydrochloride in an extemporaneously compounded topical anesthetic solution of lidocaine, recepinephrine, and tetracaine. *Am J Health Syst Pharm.* 1996; 53: 659–62.

Method 3

Allen et al. studied the stability of epinephrine hydrochloride with cefpirome using HPLC. One of two systems was utilized. One system included an Alcott model 728 Micromeritics autosampler, a Rheodyne 7010 injector with an Alcott 732 electrically actuated valve, a Waters model 501 solvent-delivery pump, a Waters model 441 UV detector, a Waters model 401 refractive index detector, and a Waters model 745 data module. The other system consisted of an Alcott model 728 Micromeritics autosampler, a Rheodyne 7010 injector with an Alcott 732 electrically actuated valve, a Shimadzu model LA-6A solvent-delivery pump, a Shimadzu model SPD-6A UV detector, and an Orion model 901 micro-processor ion analyzer. The stationary phase was a C_{18} analytical column. The mobile phase consisted of 0.05 M monobasic sodium phosphate containing 519 mg of sodium 1-octanesulfonate and 45 mg of disodium ethylenediaminetetraacetate (edetate) (pH 3.8) mixed in a ratio of 85:15 with methanol. The flow rate was 2 mL/min. UV detection was performed at 280 nm.

Samples were diluted before injection. The injection volume was 20 μL. The retention times for epinephrine and cefpirome were 3.2 and 4.7 minutes, respectively.

The analytical method was demonstrated to be stability indicating by degrading epinephrine with hydrochloric acid, sodium hydroxide, potassium chlorate, heat at 80 °C, or light (150 foot-candles from a tungsten filament source). In each case, epinephrine was suitably separated from its decomposition products on the chromatogram.

A standard curve was constructed from 10 to 100 μg/mL; the correlation coefficient was greater than 0.99. The intraday and interday coefficients of variation of the method were 2.2 and 2.0%, respectively.

References

Allen LV, Stiles ML, Prince SJ, et al. Stability of cefpirome sulfate in the presence of commonly used intensive care drugs during simulated Y-site injection. *Am J Health Syst Pharm.* 1995; 52: 2427–33.

Allen LV, Stiles ML, Wang DP, et al. Stability of bupivacaine hydrochloride, epinephrine hydrochloride, and fentanyl citrate in portable infusion-pump reservoirs. *Am J Hosp Pharm.* 1993; 50: 714–5.

Method 4

Wilson and Forde investigated the stability of epinephrine using an HPLC method. The liquid chromatograph consisted of a Varian model 5000 pump and autosampler, a Waters model 441 UV detector, a Fisher Scientific model 5000 strip-chart recorder, and a Hewlett-Packard model 3354 laboratory automation system. The stationary phase was a Whatman PXS ODS-3 stainless steel column (250 × 4.6 mm, 10-μm particle size). The mobile phase was a mixture of 800 mL of water, 125 mL of methanol, 75 mL of acetoni-trile, 5 mL of 85% phosphoric acid, and 1.08 g of sodium octanesulfonate. The flow rate was 1.5 mL/min. UV detection was performed at 254 nm and 0.01 AUFS.

Samples were diluted to about 0.02 mg/mL with mobile phase. The injection volume was 20 μL. Retention times for epinephrine and milrinone were 6.5 and 9 minutes, respectively.

The method was stated to be stability indicating.

A standard curve was constructed from 0 to 64 μg/mL of epinephrine.

Reference

Wilson TD, Forde MD. Stability of milrinone and epinephrine, atropine sulfate, lidocaine hydrochloride, or morphine sulfate injection. *Am J Hosp Pharm.* 1990; 47: 2504–7.

Method 5

Waraszkiewicz et al. developed a stability-indicating HPLC method for analysis of lidocaine hydrochloride and lidocaine hydrochloride with epinephrine injectable solutions. A Waters model ALC/GPC 204 liquid chromatograph was equipped with a Waters model U6K septumless injector, a Waters model 440 dual-channel fixed-wavelength UV detector, and a Waters μBondapak CN column (300 × 4 mm, 10-μm particle size). The mobile phase consisted of 0.01 M 1-octanesulfonic acid sodium salt, 0.1 mM edetate disodium, 2% (vol/vol) acetic acid, 2% (vol/vol) acetonitrile, and 1% (vol/vol) methanol in high-quality distilled water. The flow rate was 2.0 mL/min. UV detection was performed at 254 nm. The sample injection volume was 2 μL.

The method was determined to be stability indicating. Samples of epinephrine were spiked with the potential degradation products and chromatographed. Other samples of epinephrine were subjected to various conditions of stress such as auto-claving for 5, 10, and 20 cycles, storage for 1 month at 80 and 110 °C, exposure to atmospheric oxygen, addition of 12 ppm of aluminum, and an increase in the solution pH. Degradation products did not interfere with the analysis of the intact epinephrine.

Standard curves were constructed from ~ 80 to ~ 120% of the theoretical labeled quantity. Correlation coefficients were greater than 0.99.

Reference

Waraszkiewicz SM, Milano EA, DiRubio R. Stability-indicating high-performance liquid chromatographic analysis of lidocaine hydrochloride and lidocaine hydrochloride with epinephrine injectable solutions. *J Pharm Sci.* 1981; 70: 1215–8.

Epirubicin

Chemical Name

(8*S-cis*)-10-[(3-Amino-2,3,6-trideoxy-α-L-*arabino*-hexopyranosyl)oxy]-7,8,9,10-tetrahydro-6,8,11-trihydroxy-8-(hydroxyacetyl)-1-methoxy-5,12-naphthacenedione

Other Name
Pharmorubicin

Form

Form	Molecular Formula	MW	CAS
Epirubicin	$C_{27}H_{29}NO_{11}$	543.5	56420-45-2
Epirubicin hydrochloride	$C_{27}H_{29}NO_{11} \cdot HCl$	580.0	56390-09-1

Appearance
Epirubicin hydrochloride occurs as red-orange crystals.

Solubility
Epirubicin hydrochloride is soluble in water.

Method 1
Pujol et al. studied the stability of epirubicin in 0.9% sodium chloride injection. A Hewlett-Packard model 1090 liquid chromatograph was equipped with a diode-array detector and a Hewlett-Packard 3396D integrator. The stationary phase was a Technocroma SA Spherisorb phenyl column (250 × 4.6 mm, 5-μm particle size). The mobile phase consisted of 1 mM phosphoric acid and 2.4 mM sodium hydroxide, containing 45% sodium dodecyl sulfate buffer (1 mol/L). The pH of the buffer was adjusted to 4.6 with 50% acetonitrile and 5% methanol by volume. The flow rate was 1.8 mL/min. UV detection was performed at 254 nm. Under these conditions, the retention time for epirubicin was 6.2 minutes.

The stability-indicating nature of the method was demonstrated by accelerated degradation of epirubicin. An epirubicin solution 2 mg/mL was heated at 70 °C until more than 50% degradation of epirubicin was achieved. The pH of a second epirubicin solution was adjusted to 1.5 with 0.1 M hydrochloric acid, and another sample was adjusted to pH 11.8 with 0.1 M sodium hydroxide. Methylparaben (preservative) and degradation products did not coelute with epirubicin.

Standard curves for epirubicin were constructed from 0.2 to 2.5 mg/mL; the correlation coefficients were 0.998. Intraday and interday coefficients of variation were less than 5 and 10%, respectively.

Reference
Pujol M, Munoz M, Prat J, et al. Stability study of epirubicin in NaCl 0.9% injection. *Ann Pharmacother*. 1997; 31: 992–5.

Method 2

Wood et al. determined by HPLC the shelf life of epirubicin hydrochloride when reconstituted with water for injection, 5% dextrose injection, and 0.9% sodium chloride injection. The system included an Altex 100A pump, a Rheodyne model 7125 sample injector equipped with a 10-μL injection loop, a Pye Unicam variable-wavelength detector, and a JJ instruments CR452 chart recorder. The stationary phase was a Shandon ODS Hypersil reversed-phase stainless steel column (100 × 4.6 mm, 5-μm particle size). The mobile phase was a mixture of acetonitrile and water (50:50, vol/vol). Ten drops of diethylamine were added to each liter of mobile phase, which was then adjusted to pH 2.5 with 10% phosphoric acid. The flow rate was 1.3 mL/min. UV detection was performed at 290 nm and 0.16 AUFS.

The analytical method was determined to be stability indicating by comparison of the epirubicin hydrochloride standard solution to partially degraded and fully degraded solutions. Degradation was accelerated by subjecting solutions of epirubicin hydrochloride to extremes of temperature and pH. Degradation product peaks did not interfere with the epirubicin peak.

References

Wood MJ, Irwin WJ, Scott DK. Stability of doxorubicin, daunorubicin, and epirubicin in plastic syringes and minibags. *J Clin Pharm Ther.* 1990; 15: 279–89.

Wood MJ, Irwin WJ, Scott DK. Photodegradation of doxorubicin, daunorubicin, and epirubicin measured by high-performance liquid chromatography. *J Clin Pharm Ther.* 1990; 15: 291–300.

Method 3

Bekers et al. investigated by HPLC analysis the effect of cyclodextrins on epirubicin stability in acidic aqueous media. The liquid chromatograph consisted of a Waters model M6000 pump, a model U6K injector, a Spectroflow 757 absorbance detector, and a Spectra-Physics model 4270 integrator. The stationary phase was a Merck LiChrosorb RP8 column (300 × 3.9 mm, 10-μm particle size). The mobile phase consisted of 0.01 M sodium chloride (adjusted to pH 2.25 with perchloric acid) and acetonitrile (45:55, vol/vol). The flow rate was 1.5 mL/min. UV detection was performed at 480 nm. The sample injection volume was 20 μL.

The method was reported to be stability indicating.

Standard curves with the correlation coefficient being greater than 0.999 were generated.

Reference

Bekers O, Beijnen JH, Groot Bramel EH, et al. Effect of cyclodextrins on anthracycline stability in acidic aqueous media. *Pharm Weekbl [Sci].* 1988; 10: 207–12.

Ergotamine Tartrate

Chemical Name

(5'S)-12'-Hydroxy-2'-methyl-5'-benzylergotaman-3',6',18-trione tartrate

Other Names

Bellamine, Ergomar

Form	Molecular Formula	MW	CAS
Ergotamine tartrate	$(C_{33}H_{35}N_5O_5)_2 \cdot C_4H_6O_6$	1313.4	379-79-3

Appearance

Ergotamine tartrate occurs as colorless crystals or as a white to yellowish-white crystalline powder.

Solubility

Ergotamine tartrate is slightly soluble in water and in alcohol.

pK$_a$

Ergotamine tartrate has a pK$_a$ of 6.3.

Method

Jalal et al. described an HPLC method for the determination of ergotamine and cyclizine in commercial products. The system consisted of a DuPont model 8800 pump, a Rheodyne 7125 injector, a DuPont variable-wavelength detector, and a Spectra-Physics model SP4100 integrator. The stationary phase was a Merck reversed-phase C$_{18}$ polymeric column (250 × 4.6 mm, 5-μm particle size). The mobile phase was a mixture of acetonitrile, water, and triethylamine (35:64:1, vol/vol/vol) containing 0.01 M ammonium acetate and adjusted to pH 3.7 with glacial acetic acid. The flow rate was 1.5 mL/min. UV detection was performed at 254 nm and 0.08 AUFS. Ethylparaben 5.6 μg/mL in methanol was used as an internal standard.

A powdered sample was weighed, sonicated for 3 minutes with 20 mL of internal standard solution in a 25-mL volumetric flask, and filled to volume with internal standard. It was diluted again with internal standard solution to approximately 0.032 mg/mL of ergotamine and 0.800 mg/mL of cyclizine. The injection volume was 10 μL. Under these conditions, retention times for ergotamine, ethylparaben, and cyclizine were about 3.5, 4.1, and 4.9 minutes, respectively.

The method was stability indicating since no interference due to excipients was detected.

A calibration curve for ergotamine was constructed from 16.0 to 48.0 µg/mL. The correlation coefficient was 0.999. The limit of detection was 1.5 ng.

Reference

Jalal IM, Sasa SI, Yasin TA. Determination of ergotamine tartrate and cyclizine hydrochloride in pharmaceutical tablets by reverse phase HPLC. *Anal Lett*. 1988; 21: 1561–77.

Ertapenem

Chemical Name

(4R,5S,6S)-3-([[(3S,5S)-[(m-carboxyphenyl)carbamoyl]-3-pyrrolidinyl]thio)-6-[(1R)-1-hydroxyethyl]-4-methyl-7-oxo-1-azabicyclo[3.2.0]-hept-2-ene-2-carboxylate as the sodium salt

Other Name

Invanz

Form	Molecular Formula	MW	CAS
Ertapenem as sodium	$C_{22}H_{24}N_3NaO_7S$	497.5	153773-82-1

Method

Using an HPLC method, McQuade et al. studied the stability of ertapenem in various intravenous infusion solutions at different concentrations and multiple temperatures. An Agilent model 1090-II system included a four-solvent portioning valve solvent-delivery system, an autosampler, and a Spectra 100 UV detector. The stationary phase was a Metachem Inertsil phenyl analytical column (250 mm). UV detection was performed at 230 nm. Under these conditions, the retention time of ertapenem was 16.5 minutes.

This method was reported to be stability indicating since it allowed the separation of the peak of the intact ertapenem from the peaks of its degradation products.

The coefficient of variation of the assay was less than 1%.

Reference

McQuade MS, VanNostrand V, Schariter J, et al. Stability and compatibility of reconstituted ertapenem with commonly used i.v. infusion and coinfusion solutions. *Am J Health Syst Pharm*. 2004; 61: 38–45.

Erythromycin

Chemical Name

(2*R*,3*S*,4*S*,5*R*,6*R*,8*R*,10*R*,11*R*,12*S*,13*R*)-5-(3-Amino-3,4,6-trideoxy-*N*,*N*-dimethyl-β-D-*xylo*-hexopyranosyloxy)-3-(2,6-dideoxy-3-*C*,3-*O*-dimethyl-α-L-*ribo*-hexopyrano-syloxy)-13-ethyl-6,11,12-trihydroxy-2,4,6,8,10,12-hexamethyl-9-oxotridecan-13-olide

Other Names

ERYC, Erythrocin

Form	Molecular Formula	MW	CAS
Erythromycin A	$C_{37}H_{67}NO_{13}$	733.9	114-07-8
Erythromycin acistrate	$C_{39}H_{69}NO_{14}.C_{18}H_{36}O_2$	1060.5	96128-89-1
Erythromycin estolate	$C_{40}H_{71}NO_{14}.C_{12}H_{26}O_4S$	1056.4	3521-62-8
Erythromycin ethylsuccinate	$C_{43}H_{75}NO_{16}$	862.1	41342-53-4
Erythromycin gluceptate	$C_{37}H_{67}NO_{13}.C_7H_{14}O_8$	960.1	304-63-2
Erythromycin lactobionate	$C_{37}H_{67}NO_{13}.C_{12}H_{22}O_{12}$	1092.2	3847-29-8
Erythromycin propionate	$C_{40}H_{71}NO_{14}$	790.0	134-36-1
Erythromycin stearate	$C_{37}H_{67}NO_{13}.C_{18}H_{36}O_2$	1018.4	643-22-1

Appearance

Erythromycin occurs as white or slightly yellow crystals or powder. Erythromycin estolate is a white and odorless or almost odorless crystalline powder. Erythromycin ethylsuccinate occurs as a white or slightly yellow and odorless or almost odorless hygroscopic crystalline powder. Erythromycin gluceptate is a white and odorless or almost odorless and slightly hygroscopic powder. Erythromycin lactobionate occurs as white or slightly yellow crystals or powder. Erythromycin stearate occurs as white or slightly yellow crystals or powder.

Solubility

Erythromycin has a solubility of about 2 mg/mL in water. It is freely soluble in alcohols, acetone, chloroform, acetonitrile, and ethyl acetate. Erythromycin estolate is practically insoluble in water, freely soluble in alcohol and chloroform, and soluble in acetone. Erythromycin ethylsuccinate is practically insoluble in water and freely soluble in dehydrated alcohol, acetone, chloroform, and methanol. Erythromycin gluceptate is freely soluble in water, alcohol, and methanol; it is slightly soluble in acetone and chloroform. Erythromycin lactobionate is freely soluble in water, alcohol, and methanol; it is slightly soluble in acetone and chloroform. Erythromycin stearate is practically insoluble in water; it is soluble in acetone, dehydrated alcohol, chloroform, and methanol.

pK_a

The pK_a of erythromycin is 8.8.

Method 1

Stubbs et al. described an HPLC method for the analysis of erythromycin estolate in pharmaceutical dosage forms. A Waters liquid chromatograph included a model M6000A solvent-delivery system, a model 710B WISP autosampler, a Coulochem model 5100A electrochemical detector equipped with a model 5010 analytical cell and a model 5020 guard cell, and a Perkin-Elmer model 56 strip-chart recorder. The stationary phase was an HPLC Technology (UK) Techsil C_{18} column (250 × 3.9 mm, 10-μm particle size) with a guard column (40–60-μm Supelco glass beads). The column temperature was maintained at 35 °C. The mobile phase consisted of 350 mL of acetonitrile and 650 mL of 0.05 M phosphate buffer (pH 6.30). The flow rate was 1.3 mL/min. The upstream electrode was set at +0.70 V, the downstream electrode at +0.90 V, and the guard cell at +1.00 V. Oleandomycin phosphate 2.0 mg/mL in a mixture of acetonitrile and water (1:1) was used as an internal standard.

For tablets, six tablets were transferred into a 200-mL volumetric flask with 100 mL of water, shaken for 15 minutes, mixed with 50 mL of acetonitrile, shaken for 10 minutes, mixed with another 50 mL of acetonitrile, shaken for another 10 minutes, filled to volume with water, and sonicated for 5 minutes. One milliliter of this solution was transferred into a 10-mL volumetric flask, mixed with 1 mL of internal standard, filled to volume with acetonitrile, transferred to a Teflon centrifuge tube, and centrifuged at 12,000 rpm for 6 minutes. For capsules, 5 mg of capsule content was weighed, transferred to a 10-mL volumetric flask, mixed with 1 mL of internal standard and 5 mL of acetonitrile, vortexed for 1 minute, filled to volume with acetonitrile, sonicated for 5 minutes, and centrifuged. For suspensions, 100 μL of suspension was transferred into a 10-mL volumetric flask, mixed with 1 mL of internal standard and 4 mL of water, vortexed for 1 minute, filled to volume with acetonitrile, and centrifuged. The injection volume was 1–3 μL. Under these conditions, the retention times for the internal standard and erythromycin estolate were 5.3 and 9.5 minutes, respectively (estimated from the published chromatogram).

The method was reported to be stability indicating.

A standard curve for erythromycin estolate was constructed from 0.211 to 0.407 mg/mL. The correlation coefficient was 0.994. The coefficient of variation for the analysis of erythromycin estolate was less than 2.5%.

Reference

Stubbs C, Kanfer I. A stability-indicating high-performance liquid chromatographic assay of erythromycin estolate in pharmaceutical dosage forms. *Int J Pharm*. 1990: 63: 113–9.

Method 2

Tsuji and Kane reported a stability-indicating HPLC method for the analysis of erythromycin in solid dosage forms. The method used a modular liquid chromatograph equipped with a Laboratory Data Control model 1201 SpectroMonitor I variable-wavelength detector, a Laboratory Data Control model 196-0066-02 high-pressure pump, a Rheodyne model 7010 loop injector, and a Shimadzu Chromatopac-E1A electronic integrator. A LiChrosorb RP18 HPLC column (18–5 Å) was used and maintained at 70 °C with a circulating water bath. The mobile phase was composed of acetonitrile, methanol, 0.2 M ammonium acetate aqueous solution, and water (45:10:10:35, vol/vol/vol/vol). The flow rate was about 1.0 mL/min. UV detection was performed at 215 nm. Megestrol acetate 0.025 mg/mL was used as an internal standard. Relative retentions of erythromycin C, erythromycin A, and erythromycin B were 0.78, 1.00, and 1.46, respectively.

The assay was stated to be stability indicating since it separated erythromycin A, B, and C from degradation products and various impurities. Relative retentions were 0.73 (erythronolide B), 1.22 (8-epi-10,11-anhydroerythromycin A), 1.32 (erythralosamine), 1.38 (anhydroerythromycin A), 1.74 (dihydroerythromycin A), and 2.20 (8,9-anhydro-6,9-hemiketal erythromycin), respectively.

Calibration curves for erythromycin were constructed from 0.5 to 1.3 mg/mL; the correlation coefficients were 0.9999.

Reference
Tsuji K, Kane MP. Improved high-pressure liquid chromatographic method for the analysis of erythromycin in solid dosage forms. *J Pharm Sci.* 1982; 71: 1160–4.

Esmolol

Chemical Names
4-[2-Hydroxy-3-[(1-methylethyl)amino]propoxy]benzenepropanoic acid methyl ester
(±)-Methyl 3-[4-[2-hydroxy-3-(isopropylamino)propoxy]phenyl]propionate

Other Name
Brevibloc

Form	Molecular Formula	MW	CAS
Esmolol	$C_{16}H_{25}NO_4$	295.4	84057-94-3
Esmolol hydrochloride	$C_{16}H_{25}NO_4$.HCl	331.8	81161-17-3

Appearance
Esmolol hydrochloride is a white or off-white crystalline powder.

Solubility

Esmolol hydrochloride has solubilities of more than 650 mg/mL in water and of 350 mg/mL in alcohol at room temperature.

pK_a

Esmolol has a pK_a of 9.5.

Method 1

Using an HPLC method, Wiest et al. determined the chemical stability of esmolol hydrochloride 10, 20, and 30 mg/mL in 5% dextrose injection. The liquid chromatograph consisted of a Waters model 510 solvent-delivery system, a Waters model 712 WISP autosampler, a Waters Lambda-Max model 481 variable-wavelength UV spectrophotometer, and a Waters model 745 data module. The stationary phase was a Waters μBondapak C_{18} Radial-Pak cartridge column (100 × 8 mm) with a guard column of the same packing material. The mobile phase was a mixture of methanol and 0.01 M monobasic potassium phosphate (pH 2.69) (60:40, vol/vol) and was delivered isocratically at 2.8 mL/min. UV detection was performed at 229 nm. Methyl 4-[4-[2-hydroxy-3-[(2-methylethyl)amino]propoxy]phenyl]butyrate hydrochloride 15 μg/mL was used as the internal standard.

Each sample was diluted with HPLC-grade water. The injection volume was 15 μL. Under these conditions, retention times for esmolol hydrochloride and the internal standard were 4.76 and 8.36 minutes, respectively.

This method was determined to be stability indicating by analyzing esmolol hydrochloride reference standard after accelerated degradation. Hydrochloric acid was added to the test solution of esmolol hydrochloride to produce a final pH of 1.7; sodium hydroxide was also added to another test solution to yield a pH of 12.5. Samples were stored for 3 hours and then neutralized with potassium hydroxide or acetic acid. A third test solution was heated to 95 °C for 3 hours. Degradation product peaks did not interfere with the intact esmolol hydrochloride peak.

The standard curve was constructed by linear regression of ratios of the peak areas for esmolol to the peak areas for the internal standard and was linear from 1 to 10 μg/mL. The correlation coefficient was greater than 0.99. The intraday and interday coefficients of variation for the assay were 3.1 and 6.2%, respectively.

Reference

Wiest DB, Garner SS, Childress LM. Stability of esmolol hydrochloride in 5% dextrose injection. *Am J Health Syst Pharm.* 1995; 52: 716–8.

Method 2

Using an HPLC method, Karnatz et al. evaluated the compatibility and stability of esmolol hydrochloride and sodium nitroprusside in intravenous admixtures. The liquid chromatographic system included a Perkin-Elmer Series 2 solvent-delivery system, a Perkin-Elmer model 420 automatic sampler, a Beckman model 160 selectable fixed-wavelength UV detector, a Scientific Products model 285/mm recorder, and a Hewlett-Packard 3357 online data system. The stationary phase was a Waters μBondapak C_{18} ODS stainless steel column (300 × 3.9 mm, 10-μm particle size). The mobile phase was a mixture of methanol, acetonitrile, and buffer (20:15:64, vol/vol/vol). The buffer contained 3.0 g of monobasic potassium phosphate in 650 mL of water. The flow rate was 2 mL/min. UV detection was performed at 214 nm.

Samples were diluted with water, and the injection volume was 100 μL. Retention times for esmolol and sodium nitroprusside were 4.52 and 7.09 minutes, respectively.

The assay was determined to be stability indicating by accelerated degradation of esmolol. A solution of esmolol and sodium nitroprusside was exposed to intensive light (approximately 1200 foot-candles) for 1 week. Degradation product peaks did not interfere with the intact esmolol peak.

Similar analytical methods were described by the other researchers cited here.

References
Karnatz NN, Wong J, Baaske DM, et al. Stability of esmolol hydrochloride and sodium nitroprusside in intravenous admixtures. *Am J Hosp Pharm*. 1989; 46: 101–4.

Baaske DM, Dykstra SD, Eagenknecht DM, et al. Stability of esmolol hydrochloride in intravenous solutions. *Am J Hosp Pharm*. 1994; 51: 2693–6.

Schaaf LJ, Robinson DH, Vogel GJ, et al. Stability of esmolol hydrochloride in the presence of aminophylline, bretylium tosylate, heparin sodium, and procainamide hydrochloride. *Am J Hosp Pharm*. 1990; 47: 1567–71.

Method 3
Lee et al. reported an HPLC method for the determination of esmolol hydrochloride. A liquid chromatograph was equipped with a Perkin-Elmer model LC-15 fixed-wavelength UV absorbance detector and a Hewlett-Packard 3354 data system. The stationary phase was a Waters μBondapak CN column (300 × 3.9 mm, 10-μm particle size). The mobile phase was a mixture of 150 mL of acetonitrile, 10 mL of glacial acetic acid, and 840 mL of sodium acetate trihydrate (0.068%, wt/vol) buffer. The flow rate was 2 mL/min. UV detection was performed at 280 nm. 2-(p-Chlorophenyl)-2-methylpropanol 4 mg/mL in methanol and water (50:50, vol/vol) was used as an internal standard.

Samples were prepared in water. The injection volume was 50 μL. Under these conditions, retention times for esmolol and the internal standard were about 8.1 and 10.1 minutes, respectively (estimated from the published chromatogram).

The stability-indicating ability of the method was demonstrated by accelerated degradation of esmolol. Esmolol hydrochloride solutions were treated with 1 M hydrochloric acid, 1 M sodium hydroxide solution, or 30% hydrogen peroxide. Esmolol hydrochloride was resolved from synthetic intermediates, potential impurities, and degradation products.

A calibration curve for esmolol hydrochloride was constructed from 0 to 500 μg/mL. The limit of quantitation was about 10 μg/mL.

Reference
Lee Y-C, Baaske DM, Alam AS. High-performance liquid chromatographic method for the determination of esmolol hydrochloride. *J Pharm Sci*. 1984; 73: 1660–1.

Estradiol-3-Phosphate

Chemical Name
17β-Estra-1,3,5(10)-triene-3,17-diol-3-(dihydrogen phosphate)

Form	Molecular Formula	MW	CAS
Estradiol-3-phosphate	$C_{18}H_{25}O_5P$	352.4	13425-82-6

Method
Miller and Chen reported an HPLC method for the determination of 17β-estradiol-3-phosphate in an ophthalmic solution. The chromatographic apparatus consisted of a Waters model 600E system controller and pump, a model 712 WISP autosampler, and a model 486 variable-wavelength UV detector. The stationary phase was a Mac-Mod Zorbax SB C_{18} column (150 × 4.6 mm, 3.5-μm particle size). The mobile phase consisted of 10 mM monobasic potassium phosphate, 50 mM tetrabutylammonium chloride (adjusted to pH 3.0 with 1 N hydrochloric acid), and acetonitrile (40:15:45, vol/vol/vol). The flow rate was 0.9 mL/min. UV detection was performed at 220 nm.

Samples were diluted with water. The injection volume was 20 μL. Under these conditions, retention times of estriol, 17β-estradiol-3-phosphate, estrone-3-phosphate, 17β-estradiol, and estrone were 3.4, 6.9, 8.4, 9.6, and 11.2 minutes, respectively. The total run time was 12 minutes.

The stability-indicating nature of the method was confirmed by accelerated degradation. Solutions were exposed to a UV radiation source (200–400 nm, 40 mWatt/cm²) for 48 hours or adjusted to either pH 2 with concentrated hydrochloric acid or pH 12 with 50% sodium hydroxide and stored at 95 ºC for 48 hours. There was no interference from degradation products or excipients.

A calibration curve was constructed from 0.50 to 1.50 mg/mL. The correlation coefficient was 0.998. The limit of detection was 20 ng/mL.

Reference
Miller RB, Chen C. A stability-indicating HPLC method for the determination of 17β-estradiol-3-phosphate in an ophthalmic solution. *Chromatographia*. 1995; 40: 204–6.

Ethacrynic Acid

Chemical Name
[2,3-Dichloro-4-(2-ethylacryloyl)phenoxy]acetic acid

Other Name
Edecrin

Form	Molecular Formula	MW	CAS
Ethacrynate sodium	$C_{13}H_{11}Cl_2NaO_4$	325.1	6500-81-8
Ethacrynic acid	$C_{13}H_{12}Cl_2O_4$	303.1	58-54-8

Appearance
Ethacrynic acid is a white or almost white, odorless or almost odorless, crystalline powder.

Solubility
Ethacrynic acid is very slightly soluble in water and freely soluble in alcohol and chloroform.

pK$_a$
Ethacrynic acid has a pK$_a$ of 3.5.

Method 1
Ling and Gupta evaluated by HPLC analysis the stability of ethacrynic acid in oral liquid dosage forms. An ALC 202 system was equipped with a Rheodyne model 7125 universal injector, a Schoeffel SF 770 multiple-wavelength detector, and an Omniscribe recorder. The stationary phase was a Phenomenex Ultrasphere column (150 × 4.6 mm, 5-µm particle size). The mobile phase was 33% (vol/vol) acetonitrile in water containing 0.015 M phosphate buffer and was adjusted to pH 6.5. The flow rate was 1.2 mL/min. UV detection was performed at 278 nm and 0.1 AUFS.

An aliquot of clear decanted solution was diluted 1:25 with water. The injection volume was 80 µL. Under these conditions, the retention time for ethacrynic acid was about 4.7 minutes (estimated from the published chromatogram).

The method was reported to be stability indicating.

References
Ling J, Gupta VD. Stability of oral liquid dosage forms of ethacrynic acid. *Int J Pharm Compound.* 2001; 5: 232–3.

Ling J, Gupta VD. Stability of ethacrynate sodium after reconstitution in 0.9% sodium chloride injection and storage in polypropylene syringes for pediatric use. *Int J Pharm Compound.* 2001; 5: 73–4.

Method 2
Gupta studied ethacrynic acid using an HPLC method. An ALC 202 chromatograph was equipped with a UV detector, a Houston Omniscribe recorder, and a Spectra-Physics autolab minigrator. The stationary phase was a Waters µBondapak C$_{18}$ column (300 × 4 mm). The mobile phase was 45% methanol in 0.01 M dibasic ammonium phosphate aqueous buffer and was delivered isocratically at a flow rate of 2.0 mL/min. UV detection was performed at 254 nm and 0.16 AUFS.

For ethacrynic acid in injections and in aqueous solutions, samples were diluted with water to 50 µg/mL of ethacrynic acid. For ethacrynic acid in tablets, one tablet was ground, mixed with 3 mL of alcohol, diluted with water to the final volume of

500 mL, and filtered. The clear filtrate was collected for injection. The injection volume was 40 μL. Under these conditions, the retention time for ethacrynic acid was about 5.1 minutes (estimated from the published chromatogram).

The stability-indicating nature of the method was demonstrated by degradation of the drug at 65 °C, at various pH values, and in different vehicles.

Reference

Gupta VD. Quantitation of ethacrynic acid by high-pressure liquid chromatography. *Drug Dev Ind Pharm.* 1982; 8: 869–82.

Ethamsylate

Chemical Name

Diethylamine 2,5-dihydroxybenzenesulfonate

Other Names

Altodor, Dicynene, Dicynone, Etamsilate

Form	Molecular Formula	MW	CAS
Ethamsylate	$C_{10}H_{17}NO_5S$	263.3	2624-44-4

Appearance

Ethamsylate occurs as a white or almost white crystalline powder.

Solubility

Ethamsylate is very soluble in water, freely soluble in methanol, soluble in ethanol, and practically insoluble in methylene chloride.

Method

Kaul et al. reported an analytical method for the determination of ethamsylate as bulk material and in its dosage forms. The apparatus consisted of a Jasco model PU-1580 pump, a model AS-1555 autoinjector, and a model UV-1575 UV-visible detector. The stationary phase was a Kromasil C_{18} column (250 × 4.6 mm, 5-μm particle size). The mobile phase was a mixture of methanol and water (50:50, vol/vol). The flow rate was 0.6 mL/min. UV detection was performed at 305 nm.

A portion of powder equivalent to 1000 mg of ethamsylate from 20 tablets was accurately weighed, transferred into a 100-mL volumetric flask containing 50 mL of methanol, sonicated for 30 minutes, diluted to 100 mL with methanol, and centrifuged at 3000 rpm for 5 minutes. The supernatant was collected, diluted, and filtered through a

0.45-μm Millipore filter. The sample injection volume was 20 μL. Under these conditions, the retention time of ethamsylate was about 2.9 minutes.

To confirm the stability-indicating nature of the method, ethamsylate solutions in methanol were refluxed in 5 M hydrochloric acid, 0.1 M sodium hydroxide solution, or 6.0 and 50% hydrogen peroxide at 70 °C for 2 hours; or refluxed on a boiling water bath for 2 hours. Ethamsylate solutions were exposed to direct sunlight for 45 days, or to UV radiation for 7 days. The standard drug was also heated at 100 °C for 2 hours. In all of these cases, ethamsylate was well separated from its degradation products on its chromatogram.

A standard curve for ethamsylate was constructed from 10 to 60 μg/mL. The correlation coefficient was 0.9997. The intraday and interday coefficients of variation were 1.21 and 1.46%, respectively. The limit of detection and the limit of quantitation were 0.10 and 0.50 μg/mL, respectively.

Reference
Kaul N, Agrawal H, Kakad A, et al. Stress degradation studies on etamsylate using stability-indicating chromatographic methods. *Anal Chim Act.* 2005; 536: 49–70.

Ethinyl Estradiol

Chemical Name
(17α)-19-Norpregna-1,3,5(10)-trien-20-yne-3,17-diol

Other Names
Estinyl, Ethinyloestradiol

Form	Molecular Formula	MW	CAS
Ethinyl estradiol	$C_{20}H_{24}O_2$	296.4	57-63-6

Appearance
Ethinyl estradiol is a white to creamy white odorless crystalline powder.

Solubility
Ethinyl estradiol is insoluble in water and soluble in alcohol and vegetable oils.

Method 1
Reif et al. developed an automated stability-indicating HPLC assay for ethinyl estradiol in oral contraceptive tablets. The liquid chromatographic system consisted of a DuPont model 870 pump; two detectors connected in series, a Laboratory Data Control SpectroMonitor III UV detector and a Schoeffel FS 970 fluorescence detector set with

excitation at 210 nm and no filter in the emission path (photomultiplier window: 80% transmission at 310 nm); and a Rheodyne model 7125 fixed-loop septumless injector with a 100-μL loop. A reversed-phase DuPont Zorbax C_8 column (150 × 4.6 mm, 7-μm particle size) and also a Hypersil ODS column (100 × 4.6 mm, 3–5-μm particle size) were used with an Upchurch Scientific C-135 Uptight precolumn (40 × 4.6 mm) dry packed with silica gel 60 (230–400 mesh) from EM Reagents. The mobile phase was a mixture of acetonitrile, methanol, and water (35:15:45, vol/vol/vol). The flow rate was 1 mL/min. UV detection was performed at 215 nm. The sample injection volume was 100 μL. The retention time for ethinyl estradiol was 6.2 minutes.

The stability-indicating capability of the HPLC method was demonstrated by spiking ethinyl estradiol solution with known degradation products. Potential impurities and degradation products were well separated from the intact ethinyl estradiol peak. Retention times for degradation products were 1.7, 1.8, 3.6, and 4.1 minutes.

Standard curves were constructed from 0.006 to 0.04 mg/mL for ethinyl estradiol.

Reference
Reif VD, Eickhoff WM, Jackman JK, et al. Automated stability-indicating high-performance liquid chromatographic assay for ethinyl estradiol and (levo)norgestrel tablets. *Pharm Res.* 1987; 4: 54–8.

Method 2
Lane et al. reported an HPLC method for the simultaneous determination of norgestimate and ethinyl estradiol in oral contraceptive tablets. The liquid chromatograph included an LDC/Milton Roy model IIIG ConstaMetric pump, a model 1204A variable-wavelength detector, a Waters WISP autosampler, and a Hewlett-Packard 3357 laboratory auto-mation system. The stationary phase was an IBM reversed-phase C_{18} column (50 × 4.5 mm, 5-μm particle size). The mobile phase was a mixture of water, tetrahydrofuran, and methanol (65:25:10, vol/vol/vol). The flow rate was 2.1 mL/min. UV detection was performed at 230 nm. Dibutylphthalate 0.05 mg/mL in methanol was used as an internal standard.

Five tablets were weighed and placed into a 50-mL centrifuge tube along with two glass beads, mixed with 25 mL of internal standard, vortexed for at least 15 minutes, sonicated for 5 minutes, and filtered through a 2-μm filter. The injection volume was 25 μL. Under these conditions, retention times for ethinyl estradiol, *syn*-norgestimate, *anti*-norgestimate, and internal standard were about 3.6, 7.6, 9.3, and 13.3 minutes, respectively (estimated from the published chromatogram).

The method was evaluated to be stability indicating by stress degradation studies. Bulk norgestimate was heated at 50 °C for 90 days or exposed to 1000 foot-candles of light for 30 days. Norgestimate solutions were adjusted to pH 4.9, 7.0, and 10.0 with 0.1 M hydrochloric acid or 0.1 M sodium hydroxide and heated at 60 °C for 20 days. Bulk ethinyl estradiol was stored at 37 °C for 1 year or exposed to 80% relative humidity at 24 °C for 180 days in an open dish. There was no interference of degradation products with the intact drug.

A standard curve for ethinyl estradiol was constructed from 0.00351 to 0.0105 mg/mL. The correlation coefficient was 0.9996.

Reference
Lane PA, Mayberry DO, Young RW. Determination of norgestimate and ethinyl estradiol in tablets by high-performance liquid chromatography. *J Pharm Sci*. 1987; 76: 44–7.

Method 3
Strusiak et al. described the HPLC determination of ethinyl estradiol in solid dosage forms. The chromatographic system comprised a Waters model M6000A pump, a model 710A autosampler, a Perkin-Elmer 650-10LC fluorescence detector, and a Linear model 485 recorder. The stationary phase was a Merck LiChrosorb RP8 stainless steel column (250 × 4.6 mm, 10-μm particle size). The mobile phase consisted of 0.05 M monobasic potassium phosphate aqueous buffer and methanol (2:3, vol/vol). The flow rate was 2.0 mL/min. The fluorescence detector was set at an excitation wavelength of 280 nm and an emission wavelength of 330 nm. *o*-Phenylphenol 0.1 mg/mL in the mobile phase was used as an internal standard.

Tablets equivalent to 0.02 mg of ethinyl estradiol were added to 4 mL of 0.05 M monobasic potassium phosphate aqueous solution, rotated for 15 minutes, mixed with 2 mL of internal standard and 4 mL of methanol, rotated for another 15 minutes, and centrifuged. The supernatant was collected. The sample was extracted two more times with 5 mL of the mobile phase. The injection volume was 50 μL. Under these conditions, retention times for *o*-phenylphenol, butylparaben, and ethinyl estradiol were about 6.0, 7.3, and 8.3 minutes, respectively (estimated from the published chromatogram).

The stability-indicating nature of the assay was assessed by intentional degradation of ethinyl estradiol. Ethinyl estradiol substance was heated at 178 °C for 16 hours and ethinyl estradiol tablets were heated at 75 °C for 2 weeks. The ethinyl estradiol peak was separated from its degradation product peaks.

A standard curve for ethinyl estradiol was generated from 0.61 to 1.42 μg/mL. The correlation coefficient was 0.9995.

Reference
Strusiak SH, Hoogerheide JG, Gardner MS. Determination of ethinyl estradiol in solid dosage forms by high-performance liquid chromatography. *J Pharm Sci*. 1982; 71: 636–40.

Etoposide

Chemical Names
[5*R*-[5α,5aβ,8aα,9β(*R*∗)]]-9-[(4,6-*O*-Ethylidene-β-D-glucopyranosyl)oxy]-5,8,8a,9-tetrahydro-5-(4-hydroxy-3,5-dimethoxyphenyl)furo[3′,4′:6,7]naphtho[2,3-*d*]-1,3-dioxol-6(5a*H*)-one

4′-Demethylepipodophyllotoxin 9-[4,6-*O*-(*R*)-ethylidene-β-D-glucopyranoside]

Other Names
Etopophos, VePesid

Form

Form	Molecular Formula	MW	CAS
Etoposide	$C_{29}H_{32}O_{13}$	588.6	33419-42-0
Etoposide phosphate	$C_{29}H_{31}O_{13}.H_2PO_3$	668.5	117091-64-2

Appearance
Etoposide is a fine white or almost white crystalline powder. Etoposide phosphate occurs as a white to off-white crystalline powder.

Solubility
Etoposide is practically insoluble in water. It is slightly soluble in alcohol, chloroform, ethyl acetate, and methylene chloride. Etoposide phosphate is freely soluble in water (exceeding 100 mg/mL) and slightly soluble in alcohol.

Method 1
McLeod and Relling used an HPLC method to study the stability of etoposide solution for oral use. The chromatograph consisted of a Shimadzu model LC-6A solvent-delivery pump, a Bioanalytical Systems model LC-4B electrochemical detector, a Shimadzu model C-R6A integrator, and a Waters µBondapak phenyl column (300 × 3.9 mm, 10-µm particle size). The mobile phase consisted of deionized water, acetonitrile, and acetic acid (70:30:2, vol/vol/vol). The flow rate was 1.4 mL/min. The electrochemical detector was operated at an applied potential of 0.75 V.

Each sample was diluted 1:1000 with 0.9% sodium chloride injection. The injection volume was 100 µL. Under these conditions, the retention time for etoposide was 7.4 minutes.

The method was shown to be stability indicating by a degradation study. A test solution of etoposide was mixed 1:1 with 1 M sodium hydroxide solution and injected onto the column. The degradation product peaks did not interfere with the intact etoposide peak.

The standard curve was generated for etoposide from 5 to 20 mg/mL in 0.9% sodium chloride injection. The coefficient of variation for the assay was less than 8%.

Similar stability-indicating analytical methods were used by the other researchers cited here.

References

McLeod HL, Relling MV. Stability of etoposide solution for oral use. *Am J Hosp Pharm.* 1992; 49: 2784–5.

Joel SP, Clark PI, Slevin ML. Stability of the i.v. and oral formulations of etoposide in solution. *Cancer Chem Pharmacol.* 1995; 37: 117–24.

Barthes DMC, Rochard EB, Pouliquen IJ, et al. Stability and compatibility of etoposide in 0.9% sodium chloride injection in three containers. *Am J Hosp Pharm.* 1994; 51: 2706–9.

Shah JC, Chen JR, Chow D. Preformulation study of etoposide: Identification of physicochemical characteristics responsible for the low and erratic oral bioavailability of etoposide. *Pharm Res.* 1989; 6: 408–12.

Chow D, Shah J, Chen JR. Stability-indicating HPLC of etoposide at various pH conditions using a reversed-phase octyl column. *J Chromatogr.* 1987; 396: 217–23.

Method 2

Beijnen et al. investigated the stability of etoposide and teniposide in commonly used infusion fluids. The liquid chromatographic system consisted of a Waters model M-45 solvent-delivery system, a Waters U6K injector, a Spectra-Physics 200 programmable-wavelength detector, and a Spectra-Physics 4270 integrator. The stationary phase was a μBondapak phenyl column (300 × 3.9 mm, 10-μm particle size). The mobile phase consisted of methanol and water (50:50, wt/wt), to which 0.5% (vol/wt) 0.5 M sodium phosphate buffer (pH 6) was added. The flow rate was 1.0 mL/min. UV detection was performed at 280 nm.

The method was stated to be stability indicating because it separated the parent drug from its *cis*-fused lactone analog.

Calibration curves for etoposide were constructed from 0.1 to 0.5 mg/mL; the correlation coefficient was greater than 0.999. The coefficient of variation of the assay was 0.9%.

Reference

Beijnen JH, Beijnen-Bandhoe AU, Dubbelman AC, et al. Chemical and physical stability of etoposide and teniposide in commonly used infusion fluids. *J Parenter Sci Technol.* 1991; 45: 108–12.

Method 3

Stewart and Hampton evaluated the stability of etoposide and cisplatin in intravenous admixtures. The HPLC apparatus included an LDC Constametric III pump, a Spectra-Physics SP8300 fixed-wavelength UV detector, a Rheodyne 7125 injector with a 20-μL loop, and a Fisher Scientific Recordall series 5000 recorder. The stationary phase was a Phenomenex reversed-phase column (300 × 3.4 mm) packed with Waters μBondapak phenyl packing. The mobile phase was composed of 26% acetonitrile and 74% 0.02 M sodium acetate (pH 4.0) by volume and was delivered isocratically at 1.0 mL/min. UV detection was performed at 254 nm and 0.08 AUFS.

Samples were diluted with 0.02 M sodium acetate buffer (pH 4.0). The injection volume was 80 μL. The retention times for methyl *p*-aminobenzoic acid (the internal standard) and etoposide were 9 and 16 minutes, respectively.

The analytical method was stated to be stability indicating. Cisplatin eluted with the solvent front of the etoposide assay.

The intraday and interday coefficients of variation were less than 2 and 5%, respectively.

Reference

Stewart CF, Hampton EM. Stability of cisplatin and etoposide in intravenous admixtures. *Am J Hosp Pharm.* 1989; 46: 1400–4.

Method 4

Floor et al. developed a stability-indicating HPLC method for the analysis of etoposide and benzyl alcohol. The chromatographic system consisted of a Waters model M-45 pump, a model 710B autosampler, a model 440 fixed-wavelength detector, a Hewlett-Packard model 3354C laboratory automation system, and a model 1040A diode-array detector. The stationary phase was a Waters µBondapak phenyl column (300 × 3.9 mm, 10-µm particle size). The mobile phase consisted of 0.02 M sodium acetate buffer (pH 4.0) and acetonitrile (74:26, vol/vol). The flow rate was 1.0 mL/min. UV detection was performed at 254 nm and 0.1 AUFS. Methyl *p*-aminobenzoate 0.4 mg/mL in acetonitrile served as an internal standard.

Samples were diluted with mobile phase. The injection volume was 20 µL. Under these conditions, the retention times for benzyl alcohol, benzaldehyde, and etoposide were 6.0, 10.0, and 16.3 minutes, respectively.

The assay was found to be stability indicating by accelerated degradation of etoposide. Samples of etoposide were exposed to intense 300-nm light for 8 days, 0.1 M sodium hydroxide for 1 minute, 0.1 M hydrochloric acid for 15 minutes at 115 °C, and heat at 115 °C for 26 hours. No degradation product peaks interfered with the intact etoposide peak.

A standard curve for etoposide was constructed from 0.1 to 0.3 mg/mL; the correlation coefficient was greater than 0.999. The coefficient of variation was less than 0.2%.

A similar method was used by Mayron and Gennaro.

References

Floor BJ, Klein AE, Muhammad N, et al. Stability-indicating liquid chromatographic determination of etoposide and benzyl alcohol in injectable formulations. *J Pharm Sci.* 1985; 74: 197–200.

Mayron D, Gennaro AR. Stability and compatibility of granisetron hydrochloride in i.v. solutions and oral liquids and during simulated Y-site injection with selected drugs. *Am J Health Syst Pharm.* 1996; 53: 294–304.

Etoricoxib

Chemical Name

5-Chloro-6'-methyl-3-[4-(methylsulfonyl)phenyl]-2,3'-bipyridine

Other Name
Arcoxia

Form	Molecular Formula	MW	CAS
Etoricoxib	$C_{18}H_{15}ClN_2O_2S$	358.8	202409-33-4

Method

Hartman et al. developed and validated an HPLC method for the impurity and quantitative analysis of etoricoxib. An Agilent 1100 HPLC system consisted of an autoinjector, a quaternary pump, and a variable-wavelength detector and was also equipped with a photo-diode-array detector. The stationary phase was a YMC ODS-AQ column (150 × 4.6 mm, 3-μm particle size). Mobile phase A was 10 mM monobasic potassium phosphate aqueous buffer adjusted to pH 3.1 with 2.2 mM phosphoric acid and mobile phase B was acetonitrile. The mobile phase was delivered at 1 mL/min at 28% B for 11 minutes, linearly increased to 70% B over 19 minutes, and then to 90% B in another 5 minutes. UV detection was performed at 220 nm.

Samples of etoricoxib were prepared in water and acetonitrile (50:50, vol/vol). The injection volume was 10 μL. Under these conditions, the retention time of etoricoxib was about 11.6 minutes (estimated from the published chromatogram).

In order to demonstrate the stability-indicating nature of the method, etoricoxib samples were dissolved in 0.1 N hydrochloric acid and exposed to 1.2 million lux hours of cool white fluorescent light and 200 watt hours/square meter of near ultraviolet light. Analysis of stressed and controlled samples showed that the intact drug was well separated from the impurities and degradation products.

A standard curve for etoricoxib was constructed with a correlation coefficient of 0.9999. The limit of detection was 0.02 μg/mL and the limit of quantitation was 0.04 μg/mL.

Reference

Hartman R, Abrahim A, Clausen A, et al. Development and validation of an HPLC method for the impurity and quantitative analysis of etoricoxib. *J Liq Chromatogr Rel Technol.* 2003; 26: 2551–66.

Ezetimibe

Chemical Name
1-(4-Fluorophenyl)-(3*R*)-[3-(4-fluorophenyl)-(3S)-hydroxypropyl]-4*S*-(4-hydroxyphenyl)-
2-azetidinone

Other Names
Ezetrol, Zetia

Form	Molecular Formula	MW	CAS
Ezetimibe	$C_{24}H_{21}F_2NO_3$	409.4	163222-33-1

Method
Using an HPLC method, Singh et al. studied the degradation of ezetimibe. The Shimadzu chromatograph consisted of a model LC-10ATvp solvent-delivery module, a model SIL-10ADvp autoinjector, a model SPD-10Avp UV-visible dual-wavelength detector, a model CTO-10ASvp column oven, and a model DGU-14A on-line degasser. The stationary phase was a Merck Lichrospher C_8 column (250 × 4.0 mm, 5-μm particle size). The mobile phase consisted of 0.02 M ammonium acetate aqueous solution adjusted to pH 7.0 with ammonium hydroxide and acetonitrile. Acetonitrile was delivered linearly from 30 to 100% in 80 minutes and brought back linearly to 30% in the next 10 minutes. The flow rate was 1 mL/min. The drug was monitored at 250 nm.

The drug was dissolved in water containing 30% acetonitrile. The sample injection volume was 20 μL. Under these conditions, the retention time of ezetimibe was 26.9 minutes.

The ezetimibe solutions were treated in 0.1 and 1 M hydrochloric acid at 40 and 80 °C for 8 hours, in 0.1 M sodium hydroxide solution at 40 and 80 °C for 8 hours, and in water at 80 °C for 8 hours. The drug solutions in water and in 1 M hydrochloric acid were exposed to sunlight for 2 days. The drug powder was also dry heated at 50 °C for 45 days and at 60 °C for 7 days. In all cases, the peak of ezetimibe was well separated from peaks of its degradation products, demonstrating the stability-indicating property of the method.

A calibration curve for ezetimibe was constructed from 5 to 500 μg/mL with a correlation coefficient of 0.9997. Intraday and interday coefficients of variation were less than 1.0%.

Reference
Singh S, Singh B, Bahuguna R, et al. Stress degradation studies on ezetimibe and development of a validated stability-indicating HPLC assay. *J Pharm Biomed Anal*. 2006; 41: 1037–40.

Famotidine

Chemical Name

3-[[[2-[(Aminoiminomethyl)amino]-4-thiazolyl]methyl]thio]-N-(aminosulfonyl)-
propanimidamide

Other Name

Pepcid

Form	Molecular Formula	MW	CAS
Famotidine	$C_8H_{15}N_7O_2S_3$	337.4	76824-35-6

Appearance

Famotidine is a white to pale yellow odorless crystalline powder.

Solubility

Famotidine has the following solubilities at 20 °C (%, wt/vol): 80 in dimethylformamide, 50 in acetic acid, 0.3 in methanol, and 0.1 in water.

pK_a

Famotidine has a pK_a of 7.1 in water at 25 °C.

Method 1

Lee et al. used HPLC analysis to study the stability of famotidine with cefmetazole in 5% dextrose injection at 4 and 25 °C. The chromatograph consisted of a Shimadzu model LC-6A liquid chromatograph, a Rheodyne model 7125 single-piston 20-μL loop injector, a Shimadzu model SPD-6AV dual UV-visible detector, and a Waters model 745 data module. The stationary phase was a Waters Nova-Pak C_{18} analytical column. The mobile phase consisted of acetonitrile, 0.1% acetic acid, and 0.01 M dibasic ammonium phosphate solution at pH 7.8 (10:23:74, vol/vol/vol) and was delivered isocratically at 1.0 mL/min. UV detection was performed at 300 nm. Cefmetazole and famotidine had retention times of 8.1 and 13.9 minutes, respectively.

The assay was demonstrated to be stability indicating by analyzing a cefmetazole sodium–famotidine admixture stored at 55 °C for 15 hours. Degradation products eluted separately and did not interfere with peaks of the intact famotidine and cefmetazole.

Calibration curves were generated from 0.1 to 10 mg/mL and the correlation coefficients were greater than 0.999. The intraday and interday coefficients of variation for the assay ($n = 6$) were 0.66 and 1.03%, respectively.

References
Lee DKT, Wong C-Y, Wang D-P, et al. Stability of cefmetazole sodium and famotidine. *Am J Health Syst Pharm.* 1996; 53: 432, 442.

Wang D-P, Wang M-T, Wong C-Y, et al. Compatibility of vancomycin hydrochloride and famotidine in 5% dextrose injection. *Int J Pharm Compound.* 1997; 1: 354–6.

Method 2
Abdel-Hamid described an HPLC method for the quantitation of famotidine. The Shimadzu LC-10 chromatograph consisted of a universal injector and an SPD-M10A diode-array detector. The stationary phase was a Shimadzu GLC-ODS column (150 × 6 mm, 5-μm particle size). The mobile phase consisted of acetonitrile, methanol, and 0.1 M sodium acetate (10:10:80, vol/vol/vol), adjusted to pH 4.7 with glacial acetic acid. The flow rate was 1.5 mL/min. UV detection was performed at 275 nm. The injection volume was 50 μL. Under these conditions, the retention time for famotidine was 3.5 minutes.

The stability-indicating nature of the method was assessed by intentional degradation of famotidine. Famotidine was dissolved in 1 M hydrochloric acid and heated at 50 °C for 1 hour. The intact drug was well resolved from its degradation products. Retention times were 3.5 minutes for famotidine and 4.5, 7.0, and 9.0 minutes for its degradation products.

A calibration curve for famotidine was constructed from 5 to 20 μg/mL. The correlation coefficient was 0.9989. The limit of quantitation was 1 μg/mL.

Reference
Abdel-Hamid ME. Stability-indicating HPLC and spectrophotometric full spectrum quantitation (FSQ) for analysis of famotidine in presence of its acid-induced degradation products. *Alex J Pharm Sci.* 1997; 11: 129–34.

Method 3
Using an HPLC analytical method, Xu et al. evaluated the stability of famotidine 200 μg/mL in infusion admixtures. The chromatographic system consisted of a Waters model 501 pump, a Waters model 484 variable-wavelength UV detector, a Shimadzu model SIL-9A autoinjector, a Waters model QA-1 integrator, and a Chromatographic Sciences reversed-phase analytical column (250 × 4.6 mm) with a Waters Nova-Pak C_{18} precolumn. The mobile phase contained 7 parts of acetonitrile and 93 parts of 30 mM sodium dihydrogen phosphate adjusted to pH 2.7 with 85% phosphoric acid. The flow rate was 1.0 mL/min. UV detection was performed at 254 nm. β-Hydroxyethyltheophylline was used as the internal standard.

Samples were diluted with distilled water. The injection volume was 50 μL.

The method was determined to be stability indicating by chromatographing a sample containing famotidine, degradation products, dextrose H_4, and the internal standard. In addition, the famotidine solutions were mixed with 85% phosphoric acid or 1 M sodium hydroxide and were incubated in a water bath at 74 °C for 8 hours. In both cases, the famotidine peak was well resolved from other peaks.

Standard curves for famotidine were generated every day by linear least-squares regression and were linear from 0.125 to 25 μg/mL. The correlation coefficients were always greater than 0.999. Intraday and interday coefficients of variation for the assay were 1.1 and 1.5%, respectively.

Reference

Xu K, Gagnon N, Bisson C, et al. Stability of famotidine in polyvinylchloride minibags and polypropylene syringes and compatibility of famotidine with selected drugs. *Ann Pharmacother.* 1993; 27: 422–6.

Method 4

Bullock et al. used an HPLC method to assess the stability of famotidine in parenteral nutrition solutions. The liquid chromatograph consisted of a Waters U6K injector, a Waters M-45 pump, a Waters model 440 UV detector, and a Hewlett-Packard model 3390A integrator. The analytical column was a Waters μBondapak C_{18} cartridge. The mobile phase contained 7 parts of acetonitrile and 93 parts of 0.1 M sodium acetate with pH adjusted to 6.0. The flow rate was 1.5 mL/min. UV detection was performed at 254 nm and 0.01 AUFS.

Samples were diluted with water. The injection volume was 75 μL. Theophylline was used as the internal standard. The retention times for theophylline and famotidine were 11.2 and 15.5 minutes, respectively.

The method was determined to be stability indicating by an accelerated degradation study and by spiking the reference standard solution with known degradates. Solutions of famotidine were incubated with 2 N hydrochloric acid, 2 N sodium hydroxide, and water at 90 °C for 12 hours. The two degradation products eluted at 10.0 and 18.5 minutes and did not interfere with the famotidine peak.

Famotidine concentrations were calculated by comparing the famotidine to the theophylline area under the peak ratios to a standard curve of known famotidine concentrations. The coefficient of variation of the analysis was 5%. The intrarun and interrun coefficients of variation for the assay were less than 5%.

A similar HPLC method was used by Quercia et al.

References

Bullock L, Fitzgerald JF, Glick MR, et al. Stability of famotidine 20 and 40 mg/L and amino acids in total parenteral nutrient solutions. *Am J Hosp Pharm.* 1989; 46: 2321–5.

Quercia RA, Jay GT, Fan C, et al. Stability of famotidine in an extemporaneously prepared oral liquid. *Ann Pharmacother.* 1993; 23: 691–3.

Bullock L, Fitzgerald JF, Glick MR. Stability of famotidine 20 and 50 mg/L in total nutrient admixtures. *Am J Hosp Pharm.* 1989; 46: 2326–9.

Bullock LS, Fitzgerald JF, Glick MR. Stability of famotidine in minibags refrigerated and/or frozen. *Ann Pharmacother.* 1989; 23: 132–5.

Bullock LS, Fitzgerald JF, Mazur HI. Stability of intravenous famotidine stored in polyvinylchloride syringes. *Ann Pharmacother.* 1989; 23: 588–90.

Method 5

Montoro et al. used an HPLC method to determine the stability of famotidine 20 and 40 mg/L in total nutrient admixtures. Famotidine concentrations were analyzed by a Perkin-Elmer series-4 set with a Rheodyne-type 20-μL injector, an LC-85B UV detector, a Perkin-Elmer Sigma 15 integrator, and a Perkin-Elmer HS-3 C_{18} column (30 × 3 mm). The mobile phase was a mixture of 0.01 M heptanesulfonic acid (pH 4), acetonitrile, and water (52:38:10, vol/vol/vol) and was delivered isocratically at a rate of 1.5 mL/min. UV detection was performed at 254 nm. Ranitidine solution 100 μg/mL was used as an internal standard.

One milliliter of the nutrient solution was mixed with 1 mL of a mixture of chloroform and methanol (2:1, vol/vol) and 50 μL of internal standard, agitated for 5 minutes, and centrifuged at 3000 rpm for 5 minutes. The aqueous phase was filtered through a 0.45-μm Millipore HV filter. The injection volume was 20 μL.

To confirm the stability-indicating nature of the assay, famotidine solutions were exposed to 1 N sodium hydroxide or 1 N hydrochloric acid for 10 hours at 85 °C. In all cases, the degradation product peaks did not interfere with the determination of the intact famotidine.

A standard curve for famotidine was constructed from 10 to 50 μg/mL.

Reference

Montoro JB, Pou L, Salvador P, et al. Stability of famotidine 20 and 40 mg/L in total nutrient admixtures. *Am J Hosp Pharm.* 1989; 46: 2329–32.

Method 6

Using a stability-indicating HPLC method, Walker et al. investigated the stability of famotidine in total parenteral nutrient solutions. The HPLC system included a Spectra-Physics model 8770 dual-piston pump, a Spectra-Physics SP8780 XR autoinjector, a Kratos 783 variable-wavelength UV detector, and a Spectra-Physics SP4200 chroma-tography integrator. The stationary phase was a Beckman Ultrasphere C_{18} reversed-phase analytical column (250 × 4.6 mm, 5-μm particle size) with an RP18 precolumn (150 × 4.6 mm). The mobile phase consisted of 93 parts of water containing 0.2% triethylamine adjusted to pH 2.5 with 85% phosphoric acid and 7 parts of acetonitrile. The flow rate was 1.5 mL/min. UV detection was performed at 260 nm. The injection volume was 20 μL. Under these conditions, famotidine eluted between 4.6 and 5.5 minutes.

The assay was determined to be stability indicating by accelerated decom-position of famotidine. Standard solutions of famotidine were adjusted to pH 1.7 with 85% phosphoric acid and to pH 11 with 1 M sodium hydroxide and were then incubat-ed in a water bath at 74 °C for 5 hours. None of the degradation product peaks interfered with the intact famotidine peak.

A standard curve for famotidine was generated from 0.025 to 0.100 mg/mL. The intraday and interday coefficients of variation were 1.0 and 3.5%, respectively.

Reference

Walker SE, Iazzetta J, Lau DWC, et al. Famotidine stability in total parenteral nutrient solutions. *Can J Hosp Pharm.* 1989; 42: 97–103.

Method 7

Gupta et al. assessed the chemical stabilities of famotidine and ranitidine hydrochloride in intravenous admixtures. The HPLC system consisted of a Waters ALC 202 system, a Schoeffel SF 770 multiple-wavelength detector, and a Houston Omniscribe recorder. The stationary phase was a Waters μBondapak C_{18} nonpolar column (300 × 3.9 mm). The mobile phase consisted of 12% methanol, 2% acetonitrile, and 0.1% acetic acid in 0.01 M aqueous phosphate buffer. The flow rate was 2.0 mL/min. UV detection was performed at 268 nm and 0.04 AUFS. Sulfamerazine 0.7 mg/mL in methanol was used as the internal standard.

Samples were diluted 1:5 with water. The injection volume was 20 μL. Under these conditions, retention times for famotidine and the internal standard were about 4.6 and 8.0 minutes, respectively (estimated from the published chromatogram).

The analytical method was demonstrated to be stability indicating by accelerated decomposition of famotidine. Stock solutions of famotidine were mixed either with 1 N sulfuric acid or 1 N sodium hydroxide and were then boiled for about 10 minutes. Degradation products did not interfere with the intact famotidine peak.

Reference

Gupta VD, Parasrampuria J, Bethea C. Chemical stabilities of famotidine and ranitidine hydrochloride in intravenous admixtures. *J Clin Pharm Ther.* 1988; 13: 329–34.

Method 8

Biffar and Mazzo developed a sensitive and specific HPLC procedure for the quantification of famotidine, its potential degradation products, and preservatives. The chromatograph consisted of a Hewlett-Packard model 1090 pump module, a Kratos model 757 variable-wavelength detector, a Hewlett-Packard model 3390A integrator, and a Hewlett-Packard model 3357B laboratory automation system. The stationary phase was a Beckman Ultrasphere silica column (250 × 4.6 mm, 5-μm particle size). The mobile phase was a mixture of methanol, water, and 0.05 M sodium dihydrogen phosphate (10:74:16, vol/vol/vol) and was delivered isocratically at 1.0 mL/min. UV detection was performed at 254 nm and 0.01 AUFS. The sample injection volume was 25 μL. The retention time for famotidine was about 12.5 minutes (estimated from the published chromatogram).

The stability-indicating nature of the assay was confirmed by spiking a sample with 1.0% (wt/wt) of three possible degradation products of famotidine. The intact famotidine was well separated from its potential degradation products and preservatives.

A standard curve for famotidine was constructed from 0.02 to 0.08 mg/mL. The coefficient of variation was 0.45%.

Similar methods were used by the other researchers cited here.

References

Biffar SE, Mazzo DJ. Reversed-phase determination of famotidine, potential degradates, and preservatives in pharmaceutical formulations by high-performance liquid chromatography using silica as a stationary phase. *J Chromatogr.* 1986; 363: 243–9.

Shea BF, Souney PF. Stability of famotidine in a 3-in-1 total nutrient admixture. *Ann Pharmacother.* 1990; 24: 232–5.

Shea BF, Souney PF. Stability of famotidine frozen in polypropylene syringes. *Am J Hosp Pharm.* 1990; 47: 2073–4.

DiStefano JE, Mitrano FP, Baptista RJ, et al. Long-term stability of famotidine 20 mg/L in a total parenteral nutrient solution. *Am J Hosp Pharm.* 1989; 46: 2333–5.

Fampridine

Chemical Name
4-Aminopyridine

Other Name
Neurelan

Form	Molecular Formula	MW	CAS
Fampridine	$C_5H_6N_2$	94.1	504-24-5

Method
Trissel et al. investigated by HPLC analysis the stability of fampridine (4-aminopyridine) oral capsules. The Waters LC Module 1 Plus liquid chromatograph consisted of a multi-solvent-delivery system, an autosampler, and a UV light detector. A Waters Alliance 2960 liquid chromatograph equipped with a Waters model 996 photodiode-array detector was also used to confirm the drug peak purity. The stationary phase was a Phenomenex Prodigy C_{18} analytical column (250 × 4.6 mm, 5-μm particle size). The mobile phase was a mixture of 350 mL of acetonitrile and 1650 mL of aqueous buffer consisting of 50 mM monobasic sodium phosphate and 4 mM 1-octanesulfonic acid. The flow rate was 0.8 mL/min. UV detection was performed at 266 nm and 0.5 AUFS.

The powder contents of capsules were transferred into a glass test tube, mixed with 10 mL of water, vortexed for 30 seconds, filtered through a 0.45-μm filter, and diluted 20-fold with water. The injection volume was 10 μL. Under these conditions, the retention time of fampridine was 6.3 minutes.

The stability-indicating nature of the method was demonstrated by accelerated degradation. Sample solutions were mixed with 0.1 N sodium hydroxide solution, 0.1 N hydrochloric acid, or 3% hydrogen peroxide and heated. No degradation product peaks interfered with the peak of the intact fampridine. Photodiode-array purity analysis confirmed the peak purity of fampridine throughout the study.

A standard curve for fampridine was constructed from 5 to 40 μg/mL. The correlation coefficient was greater than 0.9993. Intraday and interday coefficients of variation were 1.2 and 0.8%, respectively.

Reference
Trissel LA, Zhang Y, Xu QA. Stability of 4-aminopyridine and 3,4-diaminopyridine oral capsules. *Int J Pharm Compound*. 2002; 6: 155–7.

Felodipine

Chemical Name
Ethyl methyl 4-(2,3-dichlorophenyl)-1,4-dihydro-2,6-dimethyl-3,5-pyridinedicarboxylate

Other Name
Plendil

Form	Molecular Formula	MW	CAS
Felodipine	$C_{18}H_{19}Cl_2NO_4$	384.3	72509-76-3

Appearance
Felodipine is a white or light yellow crystalline powder.

Solubility
Felodipine is practically insoluble in water, but it is freely soluble in acetone, absolute alcohol, methanol, and dichloromethane.

Method
Cardoza and Amin developed a reversed-phase HPLC method for the determination of felodipine as a bulk substance and in its pharmaceutical dosage forms. The apparatus consisted of a Jasco PU-980 pump, a Jasco UV-975 UV-visible detector, and a Rheodyne model 7725 injector with a 20-μL loop. The stationary phase was a Merck LiChrospher 100 RP-18 column (250 × 4 mm, 5-μm particle size). The mobile phase was a mixture of 0.01 M monobasic potassium phosphate buffer (adjusted to pH 3.5 with phosphoric acid) and methanol (25:75, vol/vol). The isocratic flow rate was 1.5 mL/min. UV detection was performed at 238 nm and 0.16 AUFS.

A powder of a ground tablet was accurately weighed and transferred into a 25-mL volumetric flask; the powder was then stirred with methanol for 30 minutes, filled to volume with methanol, and centrifuged. The clear liquid was diluted with mobile phase. Under these conditions, the retention times for the major degradation product, dehydrofelodipine, and for felodipine were about 4.4 and 6.0 minutes, respectively.

The assay was evaluated to be stability indicating by accelerated degradation of felodipine. Felodipine was degraded by heat, oxidizing agent, light, acid, or base. The intact drug was separated from its degradation products.

A standard curve for felodipine was constructed from 1 to 7 μg/mL. The correlation coefficient was 0.9994. The limit of detection and the limit of quantitation were 150 and 500 ng/mL, respectively.

Reference
Cardoza RM, Amin PD. A stability indicating LC method for felodipine. *J Pharm Biomed Anal.* 2002; 27, 711–8.

Fenoldopam Mesylate

Chemical Name
6-Chloro-2,3,4,5-tetrahydro-1-(*p*-hydroxyphenyl)-1*H*-3-benzazepine-7,8-diol methane-
   sulfonate

Other Name
Corlopam

Form	Molecular Formula	MW	CAS
Fenoldopam mesylate	$C_{17}H_{20}ClNO_6S$	401.9	67227-57-0

Appearance
Fenoldopam mesylate occurs as a white to off-white powder.

Solubility
Fenoldopam mesylate is sparingly soluble in water, ethanol, and methanol. It is soluble in propylene glycol.

Method 1
Trissel et al. evaluated by HPLC analysis the chemical stability of fenoldopam mesylate in two infusion solutions. The Waters model 2690 Alliance separation module consisted of a multisolvent-delivery system and an autosampler and was equipped with a Waters 2487 dual λ absorbance detector. The stationary phase was a Waters Symmetry C_{18} column (250 × 4.6 mm, 5-μm particle size). The mobile phase consisted of 87% 25 mM sodium phosphate in water (adjusted to pH 2.75 with phosphoric acid) and 13% acetonitrile. The flow rate was 1.0 mL/min. UV detection was performed at 225 nm.

   Samples of fenoldopam injections were diluted with the appropriate infusion solution to a nominal fenoldopam concentration of 40 μg/mL. The injection volume was 10–25 μL. Under these conditions, the retention time for fenoldopam was 6.9 minutes.

   The method was shown to be stability indicating by accelerated drug degradation. Solutions of fenoldopam mesylate were heated in water, 0.1 N hydrochloric acid, and 0.1 N sodium hydroxide at 95 °C or were exposed to 0.3% hydrogen peroxide. Degradation product peaks did not interfere with the peak of the intact fenoldopam. The retention times were 3.5, 3.8, 4.8, and 5.8 minutes for degradation products.

The standard curves for fenoldopam were constructed from 2.5 to 15 μg/mL and from 20 to 60 μg/mL. The correlation coefficient was 0.999. Intraday and interday coefficients of variation were 0.8 and 1.8%, respectively.

Reference
Trissel LA, Zhang Y, Baker MB. Stability of fenoldopam mesylate in two infusion solutions. *Am J Health Syst Pharm.* 2002; 59: 846–8.

Method 2
Thoma and Ziegler described a method for the simultaneous quantification of fenoldopam and succinic acid. The liquid chromatographic system consisted of a Waters model 6000A pump, a TSP autosampler, and a TSP fast-scanning UV detector. The stationary phase was a Nucleosil C_{18} ODS column (250 × 5.4 mm, 5-μm particle size). The mobile phase consisted of 0.025 M monobasic sodium phosphate buffer (pH 2.5) and methanol (80:20, vol/vol). The flow rate was 1 mL/min. UV detection was performed at 200–300 nm (205 nm for succinic acid and 225 nm for fenoldopam).

Pellet samples were ground, dispersed in 0.1 N hydrochloric acid, and filtered. The injection volume was 20 μL. Under these conditions, retention times for succinic acid and fenoldopam were 3.2 and 8.5 minutes, respectively.

The method was reported to be stability indicating since no interference from degradation products was observed.

Standard curves were constructed from 0.008 to 0.2 mg/mL for fenoldopam and from 0.05 to 1.35 mg/mL for succinic acid. Correlation coefficients for fenoldopam and succinic acid were 0.9993 and 0.9999, respectively. Coefficients of variation were between 0.2 and 0.9% for fenoldopam and between 0.4 and 1.8% for succinic acid.

Reference
Thoma K, Ziegler I. Simultaneous quantification of released succinic acid and a weakly basic drug compound in dissolution media. *Eur J Pharm Biopharm.* 1998; 46: 183–90.

Fenoterol

Chemical Names
5-[1-Hydroxy-2-[[2-(4-hydroxyphenyl)-1-methylethyl]amino]ethyl]-1,3-benzenediol
3,5-Dihydroxy-α-[[(*p*-hydroxy-α-methylphenethyl)amino]methyl]benzyl alcohol

Other Name
Berotec

Form

Form	Molecular Formula	MW	CAS
Fenoterol	$C_{17}H_{21}NO_4$	303.4	13392-18-2
Fenoterol hydrobromide	$C_{17}H_{21}NO_4.HBr$	384.3	1944-12-3

Appearance
Fenoterol hydrobromide occurs as a white crystalline powder.

Solubility
Fenoterol hydrobromide is soluble in water and alcohol. It is practically insoluble in ether.

Method
Jacobson and Peterson developed an HPLC assay for the simultaneous determination of ipratropium bromide, fenoterol, albuterol, and terbutaline in a nebulizer solution. The liquid chromatograph consisted of a Varian model 9010 solvent-delivery system, a Rheodyne model 7161 injector with a 10-µL external loop, a Varian model 9050 variable-wavelength UV-visible detector, and a Varian GC Star workstation. The stationary phase was a Waters Nova-Pak C_{18} Radial-Pak cartridge (100 × 8 mm, 4-µm particle size) inside a Waters RCM 8 × 10 compression module. Mobile phase A consisted of tetrahydrofuran and distilled water (40:60, vol/vol) containing 0.0025 M Waters Pic B-8 Reagent Low UV. Mobile phase B was distilled water, and mobile phase C consisted of methanol and distilled water (50:50, vol/vol). The flow rate was 2.0 mL/min and the run time was 13.0 minutes followed by a 5.0-minute equilibration time. A mixture of 50% mobile phase A and 50% mobile phase B was delivered until 7.7 minutes. The composition was then changed linearly, resulting in 60% mobile phase A, 15% mobile phase B, and 25% mobile phase C at the 13-minute endpoint. Mepivacaine hydrochloride 1% was used as an internal standard. The injection volume was 20 µL. Retention times for albuterol, terbutaline, ipratropium, mepivacaine, and fenoterol were 3.2, 4.3, 5.9, 8.2, and 12.7 minutes, respectively.

The assay was determined to be stability indicating by accelerated decomposition of fenoterol under heat, hydrogen peroxide, acid, and base. In all cases, degradation product peaks did not interfere with the intact fenoterol peak.

Reference
Jacobson GA, Peterson GM. High-performance liquid chromatographic assay for the simultaneous determination of ipratropium bromide, fenoterol, salbutamol, and terbutaline in nebulizer solution. *J Pharm Biomed Anal.* 1994; 12: 825–32.

Fenretinide

Chemical Name
4-Hydroxyphenylretinamide

Other Name
McN-R-1967

Form	Molecular Formula	MW	CAS
Fenretinide	$C_{26}H_{33}NO_2$	391.5	65646-68-6

Method

Sisco et al. reported an assay for fenretinide in soft gelatin capsules and concentrated corn oil suspensions. A DuPont model 850 liquid chromatograph was equipped with a DuPont autosampler, a fixed-wavelength detector, and a model 4100 integrator. The stationary phase was a Zorbax ODS column (250 × 4.6 mm, 5-µm particle size). The mobile phase was prepared by mixing 100 mL of water (adjusted to pH 3.0 with acetic acid) and 900 mL of acetonitrile. The flow rate was 2.0 mL/min. UV detection was performed at 254 nm.

The sample solvent was prepared by mixing 650 mL of acetonitrile with 350 mL of isopropanol. Samples were dissolved and diluted in sample solvent. The injection volume was 20 µL. Under these conditions, the retention time for fenretinide was 4.9 minutes.

The stability-indicating properties of the method were evaluated by accelerated degradation of fenretinide. Fenretinide soft gelatin capsules were stored at 100 °C overnight. The intact fenretinide was separated from its degradation products. Retention times for 4-aminophenol, 13-*cis*-fenretinide, fenretinide, and retinoic acid were 1.5, 4.5, 4.9, and 7.3 minutes, respectively.

A calibration curve for fenretinide was generated from 0 to 22 µg.

References

Sisco WR, Schrader PA, McLaughlin AM, et al. Stability-indicating reversed-phase high-performance liquid chromatographic assay for fenretinide in soft gelatin capsules and concentrated corn oil suspensions. *J Chromatogr*. 1986; 368: 184–7.

Sisco WR, DiFeo TJ. Stability-indicating reversed-phase high-performance liquid chromatographic assay for fenretinide drug substance. *J Chromatogr*. 1985; 322: 380–5.

Fentanyl

Chemical Name
N-Phenyl-*N*-[1-(2-phenylethyl)-4-piperidinyl]propanamide

Other Names
Phentanyl, Sublimaze

Form	Molecular Formula	MW	CAS
Fentanyl	$C_{22}H_{28}N_2O$	336.5	437-38-7
Fentanyl citrate	$C_{22}H_{28}N_2O.C_6H_8O_7$	528.6	990-73-8

Appearance
Fentanyl is a white crystalline powder.

Solubility
Fentanyl citrate has an aqueous solubility of about 25 mg/mL. It is soluble in methanol but sparingly soluble in chloroform.

pK$_a$
Fentanyl citrate has pK$_a$ values of 7.3 and 8.4.

Method 1
Peterson et al. used HPLC analysis to study the compatibility and stability of fentanyl in combination with scopolamine and midazolam. The liquid chromatographic system consisted of a Varian model 9010 pump, a model 9050 variable-wavelength UV-visible detector, and a Rheodyne 7161 injector. The stationary phase was a Nova-Pak C$_{18}$ column (150 × 3.9 mm, 4-μm particle size). Solvent A was acetonitrile and solvent B was 0.1 M monobasic potassium phosphate buffer (pH 7.8). The mobile phase was delivered from 25 to 60% A in 10 minutes, held at 60% A for 2 minutes, returned to 25% A in 2 minutes, and held at 25% A for 1 minute. The flow rate was 1.0 mL/min. UV detection was performed at 210 nm. Lorazepam 125 μg/mL in ethanol was used as an internal standard. The injection volume was 10 μL. Under these conditions, retention times for scopolamine, lorazepam, midazolam, and fentanyl were about 5.4, 7.4, 10.8, and 12.8 minutes, respectively.

The method was demonstrated to be stability indicating by accelerated degradation. Solutions of fentanyl were autoclaved at 121 °C for 5 hours or acidified to pH 1.0 with 0.1 M hydrochloric acid and autoclaved. None of the degradation product peaks interfered with the determination of fentanyl.

A standard curve for fentanyl was constructed from 30 to 60 µg/mL. The correlation coefficient was 0.998. The intraday and interday coefficients of variation were below 4%.

Reference

Peterson GM, Miller KA, Galloway JF, et al. Compatibility and stability of fentanyl admixtures in polypropylene syringes. *J Clin Pharm Ther*. 1998; 23: 67–72.

Method 2

Using an HPLC method, Xu et al. determined the rapid loss of fentanyl citrate admixed with fluorouracil in polyvinyl chloride containers. The liquid chromatographic system consisted of a Waters model 600E multisolvent-delivery pump, a model 490E programmable multiple-wavelength UV detector, and a model 712 WISP autosampler; the system was controlled and integrated by an NEC PowerMate SX/16 computer. The stationary phase was an Alltech Spherisorb CN column (300 × 4.6 mm, 5-µm particle size). The mobile phase was 65% acetonitrile in 0.02 M monobasic potassium phosphate buffer with the pH adjusted to 5.4 with 1 N sodium hydroxide and was run isocratically at 1.5 mL/min. UV detection was performed at 210 nm. The sample injection volume was 25 µL. Under these conditions, the retention times for fluorouracil and fentanyl were 2.4 and 6.3 minutes, respectively.

The method was determined to be stability indicating by accelerated degradation of fentanyl citrate. The addition of 1 N hydrochloric acid or 1 N sodium hydroxide resulted in a reduction in the fentanyl peak and the formation of a new peak at 2.2 minutes. Fluorouracil and the fentanyl degradation product peak did not interfere with the intact fentanyl peak.

The standard curve was constructed from a linear plot of peak area versus concentration from 2.5 to 20 µg/mL. The correlation coefficient of the standard curve was greater than 0.996. The intraday and interday coefficients of variation of the assay were 4.6 and 3.8%, respectively.

Reference

Xu QA, Trissel LA, Martinez JF. Rapid loss of fentanyl citrate admixed with fluorouracil in polyvinyl chloride containers. *Ann Pharmacother*. 1997; 31: 297–302.

Method 3

Allen et al. evaluated the stability of fentanyl citrate in portable infusion-pump reservoirs. The chromatograph consisted of a Waters model 6000A double-piston pump, a Beckman model 210A autoinjector with a 20-µL loop injector, a Beckman model 163 variable-wavelength UV detector, and a Waters Radial-Pak phenyl column (100 × 3.9 mm, 4-µm particle size). The mobile phase consisted of 65% methanol and 35% 0.005 M phosphate aqueous buffer (pH 4.8) containing 0.0014 M tetrabutyl-ammonium hydroxide. The flow rate was 3.0 mL/min. UV detection was performed at 210 nm. The injection volume was 20 µL. The retention time for fentanyl citrate was 11.6 minutes.

The HPLC method was determined to be stability indicating by accelerated degradation of fentanyl citrate. A sample of fentanyl citrate was stored at 85 °C for 60 days. The degradation product eluted separately without any interference with the intact fentanyl citrate peak.

Standard curves for fentanyl citrate were constructed from 4 to 50 µg/mL. Correlation coefficients were greater than 0.99. The intraday and interday coefficients of variation were less than 3%.

Similar methods were used by Bhatt-Mehta et al. and Tu et al.

References

Allen LV Jr, Stiles ML, Tu YH. Stability of fentanyl citrate in 0.9% sodium chloride solution in portable infusion pumps. *Am J Hosp Pharm*. 1990; 47: 1572–4.

Bhatt-Mehta V, Johnson CE, Leininger N, et al. Stability of fentanyl citrate and midazolam hydrochloride during simulated intravenous coadministration. *Am J Health Syst Pharm*. 1995; 52: 511–3.

Allen LV, Stiles ML, Wang DP, et al. Stability of bupivacaine hydrochloride, epinephrine hydrochloride, and fentanyl citrate in portable infusion-pump reservoirs. *Am J Hosp Pharm*. 1993; 50: 714–5.

Tu Y-H, Stiles ML, Allen LV Jr. Stability of fentanyl citrate and bupivacaine hydrochloride in portable pump reservoirs. *Am J Hosp Pharm*. 1990; 47: 2037–40.

Method 4

Wilson et al. reported an HPLC method for the analysis of fentanyl citrate in a parenteral dosage form. The liquid chromatograph consisted of a Waters model 510 pump, a model 712 WISP autosampler, a Kratos 757 variable-wavelength detector, and a Fisher Recordall 5000 recorder. The stationary phase was a Whatman Partisil ODS-3 column (250 × 4.6 mm, 10-µm particle size). The mobile phase consisted of water, methanol, and 85% phosphoric acid (500:500:4, vol/vol/vol). The flow rate was 1.0 mL/min. UV detection was performed at 229 nm. Under these conditions, the retention time for fentanyl citrate was about 8 minutes (estimated from the published chromatogram).

Stability-indicating capability of the method was shown by a stress study. Fentanyl samples were treated with 3 M hydrochloric acid, 3 M sodium hydroxide solution, or 15% hydrogen peroxide and heated in a 90 °C water bath for 4 hours. The degradation product did not interfere with the determination of fentanyl.

A standard curve for fentanyl was generated from 0.005 to 0.500 mg/mL. The correlation coefficient was greater than 0.9999.

Reference

Wilson TD, Maloney T, Amsden WB. High-performance liquid chromatographic determination of fentanyl citrate in a parenteral dosage form. *J Chromatogr*. 1988; 445: 299–304.

Fexofenadine Hydrochloride

Chemical Name

4-[1-Hydroxy-4-[4-(hydroxydiphenylmethyl)-1-piperidinyl]butyl]-α,α-dimethylbenzene-acetic acid hydrochloride

Other Names
Allegra, Telfast

Form	Molecular Formula	MW	CAS
Fexofenadine	$C_{32}H_{39}NO_4$	501.6	83799-24-0
Fexofenadine hydrochloride	$C_{32}H_{39}NO_4 \cdot HCl$	538.1	138452-21-8

Appearance
Fexofenadine hydrochloride is a white to off-white crystalline powder.

Solubility
Fexofenadine hydrochloride has an aqueous solubility of 2.2 mg/mL at 25 °C. It occurs as a racemic mixture and exists as a zwitterion in aqueous media at physiologic pH.

pK$_a$
Fexofenadine hydrochloride has pK$_a$ values of 4.25 and 9.53 at 25 °C.

Method
Radhakrishna and Om Reddy used an HPLC method for the simultaneous determination of fexofenadine and its related compounds. The liquid chromatograph consisted of a Waters 510 pump, a Waters 996 photodiode-array detector, and a Rheodyne injector with 10-µL sample loop. The stationary phase was an Agilent Eclipse XDB C$_8$ column (150 × 4.6 mm, 5-µm particle size). The mobile phase consisted of 1.0% triethylamine aqueous solution (adjusted to pH 3.7 with phosphoric acid), acetonitrile, and methanol (60:20:20, vol/vol/vol). The flow rate was 1.2 mL/min. UV detection was performed at 210 nm. 5-Methyl-2-nitrophenol (0.2 mg/mL) was used as an internal standard.

Twenty fexofenadine hydrochloride tablets were extracted in methanol and centrifuged. A portion of the supernatant was diluted with mobile phase before assay. Twenty fexofenadine hydrochloride capsules were treated in the same way. The injection volume was 10 µL. Under these conditions, retention times for the internal standard and fexofenadine were about 12.5 and 14 minutes, respectively.

To demonstrate that the method was stability indicating, fexofenadine samples were heated at 70 °C for 12 hours, refluxed with 0.1 N hydrochloric acid at 60 °C for 12 hours, with 0.1 N sodium hydroxide at 60 °C for 12 hours, with 3% hydrogen peroxide for 3 hours, or exposed to ultraviolet light (254 nm) for 18 hours. The peak of the intact drug was well separated from the peaks of its degradation products. Its two related compounds had retention times of about 16 and 22 minutes.

A calibration curve for fexofenadine was obtained from 60 to 750 µg/mL. The correlation coefficient was greater than 0.999.

Reference
Radhakrishna T, Om Reddy G. Simultaneous determination of fexofenadine and its related compounds by HPLC. *J Pharm Biomed Anal.* 2002; 29: 681–90.

Finasteride

Chemical Name
N-tert-Butyl-3-oxo-4-aza-5α-androst-1-ene-17β-carboxamide

Other Names
Propecia, Proscar

Form	Molecular Formula	MW	CAS
Finasteride	$C_{23}H_{36}N_2O_2$	372.5	98319-26-7

Appearance
Finasteride occurs as a white to off-white crystalline solid.

Solubility
Finasteride is very slightly soluble in water. It is freely soluble in alcohol and in chloroform.

Method
Segall et al. reported the HPLC determination of finasteride in a pharmaceutical formulation. The chromatograph included a Spectra-Physics model 150 Chrom LC pump, a Hewlett-Packard model 1050 UV-visible detector, a Rheodyne model 7125 injector, and a Hewlett-Packard series 3395 integrator. The stationary phase was a Merck LiChrospher 100 RP-18 column (250 × 4 mm, 5-µm particle size). The mobile phase was a mixture of methanol and water (70:30, vol/vol) and was delivered isocratically at 1.0 mL/min. UV detection was performed at 210 nm and 2 AUFS.

A portion of powder from 20 tablets was accurately weighed, transferred into a 25-mL volumetric flask, mixed with 20 mL of mobile phase, sonicated for 5 minutes, diluted to the volume with mobile phase, and filtered through a Whatman No. 42 paper. The injection volume was 20 µL. Under these conditions, the retention time of finasteride was about 6 minutes.

The method was demonstrated to be stability indicating by intentional degradation of finasteride. Finasteride was placed in an oven at 110 °C for 24 hours, exposed to daylight for 24 hours, and refluxed for 30 minutes in water, 1 N hydrochloric acid, 1 N sodium hydroxide, or hydrogen peroxide. Neither degradation products nor formulation ingredients interfered with the quantitation of finasteride.

A standard curve for finasteride was generated from 50 to 800 µg/mL with a correlation coefficient greater than 0.9995.

Reference
Segall AI, Vitale MF, Perez VL, et al. A stability-indicating HPLC method to determine finasteride in a tablet formulation. *J Liq Chromatogr Rel Technol*, 2002; 25: 3167–76.

Flecainide

Chemical Name
N-(2-Piperidinylmethyl)-2,5-bis(2,2,2-trifluoroethoxy)benzamide

Other Name
Tambocor

Form	Molecular Formula	MW	CAS
Flecainide	$C_{17}H_{20}F_6N_2O_3$	414.4	54143-55-4
Flecainide acetate	$C_{17}H_{20}F_6N_2O_3 \cdot C_2H_4O_2$	474.4	54143-56-5

Appearance
Flecainide acetate is a white crystalline powder.

Solubility
Flecainide acetate has a solubility of 48.4 mg/mL in water and 300 mg/mL in alcohol at 37 °C.

pK$_a$
Flecainide acetate has a pK$_a$ of 9.3.

Method 1
Allen and Erickson used HPLC to evaluate the stability of flecainide acetate 20 mg/mL in extemporaneously compounded oral liquids. A Hewlett-Packard series 1050 automated liquid chromatograph consisting of a multisolvent mixing and pumping system, an autoinjector, a diode-array detector, and a computer with Chem Station software was used. The HPLC analytical column was a Bakerbond phenylethyl column (250 × 4.6 mm, 5-µm particle size). The mobile phase was a mixture of acetonitrile and water with 0.06% phosphoric acid (40:60, vol/vol). The flow rate was 1.0 mL/min. UV detection was performed at 280 nm.

Samples were diluted 1:100. Under these conditions, the retention time for flecainide acetate was 7.7 minutes.

The HPLC assay was shown to be stability indicating. A composite chromatogram of flecainide acetate after intentional degradation by acid, base, oxidizing agent, and light showed no interference from the peaks of degradation products with the peak of the intact flecainide acetate.

Standard curves for flecainide acetate were constructed from 25 to 250 µg/mL. The intraday and interday coefficients of variation were 1.1 and 1.5%, respectively.

Reference

Allen LV, Erickson MA. Stability of baclofen, captopril, diltiazem hydrochloride, dipyridamole, and flecainide acetate in extemporaneously compounded oral liquids. *Am J Health Syst Pharm.* 1996; 53: 2179–84.

Method 2

Using an HPLC method, Wiest et al. determined the stability of flecainide acetate in an extemporaneously compounded oral suspension. The liquid chromatograph consisted of a Waters model 510 solvent-delivery system, a Waters model 712 WISP autosampler, a Perkin-Elmer model 650-15 fluorescence spectrophotometer, and a Waters model 745 data module. The stationary phase was a Waters Resolve silica column (150 × 3.9 mm, 5-µm particle size). The mobile phase consisted of 10 mM ammonium sulfate (adjusted to pH 6.8) and methanol (15:85, vol/vol) delivered isocratically at 1.0 mL/min. The fluorescence spectrophotometer was set for excitation at 225 nm and emission at 340 nm. *N*-(2-Piperidylmethyl)-2,3-bis(2,2,2-trifluoroethoxy)benzamide hydrochloride 75 µg/mL was used as the internal standard.

Sodium carbonate (0.2 M) was used as the extraction buffer and a mixture of butanol and hexane (20:80, vol/vol) served as the extraction solvent. The sample injection volume was 50 µL. Under these conditions, the retention times for flecainide acetate and the internal standard were 5.1 and 5.8 minutes, respectively.

To establish that the HPLC method was stability indicating, samples of flecainide acetate reference standard were adjusted to pH 0.5 with hydrochloric acid and to pH 13.02 with potassium hydroxide. In addition, another flecainide acetate sample was heated on a hot plate at 95 °C for 3 hours. Degradation product peaks did not interfere with the intact flecainide acetate peak.

Standard curves were constructed every day for flecainide acetate by linear regression of the peak area ratios of flecainide acetate to the internal standard against the known flecainide acetate concentrations. They were linear from 5 to 25 µg/mL and their correlation coefficients were greater than 0.97. The intraday and interday coefficients of variation for the assay were 2.50 and 4.60%, respectively.

Reference

Wiest DB, Garner SS, Pagacz LR, et al. Stability of flecainide acetate in an extemporaneously compounded oral suspension. *Am J Hosp Pharm.* 1992; 49: 1467–70.

Floxacillin

Chemical Name
(6R)-6-[3-(2-Chloro-6-fluorophenyl)-5-methylisoxazole-4-carboxamido]penicillanic acid

Other Names
Floxapen, Flucloxacillin, Staphylex

Form	Molecular Formula	MW	CAS
Floxacillin	$C_{19}H_{17}ClFN_3O_5S$	453.9	5250-39-5
Floxacillin magnesium	$(C_{19}H_{16}ClFN_3O_5S)_2Mg.8H_2O$	1074.2	58486-36-5
Floxacillin sodium	$C_{19}H_{16}ClFN_3NaO_5S.H_2O$	493.9	1847-24-1

Appearance
Floxacillin magnesium is a white or almost white powder. Floxacillin sodium is a white or almost white crystalline powder.

Solubility
Floxacillin magnesium is slightly soluble in water. It is freely soluble in methanol. Floxacillin sodium is freely soluble in water and methanol. It is soluble in alcohol.

Method
Grover et al. reported an HPLC method for the analysis of floxacillin in the presence of its degradation products. The Shimadzu model LC-10A liquid chromatograph included model LC-10AS pumps, a model SPD-10A dual-wavelength UV-visible detector, a model C-R7A data processor, and a Rheodyne model 7125 injector with a 20-μL loop. The stationary phase was a Phenomenex Resolve C_{18} column (300 × 3.9 mm, 5-μm particle size). The mobile phase was 19% acetonitrile in 20 mM monobasic potassium phosphate and 10 mM tetramethylammonium chloride in water (pH 5.0). The flow rate was 1 mL/min. UV detection was performed at 220 and 240 nm. Under these conditions, the retention time for floxacillin was 33.9 minutes.

The stability-indicating nature of the assay was demonstrated by degrading the drug. Floxacillin was degraded at pH 2.0, 7.0, and 9.0. The intact floxacillin peak was separated from its degradation product peaks.

A standard curve for floxacillin was obtained from 5 to 150 μg/mL. The correlation coefficient was greater than 0.99.

Reference
Grover M, Gulati M, Singh S. Stability-indicating analysis of isoxazolyl penicillins using dual wavelength high-performance liquid chromatography. *J Chromatogr B*. 1998; 708: 153–9.

Floxuridine

Chemical Names
2'-Deoxy-5-fluorouridine
1-(2-Deoxy-β-D-ribofuranosyl)-5-fluorouracil

Other Name
FUDR

Form	Molecular Formula	MW	CAS
Floxuridine	$C_9H_{11}FN_2O_5$	246.2	50-91-9

Appearance
Floxuridine is a white to practically white powder.

Solubility
Floxuridine is freely soluble in water and soluble in alcohol.

Method
Using an HPLC assay, Stiles et al. studied the stability of floxuridine in polypropylene infusion-pump syringes. The liquid chromatograph included a Waters model 501 pump, a Waters model 441 UV detector, a Waters model 745 data module, an Alcott model 728 autosampler, and a Rheodyne 7010 injector. The stationary phase was a Bakerbond C_{18} analytical column. The mobile phase consisted of 0.5% methanol and 99.5% water and was delivered isocratically at 1.0 mL/min. UV detection was performed at 268 nm.

Samples were diluted before injection. Under these conditions, floxuridine had a retention time of 4.6 minutes.

The stability-indicating nature of the assay was investigated by accelerated decomposition of floxuridine with sulfuric acid, 1 N sodium hydroxide, or heat at 80 °C for 24 hours. Degradation product peaks did not interfere with the intact floxuridine peak.

A standard curve was constructed for floxuridine from 0.1 to 100 μg/mL. Intraday and interday coefficients of variation of the assay were both 0.4%.

A similar method was used by Smith et al.

References
Stiles ML, Allen LV, Prince SJ. Stability of deferoxamine mesylate, floxuridine, fluorouracil, hydromorphone hydrochloride, lorazepam, and midazolam hydrochloride in polypropylene infusion-pump syringes. *Am J Health Syst Pharm.* 1996; 53: 1583–8.

Smith JA, Morris A, Duafala ME, et al. Stability of floxuridine and leucovorin calcium admixtures for intraperitoneal administration. *Am J Hosp Pharm.* 1989; 46: 985–9.

Fluconazole

Chemical Name
α-(2,4-Difluorophenyl)-α-(1*H*-1,2,4-triazol-1-ylmethyl)-1*H*-1,2,4-triazole-1-ethanol

Other Name
Diflucan

Form	Molecular Formula	MW	CAS
Fluconazole	$C_{13}H_{12}F_2N_6O$	306.3	86386-73-4

Appearance
Fluconazole is a white crystalline powder.

Solubility
Fluconazole has a solubility of 8 mg/mL in water at 37 °C and 25 mg/mL in alcohol at room temperature.

pK$_a$
Fluconazole has a pK$_a$ of 1.76 at 24 °C in 0.1 M sodium chloride.

Method 1
Hall et al. reported the compatibility of fluconazole with filgrastim during simulated Y-site administration. The liquid chromatograph consisted of a Waters model 510 pump, a model Ultra WISP 715 autosampler, a model 481 UV detector, and a Shimadzu model C-R3A integrator. The stationary phase was a Spherisorb S5 CN column. The mobile phase was a mixture of 0.02 M monobasic potassium phosphate (adjusted to pH 5.4 with 1.0 N sodium hydroxide) and acetonitrile (50:50, vol/vol). The flow rate was 1.5 mL/min. UV detection was performed at 254 nm. Under these conditions, the retention time for fluconazole was 2.5 minutes.

The stability-indicating nature of the method was tested by intentional degradation of fluconazole. Samples of fluconazole were acidified to pH 2, alkalinized to pH 8, or heated at 30 °C for 30 minutes. In all cases, degradation products did not influence the analysis of fluconazole.

A standard curve for fluconazole was constructed from 0.6 to 1.2 mg/mL. Intrarun and interrun coefficients of variation were less than 1.3%.

Reference
Hall PD, Yui D, Lyons S, et al. Compatibility of filgrastim with selected antimicrobial drugs during simulated Y-site administration. *Am J Health Syst Pharm.* 1997; 54: 185–9.

Method 2

Allen et al. studied the stability of fluconazole in the presence of cefpirome sulfate during simulated Y-site injection. The HPLC method utilized either one of two systems. One unit included an Alcott model 728 Micromeritics autosampler, a Rheodyne 7010 injector with an Alcott 732 electrically actuated valve, a Waters model 501 solvent-delivery pump, a Waters model 441 UV detector, a Waters model 401 refractive index detector, and a Waters model 745 data module. The other unit consisted of an Alcott model 728 Micromeritics autosampler, a Rheodyne 7010 injector with an Alcott 732 electrically actuated valve, a Shimadzu model LA-6A solvent-delivery pump, a Shimadzu model SPD-6A UV detector, and an Orion model 901 microprocessor ion analyzer. The stationary phase was a C_{18} analytical column. The mobile phase was a mixture of 450 mL of methanol and 550 mL of 0.025 M sodium phosphate buffer with pH adjusted to 7.0 with phosphoric acid and was delivered isocratically at 1.0 mL/min. Fluconazole was detected at 260 nm.

Samples were diluted before injection. The injection volume was 20 µL. Under these conditions, the retention times for fluconazole and cefpirome were 6.1 and 17.2 minutes, respectively.

The method was determined to be stability indicating by accelerated decomposition of fluconazole with hydrochloric acid, sodium hydroxide, potassium chlorate, heat at 80 °C, or light (150 foot-candles from a tungsten filament source). In each case, the intact fluconazole and its degradation products were well separated on the chromatogram.

The standard curve was constructed from 10 to 100 µg/mL and its correlation coefficient was greater than 0.99. The intraday and interday coefficients of variation of the method were 2.3 and 2.4%, respectively.

Reference

Allen LV, Stiles ML, Prince SJ, et al. Stability of cefpirome sulfate in the presence of commonly used intensive care drugs during simulated Y-site injection. *Am J Health Syst Pharm.* 1995; 52: 2427–33.

Method 3

Yamreudeewong et al. used an HPLC assay to investigate the stability of fluconazole in an extemporaneously prepared oral liquid. The chromatographic system included a Waters model 590 solvent-delivery pump, a Waters model 490E UV detector, and an Alltech C_{18} column (250 × 4.5 mm, 5-µm particle size). The mobile phase was a mixture of acetonitrile and water (26:74, vol/vol) and was delivered isocratically at 1 mL/min. UV detection was performed at 210 nm. Methyl *p*-hydroxybenzoate was used as an internal standard.

Samples were centrifuged for 10 minutes at 2000 × *g*. The supernatant was mixed with an equal volume of the internal standard and centrifuged for 10 seconds. The injection volume of the supernatant was 5 µL. Under these conditions, retention times for fluconazole and the internal standard were 7.0 and 9.1 minutes, respectively.

The assay was determined to be stability indicating by an accelerated degradation study. A sample of fluconazole oral liquid 1 mg/mL was added to 1 mL of 1 N nitric acid to produce a pH of 1.4; this solution was then heated at 100 °C for 30 minutes, cooled, neutralized, and assayed. Similarly, a fluconazole sample was added to 1 mL of 1 N sodium hydroxide to produce a pH of 12. In both cases, some decomposition was observed, but there were no interfering degradation peaks.

A calibration curve for fluconazole was produced from 0.1 to 1.0 mg/mL. Its correlation coefficient was 0.999. The intraday coefficient of variation for the assay was less than 2%.

Similar analytical methods were used by the other researchers cited here.

References
Yamreudeewong W, Lopez-Anaya A, Rappaport H. Stability of fluconazole in an extemporaneously prepared oral liquid. *Am J Hosp Pharm*. 1993; 50: 2366–7.

Pompilio FM, Fox JL, Inagaki K, et al. Stability of ranitidine hydrochloride with ondansetron hydrochloride or fluconazole during simulated Y-site administration. *Am J Hosp Pharm*. 1994; 51: 391–4.

Burm J-P, Choi J-S, Jhee SS, et al. Stability of paclitaxel and fluconazole during simulated Y-site administration. *Am J Hosp Pharm*. 1994; 51: 2704–6.

Inagaki K, Takagi J, Lor E, et al. Stability of fluconazole in commonly used intravenous antibiotic solutions. *Am J Hosp Pharm*. 1993; 50: 1206–8.

Method 4
Couch et al. used an HPLC assay to evaluate the stability of fluconazole in parenteral nutrient solutions. The liquid chromatograph consisted of a Waters model 501 solvent-delivery system, a Waters model 484 tunable UV detector, a Waters model U6K variable-volume injector, and a Hewlett-Packard model 3390 reporting integrator. The stationary phase was an Applied Biosystems Spheri-5 ODS analytical column (250 × 4.6 mm, 5-µm particle size). The mobile phase consisted of 30% acetonitrile and 70% 0.01 M dibasic potassium phosphate buffer, adjusted to pH 5 with 1.77% aqueous phosphoric acid solution, and was delivered isocratically at 1.0 mL/min. UV detection was performed at 260 nm and 0.005 AUFS. UK-48,134 compound (Pfizer Inc.) 21 µg/mL was used as the internal standard.

Samples were filtered through a 0.22-µm Millipore Millex GV filter and diluted with mobile phase. The injection volume was 30 µL. The retention times for fluconazole and UK-48,134 compound were about 4.2 and 5.3 minutes, respectively (estimated from the published chromatogram).

The stability-indicating capability of this assay was verified by an accelerated degradation study. Solutions of fluconazole 1.0 mg/mL were adjusted to pH 2 or 12 with 1.0 N sulfuric acid or 1.0 N sodium hydroxide and then heated to 100 °C for 30 minutes. The solutions were then cooled, neutralized, and assayed. Degradation product peaks did not interfere with the intact drug peak.

The standard curve was produced by linear regression of fluconazole concentration and the peak height ratios of fluconazole to internal standard from 4 to 10 µg/mL. Its correlation coefficient was 0.98. The intraday and interday coefficients of variation for the assay were 1.8 and 3.2%, respectively.

Similar methods were reported by the other researchers cited here.

References
Couch P, Jacobson P, Johnson CE. Stability of fluconazole and amino acids in parenteral nutrient solutions. *Am J Hosp Pharm*. 1992; 49: 1459–62.

Bosso JA, Prince RA, Fox JL. Compatibility of ondansetron hydrochloride with fluconazole, ceftazidime, aztreonam, and cefazolin sodium under simulated Y-site conditions. *Am J Hosp Pharm.* 1994; 51: 389–91.

Johnson CE, Jacobson PA, Pillen HA, et al. Stability and compatibility of fluconazole and aminophylline in intravenous admixtures. *Am J Hosp Pharm.* 1993; 50: 703–6.

Flucytosine

Chemical Names
4-Amino-5-fluoro-2(1*H*)-pyrimidinone
5-Fluorocytosine

Other Names
Alcobon, Ancobon

Form	Molecular Formula	MW	CAS
Flucytosine	$C_4H_4FN_3O$	129.1	2022-85-7

Appearance
Flucytosine is a white to off-white crystalline powder.

Solubility
Flucytosine has a solubility of 1.5 g/100 mL in water at 25 °C.

pK$_a$
Flucytosine has pK$_a$ values of 2.9 and 10.71.

Method 1
Vermes et al. investigated by HPLC the stability of 5-flucytosine in intravenous solutions. A liquid chromatograph consisted of a Shimadzu model LC-6A pump, a model C-R6A integrator, a model SPD-6A UV detector, and a Rheodyne model 7125 injector. The stationary phase was a Supelcosil LC-18-DB column (150 × 4.6 mm). The mobile phase was 0.03 M monobasic ammonium phosphate (pH 3.5) and 0.005 M 1-heptanesulfonic acid in water. The flow rate was 1 mL/min. UV detection was performed at 266 nm.

Samples were diluted with distilled water before assay. The injection volume was 5 μL. Under these conditions, retention times for 5-fluorouracil and 5-flucytosine were 3.8 and 8.0 minutes, respectively.

The stability-indicating ability of the method was demonstrated by accelerated degradation of flucytosine at 90 °C for 47 days. Flucytosine was well separated from its degradation product, fluorouracil.

A standard curve for flucytosine was generated from 20 to 100 μg/mL. The correlation coefficient was 0.9999.

Reference

Vermes A, van der Sijs H, Guchelaar H-J. An accelerated stability study of 5-flucytosine in intravenous solution. *Pharm World & Sci.* 1999; 21: 35–9.

Method 2

Wintermeyer and Nahata used an HPLC method to determine the stability of flucytosine in an extemporaneously compounded oral liquid. Flucytosine concentrations were determined using a Varian 2010 high-performance liquid chromatograph equipped with a Varian 2050 variable-wavelength detector, a Shimadzu model SIL-10A autoinjector, and a Waters μBondapak C_{18} analytical column (300 × 3.9 mm, 10-μm particle size). The mobile phase contained 750 mL of distilled water, 250 mL of acetonitrile, 2 mL of acetic acid, and 0.700 g of 1-octanesulfonic acid. The flow rate was 1.0 mL/min. UV detection was performed at 285 nm. *p*-Aminobenzoic acid 40 μg/mL was used as an internal standard. Samples and standards were maintained at 10 °C before injection.

Each sample was diluted with a solution of acetonitrile and distilled water (1:3, vol/vol). The injection volume was 5 μL. Retention times for flucytosine and the internal standard were 3.2 and 4.7 minutes, respectively.

The stability-indicating nature of the method was demonstrated by accelerated degradation. Flucytosine solution 1 mg/mL was mixed with 1 mL of 6 M hydrochloric acid in a water bath at 60 °C for 3 hours. Degradation product peaks did not interfere.

The standard curve was generated by linear regression of flucytosine concentration against the ratio of the peak height for flucytosine to that for the internal standard from 3 to 90 μg/mL. The correlation coefficient was greater than 0.999. The interday coefficient of variation was less than 0.72%.

Reference

Wintermeyer SM, Nahata MC. Stability of flucytosine in an extemporaneously compounded oral liquid. *Am J Health Syst Pharm.* 1996; 53: 407–9.

Method 3

Allen and Erickson studied the stability of flucytosine in extemporaneously compounded oral liquids. The HPLC system consisted of a Hewlett-Packard series 1050 automated high-performance liquid chromatograph, including a multisolvent mixing and pumping system, an autoinjector, a diode-array detector, and a computer with Chem Station software. The stationary phase was a Hewlett-Packard Hypersil silica column (200 × 4.6 mm, 5-μm particle size). The mobile phase was a mixture of 500 mL of methanol and 500 mL of water with 1 mL of diisopropylamine and 1 g of ammonium acetate per liter; pH was adjusted to 7.5. The flow rate was 1.0 mL/min. UV detection was performed at 280 nm.

Samples were diluted 1:200. The injection volume was 20 μL. The retention time for flucytosine was 3.3 minutes.

This assay was determined to be stability indicating by degrading samples of flucytosine in water and in commercial infusion solutions using acid, base, an oxidizing agent, heat, and light. The chromatograms showed that the presence of degradation products did not interfere with detection of the intact flucytosine.

Five-point standard curves were prepared with reference standard from 1 to 100 μg/mL. Coefficients of variation for intraday and interday assays were 1.0 and 1.2%, respectively.

Reference
Allen LV, Erickson MA. Stability of acetazolamide, allopurinol, azathioprine, clonazepam, and flucytosine in extemporaneously compounded oral liquids. *Am J Health Syst Pharm.* 1996; 53: 1944–9.

Method 4
Pramar et al. described the quantitation of 5-flucytosine by HPLC analysis. A Waters model ALC 202 chromatograph was equipped with a Rheodyne model 7125 injector, a Schoeffel SF 770 multiple-wavelength detector, and an Omniscribe recorder. The stationary phase was a Med Pharmex C_{18} column (250 × 4.6 mm). The mobile phase was 0.05 M monobasic potassium phosphate aqueous buffer (pH ~4.5) and was delivered at 1.0 mL/min. UV detection was performed at 300 nm and 0.1 AUFS. 5-Aminouracil 40 μg/mL in water was used as an internal standard.

A portion of the powder from 10 capsules equivalent to 100 mg of 5-flucytosine was mixed with 80 mL of water, stirred for 3–4 minutes, diluted to 100 mL with water, and filtered. A middle portion (2.5 mL) of the filtrate was mixed with 7.5 mL of the internal standard and then diluted to 25 mL with water. The injection volume was 25 μL. Under these conditions, retention times for 5-aminouracil, 5-flucytosine, and fluorouracil were about 4.9, 6.3, and 7.3 minutes, respectively (estimated from the published chromatogram).

The method was reported to be stability indicating since there was no interference from the major degradation product and the excipients with the 5-flucytosine peak.

Reference
Pramar Y, Das Gupta V, Bethea C. Quantitation of 5-flucytosine in capsules using high-pressure liquid chromatography. *Drug Dev Ind Pharm.* 1991; 17: 193–9.

Fludrocortisone Acetate

Chemical Name
9α-Fluoro-11β,17α,21-trihydroxypregn-4-ene-3,20-dione 21-acetate

Other Names
Blephaseptyl, Florinef Acetate, Panotile

Form	Molecular Formula	MW	CAS
Fludrocortisone acetate	$C_{23}H_{31}FO_6$	422.5	514-36-3

Appearance

Fludrocortisone acetate occurs as white or pale yellow odorless or almost odorless hygroscopic crystals or crystalline powder.

Solubility

Fludrocortisone acetate is insoluble in water and sparingly soluble in alcohol.

Method

Ast and Abdou reported the HPLC analysis of fludrocortisone acetate in solid dosage forms. The chromatographic system included a reciprocating pump, an injector, and a UV detector. The stationary phase was a Waters µBondapak C_{18} column (300 × 4 mm). The mobile phase was 42% acetonitrile in water. The flow rate was 1.8–2.0 mL/min. UV detection was performed at 254 nm. Norethindrone in acetonitrile was used as an internal standard.

A sample of ground tablets equivalent to 2.5 mg of fludrocortisone acetate was mixed with 5 mL of distilled water for 1 minute, shaken with 20 mL of internal standard for 40 minutes, and centrifuged. The clear liquid was collected for analysis. Under these conditions, retention times for fludrocortisone acetate and norethindrone were about 10.2 and 15.3 minutes, respectively (estimated from the published chromatogram).

The stability-indicating ability of the method was demonstrated. Fludrocortisone acetate was separated from its known degradation products and impurities.

A standard curve for fludrocortisone acetate was constructed from 45 to 120 µg/mL. The correlation coefficient was greater than 0.999.

Reference

Ast TM, Abdou HM. Analysis of fludrocortisone acetate and its solid dosage forms by high-performance liquid chromatography. *J Pharm Sci.* 1979; 68: 421–3.

Flumazenil

Chemical Names

8-Fluoro-5,6-dihydro-5-methyl-6-oxo-4*H*-imidazo[1,5-*a*][1,4]benzodiazepine-3-carboxylic acid ethyl ester

Ethyl 8-fluoro-5,6-dihydro-5-methyl-6-oxo-4*H*-imidazo[1,5-*a*][1,4]benzodiazepine-3-carboxylate

Other Names
Anexate, Lanexat, Romazicon

Form	Molecular Formula	MW	CAS
Flumazenil	$C_{15}H_{14}FN_3O_3$	303.3	78755-81-4

Method
Olsen et al. evaluated the stability of flumazenil with selected drugs in 5% dextrose injection. The liquid chromatograph was composed of a Shimadzu LC-6A solvent-delivery system, a Shimadzu SIL-6A autoinjector, a Shimadzu SPD-6A variable-wavelength UV spectrophotometer, a Shimadzu Chromatopac C-R5A integrator, and a Shimadzu CTO-6A column oven. The stationary phase was a Phase Separation Spherisorb C_{18} column (150 × 4.6 mm, 5-μm particle size). The mobile phase was a mixture of water, methanol, and tetrahydrofuran (60:35:5, vol/vol/vol) and was delivered at 1 mL/min. UV detection was performed at 245 nm. The injection volume was 25 μL. The temperature of the analytical column was maintained at 40 °C. The retention time for flumazenil was 4.35 minutes.

The stability-indicating nature of this assay was assessed by an accelerated degradation study. Aqueous flumazenil solutions 50 μg/mL were adjusted to pH 2 or 12 with 1.0 N hydrochloric acid or 1.0 N sodium hydroxide, respectively. These solutions were then neutralized and assayed. Degradation product peaks did not interfere with the intact flumazenil peak.

Calibration curves were generated daily and found to be linear from 1 to 50 μg/mL by regression analysis ($r = 0.9998$). The intraday and interday relative standard deviations were 1.36 and 4.65%, respectively.

Reference
Olsen KM, Gurley BJ, Davis GA, et al. Stability of flumazenil with selected drugs in 5% dextrose injection. *Am J Hosp Pharm.* 1993; 50: 1907–12.

Flunarizine Hydrochloride

Chemical Name
trans-1-Cinnamyl-4-(4,4′-difluorobenzhydryl)piperazine dihydrochloride

Other Names

Amalium, Flerudin, Sibelium

Form	Molecular Formula	MW	CAS
Flunarizine hydrochloride	$C_{26}H_{26}F_2N_2.2HCl$	477.4	30484-77-6

Method

Wahbi et al. reported the HPLC determination of flunarizine hydrochloride in the presence of its degradation product. The system consisted of a Waters model 501 pump, a model 680 automated gradient controller equipped with a U6K universal injector, a model 481 variable-wavelength UV detector, and a model 740 data module. The stationary phase was a Waters μBondapak C_{18} column (300 × 3.9 mm, 10-μm particle size). The mobile phase consisted of 78% methanol and 22% water containing 0.5% (wt/vol) sodium chloride and 0.2% (vol/vol) triethanolamine; the aqueous solution was adjusted to pH 6.6 with 30% hydrochloric acid before mixing. The flow rate was 2 mL/min. UV detection was performed at 254 nm. Clotrimazole was used as the internal standard.

An amount of powder equivalent to about 40 mg of flunarizine hydrochloride was transferred into a 100-mL flask, mixed with 50 mL of methanol, shaken for 10 minutes, filtered, and diluted to 100 mL with methanol. A 5-mL portion of this solution was added to 5 mL of internal standard solution and adjusted to 50 mL with methanol before injection. The injection volume was 20 μL. Under these conditions, retention times for the degradation product, clotrimazole, and flunarizine dihydrochloride were about 3.4, 5.1, and 6.8 minutes, respectively.

The method was stability indicating since it separated the intact drug from its degradation product.

A calibration curve for flunarizine was obtained from 0.016 to 0.064 mg/mL. The correlation coefficient was 0.9999. The limit of detection was 0.001 mg/mL.

Reference

Wahbi A, El-Walily AM, Hassan EM, et al. Liquid chromatographic determination of flunarizine dihydrochloride in the presence of its degradation product. *J Pharm Biomed Anal*. 1995; 13: 777–84.

Fluocinonide

Chemical Name

6α, 9α-Difluoro-11β,21-dihydroxy-16α,17α-isopropylidenedioxypregna-1,4-diene-3,20-dione 21-acetate

Other Names
Fluocinolide, Fluonex, Lidex, Metosyn, Topsym, Topsyn

Form

Form	Molecular Formula	MW	CAS
Fluocinonide	$C_{26}H_{32}F_2O_7$	494.5	356-12-7

Appearance
Fluocinonide occurs as a white or almost white odorless crystalline powder.

Solubility
Fluocinonide is practically insoluble in water and slightly soluble in alcohol.

Method
Shek et al. reported a reversed-phase HPLC method for a stability study of fluocinonide acetate in a cream. The instrument was a Spectra-Physics model 3500 system equipped with an oven, a Valco loop injector, and a SpectraMonitor III UV detector or a Spectra-Physics model 8300 UV detector. The stationary phase was a Spectra-Physics Spherisorb ODS stainless steel column (250 × 4.6 mm) with a Whatman Co:Pell ODS guard column (70 × 2.1 mm). The mobile phase was a mixture of tetrahydrofuran, acetonitrile, and water (1:3:6, vol/vol/vol) and was delivered isocratically at 1.2 mL/min. UV detection was performed at 254 nm and 0.02 AUFS. The column temperature was 45 °C. Flucloronide 190 μg/mL in acetonitrile was used as an internal standard.

Two grams of cream was transferred to a mixture of 75 mL of isooctane and 25 mL of acetonitrile containing 1 mL of the internal standard, and the mixture was shaken vigorously for 1 minute. The lower acetonitrile layer was collected. This extraction was repeated three or four times. The combined acetonitrile solution was evaporated to dryness using a steam bath and a stream of nitrogen; the residue was redissolved in 10 mL of acetonitrile, diluted with 10 mL of water, and centrifuged at 3000 rpm for 10 minutes. The clear solution was collected and assayed. The injection volume was 50 μL. Under these conditions, retention times for the major degradation product (fluocinolone acetonide), the internal standard, and fluocinonide were about 7.2, 13.3, and 16.7 minutes, respectively (estimated from the published chromatogram).

The stability-indicating nature of the method was demonstrated by assaying a mixture of the drug and its possible degradation products. The chromatogram showed excellent separation of the intact drug from all degradation products.

Reference
Shek E, Bragonje J, Benjamin EJ, et al. A stability indicating high-performance liquid chromatography determination of triple corticoid integrated system in a cream. *Int J Pharm.* 1982; 11: 257–69.

Fluorouracil

Chemical Names
5-Fluoro-2,4(1*H*,3*H*)-pyrimidinedione
2,4-Dioxo-5-fluoropyrimidine

Other Name
Adrucil

Form	Molecular Formula	MW	CAS
Fluorouracil	$C_4H_3FN_2O_2$	130.1	51-21-8

Appearance
Fluorouracil is a white to practically white crystalline powder.

Solubility
Fluorouracil is sparingly soluble in water and slightly soluble in alcohol.

pK$_a$
Fluorouracil has a pK$_a$ of 7.71 at 25 °C.

Method 1
Xu et al. used an HPLC method to study the stability of fluorouracil admixed with fentanyl in polyvinyl chloride containers. A Waters LC-Module 1 consisted of a multisolvent-delivery pump, a multiple-wavelength UV detector, and an autosampler. This system was controlled and integrated by an NEC PowerMate SX/16 computer. The stationary phase was a Vydac C$_{18}$ column (250 × 4.6 mm, 5-µm particle size) with a guard column of the same material. The mobile phase consisted of 2% methanol in water. The flow rate was 0.7 mL/min. UV detection was performed at 254 nm.

Samples were diluted with the appropriate infusion solution. The injection volume was 10 µL. Under these conditions, the retention time for fluorouracil was about 7 minutes.

The method was determined to be stability indicating by accelerated degradation of fluorouracil. The addition of 1 N sodium hydroxide and 1 N hydrochloric acid yielded a reduction in the intact fluorouracil peak and the formation of a new peak at 4.4 minutes. Fentanyl citrate did not elute.

The calibration curve was constructed from 0.025 to 0.150 mg/mL; the correlation coefficient was greater than 0.9999. The intraday and interday coefficients of variation were 1.4 and 0.9%, respectively.

Reference
Xu QA, Trissel LA, Martinez JF. Rapid loss of fentanyl citrate admixed with fluorouracil in polyvinyl chloride containers. *Ann Pharmacother*. 1997; 31: 297–302.

Method 2

Mayron and Gennaro assessed the stability of fluorouracil with granisetron hydrochloride during simulated Y-site administration. The liquid chromatograph consisted of a piston pump with a pulse dampener, a rotary injection port with a 20-μL loop, a variable-wavelength UV detector, and an integrator. The stationary phase was a C_{18} reversed-phase column (250 × 4.6 mm, 5-μm particle size). The mobile phase was a mixture of acetonitrile and 0.1 M monobasic sodium phosphate dihydrate (20:80, vol/vol) adjusted to pH 4.2 with phosphoric acid. The flow rate was 2.5 mL/min. UV detection was performed at 300 nm.

Samples were diluted with mobile phase. The retention times for fluorouracil and granisetron were 1.58 and 5.15 minutes, respectively.

The stability-indicating capability of the analytical method was demonstrated by an accelerated degradation study. Solutions of fluorouracil and granisetron hydrochloride were adjusted to pH 2 or 11, boiled for 1 hour, readjusted to pH 5, diluted with the mobile phase, and analyzed. The degradation product peaks did not interfere with the intact fluorouracil or granisetron peaks.

Reference

Mayron D, Gennaro AR. Stability and compatibility of granisetron hydrochloride in i.v. solutions and oral liquids and during simulated Y-site injection with selected drugs. *Am J Health Syst Pharm.* 1996; 53: 294–304.

Method 3

Wang et al. used an HPLC method to study the stability of a fluorouracil–metoclopramide hydrochloride admixture. The liquid chromatographic system consisted of a Shimadzu model LC-6A single-piston pump, a Shimadzu model SPD-6AV dual UV-visible detector, a Waters model 745 electronic integrator, and a Waters μBondapak phenyl column (150 × 4.6 mm). The mobile phase was a mixture of acetonitrile, methanol, 0.01% acetic acid, and 0.005 N sulfonic acid solution (20:15:40:25, vol/vol/vol/vol) and was delivered isocratically at 1.0 mL/min. UV detection was performed at 230 nm.

Samples were diluted 1:10 with mobile phase. Under these conditions, retention times for fluorouracil and metoclopramide were 3.5 and 10.0 minutes, respectively.

The stability-indicating capability of the assay was shown by analyzing a fluorouracil–metoclopramide hydrochloride admixture that had been stored at 50 °C for 10 days. No degradation product peak interfered with the intact fluorouracil peak.

Standard curves were constructed from 0.01 to 0.25 mg/mL. Correlation coefficients were greater than 0.99. Intraday and interday coefficients of variation for the assay were 0.66 and 0.93%, respectively.

Reference

Wang D-P, Chang L-C, Lee DKT, et al. Stability of fluorouracil–metoclopramide hydrochloride admixture. *Am J Hosp Pharm.* 1995; 52: 98–9.

Method 4

Rochard et al. determined the stability of fluorouracil in ethylene vinyl acetate portable infusion-pump reservoirs. The chromatograph included a Waters model 510 reciprocating piston pump, a Waters model 710A WISP autosampler, a Waters model 484 variable-wavelength UV detector, and a Waters model 810 computing integrator. The stationary phase was a Nucleosil C_{18} column (250 × 4.6 mm, 5-μm particle size). The mobile phase was 0.05 M ammonium dihydrogen phosphate and was delivered at 1.2 mL/min. UV detection was performed at 266 nm. The retention time for fluorouracil was 5.3 minutes.

The stability-indicating nature of the method was verified by using solutions of fluorouracil partially decomposed by heat (70 °C), 1 N hydrochloric acid, 1 N sodium hydroxide, or 1% hydrogen peroxide. Degradation product peaks did not interfere with the intact fluorouracil peak.

Standard curves for fluorouracil were constructed each day for calibration from 5 to 100 μg/mL. Correlation coefficients were greater than 0.990. Intraday and interday coefficients of variation of the assay were 0.50 and 0.64%, respectively.

Reference

Rochard EB, Barthes DMC, Courtois PY. Stability of fluorouracil, cytarabine, or doxorubicin hydrochloride in ethylene vinylacetate portable infusion-pump reservoirs. *Am J Hosp Pharm.* 1992; 49: 619–23.

Method 5

Woloschuk et al. assessed the chemical stability of fluorouracil during simulated Y-site administration with mannitol. The HPLC system consisted of a Waters model M45 pump, a Waters model U6K injector, a Waters model 481 absorbance detector, an Omniscribe strip-chart recorder, and a Waters Deltapak C_{18} column (300 × 3.9 mm). The mobile phase was 0.5 M ammonium acetate (pH 6.5) and was delivered at 1.5 mL/min. UV detection was performed at 254 nm and 0.2 AUFS. 5-Chlorouracil was used as an internal standard.

Samples were diluted with distilled water. The injection volume was 10 μL. Under these conditions, retention times were 3.1 and 4.8 minutes for fluorouracil and the internal standard, respectively.

The HPLC assay was determined to be stability indicating by assaying a sample containing fluorouracil and its related compounds (5-chlorouracil, 5-bromouracil, uracil, and cytosine). The related compound peaks did not interfere with the intact fluorouracil peak.

The amount of fluorouracil was determined from the peak height ratio of fluorouracil to 5-chlorouracil.

Reference

Woloschuk DMM, Wermeling JR, Pruemer JM. Stability and compatibility of fluorouracil and mannitol during simulated Y-site administration. *Am J Hosp Pharm.* 1991; 48: 2158–60.

Method 6

Northcott et al. evaluated the chemical stability of fluorouracil infusion 25 mg/mL in an ambulatory pump. The HPLC method used a Constametric 3000 pump, a SpectroMonitor 3100 UV-visible variable-wavelength detector, a Rheodyne 7125 loop valve fitted with a 20-μL loop, an LDC-Milton Roy CI-10B integrator/printer plotter, and a Kontron S5ODS1 analytical column (250 × 4.8 mm). The mobile phase was a mixture of methanol and 0.01 M acetate buffer (5:95, vol/vol) with a pH of 4.0 and was delivered isocratically at 1.0 mL/min. UV detection was performed at 270 nm.

Samples were diluted 1:250 with water. The injection volume was 20 μL.

The stability-indicating nature of the HPLC assay was determined using an accelerated degradation study. Fluorouracil sample solutions were subjected to 1 N sodium hydroxide, 1 M hydrochloric acid, hydrogen peroxide, or heat. Each of the stress treatments decreased the intact fluorouracil peak height. In all cases, the intact fluorouracil peak was stated to be well resolved from the degradation product peaks.

Calibration curves for fluorouracil were constructed from the peak height versus drug concentration from 25 to 300 μg/mL; the correlation coefficient was 0.999. The coefficient of variation of the peak height was 0.42% ($n = 6$).

Reference

Northcott M, Allsopp MA, Powell H, et al. The stability of carboplatin, diamorphine, 5-fluorouracil and mitozantrone infusions in an ambulatory pump under storage and prolonged 'in-use' conditions. *J Clin Pharm Ther*. 1991; 16: 123–9.

Method 7

Quebbeman et al. evaluated the stability of fluorouracil in plastic containers used for continuous infusion at home. The high-performance liquid chromatograph consisted of a Waters M6000A pump, a Waters U6K injector, a Waters model 440 UV detector, and a Waters model 730 data module. Three columns were tested and used: a Waters μBondapak C_{18} steel column (300 × 3.9 mm), a Waters radial compression module (100 × 8 mm), and an Alltech C_{18} column (250 × 4.6 mm). The mobile phase was 0.16 mM phosphate buffer at pH 6.70 and was delivered isocratically at 3 mL/min. UV detection was performed at 254 nm and 0.005 AUFS. 5-Bromouracil 402 μg/mL was used as an internal standard.

Samples were diluted with distilled water. The injection volume was 10 μL. Under these conditions, retention times for fluorouracil and the internal standard were approximately 2.5 and 6.0 minutes, respectively.

The analytical method was determined to be stability indicating. One fluorouracil sample was mixed with 1 M hydrochloric acid and the other was mixed with 1 M sodium hydroxide, and both were heated in a boiling water bath (100 °C) for 3.5 hours. These two samples were then cooled, neutralized, and chromatographed. Degradation product peaks did not interfere with the fluorouracil peak.

Reference

Quebbeman EJ, Hamid AAR, Hoffman NE, et al. Stability of fluorouracil in plastic containers used for continuous infusion at home. *Am J Hosp Pharm*. 1984; 41: 1153–6.

Method 8

Stiles et al. determined the stability of fluorouracil in polypropylene infusion-pump syringes. The liquid chromatograph included a Waters model 501 pump, a Waters model 441 UV detector, a Waters model 745 data module, an Alcott model 728 autosampler, and a Rheodyne 7010 injector. The stationary phase was a Bakerbond C_{18} analytical column. The mobile phase was 0.005 M dibasic potassium phosphate buffer (pH 7.8) and was delivered isocratically at 0.5 mL/min. UV detection was performed at 214 nm.

Samples were diluted 1:100 before injection. Under these conditions, the retention time for fluorouracil was 6.6 minutes.

The analytical method was stated to be stability indicating.

The standard curve was constructed for fluorouracil from 100 to 1000 μg/mL. Intraday and interday coefficients of variation of the assay were 1.2 and 1.3%, respectively.

References

Stiles ML, Allen LV Jr, Prince SJ. Stability of deferoxamine mesylate, floxuridine, fluorouracil, hydromorphone hydrochloride, lorazepam, and midazolam hydrochloride in polypropylene infusion-pump syringes. *Am J Health Syst Pharm.* 1996; 53: 1583–8.

Stiles ML, Allen LV Jr, Tu YH. Stability of fluorouracil administered through four portable infusion pumps. *Am J Hosp Pharm.* 1989; 46: 2036–40.

Allen LV, Stiles ML, Prince SJ, et al. Stability of 14 drugs in the latex reservoir of an elastomeric infusion device. *Am J Health Syst Pharm.* 1996; 53: 2740–3.

Fluorouridine

Chemical Name
5-Fluorouridine

Form	Molecular Formula	MW	CAS
Fluorouridine	$C_9H_{11}FN_2O_6$	262.2	316-46-1

Method
Dorta et al. developed an HPLC method for 5-fluorouridine in aqueous solution. A Waters system was equipped with a model 600E pump and a model 490E programmable multiwavelength detector. The stationary phase was a Waters reversed-phase Resolve C_{18} column (100 × 8 mm). The mobile phase consisted of 50 mM monobasic ammonium phosphate (adjusted to pH 3.5 with phosphoric acid) and acetonitrile (96:4, vol/vol). The flow rate was 1.7 mL/min. UV detection was performed at 268 nm. Under these conditions, retention times for 5-fluorouracil and 5-fluorouridine were about 2.7 and 4.2 minutes, respectively (estimated from the published chromatogram).

The stability-indicating ability of the assay was demonstrated by thermal degradation of 5-fluorouridine. 5-Fluorouridine was separated from its degradation product, 5-fluorouracil.

A standard curve for 5-fluorouridine was constructed from 0.4 to 5.0 μg/mL.

Reference

Dorta MJ, Munguia O, Farina JB, et al. Stability indicating high performance liquid chromatography methods for 5-fluorouridine in aqueous solution. *Arzneim-Forsch/Drug Res.* 1997; 47: 1388–92.

Fluoxetine

Chemical Names
(±)-*N*-Methyl-γ-[4-(trifluoromethyl)phenoxy]benzenepropanamine
(±)-*N*-Methyl-3-phenyl-3-[(α,α,α-trifluoro-*p*-tolyl)oxy]propylamine

Other Name
Prozac

Form	Molecular Formula	MW	CAS
Fluoxetine	$C_{17}H_{18}F_3NO$	309.3	54910-89-3
Fluoxetine hydrochloride	$C_{17}H_{18}F_3NO \cdot HCl$	345.8	59333-67-4

Appearance
Fluoxetine hydrochloride occurs as a white to off-white crystalline solid.

Solubility
Fluoxetine hydrochloride has a solubility of 50 mg/mL in water.

Method 1
Pramar et al. developed an assay method for the determination of fluoxetine hydrochloride. A Waters ALC202 liquid chromatograph was equipped with a Rheodyne model 7125 injector, a Schoeffel SF770 multiple-wavelength detector, and a Houston Omniscribe recorder. The stationary phase was a Waters μBondapak C_{18} column (300 × 3.9 mm). The mobile phase consisted of acetonitrile and 0.01 M monobasic potassium phosphate aqueous buffer (45:55, vol/vol). The flow rate was 2.0 mL/min. UV detection was carried out at 234 nm and 0.1 AUFS. Methyltestosterone 50 μg/mL in methanol was used as an internal standard.

A portion of powder from 10 capsules, equivalent to 20 mg of fluoxetine hydrochloride, was weighed, mixed with 0.5 mL of 0.1 N hydrochloric acid and 20 mL of water, stirred for 3 minutes, brought to 25 mL with water, and filtered. The filtrate was mixed with internal standard solution and diluted with water. The injection volume was 20 μL. Under these conditions, retention times of fluoxetine and methyltestosterone were about 9.3 and 10.9 minutes, respectively (estimated from the published chromatogram).

Fluoxetine solutions were mixed with 1 N sulfuric acid or 0.1 N sodium hydroxide solution and heated to boiling for 15 minutes. Degradation product peaks did not interfere with the peak of intact drug, confirming the stability-indicating ability of the analytical method.

Reference
Pramar Y, Gupta VD, Bethea C. Quantitation of fluoxetine hydrochloride in capsules using high-performance liquid chromatography. *Drug Develop Ind Pharm.* 1992; 18: 257–64.

Method 2

Peterson et al. studied the stability of fluoxetine hydrochloride in common pharmaceutical diluents. Fluoxetine samples were assayed on a Hewlett-Packard 1050 series HPLC system including a variable-wavelength UV detector, an autosampler, a gradient pump, and a DuPont Zorbax cyano special column (250 × 4.6 mm). The mobile phase consisted of acetonitrile and a triethylamine solution. Triethylamine solution contained 1% triethylamine and 99% water with the pH adjusted to 6 with concentrated phosphoric acid. The flow rate was 1.0 mL/min. UV detection was performed at 215 nm.

Each sample was diluted with mobile phase. The injection volume was 20 μL. Under these conditions, fluoxetine had a retention time of about 650 seconds.

The assay was determined to be stability indicating by accelerated degradation of fluoxetine. Fluoxetine was intentionally degraded by acid, base, oxidizing agent, heat, light, and humidity. The fluoxetine assay separated all known related substances and degradation product peaks from the intact fluoxetine peak.

A standard curve was generated by least-squares regression analysis of peak area versus concentration from 30 to 60 μg/mL. The correlation coefficient was greater than 0.9992. The coefficient of variation of the assay was 0.4%.

Reference

Peterson JA, Risley DS, Anderson PN, et al. Stability of fluoxetine hydrochloride in fluoxetine solution diluted with common pharmaceutical diluents. *Am J Hosp Pharm.* 1994; 51: 1342–5.

Fluphenazine

Chemical Name

4-[3-[2-(Trifluoromethyl)-10H-phenothiazin-10-yl]propyl]-1-piperazineethanol

Other Names

Dapotum, Modecate, Permitil, Prolixin

Form	Molecular Formula	MW	CAS
Fluphenazine	$C_{22}H_{26}F_3N_3OS$	437.5	69-23-8
Fluphenazine enanthate	$C_{29}H_{38}F_3N_3O_2S$	549.7	2746-81-8
Fluphenazine hydrochloride	$C_{22}H_{26}F_3N_3OS.2HCl$	510.4	146-56-5
Fluphenazine decanoate	$C_{32}H_{44}F_3N_3OS$	591.8	5002-47-1

Appearance

Fluphenazine enanthate is a pale yellow to yellow-orange and clear to slightly turbid viscous liquid. Fluphenazine hydrochloride occurs as a white or almost white and odorless crystalline powder. Fluphenazine decanoate occurs as a pale yellow viscous liquid or a yellow crystalline oily solid.

Solubility

Fluphenazine enanthate is insoluble in water. It is very soluble in dehydrated alcohol and in dichloromethane. It is freely soluble in methanol. Fluphenazine hydrochloride is freely soluble in water. It is slightly soluble in alcohol. Fluphenazine decanoate is insoluble in water. It is freely soluble in alcohol.

Method

El-Ragehy et al. developed a stability-indicating HPLC method for the analysis of fluphenazine hydrochloride in commercial tablets and in the presence of its degradates. A Shimadzu liquid chromatograph consisted of a model LC-10AD pump, a model SPD-10A UV-visible detector, and a model C-R6A Chromatopac recorder. The stationary phase was a Nova-Pak C_{18} column (150 × 3.9 mm, 4-μm particle size). The mobile phase consisted of 0.05 M ammonium acetate aqueous solution, methanol, and acetonitrile (4:1:5 vol/vol/vol); the pH was adjusted to 7.5. The flow rate was 1 mL/min. UV detection was performed at 254 nm. The injection volume was 20 μL. Under these conditions, the retention time of fluphenazine was about 8.1 minutes.

The method was shown to be stability indicating by intentional degradation of the drug. Fluphenazine solution was treated with 30% hydrogen peroxide and heated. The retention time of its degradation product was about 4.8 minutes. The intact drug peak and its degradation product peak were well resolved on the chromatogram.

A standard curve for fluphenazine was constructed from 1.2 to 4.2 μg/mL. The correlation coefficient was greater than 0.9999. The limit of detection and the limit of quantitation were 0.60 and 1.00 μg/mL, respectively.

Reference

El-Ragehy NA, Abbas SS, El-Khateeb SZ. Spectrophotometric and stability indicating high performance liquid chromatographic determination of nortriptyline hydrochloride and fluphenazine hydrochloride. *Anal Lett.* 2002; 35: 1171–91.

Flurbiprofen

Chemical Name

2-(2-Fluorobiphenyl-4-yl)propionic acid

Other Names
Ansaid, Froben, Ocufen

Form	Molecular Formula	MW	CAS
Flurbiprofen	$C_{15}H_{13}FO_2$	244.3	5104-49-4
Flurbiprofen sodium	$C_{15}H_{12}FNaO_2.2H_2O$	302.3	56767-76-1

Appearance
Flurbiprofen is a white or almost white crystalline powder. Flurbiprofen sodium is a white to creamy white crystalline powder.

Solubility
Flurbiprofen is practically insoluble in water. It is freely soluble in alcohol, acetone, chloroform, ether, and methanol. Flurbiprofen sodium has a solubility of 4 mg/mL in water (pH 7) at 26 °C.

pK_a
Flurbiprofen has a pK_a of 4.22.

Method
Mathew et al. reported the quantitation of flurbiprofen in tablets using high-performance liquid chromatography. A Waters ALC 202 chromatograph was equipped with a Rheodyne model 7125 universal injector, a Schoeffel SF 770 multiple-wavelength detector, and a Houston Omniscribe recorder. The stationary phase was a Waters micro C_{18} column (300 × 3.9 mm). The mobile phase was 48% acetonitrile in 0.01 M monobasic potassium phosphate aqueous buffer and was delivered isocratically at 2.2 mL/min. UV detection was performed at 234 nm and 0.1 AUFS. Ibuprofen 0.32 mg/mL in methanol was used as an internal standard.

Five flurbiprofen tablets were mixed with 40 mL of methanol, stirred for 5 minutes, and brought to volume (50 mL) with methanol. The mixture was filtered and the middle portion of the filtrate was collected. The filtrate was appropriately diluted with 40% methanol in 0.02 M monobasic potassium phosphate aqueous buffer before injection. The injection volume was 20 µL. Under these conditions, retention times for flurbiprofen and ibuprofen were about 4.6 and 8.0 minutes, respectively (estimated from the published chromatogram).

The stability-indicating nature of the method was demonstrated by accelerated degradation of flurbiprofen. Solutions of flurbiprofen were mixed with 1 N sulfuric acid or 1 N sodium hydroxide solution and heated to boiling for 5 minutes. Degradation products and excipients in the dosage forms did not interfere with the determination of flurbiprofen.

Reference
Mathew M, Das Gupta V, Bethea C. Quantitation of flurbiprofen in tablets using high performance liquid chromatography. *Drug Dev Ind Pharm*. 1993; 19: 493–8.

Flurogestone Acetate

Chemical Name
9α-Fluoro-11β,17α-dihydroxypregn-4-ene-3,20-dione 17-acetate

Other Names
Flugestone Acetate, Progestin

Form	Molecular Formula	MW	CAS
Flurogestone acetate	C$_{23}$H$_{31}$FO$_5$	406.5	2529-45-5

Method
Kabadi et al. described an HPLC method for the determination of flurogestone in the presence of its degradation products. The Waters liquid chromatograph included a model 6000A pump, a model U6K injector, a model 440 UV detector, and an Omniscribe recorder. The stationary phase was a Waters μBondapak C$_{18}$ column (300 × 3.9 mm, 10-μm particle size). The mobile phase consisted of methanol and water (50:50, vol/vol) and was delivered at 2 mL/min. UV detection was performed at 254 nm. Testosterone was used as an internal standard. The injection volume was 100 μL. Under these conditions, retention times for flurogestone acetate and testosterone were 21.2 and 31.5 minutes, respectively.

The stability-indicating nature of the method was established by accelerated degradation of flurogestone. Flurogestone acetate was stored in 0.1 M hydrochloric acid at 94 °C for 95 hours. The flurogestone acetate peak was well separated from the peaks of its degradation products.

Standard curves for flurogestone acetate were obtained from 0 to 10 and from 10 to 40 μg/mL. The correlation coefficients were greater than 0.99.

References
Kabadi MB, Valia KH, Chien YW. Intravaginal controlled administration of flurogestone acetate I: Development of a stability-indicating liquid chromatographic method and stability kinetics of flurogestone acetate. *J Pharm Sci.* 1984; 73: 1461–4.

Kabadi MB, Chien YW. Intravaginal controlled administration of flurogestone acetate II: Development of an *in vitro* system for studying the intravaginal release and permeation of flurogestone acetate. *J Pharm Sci.* 1984; 73: 1464–8.

Folic Acid (Vitamin B$_9$)

Chemical Name
N-[4-[[(2-Amino-1,4-dihydro-4-oxo-6-pteridinyl)methyl]amino]benzoyl]-L-glutamic acid

Other Name
Folicet

Form

Form	Molecular Formula	MW	CAS
Folic acid	C$_{19}$H$_{19}$N$_7$O$_6$	441.4	59-30-3

Appearance
Folic acid is a yellow or yellowish-orange crystalline powder.

Solubility
Folic acid has an aqueous solubility of 0.0016 mg/mL at 25 °C. It is insoluble in ethanol.

Method 1
Van der Horst et al. described the HPLC analysis of folic acid in total parenteral nutrition solution. The liquid chromatographic system consisted of a Waters model M6000A pump, a model WISP autosampler, a model 481 variable-wavelength detector, a Perkin-Elmer model 204-A fluorescence spectrophotometer, and a Shimadzu Chromatopac C-R3A integrator. The stationary phase was a LiChrosorb RP8 analytical column (250 × 4.6 mm, 10-μm particle size) with a RCSS Guard-Pak C$_{18}$ precolumn (4 × 6 mm). The mobile phase was a mixture of 0.1% triethylamine in 0.2 M monobasic potassium phosphate buffer (pH 5.75) and acetonitrile (100:4, vol/vol). The flow rate was 3 mL/min. UV detection was performed at 314 nm.

An aliquot of 400 μL of the sample was injected onto the precolumn, washed with 20 mL of 0.030 M monobasic potassium phosphate buffer (pH 2.5), and switched to the analytical column. Under these conditions, the retention time for folic acid was about 4.0 minutes (estimated from the published chromatogram).

The method was demonstrated to be stability indicating by accelerated degradation of folic acid. Folic acid solutions were treated with 1 M sodium hydroxide solution or 1 M hydrochloric acid and then heated to boiling for about 30 minutes. Degradation product peaks were separated from the peak of folic acid.

A standard curve for folic acid was generated from 0.01 to 0.32 mg/L. The correlation coefficient was 0.9995. The limit of detection was 0.006 mg/L.

Reference
Van der Horst A, Martens HJM, de Goede PNFC. Analysis of water-soluble vitamins in total parenteral nutrition solution by high pressure liquid chromatography. *Pharm Weekbl [Sci]*. 1989; 11: 169–74.

Method 2

Paveenbampen et al. reported a stability-indicating HPLC method for the determination of folic acid in multivitamin preparations. The liquid chromatographic system consisted of a Waters 6000A constant-flow pump, a Waters 710B WISP autosampler injector, a Kratos Spectroflow 773 variable-wavelength detector, a Houston Omniscribe 10-mV recorder, and a Hewlett-Packard LAS 3357 computer. The stationary phase was a Waters μBondapak C_{18} analytical column (300 × 3.9 mm) connected to a precolumn (70 × 2 mm) packed with Waters Corasil C_{18} (37–50 μm). The mobile phase was a mixture of methanol, acetonitrile, and 0.01 M sodium acetate buffer (6:4:90, vol/vol/vol). The mobile phase was adjusted to pH 4.5 with acetic acid. The flow rate was 2.5 mL/min. UV detection was performed at 280 nm and 0.02 AUFS.

Samples were diluted with mobile phase. The injection volume was 25 μL. The retention time for folic acid was about 10 minutes.

The HPLC method was demonstrated to be stability indicating by spiking folic acid samples with 12 other vitamin components, the formulation excipients, potential impurities, or potential degradation products of folic acid. Also, folic acid was heated using a steam bath for 0.5 hour and placed at a distance of 5 cm from a 254-nm lamp for 53 hours. No interference from any of these compounds or decomposition products with folic acid was observed.

A standard curve was constructed from 0.1 to 15.0 μg/mL and had a correlation coefficient of 0.999. The coefficients of variation for the intraday precision study at 0.1, 5.0, 10.0, and 15.0 μg/mL were 5.6, 0.6, 0.3, and 0.3%, respectively. The coefficients of variation for the interday precision study were 7.2, 0.8, 0.9, and 1.0%, respectively.

Reference

Paveenbampen C, Lamontanaro D, Moody J, et al. Liquid chromatographic determination of folic acid in multivitamin preparations. *J Pharm Sci.* 1986; 75: 1192–4.

Formoterol Fumarate

Chemical Name

(±)-2'-Hydroxy-5'-[(R*)-1-hydroxy-2-[[(R*)-p-methoxy-α-methylphenethyl]amino]-ethyl]formanilide fumarate (2:1)

Other Names

Atock, Foradil, Oxis

Form	Molecular Formula	MW	CAS
Formoterol	$C_{19}H_{24}N_2O_4$	344.4	73573-87-2
Formoterol fumarate	$(C_{19}H_{24}N_2O_4)_2.C_4H_4O_4$	804.9	43229-80-7

Appearance

Formoterol fumarate is a white or almost white or slightly yellow powder.

Solubility

Formoterol fumarate is slightly soluble in water and isopropyl alcohol. It is practically insoluble in acetonitrile. It is soluble in methanol.

Method

Akapo and Asif reported an HPLC method for the determination of formoterol fumarate and its related substances in the bulk drug. A Hitachi liquid chromatograph consisting of a model L-7100 pump, a model L-7300 oven, a model L-7200 autosampler, and a model L-7450 photodiode-array detector was used. The stationary phase was an Alltech Alltima C_{18} silica column (150 × 4.6 mm, 5-μm particle size). The mobile phase consisted of 50 mM ammonium acetate (pH 5.0) and methanol (65:35, vol/vol). The flow rate was 1.0 mL/min. UV detection was performed at 242 nm. The run time was 20 minutes. The injection volume was 20 μL. Under these conditions, the retention time of formoterol fumarate was about 10.2 minutes (estimated from the published chromatogram).

Formoterol fumarate samples were exposed to 0.1 N hydrochloric acid, 0.1 N sodium hydroxide solution, or 5% hydrogen peroxide at room temperature for 5 hours or at 60 °C for 1 hour. The drug solid was also exposed to heat at 60 °C for 72 hours and to light for 7 days. The intact drug was well separated from its degradation product and its related substances, indicating that the method was stability indicating.

A calibration curve for formoterol fumarate was constructed from 0.03 to 255 μg/mL. The correlation coefficient was 1.000. The limit of detection and the limit of quantitation were 0.03 and 0.08 μg/mL, respectively. The intraday coefficient of variation was 0.3%.

Reference

Akapo SO, Asif M. Validation of a RP-HPLC method for the assay of formoterol and its related substances in formoterol fumarate dihydrate drug substance. *J Pharm Biomed Anal.* 2003; 33: 935–45.

Foscarnet Sodium

Chemical Names

Dihydroxyphosphinecarboxylic acid oxide trisodium salt
Trisodium phosphonoformate

Other Name
Foscavir

Form

Form	Molecular Formula	MW	CAS
Foscarnet sodium	$CNa_3O_5P \cdot 6H_2O$	300.1	34156-56-4

Appearance
Foscarnet sodium is a white or almost white crystalline powder.

Solubility
Foscarnet sodium is soluble in water. It is practically insoluble in alcohol.

Method
Woods et al. evaluated the stability of foscarnet sodium 12 mg/mL in 0.9% sodium chloride injection. The liquid chromatograph consisted of a Waters model 440 ABS detector, a Waters model 740 data module, a Rheodyne model 7010 injector, and a Waters C_{18} reversed-phase Nova-Pak column (150 × 3.9 mm). The mobile phase consisted of 5% methanol, 0.005 M sulfuric acid, and 0.904 g of tetrahexylammonium hydrogen sulfate in a total volume of 1000 mL. The flow rate was 1.5 mL/min. UV detection was performed at 254 nm.

Samples were diluted in an isotonic vehicle (pH 7.6); injection volume was 20 μL.

Foscarnet sodium in a solution was degraded with 10 M sulfuric acid. The chromatograms showed an absence of interference from degradation products with the determination of foscarnet.

The coefficient of variation of the assay was less than 1.5%.

Similar methods were used by the other researchers cited here.

References

Woods K, Steinmann W, Bruns L, et al. Stability of foscarnet sodium in 0.9% sodium chloride injection. *Am J Hosp Pharm.* 1994; 51: 88–90.

Mathew M, Gupta VD, Bethea C. Stability of foscarnet sodium in 5% dextrose and 0.9% sodium chloride injections. *J Clin Pharm Ther.* 1994; 19: 35–6.

Bundgaard H, Mork N. Kinetics of the decarboxylation of foscarnet in acidic aqueous solution and its implication in its oral absorption. *Int J Pharm.* 1990; 63: 213–8.

Fosphenytoin Sodium

Chemical Name
2,5-Dioxo-4,4-diphenylimidazolidin-1-yl-methyl phosphate disodium

Other Names
Cerebyx, Cetebyx

Form	Molecular Formula	MW	CAS
Fosphenytoin sodium	$C_{16}H_{13}N_2Na_2O_6P$	406.2	92134-98-0

Appearance
Fosphenytoin sodium occurs as a white to pale yellow solid.

Solubility
Fosphenytoin sodium is freely soluble in water.

Method
Herbranson and Kriss-Danziger developed an analytical method for fosphenytoin in solutions, parenteral formulations, and active drug substance. The chromatograph consisted of a Perkin-Elmer series 10 pump, a model LC600 autosampler, a Beckman model 160 selectable-wavelength detector, and a Linear Instruments model 555 recorder. The stationary phase was a Waters Nova-Pak phenyl column (150 × 3.9 mm, 5-μm particle size). The mobile phase consisted of 650 mL of water containing 2.21 g of monobasic potassium phosphate and 7.16 g of 1-heptanesulfonic acid sodium salt (adjusted to pH 4.1 with 85% phosphoric acid) and 350 mL of methanol. The flow rate was 1.0 mL/min. UV detection was performed at 214 nm.

Samples were diluted in a diluent composed of 650 mL of water containing 2.21 g of monobasic potassium phosphate, adjusted to pH 4.1 with 85% phosphoric acid and 350 mL of water. The injection volume was 50 μL. Under these conditions, the retention time for fosphenytoin was 3–4 minutes.

The procedure was determined to be stability indicating by thermal degradation of fosphenytoin at 75 °C for up to 4 months. All anticipated and potential degradation products and excipients were well resolved from fosphenytoin.

Standard curves were constructed from 5 to 50 μg/mL. Correlation coefficients were greater than 0.999. The limits of detection and quantitation were 0.1 and 5 μg/mL, respectively.

Reference
Herbranson DE, Kriss-Danziger P. Development and validation of a high performance liquid chromatographic (HPLC) method for the determination of phenytoin prodrug (fosphenytoin) in solutions, parenteral formulations, and active drug substance. *J Liq Chromatogr*. 1993; 16: 1143–61.

Fotemustine

Chemical Name
(±)-Diethyl-[1-[3-(2-chloroethyl)-3-nitrosoureido]ethyl]phosphonate

Other Names
Muphoran, Mustoforan

Form	Molecular Formula	MW	CAS
Fotemustine	$C_9H_{19}ClN_3O_5P$	315.7	92118-27-9

Method
Dine et al. evaluated by HPLC analysis the stability of fotemustine in polyvinyl chloride infusion bags and sets under various conditions. A Hewlett-Packard 1090M system was equipped with a variable-volume injector, an autosampler, a model 79994 photodiode-array detector, and a model 300 integrator. The stationary phase was a Kromasil BDS C_{18} column (150 × 4.6 mm, 5-μm particle size). The mobile phase consisted of 0.05 M ammonium acetate buffer (adjusted to pH 4.5 with acetic acid) and acetonitrile (70:30, vol/vol). The flow rate was 1 mL/min. UV detection was performed at 230 nm. Phenobarbital sodium was used as an internal standard.

Samples were diluted with mobile phase. The injection volume was 10 μL. Under these conditions, retention times for phenobarbital sodium and fotemustine were about 2.7 and 5.4 minutes, respectively.

The method was evaluated to be stability indicating by accelerated degradation of fotemustine. Samples of fotemustine were treated with concentrated hydrochloric acid, 1 N sodium hydroxide solution, or 1% hydrogen peroxide. No degradation products interfered with the analysis of fotemustine. The peak purity of fotemustine was also confirmed by mass spectrometry.

A calibration curve for fotemustine was constructed from 5 to 20 μg/mL. The correlation coefficient was greater than 0.999. Intraassay and interassay coefficients of variation were less than 1.60 and 2.36%, respectively.

Reference
Dine T, Khalfi F, Gressier B, et al. Stability study of fotemustine in PVC infusion bags and sets under various conditions using a stability-indicating high-performance liquid chromatographic assay. *J Pharm Biomed Anal.* 1998; 18: 373–81.

Fumagillin

Chemical Name
4-(1,2-Epoxy-1,6-dimethylhex-4-enyl)-5-methoxy-1-oxaspiro[2.5]oct-6-yl hydrogen deca-2,4,6,8-tetraenedioate

Other Name
Fumidil

Form
Fumagillin

Molecular Formula
$C_{26}H_{34}O_7$

MW
458.5

CAS
23110-15-8

Method
Brackett et al. reported the determination of fumagillin by HPLC. One system included a Waters model M45 pump, a Beckman model 501 autosampler, a Schoeffel model SF 770 UV detector, and a Perkin-Elmer LCI 100 integrator. The other system consisted of a Waters model M6000 pump, a model 710B WISP autosampler, a Kratos model SF 769Z UV detector, and a Waters data module. The stationary phase was an IBM C_{18} column (150 × 4 mm, 5-μm particle size). The mobile phase consisted of acetonitrile, water, and glacial acetic acid (500:500:1.5, vol/vol/vol). UV detection was performed at 351 nm.

Samples were dissolved and diluted with acetonitrile. Under these conditions, the retention time for fumagillin was about 10.6 minutes.

The method was evaluated to be stability indicating by accelerated degradation of fumagillin. Fumagillin was refluxed in distilled water, 1 N hydrochloric acid, or 1N sodium hydroxide solution for 3 hours; it was exposed to UV light at 254 or 366 nm for 3 hours; or it was stored in dry heat at 105 °C for 1 week. Degradation products did not interfere with the analysis of fumagillin.

A calibration curve for fumagillin was generated from 0 to 0.035 mg/mL. The correlation coefficient was 0.9999.

Reference
Brackett JM, Arguello MD, Schaar JC. Determination of fumagillin by high-performance liquid chromatography. *J Agric Food Chem.* 1988; 36: 762–4.

Furazolidone

Chemical Name
3-[(5-Nitrofurfurylidene)amino]-2-oxazolidinone

Other Names
Furoxone, Nifuran

Form	Molecular Formula	MW	CAS
Furazolidone	$C_8H_7N_3O_5$	225.2	67-45-8

Appearance
Furazolidone is a yellow crystalline powder with a bitter aftertaste.

Solubility
Furazolidone is practically insoluble in water and in alcohol.

Method
Hassan et al. described an analytical method for furazolidone. The system included an LKB.2150 pump, a 50-μL loop injector, an LKB.2151 variable-wavelength detector, and an LKB.2220 integrator. The stationary phase was a LiChrosorb RP18 column (250×4 mm, 5-μm particle size). The mobile phase was a mixture of methanol, water, and Britton Robinson buffer (pH 3) (40:55:5, vol/vol/vol). The flow rate was 1 mL/min. UV detection was performed at 365 nm.

A portion of the ground tablets containing about 50 mg of furazolidone was accurately weighed, extracted with five 15-mL portions of water, and filtered. The residue was collected and extracted with five 10-mL portions of dimethylformamide. The extract was transferred to a 100-mL flask, filled to volume with water, filtered, and diluted by a factor of five with water. Under these conditions, the retention time of furazolidone was about 4.9 minutes.

The method was evaluated to be stability indicating. Samples with different concentrations of the drug were exposed to sunlight for up to 24 hours. None of the degradation products interfered with the analysis of the intact drug.

A standard curve for furazolidone was constructed from 3 to 18 μg/mL. The correlation coefficient was 0.9992.

Reference
Hassan SM, Ibrahim FA, El-Din MS, et al. A stability-indicating high-performance liquid chromatographic assay for the determination of some pharmaceutically important nitrocompounds. *Chromatographia*. 1990; 30: 176–80.

Furosemide

Chemical Names
5-(Aminosulfonyl)-4-chloro-2-[(2-furanylmethyl)amino]benzoic acid
4-Chloro-*N*-furfuryl-5-sulfamoylanthranilic acid

Other Names
Frusemide, Lasix

Form	Molecular Formula	MW	CAS
Furosemide	$C_{12}H_{11}ClN_2O_5S$	330.8	54-31-9

Appearance
Furosemide is a white to slightly yellow odorless crystalline powder.

Solubility
Furosemide is practically insoluble in water and sparingly soluble in alcohol but is freely soluble in alkali hydroxides.

pK$_a$
Furosemide has a pK$_a$ of 3.9.

Method 1
Yang et al. used HPLC to study the stability of furosemide in extemporaneously compounded powder packets from tablets. The liquid chromatograph consisted of a Waters model 510 pump, a model 717 plus autosampler, and a model 486 tunable UV detector. A Waters model 2690 separation module and a model 996 photodiode-array detector were also used in validating the method. The stationary phase was a Waters Spherisorb S5 ODS2 column (250 × 4.6 mm). The mobile phase was 0.01 M dibasic ammonium phosphate in water containing 30% (vol/vol) methanol. The flow rate was 1 mL/min. UV detection was performed at 254 nm.

Furosemide powder from comminuted tablets was mixed and diluted to 100 mL with a mixture of methanol and water (1:1), sonicated for 10 minutes, and centrifuged. The supernatant liquid was collected. The injection volume was 10 μL. Under these conditions, the retention time for furosemide was about 7 minutes.

The method was shown to be stability indicating by accelerated decomposition of furosemide. The drug was prepared in 0.1 N hydrochloric acid, 1 N sodium hydroxide solution, or water and stored at 70 °C for 1 month. The degradation product peaks did not interfere with the peak of the intact drug.

A standard curve for furosemide was constructed from 20 to 60 μg/mL.

References

Yang Y-HK, Lin T-R, Huang Y-F, et al. Stability of furosemide, nadolol and propranolol hydrochloride in extemporaneously compounded powder packets from tablets. *Chin Pharm J.* 2000; 52: 51–8.

Lin T-R, Yang Y-HK, Huang Y-F, et al. Content uniformity of captopril, furosemide, nadolol and propranolol hydrochloride powder packets extemporaneously compounded from tablets. *Chin Pharm J.* 2000; 52: 59–68.

Method 2

Using an HPLC method, Mayron and Gennaro assessed the stability of furosemide with granisetron in intravenous fluids and oral liquids. The liquid chromatograph consisted of a piston pump with a pulse dampener, a rotary injection port with a 20-μL loop, a variable-wavelength UV detector, and an integrator. The stationary phase was a C_{18} reversed-phase column (300 × 4.6 mm, 5-μm particle size). The mobile phase was a mixture of acetonitrile and 0.1 M monobasic sodium phosphate dihydrate (20:80, vol/vol) adjusted to pH 4.2 with phosphoric acid. The flow rate was 1.2 mL/min. UV detection was performed at 228 nm.

Samples were diluted with mobile phase. The retention times for furosemide and granisetron were 4.52 and 6.18 minutes, respectively.

The stability-indicating capability of the analytical method was demonstrated by an accelerated degradation study. Solutions of furosemide were adjusted to pH 2 or 11, boiled for 1 hour, readjusted to pH 5, diluted with the mobile phase, and analyzed. The degradation product peaks did not interfere with the intact furosemide peak.

Reference

Mayron D, Gennaro AR. Stability and compatibility of granisetron hydrochloride in i.v. solutions and oral liquids and during simulated Y-site injection with selected drugs. *Am J Health Syst Pharm.* 1996; 53: 294–304.

Method 3

Neil et al. developed a rapid, flexible, stability-indicating, ion-pair reversed-phase HPLC method for the separation and quantitation of furosemide and its principal hydrolysis product, saluamine, in the presence of potential contaminants, photolytic products, and other degradants. The liquid chromatograph consisted of a Perkin-Elmer series 1 pump, a Pye Unicam variable-wavelength UV monitor with an 8-μL flow cell, a Rheodyne model 7125 20-μL loop valve injector, and a Perkin-Elmer type 56 recorder. The stationary phase was a laboratory-packed Shandon SAS-Hypersil reversed-phase stainless steel column (100 × 5 mm, 5-μm particle size). The mobile phase was *n*-propanol and 20 mM monobasic potassium phosphate buffer (pH 7.0) (25:75, vol/vol) containing 0.25% cetrimide (wt/vol). The flow rate was 1.0 mL/min. UV detection was performed at 273 nm and 0.32 AUFS. The sample injection volume was 4 μL. Under these conditions, the retention time for furosemide was 7.5 minutes.

The method was determined to be stability indicating by accelerated degradation. Furosemide samples were heated for 3 hours at 118 °C in an autoclave or irradiated at 254 nm in a thin-layer chromatography viewing cabinet for 3 hours. The intact furosemide peak was well resolved from other peaks.

A standard curve for furosemide was constructed from 0 to 50 μg/mL with a correlation coefficient of 0.9999.

Reference

Neil JM, Fell AF, Smith G. Evaluation of the stability of frusemide in intravenous infusions by reversed-phase high-performance liquid chromatography. *Int J Pharm.* 1984; 22: 105–26.

Fusidic Acid

Chemical Names

$(3\alpha,4\alpha,8\alpha,9\beta,11\alpha,13\alpha,14\beta,16\beta,17Z)$-16-(Acetyloxy)-3,11-dihydroxy-29-nordammara-17(20),24-dien-21-oic acid

$3\alpha,11\alpha,16\beta$-Trihydroxy-29-nor-$8\alpha,9\beta,13\alpha,14\beta$-dammara-17(20),24-dien-21-oic acid 16-acetate

Other Name

Fucidin

Form	Molecular Formula	MW	CAS
Fusidic acid	$C_{31}H_{48}O_6$	516.7	6990-06-3
Fusidate diethanolamine	$C_{35}H_{59}NO_8$	621.9	16391-75-6
Fusidate sodium	$C_{31}H_{47}NaO_6$	538.7	751-94-0

Appearance

Fusidic acid is a white crystalline powder. Fusidate sodium is a white or almost white and slightly hygroscopic crystalline powder.

Solubility

Fusidic acid is soluble in alcohol, acetone, chloroform, pyridine, and dioxane. It is sparingly soluble in water, ether, and hexane. Fusidate sodium is freely soluble in water and alcohol; it is slightly soluble in chloroform.

Method

Using an HPLC method, McLaughlin and Simpson investigated the stability of reconstituted fusidate diethanolamine in a 5% dextrose infusion. The liquid chromatograph consisted of a Merck/Hitachi L5000 LC controller with a 655A-12 pump, a 655A variable-wavelength UV detector, a D-2000 integrator, and a Rheodyne injection valve with a 20-µL fixed injection loop. The stationary phase was a Jones Chromatography reversed-phase Apex 1 C_{18} column (50 × 4 mm, 5-µm particle size). The mobile phase was a mixture of buffer, methanol, and water (62:5:33, vol/vol/vol). The buffer contained 1.5 mL of acetic acid, 148.5 mL of water, and 350 mL of acetonitrile. The flow rate was 1.8 mL/min. UV detection was performed at 235 nm. The injection volume was 20 µL. The retention time for fusidic acid was 6.1 minutes.

The method was stability indicating since it separated the fusidic acid peak from its degradation product peak. Retention times for the principal degradation product and fusidic acid were 5.1 and 6.1 minutes, respectively.

A standard curve was generated by a plot of concentration against peak area over a fusidate diethanolamine concentration range from 0.200 to 0.600 mg/mL. The correlation coefficient was 0.9997. The coefficient of variation was 0.61%.

Reference

McLaughlin JP, Simpson C. The stability of reconstituted diethanolamine fusidate in a 5% dextrose infusion. *Hosp Pharm Prac.* 1992; 2: 59–62.

Gabapentin

Chemical Name

1-(Aminomethyl)cyclohexaneacetic acid

Other Name

Neurontin

Form	Molecular Formula	MW	CAS
Gabapentin	$C_9H_{17}NO_2$	171.2	60142-96-3

Appearance

Gabapentin is a white to off-white crystalline solid.

Solubility

Gabapentin is freely soluble in water.

Method

Zour et al. reported stability studies of gabapentin in aqueous solution. All samples were assayed on a Hewlett-Packard 1090 series L system fitted with a model 1090L diode-array detector and a Beckman C_{18} reversed-phase Ultrasphere ODS column (250 × 4.6 mm, 5-µm particle size). The mobile phase consisted of water, methanol, and acetonitrile (55:35:10, vol/vol/vol) and the flow rate was 1.0 mL/min. UV detection was performed at 210 nm.

All samples were diluted 1:10. The injection volume was 50 µL.

The assay was stability indicating because it could separate the intact gabapentin from its degradation products. The retention times for gabapentin and the only known degradation product (3,3-pentamethylene-4-butyrolactam) were 3.1 and 13.3 minutes, respectively.

The assay was linear from 0.1 to 6.32 mg/mL. The correlation coefficient was 0.9999. The coefficient of variation was less than 1.0%.

Reference

Zour E, Lodhi SA, Nesbitt RU, et al. Stability studies of gabapentin in aqueous solutions. *Pharm Res.* 1992; 9: 595–600.

Galactose

Chemical Name

D-Galactose

Form	Molecular Formula	MW	CAS
Galactose	$C_6H_{12}O_6$	180.2	59-23-4

Appearance

Galactose occurs as crystalline prisms.

Solubility

Galactose is soluble in about 0.5 part water and freely soluble in hot water. It is also soluble in pyridine but is only slightly soluble in alcohol.

Method

Bhargava et al. investigated the stability of galactose in aqueous solutions. The HPLC system consisted of a Laboratory Data Control Constametric III dual reciprocating pump, a Rheodyne model 7125 20-μL loop sample injector, a Bio-Rad model 1750A refractive index monitor, a Bio-Rad column heater, and a Coleman 550 10-inch strip-chart recorder. The stationary phase was a Bio-Rad (HPX-87P) bonded resin chromatography column (300 × 7.8 mm). The mobile phase was deionized water and was delivered isocratically at 0.6 mL/min. The column heater was set at 85 °C, and the refractive index detector was set at a sensitivity value of 16. Cellobiose 0.25% was used as the internal standard.

Samples were diluted to a nominal concentration of 0.5% (wt/vol). Under these conditions, the retention times for cellobiose and galactose were 11 and 15 minutes, respectively.

The assay was stability indicating because common galactose contaminants, including arabinose, glucose, mannose, xylose, and the initial degradation product, 5-hydroxymethyl-furfural, did not interfere with the intact galactose peak.

The assay was found to be linear by plotting the ratios of peak height for galactose to that for cellobiose versus the galactose concentrations from 0.05 to 0.75% galactose. The minimum correlation coefficient was 0.996. The intraday and interday coefficients of variation were less than 2.5 and 2.7%, respectively.

Reference
Bhargava VO, Rahman S, Newton DW. Stability of galactose in aqueous solutions. *Am J Hosp Pharm.* 1989; 46:104–8.

Ganciclovir

Chemical Names
2-Amino-1,9-dihydro-9-[[2-hydroxy-1-(hydroxymethyl)ethoxy]methyl]-6*H*-purin-6-one
9-[(1,3-Dihydroxy-2-propoxy)methyl]guanine

Other Names
Cytovene, DHPG

Forms	Molecular Formula	MW	CAS
Ganciclovir	$C_9H_{13}N_5O_4$	255.2	82410-32-0
Ganciclovir sodium	$C_9H_{12}N_5NaO_4$	277.2	107910-75-8

Appearance
Ganciclovir sodium is a white to off-white solid.

Solubility
Ganciclovir sodium has a solubility of 4.3 mg/mL in water at pH 7 at 25 °C.

pK$_a$
Ganciclovir has pK$_a$ values of 2.2 and 9.4.

Method 1
Anaizi et al. evaluated by HPLC the stability of ganciclovir in extemporaneously compounded oral liquids. The stationary phase was a MetaChem Inertsil ODS-3

reversed-phase column (100 × 4.6 mm, 5-μm particle size). The mobile phase was 2.5% (vol/vol) acetonitrile in 25 mM phosphate aqueous buffer (adjusted to pH 2.5 with phosphoric acid). The flow rate was 1.5 mL/min. UV detection was performed at 254 nm and 0.01 AUFS. Hypoxanthine 0.4 mg/mL in water was used as an internal standard.

Samples were filtered through a 0.2-μm filter before assay. The injection volume was 10 μL. Under these conditions, retention times for hypoxanthine and ganciclovir were about 1.8 and 2.4 minutes, respectively.

The stability-indicating nature of the method was demonstrated. Ganciclovir samples were adjusted to pH 1.2 with 10 N hydrochloric acid and to pH 12.5 with 10 N sodium hydroxide and kept at 65 °C for 12 days. Degradation products did not interfere with the analysis of ganciclovir.

A calibration curve for ganciclovir was constructed from 1.25 to 8.75 μg/mL. The correlation coefficient was 1.0000. The intraday and interday coefficients of variation were 0.35 and 1.32%, respectively.

Reference
Anaizi NH, Swenson CF, Dentinger PJ. Stability of ganciclovir in extemporaneously compounded oral liquids. *Am J Health Syst Pharm.* 1999; 56: 1738–41.

Method 2
Silvestri et al. assessed the stability of ganciclovir sodium at concentrations of 1, 5, and 10 mg/mL in polyvinyl chloride minibags containing 5% dextrose injection when stored under refrigeration and in the dark for 35 days. The sample analyses were conducted with a Spectra-Physics Chromstation including a model 8800 ternary pump, a model 8780 autosampler, a model 8490 variable-wavelength UV detector, a model 8790 column heater, and a model 4270 integrator and plotter. The stationary phase was a Brownlee Laboratories Spheri-5 RP18 cartridge (200 × 4.6 mm). The mobile phase was 6 mM ammonium phosphate in water, adjusted to pH 2.5 with phosphoric acid. The flow rate was 1.5 mL/min. UV detection was performed at 254 nm. Hypoxanthine 0.4 mg/mL was used as the internal standard.

Samples were diluted with HPLC-grade water. The injection volume was 5 μL. Retention times for hypoxanthine and ganciclovir were 3.4 and 6.3 minutes, respectively.

The assay was confirmed to be stability indicating by accelerated degradation of ganciclovir. A solution of ganciclovir was stored at 80 °C for 3 days and then was chromatographed. The degradation products did not interfere with the determination of the intact ganciclovir.

A calibration curve was plotted as the peak area ratios of ganciclovir to that of the internal standard against the ganciclovir concentrations from 0.1 to 0.3 mg/mL. The correlation coefficient was 0.9998. The intraday and interday coefficients of variation were less than 2%.

Reference
Silvestri AP, Mitrano FP, Baptista RJ, et al. Stability and compatibility of ganciclovir sodium in 5% dextrose injection over 35 days. *Am J Hosp Pharm.* 1991; 48: 2641–3.

Method 3
Visor et al. evaluated the stability of ganciclovir sodium in 0.9% sodium chloride injection and 5% dextrose injection. The liquid chromatograph consisted of a Waters model 710B WISP autosampler, a Laboratory Data Control Constametric III pump, a Waters model 480 Lambda-

Max detector, and a Spectra-Physics SP4000 integrator. The stationary phase was a Whatman C_{18} column (250 mm, 10-µm particle size). The mobile phase consisted of 12 mM ammonium phosphate and 0.1% phosphoric acid (pH 2.5) and was delivered at 1.2 mL/min. UV detection was performed at 254 nm and 0.05 AUFS. Hypoxanthine was used as the internal standard.

Samples were diluted before injection. The injection volume was 10 µL. Retention times for hypoxanthine and ganciclovir were 8.6 and 13.7 minutes, respectively.

The HPLC method was demonstrated to be stability indicating by an accelerated degradation study. The ganciclovir sample was stored at 80 °C for 6 weeks and its chromatogram showed no interference from the degradation product peak. This finding was further confirmed with a peak-spectra purity study using a Hewlett-Packard model 1040A rapid scanner linear diode-array spectrophotometer.

The coefficient of variation was less than 1.5%.

Similar methods were reported by the other investigators cited here.

References
Visor GC, Lin L-H, Jackson SE, et al. Stability of ganciclovir sodium (DHPG sodium) in 5% dextrose or 0.9% sodium chloride injections. *Am J Hosp Pharm.* 1986; 43: 2810–2.

Mulye NV, Turco SJ, Speaker TJ. Stability of ganciclovir sodium in an infusion-pump syringe. *Am J Hosp Pharm.* 1994; 51: 1348–9.

Parasrampuria J, Li LC, Stelmach AH, et al. Stability of ganciclovir sodium in 5% dextrose injection and in 0.9% sodium chloride injection over 35 days. *Am J Hosp Pharm.* 1992; 49: 116–8.

Gentamicin

Chemical Names
Gentamicin A: *O*-2-Amino-2-deoxy-α-D-glucopyranosyl-(1→ 4)-*O*-[3-deoxy-3-(methyl-amino)-α-D-xylopyranosyl-(1→6)]-2-deoxy-D-streptamine

Gentamicin C_{1A}: *O*-3-deoxy-4-*C*-methyl-3-(methylamino)-β-L-arabinopyranosyl-(1→6)-*O*-[2,6-diamino-2,3,4,6-tetradeoxy-α-D-*erythro*-hexopyranosyl-(1→4)]-2-deoxy-D-streptamine

Gentamicin C_1 $R_1 = R_2 = CH_3$
Gentamicin C_2 $R_1 = CH_3$, $R_2 = H$
Gentamicin C_{1a} $R_1 = R_2 = H$

Other Names

Garamycin, Gentamycin

Form	Molecular Formula	MW	CAS
Gentamicin A	$C_{18}H_{36}N_4O_{10}$	468.5	13291-74-2
Gentamicin C_1	$C_{21}H_{43}N_5O_7$	477.6	1403-66-3
Gentamicin C_{1A}	$C_{19}H_{39}N_5O_7$	449.4	26098-04-4
Gentamicin C_2	$C_{20}H_{41}N_5O_7$	463.5	25876-11-3

Appearance

Gentamicin is a white amorphous powder. Gentamicin sulfate is a mixture of the sulfates of gentamicin C_1, gentamicin C_{1A}, and gentamicin C_2 and occurs as a white to buff powder.

Solubility

Gentamicin is freely soluble in water, soluble in pyridine and dimethylformamide, and moderately soluble in methanol, ethanol, and acetone. Gentamicin sulfate is freely soluble in water and practically insoluble in alcohol, acetone, chloroform, and ether.

Method 1

Allen et al. used HPLC to evaluate the stability of gentamicin sulfate 1 mg/mL in the presence of cefpirome sulfate during simulated Y-site injection. The method utilized one of two systems. One unit included an Alcott model 728 Micromeritics autosampler, a Rheodyne 7010 injector with an Alcott 732 electrically actuated valve, a Waters model 501 solvent-delivery pump, a Waters model 441 UV detector, a Waters model 401 refractive index detector, and a Waters model 745 data module. The other unit consisted of an Alcott model 728 Micromeritics autosampler, a Rheodyne 7010 injector with an Alcott 732 electrically actuated valve, a Shimadzu model LA-6A solvent-delivery pump, a Shimadzu model SPD-6A UV detector, and an Orion model 901 microprocessor ion analyzer. The stationary phase was a C_{18} analytical column. The mobile phase consisted of 700 mL of methanol, 250 mL of water, 50 mL of acetic acid, and 5 g of sodium 1-heptanesulfonate and was delivered isocratically at 1.5 mL/min. Gentamicin was monitored at 330 nm.

The derivatizing solution was prepared by dissolving 1 g of o-phthalaldehyde in 5 mL of methanol with 95 mL of 0.4 M boric acid previously adjusted to pH 10.4 with 8 N potassium hydroxide and with 2 mL of thioglycolic acid added. The pH of the mixture was again adjusted to 10.4 with 8 N potassium hydroxide. To 10 mL of a gentamicin sample solution, 5 mL of isopropanol and 4 mL of the derivatizing solution were added. The solution was brought to a volume of 25 mL with additional isopropanol. The mixture was heated at 60 °C using a water bath for 15 minutes and then cooled. The injection volume was 20 μL. Under these conditions, retention times for different forms of derivatized gentamicin were 3.1, 5.1, 7.4, and 8.9 minutes. Cefpirome had a retention time of 2.0 minutes.

The stability-indicating nature of the assay was determined by accelerated degradation of gentamicin with hydrochloric acid, sodium hydroxide, potassium chlorate, heat at 80 °C, or light (150 foot-candles from a tungsten filament source). In each case, gentamicin was separated from its degradation products and from cefpirome.

A standard curve was constructed from 10 to 100 μg/mL; the correlation coefficient was greater than 0.99. The intraday and interday coefficients of variation of the assay were 3.8 and 3.2%, respectively.

Similar methods were used by the other researchers cited here.

References

Allen LV, Stiles ML, Prince SJ, et al. Stability of cefpirome sulfate in the presence of commonly used intensive care drugs during simulated Y-site injection. *Am J Health Syst Pharm.* 1995; 52: 2427–33.

Xu QA, Trissel LA, Saenz CA, et al. Stability of gentamicin sulfate and tobramycin sulfate in AutoDose infusion system bags. *Int J Pharm Compound.* 2002; 6: 152–4.

Xu QA, Trissel LA, Zhang Y, et al. Compatibility and stability of linezolid injection admixed with gentamicin sulfate and tobramycin sulfate. *Int J Pharm Compound.* 2000; 4: 476–9.

Method 2

Tu et al. investigated the stability of gentamicin sulfate in prefilled drug reservoirs. Samples were analyzed on an HPLC system including a Waters model 501 pump, a model U6K injector, a model 441 UV detector, a model 745 data module, and a Waters Nova-Pak C_{18} column (150 × 3.9 mm, 5-μm particle size). The mobile phase was a mixture of 30% acetonitrile and 70% distilled water and was delivered isocratically at 1.0 mL/min. UV detection was performed at 300 nm.

Gentamicin was derivatized as described in Method 1. The retention time for the gentamicin derivative was 2.9 minutes.

This method was determined to be stability indicating by accelerated degradation of the gentamicin sample at 90 °C over 24 hours. The intact gentamicin derivative peak was well separated from its degradation product and from excipients used in the formulation.

A standard curve was constructed each day from 1.0 to 5.5 mg/mL. The correlation coefficient was greater than 0.99.

Reference

Tu Y-H, Stiles ML, Allen LV Jr, et al. Stability study of gentamicin sulfate administered via Pharmacia Deltec CADD-VT® pump. *Hosp Pharm.* 1990; 25: 843–5.

Glucosamine

Chemical Names

2-Amino-2-deoxy-D-glucose
2-Amino-2-deoxy-β-D-glucopyranose

Other Names
Anartril, Dona, Dorofen

Form

Form	Molecular Formula	MW	CAS
Glucosamine	$C_6H_{13}NO_5$	179.2	3416-24-8
Glucosamine hydrochloride	$C_6H_{13}NO_5.HCl$	215.7	66-84-2
Glucosamine sulfate potassium	$(C_6H_{14}NO_5)_2.SO_4.2KCl$	605.5	38899-05-7

Solubility
Glucosamine is very soluble in water. It is practically insoluble in ether and chloroform.

Method
Shao et al. developed and validated a stability-indicating HPLC method for the determination of glucosamine in bulk materials and pharmaceutical formulations. Two HPLC systems were used. One was a Waters Alliance 2690D system equipped with a quaternary pump, a 2487 dual λ absorbance detector, and a column oven. The other was an Agilent 1100 system equipped with a quaternary pump, an autosampler, a thermostated column heater, and a diode-array detector. The stationary phase was a Phenomenex Luna amino column (150 × 4.6 mm, 5-μm particle size). The column temperature was 35 °C. The mobile phase consisted of 0.020 M monobasic potassium phosphate buffer (adjusted to pH 7.5 with phosphoric acid) and acetonitrile (75:25, vol/vol). The flow rate was 1.5 mL/min. UV detection was performed at 195 nm.

A portion of the powder (20 tablets) equivalent to the average tablet weight was accurately weighed, transferred into a 100-mL volumetric flask, mixed with 50 mL of sample preparation solvent (500 mL of acetonitrile and 500 mL of water), shaken for about 30 minutes to dissolve, filled to volume with sample preparation solvent, mixed well, and filtered through a 0.45-μm nylon syringe filter. The injection volume was 10 μL. Under these conditions, the retention time of glucosamine was about 10.5 minutes (estimated from the published chromatogram).

The accelerated degradation study of glucosamine was conducted by subjecting standard, placebo, and tablet formulation samples to heat, light, oxidation, acid, and base. Samples were stored at 60 °C for 14 days or treated with 3% hydrogen peroxide at room temperature for 7 days. Degradation product peaks did not interfere with the glucosamine peak. Therefore, the method was stability indicating.

A standard curve was constructed from 1.88 to 5.62 mg/mL, with a correlation coefficient of 0.9998. The limit of detection and limit of quantitation values for glucosamine were 37.0 and 149.0 μg/mL, respectively.

Reference
Shao Y, Alluri R, Mummert M, et al. A stability-indicating HPLC method for the determination of glucosamine in pharmaceutical formulations. *J Pharm Biomed Anal.* 2004; 35: 625–31.

Glycolic Acid

Chemical Name
Hydroxyethanoic acid

Other Name
Neostrata

Form	Molecular Formula	MW	CAS
Glycolic acid	$C_2H_4O_3$	76.1	79-14-1

Method
De Villiers et al. described an HPLC assay for lactic acid and glycolic acid in nonionic creams. A Shimadzu LC-6A system was equipped with a model C-3RA integrator. The stationary phase was a Waters μBondapak C_{18} column (250 × 3.9 mm, 5-μm particle size). The mobile phase consisted of water and acetonitrile (60:40, vol/vol) containing 0.02 mM monobasic potassium phosphate adjusted to pH 2.25 with phosphoric acid. The flow rate was 1 mL/min. UV detection was performed at 214 nm.

A sample equivalent to 100 mg of lactic acid and glycolic acid was weighed and placed into a 150-mL beaker. It was then mixed with 20 mL of 0.1 M sodium hydroxide solution, boiled for 20 minutes, cooled to room temperature, mixed with 20 mL of water, adjusted to pH 6 with diluted phosphoric acid, diluted with more water, filtered, and diluted with the mobile phase. The injection volume was 20 μL. Under these conditions, retention times for lactic acid and glycolic acid were about 8.8 and 10.8 minutes, respectively (estimated from the published chromatogram).

The method was shown to be stability indicating by accelerated degradation of glycolic and lactic acids. The acids were degraded in an oven at 120 °C for 5 days. Degradation products of lactic and glycolic acids did not interfere with the determination of acids.

A standard curve was constructed. The correlation coefficient was 0.9997.

Reference
De Villiers MM, Wurster DE, Bergh T, et al. Stability indicating HPLC assay of the α-hydroxy acids lactic acid and glycolic acid in non-ionic creams. *Pharmazie*. 1998; 53: 204–5.

Glycopyrrolate

Chemical Name
3-[(Cyclopentylhydroxyphenylacetyl)oxy]-1,1-dimethylpyrrolidinium bromide

Other Name
Robinul

Form	Molecular Formula	MW	CAS
Glycopyrrolate	$C_{19}H_{28}BrNO_3$	398.3	596-51-0

Appearance
Glycopyrrolate occurs as a white odorless crystalline powder.

Solubility
Glycopyrrolate is freely soluble in water and soluble in alcohol. It is slightly soluble in chloroform.

Method
Using HPLC, Storms et al. studied the stability of glycopyrrolate injection at 4 and 23 °C in polypropylene syringes. The instrument consisted of a Micromeritics model 760 pump, a model 728 autosampler with a 50-μL loop, a Kratos model 757 ultraviolet variable-wavelength detector, and a Hewlett-Packard model 3395 integrator. The stationary phase was a Phenomenex silica column (250 × 4.6 mm, 3-μm particle size). The mobile phase consisted of methanol and 10 mM monobasic potassium phosphate buffer at pH 3.0 (50:50, vol/vol). The flow rate was 0.85 mL/min. UV detection was performed at 222 nm.

Samples were diluted 1:4 with mobile phase. The injection volume was 50 μL. Under these conditions, retention times for glycopyrrolate and benzyl alcohol were 4.2 and 12.3 minutes, respectively.

To ensure that the assay method was stability indicating, glycopyrrolate solutions 1 mg/mL were degraded with 0.1 N hydrochloric acid, 0.1 N sodium hydroxide, 3% hydrogen peroxide, and heat.

Reference
Storms ML, Stewart JT, Warren FW. Stability of glycopyrrolate injection at ambient temperature and 4 °C in polypropylene syringes. *Int J Pharm Comp.* 2003; 7: 65–7.

Glycyrrhizinic Acid

Chemical Name
(3β,20β)-20-Carboxy-11-oxo-30-norolean-12-en-3-yl-2-*O*-β-D-glucopyranuronosyl-α-D-glucopyranosiduronic acid

Other Names
Glycyrrhetinic Acid, Glycyrrhizic Acid, Glycyrrhizin

Form	Molecular Formula	MW	CAS
Glycyrrhizinic acid	$C_{42}H_{62}O_{16}$	822.94	1405-86-3

Appearance
Glycyrrhizinic acid occurs as crystals with an intensely sweet taste.

Solubility
Glycyrrhizinic acid is freely soluble in hot water and alcohol. It is practically insoluble in ether.

Method
Hansen et al. reported an HPLC method for the determination of glycyrrhizinic acid in liquoric extract and three cough mixtures. A liquid chromatographic system comprised a Shimadzu model LC-10AD pump, a model SIL-10A autoinjector, a model SPD-10A UV detector, and an integrator. The stationary phase was a Waters Symmetry Shield RP8 column (150×3.9 mm, 5-μm particle size). The column oven was maintained at 40 °C. The mobile phase consisted of methanol, 0.2 M potassium phosphate (pH 7.0), and water (47:10:43, vol/vol/vol) containing 5 mM tetrabutylammonium bromide. The flow rate was 1.0 mL/min. UV detection was performed at 248 nm.

Samples of cough mixtures were appropriately diluted with mobile phase, and samples of liquoric extract were diluted with water and then mobile phase. The injection volume was 20 μL. Under these conditions, the retention time for glycyrrhizinic acid was about 7.3 minutes (estimated from the published chromatogram).

The method was reported to be stability indicating.

A standard curve for glycyrrhizinic acid was generated from 1.36 to 13.63 μg/mL with a correlation coefficient of 0.9999.

Reference
Hansen HK, Hansen SH, Kraunsoe M, et al. Comparison of high-performance liquid chromatography and capillary electrophoresis methods for quantitative determination of glycyrrhizinic acid in pharmaceutical preparations. *Eur J Pharm Sci.* 1999; 9: 41–6.

Gonadorelin

Chemical Name
5-Oxo-L-prolyl-L-histidyl-L-tryptophyl-L-seryl-L-tyrosylglycyl-L-leucyl-L-arginyl-L-prolylglycinamide

Other Name
Cryptocur

Form

Form	Molecular Formula	MW	CAS
Gonadorelin	$C_{55}H_{75}N_{17}O_{13}$	1182.3	33515-09-2
Gonadorelin acetate	$C_{55}H_{75}N_{17}O_{13}.C_2H_4O_2$	1242.4	34973-08-5
Gonadorelin hydrochloride	$C_{55}H_{75}N_{17}O_{13}.HCl$	1218.8	51952-41-1

Appearance
Gonadorelin is a white or faintly yellowish-white powder.

Solubility
Gonadorelin is soluble in water and 1% (vol/vol) acetic acid and is sparingly soluble in methanol.

Method 1
Hoitink et al. used a stability-indicating HPLC method to study the degradation kinetics of gonadorelin in aqueous solution. The liquid chromatographic system was composed of a Waters model 510 pump, a Gilson 231 sample injector with a 20-μL loop, a Gilson 401 dilutor, an Applied Biosystems 785A programmable absorbance detector, and a Kip & Zonen BD 40 recorder. The stationary phase was a Merck LiChrospher 100 RP18 column (125 × 4 mm, 5-μm particle size). The mobile phase consisted of 16% acetonitrile (wt/wt) with 0.1% trifluoroacetic acid (vol/vol). The flow rate was 1 mL/min. UV detection was performed at 214 nm. The retention time for gonadorelin was about 9.7 minutes (estimated from the published chromatogram).

The assay was determined to be stability indicating by accelerated degradation of gonadorelin at pH 2 and 9. The degradation product peaks were separated from the parent peak on the chromatogram.

A calibration curve was constructed for gonadorelin and was linear ($r > 0.99$) from 5 to 25 µg/mL.

Reference

Hoitink MA, Beijnen JH, Bult A, et al. Degradation kinetics of gonadorelin in aqueous solution. *J Pharm Sci.* 1996; 85: 1053–9.

Method 2

Helm and Muller investigated the stability of gonadorelin in aqueous solution using an HPLC method. The system consisted of a Pye Unicam model PU 4011 dual-piston pump, a model PU 4020 variable-wavelength UV absorbance detector, a model PU4810 integrator, and a Rheodyne 20-µL loop injector. The stationary phase was a Vydac C_{18} protein and peptide column (250×4.6 mm, 5-µm particle size). The mobile phase was a mixture of 0.2 M triethylammonium phosphate buffer (pH 2.0) and tetrahydrofuran (90:10, vol/vol). The flow rate was 1 mL/min. UV detection was performed at 210 nm. Under these conditions, the retention time of gonadorelin was about 14.7 minutes (estimated from the published chromatogram).

The method was stability indicating since the degradation products generated by storing the drug at pH 8.0 at 50 °C for 28 days were well resolved from the intact drug.

A calibration curve for gonadorelin was generated from 100 to 300 µg/mL. The correlation coefficient was 0.9998.

Reference

Helm VJ, Muller BW. Stability of gonadorelin and triptorelin in aqueous solution. *Pharm Res.* 1990; 7: 1253–6.

Granisetron

Chemical Name

endo-1-Methyl-*N*-(9-methyl-9-azabicyclo[3.3.1]non-3-yl)-1*H*-indazole-3-carboxamide

Other Name

Kytril

Form	Molecular Formula	MW	CAS
Granisetron	$C_{18}H_{24}N_4O$	312.4	109889-09-0
Granisetron hydrochloride	$C_{18}H_{24}N_4O.HCl$	348.9	107007-99-8

Appearance

Granisetron hydrochloride is a white to off-white solid.

Solubility

Granisetron hydrochloride is readily soluble in water and in normal saline at 20 °C.

Method 1

Mayron and Gennaro assessed the stability of granisetron hydrochloride in oral liquids and intravenous fluids using an HPLC method. The liquid chromatograph consisted of a piston pump with a pulse dampener, a rotary injection port with a 20-μL loop, a variable-wavelength spectrometric detector, and an integrator. The stationary phase was a cyano reversed-phase column (250 × 4.6 mm, 10-μm particle size). The mobile phase was a mixture of 0.1 M monobasic sodium phosphate dihydrate and acetonitrile adjusted to pH 4.2 with phosphoric acid (80:20, vol/vol). The flow rate was 2.0 mL/min. UV detection was performed at 300 nm.

Samples were diluted with mobile phase. The retention time for granisetron was 5.60 minutes.

The stability-indicating capability of the analytical method was demonstrated by separating synthetic intermediates and degradation products of granisetron hydrochloride supplied by SmithKline Beecham Pharmaceuticals from granisetron hydrochloride.

Reference

Mayron D, Gennaro AR. Stability and compatibility of granisetron hydrochloride in i.v. solutions and oral liquids and during simulated Y-site injection with selected drugs. *Am J Health Syst Pharm.* 1996; 53: 294–304.

Method 2

Mayron and Gennaro investigated the stability of granisetron hydrochloride with carboplatin, gentamicin sulfate, paclitaxel, and vincristine sulfate separately during simulated Y-site administration. The liquid chromatograph consisted of a piston pump with a pulse dampener, a rotary injection port with a 20-μL loop, a variable-wavelength UV detector, and an integrator. The stationary phase was a C_{18} reversed-phase column (300 × 4.6 mm, 5-μm particle size). The mobile phase was a mixture of 0.1 M monobasic sodium phosphate dihydrate and acetonitrile adjusted to pH 4.2 with phosphoric acid (80:20, vol/vol). UV detection was performed at 300 nm.

Samples were diluted with the mobile phase before injection. Retention times for granisetron were 4.56 minutes at 2.50 mL/min with carboplatin, 5.28 minutes at 2.00 mL/min with gentamicin sulfate, 3.95 minutes at 2.25 mL/min with paclitaxel, and 5.65 minutes at 1.25 mL/min with vincristine sulfate.

The analytical method was determined to be stability indicating by accelerated degradation studies. Solutions of granisetron hydrochloride with carboplatin, gentamicin sulfate, paclitaxel, or vincristine sulfate were adjusted to pH 2 or 11, boiled for 1 hour, readjusted to pH 5, diluted with the mobile phase, and analyzed. The degradation product peaks did not interfere with the intact granisetron peak.

Mayron and Gennaro also studied the stability of granisetron hydrochloride during simulated Y-site administration with each of the following drugs: ceftazidime, cimetidine hydrochloride, cisplatin, cyclophosphamide, cytarabine, dacarbazine, dexamethasone sodium phosphate, diphenhydramine hydrochloride, doxorubicin hydrochloride, etoposide, fluorouracil, furosemide, hydromorphone hydrochloride, ifosfamide, lorazepam, mechlorethamine hydrochloride, mesna, methotrexate sodium, mezlocillin sodium, morphine sulfate, streptozocin, and ticarcillin disodium–clavulanate potassium. The analytical methods that were used are presented in the monographs for each of these drugs.

Reference
Mayron D, Gennaro AR. Stability and compatibility of granisetron hydrochloride in i.v. solutions and oral liquids and during simulated Y-site injection with selected drugs. *Am J Health Syst Pharm.* 1996; 53: 294–304.

Method 3
Quercia et al. studied the stability of granisetron in 0.9% sodium chloride injection and 5% dextrose injection when stored in plastic syringes at 5 and 24 °C for up to 14 days. The liquid chromatograph consisted of a Waters model 6000A pump, a Waters model 717-plus autosampler, a Milton Roy model SpectroMonitor 3000 variable-wavelength UV detector, a Hewlett-Packard model 3396 series II integrator, and a PhaseSep Spherisorb cyano column (250 × 4.6 mm, 5-μm particle size). The mobile phase consisted of acetonitrile and 0.1 M sodium dihydrogen phosphate (15:85, vol/vol) and was delivered isocratically at 2 mL/min. UV detection was performed at 302 nm. The injection volume was 20 μL.

The assay was determined to be stability indicating by an accelerated decomposition study. A sample of granisetron was mixed with 6 N hydrochloric acid and was heated at 100 °C for 5 hours. No degradation product peaks interfered with the intact granisetron peak.

Calibration curves were generated from 0.04 to 0.12 mg/mL. The correlation coefficient was greater than 0.999. The intraday and interday coefficients of variation were 0.25–0.46% and 1.04–2.18%, respectively.

References
Quercia RA, Zhang J, Fan C, et al. Stability of granisetron hydrochloride in polypropylene syringes. *Am J Health Syst Pharm.* 1996; 53: 2744–6.

Quercia RA, Zhang J, Fan CC, et al. Stability of granisetron hydrochloride in an extemporaneously prepared oral liquid. *Am J Health Syst Pharm.* 1997; 54: 1404–6.

Method 4
Chung et al. evaluated the stability of granisetron hydrochloride in 5% dextrose injection or 0.9% sodium chloride injection when stored in a disposable elastomeric infusion device. The chromatograph included a Hitachi model L-6200 Intelligent pump, a model L-4200 UV-visible detector, a model AS 2000 autosampler, a model D-2500 chromatointegrator, and a J&W Scientific Accubond CN column (250 × 4.6 mm, 5-μm particle size). The mobile phase was a mixture of acetonitrile and a buffer solution (60:40, vol/vol). The buffer solution contained 0.02 M monobasic potassium phosphate and 5 mM octanesulfonic acid adjusted to pH 6.0 with 1 N sodium hydroxide. The flow rate was 1.5 mL/min. UV detection was performed at 305 nm. The sample injection volume was 50 μL. Under these conditions, the retention time for granisetron was about 7.5 minutes.

The stability-indicating capability of the assay was demonstrated by accelerated degradation of granisetron. Samples were subjected to the addition of 1 N hydrochloric acid, 1 N sodium hydroxide, or 1% hydrogen peroxide solution and were heated at 58 °C for 4 hours. The resulting solutions were then neutralized and assayed. Degradation product peaks did not interfere with the intact granisetron peak.

A standard curve for granisetron was constructed daily from 5 to 75 μg/mL. The correlation coefficients for the standard curves were all greater than 0.999. The intraday and interday coefficients of variation were less than 0.7 and 3.1%, respectively.

A similar stability-indicating assay was reported by Chin et al.

References

Chung KC, Chin A, Gill MA. Stability of granisetron hydrochloride in a disposable elastomeric infusion device. *Am J Health Syst Pharm*. 1995; 52: 1541–3.

Chin A, Moon YSK, Chung KC, et al. Stability of granisetron hydrochloride with dexamethasone sodium phosphate for 14 days. *Am J Health Syst Pharm*. 1996; 53: 1174–6.

Guaifenesin

Chemical Name
(*RS*)-3-(2-Methoxyphenoxy)propane-1,2-diol

Other Name
Amonidrin

Form	Molecular Formula	MW	CAS
Guaifenesin	$C_{10}H_{14}O_4$	198.2	93-14-1

Appearance
Guaifenesin is a white to slightly gray crystalline powder with a bitter taste.

Solubility
Guaifenesin is soluble in water and in alcohol.

Method 1
Grosa et al. reported a method for the simultaneous determination of guaifenesin, methyl *p*-hydroxybenzoate, and propyl *p*-hydroxybenzoate in a commercial cough syrup dosage form. The Shimadzu HPLC system included two LC-10ADVp module pumps, a DGU-14-A online degasser, a SIL-10ADVp autosampler, and an SPD-M10Avp photodiode-array detector. The stationary phase was a Phenomenex Synergi Polar-RP column

(150 × 4.6 mm, 5-μm particle size). The mobile phase consisted of 10 mM monobasic potassium phosphate aqueous buffer (pH 3.0), acetonitrile, and methanol (21:64:15, vol/vol/vol). The flow rate was 1.0 mL/min. UV detection was performed at 254 nm.

Five milliliters of cough syrup was transferred into a 100-mL volumetric flask, filled to volume with water and acetonitrile (75:25, vol/vol), vortexed for 2 minutes, diluted 1:5 with the same solvent, filtered through a 0.2-μm polytetrafluoroethylene (PTFE) membrane filter, and assayed. The injection volume was 20 μL. Under these conditions, the retention time for guaifenesin was about 3.9 minutes (estimated from the published chromatogram).

The stability-indicating nature of the method was demonstrated by assaying a mixture of guaifenesin and its excipients and degradation products. Retention times for 4-hydroxybenzoic acid, guaifenesin, guaiacol, methyl *p*-hydroxybenzoate, and propyl *p*-hydroxybenzoate were about 3.1, 3.9, 5.8, 6.5, and 16 minutes, respectively (estimated from the published chromatogram).

A standard curve for guaifenesin was constructed from 95 to 285 μg/mL. The correlation coefficient was 0.9995. Intraday and interday coefficients of variation were 0.60 and 0.92%, respectively.

Reference
Grosa G, Grosso ED, Russo R, et al. Simultaneous, stability indicating, HPLC-DAD determination of guaifenesin and methyl and propyl-parabens in cough syrup. *J Pharm Biomed Anal.* 2006; 41: 798–803.

Method 2
Hewala described an HPLC method for paracetamol (acetaminophen), guaifenesin, sodium benzoate, and oxomemazine in the presence of degradation products. The Beckman Gold system consisted of a model 125 programmable pump, a model 166 programmable UV detector, and a Rheodyne 20-μL loop injector. The stationary phase was a C_{18} ODS stainless steel column (250 × 4.6 mm, 5-μm particle size) with a guard column (50 × 4.6 mm) of the same packing material. Mobile phase A was a mixture of methanol and water (18:82, vol/vol) adjusted to pH 3.9 with phosphoric acid, and mobile phase B was a mixture of methanol and water (80:20, vol/vol) adjusted to pH 3.9 with phosphoric acid. Mobile phase A was delivered from 0 to 12 minutes and mobile phase B from 12 to 22 minutes. The flow rate was 1.5 mL/min. UV detection was performed at 235 nm and 0.05 AUFS. Metronidazole was used as an internal standard.

A 10-mL portion of the cough syrup was diluted to 100 mL with methanol, mixed with internal standard solution, and further diluted with methanol to 20 μg/mL of metronidazole in the final solution. Under these conditions, retention times for paracetamol, metronidazole, sodium benzoate, guaifenesin, and oxomemazine were about 2.6, 5.5, 8.6, 10.7, and 16.8 minutes, respectively (estimated from the published chromatogram).

The method was evaluated to be stability indicating by assaying a mixture of active compounds and possible degradation products. Retention times for possible degradants 4-aminophenol and guaicol were about 3.7 and 7.0 minutes, respectively (estimated from the published chromatogram).

A standard curve for guaifenesin was generated from 5.0 to 30.0 μg/mL. The correlation coefficient was 0.9999. Intraday and interday coefficients of variation were 1.71 and 2.31%, respectively.

Reference

Hewala II. Stability-indicating HPLC assay for paracetamol, guaiphenesin, sodium benzoate and oxomemazine in cough syrup. *Anal Lett.* 1994; 27: 71–93.

Method 3

Heidemann et al. reported an HPLC analysis of guaifenesin in cough-cold products. The apparatus consisted of a Waters model 510 dual-piston pump, a model 481 variable-wavelength detector, and a model WISP 710B autosampler. The stationary phase was a Brownlee Spheri-5 RP18 column (30 × 2.1 mm, 5-μm particle size). The mobile phase was 1 mM citrate acid buffer (adjusted to pH 6.0 with 1 N sodium hydroxide). The flow rate was 1 mL/min. UV detection was performed at 250 nm.

The syrup sample was diluted appropriately with water. Under these conditions, the retention time of guaifenesin was about 1.5 minutes (estimated from the published chromatogram).

The method was verified to be stability indicating by the analysis of degraded samples of the different formulations. Degradation products did not interfere with the determination of the drug.

A standard curve for guaifenesin was constructed from 80 to 120% for the labeled concentrations. The correlation coefficient was 1.000.

Reference

Heidemann DR, Groon KS, Smith JM. HPLC determination of cough-cold products. *LC-GC.* 1987; 5: 422–6.

Method 4

Schieffer et al. presented a simultaneous determination of phenylephrine hydrochloride, phenylpropanolamine hydrochloride, guaifenesin, and sodium benzoate in dosage forms. The Waters model ALC 204 system was equipped with a Waters model 6000A reciprocating pump, a model 440 absorbance detector, a Rheodyne model 7125 injector, and a Hewlett-Packard model 3352B integrator. The stationary phase was a Whatman Partisil-10 C_8 column. The mobile phase consisted of 50 mL of methanol, 170 mL of acetonitrile, 755 mL of water, and 25 mL of pentanesulfonic acid sodium salt in glacial acetic acid. The flow rate was 2.0 mL/min. UV detection was performed at 254 nm.

For liquid forms, samples were diluted with water. For solid forms, samples were leached with water or mobile phase and then diluted. Under these conditions, retention times for phenylephrine hydrochloride, phenylpropanolamine hydrochloride, guaifenesin, and benzoic acid were about 2.8, 4.4, 7.4, and 14.4 minutes, respectively (estimated from the published chromatogram).

The stability-indicating nature of the method was demonstrated by spiking the sample solution with possible impurities and degradation products. Phenylephrine hydrochloride, phenylpropanolamine hydrochloride, guaifenesin, and benzoic acid were separated from α-aminopropiophenone, 2-(2-methoxyphenoxy)-1,3-propanediol, *m*-hydroxybenzaldehyde, guaicol, and benzaldehyde.

A standard curve for guaifenesin was constructed from 0 to 0.1 μg/mL. The correlation coefficient was 0.9999.

References

Schieffer GW, Smith WO, Lubey GS, et al. Determination of the structure of a synthetic impurity in guaifenesin: Modification of a high-performance liquid chromatographic method for phenylephrine hydrochloride, phenylpropanolamine hydrochloride, guaifenesin, and sodium benzoate in dosage forms. *J Pharm Sci*. 1984; 73: 1856–8.

Schieffer GW, Hughes DE. Simultaneous stability-indicating determination of phenylephrine hydrochloride, phenylpropanolamine hydrochloride, and guaifenesin in dosage forms by reversed-phase paired-ion high-performance liquid chromatography. *J Pharm Sci*. 1983; 72: 55–9.

Method 5

Heidemann reported an HPLC method for the determination of theophylline, guaifenesin, and benzoic acid in liquid and solid pharmaceutical dosage forms. A DuPont model 830 liquid chromatograph was equipped with a DuPont model 837 variable-wavelength detector, a Rheodyne model 7105 injector, and a Spectra-Physics autolab system 1V integrator. The stationary phase was a Whatman Partisil-10-ODS reversed-phase column (250 × 4.6 mm, 10-μm particle size). The mobile phase was a mixture of 0.001 M sodium citrate–citric acid buffer (adjusted to pH 4.15) and acetonitrile (9:1, vol/vol). The flow rate was 2 mL/min. UV detection was performed at 230 nm. Methylparaben was used as an internal standard.

Liquid samples were diluted with water and the internal standard. Tablets were ground, wetted with alcohol, diluted with water, stirred for 1 hour, and filtered. The filtrate was further diluted with water and the internal standard. The injection volume was 20 μL. Under these conditions, retention times for benzoic acid, theophylline sodium glycinate, guaifenesin, and methylparaben were about 6.8, 9.4, 11.6, and 16.0 minutes, respectively.

The method was reported to be stability indicating.

Reference

Heidemann DR. Rapid, stability-indicating, high-pressure liquid chromatographic determination of theophylline, guaifenesin, and benzoic acid in liquid and solid pharmaceutical dosage forms. *J Pharm Sci*. 1979; 68: 530–2.

Halofantrine

Chemical Name

1,3-Dichloro-α-[2-(dibutylamino)ethyl]-6-(trifluoromethyl)-9-phenanthrenemethanol-
hydrochloride

Other Name
Halfan

Form

Form	Molecular Formula	MW	CAS
Halofantrine hydrochloride	$C_{26}H_{30}Cl_2F_3NO.HCl$	536.9	36167-63-2

Appearance
Halofantrine hydrochloride occurs as a white or almost white powder.

Solubility
Halofantrine hydrochloride is practically insoluble in water. It is sparingly soluble in ethanol, but it is freely soluble in methanol.

Method
Abdul-Fattah and Bhargava reported an HPLC method for the analysis of halofantrine in pharmaceuticals. The Hewlett-Packard chromatograph 1100 included a multiple-wavelength UV detector and a model 3395 integrator. The stationary phase was a MAC-MOD Zorbax SB-CN column (150 × 4.6 mm, 5-μm particle size). The mobile phase consisted of 0.025 M monobasic potassium phosphate aqueous buffer (adjusted to pH 5.0 with 0.1 N sodium hydroxide solution) and acetonitrile (20:80, vol/vol). The flow rate was 1 mL/min. UV detection was performed at 259 nm.

Ten milligrams of solid sample was accurately weighed, dissolved in 50 mL of mobile phase, and diluted by a factor of 100 with mobile phase. The injection volume was 100 μL. Under these conditions, the retention time for halofantrine was about 7.2 minutes.

The assay was shown to be stability indicating by accelerated degradation of halofantrine. Halofantrine solutions were treated with 1 N sodium hydroxide, 1 N hydrochloric acid, or 3% hydrogen peroxide and heated at 80 °C for 1 hour. The degradation products did not interfere with the determination of halofantrine.

A standard curve for halofantrine was constructed from 0.1 to 5 μg/mL. The correlation coefficient was greater than 0.9998. The limit of detection was 1 ng/mL and the limit of quantitation was 10 ng/mL.

Reference
Abdul-Fattah AM, Bhargava HN. A new high-performance liquid chromatography (HPLC) method for the analysis of halofantrine (HF) in pharmaceuticals. *J Pharm Biomed Anal.* 2002; 29: 901–8.

Haloperidol

Chemical Name
4-[4-(4-Chlorophenyl)-4-hydroxypiperidino]-4′-fluorobutyrophenone

Other Names
Dozic, Haldol, Halperon

Form	Molecular Formula	MW	CAS
Haloperidol	$C_{21}H_{23}ClFNO_2$	375.9	52-86-8
Haloperidol decanoate	$C_{31}H_{41}ClFNO_3$	530.1	74050-97-8

Appearance
Haloperidol is a white or faintly yellowish amorphous or microcrystalline powder. Haloperidol decanoate is a clear light amber oily liquid.

Solubility
Haloperidol has solubilities of less than 0.1 mg/mL in water and of approximately 16.7 mg/mL in alcohol at 25 °C. Haloperidol decanoate is soluble in fixed oil and in most organic solvents.

pK$_a$
Haloperidol has a pK$_a$ of 8.3.

Method
Driouich et al. developed an HPLC assay for haloperidol syrup. The instrument consisted of a Perkin Elmer 200-LC pump equipped with a 20-µL manual injector, a Perkin-Elmer LC-295 UV detector, and a Shimadzu C-R6A integrator. The stationary phase was an Applied Biosystems LiChrosorb C_{18} column (250 × 4.6 mm, 5-µm particle size). The mobile phase consisted of 50% methanol, 50% phosphate buffer (pH 2.0), and 0.2% triethylamine and was delivered isocratically at 1.2 mL/min. UV detection was performed at 254 nm.

Samples were diluted 1:50 with methanol before injection. The injection volume was 20 µL. Under these conditions, retention times for 4-hydroxybenzoic acid, methylparaben, 4-fluorobenzoic acid, haloperidol, and propylparaben were about 3.0, 5.5, 6.0, 9.5, and 13.5 minutes, respectively (estimated from the published chromatogram).

The method was proven to be stability indicating by accelerated decomposition of the drug. The drug was directly exposed to daylight. The drug samples were also stored in the dark at 36 and 48 °C. The degradation products did not interfere with the analysis of haloperidol.

A linear calibration curve for haloperidol was constructed. The correlation coefficient was 0.999.

Reference

Driouich R, Trabelsi H, Bouzouita KA. Stability-indicating assay for haloperidol syrup by high-performance liquid chromatography. *Chromatographia*. 2001; 53: 629–34.

Homoharringtonine

Chemical Name

Cephalotaxine 2-(methoxycarbonylmethyl)-2,6-dihydroxy-5-methylheptanoate

Form	Molecular Formula	MW	CAS
Homoharringtonine	$C_{29}H_{39}NO_9$	545.6	26833-87-4

Method

He et al. presented an HPLC assay for the determination of homoharringtonine. The liquid chromatograph was a Hewlett-Packard 1050 system. The stationary phase was a Phenomenex IB-SIL C_{18} stainless steel column (250 × 4.6 mm, 5-μm particle size). The mobile phase consisted of water containing 0.2% acetic acid (adjusted to pH 6.5 with triethylamine) (76:24, vol/vol). The flow rate was 1.0 mL/min. UV detection was performed at 280 nm. *o*-Nitroaniline 0.1 mg/mL in acetonitrile was used as an internal standard. The injection volume was 10 μL. Under these conditions, the retention time for homoharringtonine was about 10.4 minutes (estimated from the published chromatogram).

The stability-indicating ability of the method was demonstrated by hydrolysis and thermal degradation of homoharringtonine. Homoharringtonine solution 2 mg/mL was diluted with an equal volume of acetonitrile, water, 0.1 N and 1.0 N hydrochloric acid, and 0.1 M ammonium hydroxide and heated at 75 °C for 2 hours. A solid sample of homoharringtonine was kept in a 60 °C oven with 60% relative humidity for 30 days. The intact homoharringtonine peak was resolved from the degradation product peaks.

A standard curve for homoharringtonine was constructed from 4 to 12 μg/mL. The correlation coefficient was 0.9999. The limit of quantitation was 27 ng.

Reference

He J, Cheung AP, Wang E, et al. Stability-indicating LC assay of and impurity identification in homoharringtonine samples. *J Pharm Biomed Anal*. 2000; 22: 541–54.

Hydralazine

Chemical Names

1(2*H*)-Phthalazinone hydrazone
1-Hydrazinophthalazine

Other Names
Apresoline, Hydrallazine

Form

Form	Molecular Formula	MW	CAS
Hydralazine	$C_8H_8N_4$	160.2	86-54-4
Hydralazine hydrochloride	$C_8H_8N_4 \cdot HCl$	196.6	304-20-1

Appearance
Hydralazine hydrochloride is a white to off-white crystalline powder.

Solubility
Hydralazine hydrochloride has solubilities of approximately 40 mg/mL in water and 2 mg/mL in alcohol at 25 °C.

pK_a
Hydralazine hydrochloride has a pK_a of 7.3.

Method
Alexander et al. studied the stability of hydralazine hydrochloride syrup compounded from tablets. The liquid chromatograph was composed of a Beckman model 110B solvent-delivery system, a Beckman system organizer injector fitted with a 20-µL loop, a Beckman model 420 system controller programmer, a Knauer model 731.7100000 spectrophotometric UV detector, and a Shimadzu model C-R3A Chromatopac integrator. The stationary phase was a Supelco Partisil ODS-3 C_{18} column (250 × 4.6 mm). The mobile phase was prepared by dissolving 2.3 g of tetraethylammonium perchlorate in 900 mL of distilled water, adding 2 mL of acetic acid to the mixture, and bringing the final volume to 1 L with methanol. The pH of the mobile phase was adjusted to 2.7 with acetic acid. The flow rate was 1.4 mL/min. UV detection was performed at 254 nm and 0.04 AUFS. Procaine hydrochloride 1% in distilled water was used as the internal standard.

Samples were diluted 1:250 with 0.1 N hydrochloric acid. The injection volume was 50 µL. Under these conditions, retention times for hydralazine hydrochloride and procaine hydrochloride were 5.1 and 10.4 minutes, respectively.

This assay was determined to be stability indicating by accelerated degradation. Samples of hydralazine hydrochloride syrup were heated for 2 hours under acidic and alkaline conditions. The resulting solutions were assayed. No degradation products interfered with either the intact hydralazine hydrochloride peak or the procaine hydrochloride peak.

A calibration curve was generated for hydralazine hydrochloride from 0.0025 to 0.0125 mg/mL by regression analysis of peak-area ratio versus hydralazine concentration. The associated correlation coefficient was 0.999.

Reference

Alexander KS, Pudipeddi M, Parker GA. Stability of hydralazine hydrochloride syrup compounded from tablets. *Am J Hosp Pharm.* 1993; 50: 683–6.

Hydrochlorothiazide

Chemical Name

6-Chloro-3,4-dihydro-2*H*-1,2,4-benzothiadiazine-7-sulfonamide 1,1-dioxide

Other Names

HydroDIURIL, Oretic

Form	Molecular Formula	MW	CAS
Hydrochlorothiazide	$C_7H_8ClN_3O_4S_2$	297.7	58-93-5

Appearance

Hydrochlorothiazide is a white or practically white crystalline powder.

Solubility

Hydrochlorothiazide is soluble in diluted ammonia or sodium hydroxide and also in methanol, ethanol, and acetone but is practically insoluble in water.

pK$_a$

Hydrochlorothiazide has pK$_a$ values of 7.9 and 9.2.

Method 1

Allen and Erickson used HPLC to assess the stability of hydrochlorothiazide 5 mg/mL in extemporaneously compounded oral liquids. The liquid chromatograph was an automated Hewlett-Packard series 1050 system with a multisolvent mixing and pumping system, an autoinjector, a diode-array detector, and a computer with Chem Station data-handling software. The stationary phase was a Bakerbond C_{18} analytical column (250 × 4.6 mm, 5-μm particle size). The mobile phase was 70% methanol in water and was delivered isocratically at 1.0 mL/min. UV detection was performed at 254 nm.

Samples were diluted 1:50. Under these conditions, hydrochlorothiazide eluted in 3.5 minutes.

The stability-indicating capability of the HPLC assay was demonstrated by accelerated degradation of hydrochlorothiazide. A composite chromatogram of hydrochlorothiazide after intentional degradation by acid, base, heat, oxidizing agent, and light did not show any interference of the decomposition product peaks with the intact hydrochlorothiazide peak.

Calibration curves for hydrochlorothiazide were constructed from a linear plot of peak areas versus concentration from 10 to 150 µg/mL. The intraday and interday coefficients of variation were 1.4 and 1.8%, respectively.

Reference

Allen LV, Erickson MA. Stability of labetalol hydrochloride, metoprolol tartrate, verapamil hydrochloride, and spironolactone with hydrochlorothiazide in extemporaneously compounded oral liquids. *Am J Health Syst Pharm.* 1996; 53: 2304–9.

Method 2

Sasa et al. used reversed-phase HPLC to analyze atenolol in combination with hydrochlorothiazide and chlorthalidone in tablet formulations. The chromatographic apparatus included a Varian model 2010 pump, a Rheodyne model 7125 10-µL loop injector, a Varian model 2050 spectrophotometric detector, and a Varian model 4290 integrator. The stationary phase was a Supelco Supelcosil LC-8-DB reversed-phase column (250 × 4.6 mm, 5-µm particle size). The mobile phase was a mixture of 1.0 mM ammonium acetate and 2.0 mM sodium octanesulfonate in acetonitrile–water (25:75, vol/vol) that was adjusted to pH 3.5 with glacial acetic acid. The flow rate was 1.5 mL/min. UV detection was performed at 254 nm and 0.2 AUFS. Methyl *p*-hydroxy-benzoate 10 µg/mL in methanol was used as the internal standard.

Twenty tablets were ground. A portion of this powder equivalent to one tablet was weighed, transferred into a 25-mL volumetric flask, sonicated with 20 mL of the internal standard for 3 minutes, diluted to 1.00 mg/mL of atenolol with internal standard, and filtered through a 0.45-µm membrane filter. The injection volume was 10 µL. Under these conditions, the retention times of hydrochlorothiazide, atenolol, chlorthalidone, and the internal standard were 3.5, 5.2, 5.9, and 8.4 minutes, respectively.

The method was verified to be stability indicating by studying a synthetic mixture of the drugs and their degradation products. All compounds were separated. No interference from excipients, other drugs, or the degradation products was observed.

A standard curve for hydrochlorothiazide was generated from 1.25 to 3.75 µg/mL. The correlation coefficient was greater than 0.999. The limit of detection was 25 pg.

Reference

Sasa SI, Jalal IM, Khalil HS. Determination of atenolol combinations with hydrochlorothiazide and chlorthalidone in tablet formulations by reverse phase HPLC. *J Liq Chromatogr.* 1988; 11: 1673–96.

Method 3

Hitscherich et al. presented a simple analytical method for the simultaneous determination of hydrochlorothiazide and propranolol hydrochloride in tablets. The liquid chromatograph included a Waters model 501 pump, a Perkin-Elmer model LC-420B autoinjector, a Perkin-Elmer model LC-95 variable-wavelength detector, and a Spectra-Physics model SP-4270 integrator. The stationary phase was an Altex Ultrasphere cyano column (250 × 4.6 mm, 5-µm particle size). The mobile phase was prepared by mixing 15 mL of acetonitrile with 85 mL of 0.05 M monobasic ammonium phosphate. The pH was adjusted to 3.0 with phosphoric acid. The flow rate was 2 mL/min. UV detection was carried out at 290 nm.

A portion of powder from 20 tablets, equivalent to 25 mg of hydrochlorothiazide, was accurately weighed, transferred to a 100-mL volumetric flask, mixed with 5 mL of 0.1 N hydrochloric acid, diluted with 50 mL of methanol, sonicated for 10 minutes, cooled down, filled to volume with methanol, mixed well, filtered through a 0.45-µm solvent-resistant filter, diluted with mobile phase, and assayed. The injection volume was 20 µL. Under these conditions, retention times for hydrochlorothiazide and propranolol hydrochloride were 4.5 and 14.8 minutes, respectively.

The method was stability indicating since the intact drug was well separated from its hydrolysis product and synthesis precursor 4-amino-6-chloro-1,3-benzenedisulfon-amide (3.6 minutes) and from its synthesis impurity chlorothiazide (4.0 minutes).

A standard curve was constructed from 0 to 92.4 µg/mL, with a correlation coefficient of 0.9999. The average recovery for hydrochlorothiazide was 100.5%.

Reference

Hitscherich ME, Rydberg EM, Tsilifonis DC, et al. Simultaneous determination of hydrochlorothiazide and propranolol hydrochloride in tablets by high performance liquid chromatography. *J Liq Chromatogr.* 1987; 10: 1011–21.

Method 4

Menon and White described a simultaneous determination by HPLC of hydrochlorothiazide and triamterene in capsules. The instrument consisted of a Waters model 6000A dual-piston reciprocating pump, a model U6K universal injector, a Schoeffel model SF 770 variable-wavelength detector, and a Hewlett-Packard model 3385A automation system. The stationary phase was a Waters µBondapak C_{18} column (300 × 4 mm, 10-µm particle size). The mobile phase was a mixture of 20 mL of 0.2 M sodium acetate (adjusted to pH 5.0 with acetic acid), 780 mL of water, 150 mL of acetonitrile, and 50 mL of methanol. The flow rate was 2 mL/min. UV detection was performed at 273 nm and 0.1 AUFS. *m*-Hydroxyacetophenone 3 mg/mL in acetonitrile–water (1:1, vol/vol) was used as an internal standard.

A portion of powder from 10 capsules equivalent to 50 mg of hydrochlorothiazide and 100 mg of triamterene was prepared as a slurry with 25 mL of acetonitrile, followed by 4 mL of acetic acid. The slurry was shaken well for 5 minutes, dissolved with 10 mL of internal standard solution, diluted to 100 mL with water, and filtered through a Nuclepore polycarbonate membrane. The injection volume was 40 µL. Under these conditions, retention times for hydrochlorothiazide, *m*-hydroxyacetophenone, and triamterene were about 4.0, 7.1, and 10.0 minutes, respectively (estimated from the published chromatogram).

The method was demonstrated to be stability indicating by accelerated degradation of the drugs. Samples were refluxed in 1 N hydrochloric acid, 1 N sodium hydroxide solution, or water. The intact drugs were resolved from their degradation products.

A standard curve for hydrochlorothiazide was obtained from 15 to 30 mg/capsule. The correlation coefficient was 1.000. The coefficient of variation for the analysis of hydrochlorothiazide was 1.2%.

Reference

Menon GN, White LB. Simultaneous determination of hydrochlorothiazide and triamterene in capsule formulations by high-performance liquid chromatography. *J Pharm Sci.* 1981; 70: 1083–5.

Method 5

Daniels and Vanderwielen described a reversed-phase HPLC method for hydro-chlorothiazide. A Waters model ALC 202 liquid chromatograph was equipped with a low-volume septumless injector and a Waters model 440 fixed-wavelength detector. The stationary phase was a Waters μBondapak C_{18} column (10-μm particle size). The mobile phase was 5% methanol in double-distilled water adjusted to pH 4.5–5.0 with 0.1 M acetic acid. The flow rate was 2.0 mL/min. UV detection was performed at 254 nm and 0.2–0.5 AUFS. Sulfadiazine 0.7 mg/mL in methanol was used as an internal standard.

A portion of powder prepared from ground tablets equivalent to 20 mg of hydrochlorothiazide was weighed, mixed with 10 mL of the internal standard solution, shaken for 35–40 minutes, and filtered or centrifuged. The injection volume was 6–10 μL. Under these conditions, the retention time of hydrochlorothiazide was 13 minutes.

The method was evaluated to be stability indicating by spiking a hydrochlor-othiazide solution with its degradation product and impurities. Hydrochlorothiazide was well separated from these compounds.

Reference

Daniels SL, Vanderwielen AJ. Stability-indicating assay of hydrochlorothiazide. *J Pharm Sci*. 1981; 70: 211–5.

Hydrocodone

Chemical Names

4,5-Epoxy-3-methoxy-17-methylmorphinan-6-one
Dihydrocodeinone

Other Names

Dicodid, component of Vicodin

Form	Molecular Formula	MW	CAS
Hydrocodone	$C_{18}H_{21}NO_3$	299.4	125-29-1
Hydrocodone bitartrate	$C_{18}H_{21}NO_3.C_4H_6O_6.2\frac{1}{2}H_2O$	494.5	34195-34-1
Hydrocodone hydrochloride	$C_{18}H_{21}NO_3.HCl.2\frac{1}{2}H_2O$	380.9	25968-91-6

Appearance
Hydrocodone bitartrate occurs as fine white crystals or crystalline powder.

Solubility
Hydrocodone bitartrate is soluble in water and slightly soluble in alcohol.

Method 1
Alvi and Castro reported a simultaneous HPLC analysis of acetaminophen and hydrocodone bitartrate in a tablet formulation. The liquid chromatograph consisted of a Waters model M6000A pump, a model 710B/712B WISP autosampler, a model 490 variable-wavelength detector, and a model 730 data module. The stationary phase was a Waters Nova-Pak C_{18} Radial-Pak cartridge column (100 × 8 mm, 4-μm particle size) with a Waters C_{18} guard column. The column temperature was 30 °C. The mobile phase consisted of an aqueous phosphate buffer and acetonitrile (84:16, vol/vol); the phosphate buffer was 0.02 M monobasic potassium phosphate containing 0.2 mL of triethylamine and 0.2 mL of phosphoric acid and adjusted to pH 3.3 with 3 N phosphoric acid. The flow rate was 2.0 mL/min. UV detection was performed at 215 nm and 0.02 AUFS.

A portion of powder prepared from ground tablets equivalent to one tablet was weighed, transferred into a 100-mL volumetric flask, mixed with 50 mL of the diluent (0.5 mL of 3 N phosphoric acid in 1000 mL of water), and shaken for 45 minutes. The mixture was diluted to volume with the diluent, mixed, filtered, and further diluted with the diluent. The injection volume was 10 μL. Under these conditions, the retention times of acetaminophen and hydrocodone bitartrate were 2.6 and 5.0 minutes, respectively.

The method was evaluated to be stability indicating by assaying a synthetic mixture of drugs and their known degradation products. Retention times for p-aminophenol, hydromorphone hydrochloride, codeine sulfate, and p-chloroacetanilide were 1.6, 2.3, 3.3, and 34.6 minutes, respectively. Samples were also refluxed in 1 N hydrochloric acid, 1 N sodium hydroxide solution, or 10% hydrogen peroxide for 4 hours. The decomposition products were separated from the intact drugs.

A standard curve for hydrocodone bitartrate was obtained from 0.0031 to 0.0072 mg/mL. The correlation coefficient was 0.9993.

Reference
Alvi SU, Castro F. A stability-indicating simultaneous analysis of acetaminophen and hydrocodone bitartrate in tablets formulation by HPLC. *J Liq Chromatogr*. 1987; 10: 3413–26.

Method 2
Wallo and D'Adamo developed an HPLC method for the simultaneous assay of hydrocodone bitartrate and acetaminophen in a tablet formulation. A Waters model ALC 204 liquid chromatograph was equipped with a Waters model 710B WISP autosampler, a Schoeffel SF 770 variable-wavelength UV detector, and a Waters data module. The stationary phase was a Waters μBondapak reversed-phase C_{18} column. The mobile phase consisted of 25% methanol and 75% aqueous solution containing 0.01 N monobasic potassium phosphate and 0.05 N potassium nitrate, adjusted to a pH of about 4.5 with 3 N phosphoric acid solution. The flow rate was about 1.1 mL/min. UV detection was performed at 283 nm and 0.01 AUFS. The injection volume was 13 μL.

The assay was determined to be stability indicating by accelerated degradation of hydrocodone bitartrate. Hydrocodone bitartrate powder was heated at 105 °C for 1 week. Hydrocodone bitartrate 1 mg/mL was refluxed in 0.1 N sodium hydroxide for 24 hours or in 1.0 N hydrochloric acid for 48 hours. Hydrocodone bitartrate 1 mg/mL in water was irradiated with a 275-watt sunlamp for 2 hours. Hydrocodone bitartrate 1 mg/mL was dissolved in 3% hydrogen peroxide and stored for 1 month. In each case, degradation product peaks did not interfere with the hydrocodone peak. Retention times in minutes were 3.4 (*p*-aminophenol), 5.2 (hydromorphone), 5.9 (acetaminophen), 7.3 (codeine), 10.0 (hydrocodone), and 43.3 (*p*-chloroacetanilide).

Standard curves were constructed from 0.035 to 0.065 mg/mL. The correlation coefficient was 0.996.

Reference

Wallo WE, D'Adamo A. Simultaneous assay of hydrocodone bitartrate and acetaminophen in a tablet formulation. *J Pharm Sci.* 1982; 71: 1115–8.

Hydrocortisone

Chemical Names

(11β)-11,17,21-Trihydroxypregn-4-ene-3,20-dione
4-Pregnene-11β,17α,21-triol-3,20-dione

Other Names

Cortef, Hydrocortone, Solu-Cortef

Form	Molecular Formula	MW	CAS
Hydrocortisone	$C_{21}H_{30}O_5$	362.5	50-23-7
Hydrocortisone acetate	$C_{23}H_{32}O_6$	404.5	50-03-3
Hydrocortisone sodium phosphate	$C_{21}H_{29}Na_2O_8P$	486.4	6000-74-4
Hydrocortisone sodium succinate	$C_{25}H_{33}NaO_8$	484.5	125-04-2

Appearance

Hydrocortisone and hydrocortisone acetate occur as white or practically white odorless crystalline powders. Hydrocortisone sodium phosphate is a white or light yellow and odorless or almost odorless hygroscopic powder. Hydrocortisone sodium succinate is a white or nearly white amorphous solid.

Solubility

Hydrocortisone is very slightly soluble in water and sparingly soluble in alcohol. Hydrocortisone acetate is insoluble in water and slightly soluble in alcohol. Hydrocortisone sodium phosphate is freely soluble in water and slightly soluble in alcohol. Hydrocortisone sodium succinate is very soluble in water and alcohol; it is very slightly soluble in acetone and practically insoluble in chloroform.

pK$_a$

The apparent pK$_a$ by titration with sodium hydroxide is 11.05 at 25 °C.

Method 1

Zhang et al. reported the chemical stability of hydrocortisone 2 mg/mL (as the sodium succinate) in Elliott's B solution, an artificial cerebrospinal fluid. The liquid chromatograph consisted of a Waters model 600E multisolvent-delivery pump, a Waters 490E programmable multiple-wavelength UV detector, and a Waters model 710 WISP autosampler. The system was controlled and integrated by an NEC PowerMate SX/16 personal computer utilizing Waters Maxima 820 chromatography software. The stationary phase was a Waters µBondapak reversed-phase C$_{18}$ analytical column (150 × 3.9 mm, 10-µm particle size). The mobile phase consisted of 33.5% acetonitrile in 1% acetic acid in water. The flow rate was 1.2 mL/min. UV detection was performed at 240 nm.

Samples were diluted with mobile phase. The injection volume was 15 µL. Under these conditions, the retention time for hydrocortisone sodium succinate was about 5.6 minutes.

The HPLC method was evaluated to be stability indicating by accelerated hydrocortisone 21-hemisuccinate decomposition. Boiling a solution of hydrocortisone 21-hemisuccinate for 2 hours yielded a reduction in the intact drug peak and the formation of two new peaks at 3.3 and 3.6 minutes. Degradation product peaks did not interfere with the intact hydrocortisone 21-hemisuccinate peak.

Calibration curves were constructed from a linear plot of peak area versus concentration for hydrocortisone from 25 to 150 µg/mL. The correlation coefficient was greater than 0.9997. The intraday and interday coefficients of variation were 1.7 and 1.1%, respectively.

A similar method was used by Cheung et al.

References

Zhang Y, Xu QA, Trissel LA, et al. Physical and chemical stability of methotrexate sodium, cytarabine, and hydrocortisone sodium succinate in Elliott's B solution. *Hosp Pharm.* 1996; 31: 965–70.

Cheung Y-W, Vishnuvajjala BR, Flora KP. Stability of cytarabine, methotrexate sodium, and hydrocortisone sodium succinate admixtures. *Am J Hosp Pharm.* 1984; 41: 1802–6.

Method 2

Gupta and Stewart used a stability-indicating HPLC method to investigate the chemical stability of hydrocortisone sodium succinate when mixed with metronidazole injection for intravenous infusion. A Waters ALC 202 high-performance liquid chromatograph was equipped with a Schoeffel SF 770 multiple-wavelength detector, an Omniscribe recorder, and a Waters μBondapak C_{18} column. The mobile phase consisted of 32% acetonitrile and 68% 0.02 M monobasic potassium phosphate in water. The flow rate was 2.5 mL/min. UV detection was performed at 254 nm and 0.2 AUFS. A solution containing approximately 8 μg of hydrocortisone was injected. The retention time for hydrocortisone was about 3.5 minutes (estimated from the published chromatogram).

The method was stated to be stability indicating because a chromatogram of a 5-day-old solution of hydrocortisone stored at 25 °C showed that degradation product peaks did not interfere with the hydrocortisone peak.

Reference
Gupta VD, Stewart KR. Chemical stability of hydrocortisone sodium succinate and several antibiotics when mixed with metronidazole injection for intravenous infusion. *J Parenter Sci Technol.* 1985; 39: 145–8.

Hydromorphone

Chemical Name
4,5-Epoxy-3-hydroxy-17-methylmorphinan-6-one

Other Names
Dihydromorphinone, Dilaudid

Form	Molecular Formula	MW	CAS
Hydromorphone	$C_{17}H_{19}NO_3$	285.3	466-99-9
Hydromorphone hydrochloride	$C_{17}H_{19}NO_3 \cdot HCl$	321.8	71-68-1

Appearance
Hydromorphone hydrochloride is a fine white or practically white crystalline powder.

Solubility
Hydromorphone hydrochloride is freely soluble in water but sparingly soluble in alcohol.

Method 1
Christen et al. evaluated the stability of hydromorphone hydrochloride 20 and 100 μg/mL with bupivacaine hydrochloride in 0.9% sodium chloride injection during simulated

epidural coadministration. The liquid chromatograph consisted of a Waters model 501 pump, a Waters model U6K variable-volume injector, a Waters model 486 variable-wavelength UV detector, and a Hewlett-Packard model 3390 integrator and recorder. The stationary phase was an Alltech Hypersil ODS C_{18} analytical column (150×4.6 mm, 5-µm particle size). The mobile phase consisted of 40% acetonitrile, 20 mL of acetic acid, and 10 g of sodium lauryl sulfate in water to make a total volume of 2000 mL. The flow rate was 1 mL/min. UV detection was performed at 280 nm and 0.01 AUFS. The sample injection volume was 10 µL. Under these conditions, the retention time for hydromorphone was 4.6 minutes.

The stability-indicating capability of this assay was demonstrated by accelerated decomposition of hydromorphone. Addition of acid to the sample solution resulted in degradation products eluting at 7.2 minutes; addition of base to the sample solution resulted in decomposition products eluting at 3.0, 3.5, and 3.9 minutes. No peak interfered with the intact hydromorphone peak.

Standard curves for hydromorphone were constructed based on the peak height versus concentration from 15 to 25 µg/mL and from 90 to 110 µg/mL. Correlation coefficients were greater than 0.999. The intraday and interday coefficients of variation were 1.2 and 2.4%, respectively.

Reference

Christen C, Johnson CE, Walters JR. Stability of bupivacaine hydrochloride and hydromorphone hydrochloride during simulated epidural coadministration. *Am J Health Syst Pharm.* 1996; 53: 170–3.

Method 2

Mayron and Gennaro used an HPLC method to assess the stability of hydromorphone hydrochloride with granisetron hydrochloride during simulated Y-site administration. The liquid chromatograph consisted of a piston pump with a pulse dampener, a rotary injection port with a 20-µL loop, a variable-wavelength UV detector, and an integrator. The stationary phase was a cyano reversed-phase column (250×4.6 mm, 5-µm particle size). The mobile phase was a mixture of acetonitrile and 0.1 M monobasic sodium phosphate dihydrate (20:80, vol/vol) adjusted to pH 4.2 with phosphoric acid. The flow rate was 1.25 mL/min. UV detection was performed at 290 nm.

Samples were diluted with mobile phase. The retention times for hydromorphone and granisetron were 3.72 and 6.96 minutes, respectively.

The stability-indicating capability of the analytical method was demonstrated by an accelerated degradation study. Solutions of hydromorphone hydrochloride and granisetron hydrochloride were adjusted to pH 2 and 11, boiled for 1 hour, readjusted to pH 5, diluted with the mobile phase, and analyzed. The degradation product peaks did not interfere with the intact hydromorphone or granisetron peak.

Reference

Mayron D, Gennaro AR. Stability and compatibility of granisetron hydrochloride in i.v. solutions and oral liquids and during simulated Y-site injection with selected drugs. *Am J Health Syst Pharm.* 1996; 53: 294–304.

Method 3

Using an HPLC assay, Stiles et al. studied the stability of hydromorphone hydrochloride in polypropylene infusion-pump syringes. The liquid chromatograph included a Waters model 501 pump, a Waters model 441 UV detector, a Waters model 745 data module, an Alcott model 728 autosampler, and a Rheodyne 7010 injector. The stationary phase was a Bakerbond C_{18} analytical column. The mobile phase consisted of 15% methanol and 85% 15 mM sodium dihydrogen phosphate buffer containing 3 mM 1-heptanesulfonic acid (adjusted to pH 3.5 with phosphoric acid). The flow rate was 0.8 mL/min. UV detection was performed at 230 nm.

Samples were diluted 1:100 before injection. Under these conditions, the retention time for hydromorphone hydrochloride was 6.0 minutes.

The stability-indicating nature of the assay was investigated by accelerated degradation of hydromorphone hydrochloride with sulfuric acid, 1 N sodium hydroxide, and heat at 80 °C for 24 hours. Degradation product peaks did not interfere with the intact hydromorphone peak.

A standard curve was constructed from 1 to 200 µg/mL. Intraday and interday coefficients of variation of the assay were 0.2 and 0.3%, respectively.

Reference
Stiles ML, Allen LV, Prince SJ. Stability of deferoxamine mesylate, floxuridine, fluorouracil, hydromorphone hydrochloride, lorazepam, and midazolam hydrochloride in polypropylene infusion-pump syringes. *Am J Health Syst Pharm.* 1996; 53: 1583–8.

Method 4
Trissel et al. determined the stability of hydromorphone hydrochloride 0.5 mg/mL when admixed with ondansetron hydrochloride in 0.9% sodium chloride injection and stored at 32 °C for up to 7 days and at 4 and 22 °C for up to 31 days. The liquid chromatograph consisted of a Waters model 600E multisolvent-delivery system, a Waters model 490E programmable multiple-wavelength UV detector, a Waters model 712 WISP autosampler, and an Alltech Spherisorb cyano column (300 × 4.6 mm, 5-µm particle size). The system was controlled and the peak area was integrated by an NEC PowerMate SX/16 personal computer. The mobile phase was 50% acetonitrile in 0.02 M monobasic potassium phosphate with the pH adjusted to 5.40 with 1 N sodium hydroxide; it was delivered isocratically at 1.5 mL/min. UV detection was performed at 216 nm.

Samples were diluted 1:5 before injection. The injection volume was 20 µL. Under these conditions, the retention time for hydromorphone was 7.5 minutes. The retention time for ondansetron was 8.8–10 minutes.

The assay was demonstrated to be stability indicating by accelerated degradation of hydromorphone hydrochloride. Heating hydromorphone hydrochloride solution at 95 °C for 4 hours yielded a reduction in the parent hydromorphone peak and the formation of new peaks eluting around the solvent front and at 5–6 minutes. Neither the ondansetron nor the degradation product peaks interfered with the parent hydromorphone peak.

Standard curves were constructed from a linear plot of peak area versus concentration from 0.01 to 0.15 mg/mL. The correlation coefficient was greater than 0.9993. The intraday and interday coefficients of variation of the assay were 1.7 and 2.2%, respectively.

References
Trissel LA, Xu Q, Martinez JF, et al. Compatibility and stability of ondansetron hydrochloride with morphine sulfate and with hydromorphone hydrochloride in 0.9% sodium chloride injection at 4, 22, and 32 °C. *Am J Hosp Pharm.* 1994; 51: 2138–42.

Trissel LA, Xu QA, Pham L. Physical and chemical stability of hydromorphone hydrochloride 1.5 and 80 mg/mL packaged in plastic syringes. *Int J Pharm Compound.* 2002; 6: 74–6.

Method 5

Walker et al. used an HPLC assay to evaluate the stability of hydromorphone hydrochloride. The chromatograph consisted of a Schoeffel SF 770 UV detector, a Spectra-Physics SP4200 chromatographic integrator, and a Beckman Ultrasphere C_{18} reversed-phase column (250 × 4.2 mm, 5-μm particle size). The mobile phase consisted of acetonitrile and phosphate buffer containing 1 mg/mL of heptanesulfonic acid and was pumped gradiently at 2.0 mL/min from 19 to 50% acetonitrile in 15 minutes. UV detection was performed at 230 nm.

The analytical method was determined to be stability indicating by accelerated degradation of hydromorphone hydrochloride using acid, base, and heat, with inspection of the chromatograms for the appearance of additional peaks, changes in retention times, and peak shapes. Intraday and interday coefficients of variation for the assay were 1.40 and 2.22%, respectively.

References

Walker SE, Iazzetta J, DeAngelis C, et al. Stability and compatibility of combinations of hydromorphone and dimenhydrinate, lorazepam or prochlorperazine. *Can J Hosp Pharm.* 1993; 46: 61–5.

Walker SE, DeAngelis C, Iazzetta J, et al. Compatibility of dexamethasone sodium phosphate with hydromorphone hydrochloride or diphenhydramine hydrochloride. *Am J Hosp Pharm.* 1991; 48: 2161–6.

Walker SE, DeAngelis C, Iazzetta J. Stability and compatibility of combinations of hydromorphone and a second drug. *Can J Hosp Pharm.* 1991; 44: 289–94.

Method 6

Walker et al. investigated the stability of hydromorphone hydrochloride in portable infusion pump cassettes and minibags. The liquid chromatographic system consisted of a Spectra-Physics model SP8770 isocratic solvent-delivery pump, a Schoeffel SF 770 variable-wavelength UV detector, a Spectra-Physics model SP4270 integrator, and a Hamilton PRP-1 rigid macroporous poly(styrene-divinylbenzene) copolymer column (10-μm particle size). The mobile phase consisted of 7% acetonitrile and 93% 0.01 M dibasic sodium phosphate adjusted to pH 11 with 10 M sodium hydroxide. The flow rate was 2 mL/min. UV detection was performed at 230 nm. The sample injection volume was 5 μL. The retention time for hydromorphone was about 3.5 minutes (estimated from the published chromatogram).

The analytical method was determined to be stability indicating by an accelerated degradation study. Hydromorphone hydrochloride in 0.9% sodium chloride injection was adjusted to pH 11 using sodium hydroxide and stored at 66 °C for 125 hours. The degradation product peak did not interfere with hydromorphone quantitation.

Standard curves were prepared daily.

Reference
Walker SE, Coons C, Matte D, et al. Hydromorphone and morphine stability in portable infusion pump cassettes and minibags. *Can J Hosp Pharm.* 1988; 41: 177–82.

Hydroxyzine Hydrochloride

Chemical Name
2-[2-[4-[(4-Chlorophenyl)phenylmethyl]-1-piperazinyl]ethoxy]ethanol dihydrochloride

• 2HCl

Other Names
Anxanil, Atarax, Vistaril

Form	Molecular Formula	MW	CAS
Hydroxyzine	$C_{21}H_{27}ClN_2O_2$	374.9	68-88-2
Hydroxyzine dihydrochloride	$C_{21}H_{27}ClN_2O_2 \cdot 2HCl$	447.8	2192-20-3
Hydroxyzine pamoate	$C_{44}H_{43}ClN_2O_8$	763.3	10246-75-0

Appearance
Hydroxyzine hydrochloride occurs as a white or almost white and hygroscopic odorless crystalline powder. Hydroxyzine pamoate occurs as a light yellow practically odorless powder.

Solubility
Hydroxyzine hydrochloride is very soluble in water and freely soluble in alcohol. Hydroxyzine pamoate is practically insoluble in water and in alcohol.

pK_a
Hydroxyzine has pK_a values of 2.6 and 7.0.

Method
Menon and Norris reported a simultaneous HPLC determination of hydroxyzine hydrochloride and benzyl alcohol in injection solutions. The liquid chromatographic system consisted of a Waters model 6000A dual-piston reciprocating pump, a model U6K universal injector, and a Perkin-Elmer model LC-55 or a Schoeffel model SF 770 Spectroflow variable-wavelength UV detector. The stationary phase was a Waters μBondapak C_{18} column (300 × 4 mm, 10-μm particle size). The mobile phase was a mixture of water, acetonitrile, and methanol (60:25:15, vol/vol/vol, pH 2.6) containing 0.06% (vol/vol) sulfuric acid, 0.5% (wt/vol) sodium sulfate, and 0.02% (wt/vol) heptane-

sulfonic acid sodium salt. The flow rate was 2 mL/min. UV detection was performed at 257 nm and at 0.2 AUFS. *p*-Nitroacetophenone 0.2 mg/mL and isobutyrophenone 2.5 mg/mL in methanol were used as internal standards.

Five-milliliter samples were mixed with 5 mL of the internal standard and diluted to 50 mL with methanol. The injection volume was 10 µL. Under these conditions, retention times for benzyl alcohol, *p*-nitroacetophenone, hydroxyzine hydrochloride, and isobutyrophenone were about 3.1, 5.6, 9.3, and 12.5 minutes, respectively (estimated from the published chromatogram).

The stability-indicating ability of the method was demonstrated by assaying a synthetic mixture of benzyl alcohol, hydroxyzine hydrochloride, and their potential degradation products. Benzyl alcohol and hydroxyzine hydrochloride were separated from their degradation products. Retention times in minutes were about 3.1 for benzyl alcohol, 3.5 for benzoic acid, 4.0 for benzaldehyde, 5.6 for *p*-nitroacetophenone, 6.7 for *p*-chlorobenzoic acid, 7.6 for *p*-chlorobenzaldehyde, 9.3 for hydroxyzine hydrochloride, 12.5 for isobutyrophenone, and 46.0 for *p*-chlorobenzophenone (estimated from the published chromatogram).

A standard curve for hydroxyzine hydrochloride was constructed from 3 to 10 mg/mL. The correlation coefficient was 0.9999.

Reference

Menon CN, Norris BJ. Simultaneous determination of hydroxyzine hydrochloride and benzyl alcohol in injection solutions by high-performance liquid chromatography. *J Pharm Sci.* 1981; 70: 697–8.

Ibuprofen

Chemical Names

α-Methyl-4-(2-methylpropyl)benzeneacetic acid
p-Isobutylhydratropic acid

Other Names

Advil, Motrin

Form	Molecular Formula	MW	CAS
Ibuprofen	$C_{13}H_{18}O_2$	206.3	15687-27-1

Appearance

Ibuprofen is a white to off-white crystalline powder.

Solubility
Ibuprofen is practically insoluble in water but is very soluble in most alcohols.

pK_a
Ibuprofen has a pK_a of 4.4.

Method 1
Farmer et al. evaluated the assay method for the determination of ibuprofen in bulk drug and in tablets. An Agilent HPLC system was used. The stationary phase was a Zorbax ODS column (250×4.6 mm). The mobile phase consisted of 400 mL of 1% chloroacetic acid and 600 mL of acetonitrile. The flow rate was 2 mL/min. UV detection was performed at 254 nm. An Agilent photodiode-array detector was used to determine the peak purity. Valerophenone was used as an internal standard.

The extraction solvent was 400 mL of 1% chloroacetic acid solution (adjusted to pH 3.0 with ammonium hydroxide), 600 mL of acetonitrile, and 3 mL of internal standard. One tablet or 0.8 g of ibuprofen bulk substance was added to 65 mL of extraction solvent, shaken for 1 hour, and filtered through a Whatman 0.45-μm filter into autosampler vials for assay. The retention times of ibuprofen and valerophenone were 4.0 and 5.3 minutes, respectively.

Ibuprofen samples were stressed with hydrochloric acid for 240 hours, light for 380 hours, temperature (at 80 °C) for 244 hours, and 30% hydrogen peroxide for 216 hours. The degradation products did not interfere with the quantitation of ibuprofen, demonstrating the stability-indicating nature of the assay.

Reference
Farmer S, Anderson P, Burns P, et al. Forced degradation of ibuprofen in bulk drug and tablets and determination of specificity, selectivity, and the stability-indicating nature of the USP ibuprofen assay method. *Pharm Technol*. 2002; 28: 42.

Method 2
Nada and Walily reported the formulation and stability of directly compressed ibuprofen tablets. A Hewlett-Packard 1090 liquid chromatograph was used; it included a binary solvent-delivery system, an autosampler, a column temperature controller, a photodiode-array detector, and an HP 3392A integrator. The stationary phase was a LiChrosorb RP18 column (250×4.0 mm, 5-μm particle size). The mobile phase was a mixture of methanol and water (80:20, vol/vol). The flow rate was 1 mL/min. UV detection was performed at 230 nm. Acetaminophen was used as an internal standard.

A sample of ground tablets equivalent to one tablet was diluted with 50 mL of methanol in a 100-mL volumetric flask, sonicated for 10 minutes, and further diluted with internal standard solution and methanol. The sample injection volume was 10 μL. Retention times for acetaminophen and ibuprofen were 2.08 and 5.76 minutes, respectively.

The method was reported to be stability indicating since it resolved ibuprofen from its main impurities and from the internal standard.

A standard curve for ibuprofen was constructed from 2.5 to 12.5 mg/mL. The correlation coefficient was 0.9996.

Reference
Nada AH, Walily AFE. Formulation and stability of directly compressed ibuprofen tablets. *Mans J Pharm Sci*. 1997; 13: 1–29.

Method 3

Haikala et al. developed an HPLC method for the determination of ibuprofen in ointments. The HPLC system included a Pye Unicam PU4010 pump, an LC-XP gradient programmer, a PU4020 detector, a CDP1 computing integrator, and a Rheodyne model 7125 injector equipped with a 10-µL sample loop. The stationary phase was a Pye Unicam RP18 analytical column (100 × 4.6 mm, 10-µm particle size) with a Pye Unicam RP18 precolumn (30 × 4.6 mm, 10-µm particle size). The mobile phase was 45% tetrahydrofuran in 0.02 M sodium dihydrogen phosphate aqueous buffer with the pH adjusted to 4.0. The flow rate was 1.0 mL/min. UV detection was performed at 219 nm and 0.008 AUFS.

Ibuprofen samples were dissolved in tetrahydrofuran, diluted with 0.02 M sodium dihydrogen phosphate (pH 4.0), and filtered. The injection volume was 10 µL. Under these conditions, the retention time for ibuprofen was 6.58 minutes.

The HPLC method was determined to be stability indicating. Ibuprofen solutions in 0.1 M hydrochloric acid and 1 M sodium hydroxide were boiled for 48 hours. In both cases, one degradation peak was obtained just before the intact ibuprofen peak.

Calibration curves for ibuprofen were constructed from 125 to 750 ng. The correlation coefficient was 1.000.

Reference

Haikala VE, Heimonen IK, Vuorela HJ. Determination of ibuprofen in ointments by reversed-phase liquid chromatography. *J Pharm Sci*. 1991; 80: 456–8.

Method 4

George and Contario described an HPLC method for the determination of terfenadine, pseudoephedrine hydrochloride, and ibuprofen in a liquid formulation. The liquid chromatographic system comprised a Waters model M6000A pump, a model 710B WISP autosampler, a Spectroflow 757 variable-wavelength detector, and a Beckman CALS data system. The stationary phase was a Phase Separations Spherisorb ODS-2 column (100 × 4.6 mm, 3-µm particle size). The mobile phase consisted of 400 mL of water, 1.0 g of monobasic sodium phosphate monohydrate, 0.5 g of phosphoric acid (85%), 3.0 g of sodium perchlorate monohydrate, and 600 mL of acetonitrile. The flow rate was 1.5 mL/min. UV detection was performed at 210 nm and 0.32 AUFS.

Samples were diluted with methanol and then with the mobile phase. The injection volume was 20 µL. Under these conditions, retention times for pseudoephedrine, ibuprofen, and terfenadine were about 1.1, 2.4, and 4.4 minutes, respectively (estimated from the published chromatogram).

The stability-indicating nature of the method was established by intentional degradation of ibuprofen. Samples of ibuprofen were treated with 0.1 M hydrochloric acid at 90 °C for up to 72 hours. The intact ibuprofen was well resolved from its degradation products.

Reference

George RC, Contario JJ. Quantitation of terfenadine, pseudoephedrine hydrochloride, and ibuprofen in a liquid animal dosing formulation using high performance liquid chromatography. *J Liq Chromatogr*. 1988; 11: 475–88.

Ifosfamide

Chemical Names

N,3-Bis(2-chloroethyl)tetrahydro-2*H*-1,3,2-oxazaphosphorin-2-amine 2-oxide

3-(2-Chloroethyl)-2-[(2-chloroethyl)amino]tetrahydro-2*H*-1,3,2-oxazaphosphorin-2-oxide

Other Names

Ifex, Isophosphamide

Form	Molecular Formula	MW	CAS
Ifosfamide	$C_7H_{15}Cl_2N_2O_2P$	261.1	3778-73-2

Appearance

Ifosfamide is an off-white to white crystalline powder.

Solubility

Ifosfamide is soluble in water.

Method 1

Mayron and Gennaro used an HPLC method to assess the stability of ifosfamide with granisetron hydrochloride during simulated Y-site administration. The liquid chromatograph consisted of a piston pump with a pulse dampener, a rotary injection port with a 20-μL loop, a variable-wavelength UV detector, and an integrator. The stationary phase was a C_{18} reversed-phase column (300 × 4.6 mm, 5-μm particle size). The mobile phase was a mixture of acetonitrile and 0.1 M monobasic sodium phosphate dihydrate (20:80, vol/vol) adjusted to pH 4.2 with phosphoric acid. The flow rate was 1.80 mL/min. UV detection was performed at 198 nm.

Samples were diluted with mobile phase. The retention times for ifosfamide and granisetron were 2.89 and 4.86 minutes, respectively.

The stability-indicating capability of the analytical method was demonstrated by an accelerated degradation study. Solutions of ifosfamide and granisetron hydrochloride were adjusted to pH 2 and 11, boiled for 1 hour, readjusted to pH 5, diluted with the mobile phase, and analyzed. The degradation product peaks did not interfere with the intact ifosfamide or granisetron peaks.

Reference

Mayron D, Gennaro AR. Stability and compatibility of granisetron hydrochloride in i.v. solutions and oral liquids and during simulated Y-site injection with selected drugs. *Am J Health Syst Pharm.* 1996; 53: 294–304.

Method 2

Munoz et al. determined the stability of ifosfamide in 0.9% sodium chloride solution and water for injection in a portable intravenous pump cassette. The HPLC system consisted of an isocratic pump, a Hewlett-Packard model 1090 diode-array UV detector, and a Hewlett-Packard model 3396D integrator. The stationary phase was a Hewlett-Packard Hypersil ODS column (100 × 4.6 mm, 5-μm particle size). The mobile phase consisted of 30% acetonitrile and 70% double-distilled water. The flow rate was 1.5 mL/min. UV detection was performed at 210 nm.

Samples were diluted with distilled water before injection. Under these conditions, ifosfamide had a retention time of 1.5 minutes.

The stability-indicating nature of the assay was verified by accelerated decomposition of ifosfamide. One ifosfamide sample was heated at 50 °C for 6 days. Another sample was adjusted to pH 1.5 with 0.1 M hydrochloric acid, and the third sample was adjusted to pH 11.8 with 0.1 M sodium hydroxide. In all cases, the degradation product peaks were resolved from the intact ifosfamide peak.

Standard curves were constructed and were linear ($r > 0.99$) from 1 to 10 mg/mL. The intraday and interday coefficients of variation were less than 3%.

A similar analytical method was used by Radford et al.

References

Munoz M, Girona V, Pujol M, et al. Stability of ifosfamide in 0.9% sodium chloride solution or water for injection in a portable i.v. pump cassette. *Am J Hosp Pharm.* 1992; 49: 1137–9.

Radford JA, Margison JM, Swindell R, et al. The stability of ifosfamide in aqueous solution and its suitability for continuous 7-day infusion by ambulatory pump. *J Cancer Res Clin Oncol.* 1991; 117: S154–6.

Imexon

Chemical Name

4-Imino-1,3-diazabicyclo[3.1.0]hexan-2-one

Form	Molecular Formula	MW	CAS
Imexon	$C_4H_5N_3O$	111.1	59643-91-3

Method

Den Brok et al. reported the development and validation of an assay for the investigational anticancer agent imexon and identification of its degradation products. The liquid chromatograph consisted of an Agilent 1100 series binary pump, a Spectra series AS3000 autosampler with a 100-µL loop, and a Waters 996 photodiode-array detector. The stationary phase was a Rockland Technologies Zorbax Bonus RP analytical column (150 × 4.6 mm, 5-µm particle size) coupled with a guard column with a reversed-phase Chrompack material. The mobile phase was 50 mM sodium phosphate buffer (pH 6). The flow rate was 0.8 mL/min. UV detection was carried out at 230 nm.

Imexon standard solution was prepared in 20 mM sodium phosphate buffer (pH 7.4). For the dosage form, imexon lyophilized powder for intravenous administration was also dissolved in 20 mM phosphate buffer (pH 7.4). The injection volume was 10 µL. Under these conditions, the retention time of imexon was about 3.4 minutes.

To test the stability-indicating nature of the method, imexon stock solutions were exposed to heat at 100 °C, 30% hydrogen peroxide solution, 1 M hydrochloric acid, and 1 M sodium hydroxide solution. Degradation products did not interfere with the analysis of imexon. LC–MS also confirmed the specificity of the method.

A linear standard curve was constructed from 1.0 to 25 µg/mL. The correlation coefficient was greater than 0.9999. Within-run precision and between-run precision were less than 1.7 and 1.3%, respectively.

Reference

Den Brok MWJ, Nuijen B, Hillebrand MJX, et al. LC-UV method development and validation for the investigational anticancer agent imexon and identification of its degradation products. *J Pharm Biomed Anal.* 2005; 38: 686–94.

Imipenem

Chemical Name

[5*R*-[5α,6α(*R**)]]-6-(1-Hydroxyethyl)-3-[[2-[(iminomethyl)amino]ethyl]thio]-7-oxo-1-azabicyclo[3.2.0]hept-2-ene-2-carboxylic acid monohydrate

Other Name

Component of Primaxin

Form	Molecular Formula	MW	CAS
Imipenem	$C_{12}H_{17}N_3O_4S.H_2O$	317.4	74431-23-5

Appearance

Imipenem monohydrate is a white or off-white and nonhygroscopic crystalline compound.

Solubility

Imipenem has the following solubilities, given in milligrams per milliliter in parentheses: water (10), methanol (5), ethanol (0.2), acetone (<0.1), dimethylformamide (<0.1), and dimethyl sulfoxide (0.3).

Method 1

Using an HPLC method, Trissel and Xu evaluated the stability of imipenem in imipenem–cilastatin admixtures in AutoDose infusion system bags. A Waters Alliance 2690 separation model was coupled with a Waters model 996 photodiode-array detector to perform the analysis. The stationary phase was an Alltech Spherisorb CN column (250 × 4.6 mm, 5-μm particle size) coupled with a guard column of the same material. The mobile phase was a 0.010 M monobasic potassium phosphate aqueous solution. It was delivered isocratically at 0.8 mL/min. UV detection was performed at 300 nm and 0.5 AUFS.

Samples were diluted appropriately with mobile phase to a nominal imipenem concentration of 0.1 mg/mL for analysis. Injection volume was 10 μL. Under these conditions, the principal imipenem peak eluted at 6 minutes. Cilastatin had no UV absorption at 300 nm.

To demonstrate the stability-indicating nature of the assay, imipenem solution was mixed with 1 N sodium hydroxide solution, 1 N hydrochloric acid, and 3% hydrogen peroxide and subjected to heating. Loss of the intact imipenem was observed and there was no interference by the degradation product peaks with the principal imipenem peak.

A standard curve for imipenem was constructed from 25 to 150 μg/mL; the correlation coefficient was greater than 0.9997. The intraday and interday coefficients of variation were 0.4 and 3.3%, respectively.

Reference

Trissel LA, Xu QA. Stability of imipenem–cilastatin sodium in AutoDose infusion system bags. *Hosp Pharm.* 2003; 38: 130–4.

Method 2

Hall et al. studied the compatibility of imipenem and cilastatin with filgrastim during simulated Y-site administration. The liquid chromatograph consisted of a Waters model 510 pump, a model Ultra WISP 715 autosampler, a model 481 UV detector, and a Shimadzu model C-R3A integrator. The stationary phase was a Waters μBondapak C_{18} column. The mobile phase was 8% acetonitrile in 0.02 M dibasic potassium phosphate (adjusted to pH 7.0 with 10% phosphoric acid). The flow rate was 2 mL/min. UV detection was performed at 254 nm. Under these conditions, the retention times for imipenem and cilastatin were 2.0 and 6.7 minutes, respectively.

The stability-indicating properties of the method were demonstrated by intentional degradation of drugs. Samples of imipenem and cilastatin were acidified to pH 2, alkalinized to pH 8, or heated at 30 °C for 30 minutes. In all cases, degradation products did not influence the determination of imipenem.

A standard curve for imipenem was constructed from 1.0 to 3.0 mg/mL. Intrarun and interrun coefficients of variation were less than 3.1%.

Reference

Hall PD, Yui D, Lyons S, et al. Compatibility of filgrastim with selected antimicrobial drugs during simulated Y-site administration. *Am J Health Syst Pharm.* 1997; 54: 185–9.

Method 3

Using one of two HPLC systems, Allen et al. determined the stability of imipenem in the latex reservoir of an elastomeric infusion device. One system consisted of a Waters model 501 pump, a Waters model 441 UV detector, a Waters model 745 data module, and a Rheodyne 7010 injector. The other system was composed of a Shimadzu model LC-6A pump, a Shimadzu model SPD-6A UV detector, a Shimadzu model CR-601 data module, a Shimadzu model SIL-9A autosampler, and a Rheodyne 7010 injector. The stationary phase was a J.T. Baker C_{18} column. The mobile phase contained 0.54 g of monobasic potassium phosphate in 3600 mL of water; it was adjusted with 0.5 N sodium hydroxide or 0.5 M phosphoric acid to a pH of 6.8 and diluted with water to make 4000 mL of solution. The flow rate was 1.5 mL/min. UV detection was performed at 300 nm.

Samples were diluted 1:50 with mobile phase. Under these conditions, the retention time for imipenem was 5.5 minutes.

To show the stability-indicating nature of the assay, imipenem was degraded with heat (to boiling) or acid (1 N sulfuric acid to pH 1.5–2.0). Degradation product peaks did not interfere with the intact drug peak. Retention times for the degradation products were 1.3 and 3.0 minutes.

A standard curve was constructed from 25 to 125 µg/mL; the correlation coefficient was greater than 0.99. Intraday and interday coefficients of variation for the assay were 1.1%.

Reference

Allen LV Jr, Stiles ML, Prince SJ, et al. Stability of 14 drugs in the latex reservoir of an elastomeric infusion device. *Am J Health Syst Pharm.* 1996; 53: 2740–3.

Method 4

Zaccardelli et al. used an HPLC method to study the stability of imipenem and cilastatin sodium in total parenteral nutrient solution. The liquid chromatograph consisted of a Rheodyne model 7125 injector, a Waters model 6000A solvent-delivery system, a Waters model Lambda-Max 481 UV detector, and a Hewlett-Packard model 3390A integrator. The stationary phase was a Hewlett-Packard RP8 column (200 × 4.6 mm). The mobile phase consisted of 0.0004 M 3-(N-morpholino)propanesulfonic acid buffer with sodium hexane sulfate (2 g), acetonitrile (4 mL), and methanol (5 mL) per liter. The mobile phase was adjusted to pH 7.00 with sodium hydroxide. The flow rate was 1.8 mL/min. UV detection was performed at 250 nm and 0.1 AUFS.

Samples were diluted 1:100 with mobile phase. The injection volume was 30 µL. Under these conditions, retention times for imipenem and cilastatin were 5.5 and 8.8 minutes, respectively.

The method was determined to be stability indicating by accelerated degradation of imipenem–cilastatin sodium. Standard solutions were adjusted to pH 1.8 with phosphoric acid, evaporated on a hot plate, allowed to dry at excessive heat before reconstitution to the initial volume with water, and heated on a boiling water bath for 15 minutes. Degradation product peaks did not interfere with the intact drug peaks.

A standard curve for imipenem was constructed each day from 0 to 86.4 µg/mL; the correlation coefficient was 0.9995.

Reference

Zaccardelli DS, Krcmarik CS, Wolk R, et al. Stability of imipenem and cilastatin sodium in total parenteral nutrient solution. *J Parenter Enter Nutr.* 1990; 14: 306–9.

Inamrinone

Chemical Names

5-Amino-[3,4'-bipyridin]-6(1*H*)-one

3-Amino-5-(4-pyridinyl)-2(1*H*)-pyridinone

Other Names

Inocar, Inocor

Form	Molecular Formula	MW	CAS
Inamrinone	$C_{10}H_9N_3O$	187.2	60719-84-8
Inamrinone lactate	$C_{10}H_9N_3O.C_3H_6O_3$	277.3	75898-90-7

Appearance

Inamrinone is a pale yellow crystalline powder. Inamrinone lactate injection occurs as a clear yellow solution.

Solubility

Inamrinone is insoluble in water and slightly soluble in alcohol.

Method

Using a stability-indicating HPLC method, Riley and Junkin studied the stability of inamrinone (amrinone) in intravenous admixtures. The system consisted of a Kratos Spectroflow 400 solvent-delivery system, a Waters model 450 variable-wavelength UV detector, a Micromeritics model 728 autosampler, a Valco VICI six-port electrically actuated injection valve fitted with a 20-μL loop, a Fisher Scientific series 5000 Recordall strip-chart recorder, and a Shimazdu model C-R3A electronic integrator.

Two chromatographic systems were used. For the first, the stationary phase was an Alltech Partisil ODS III analytical column (250 × 4.6 mm, 10-μm particle size). The mobile phase was a mixture of methanol, water, and 0.5 M borate (40:58:2, vol/vol/vol) and had a pH of 7.0. The flow rate was 2 mL/min. UV detection was performed at 254 nm.

The other chromatographic system consisted of a Phase Separations Spherisorb phenyl column (150 × 4.6 mm, 5-μm particle size). The mobile phase was a mixture of acetonitrile, water, 100 mM tetrabutylammonium hydrogen sulfate, and 0.5 M potassium dihydrogen phosphate (15:50:25:10, vol/vol/vol/vol) and had a pH of 5.1. The flow rate was 2 mL/min. UV detection was performed at 268 nm. Samples were diluted 1:10 with the mobile phase.

These methods were determined to be stability indicating and were free from interference by formulatory excipients and degradation products.

The peak areas were linearly related to the concentration from 0.1 to 3.75 mg/mL. The coefficient of variation was less than 1%.

Reference

Riley CM, Junkin P. Stability of amrinone and digoxin, procainamide hydrochloride, propranolol hydrochloride, sodium bicarbonate, potassium chloride, or verapamil hydrochloride in intravenous admixtures. *Am J Hosp Pharm.* 1991; 48: 1245–52.

Indapamide

Chemical Names

3-(Aminosulfonyl)-4-chloro-*N*-(2,3-dihydro-2-methyl-1*H*-indol-1-yl)benzamide

4-Chloro-*N*-(2-methyl-1-indolinyl)-3-sulfamoylbenzamide

Other Name

Lozol

Form	Molecular Formula	MW	CAS
Indapamide	$C_{16}H_{16}ClN_3O_3S$	365.8	26807-65-8

Appearance

Indapamide is a white to yellow-white crystalline powder.

Solubility

Indapamide is practically insoluble in water. It is soluble in alcohol, methanol, acetonitrile, acetic acid, and ethyl acetate and is very slightly soluble in ether and chloroform.

pK$_a$

Indapamide has a pK$_a$ of 8.8.

Method

Padval and Bhargava reported the liquid chromatographic determination of indapamide in the presence of its degradation products. The HPLC system consisted of a Waters model 6000A pump, a Waters model 710 WISP autosampler, a Waters model 450 variable-wavelength UV detector, and a Hewlett-Packard model 3390A integrator-plotter. The stationary phase was an Alltech LiChrosorb RP18 column (250 × 4.6 mm, 5-μm particle size). The mobile phase consisted of methanol and aqueous acetic acid (1%) with triethylamine (0.2%) (50:50, vol/vol). The flow rate was 1.0 mL/min. UV detection was performed at 250 nm and at 0.02 AUFS.

Samples were diluted 1:100 with mobile phase. The injection volume was 20 μL. Under these conditions, the retention time for indapamide was 10.99 minutes.

The analytical method was determined to be stability indicating by accelerated decomposition of indapamide. Indapamide was degraded with 0.01 N hydrochloric acid and 0.01 N sodium hydroxide at 55 °C for 6 months. Degradation products were *N*-amino-2-methyl-indoline and 4-chloro-3-sulfamoylbenzoic acid, which eluted at 3.40 and 14.16 minutes, respectively.

A standard curve for indapamide was constructed from 1 to 50 μg/mL; the correlation coefficient was 0.9999. The coefficient of variation was 1.21%. The limit of detection was 0.5 μg/mL.

Reference
Padval MV, Bhargava HN. Liquid chromatographic determination of indapamide in the presence of its degradation products. *J Pharm Biomed Anal*. 1993; 11: 1033–6.

Indomethacin

Chemical Name
1-(4-Chlorobenzoyl)-5-methoxy-2-methyl-1*H*-indole-3-acetic acid

Other Names
Indocin, Indometacin

Form	Molecular Formula	MW	CAS
Indomethacin	$C_{19}H_{16}ClNO_4$	357.8	53-86-1
Indomethacin sodium trihydrate	$C_{19}H_{15}ClNNaO_4 \cdot 3H_2O$	433.8	74252-25-8

Appearance
Indomethacin occurs as a pale yellow to yellow-tan crystalline powder with a slight odor. Indomethacin sodium trihydrate occurs as a lyophilized white to yellow powder or plug.

Solubility
Indomethacin is practically insoluble in water and sparingly soluble in alcohol. Indomethacin sodium trihydrate is soluble in water and in alcohol.

pK$_a$

Indomethacin has a pK$_a$ of 4.5.

Method

Walker et al. studied the stability of reconstituted indomethacin sodium trihydrate in original vials and in polypropylene syringes. The liquid chromatograph consisted of a Spectra-Physics model P100 pump, a Waters model 715 WISP autoinjector, a Hewlett-Packard model 1050 variable-wavelength UV detector, and a Spectra-Physics model 4240 chromatographic integrator. The stationary phase was a Beckman Ultrasphere ODS C$_{18}$ column (250 × 4.2 mm, 5-µm particle size). The mobile phase consisted of acetonitrile and 0.05 M phosphoric acid (60:40, vol/vol). The flow rate was 1.0 mL/min. UV detection was performed at 260 nm. The sample injection volume was 5 µL. Under these conditions, the retention time for indomethacin was 6.5–7 minutes.

The HPLC method was determined to be stability indicating by accelerated decomposition of indomethacin. At pH 2.2 (adjusted with 0.05 M phosphoric acid) and 79–80 °C, the indomethacin concentration decreased to 38% of the initial concentration. At least three degradation products appeared in the first 5 minutes. At a pH greater than 12 (adjusted with 1.0 or 0.1 M sodium hydroxide) and room temperature, indomethacin completely degraded in 14 minutes. In each case, degradation product peaks did not interfere with the intact indomethacin peak.

A standard curve for indomethacin was constructed from 0.125 to 0.75 mg/mL. Intraday and interday coefficients of variation averaged 1.46 and 4.3%, respectively.

Reference

Walker SE, Gray S, Schmidt B. Stability of reconstituted indomethacin sodium trihydrate in original vials and polypropylene syringes. *Am J Health Syst Pharm.* 1998; 55: 154–8.

Insulin

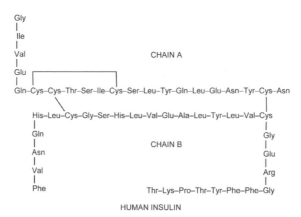

HUMAN INSULIN

Other Names

Actrapid, Humalin BR, Humalog

Form	Molecular Formula	MW	CAS
Insulin (bovine)	$C_{254}H_{377}N_{65}O_{75}S_6$	5733.6	11070-73-8
Insulin human	$C_{257}H_{383}N_{65}O_{77}S_6$	5807.7	11061-68-0
Insulin (porcine)	$C_{256}H_{381}N_{65}O_{76}S_6$	5777.6	12584-58-6

Method 1

Oliva et al. described an analysis of insulin by reversed-phase HPLC. The chromatographic apparatus consisted of a Waters model 600E multisolvent-delivery system, a model 490E programmable multiwavelength detector, and Maxima 820 data acquisition software. The stationary phase was a Waters Delta Pack reversed-phase C_8 column (100 × 8 mm, 300-Å pore size). The column temperature was 40 °C. The mobile phase consisted of 25% acetonitrile and 75% 0.25 M sulfate buffer (pH 2.3). The flow rate was 1.0 mL/min. UV detection was performed at 214 nm.

Samples of insulin in solution were precipitated by 300 µL of 0.01 M zinc acetate and stored at 4 °C overnight. Supernatants were discarded and the residues were digested. Under these conditions, the retention time for human insulin was 54 minutes.

The analytical method was reported to be stability indicating.

Reference

Oliva A, Farina J, Llabres M. Fingerprint analysis of insulin: Application in stability studies of pharmaceutical preparations. *Drug Dev Ind Pharm*. 1997; 23: 127–32.

Method 2

Hoyer et al. presented an HPLC method for separating recombinant human regular insulin from insulin breakdown and transformation products. The liquid chromatograph consisted of a Spectra-Physics Isochrom LC pump, a detector, a model SP8880 autosampler, and a ChromJet integrator. The stationary phase was a Vydac ODS silica column (250 × 4.6 mm, 5-µm particle size) with 300-Å pore size for the separation of proteins and peptides. The mobile phase consisted of 0.05 M monobasic potassium phosphate (pH 2.4) and acetonitrile (75:25, vol/vol) and was isocratically delivered at 1 mL/min. UV detection was performed at 230 nm. Benzoic acid 50 µg/mL in water was used as an internal standard.

Samples were mixed with the internal standard and diluted with water. The injection volume was 50 µL. Under these conditions, the retention time for insulin was 11.2 minutes.

The stability-indicating nature of the method was demonstrated by accelerated degradation of insulin. Solutions of insulin 0.5 unit/mL were exposed to each one of the following seven conditions for 72 hours: freezing at −15 °C, heat at 75 °C, three cycles of freezing and heating to 75 °C at 24 hours per cycle, pH < 1, pH > 13, room temperature in darkness for 10 hours per day, and room temperature under fluorescent light for 10 hours per day. The intact insulin peak was separated from its degradation and transformation product peaks on its chromatogram.

A standard curve for insulin was constructed from 0.2 to 2.5 units/mL. Intraday and interday coefficients of variation were 4.2 and 6.4%, respectively.

Reference

Hoyer GL, Nolan PE Jr, LeDoux JH, et al. Selective stability-indicating high-performance liquid chromatographic assay for recombinant human regular insulin. *J Chromatogr A*. 1995; 699: 383–8.

Method 3

Adams and Haines-Nutt used HPLC to evaluate the stability of insulins stored in syringes. The liquid chromatographic system consisted of an ACS model 740 pump, a Perkin-Elmer model LC75 variable-wavelength detector, a Rheodyne model 7125 injector, and a Hewlett-Packard 3900A integrator. The stationary phase was a Perkin-Elmer ODS column (30 × 5.2 mm, 3-μm particle size). The mobile phase consisted of a buffer and acetonitrile (78:27.5, vol/vol). The buffer was made of 0.1 M ammonium sulfate and 0.005 M tartaric acid, adjusted to pH 3.5 with sulfuric acid. The mobile phase also contained 0.002% cetrimide. The flow rate was 2 mL/min. UV detection was performed at 215 nm.

Soluble insulins were appropriately diluted with water for injection and insoluble insulins were dissolved in and appropriately diluted with 0.1 M hydrochloric acid. The injection volume was 20 μL. Under these conditions, retention times for human and porcine insulin were about 4.5 and 5 minutes, respectively.

The method was reported to be stability indicating since it separated insulin from its deamidation product, desamidoinsulin, when subjected to accelerated degradation by storage at 55 °C.

Standard curves for insulins were constructed from 0 to 80 units/mL. The correlation coefficients were 0.98.

Reference

Adams PS, Haines-Nutt RF. Analysis of bovine, porcine and human insulins in pharmaceutical dosage forms and drug delivery systems. *J Chromatogr.* 1986; 351: 574–9.

Ipratropium Bromide

Chemical Names

(*endo,syn*)-(±)-3-(3-Hydroxy-1-oxo-2-phenylpropoxy)-8-methyl-8-(1-methylethyl)-8-azoniabicyclo[3.2.1]octane bromide

(8*r*)-3α-Hydroxy-8-isopropyl-1α*H*,5α*H*-tropanium bromide (±)-tropate

Other Name
Atrovent

Form	Molecular Formula	MW	CAS
Ipratropium bromide	$C_{20}H_{30}BrNO_3$	412.4	22254-24-6

Appearance
Ipratropium bromide is a white bitter-tasting crystalline powder.

Solubility
Ipratropium bromide has solubilities of 90 mg/mL in water and 28 mg/mL in alcohol.

Method 1
Simms et al. reported the separation of ipratropium bromide and its related compounds by HPLC analysis. The Hitachi liquid chromatograph consisted of a model L-7100 low-pressure gradient pump, a model L-7200 sequential autosampler with a 100-µL loop, a model L-7300 column oven, and a model L-7400 programmable UV detector or model L-7450 photodiode-array detector. The stationary phase was an Alltech Alltima C_{18} column (250 × 4.6 mm, 5-µm particle size). The column temperature was maintained at 35 °C. Mobile phase A consisted of 800 mL of 100 mM monobasic potassium phosphate buffer (pH 4.0) and 200 mL of acetonitrile. Mobile phase B was a mixture of 550 mL of 100 mM monobasic potassium phosphate buffer (pH 4.0) and 450 mL of acetonitrile. The mobile phase was gradiently delivered as follows:

Time, minutes	A, %	B, %	Flow Rate, mL/min
0	100	0	1
5	100	0	1
8	0	100	1
13	0	100	1
14	100	0	2
20	100	0	2

UV detection was performed at 210 nm.

Each sample was diluted with 1×10^{-4} N hydrochloric acid, sonicated, and filtered through a 0.2-µm nylon filter. The injection volume was 20 µL. Under these conditions, the retention time for ipratropium was about 7.9 minutes (estimated from the published chromatogram).

The method was evaluated to be stability indicating by accelerated degradation of ipratropium. A sample solution was treated with hydrochloric acid and heated at 95 °C for 5 days. The degradation products did not interfere with the analysis of ipratropium.

A standard curve for ipratropium was constructed from 10 to 1000 µg/mL. The correlation coefficient was 0.9992. The coefficient of variation for the analysis of ipratropium was less than 2.0%. The limit of detection was 60 ng/mL.

Reference
Simms PJ, Towne RW, Gross CS, et al. The separation of ipratropium bromide and its related compounds. *J Pharm Biomed Anal*. 1998; 17: 841–9.

Method 2

Jacobson and Peterson developed an HPLC assay for the simultaneous determination of ipratropium bromide, fenoterol, albuterol, and terbutaline in nebulizer solutions. The liquid chromatograph consisted of a Varian model 9010 solvent-delivery system, a Rheodyne model 7161 injector with a 10-µL external loop, a Varian model 9050 variable-wavelength UV-visible detector, and a Varian GC Star workstation. The stationary phase was a Waters Nova-Pak C_{18} Radial-Pak cartridge (100 × 8 mm, 4-µm particle size) inside a Waters RCM 8 × 10 compression module. Mobile phase A consisted of tetrahydrofuran and distilled water (40:60, vol/vol) containing 0.0025 M Waters Pic B-8 Reagent Low UV. Mobile phase B was distilled water and mobile phase C consisted of methanol and distilled water (50:50, vol/vol). The flow rate was 2.0 mL/min. A mixture of 50% mobile phase A and 50% mobile phase B was delivered up to 7.7 minutes; then the composition of the mixture was changed linearly to 60% mobile phase A, 15% mobile phase B, and 25% mobile phase C at 13.0 minutes. The run time was 13.0 minutes with a 5.0-minute equilibration time. Mepivacaine hydrochloride 1% was used as an internal standard. The injection volume was 20 µL. Retention times for albuterol, terbutaline, ipratropium, mepivacaine, and fenoterol were 3.2, 4.3, 5.9, 8.2, and 12.7 minutes, respectively.

  The assay was determined to be stability indicating by accelerated decomposition of ipratropium with heat, hydrogen peroxide, acid, and base. In all cases, degradation product peaks did not interfere with the intact ipratropium peak.

Reference

Jacobson GA, Peterson GM. High-performance liquid chromatographic assay for the simultaneous determination of ipratropium bromide, fenoterol, salbutamol, and terbutaline in nebulizer solution. *J Pharm Biomed Anal.* 1994; 12: 825–32.

Irinotecan Hydrochloride

Chemical Name

(*S*)-4,11-Diethyl-3,4,12,14-tetrahydro-4-hydroxy-3,14-dioxo-1*H*-pyrano[3′,4′:6,7]-indolizino[1,2-*b*]quinolin-9-yl [1,4′-dipiperidine]-1′-carboxylate monohydrochloride trihydrate

• HCl • 3H₂O

Other Names
Campto, Camptosar

Form	Molecular Formula	MW	CAS
Irinotecan hydrochloride trihydrate	$C_{33}H_{38}N_4O_6 \cdot HCl \cdot 3H_2O$	677.2	136572-09-3

Appearance
Irinotecan hydrochloride occurs as a pale yellow to yellow crystalline powder.

Solubility
Irinotecan hydrochloride is slightly soluble in water and organic solvents.

Method
Thiesen and Kramer evaluated the physicochemical stability of irinotecan injection concentrate and diluted infusion solutions in polyvinyl chloride bags. The liquid chromatograph was composed of a Waters model 717 plus autosampler, a model 510 pump, and a model 996 photodiode-array detector. The stationary phase was a Supelco Suplex C_{18} reversed-phase column (250 × 4.6 mm, 5-μm particle size). The mobile phase was a mixture of 0.01 M monobasic potassium phosphate, methanol, and acetonitrile (54:27:19, vol/vol/vol) containing 1.22 g of sodium 1-decasulfonate per liter. The flow rate was 1.2 mL/min. UV detection was performed at 254 nm. The run time was 25 minutes.

Samples were diluted appropriately before assaying. The injection volume was 10 μL. Under these conditions, the retention time for irinotecan was about 15 minutes.

The method was shown to be stability indicating by accelerated degradation of irinotecan. Irinotecan samples were treated with acid (pH 1) for 3 hours, with base (pH 14) for 3 hours, or exposed to daylight and fluorescent light for 14 days. Degradation products did not interfere with the determination of irinotecan.

A calibration curve for irinotecan was generated from 0.175 to 1.225 mg/mL. The correlation coefficient was 0.9999.

Reference
Thiesen J, Kramer I. Physicochemical stability of irinotecan injection concentrate and diluted infusion solutions in PVC bags. *J Oncol Pharm Practice*. 2000; 6: 115–21.

Isoniazid

Chemical Names
Isonicotinic acid hydrazide
4-Pyridinecarboxylic acid hydrazide

Other Names
Dipasic, Inapsade, Laniazid, Nydrazid, Rimifon

Form	Molecular Formula	MW	CAS
Isoniazid	$C_6H_7N_3O$	137	54-85-3

Appearance
Isoniazid occurs as colorless odorless crystals or as a white crystalline powder.

Solubility
Isoniazid is freely soluble in water, sparingly soluble in alcohol, slightly soluble in chloroform, and very slightly soluble in ether.

Method 1
Glass et al. optimized an assay for the simultaneous determination of isoniazid, rifampicin, and pyrazinamide in a fixed-dose combination. The Waters liquid chromatograph included a WISP model 6000A solvent-delivery system, a WISP model 710B autosampler, a model 481 Lambda Max UV detector, and a data module. The stationary phase was a Waters µBondapak analytical column (250 × 4.6 mm, 10-µm particle size). The mobile phase consisted of acetonitrile and 0.0002 M tetrabutylammonium hydroxide solution (42.5:57.5, vol/vol) (pH 3.1). The flow rate was 1.0 mL/min. The eluent was monitored at 260 nm.

Isoniazid, rifampicin, and pyrazinamide standard solutions were prepared in water. The injection volume was 20 µL. Under these conditions, the retention times of rifampicin, isoniazid, and pyrazinamide were 2.85, 3.54, and 10.97 minutes, respectively.

The method was stated to be stability indicating.

A standard curve was obtained with a correlation coefficient of greater than 0.9998. The limit of detection was 0.150 µg/mL.

Reference
Glass BD, Agatonovic-Kustrin S, Chen Y-J., et al. Optimization of a stability indicating HPLC method for the simultaneous determination of rifampicin, isoniazid, and pyrazinamide in fixed-dose combination using artificial neural networks. *J Chromatogr Sci.* 2007; 45: 38–44.

Method 2
Gupta and Sood developed an HPLC method for the quantitation of isoniazid, and they studied the stability of isoniazid in an oral liquid dosage form. The liquid chromatographic system consisted of a Waters model ALC202 pump, a model 484 multiple-wavelength detector, a Rheodyne model 7125 injector, and an Omniscribe recorder. The stationary phase was a Beckman Ultrasphere C_{18} column (150 × 4.6 mm, 5-µm particle size). The mobile phase consisted of acetonitrile and 0.02 M potassium dihydrogen phosphate aqueous buffer (3:97, vol/vol), pH 4.9. The flow rate was 1.0 mL/min. UV detection was carried out at 265 nm and 0.28 AUFS.

An oral sample of isoniazid was diluted with water to about 50 µg/mL. The sample injection volume was 80 µL. Under these conditions, the retention time of isoniazid was about 3.0 minutes (estimated from the published chromatogram).

To confirm the stability-indicating nature of the method, isoniazid solutions were boiled in 1 N sodium hydroxide solution or 1 N sulfuric acid for 5 minutes, cooled down, neutralized, and assayed. The degradation products of isoniazid did not interfere with the determination of the intact drug.

Reference

Gupta VD, Sood A. Chemical stability of isoniazid in an oral liquid dosage form. *Int J Pharm Compound.* 2005; 9: 165–6.

Isoproterenol

Chemical Names

4-[1-Hydroxy-2-[(1-methylethyl)amino]ethyl]-1,2-benzenediol
3,4-Dihydroxy-α-[(isopropylamino)methyl]benzyl alcohol

Other Names

Isoprenaline, Isuprel

Form	Molecular Formula	MW	CAS
Isoproterenol	$C_{11}H_{17}NO_3$	211.3	7683-59-2
Isoproterenol hydrochloride	$C_{11}H_{17}NO_3.HCl$	247.7	51-30-9
Isoproterenol sulfate	$(C_{11}H_{17}NO_3)_2.H_2SO_4.2H_2O$	556.6	6700-39-6

Appearance

Isoproterenol hydrochloride and isoproterenol sulfate occur as white to practically white crystalline powders.

Solubility

Isoproterenol hydrochloride and isoproterenol sulfate are freely soluble in water. Isoproterenol hydrochloride is sparingly soluble in alcohol and isoproterenol sulfate is very slightly soluble in alcohol.

Method 1

Smith et al. investigated the effects of ascorbic acid and edetate disodium, alone and in combination, on the stability of isoproterenol hydrochloride in injections after sterilization by autoclaving or filtration and subsequent storage for 1 year. The chromatograph consisted of a Milton-Roy model 110 double reciprocating pump, a septum injection, a Cecil Instruments model 212 variable-wavelength UV detector, a Servoscribe RES 11-20 linear potentiometric recorder, and a Belmont Instruments model 308 computer integrator. The stationary phase was a Shandon ODS Hypersil-5 column (100 × 4.5 mm). The mobile phase consisted of methanol, acetic acid, sodium lauryl sulfate, and water

(30:2:0.002:68, vol/vol/vol/vol). The flow rate was 1.4 mL/min. The injection volume was 5 μL. UV detection was performed at 280 nm. Epinephrine bitartrate solution 0.1% was used as an internal standard.

The method was stated to be stability indicating since it separated the intact isoproterenol from its degradation products (*N*-isopropylnoradrenochrome and *N*-isopropylnoradrenolutin), an impurity (isoproterenone), antioxidants (ascorbic acid and sodium metabisulfite), and an antioxidant synergist (edetate disodium).

The peak area ratio of isoproterenol to that of epinephrine was related rectilinearly to the concentration of isoproterenol hydrochloride from 0.1 to 0.8 mg/mL; the coefficient of variation of the 0.5-mg/mL solution was 0.6%.

Reference
Smith G, Hasson K, Clements JA. Effects of ascorbic acid and disodium edetate on the stability of isoprenaline hydrochloride injection. *J Clin Hosp Pharm.* 1984; 9: 209–15.

Method 2
Williams et al. described a stability-indicating ion-pair HPLC method for isoproterenol hydrochloride in the presence of aminophylline in intravenous solutions. The HPLC system included a Waters model M6000A solvent-delivery system, a Waters model U6K injector, and a Perkin-Elmer model 204A spectrofluorometer or model LC-55 spectro-photometer. The stationary phase was a Waters C_{18} column. The mobile phase consisted of 0.005 M 1-heptanesulfonate sodium and 0.35 M acetic acid in 35% (vol/vol) methanol. The flow rate was 1.6 mL/min. The fluorometric detector was set at an excitation wavelength of 285 nm and an emission wavelength of 315 nm. The injection volume was 20 μL. The retention time for isoproterenol hydrochloride was 4.33 minutes.

The stability-indicating capability of the assay was demonstrated by an accelerated degradation study. A solution of isoproterenol 2 μg/mL and aminophylline 500 μg/mL was exposed to fluorescent light for 17 hours. The degradation product did not interfere with the analysis of the parent isoproterenol hydrochloride.

A standard curve was constructed by plotting drug concentrations versus the peak areas from 0.2 to 2.4 μg/mL. The correlation coefficient was greater than 0.997.

Reference
Williams DA, Fung EYY, Newton DW. Ion-pair high-performance liquid chroma-tography of terbutaline and catecholamines with aminophylline in intravenous solutions. *J Pharm Sci.* 1982; 71: 956–8.

Isoxicam

Chemical Name
4-Hydroxy-2-methyl-*N*-(5-methyl-3-isoxazolyl)-2*H*-1,2-benzothiazine-3-carboxamide 1,1-dioxide

Other Names
Flexicam, Maxicam

Form	Molecular Formula	MW	CAS
Isoxicam	$C_{14}H_{13}N_3O_5S$	335.3	34552-84-6

Method

Bartsch et al. investigated the stability of isoxicam by HPLC. A Shimadzu HPLC system consisted of a model LC-10AS pump, a model SPD-M10A diode-array detector, a model CTO-10AC column oven, and a Rheodyne injector. The stationary phase was a Merck LiChrospher 100 RP18 endcapped column (119 × 3 mm, 5-μm particle size). The mobile phase consisted of 0.4 M acetate buffer (pH 4.6) and methanol (60:40, vol/vol). The flow rate was 0.8 mL/min. UV detection was performed at 280 nm.

Samples were diluted with mobile phase. Under these conditions, the retention time for isoxicam was about 6.6 minutes (estimated from the published chromatogram).

The stability-indicating ability of the method was proven by accelerated degradation of isoxicam. Samples were exposed to irradiation from a xenon source in a Suntest. Degradation products did not influence the determination of isoxicam.

Standard curves for isoxicam were constructed. Correlation coefficients were greater than 0.999. Intraday and interday coefficients of variation were less than 2.57 and 2.78%, respectively. The limit of detection and limit of quantitation were 0.32 and 1.25 μg/mL, respectively.

Reference

Bartsch H, Eiper A, Habiger K, et al. Comparison of analytical methods for investigating the photostability of isoxicam. *J Chromatogr A*. 1999; 846: 207–16.

Isoxsuprine Hydrochloride

Chemical Name

1-(4-Hydroxyphenyl)-2-(1-methyl-2-phenoxyethylamino)propan-1-ol hydrochloride

Other Names

Duvadilan, Vasodilan, Voxsuprine, Xuprin

Form	Molecular Formula	MW	CAS
Isoxsuprine hydrochloride	$C_{18}H_{23}NO_3 \cdot HCl$	337.8	579-56-6

Appearance

Isoxsuprine hydrochloride is a white or almost white odorless or almost odorless crystalline powder.

Solubility

Isoxsuprine hydrochloride is slightly soluble in water and sparingly soluble in alcohol.

Method

Belal et al. described an HPLC method for the determination of isoxsuprine in dosage forms. The HPLC system consisted of a Waters model 600E system controller, a model U6K injector, a model 486 tunable absorbance detector, and a model 746 data module. The stationary phase was a Waters μBondapak C_{18} stainless steel column (150 × 3.9 mm). The mobile phase consisted of 0.01 M monobasic potassium phosphate (adjusted to pH 2.2 with phosphoric acid) and acetonitrile (82:18, vol/vol). The flow rate was 2.5 mL/min. UV detection was performed at 275 nm. Methyl *p*-hydroxybenzoate 8.0 μg/mL in the mobile phase was used as an internal standard.

A portion of 10 ground tablets equivalent to 20 mg of the drug was accurately weighed and placed into a 100-mL volumetric flask. The sample was mixed with 50 mL of the mobile phase, heated in a boiling water bath for 30 minutes, cooled, diluted to volume with the mobile phase, and filtered. The filtrate was mixed with internal standard and diluted with the mobile phase. The injection volume was 20 μL. Under these conditions, retention times for the internal standard and isoxsuprine were 4.8 and 5.5 minutes, respectively.

The stability-indicating nature of the assay was tested by accelerated degradation of the drug. Isoxsuprine solutions were treated with 2 M sodium hydroxide solution at 80 °C for 4 hours or exposed to UV irradiation. Degradation products did not interfere with the analysis of isoxsuprine. Retention times for 4-hydroxybenzyl alcohol, 4-hydroxybenzoic acid, 4-hydroxybenzaldehyde, 4-hydroxyacetophenone, and isoxsuprine were about 1.1, 1.6, 2.2, 2.5, and 5.5 minutes, respectively.

A standard curve for isoxsuprine was generated from 2 to 40 μg/mL. The correlation coefficient was 0.9998. The limit of detection was 0.2 μg/mL.

Reference

Belal F, Al-Malaq HA, Al-Majed AA, et al. A stability-indicating HPLC method for the determination of isoxsuprine in dosage forms. Application to kinetic studies of isoxsuprine. *J Liq Chromatogr Rel Technol.* 2000; 23: 3175–89.

Isradipine

Chemical Name

Isopropyl methyl 4-(2,1,3-benzoxadiazol-4-yl)-1,4-dihydro-2,6-dimethylpyridine-3,5-dicarboxylate

Other Names
Clivoten, Dyna Circ, Lomir, Prescal

Form	Molecular Formula	MW	CAS
Isradipine	$C_{19}H_{21}N_3O_5$	371.4	75695-93-1

Appearance
Isradipine is a yellow fine crystalline powder.

Solubility
Isradipine is practically insoluble in water. It is soluble in methanol and freely soluble in acetone.

Method
Elghany et al. described a method for the determination of isradipine in tablets and capsules. The HPLC system consisted of a Beckman model 110B solvent-delivery module, a Rheodyne model 7125 injector with a 20-µL loop, a Waters model 481 UV-visible detector, and a Spectra-Physics model 4290 integrator. The stationary phase was a Beckman Ultrasphere ODS column (150 × 4.6 mm, 5-µm particle size). The mobile phase consisted of methanol and water (60:40, vol/vol). The flow rate was 1.0 mL/min. UV detection was performed at 325 nm.

A portion of powder equivalent to 25 mg of isradipine prepared from 20 capsules or tablets was weighed into a 25-mL volumetric flask, dissolved in 10 mL of methanol, filtered, and diluted with methanol and then with mobile phase to approximately 35 µg/mL of isradipine. Under these conditions, the retention time for isradipine was 13 minutes.

The stability-indicating capability of the method was verified by accelerated degradation of isradipine. Isradipine solutions were treated with 0.1 N hydrochloric acid and stored at 45 and 60 °C for 1–6 hours, with 0.1 N sodium hydroxide solution and kept at 45 and 60 °C, or with 30% hydrogen peroxide and kept at room temperature for 1–6 hours. The intact isradipine was separated from its degradation products.

A calibration curve for isradipine was constructed from 10 to 60 µg/mL. The intraday and interday coefficients of variation were 2.47 and 1.76%, respectively. The limit of detection was 0.25 µg/mL.

Reference
Elghany MFA, Elzeany BE, Elkawy MA, et al. A stability indicating high performance liquid chromatographic assay of isradipine in pharmaceutical preparations. *Anal Lett.* 1996; 29: 1157–65.

Itraconazole

Chemical Name
4-[4-[4-[4-[[2-(2,4-Dichlorophenyl)-2-(1*H*-1,2,4-triazol-1-ylmethyl)-1,3-dioxolan-4-yl]-methoxy]phenyl]-1-piperazinyl]phenyl]-2,4-dihydro-2-(1-methylpropyl)-3*H*-1,2,4-triazol-3-one

Other Name
Sporanox

Form	Molecular Formula	MW	CAS
Itraconazole	$C_{35}H_{38}Cl_2N_8O_4$	705.6	84625-61-6

Appearance
Itraconazole is a white to slightly yellowish powder.

Solubility
Itraconazole is practically insoluble in water and very slightly soluble in alcohols. It is freely soluble in dichloromethane.

Method
Jacobson et al. used an HPLC method to determine the stability of itraconazole in an extemporaneously compounded oral liquid. The liquid chromatograph consisted of a Waters model 501 constant-flow solvent-delivery system, a Waters model U6K variable-volume injector, a Waters model 486 tunable UV detector, and a Hewlett-Packard model 3394 integrator-recorder. The stationary phase was an Applied Biosystems Spheri-5 ODS column (250 × 4.6 mm, 5-µm particle size). The mobile phase was a mixture of water, acetonitrile, and diethylamine (40:60:0.05, vol/vol/vol). The flow rate was 1.0 mL/min. UV detection was performed at 263 nm and 0.1 AUFS. Janssen R5 1012 1 mg/mL in methanol was used as an internal standard.

The itraconazole suspension was dissolved in dimethylformamide and diluted with mobile phase. The injection volume was 5 µL. Retention times for itraconazole and the internal standard were about 7.1 and 10.5 minutes (estimated from the published chromatogram).

The stability-indicating nature of the assay was determined by an accelerated degradation study of itraconazole. Solutions of itraconazole 0.5 mg/mL were heated to 100 °C for 1 hour after adjustment to pH 2 with 1 N sulfuric acid and to pH 12 with 1 N sodium hydroxide. These solutions were exposed to direct sunlight at room temperature for 35 days. Degradation product peaks did not interfere with the intact itraconazole peak.

A standard curve was constructed each day by linear regression of the peak height ratio of itraconazole to internal standard versus the itraconazole concentration from 15 to 35 µg/mL. The correlation coefficient was greater than 0.9997. The intraday and interday coefficients of variation of the assay were less than 2%.

Reference
Jacobson PA, Johnson CE, Walters JR. Stability of itraconazole in an extemporaneously compounded oral liquid. *Am J Health Syst Pharm.* 1995; 52: 189–91.

Ketamine Hydrochloride

Chemical Name
(±)-2-(2-Chlorophenyl)-2-(methylamino)cyclohexanone hydrochloride

Other Names
Ketaject, Ketalar, Ketaset, Vetalar

Form	Molecular Formula	MW	CAS
Ketamine hydrochloride	$C_{13}H_{16}ClNO.HCl$	274.2	1867-66-9

Appearance
Ketamine hydrochloride occurs as a white crystalline powder with a slight characteristic odor.

Solubility
Ketamine hydrochloride is freely soluble in water (to about 200 mg/mL) and methanol. It is soluble in ethanol.

Method 1
Gupta developed an analytical method for the analysis of ketamine hydrochloride after reconstitution in water for injection. The Waters ALC202 liquid chromatograph was equipped with a Rheodyne model 7125 universal injector, a Schoeffel SF770 multiple-wavelength detector, and a Houston Omniscribe recorder. The stationary phase was a Beckman Ultrasphere column (150 × 4.6 mm, 5-μm particle size). The mobile phase consisted of 0.01 M monobasic potassium phosphate buffer and acetonitrile (77:23, vol/vol); the pH was adjusted to 3.9. The flow rate was 1.6 mL/min. UV detection was carried out at 269 nm and 0.1 AUFS.

Samples were diluted with water. The injection volume was 80 μL. Under these conditions, the retention time for ketamine was about 2.6 minutes (estimated from the published chromatogram).

To show the stability-indicating nature of the method, ketamine hydrochloride solutions were heated to boiling in water, 0.5 N sodium hydroxide, and 0.5 N sulfuric acid. There was no interference with the analysis of the drug from its decomposition products.

Reference
Gupta VD. Stability of ketamine hydrochloride injection after reconstitution in water for injection and storage in 1-mL tuberculin polypropylene syringes for pediatric use. *Int J Pharm Compound.* 2002 (July/Aug); 316–7.

Method 2

Walker et al. used an HPLC method to study the stability of ketamine and hydromorphone in normal saline. The Waters 600E system was coupled with a Thermo Separation model UV6000LP scanning variable-wavelength UV detector. The stationary phase was a Beckman Ultrasphere ODS column (250 × 4.2 mm, 5-μm particle size). The mobile phase consisted of a phosphate buffer and acetonitrile (50:50, vol/vol); the pH was adjusted to 7.3 with phosphoric acid. The phosphate buffer was prepared by dissolving 10.7 g of dibasic sodium phosphate heptahydrate in 4 liters of water. The flow rate was 1 mL/min. The run time was 12 minutes. UV detection was performed at 280 nm. Under these conditions, the retention time for ketamine was about 9.0 minutes (estimated from the published chromatogram).

This method was demonstrated to be stability indicating by accelerated studies. Ketamine solutions were adjusted to pH 1.5 with hydrochloric acid or to pH 12.7 with sodium hydroxide solution. Degradation products did not interfere with the determination of ketamine.

A standard curve for ketamine was constructed from 0.3 to 50.0 mg/mL.

Reference

Walker SE, Law S, DeAngelis C. Stability and compatibility of hydromorphone and ketamine in normal saline. *Can J Hosp Pharm.* 2001; 54: 191–9.

Ketoconazole

Chemical Name

cis-1-Acetyl-4-[4-[[2-(2,4-dichlorophenyl)-2-(1*H*-imidazol-1-ylmethyl)-1,3-dioxolan-4-yl]methoxy]phenyl]piperazine

Other Name

Nizoral

Form	Molecular Formula	MW	CAS
Ketoconazole	$C_{26}H_{28}Cl_2N_4O_4$	531.4	65277-42-1

Appearance

Ketoconazole is a white to slightly beige powder.

Solubility

Ketoconazole has a solubility of 40 μg/mL in water at 23 °C and is relatively insoluble in alcohol at 23 °C.

pK$_a$

Ketoconazole has pK$_a$ values of 2.9 and 6.5.

Method

Allen and Erickson used HPLC to study the stability of ketoconazole in extemporaneously compounded oral liquids. The chromatograph was composed of a Hewlett-Packard series 1050 automated high-performance liquid chromatograph, including a multisolvent mixing and pumping system, an autoinjector, a diode-array detector, and a computer with Chem Station data-handling software. The stationary phase was a Bakerbond C$_{18}$ analytical column (250 × 4.6 mm, 5-μm particle size). The mobile phase was a mixture of 7 parts of diisopropylamine in methanol (1:500, vol/vol) and 3 parts of ammonium acetate solution (1:200, vol/vol). The flow rate was 3.0 mL/min. UV detection was performed at 225 nm.

Samples were diluted 1:50 before injection. Under these conditions, ketoconazole eluted with a retention time of 3.1 minutes.

The method was evaluated to be stability indicating. A composite chromatogram of ketoconazole after accelerated decomposition by heat, acid, base, oxidizing agent, and light did not show any interference from the degradation products with the peak of the intact ketoconazole.

Standard curves were constructed for ketoconazole from 50 to 500 μg/mL. Coefficients of variation for intraday and interday assay were 1.9 and 2.0%, respectively.

Reference

Allen LV, Erickson MA. Stability of ketoconazole, metolazone, metronidazole, procainamide hydrochloride, and spironolactone in extemporaneously compounded oral liquids. *Am J Health Syst Pharm.* 1996; 53: 2073–8.

Ketoprofen

Chemical Name

(*RS*)-2-(3-Benzoylphenyl)propionic acid

Other Names
Actron, Dexal, Fenoket, Orudis

Form	Molecular Formula	MW	CAS
Ketoprofen	$C_{16}H_{14}O_3$	254.3	922071-15-4

Appearance
Ketoprofen is a white or almost white crystalline powder.

Solubility
Ketoprofen is practically insoluble in water but is freely soluble in alcohol.

pK_a
Ketoprofen has a pK_a of 5.9 in a 3:1 solution of methanol and water.

Method 1
Bempong and Bhattacharyya reported a validated method for the determination of keto-profen in a topical gel. The Hewlett-Packard 1100 series system included a model G1322A degasser, a model G1311A quaternary pump, a model 1313A autosampler, a model G1314A UV detector, and a model G1316A column oven. The stationary phase was a YMC ODS-AQ reversed-phase analytical column (150 × 4.6 mm, 5-μm particle size) with an Econosphere C_{18} guard column (5 × 4.6 mm, 5-μm particle size). The mobile phase consisted of a phosphate buffer (pH 3.5), water, and acetonitrile (8:43:49, vol/vol/vol). The buffer was prepared by dissolving 3.4 g of monobasic potassium phosphate in water, adjusting the pH to 3.5 with phosphoric acid, and diluting with water to 1 L. The flow rate was 1.0 mL/min. UV detection was carried out at 233 nm.

A portion of the topical gel was weighed, dispersed by sonication for 10 minutes in 50 mL of mobile phase, diluted to 100 mL with mobile phase, filtered through a 0.45-μm Millex-HV membrane filter, and assayed. Under these conditions, the retention time of ketoprofen was about 6.5 minutes.

The method was verified to be stability indicating by forced degradation studies. Ketoprofen topical gel was subjected to UV radiation at 254 and 365 nm for up to 2 hours, heat at 90 °C for 4 hours, 3% hydrogen peroxide for up to 26 hours, dilute hydrochloric acid for up to 26 hours, and 0.5% sodium hydroxide solution for up to 26 hours. The degradation products and excipients did not interfere with the analysis of ketoprofen.

A standard curve for ketoprofen was obtained from 0.04 to 100 μg/mL. The correlation coefficient was 1.0000. Intraday and interday coefficients of variation were less than 0.6 and 0.5%, respectively.

Reference
Bempong DK, Bhattacharyya L. Development and validation of a stability-indicating high-performance liquid chromatographic assay for ketoprofen topical penetrating gel. *J Chromatogr A*. 2005; 1073: 341–6.

Method 2
Proniuk et al. reported an HPLC assay for ketoprofen in isopropyl myristate. The liquid chromatographic system comprised a Thermo Separation Products model 8815 Isochrom pump, a Rheodyne model 9125 20-μL loop injector, a Spectra-Physics model

100 variable-wavelength detector, and a Thermo Separation Products model 4290 integrator. The stationary phase was a Phenomenex Luna C_{18} column (250 × 4.6 mm, 5-μm particle size) with a Luna guard column (30 × 4.6 mm, 5-μm particle size). The mobile phase consisted of acetonitrile, methanol, and water (36:54:10, vol/vol/vol). The flow rate was 1.2 mL/min. UV detection was performed at 265 nm. Propylparaben 200 μg/mL in isopropyl myristate was used as an internal standard.

Samples were properly diluted for testing. The injection volume was 20 μL. Under these conditions, retention times for ketoprofen and propylparaben were about 2.4 and 3.3 minutes, respectively.

The stability-indicating nature of this method was verified using electrospray mass spectrometry.

Standard curves for ketoprofen were constructed from 0.625 to 10 μg/mL and from 6.25 to 100 μg/mL. The correlation coefficients were greater than 0.988. Intraday and interday coefficients of variation were 2.91 and 3.32%, respectively. The limit of quantitation was approximately 0.625 μg/mL.

Reference

Proniuk S, Lerkpulsawad S, Blanchard J. A simplified and rapid high-performance liquid chromatographic assay for ketoprofen in isopropyl myristate. *J Chromatogr Sci.* 1998; 36: 495–8.

Method 3

Hsu et al. reported an HPLC method for the determination of ketoprofen in its dosage forms. The Shimadzu chromatograph consisted of a model C-R6A dual pump, a model SCL-6A system controller, a model SIL-A autosampler, a model SPD-6A UV detector, and a model C-R4A integrator. The stationary phase was a Zorbax ODS column (150 × 4.6 mm). The mobile phase consisted of 0.05 M monobasic potassium phosphate buffer (adjusted to pH 2.5 with glacial acetic acid) and acetonitrile (55:45, vol/vol). The flow rate was 1.2 mL/min. UV detection was performed at 265 nm and 0.02 AUFS. Flurbiprofen was used as an internal standard.

For injection, the sample was diluted with methanol and then with the mobile phase to approximately 5 μg/mL of ketoprofen. For capsules, the sample was mixed with methanol, sonicated for 15 minutes, and filtered. The filtrate was diluted to approximately 5 μg/mL of ketoprofen with the mobile phase. The injection volume was 20 μL. Under these conditions, retention times for ketoprofen and flurbiprofen were about 5.5 and 11.5 minutes, respectively.

The stability-indicating nature of the assay was established by accelerated decomposition of ketoprofen. Ketoprofen solutions were treated with 0.1 N hydrochloric acid or 0.1 N sodium hydroxide solution and kept at 60 °C for 48 hours or were exposed to sunlight for 16 hours. The degradation products did not interfere with the determination of ketoprofen.

A calibration curve for ketoprofen was constructed from 1.25 to 20 μg/mL. The correlation coefficient was 0.9999. Intraday and interday coefficients of variation were less than 0.01%. The limit of detection was 0.023 μg/mL. The limit of quantitation was 0.076 μg/mL.

Reference

Hsu S-Y, Shaw C-Y, Chang B-L. A stability-indicating HPLC method for ketoprofen. *J Food & Drug Anal.* 1995; 3: 275–85.

Ketorolac Tromethamine

Chemical Name

(±)-5-Benzoyl-2,3-dihydro-1*H*-pyrrolizine-1-carboxylic acid compound with 2-amino-2-(hydroxymethyl)-1,3-propanediol (1:1)

Other Names

Acular, Ketorolac Trometamol, Taradyl, Toradol

Form	Molecular Formula	MW	CAS
Ketorolac tromethamine	$C_{15}H_{13}NO_3.C_4H_{11}NO_3$	376.4	74103-07-4

Appearance

Ketorolac tromethamine is a white to off-white crystalline powder.

Solubility

Ketorolac tromethamine is freely soluble in water and in methanol.

pK$_a$

Ketorolac tromethamine has a pK$_a$ of 3.54.

Method

Kumar et al. described an HPLC determination of ketorolac tromethamine in an ophthalmic formulation. A Jasco instrument was equipped with a model 880PU reciprocating pump, a model 875UV variable-wavelength detector, and a Rheodyne model 7125 injector with a 20-µL loop. The stationary phase was a Finepak C_8 analytical column (250 × 4.6 mm, 10-µm particle size) with a Corasil C_{18} guard column (30 mm, 35–45-µm particle size). The mobile phase consisted of acetonitrile, methanol, and 0.01 M monobasic potassium phosphate buffer (pH 4.2) (25:25:50, vol/vol/vol) and was delivered isocratically at 1 mL/min. UV detection was performed at 319 nm.

Samples were diluted 1:100 with double-distilled water. The sample injection volume was 20 µL. Under these conditions, the retention time for ketorolac tromethamine was 6.1 minutes.

The method was evaluated to be stability indicating by assaying the authentic degradation products and a sample of ketorolac tromethamine kept at 60 °C for 4 months. The intact drug peak was well resolved from all degradation product peaks.

A calibration curve of ketorolac was constructed from 20 to 70 µg/mL. The correlation coefficient was 0.9996. The intraday and interday coefficients of variation were less than 1.56 and 2.02%, respectively.

Reference

Kumar STR, Shedbalkar VP, Bhalla HL. High performance liquid chromatographic determination of ketorolac tromethamine in ophthalmic formulation. *Indian Drugs*. 1997; 34: 532–5.

Ketotifen Fumarate

Chemical Name

4-(1-Methylpiperidin-4-ylidene)-4*H*-benzo[4,5]cyclohepta[1,2-*b*]thiophen-10(9*H*)-one hydrogen fumarate

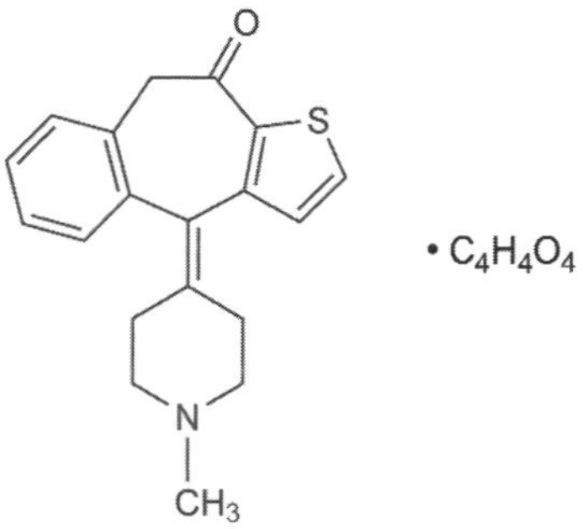

$\cdot C_4H_4O_4$

Other Names

Zaditen, Zaditor

Form	Molecular Formula	MW	CAS
Ketotifen fumarate	$C_{19}H_{19}NOS.C_4H_4O_4$	425.5	34580-14-8

Appearance

Ketotifen fumarate occurs as a fine crystalline powder.

Solubility

Ketotifen is readily soluble in water.

Method 1

Nnane et al. developed an assay for ketotifen in aqueous and silicon oil formulations. The liquid chromatograph comprised a ConstaMetric 3000 solvent-delivery system, an LDC SpectroMonitor 3100 UV detector, an LDC CI4000 integrator, and an LKB 2157 autoinjector. The stationary phase was a reversed-phase µBondapak C_{18} column (300 × 3.9 mm, 5-µm particle size) with a pellicular ODS guard column (50 × 2 mm, 10-µm particle size). The mobile phase consisted of 0.001 M phosphate buffer, methanol, acetonitrile, and trimethylamine (29.8:45:25:0.2, vol/vol/vol/vol). The flow rate was 1 mL/min. UV detection was performed at 299 nm. Imipramine hydrochloride (20 mg/mL free base) in ethanol was used as an internal standard.

A ketotifen suspension (250 µL) equivalent to 600 µg/mL free base was transferred into a microfuge tube, mixed with 20 µL of internal standard followed by 250 µL of 0.05 M hydrochloric acid, mixed for 1 minute, and centrifuged at 3000 rpm for 5 minutes. The supernatant was diluted 1:10 with mobile phase. The injection volume was 25 μL. Under these conditions, retention times for ketotifen and imipramine were 7.7 and 15.4 minutes, respectively.

The method was evaluated to be stability indicating. Ketotifen samples were stored at −20 °C for 8 hours, autoclaved at 100 °C for 1 and 4 hours, exposed to UV irradiation (254 and 360 nm) for 8 hours, or treated with 30% hydrogen peroxide for 8 hours. The intact ketotifen peak was resolved from its degradation product peaks.

A calibration curve for ketotifen in 0.001 M phosphate buffer (pH 7.4) was constructed from 1.0 to 25 μg/mL. The correlation coefficient was greater than 0.992. Intraday and interday coefficients of variation were 1.96 and 7.40%, respectively. The limit of quantitation was 1 μg/mL. Another standard curve for ketotifen in silicon oil was generated from 50 to 600 μg/mL. The correlation coefficient was greater than 0.998. Intraday and interday coefficients of variation were 4.85 and 5.85%, respectively. The limit of quantitation was 50 μg/mL.

Reference

Nnane IP, Damani LA, Hutt AJ. Development and validation of stability indicating high-performance liquid chromatographic assays for ketotifen in aqueous and silicon oil formulations. *Chromatographia*. 1998; 48: 797–805.

Method 2

Elsayed developed and validated a simple method for the determination of ketotifen in raw materials and pharmaceutical formulations. A Perkin-Elmer series 200 system was equipped with a pump, a vacuum degasser, an autosampler, and a UV-visible detector. The stationary phase was a Perkin-Elmer Spheri-5 RP-8 column (250 × 4.6 mm, 5-μm particle size). The mobile phase consisted of methanol, 0.04 M triethylamine phosphate buffer (pH 2.8), and tetrahydrofuran (43:55:2, vol/vol/vol). The buffer was prepared by dissolving 4.61 g of phosphoric acid (85%, wt/wt) and 3.57 g of triethylamine in 900 mL of water; the pH was adjusted to 2.8 with phosphoric acid, and the flask was filled to 1000 mL with water. The flow rate was 1.2 mL/min. UV detection was performed at 297 nm.

For tablets, a portion of powder from 10 tablets equivalent to 1 mg of ketotifen base was transferred to a 50-mL volumetric flask, mixed with 40 mL of a diluent (methanol and 0.04 M triethylamine phosphate buffer, 40:60, vol/vol), sonicated for 5 minutes, filled to volume with the diluent, and mixed well. For syrup, a portion of syrup equivalent to 1 mg of ketotifen base was diluted to 50 mL with the diluent and mixed well. Before injection, all samples were filtered through a 0.45-μm membrane filter. The injection volume was 50 μL. Under these conditions, the retention time for ketotifen was about 4.9 minutes.

The method was demonstrated to be stability indicating by forced degradation studies. Ketotifen was treated in 1 M hydrochloric acid at 70 °C for 3 hours and 1 M sodium hydroxide solution at 70 °C for 3 hours. The degradation product did not interfere with the analysis of ketotifen.

A standard curve was obtained from 0.73 to 145.43 μg/mL. The correlation coefficient was 0.9999. The coefficient of variation of the assay was less than 2.37%. The limit of quantitation was 0.60 μg/mL.

Reference
Elsayed MA. Development and validation of a rapid HPLC method for the determination of ketotifen in pharmaceuticals. *Drug Develop Ind Pharm.* 2006; 32: 457–61.

Labetalol

Chemical Name
2-Hydroxy-5-[1-hydroxy-2-[(1-methyl-3-phenylpropyl)amino]ethyl]benzamide

Other Names
Normodyne, Trandate

Form	Molecular Formula	MW	CAS
Labetalol	$C_{19}H_{24}N_2O_3$	328.4	36894-69-6
Labetalol hydrochloride	$C_{19}H_{24}N_2O_3 \cdot HCl$	364.9	32780-64-6

Appearance
Labetalol hydrochloride occurs as a white or off-white crystalline powder.

Solubility
Labetalol hydrochloride is soluble in water and ethanol. It is insoluble in ether and chloroform.

pK$_a$
Labetalol hydrochloride has a pK$_a$ of 9.3.

Method 1
Allen and Erickson used HPLC to study the stability of labetalol hydrochloride 40 mg/mL in extemporaneously compounded oral liquids. A Hewlett-Packard series 1050 automated high-performance liquid chromatograph consisting of a multisolvent mixing and pumping system, an autoinjector, a diode-array detector, and a computer with Chem Station data-handling software was used. The stationary phase was a Bakerbond C_{18} analytical column (250 × 4.6 mm, 5-µm particle size). The mobile phase was 45% methanol in 0.1 M monobasic sodium phosphate buffer solution and was delivered isocratically at 1.3 mL/min. UV detection was performed at 230 nm.

Samples were diluted 1:100. Under these conditions, labetalol eluted in 7.5 minutes.

The stability-indicating nature of the assay was demonstrated by accelerated degradation of labetalol hydrochloride. A composite chromatogram of labetalol after accelerated decomposition by acid, base, oxidizing agent, heat, and light showed that the intact labetalol peak was well separated from two major degradation product peaks.

Calibration curves for labetalol were constructed from 100 to 500 μg/mL. The intraday and interday coefficients of variation were 1.6 and 1.9%, respectively.

Reference

Allen LV, Erickson MA. Stability of labetalol hydrochloride, metoprolol tartrate, verapamil hydrochloride, and spironolactone with hydrochlorothiazide in extemporaneously compounded oral liquids. *Am J Health Syst Pharm*. 1996; 53: 2304–9.

Method 2

Hassan et al. determined the compatibility and stability of labetalol hydrochloride in combination with dobutamine, dopamine, morphine, nitroglycerin, and ranitidine as used in the intensive care unit in a Y-site administration set. The HPLC system consisted of a Waters model 6000A pump, a Waters model 710B WISP autosampler, a Waters model 480 UV detector, a Hewlett-Packard model 3390A integrator, and a Phenomenex Bondex column (300 × 3.9 mm, 10-μm particle size). The mobile phase consisted of 0.004 M sodium 1-decanesulfonate and 1% acetic acid in methanol and water (60:40, vol/vol). UV detection was performed at 254 nm. The sample injection volume was 20 μL.

The stability-indicating nature of the analytical method was established by an intentional degradation study. Solutions of labetalol were alkalinized by dropwise addition of 1 N sodium hydroxide to pH 12 or acidified by the dropwise addition of 1 N hydrochloric acid to pH 2 and were then stored at room temperature in full fluorescent light for at least 2 weeks. No degradation products interfered with the separation of the intact labetalol.

Standard curves were constructed from 0.50 to 1.50 mg/mL. Correlation coefficients were greater than 0.99. The coefficient of variation was less than 1.0%.

A similar analytical method was used by Yuen et al.

References

Hassan E, Leslie J, Marter-Herrero ML. Stability of labetalol hydrochloride with selected critical care drugs during simulated Y-site injection. *Am J Hosp Pharm*. 1994; 51: 2143–5.

Yuen P-H C, Taddei CR, Wyka BE, et al. Compatibility and stability of labetalol hydrochloride in commonly used intravenous solutions. *Am J Hosp Pharm*. 1983; 40: 1007–9.

Lactic Acid

Chemical Name

2-Hydroxypropionic acid

Other Names
Calmuril, Lactinol, Lactisan, Unigyn

Form	Molecular Formula	MW	CAS
Lactic acid	$C_3H_6O_3$	90.1	50-21-5

Appearance
Lactic acid is a colorless or slightly yellow syrupy hygroscopic and practically odorless liquid.

Solubility
Lactic acid is miscible with water, alcohol, and ether.

Method
De Villiers et al. developed an HPLC assay for lactic acid and glycolic acid in nonionic creams. A Shimadzu LC-6A system was equipped with a model C-3RA integrator. The stationary phase was a Waters μBondapak C_{18} column (250 × 3.9 mm, 5-μm particle size). The mobile phase consisted of water and acetonitrile (60:40, vol/vol) containing 0.02 mM monobasic potassium phosphate adjusted to pH 2.25 with phosphoric acid. The flow rate was 1 mL/min. UV detection was performed at 214 nm.

A sample equivalent to 100 mg of lactic acid and glycolic acid was weighed and placed in a 150-mL beaker, mixed with 20 mL of 0.1 M sodium hydroxide solution, boiled for 20 minutes, and cooled to room temperature. It was then mixed with 20 mL of water, adjusted to pH 6 with dilute phosphoric acid, diluted with water, filtered, and diluted with the mobile phase. The injection volume was 20 μL. Under these conditions, retention times for lactic acid and glycolic acid were about 8.8 and 10.8 minutes, respectively (estimated from the published chromatogram).

The method was shown to be stability indicating by accelerated degradation of glycolic and lactic acids. The acids were degraded in an oven at 120 °C for 5 days. Degradation products of lactic and glycolic acid did not interfere with the analysis of the acids.

A standard curve was constructed. The correlation coefficient was 0.9997.

A similar method for lactic acid in lotions was described by Cheng and Gadde.

References
De Villiers MM, Wurster DE, Bergh T, et al. Stability indicating HPLC assay of the α-hydroxy acids lactic acid and glycolic acid in non-ionic creams. *Pharmazie*. 1998; 53: 204–5.

Cheng H, Gadde RR. Stability-indicating high-performance liquid chromatographic assay for lactic acid in lotions. *J Chromatogr*. 1986; 355: 399–406.

Lansoprazole

Chemical Name

2-[[[3-Methyl-4-(2,2,2-trifluoroethoxy)-2-pyridinyl]methyl]sulfinyl]-1*H*-benzimidazole

Other Names

Lanzo, Lanzor, Prevacid, Zoton

Form	Molecular Formula	MW	CAS
Lansoprazole	$C_{16}H_{14}F_3N_3O_2S$	369.4	103577-45-3

Appearance

Lansoprazole occurs as a white to brownish-white odorless crystalline powder.

Solubility

Lansoprazole is practically insoluble in water and sparingly soluble in ethanol. It is freely soluble in dimethylformamide.

Method

DiGiacinto et al. investigated the stability of lansoprazole prepared in suspensions and stored in amber-colored plastic oral syringes at 22 and 4 °C for up to 60 days. The HPLC apparatus comprised a Shimadzu model SPD-10A UV-visible detector, a model SIL-10A autoinjector, a model LC-1AS pump, a model SCL-10A system controller, and a model CTO-10A column oven. The stationary phase was a Phenomenex Maxsil C_{18} column (250 × 4.6 mm, 5-μm particle size). The column oven temperature was kept at 35 °C. The mobile phase consisted of acetonitrile and water (45:55, vol/vol) adjusted to pH 7.5 with 1 M monobasic sodium phosphate. The flow rate was 1 mL/min. UV detection was performed at 285 nm.

Samples were diluted with a mixture of acetonitrile and water (45:55, vol/vol) and vortexed for 1 minute. The injection volume was 150 μL. Under these conditions, the retention time for lansoprazole was about 6.4 minutes.

The stability-indicating nature of the assay was assessed by accelerated degradation of lansoprazole. Stock solutions of lansoprazole were adjusted to pH 2 with 1 N hydrochloric acid and stood for 15 minutes, were adjusted to pH 13 with 1 M sodium hydroxide solution and heated to 100 °C for 20 minutes, or were stored at room temperature for 72 hours. In all cases, degradation products did not interfere with the determination of lansoprazole.

A standard curve for lansoprazole was constructed from 0 to 200 μg/mL. The correlation coefficient was 0.9995. Intraday and interday coefficients of variation were 1.15 and 7.79%, respectively.

Reference

DiGiacinto JL, Olsen KM, Bergman KL, et al. Stability of suspension formulations of lansoprazole and omeprazole stored in amber-colored plastic oral syringes. *Ann Pharmacother.* 2000; 34: 600–5.

Lercanidipine

Chemical Name

(±)-2-[(3,3-Diphenylpropyl)methylamino]-1,1-dimethylethyl methyl 1,4-dihydro-2,6-dimethyl-4-(*m*-nitrophenyl)-3,5-pyridinedicarboxylate

Form

Form	Molecular Formula	MW	CAS
Lercanidipine	$C_{36}H_{41}N_3O_6$	611.7	100427-26-7
Lercanidipine hydrochloride	$C_{36}H_{41}N_3O_6.HCl$	648.2	132866-11-6

Method

Using HPLC, Fiori et al. investigated the photochemical stability of lercanidipine. A Jasco PU-1585 liquid chromatograph was equipped with a Jasco 1575 UV-visible detector and a Rheodyne model 7725i injector. The stationary phase was a Phenomenex Luna C_{18} reversed-phase column (150 × 2.0 mm, 3.5-μm particle size). The mobile phase consisted of methanol and 0.01 M triethylamine (60:40, vol/vol); the pH was adjusted to 4 with acetic acid. The flow rate was 250 μL/min. UV detection was performed at 265 nm.

Powder equivalent to about 100 mg from five tablets was accurately weighed, treated with 50 mL of ethanol, sonicated for 15 minutes, filtered, diluted 1:10 with mobile phase to about 20 μg/mL of lercanidipine, and assayed. Under these conditions, the retention time of lercanidipine was about 8.4 minutes (estimated from the published chromatogram).

Lercanidipine solutions in ethanol and in an equal parts mixture of ethanol and PBS (50:50, vol/vol), lercanidipine pure powder, and commercial tablets were exposed to UV-A radiations (Xe arc lamp) at ambient temperature. The peak of the intact lercanidipine was separated from peaks of its degradation products on its chromatogram, demonstrating that the analytical method was stability indicating.

A standard curve of lercanidipine was constructed from 5.0 to 40.0 μg/mL. The correlation coefficient was 0.9994. The limit of detection and the limit of quantitation were 1.49 and 4.96 μg/mL, respectively.

Reference

Fiori J, Gotti R, Bertucci C, et al. Investigation on the photochemical stability of lercanidipine and its determination in tablets by HPLC-UV and LC-ESI-MS/MS. *J Pharm Biomed Anal.* 2006; 41: 176–81.

Leucovorin Calcium

Chemical Name

Calcium N-[4-[[(2-amino-5-formyl-5,6,7,8-tetrahydro-4-hydroxy-6-pteridinyl)-methyl]-amino]benzoyl]-L-glutamate

Other Names

Calcium Folinate, Wellcovorin

Form	Molecular Formula	MW	CAS
Leucovorin calcium	$C_{20}H_{21}CaN_7O_7$	511.5	1492-18-8

Appearance

Leucovorin calcium is an off-white to light beige amorphous odorless powder.

Solubility

Leucovorin calcium has solubilities of more than 500 mg/mL in water and less than 1 mg/mL in alcohol.

pK$_a$

Leucovorin calcium has pK$_a$ values of 3.1, 4.8, and 10.4.

Method 1

Lecompte et al. determined the stability of reconstituted and diluted solutions of calcium folinate using an HPLC method. The liquid chromatograph consisted of two Waters model 510 pumps, a Waters model 712 WISP autosampler, a Waters TCM oven, a Waters 490E programmable multiple-wavelength detector, and an NEC computer as a chromatographic workstation equipped with a Waters Maxima 820 manager. The stationary phase was an Interchim Hypersil C_{18} stainless steel column (250 × 4 mm, 5-µm particle size). The injection volume was 20 µL.

The analytical method was stated to be stability indicating since it separated folinic acid from its degradation products. The retention time for folinic acid was 11 minutes, and retention times for the related compounds N-(p-aminobenzoyl)glutamic acid, N^{10}- formylfolic acid, and folic acid were 8, 20, and 24 minutes, respectively.

A standard curve for folinic acid was constructed from 20 to 200 µg/mL. The correlation coefficient was 0.9999. The coefficient of variation was 2.9%.

Reference

Lecompte D, Bousselet M, Gayrard D, et al. Stability study of reconstituted and diluted solutions of calcium folinate. *Pharm Ind.* 1991; 53: 90–4.

Method 2

Smith et al. used an HPLC method to study the stability of leucovorin calcium and floxuridine of various concentrations in admixtures. The liquid chromatograph consisted of a Perkin-Elmer model series 500 pump, a Perkin-Elmer model ISS 100 autosampler, a Perkin-Elmer model LCI 100 programmable integrator, and a Kratos model 757 autogain variable-wavelength detector. The stationary phase was a μBondapak C_{18} column (10-μm particle size). The mobile phase consisted of 210 mL of methanol and 790 mL of aqueous buffer. The buffer was prepared with 15 mL of 1 M tetrabutylammonium hydroxide in methanol and 850 mL of sterile water for injection; the pH was adjusted to 7.5 with 2 N monobasic sodium phosphate, giving a final volume of 875 mL. The flow rate was 1.5 mL/min. UV detection was performed at 254 nm.

Concentrated samples were diluted before injection, and the injection volume was 20 μL. The retention times for floxuridine and leucovorin calcium were about 2.0 and 4.0 minutes, respectively (estimated from the published chromatogram).

To determine the stability-indicating capability of the method, solutions of leucovorin calcium were exposed to extreme conditions of heat (80 °C for 5 minutes) and acid (pH 3.0–3.5 with 1 N hydrochloric acid) and then chromatographed. Decomposition product peaks did not interfere with the parent drug peak.

The standard curve for leucovorin calcium was linear from 0.03 to 0.96 mg/mL. The correlation coefficient was greater than 0.998.

Reference

Smith JA, Morris A, Duafala ME, et al. Stability of floxuridine and leucovorin calcium admixtures for intraperitoneal administration. *Am J Hosp Pharm.* 1989; 46: 985–9.

Leuprolide

Chemical Name

(D-Leu[6])-des-Gly[10]-LH-RH-ethylamide

Other Names

Leuprorelin, Lupron

Form	Molecular Formula	MW	CAS
Leuprolide	$C_{59}H_{84}N_{16}O_{12}$	1209.4	53714-56-0
Leuprolide acetate	$C_{59}H_{84}N_{16}O_{12}.C_2H_4O_2$	1269.5	74381-53-6

Appearance
Leuprolide acetate is a white to off-white powder.

Solubility
Leuprolide acetate has solubilities of greater than 250 mg/mL in water and greater than 1 g/mL in alcohol at 25 °C.

Method
Sutherland and Menon reported a stability-indicating HPLC method for the determination of leuprolide acetate in an injectable formulation. The liquid chromatograph consisted of a Waters 6000A pump, a Waters WISP injector, a Waters multiple-wavelength detector, and a Spectra-Physics SP4100 electronic integrator/recorder. The stationary phase was an IBM Nucleosil ODS column (150 × 4.6 mm, 5-μm particle size). The mobile phase consisted of 0.087 M aqueous monobasic ammonium phosphate solution (adjusted to pH 6.5 with ammonium hydroxide) and acetonitrile (77:23, vol/vol). The flow rate was 2.0 mL/min. UV detection was performed at 220 nm and 0.2 AUFS. Ethyl p-hydroxybenzoate 2 mg/mL in methanol was used as an internal standard.

Samples and internal standard solution were diluted to concentrations of 0.1 mg/mL leuprolide acetate and 0.15 mg/mL ethyl p-hydroxybenzoate, respectively, with 0.9% sodium chloride injection. The injection volume was 20 μL.

The analytical method was determined to be stability indicating by accelerated degradation of leuprolide. At pH 10.3, 70% of leuprolide was lost when heated at 100 °C for 30–40 minutes, producing two major racemization products, D-Ser[4] and D-His[2]. Following are these compounds and others, with their retention times in minutes shown in parentheses: p-hydroxybenzoic acid (3.3), ethyl p-hydroxybenzoate (6.7), D-Ser[4] analog (17.0), leuprolide (18.7), and D-His[2] analog (21.3) (estimated from the published chromatogram).

Reference
Sutherland JW, Menon GN. HPLC of leuprolide acetate in injectable solutions. *J Liq Chromatogr.* 1987; 10: 2281–9.

Levamisole Hydrochloride

Chemical Name
(−)-(S)-2,3,5,6-Tetrahydro-6-phenylimidazo[2,1-b]thiazole hydrochloride

Other Names
Ergamisol, Immunol, Levadin, Levasole, Solaskil

Form	Molecular Formula	MW	CAS
Levamisole hydrochloride	$C_{11}H_{12}N_2S.HCl$	240.8	16595-80-5

Appearance
Levamisole hydrochloride occurs as a white to almost white crystalline powder.

Solubility
Levamisole hydrochloride is freely soluble in water, soluble in alcohol, and practically insoluble in ether.

Method
Using HPLC, Chiadmi et al. investigated the chemical stability of levamisole in solutions prepared from tablets and powder. The liquid chromatograph consisted of a Shimadzu model LC-6A isocratic pump, a model SPD-6A UV detector, a model C-R6A integrator, and a Rheodyne model 7125 injector with a 20-µL loop. The stationary phase was a Nucleosil C_{18} column (250 × 4.6 mm). The mobile phase was 0.05 M potassium hydrogen phosphate and acetonitrile (85:15, vol/vol). The flow rate was 1 mL/min. UV detection was carried out at 235 nm.

Sample solutions were diluted with water, mixed with internal standard solution, further diluted with water, and assayed. Quinine in water 1.5 mg/mL was used as an internal standard. Under these conditions, the retention times of levamisole and quinine were 8 and 15 minutes, respectively. The run time of a sample injection was 20 minutes.

This method was shown to be stability indicating by accelerated degradation studies of levamisole. The drug solutions were adjusted to pH 1 with 1.0 M hydrochloric acid or to pH 12 with 1.0 M sodium hydroxide solution or were oxidized with hydrogen peroxide. The oral solution was also incubated at 60 °C. Degradation products did not interfere with the analysis of levamisole.

A calibration curve of levamisole hydrochloride was obtained from 50 to 500 µg/mL with a correlation coefficient of 0.999. Intraday and interday coefficients of variation of the analysis were less than 3.79 and 3.69%, respectively. The limit of detection and the limit of quantitation were 0.36 and 1 µg/mL, respectively.

Reference
Chiadmi F, Lyer A, Cisternino S, et al. Stability of levamisole oral solutions prepared from tablets and powder. *J Pharm Pharm Sci.* 2005; 8: 322–5.

Levodopa

Chemical Names
3-Hydroxy-L-tyrosine
(–)-3-(3,4-Dihydroxyphenyl)-L-alanine

Other Names

Dopar, Larodopa, Levopa

Form	Molecular Formula	MW	CAS
Levodopa	$C_9H_{11}NO_4$	197.2	59-92-7

Appearance

Levodopa occurs as white to off-white odorless crystals or crystalline powder.

Solubility

Levodopa has a solubility of 66 mg in 40 mL of water. It is readily soluble in diluted hydrochloric acid and formic acid, but it is practically insoluble in ethanol, benzene, chloroform, and ethyl acetate.

Method 1

Kafil and Dhingra reported a method for the separation of levodopa, carbidopa, and the related impurities. The liquid chromatograph was equipped with a Waters model 510 pump, a model 710B WISP autosampler, and an Environmental Sciences model 5100A Coulochem dual-electrode coulimetric detector. The stationary phase was a Waters μBondapak C_{18} reversed-phase column (250 × 4.6 mm, 10-μm particle size) with a guard column containing a 0.2-μm filter. The mobile phase was 0.05 M ammonium acetate with 0–2% methanol adjusted to pH 4.1 with 0.6 M acetic acid. The flow rate was 0.9 mL/min. The electrochemical detector was set at applied potentials of +0.3 V for the first (screen) electrode and +0.6 V for the second (sample) electrode.

Samples were prepared by dissolving the compound in the mobile phase and filtering it. Under these conditions, retention times for levodopa and carbidopa were 6.35 and 11.41 minutes, respectively.

The stability-indicating capability of the assay was shown by assaying a synthetic mixture of levodopa, carbidopa, and their potential degradation products. Retention times in minutes were 5.49 for 6-hydroxydopa, 6.35 for levodopa, 8.59 for methyldopa, 11.41 for carbidopa, 14.31 for 3-methoxytyrosine, and 22.68 for 3-*o*-methylcarbidopa.

Intraday and interday coefficients of variation were both 2.0%. The limit of detection was 20 ng/mL.

Reference

Kafil JB, Dhingra BS. Stability-indicating method for the determination of levodopa, levodopa–carbidopa and related impurities. *J Chromatogr A*. 1994; 667: 175–81.

Method 2

Schieffer reported a reversed-phase HPLC procedure for levodopa determination in pharmaceutical preparations. A Waters model ALC/GPC 202 chromatograph was equipped with an M6000 pump, a U6K injector, and a Schoeffel model 770 variable-wavelength detector. The stationary phase was a Waters μBondapak C_{18} column (300 × 3.9 mm, 10-μm particle size). The mobile phase was 0.01 M monobasic sodium phosphate solution adjusted to pH 3.5 with phosphoric acid. The flow rate was 2.0 mL/min. UV detection was performed at 280 and 485 nm. Hypoxanthine 0.15 mg/mL was used as an internal standard.

The assay was determined to be stability indicating by spiking the levodopa solution with the internal standard, five postulated degradation products, and five impurities. The intact levodopa peak was well separated from peaks of hypoxanthine, 5-hydroxydopa, 6-hydroxydopa, dopamine, tyrosine, 3-*o*-methyldopa, 3-(3-hydroxy-4-methoxyphenyl)alanine, 5,6-dihydroxyindole, and 5,6-dihydroxyindole-2-carboxylic acid.

A standard curve for levodopa was constructed from 0.02 to 0.2 mg/mL; the correlation coefficient was 0.9999.

A similar method was used by Stennett et al.

References

Schieffer GW. Reversed-phase high-performance liquid chromatographic investigation of levodopa preparations I: Amino acid impurities. *J Pharm Sci.* 1979; 68: 1296–8.

Schieffer GW. Reversed-phase high-performance liquid chromatographic investigation of levodopa preparations II: Levodopa determination. *J Pharm Sci.* 1979; 68: 1299–1301.

Stennett DJ, Christensen JM, Anderson JL, et al. Stability of levodopa in 5% dextrose injection at pH 5 or 6. *Am J Hosp Pharm.* 1986; 43: 1726–8.

Levofloxacin

Chemical Name
(–)-(*S*)-9-Fluoro-2,3-dihydro-3-methyl-10-(4-methyl-1-piperazinyl)-7-oxo-7*H*-pyrido[1,2,3-*de*]-1,4-benzoxazine-6-carboxylic acid

Other Name
Levaquin

Form	Molecular Formula	MW	CAS
Levofloxacin	$C_{18}H_{20}FN_3O_4$	361.4	100986-85-4

Method
Williams et al. studied the stability of levofloxacin diluted in some commonly used infusion fluids. The HPLC system consisted of Hewlett-Packard models 1050 and 1090 and Perkin-Elmer model 4000 high-performance liquid chromatographs, including automatic injectors, which were capable of injecting 25-µL samples, and variable-

wavelength UV detectors. The stationary phase was a Zorbax SB-phenyl column (250 × 4.6 mm, 5-µm particle size). The mobile phase consisted of 0.094 M aqueous monobasic potassium phosphate, acetonitrile, methanol, and trifluoroacetic acid (80:15:5:0.3, vol/vol/vol/vol). The flow rate was 1.0 mL/min. The column temperature was 45 °C. UV detection was performed at 294 nm. The typical run time was 30 minutes.

Samples were diluted with a sample diluent (water, acetonitrile, and methanol; 80:15:5, vol/vol/vol). The retention time for levofloxacin was about 14.8 minutes (estimated from the published chromatogram).

The analytical method was demonstrated to be stability indicating. Levofloxacin solutions were subjected to heat and light. They were also spiked with all known synthesis-related impurities and degradation products. Degradation products, impurities, and formulation excipients did not interfere with the intact levofloxacin peak. Retention times for desfluoro-levofloxacin, the diamine derivative of levofloxacin, *N*-desmethyl-levofloxacin, and levofloxacin-*N*-oxide were about 9.1, 10.7, 12.6, and 20.5 minutes, respectively (estimated from the published chromatogram).

A standard curve for levofloxacin was constructed from 0.1 to 150 µg/mL. The correlation coefficient was greater than 0.9999. The coefficient of variation ranged from 0.19 to 0.83%.

Reference
Williams NA, Bornstein M, Johnson K. Stability of levofloxacin in intravenous solutions in polyvinyl chloride bags. *Am J Health Syst Pharm.* 1996; 53: 2309–13.

Levonordefrin

Chemical Name
(−)-2-Amino-1-(3,4-dihydroxyphenyl)propan-1-ol

Other Names
Corbadrine

Form	Molecular Formula	MW	CAS
Levonordefrin	$C_9H_{13}NO_3$	183.2	829-74-3

Appearance
Levonordefrin occurs as a white to buff-colored and odorless crystalline solid.

Solubility
Levonordefrin is practically insoluble in water and slightly soluble in alcohol, acetone, chloroform, and ether. It is freely soluble in aqueous solutions of mineral acids.

Method
Storms and Stewart reported an HPLC method for the determination of a levonordefrin, tetracaine, and procaine hydrochloride drug combination. The liquid chromatograph consisted of a Beckman model 110-B pump, a Rheodyne model 7125 injector with a

20-μL loop, a Waters model 486 UV-visible detector, and a Shimadzu model C-R3A integrator. The stationary phase was a Waters μBondapak phenyl column (300 × 3.9 mm). The mobile phase consisted of an ion-pair aqueous buffer and acetonitrile (80:20, vol/vol). The ion-pair aqueous buffer was 25 mM monobasic potassium phosphate in water (pH 3.0) with 50 mM heptanesulfonic acid sodium salt. The flow rate was 1 mL/min. UV detection was performed at 254 nm. The injection volume was 20 μL. Under these conditions, retention times of levonordefrin, tetracaine, and procaine hydrochloride were 4.4, 8.7, and 12.3 minutes, respectively.

The method was demonstrated to be stability indicating by accelerated degradation of levonordefrin, tetracaine, and procaine hydrochloride. At ambient temperature and 90 °C, each drug solution was separately mixed with 0.1 N hydrochloric acid, with 0.1 N sodium hydroxide solution, and with 3% hydrogen peroxide. When present, degradation product peaks did not interfere with the analysis of the intact drugs.

Reference

Storms ML, Stewart JT. Stability-indicating HPLC assays for the determination of prilocaine and procaine drug combinations. *J Pharm Biomed Anal.* 2002; 30: 49–58.

Levothyroxine Sodium

Chemical Names

O-(4-Hydroxy-3,5-diiodophenyl)-3,5-diiodo-L-tyrosine monosodium salt
L-Thyroxine sodium salt

Other Names

Levothroid, Levoxyl, Synthroid

Form	Molecular Formula	MW	CAS
Levothyroxine sodium	$C_{15}H_{10}I_4NNaO_4$	798.9	55-03-8

Appearance

Levothyroxine sodium occurs as a light yellow to buff odorless and tasteless hygroscopic powder.

Solubility

Levothyroxine sodium has a solubility of 15 mg/100 mL in water at 25 °C. It is soluble in mineral acids and solutions of alkali hydroxides and carbonates and more soluble in alcohol; it is very slightly soluble in chloroform and ether.

Method 1

Alexander et al. used an HPLC method to evaluate the stability of an extemporaneously formulated levothyroxine sodium syrup compounded from commercial tablets. The liquid chromatograph consisted of a Waters model 501 solvent-delivery system, a Waters model U6K universal liquid chromatography injector with a 2-mL loop, a Waters 484 tunable UV-visible single-channel detector, and an NEC APC IV personal computer with a Baseline 810 chromatography workstation. The stationary phase was a Supelco C_{18} 5-μm Supelcosil LC 18-DB analytical column. The mobile phase was a mixture of 1 part of buffer solution and 3 parts of acetonitrile. The buffer solution was 12.5 mL of 25% acetic acid diluted to volume with deionized water. The mobile phase was filtered through a Millipore 0.45-μm membrane filtration system, and the pH was adjusted to 7.3 with 4 N sodium hydroxide. The flow rate was 1.2 mL/min. UV detection was performed at 229 nm and 0.04 AUFS. Testosterone propionate 1000 μg/mL in methanol was used as the internal standard. The injection volume was 100 μL. Under these conditions, retention times for testosterone propionate and levothyroxine were 9.23 and 12.52 minutes, respectively.

The stability-indicating nature of the assay was demonstrated by accelerated degradation of a levothyroxine sample. A solution of levothyroxine was mixed with 1 N hydrochloric acid, and another solution was mixed with 1 N sodium hydroxide. These two solutions were heated to boiling for 90 minutes and cooled. Their chromatograms showed that the intact levothyroxine was well separated from degradation product peaks.

A standard curve for levothyroxine was constructed from 8 to 24 μg/mL. The intraday and interday coefficients of variation of the assay were less than 2%.

Reference
Alexander KS, Kothapalli MR, Dolimor D. Stability of an extemporaneously formulated levothyroxine sodium syrup compounded from commercial tablets. *Int J Pharm Compound.* 1997; 1: 60–4.

Method 2
Boulton et al. determined the stability of levothyroxine sodium in extemporaneously prepared oral liquid formulations made from commercially available tablets and from pure powder. The liquid chromatograph consisted of a Japan Spectroscopic model 880-PU solvent-delivery pump, a Shimadzu model SIL9A autosampler, a Japan Spectroscopic model 875 variable-wavelength UV detector, and a Digital Solutions Pty Delta V4.01 chromatography data analysis system. The stationary phase was a Hichrom Spherisorb CN stainless steel analytical column with 5-μm packing materials. The mobile phase was a mixture of acetonitrile, water, and phosphoric acid (35:65:1, vol/vol/vol). The flow rate was 1.0 mL/min. UV detection was performed at 225 nm.

Samples were centrifuged at 2500 rpm for 5 minutes. The injection volume was 50 μL. Under these conditions, the retention times for methylparaben, liothyronine, and levothyroxine were 2.3, 6.2, and 7.6 minutes, respectively.

The stability-indicating capability of the assay was shown by accelerated decomposition of levothyroxine. Samples of levothyroxine were acidified with 0.1 M phosphoric acid, sealed in glass tubes, and placed in an oven at 70 °C for 14 days. Degradation product peaks were well separated from the intact levothyroxine peak.

Calibration curves were generated over the levothyroxine concentration range of 1 to 50 μg/mL. The correlation coefficient was greater than 0.99. The intraday and interday coefficients of variation for the assay were less than 5 and 6.4%, respectively.

Similar stability-indicating methods were used by Garnick et al. and Gupta et al.

References

Boulton DW, Fawcett JP, Woods DJ. Stability of an extemporaneously compounded levothyroxine sodium oral liquid. *Am J Health Syst Pharm.* 1996; 53: 1157–61.

Gupta VD, Odom C, Bethea C, et al. Effect of excipients on the stability of levothyroxine sodium tablets. *J Clin Pharm Ther.* 1990; 15: 331–6.

Garnick RL, Burt GF, Long DA, et al. High-performance liquid chromatographic assay for sodium levothyroxine in tablet formulations: Content uniformity applications. *J Pharm Sci.* 1984; 73: 75–7.

Method 3

Richheimer and Amer developed a stability-indicating assay for analysis of sodium levothyroxine in tablets. The high-performance liquid chromatograph consisted of a Varian model 5020 system, a Varian model UV-50 variable-wavelength detector, a Beckman model 1005 strip-chart recorder, a Varian model CDS-111L electronic integrator, and a Valco model CV-6-UHPa-N60 200-μL loop injector. The stationary phase was a Waters μBondapak C_{18} column (300 × 4.0 mm, 10-μm particle size). The guard column was a Vydac C_{18} stainless steel column (40 × 4 mm, 40-μm particle size). The mobile phase consisted of 60% acetonitrile and 40% aqueous buffer. The buffer contained 0.005 M 1-octanesulfonic acid and 0.005 M tetramethylammonium chloride (pH 3.0). The flow rate was 2.0 mL/min. UV detection was performed at 230 nm and 0.08 AUFS. The injection volume was 200 μL. The retention times for 3,5-diiodo-L-thyronine, liothyronine, and levothyroxine were about 2.0, 3.0, and 4.2 minutes, respectively (estimated from the published chromatogram).

The method was determined to be stability indicating by accelerated degradation of levothyroxine. Samples of levothyroxine were exposed to short wavelength light for 24 hours, refluxed with 0.1 N hydrochloric acid for 16 hours, refluxed with 0.1 N sodium hydroxide for 16 hours, oxidized with 0.01 M hydrogen peroxide and 0.01 M ferric ion for 16 hours, and heated at 80 °C for 24 hours. In every case, degradation product peaks did not interfere with the intact levothyroxine peak.

Standard curves for levothyroxine were constructed from 5 to 85 μg/mL. The correlation coefficients were 0.9999.

Reference

Richheimer SL, Amer TM. Stability-indicating assay, dissolution, and content uniformity of sodium levothyroxine in tablets. *J Pharm Sci.* 1983; 72: 1349–51.

Lidocaine

Chemical Names

2-(Diethylamino)-*N*-(2,6-dimethylphenyl)acetamide
2-Diethylamino-2′,6′-acetoxylidide

Other Names
Lignocaine, Xylocaine

Form	Molecular Formula	MW	CAS
Lidocaine	$C_{14}H_{22}N_2O$	234.3	137-58-6
Lidocaine hydrochloride	$C_{14}H_{22}N_2O.HCl.H_2O$	288.8	6108-05-0

Appearance
Lidocaine is a white to slightly yellow crystalline powder. Lidocaine hydrochloride is a white odorless crystalline powder.

Solubility
Lidocaine is practically insoluble in water; it is very soluble in alcohol, chloroform, and methylene chloride. It is freely soluble in ether. Lidocaine hydrochloride is very soluble in water and alcohol and soluble in chloroform, but it is insoluble in ether.

pK_a
Lidocaine hydrochloride has a pK_a of 7.86.

Method 1
Gebauer et al. developed a method for the study of lidocaine and oxycodone stability in a rectal gel. The HPLC instrument consisted of a Waters model 510 pump, a model W717 autoinjector, and a model 996 photodiode-array detector and was managed by Waters Millenium 32 software. The stationary phase was a Hewlett-Packard Zorbax SB-C_8 reversed-phase column (250 × 4.6 mm, 5-μm particle size). The mobile phase was a mixture of 350 mL of methanol, 150 mL of water, 10 mL of acetic acid, and 1.60 g of sodium dodecyl sulfate. The flow rate was 1.5 mL/min. The eluant was monitored by a photodiode-array detector from 250 to 300 nm. A chromatogram was extracted from the contour plot at 264 nm for lidocaine.

A 0.5-mL sample of the gel was placed in a disposable filtration tube and centrifuged at 2600 × g for 30 min at 10 °C. The filtrate was collected and assayed. Under these conditions, retention times for the preservative methyl 4-hydroxybenzoate and for oxycodone and lidocaine were about 2.8, 5.1, and 8.0 minutes, respectively.

The method was demonstrated to be stability indicating by intentional degradation of lidocaine. Aqueous solutions of lidocaine hydrochloride were heated to dryness or mixed with hydrogen peroxide. The peak of the intact lidocaine hydrochloride was resolved from its degradation product peaks on the chromatogram.

A calibration curve for lidocaine was constructed in the concentration range from 0.1 to 5.0% (wt/wt). The correlation coefficient was greater than 0.998. The interday coefficient of variation for the analysis was 3.0%.

Reference

Gebauer MG, McClure AF, Vlahakis TL. Stability indicating HPLC method for the estimation of oxycodone and lidocaine in rectal gel. *Int J Pharm*. 2001; 223: 49–54.

Method 2

Wilson and Forde determined the stability of lidocaine hydrochloride using an HPLC method. The liquid chromatograph consisted of a Varian model 5000 pump and an autosampler, a Waters model 441 UV detector, a Fisher Scientific model 5000 strip-chart recorder, and a Hewlett-Packard model 3354 laboratory automation system. The stationary phase was a Whatman PXS ODS-3 stainless steel column (250 × 4.6 mm, 10-μm particle size). The mobile phase consisted of 5% acetic acid in water adjusted to pH 3.0 with 1 N sodium hydroxide and acetonitrile (800:200, vol/vol). The flow rate was 2.0 mL/min. UV detection was performed at 254 nm and 0.1 AUFS.

Samples were diluted to about 0.8 mg/mL with a mixture of water, methanol, and 0.5 M sodium borate (pH 7.0) (650:350:20, vol/vol/vol). The injection volume was 20 μL. Retention times for milrinone and lidocaine were about 3 and 5 minutes, respectively.

The method was stated to be stability indicating.

A standard curve for lidocaine was constructed from 0 to 7.668 mg/mL.

Reference

Wilson TD, Forde MD. Stability of milrinone and epinephrine, atropine sulfate, lidocaine hydrochloride, or morphine sulfate injection. *Am J Hosp Pharm*. 1990; 47: 2504–7.

Method 3

Gupta and Stewart investigated the stabilities of lidocaine hydrochloride and phenyl-ephrine hydrochloride in combination in the most commonly used concentrations. A Waters ALC 202 high-pressure liquid chromatograph was equipped with a Kratos SF 770 multiple-wavelength detector, a Houston Omniscribe recorder, and a Waters μBondapak C_{18} nonpolar column (300 × 3.9 mm). The mobile phase was an aqueous solution of 54% (vol/vol) acetonitrile containing 0.01 M monobasic potassium phosphate, adjusted to pH 7.05 ± 0.05 with phosphoric acid. The flow rate was 2.0 mL/min. UV detection was performed at 261 nm and 0.04 AUFS. An aqueous solution of verapamil hydro-chloride 0.16% was used as the internal standard.

Samples were diluted with water. The injection volume was 20 μL. The retention time for lidocaine was about 3.3 minutes (estimated from the published chromatogram).

The analytical method was stated to be stability indicating.

Reference

Gupta VD, Stewart KR. Chemical stabilities of lignocaine hydrochloride and phenyl-ephrine hydrochloride in aqueous solution. *J Clin Hosp Pharm*. 1986; 11: 449–52.

Method 4

Waraszkiewicz et al. developed a stability-indicating HPLC method for analysis of lidocaine hydrochloride and lidocaine hydrochloride with epinephrine injectable solutions. A Waters model ALC/GPC 204 liquid chromatograph was equipped with a

Waters model U6K septumless injector, a Waters model 440 dual-channel fixed-wave-length UV detector, and a Waters μBondapak CN column (300 × 4 mm, 10-μm particle size). The mobile phase consisted of 0.01 M 1-octanesulfonic acid sodium salt, 0.1 mM edetate disodium, 2% (vol/vol) acetic acid, 2% (vol/vol) acetonitrile, and 1% (vol/vol) methanol in high-quality distilled water. The flow rate was 2.0 mL/min. UV detection was performed at 254 nm. The sample injection volume was 2 μL. Under these conditions, the retention time for lidocaine was 6.8 minutes.

The assay was determined to be stability indicating. Samples of lidocaine hydrochloride were spiked with the potential degradation products and chromatographed. Other samples of lidocaine with epinephrine were subjected to various conditions of stress such as autoclaving for 5, 10, and 20 cycles, storage for 1 month at 80 and 110 °C, exposure to atmospheric oxygen, addition of 12 ppm of aluminum, and an increase in the pH of the solution. Degradation products did not interfere with the analysis of the intact lidocaine. These along with other compounds follow, with their retention times in parentheses in minutes: epinephrine sulfonic acid (1.8), adrenochrome (2.2), epinephrine (3.0), p-hydroxybenzoic acid (3.2), methylparaben (4.4), and 2,6-xylidine (5.0).

Standard curves were constructed from 5 to 20 mg/mL. Correlation coefficients were greater than 0.99.

Reference
Waraszkiewicz SM, Milano EA, DiRubio R. Stability-indicating high-performance liquid chromatographic analysis of lidocaine hydrochloride and lidocaine hydrochloride with epinephrine injectable solutions. *J Pharm Sci.* 1981; 70: 1215–8.

Linezolid

Chemical Name
(*S*)-*N*-[3-[3-Fluoro-4-(4-morpholinyl)phenyl-2-oxo-5-oxazolidinyl]methyl]acetamide

Other Name
Zyvox

Form	Molecular Formula	MW	CAS
Linezolid	$C_{16}H_{20}FN_3O_4$	337.4	165800-03-3

Appearance
Linezolid occurs as white crystals.

Method

Xu et al. used HPLC to evaluate the chemical stability of linezolid injection at 2 mg/mL admixed with three common cephalosporin antibiotics and stored for 7 days at 4 and 23 °C. The Waters liquid chromatograph consisted of a model 600E multisolvent-delivery pump, a model 996 photodiode-array detector, and a model 712 WISP auto-sampler. The stationary phase was a YMC-Pack ODS-AM column (150 × 4.6 mm, 5-μm particle size). The mobile phase consisted of 820 mL of water, 180 mL of acetonitrile, 1 mL of trifluoroacetic acid, and 1 mL of triethylamine. The flow rate was 1.5 mL/min. UV detection was performed at 254 nm and 0.5 AUFS.

Samples were diluted 1:20 with 10% acetonitrile in water. The injection volume was 10 μL. Under these conditions, retention times for ceftazidime, cefazolin, and linezolid were 2.0, 6.1, and 10.9 minutes, respectively.

The method was shown to be stability indicating by accelerated degradation. Sample solutions were treated with 1 N sodium hydroxide, 1 N hydrochloric acid, or 3% hydrogen peroxide, or were subjected to heating. There was no interference of the degradation product peaks with the intact drug peak.

A standard curve for linezolid was constructed from 25 to 150 μg/mL. The correlation coefficient was greater than 0.9999. Intraday and interday coefficients of variation were both 0.1%.

Similar methods were used by the authors in additional studies.

References

Xu QA, Trissel LA, Williams KY. Compatibility and stability of linezolid injection admixed with three cephalosporin antibiotics. *J Am Pharm Assoc*. 2000; 40: 509–14.

Zhang Y, Xu QA, Trissel LA, et al. Compatibility and stability of linezolid injection admixed with three quinolone antibiotics. *Ann Pharmacother*. 2000; 34: 996–1001.

Zhang Y, Xu QA, Trissel LA, et al. Compatibility and stability of linezolid injection admixed with aztreonam or piperacillin sodium. *J Am Pharm Assoc*. 2000; 40: 520–4.

Liothyronine Sodium

Chemical Name
Sodium 4-*O*-(4-hydroxy-3-iodophenyl)-3,5-diiodo-L-tyrosine

Other Names
Cynomel, Cytomel, Thybon, Triostat

Form	Molecular Formula	MW	CAS
Liothyronine sodium	$C_{15}H_{11}I_3NNaO_4$	673.0	55-06-1

Appearance
Liothyronine sodium is a white to light tan odorless crystalline powder.

Solubility

Liothyronine sodium is very slightly soluble to practically insoluble in water and slightly soluble in alcohol.

Method

Garnick et al. described a method for the quality control of liothyronine sodium and levothyroxine sodium in tablet formulations. The liquid chromatograph comprised a pump, an autosampler, a UV detector, and a plotter. The stationary phase was a DuPont Zorbax CN reversed-phase column. The mobile phase consisted of water and acetonitrile (40:60, vol/vol) containing 0.05% orthophosphoric acid. UV detection was performed at 225 nm.

A sample of ground tablets was mixed with 10 mL of mobile phase, vortexed for 3 minutes, and centrifuged. The clear supernatant liquid was collected and assayed. Under these conditions, retention times for liothyronine sodium and levothyroxine sodium were about 10.8 and 15.9 minutes, respectively (estimated from the published chromatogram).

The stability-indicating nature of the method was demonstrated by subjecting tablets to extreme stress conditions. Samples were treated with 0.1 N hydrochloric acid, 0.1 N sodium hydroxide solution, or 3% *tert*-butyl alcohol peroxide or were exposed to UV light. The degradation products did not interfere with the analysis of the drug.

A calibration curve for liothyronine sodium was constructed from 1 to 13 µg/mL. The correlation coefficient was 0.993.

Reference

Garnick RL, Burt GF, Borger FR, et al. Stability indicating high-pressure liquid chromatographic method for quality control of sodium liothyronine and sodium levothyroxine in tablet formulations. *Hormone Drugs: Proceedings of the FDA–USP Workshop on Drug and Reference Standards for Insulins, Somatropins, and Thyroid-axis Hormones.* Bethesda, MD, 1982; 504–16.

Lisinopril

Chemical Name

(*S*)-1-[*N*^{2}-(1-Carboxy-3-phenylpropyl)-L-lysyl]-L-proline dihydrate

Other Names
Prinivil, Zestril

Form	Molecular Formula	MW	CAS
Lisinopril	$C_{21}H_{31}N_3O_5.2H_2O$	441.5	83915-83-7

Appearance
Lisinopril is a white crystalline powder.

Solubility
Lisinopril is soluble 1 in 10 of water and 1 in 70 of methanol. It is practically insoluble in alcohol, acetone, chloroform, and ether.

Method 1
Nahata and Morosco used an HPLC method to study the stability of lisinopril in two liquid dosage forms. The HP 1050 instrumentation included a pump, an autosampler, a variable-wavelength detector, and a model 3396 integrator. The stationary phase was a MAC-MOD Zorbax C_{18} column (150 × 3.0 mm). The mobile phase consisted of 0.8% diethylamine in water and acetonitrile (43:57, vol/vol). The flow rate was 0.4 mL/min. UV detection was carried out at 220 nm.

Lisinopril stock solutions were prepared in methanol, mixed with mobile phase, and centrifuged, and the supernatant was then assayed. The injection volume was 10 μL. Under these conditions, the retention time for lisinopril was about 3.8 minutes. The run time was 10 minutes.

The method was established to be stability indicating by forced degradation studies of lisinopril. Lisinopril samples were treated with 2.0 M hydrochloric acid or 2.0 M sodium hydroxide solution, oxidized with 0.3% hydrogen peroxide, or heated at 80 °C. The analysis of lisinopril was not influenced by its degradation products.

A standard curve for lisinopril was constructed from 0.1 to 1.5 mg/mL. The correlation coefficient was greater than 0.999. Intraday and interday coefficients of variation of the analysis were less than 2.7 and 3.3%, respectively.

Reference
Nahata MC, Morosco RS. Stability of lisinopril in two liquid dosage forms. *Ann Pharmacother.* 2004; 38: 396–9.

Method 2
Webster et al. investigated the stability of lisinopril as an extemporaneous syrup. The liquid chromatograph consisted of a Waters model 510 pump, a Waters model 712 WISP refrigerated autosampler, and a Waters model 490E programmable multiple-wavelength detector. The stationary phase was a Supelco Supelcosil C_8 analytical column (250 × 4.6 mm, 5-μm particle size). The mobile phase consisted of 0.03 M monobasic potassium phosphate adjusted to pH 4.1 with phosphoric acid and acetonitrile (80:20, vol/vol) with 0.004 M 1-octanesulfonic acid sodium salt. The flow rate was 1.5 mL/min. UV detection was performed at 215 nm.

Each sample was diluted with a mixture of water and methanol (4:1, vol/vol) to a 0.2-mg/mL concentration of lisinopril. The retention time for lisinopril was about 4.6 minutes (estimated from the published chromatogram).

The stability-indicating capability of the method was demonstrated by accelerated decomposition of lisinopril. Solutions of lisinopril 2 mg/mL in distilled water and lisinopril 2 mg/mL in syrup were diluted 1:5 with 0.1 M sodium hydroxide, then diluted 1:5 with 0.05 M sulfuric acid, and then refluxed for 1 hour. In all cases, degradation product peaks did not interfere with the intact lisinopril peak.

A standard curve for lisinopril was constructed daily from 0.05 to 0.4 mg/mL; the correlation coefficient was 0.999. The coefficient of variation was less than 4%.

A similar method was described by Beasley et al.

References

Webster AA, English BA, Rose DJ. The stability of lisinopril as an extemporaneous syrup. *Int J Pharm Compound.* 1997; 1: 352–3.

Beasley CA, Shaw J, Zhao, Z, et al. Development and validation of a stability indicating HPLC method for determination of lisinopril, lisinopril degradation product and parabens in the lisinopril extemporaneous formulation. *J Pharm Biomed Anal.* 2005; 37: 559–67.

Lithospermic Acid B

Form	Molecular Formula	MW	CAS
Lithospermic acid B	$C_{36}H_{30}O_{16}$	718	121521-90-2

Method

Using an HPLC method, Guo et al. studied the kinetics and mechanism of degradation of lithospermic acid B in aqueous solution. The Agilent 1100 series included a model G1314A UV-visible detector. The stationary phase was an Agilent RP-8 column (200 × 4.6 mm, 5-µm particle size). The column temperature was maintained at 30 °C. Solvent A was 1% (vol/vol) acetic acid in water and solvent B was a mixture of acetonitrile and methanol (3:2, vol/vol). The mobile phase was delivered in a linear gradient mode at 1 mL/min:

Time, minutes	Solvent A, %	Solvent B, %
0	80	20
15	70	30
20	70	30
22	80	20
30	80	20

UV detection was performed at 288 nm.

Lithospermic acid B stock solution was prepared in double-distilled water. Under these conditions, the retention time of lithospermic acid B was about 17.8 minutes (estimated from the published chromatogram).

The analytical method was reported to be stability indicating.

Reference
Guo Y-X, Xiu Z-L, Zhang D-J, et al. Kinetics and mechanism of degradation of litho-spermic acid B in aqueous solution. *J Pharm Biomed Anal.* 2006; 43: 1249–55.

Loperamide Hydrochloride

Chemical Name
4-(4-*p*-Chlorophenyl-4-hydroxypiperidino)-*N,N*-dimethyl-2,2-diphenylbutyramide hydrochloride

Other Names
Arestal, Arret, Blox, Imodium

Form	Molecular Formula	MW	CAS
Loperamide hydrochloride	$C_{29}H_{33}ClN_2O_2 \cdot HCl$	513.5	34552-83-5

Appearance
Loperamide hydrochloride is a white to faintly yellow amorphous or microcrystalline powder.

Solubility
Loperamide hydrochloride is slightly soluble in water and soluble in alcohol.

pK$_a$
Loperamide hydrochloride has a pK$_a$ of 8.6.

Method
Tu et al. evaluated by HPLC the stability of loperamide hydrochloride in aqueous solutions. The chromatograph included a Waters model 6000A dual-piston pump, a model 440 dual-wavelength UV absorbance detector, and a Houston Omniscribe strip-chart recorder. The stationary phase was a Waters μBondapak C_{18} column (300 × 3.9 mm, 10-μm particle size). The mobile phase consisted of methanol, acetonitrile, and 0.2 M phosphate buffer (pH 3.0) containing 0.005 M heptanesulfonic acid sodium salt (12:1:10, vol/vol/vol). The flow rate was 3.0 mL/min. UV detection was performed at 201 nm. Under these conditions, the retention time for loperamide hydrochloride was 4.4 minutes.

The stability-indicating capability of the assay was demonstrated by accel-erated decomposition of a sample of loperamide hydrochloride solution at pH 8.0 and

90 °C for 85 days. The decomposition products were well separated from the intact loperamide hydrochloride on the chromatogram.

A standard curve for loperamide hydrochloride was generated from 2 to 10 µg/mL. The correlation coefficient was greater than 0.99.

Reference

Tu Y-H, Allen LV Jr, Wang D-P. Stability of loperamide hydrochloride in aqueous solutions as determined by high performance liquid chromatography. *Int J Pharm.* 1989; 51: 157–60.

Loratadine

Chemical Name

Ethyl 4-(8-chloro-5,6-dihydro-11*H*-benzo[5,6]cyclohepta[1,2-*b*]pyridin-11-ylidine)-1-piperidinecarboxylate

Other Name

Claritin

Form	Molecular Formula	MW	CAS
Loratadine	$C_{22}H_{23}ClN_2O_2$	382.9	79794-75-5

Appearance

Loratadine is a white to off-white powder.

Solubility

Loratadine is insoluble in water. It is very soluble in ethanol, acetone, and chloroform.

Method

El-Ragehy et al. described an HPLC method for the determination of loratadine in dosage forms and in the presence of its degradate. The Shimadzu liquid chromatograph included an LC-10 AD pump and an SPD-10A UV-visible detector. The stationary phase was a µBondapak C_{18} column (250 × 4.6 mm, 10-µm particle size). The mobile phase consisted of acetonitrile and phosphoric acid (35:65, vol/vol). The flow rate was 2.0 mL/min. UV detection was performed at 250 nm. Benzophenone 0.1 mg/mL in methanol was used as an internal standard.

For tablets, a portion of powder from 20 tablets equivalent to 10 mg of loratadine was accurately weighed into a beaker, extracted with 80 mL of methanol, filtered into a 100-mL volumetric flask, and filled to volume with methanol. A portion of the solution was mixed with internal standard and properly diluted with methanol. For syrup, 10 mL of loratadine (Claritin) syrup was extracted three times with 10 mL of chloroform, evaporated to dryness, and reconstituted with 100 mL of methanol. An aliquot of this solution was mixed with internal standard and properly diluted with methanol. The injection volume was 20 μL. Under these conditions, retention times for loratadine and benzophenone were about 4.6 and 6.4 minutes, respectively.

To show the stability-indicating nature of this method, loratadine was assayed in the presence of its degradation product. The retention time of its degradation product was about 2.0 minutes.

A calibration curve for loratadine was constructed from 5.0 to 50.0 μg/mL. The limit of detection and the limit of quantitation of the analysis were 4.1 and 5.0 μg/mL, respectively.

Reference
El-Ragehy NA, Badawey AM, El-Khateeb SZ. Stability indicating method for the determination of loratadine in the presence of its degradation product. *J Pharm Biomed Anal.* 2002; 28: 1041–53.

Lorazepam

Chemical Name
7-Chloro-5-(2-chlorophenyl)-1,3-dihydro-3-hydroxy-2*H*-1,4-benzodiazepin-2-one

Other Name
Ativan

Form	Molecular Formula	MW	CAS
Lorazepam	$C_{15}H_{10}Cl_2N_2O_2$	321.2	846-49-1

Appearance
Lorazepam is a white or almost white crystalline powder.

Solubility
Lorazepam has the following solubilities: 0.08 mg/mL in water, 3 mg/mL in chloroform, 14 mg/mL in alcohol, 16 mg/mL in propylene glycol, and 30 mg/mL in ethyl acetate.

pK$_a$

Lorazepam has pK$_a$ values of 1.3 and 11.5.

Method 1

Mayron and Gennaro used an HPLC method to assess the stability of lorazepam with granisetron hydrochloride during simulated Y-site administration. The liquid chromatograph consisted of a piston pump with a pulse dampener, a rotary injection port with a 20-µL loop, a variable-wavelength spectrometric detector, and an integrator. The stationary phase was a cyano reversed-phase column (250 × 4.6 mm, 5-µm particle size). The mobile phase was a mixture of acetonitrile and 0.1 M monobasic sodium phosphate dihydrate (20:80, vol/vol) adjusted to pH 4.2 with phosphoric acid. The flow rate was 1.5 mL/min. UV detection was performed at 220 nm.

Samples were diluted with mobile phase. The retention times for granisetron and lorazepam were 4.62 and 8.72 minutes, respectively.

The stability-indicating capability of the analytical method was demonstrated by an accelerated degradation study. Solutions of lorazepam and granisetron hydrochloride were adjusted to pH 2 and 11, boiled for 1 hour, readjusted to pH 5, diluted with the mobile phase, and analyzed. The degradation product peaks did not interfere with the intact lorazepam and granisetron peaks.

Reference

Mayron D, Gennaro AR. Stability and compatibility of granisetron hydrochloride in i.v. solutions and oral liquids and during simulated Y-site injection with selected drugs. *Am J Health Syst Pharm.* 1996; 53: 294–304.

Method 2

Stiles et al. evaluated the stability of lorazepam in polypropylene infusion-pump syringes. The liquid chromatograph included a Waters model 501 pump, a Waters model 441 UV detector, a Waters model 745 data module, an Alcott model 728 autosampler, and a Rheodyne 7010 injector. The stationary phase was a Bakerbond C$_{18}$ analytical column. The mobile phase consisted of 46% acetonitrile and 54% 0.08 M phosphoric acid in water (pH 2.0). It was delivered isocratically at 1.2 mL/min. UV detection was performed at 240 nm.

Samples were diluted 1:100 before injection. Under these conditions, the retention time for lorazepam was 6.6 minutes.

The assay was stated to be stability indicating and was originally published by Walmsley and Chasseaud.

A standard curve for lorazepam was constructed from 5 to 25 µg/mL. Intraday and interday coefficients of variation of the assay were 1.4 and 1.9%, respectively.

References

Stiles ML, Allen LV, Prince SJ. Stability of deferoxamine mesylate, floxuridine, fluorouracil, hydromorphone hydrochloride, lorazepam, and midazolam hydrochloride in polypropylene infusion-pump syringes. *Am J Health Syst Pharm.* 1996; 53: 1583–8.

Walmsley LM, Chasseaud LF. High performance liquid chromatographic determination of lorazepam in monkey plasma. *J Chromatogr.* 1981; 226: 155–63.

Method 3

Trissel and Pearson used an HPLC method to determine the stability of lorazepam injection at 0.1 mg/mL in intravenous infusion solutions. The chromatograph consisted of a Waters model 600E multisolvent-delivery pump, a Waters model 490E multiple-wavelength UV detector, a Waters model 710B WISP autosampler, and a Vydac C_8 reversed-phase HPLC column (250 × 4.6 mm, 5-μm particle size). The system was controlled and integrated by an NEC PowerMate SX/16 personal computer running Waters chromatography manager Maxima 820. The mobile phase consisted of 57% methanol in 50 mM monobasic ammonium phosphate with the pH adjusted to 6.5 with ammonium hydroxide. The mobile phase was delivered isocratically at 1 mL/min. UV detection was performed at 254 nm. Under these conditions, the retention time of lorazepam was 6.1 minutes.

The stability-indicating nature of the method was established by accelerated degradation of lorazepam. A lorazepam solution was heated at 90 °C for 3 hours. A 67% reduction in the peak area for intact lorazepam and the formation of a new peak eluting at 9.6 minutes were observed. The degradation product did not interfere with the analysis of lorazepam.

Standard curves were constructed from a linear plot of peak area versus concentration of lorazepam from 25 to 150 μg/mL. The correlation coefficients were greater than 0.9999.

Reference

Trissel LA, Pearson SD. Storage of lorazepam in three injectable solutions in polyvinyl chloride and polyolefin bags. *Am J Hosp Pharm.* 1994; 51: 368–72.

Method 4

McGuire et al. evaluated the compatibility and stability of lorazepam, dexamethasone sodium phosphate, and ondansetron hydrochloride in injectable solutions. The liquid chromatograph consisted of a Shimadzu LC-6A single-piston low-pulsatile-flow pump, a Shimadzu SPD-6AV UV-visible spectrophotometer, and a Shimadzu SCL-6B controller. The stationary phase was a Phenomenex C_{18} column (150 × 4.6 mm, 5-μm particle size). The mobile phase consisted of 0.02 M dibasic sodium phosphate buffer and acetonitrile (56:44, vol/vol) adjusted to pH 7.5. The flow rate was 0.9 mL/min. The sample injection volume was 50 μL. Retention times for dexamethasone, lorazepam, and ondansetron were 2.7, 6.8, and 16.5 minutes, respectively.

The HPLC method was evaluated to be stability indicating by accelerated decomposition of lorazepam, dexamethasone sodium phosphate, and ondansetron hydrochloride. Solutions of these drugs were treated with 6 N hydrochloric acid at 90 °C for 6 hours. Degradation product peaks could be distinguished from their parent drug peaks.

A standard curve for lorazepam was constructed from 0 to 0.1 mg/mL; the correlation coefficient was 0.999. Intraday and interday coefficients of variation were less than 5%.

Reference

McGuire TR, Narducci WA, Fox JL. Compatibility and stability of ondansetron hydrochloride, dexamethasone, and lorazepam in injectable solutions. *Am J Hosp Pharm.* 1993; 50: 1410–4.

Method 5

Nahata et al. reported the stability of lorazepam injection diluted from 4 to 1 mg/mL in bacteriostatic water for injection and stored in glass vials. The liquid chromatograph consisted of a Hewlett-Packard 1050 series pump, an autosampler, a variable-wavelength detector, and a Hewlett-Packard 3394A integrator. The stationary phase was a Beckman Ultrasphere ODS column (250 × 4.6 mm, 5-μm particle size). The mobile phase was 60% methanol in deionized water and was delivered isocratically at 1.0 mL/min. UV detection was performed at 230 nm. Acetophenone 2.5 μg/mL was used as the internal standard. The injection volume was 25 μL. Under these conditions, retention times for acetophenone and lorazepam were 4.3 and 6.4 minutes, respectively.

The analytical method was determined to be stability indicating. One milliliter of lorazepam solution 1 mg/mL was mixed with 1 mL of 1.0 M sodium hydroxide and heated to 60 °C for 30 minutes. The determination of lorazepam was not influenced by degradation products.

Standard curves were generated from 0.50 to 1.50 mg/mL; the correlation coefficient was greater than 0.999. The coefficient of variation was less than 1%.

Reference

Nahata MC, Morosco RS, Hipple TF. Stability of lorazepam diluted in bacteriostatic water for injection at two temperatures. *J Clin Pharm Ther*. 1993; 18: 69–71.

Method 6

Walker et al. studied the stability of lorazepam using an HPLC assay. The chromatograph consisted of a Schoeffel SF 770 UV detector, a Spectra-Physics SP4200 chromatographic integrator, and a Beckman Ultrasphere C_{18} reversed-phase column (250 × 4.2 mm, 5-μm particle size). The mobile phase consisted of acetonitrile and phosphate buffer containing 1 mg/mL of heptanesulfonic acid and was pumped gradiently at 2.0 mL/min from 19 to 50% acetonitrile in 15 minutes. UV detection was performed at 230 nm.

The analytical method was determined to be stability indicating by accelerated degradation of lorazepam using acid, base, and heat and by inspection of the chromatograms for the appearance of additional peaks and changes in retention times and peak shapes. Intraday and interday coefficients of variation for the assay were 2.71 and 1.53%, respectively.

Reference

Walker SE, Iazzetta J, DeAngelis C, et al. Stability and compatibility of combination of hydromorphone and dimenhydrinate, lorazepam or prochlorperazine. *Can J Hosp Pharm*. 1993; 46: 61–5.

Lornoxicam

Chemical Name

6-Chloro-4-hydroxy-2-methyl-*N*-2-pyridinyl-2*H*-thieno[2,3-*e*][1,2]-thiazine-3-carboxamide 1,1-dioxide

Other Names
Acabel, Artok, Xefo

Form	Molecular Formula	MW	CAS
Lornoxicam	$C_{13}H_{10}ClN_3O_4S_2$	371.8	70374-39-9

Appearance
Lornoxicam occurs as orange to yellow crystals.

Method
Taha et al. developed an HPLC method for the determination of lornoxicam in the presence of its alkaline degradation products. The Waters chromatograph consisted of a model 600 LC series pump, a model 600 controller unit, a model 486 tunable absorbance detector, an injector with a 20-μL loop, and a model 746 data module. The stationary phase was a Nova-Pak C_{18} column (150 × 3.9 mm, 4-μm particle size). The mobile phase was a mixture of methanol, acetonitrile, and acetate buffer (pH 4.6) (4.5:0.5:5.0, vol/vol/vol). The flow rate was 0.8 mL/min. The detector was set at 280 nm.

A portion of powder from 10 tablets equivalent to 25 mg of lornoxicam was accurately weighed, mixed in 0.3 mL of 1 N sodium hydroxide solution and 15 mL of methanol, stirred for 30 minutes, further diluted with methanol, and filtered. The sample injection volume was 20 μL. Under these conditions, the retention time for lornoxicam was about 4.0 minutes.

Lornoxicam was heated in 1 N sodium hydroxide solution for 3 hours. Its degradation products did not interfere with the determination of the intact lornoxicam, demonstrating the stability-indicating nature of the method.

A calibration curve for lornoxicam was constructed from 0.5 to 20 μg/mL. The correlation coefficient was 0.9998. Intraday and interday coefficients of variation of the analysis were 1.76 and 0.97%, respectively. The limit of detection and the limit of quantitation were 0.01 and 0.04 μg/mL, respectively.

Reference
Taha EA, Salama NN, Fattah LEA. Stability-indicating chromatographic methods for the determination of some oxicams. *J AOAC Int.* 2004; 87: 366–73.

Losartan Potassium

Chemical Name
2-Butyl-4-chloro-1-[p-(1H-tetrazol-5-ylphenyl)benzyl]imidazole-5-methanol potassium

Other Names
Cozaar, Lortaan

Form	Molecular Formula	MW	CAS
Losartan potassium	$C_{22}H_{22}ClKN_6O$	461.0	124750-99-8

Appearance
Losartan potassium is a white to off-white free-flowing crystalline powder.

Solubility
Losartan potassium is freely soluble in water and soluble in alcohol.

Method 1
Hertzog et al. reported the development and validation of a method for the simultaneous determination of losartan potassium, hydrochlorothiazide, and their degradation products. The Thermo Separations liquid chromatographic system consisted of a model P4000 pump, a model AS3000 injector, a SpectraFOCUS variable-wavelength UV detector, and a model SN4000 controller. The stationary phase was a Waters Symmetry C_8 column (150 × 3.9 mm, 5-μm particle size). Mobile phase A consisted of a phosphate buffer and acetonitrile (85:15, vol/vol). The phosphate buffer was a solution of monobasic potassium phosphate and dibasic sodium phosphate (0.02 M, pH 7.0). Mobile phase B was acetonitrile. The mobile phase was delivered in a gradient mode from 80% A to 40% A in 10 minutes and back to 80% A for equilibration prior to the next injection, for a total run time of 15 minutes. The flow rate was 1.0 mL/min. UV detection was carried out at 250 nm.

Losartan potassium reference standard was prepared in mobile phase A. Losartan tablets were dissolved in mobile phase A with sonication and shaking and further dilution. Standards and samples were filtered through a Gelman 0.45-μm Acrodisc CR polytetrafluoroethylene (PTFE)-membrane filter. The injection volume was 10 μL. Under these conditions, the retention time of losartan was about 2.5 minutes (estimated from the published chromatogram).

This analytical method was verified to be stability indicating by assaying the worst-case placebo mixture containing excipients from all losartan tablet formulations. The peak of losartan was well separated from the peaks of other materials.

Standard curves were constructed from 50 to 150 μg/mL. The correlation coefficient was 0.9999.

Reference
Hertzog DL, McCafferty JF, Fang X, et al. Development and validation of a stability-indicating HPLC method for the simultaneous determination of losartan potassium, hydrochlorothiazide, and their degradation products. *J Pharm Biomed Anal.* 2002; 30: 747–60.

Method 2

Zhao et al. described an HPLC analytical method for losartan in stressed tablets. The liquid chromatograph consisted of a Dionex GP40 gradient pump, a Thermal Separation AS3500 autosampler, and a Liner UVIS-205 UV detector. The stationary phase was a Spherisorb C_8 column (250 × 4.6 mm, 5-μm particle size). The mobile phase consisted of 0.001 M phosphate buffer (pH 2.3) and acetonitrile and was delivered from 40 to 75% acetonitrile in 25 minutes and held for 5 minutes at 75% acetonitrile. The flow rate was 1.5 mL/min. UV detection was performed at 230 nm.

A tablet equivalent to 25 mg of losartan was transferred into a 250-mL volumetric flask, filled to volume with acetonitrile–water (1:1, vol/vol), vigorously stirred for 30 minutes, and centrifuged. The clear supernatant was diluted twofold prior to analysis. The injection volume was 50 μL. The total run time was 30 minutes.

The method was reported to be stability indicating. The intact losartan peak was separated from its three degradation product peaks.

Reference

Zhao ZZ, Wang Q, Tsai EW, et al. Identification of losartan degradates in stressed tablets by LC-MS and LC-MS/MS. *J Pharm Biomed Anal.* 1999; 20: 129–36.

Lovastatin

Chemical Name

(3*R*,5*R*)-7-{(1*S*,2*S*,6*R*,8*S*,8a*R*)-1,2,6,7,8,8a-Hexahydro-2,6-dimethyl-8-[(*S*)-2-methylbu-tyryloxy]-1-naphthyl}-3-hydroxyheptan-5-olide

Other Names

Lipofren, Mevacor, Mevinacor

Form	Molecular Formula	MW	CAS
Lovastatin	$C_{24}H_{36}O_5$	404.5	75330-75-5

Appearance

Lovastatin occurs as a white to off-white crystalline powder.

Solubility

Lovastatin is practically insoluble in water; freely soluble in chloroform; and soluble in acetone, acetonitrile, and methyl alcohol. It is sparingly soluble in alcohol.

Method

Alvarez-Lueje et al. developed an HPLC method to study the hydrolytic degradation of lovastatin in different media. The liquid chromatograph included a Waters model 600 controller pump, a model 996 photodiode-array detector, and a Rheodyne model 7125 injector with a 20-μL loop. The stationary phase was a μBondapak/Porasil C_{18} column (150 × 3.9 mm, 10-μm particle size) with a μBondapak C_{18} guard column (30 × 4.6 mm). The column temperature was maintained at 45 °C. The mobile phase consisted of acetonitrile, methanol, and 0.05 M phosphate buffer (pH 4) (32:33:35, vol/vol/vol). The flow rate was 1.5 mL/min. UV detection was performed at 238 nm. Under these conditions, the retention time of lovastatin was 8.3 minutes.

Lovastatin was hydrolyzed in 0.1 N hydrochloric acid or phosphate buffer (pH 7.4) at 37, 60, and 80 °C for up to 20 minutes. The peak of lovastatin was separated from peaks of its degradation products, demonstrating the stability-indicating property of the method.

A standard curve for lovastatin was obtained from 1.0×10^{-6} to 1.0×10^{-3} M. The correlation coefficient was 0.9997. The limit of detection and the limit of quantitation were 9.1×10^{-7} and 2.8×10^{-6} M, respectively.

Reference

Alvarez-Lueje A, Pastine J, Squella JA, et al. Assessment of the hydrolytic degradation of lovastatin by HPLC. *J Chil Chem Soc.* 2005; 50: 639–46.

Lysine

Other Name
Enisyl

Form	Molecular Formula	MW	CAS
Lysine	$C_6H_{14}N_2O_2$	146.2	56-87-1
Lysine acetate	$C_6H_{14}N_2O_2 \cdot C_2H_4O_2$	206.2	57282-49-2
Lysine hydrochloride	$C_6H_{14}N_2O_2 \cdot HCl$	182.7	657-27-2

Appearance
Lysine hydrochloride occurs as a white odorless powder or as colorless crystals.

Solubility
Lysine hydrochloride is freely soluble in water and slightly soluble in alcohol. It is practically insoluble in ether.

Method

Muhammad and Bodnar reported an assay for the quantitative determination of L-lysine in its raw materials or stability samples. The apparatus consisted of a Waters model 6000A pump, a model 440 UV detector, a Valco injector with a 20-µL loop, and a Hewlett-Packard 7130A recorder. The stationary phase was a reversed-phase LiChrosorb RP18 column (250 × 4.6 mm, 10-µm particle size). The mobile phase was prepared by mixing 600 mL of methanol, 400 mL of water, and 2.5 mL of glacial acetic acid (pH ~ 3.0). The flow rate was 1.5 mL/min.

L-Lysine was derivatized prior to analysis. Tris(hydroxymethyl)aminoethane (THAM) solution was prepared by dissolving 1.44 mg of THAM in a 100-mL volumetric flask and filling to volume with distilled water. 2,4-Dinitrofluorobenzene solution was prepared by dissolving 760 mg of 2,4-dinitrofluorobenzene in a 50-mL volumetric flask and diluting to volume with absolute alcohol. The internal standard solution was prepared by dissolving 27 mg of 2-naphthoic acid in a 25-mL volumetric flask by sonicating and diluting to volume with THAM solution. Thirty-six milligrams of sample was accurately weighed, placed in a 50-mL volumetric flask, and then dissolved in and diluted to volume with distilled water. Two milliliters of this solution was transferred to a 10-mL volumetric flask, mixed with 2 mL of THAM solution and 2 mL of 2,4-dinitrofluorobenzene solution, suspended in a 50 °C water bath for 30 minutes, cooled to room temperature, mixed with 2 mL of internal standard solution, and diluted to volume with methanol. The injection volume was 20 µL. Under these conditions, retention times for 2-naphthoic acid and L-lysine derivative were about 8.6 and 11.7 minutes, respectively (estimated from the published chromatogram).

The assay was reported to be stability indicating by accelerated degradation of L-lysine. Aqueous L-lysine solutions were mixed with 1 N hydrochloric acid and 1 N sodium hydroxide solution and refluxed for 30 minutes. Another L-lysine solid sample was irradiated with 2500 Å light for 6 days. All stressed samples were derivatized and assayed. The degradation product did not interfere with the determination of L-lysine.

Reference

Muhammad N, Bodnar JA. Stability-indicating high-pressure liquid chromatographic assay for L-lysine. *J Liq Chromatogr.* 1980; 3: 529–36.

Malonic Acid

Chemical Name
2,2-Dimethyl-1,3-dioxane-4,6-dione

Other Name
Meldrum's Acid

Form	Molecular Formula	MW	CAS
Malonic acid	$C_6H_8O_4$	144.1	2033-24-1

Appearance
Malonic acid occurs as small crystals.

Solubility
Malonic acid has solubilities of 1 g in 0.65 mL of water, in about 2 mL of alcohol, in 1.1 mL of methanol, and in 13 mL of ether.

pK$_a$
Meldrum's acid has a pK$_a$ of 5.1.

Method
Kaczvinsky and Read developed an assay for malonic acid. The chromatographic apparatus consisted of a Waters model 510 pump, a model 840 datastation, a model 712B WISP autosampler, and a model 490 multiwavelength UV-visible detector. The stationary phase was a Supelco LC-18 column (250 × 4.6 mm, 5-µm particle size). The mobile phase consisted of 5% acetonitrile and 5% phosphoric acid in water. The flow rate was 1 mL/min. UV detection was performed at 210 nm.

The sample was diluted 1:10 with water (pH 7.5) prior to assay if necessary. Under these conditions, the retention time for malonic acid was about 11 minutes (estimated from the published chromatogram).

The assay was reported to be stability indicating since it separated the intact malonic acid from its precursors and potential degradation products. Retention times for malonic acid, acetone, and an unknown compound were about 4.0, 5.8, and 14 minutes, respectively (estimated from the published chromatogram).

Standard curves for malonic acid were constructed from 0.1 to 10 mg/mL. The mean correlation coefficient was more than 0.999.

Reference
Kaczvinsky JR Jr, Read SA. Development and use of a stability-indicating high-performance liquid chromatographic assay for Meldrum's acid. *J Chromatogr.* 1992; 575: 177–81.

Mebendazole

Chemical Name
Methyl 5-benzoyl-1*H*-benzimidazol-2-ylcarbamate

Other Names
Madicure, Ovex, Vermox

Form	Molecular Formula	MW	CAS
Mebendazole	$C_{16}H_{13}N_3O_3$	295.3	31431-39-7

Appearance
Mebendazole is a white to slightly yellow almost odorless powder.

Solubility
Mebendazole is practically insoluble in water, alcohol, chloroform, and ether. It is freely soluble in formic acid.

Method 1
Al-Kurdi et al. investigated mebendazole in the presence of its major degradation products in pharmaceutical formulations. The Beckman Gold system included a model 116 pump, a model 166 UV spectrophotometric detector, and a Rheodyne model 7010 injector with a 20-µL loop. The stationary phase was a Waters Spherisorb S5 ODS1 column (250 × 4.6 mm, 5-µm particle size). The mobile phase consisted of 0.05 M monobasic potassium phosphate, methanol, and acetonitrile (5:3:2, vol/vol/vol). The flow rate was 1 mL/min. UV detection was performed at 290 nm.

For tablets, a portion of ground tablets equivalent to 100 mg of mebendazole was weighed, dissolved in 100 mL of 0.1 M methanolic hydrochloric acid, shaken for 15 minutes, and centrifuged. For suspensions, an amount of suspension equivalent to 100 mg of mebendazole was dissolved in 100 mL of 0.1 M methanolic hydrochloric acid, shaken for 10 minutes, and centrifuged for 15 minutes. In both cases, 5 mL of the supernatant was diluted to 50 mL with the mobile phase. The injection volume was 20 µL. Under these conditions, the retention time for mebendazole was 7.2 minutes.

The stability-indicating nature of the method was shown by accelerated degradation of mebendazole. Samples were treated with 1 M sodium hydroxide solution, 1 M hydrochloric acid, or 3.3% hydrogen peroxide and refluxed for 30 minutes. The degradation product eluted at 4.6 minutes.

A standard curve for mebendazole in 0.1 M methanolic hydrochloric acid was generated from 4 to 16 µg/mL. The correlation coefficient was 0.9999. The limit of detection and the limit of quantitation were 0.243 and 0.810 µg/mL, respectively.

Reference
Al-Kurdi Z, Al-Jallad T, Badwan A, et al. High performance liquid chromatography method for determination of methyl-5-benzoyl-2-benzimidazole carbamate (mebendazole) and its main degradation product in pharmaceutical dosage forms. *Talanta*. 1999; 50: 1089–97.

Method 2
Argekar et al. developed a simple, precise, and rapid HPLC method for the simultaneous determination of mebendazole and pyrantel pamoate from tablets. The chromatograph consisted of a Tosho CCPE pump equipped with a universal injector and an Oracle 2 integrator. The stationary phase was a reversed-phase Shodex C$_8$ column (250 × 3.9 mm, 5-µm particle size). The mobile phase was a mixture of 0.05 M monobasic sodium

phosphate, acetonitrile, and triethylamine (60:40:1.5, vol/vol/vol) and was adjusted to pH 6.8 with dilute phosphoric acid. UV detection was performed at 290 nm.

Tablets were weighed and finely ground. An amount of powder equivalent to 75 mg of mebendazole and 50 mg of pyrantel pamoate was mixed with 20 mL of glacial acetic acid, ultrasonicated for 30 minutes, kept in a water bath for 20 minutes, cooled, and diluted with 65 ml of a mixture of acetonitrile, diethylamine, and acetic acid (90:10:10, vol/vol/vol) (diluent A). The mixture was sonicated for another 20 minutes and was brought to a final volume of 100 mL with diluent A. The resulting solution was filtered through Whatman No. 42 filter paper. Six milliliters of the filtrate was diluted to 50 mL with the mobile phase and used for the assay. Under these conditions, retention times for pamoic acid, pyrantel, and mebendazole were about 3.6, 6.4, and 12.8 minutes, respectively.

The HPLC method was demonstrated to be stability indicating by accelerated decomposition of mebendazole. Tablets were kept in an oven at 105 °C for 1 month and the solution of tablets was kept in sunlight for 1 hour. The test samples were analyzed by the HPLC method. Degradation products of mebendazole and pyrantel pamoate did not interfere with the intact mebendazole.

A standard curve for mebendazole was constructed from 30 to 240 µg. The correlation coefficient was 0.999. The limit of detection and the limit of quantification for mebendazole were 15 and 60 µg, respectively.

Reference

Argekar AP, Raj SV, Kapadia SU. Simultaneous determination of mebendazole and pyrantel pamoate from tablets by high performance liquid chromatography–reverse phase (RP-HPLC). *Talanta*. 1997; 44: 1959–65.

Mechlorethamine

Chemical Names

2-Chloro-*N*-(2-chloroethyl)-*N*-methylethanamine
2,2′-Dichloro-*N*-methyldiethylamine

Other Names

Mustargen, Mustine

Form	Molecular Formula	MW	CAS
Mechlorethamine	$C_5H_{11}Cl_2N$	156.1	51-75-2
Mechlorethamine hydrochloride	$C_5H_{11}Cl_2N.HCl$	192.5	55-86-7

Appearance

Mechlorethamine hydrochloride is a hygroscopic light yellow to brown crystalline powder.

Solubility

Mechlorethamine hydrochloride is very soluble in water. It is also soluble in alcohol.

pK$_a$

Mechlorethamine hydrochloride has an approximate pK$_a$ of 6.1.

Method 1

Zhang et al. investigated the stability of mechlorethamine hydrochloride 0.01% ointment in Aquaphor base. The Waters LC Module-1 HPLC liquid chromatograph consisted of a multisolvent-delivery pump, a UV detector, and an autosampler in one unit. The system was controlled and integrated by a personal computer with Waters Millennium 2010 chromatography management software. The stationary phase was a Waters μBondapak C$_{18}$ reversed-phase analytical column (300 × 3.9 mm, 10-μm particle size) with a guard column of the same packing material. The mobile phase consisted of 5 mM phosphoric acid (pH 3.0) and acetonitrile by gradient delivery at 1.0 mL/min. For the first 3 minutes, the mobile phase was composed of 70% phosphoric acid and 30% acetonitrile. Between 3 and 14 minutes, acetonitrile was increased linearly from 30 to 100%. From 14 to 21 minutes, the acetonitrile was ramped back down linearly to 30%. From 21 to 26 minutes, the mobile phase was composed of 70% phosphoric acid and 30% acetonitrile and was delivered isocratically. UV detection was performed at 276 nm and 0.5 AUFS.

A mechlorethamine hydrochloride ointment aliquot was dissolved in a solution of 4 mL of chloroform, 0.9 mL of 2-propanol, and 0.1 mL of water. Mechlorethamine was then extracted into 5 mL of 10 mM hydrochloric acid and centrifuged at 3500 rpm for 10 minutes. One milliliter of the resulting liquid phase was neutralized with 0.1 N sodium hydroxide. To this solution 0.1 mL of a diethyldithiocarbamate (DDTC) sodium solution was added. It was incubated for 1 hour at 37 °C. The sample was then mixed with 1 mL of methanol, and 50 μL was analyzed by HPLC. Under these conditions, the retention time for the mechlorethamine–DDTC adduct was about 13.8 minutes.

The HPLC method was determined to be stability indicating by accelerated decomposition of mechlorethamine hydrochloride. Aqueous solutions of mechlorethamine hydrochloride 100 μg/mL were allowed to stand at room temperature. Approximately 35% loss of peak area for mechlorethamine–DDTC occurred, and new peaks formed at a retention time of about 17.4 minutes. The degradation product peaks did not interfere with the peak for the mechlorethamine–DDTC adduct.

Standard curves were constructed from a linear plot of peak area versus concentration of mechlorethamine hydrochloride reference standard from 0.5 to 10 μg/mL. The correlation coefficient was greater than 0.9999. The intraday and interday coefficients of variation of the assay were 4.1 and 1.9%, respectively. The extraction recovery efficiency was 84.5 ± 1.6%.

Reference

Zhang Y, Trissel LA, Johansen JF, et al. Stability of mechlorethamine hydrochloride 0.01% ointment in Aquaphor® base. *Int J Pharm Compound*. 1998; 2: 89–91.

Method 2

Mayron and Gennaro used an HPLC method to assess the stability of mechlorethamine hydrochloride with granisetron hydrochloride during Y-site administration. The liquid chromatographic system consisted of a piston pump with a pulse dampener, a rotary injection port with a 20-μL loop, a variable-wavelength spectrometric detector, and an integrator. The stationary phase was a cyano reversed-phase column (250 × 4.6 mm, 5-μm particle size). The mobile phase was a mixture of acetonitrile and 0.1 M monobasic sodium phosphate (20:80, vol/vol) adjusted to pH 4.2 with phosphoric acid. The flow rate was 1.0 mL/min. UV detection was performed at 200 nm. Samples were diluted with the mobile phase. The retention times for mechlorethamine and granisetron were 2.24 and 5.60 minutes, respectively.

The stability-indicating capability of the analytical method was demonstrated by an accelerated degradation study. Solutions of mechlorethamine hydrochloride and granisetron hydrochloride were adjusted to pH 2 and 11, boiled for 1 hour, and then readjusted to pH 5, diluted with the mobile phase, and assayed. Degradation product peaks did not interfere with the intact mechlorethamine or granisetron peaks.

Reference

Mayron D, Gennaro AR. Stability and compatibility of granisetron hydrochloride in i.v. solutions and oral liquids and during simulated Y-site injection with selected drugs. *Am J Health Syst Pharm.* 1996; 53: 294–304.

Mecillinam

Chemical Name

(6*R*)-6-(Perhydroazepin-1-ylmethyleneamino)penicillanic acid

Other Names

Amdinocillin, Selexid, Selexidin

Form	Molecular Formula	MW	CAS
Mecillinam	$C_{15}H_{23}N_3O_3S$	325.4	32887-01-7

Solubility

Mecillinam is freely soluble in water and methanol.

Method 1

Grover et al. evaluated by HPLC analysis the degradation rate of mecillinam at 35 °C in borate buffer (pH 9.2). The Shimadzu LC-10A liquid chromatograph included model

LC-10AS pumps, a model SPD-10A dual-wavelength detector, a model C-R7A data processor, and a Rheodyne model 7125 injector with a 20-µL loop. The stationary phase was a Phenomenex C_{18} column (300 × 3.9 mm, 5-µm particle size). The mobile phase was a mixture of 88% water containing 20 mM monobasic potassium phosphate and 10 mM tetramethylammonium chloride (pH 5.0) and 12% acetonitrile. The flow rate was 1.0 mL/min. UV detection was performed at 220 and 240 nm.

Samples were incubated at 35 °C and assayed. Under these conditions, the retention time for mecillinam was 13.9 minutes.

The method was reported to be stability indicating.

A standard curve for mecillinam was generated from 5 to 150 µg/mL. The correlation coefficient was greater than 0.99. The coefficient of variation for the analysis of mecillinam was less than 1%.

Reference
Grover M, Gulati M, Singh B, et al. Correlation of penicillin structure with rate constants for basic hydrolysis. *Pharm Pharmacol Commun*. 2000; 6: 355–63.

Method 2
Hagel et al. developed an HPLC assay for mecillinam. The liquid chromatograph consisted of an LDC model 396 Mini pump or a Waters model 6000A pump, a Waters model U6K injector or a Rheodyne model 7120 injector, and a Tracor model 970A variable-wavelength detector or an LDC model 1202 variable-wavelength detector. The stationary phase was a Waters µBondapak C_{18} column (300 × 3.9 mm). The mobile phase consisted of 0.01 M sodium phosphate (pH 5.0) and acetonitrile (85:15, vol/vol). The flow rate was 1.0 mL/min. The UV detector was set at 220 nm and 0.32 AUFS.

Samples were dissolved in distilled water and diluted appropriately. Under these conditions, the retention time for mecillinam was about 9 minutes (estimated from the published chromatogram).

The stability-indicating nature of the assay was verified by four experiments. Mecillinam standard samples were spiked with its potential degradation products, stored at 55 °C for 3 months, degraded enzymatically by reaction with penicillinase derived from *Bacillus cereus*, and heated at 110 °C for 13 hours. The intact mecillinam peak was separated from its degradation product peaks.

A standard curve for mecillinam was constructed from 0.500 to 1.500 µg. The correlation coefficient was 0.9999. The limit of detection was 2 ng.

Reference
Hagel RB, Waysek EH, Humphrey D. Stability-indicating assay for mecillinam using high-pressure liquid chromatography. *J Chromatogr*. 1979; 170: 391–8.

Meclizine Hydrochloride

Chemical Name
1-(4-Chlorobenzhydryl)-4-(3-methylbenzyl)piperazine dihydrochloride

Other Names
Antivert, Bonamine, Meclozine

Form
Meclizine hydrochloride

Molecular Formula
$C_{25}H_{27}ClN_2.2HCl$

MW
463.9

CAS
1104-22-9

Appearance
Meclizine hydrochloride is a white to slightly yellowish crystalline powder with a slight odor.

Solubility
Meclizine hydrochloride is practically insoluble in water and slightly soluble in alcohol.

Method
Al-Jallad et al. developed an HPLC method for the simultaneous determination of pyridoxine hydrochloride and meclizine hydrochloride in tablets. The Beckman Gold system comprised a model 116 pump, a model 116 UV spectrophotometric detector, and a Rheodyne model 7725 20-µL loop injector. The stationary phase was a Waters Symmetry C_{18} silica column (250 × 4.6 mm, 5-µm particle size). The mobile phase was a mixture of 200 mL of water containing 1.5 g of dodecyl sulfate sodium salt, 500 mL of acetonitrile, 250 mL of methanol, 15 mL of glacial acetic acid, and 15 mL of tetrahydrofuran. The flow rate was 1.5 mL/min. UV detection was performed at 254 nm.

A sample equivalent to 50 mg of pyridoxine hydrochloride and 75 mg of meclizine hydrochloride was accurately weighed, transferred to a 100-mL volumetric flask, mixed with mobile phase, shaken for 15 minutes, brought to volume with the mobile phase, and centrifuged. The injection volume was 20 µL. Under these conditions, retention times for pyridoxine hydrochloride and meclizine hydrochloride were about 2.2 and 7.4 minutes, respectively (estimated from the published chromatogram).

The stability-indicating capability of the assay was demonstrated by accelerated decomposition of meclizine hydrochloride. The drug was treated with 1 M sodium hydroxide solution, 1 M sulfuric acid, or 3.3% hydrogen peroxide. No interference from degradation products with the determination of meclizine hydrochloride was observed.

A calibration curve for meclizine hydrochloride was obtained from 0.13 to 0.39 mg/mL. The correlation coefficient was 0.9998. The limit of detection was 7.65 µg/mL.

Reference
Al-Jallad T, Al-Kurdi Z, Badwan A, et al. Simultaneous determination of pyridoxine hydrochloride and meclizine hydrochloride in tablet formulations by HPLC. *Pharm Pharmacol Commun.* 1999; 5: 479–83.

Medroxyprogesterone

Chemical Names
(6α)-17-Hydroxy-6-methylpregn-4-ene-3,20-dione
17α-Hydroxy-6α-methylprogesterone

Other Names
Amen, Provera

Form	Molecular Formula	MW	CAS
Medroxyprogesterone	$C_{22}H_{32}O_3$	344.5	520-85-4
Medroxyprogesterone acetate	$C_{24}H_{34}O_4$	386.5	71-58-9

Appearance
Medroxyprogesterone acetate occurs as a white to off-white odorless crystalline powder.

Solubility
Medroxyprogesterone acetate is insoluble in water and sparingly soluble in alcohol and methanol. It is slightly soluble in ether, freely soluble in chloroform, and soluble in acetone and dioxane.

Method
Fatmi et al. developed a reversed-phase liquid chromatographic method for the analysis of medroxyprogesterone acetate in tablets. The liquid chromatograph consisted of a Waters model 510 solvent pump, a Waters model 481 variable-wavelength detector, a Waters model 710B WISP autosampler, and a Hewlett-Packard model 3390A integrator. The stationary phase was a Waters Nova-Pak C_{18} stainless steel column (150 × 3.9 mm, 5-μm particle size). The mobile phase consisted of methanol and 0.01 M ammonium phosphate (80:20, vol/vol) adjusted to pH 7.2 ± 0.1 with 85% phosphoric acid. The flow rate was 1.0 mL/min. UV detection was performed at 254 nm.

Samples were extracted with methanol. The injection volume was 20 μL.

The method was determined to be stability indicating by accelerated decomposition of medroxyprogesterone acetate. One tablet was dissolved in methanol and mixed with 1 N sodium hydroxide. The sample was also incubated in a 50 °C water bath for 4 hours. Degradation product peaks did not interfere with the intact medroxyprogesterone acetate peak.

A standard curve for medroxyprogesterone acetate was constructed from 50 to 150 μg/mL. The coefficient of variation was less than 2.0%.

Reference
Fatmi AA, Williams GV, Hickson EA. Liquid chromatographic determination of medroxyprogesterone acetate in tablets. *J Assoc Off Anal Chem.* 1988; 71: 528–30.

Mefenamic Acid

Chemical Name
N-(2,3-Xylyl)anthranilic acid

Other Names
Contraflam, Ponstan, Ponstel

Form	Molecular Formula	MW	CAS
Mefenamic acid	$C_{15}H_{15}NO_2$	241.3	61-68-7

Appearance
Mefenamic acid is a white powder.

Solubility
Mefenamic acid is insoluble in water and slightly soluble in alcohol.

pK$_a$
Mefenamic acid has an apparent pK$_a$ of 4.2.

Method
Maron and Wright reported an HPLC analytical method for the determination of mefenamic acid. The chromatographic system consisted of a Waters model 6000A pump, a model 710B WISP autosampler, a Varian Polychrom 9060 photodiode-array detector, and a Hewlett-Packard 3390 integrator. The stationary phase was a Waters Nova-Pak C_{18} reversed-phase column (150 × 3.9 mm, 5-μm particle size) with a Waters μBondapak C_{18} guard column. The mobile phase was a mixture of acetonitrile, tetrahydrofuran, water, and glacial acetic acid (15:40:45:2, vol/vol/vol/vol). The flow rate was 1 mL/min. UV detection was performed at 278 nm. The injection volume was 20 μL. Under these conditions, the retention time for mefenamic acid was about 4 minutes.

The stability-indicating nature of the method was investigated by accelerated degradation of mefenamic acid. Mefenamic acid was refluxed with 0.5 N sulfuric acid for 48 hours, with 0.1 N sodium hydroxide solution for 48 hours, or with 30% hydrogen peroxide and 0.03 N hydrochloric acid for 24 hours. In all cases, the intact mefenamic acid was well separated from its degradation products. The placebo ingredients did not interfere with the analysis of mefenamic acid.

A calibration curve for mefenamic acid was generated from 25 to 150 μg/mL with a correlation coefficient of 0.9993.

Reference

Maron N, Wright G. Application of photodiode-array UV detection in the development of stability-indicating LC methods: Determination of mefenamic acid. *J Pharm Biomed Anal.* 1990; 8: 101–5.

Melanotropin

Chemical Name

N-Acetyl-L-norleucyl-L-α-aspartyl-L-histidyl-D-phenylalanyl-L-arginyl-L-tryptophyl-L-lysinamide

Other Name

MT II

Form	Molecular Formula	MW	CAS
Melanotropin	$C_{50}H_{69}N_{15}O_9$	1024.2	121062-08-6

Method

Ugwu et al. evaluated by HPLC analysis the influence of pH, phosphate buffer concentration, temperature, and ionic strength on the degradation rate of α-melanotropin. The chromatographic apparatus was composed of a Spectra-Physics Isochrom pump, a Rheodyne model 7125 injector with a 50-μL loop, a Hitachi/Spectra-Physics model 100-30 variable-wavelength UV detector, and a Spectra-Physics model 4400 integrator. The stationary phase was a DuPont Zorbax C_8 column (150 × 4.6 mm, 10-μm particle size) with a Whatman C_{18} guard column (10 × 4.6 mm, 30-μm particle size). The mobile phase consisted of 0.1 M dibasic potassium phosphate buffer and acetonitrile (73:27, vol/vol) containing 18 μL of 99% (vol/vol) triethylamine per liter of the mobile phase (pH 2.5). The flow rate was 1.0 mL/min. UV detection was performed at 214 nm.

Samples were prepared in water. The injection volume was 100 μL. Under these conditions, the retention time for α-melanotropin was about 9.2 minutes (estimated from the published chromatogram).

The method was shown to be stability indicating. Samples in pH 2.4 and 7.0 buffer solutions were stored at 4 °C for 300 days. Degradation products did not interfere with the analysis of α-melanotropin.

A standard curve for α-melanotropin was constructed from 1 to 10 μg/mL. The correlation coefficient was 0.9997.

Reference

Ugwu SO, Lan E-L, Sharma S, et al. Kinetics of degradation of a cyclic lactam analog of α-melanotropin (MT-II) in aqueous solution. *Int J Pharm.* 1994; 102: 193–9.

Meloxicam

Chemical Name
4-Hydroxy-2-methyl-*N*-(5-methyl-2-thiazolyl)-2*H*-1,2-benzothiazine-3-carboxamide 1,1-dioxide

Other Names
Coxflam, Metacam, Mobic, Mobicox, Zilutrol

Form	Molecular Formula	MW	CAS
Meloxicam	$C_{14}H_{13}N_3O_4S_2$	351.4	71125-38-7

Appearance
Meloxicam occurs as a pale yellow powder.

Solubility
Meloxicam is practically insoluble in water and very slightly soluble in ethanol and methanol.

Method 1
Taha et al. developed an HPLC method for the determination of meloxicam in the presence of its alkaline degradation products. The Waters chromatograph consisted of a model 600 LC series pump, a model 600 controller unit, a model 486 tunable absorbance detector, an injector with a 20-µL loop, and a model 746 data module. The stationary phase was a Nova-Pak C_{18} column (150 × 3.9 mm, 4-µm particle size). The mobile phase was a mixture of methanol, acetonitrile, and acetate buffer (pH 4.6) (4.5:0.5:5.0, vol/vol/vol). The flow rate was 0.8 mL/min. The detector was set at 280 nm.

A portion of powder from 10 tablets equivalent to 25 mg of meloxicam was accurately weighed, mixed in 0.3 mL of 1 N sodium hydroxide solution and 15 mL of methanol, stirred for 30 minutes, further diluted with methanol, and filtered. The sample injection volume was 20 µL. Under these conditions, the retention time for meloxicam was about 3.1 minutes.

Meloxicam was heated in 1 N sodium hydroxide solution for 4 hours. Its degradation products did not interfere with the determination of the intact meloxicam, demonstrating the stability-indicating nature of the method.

A calibration curve for meloxicam was constructed from 1.25 to 50 µg/mL. The correlation coefficient was 0.9996. Intraday and interday coefficients of variation of the analysis were 0.76 and 1.12%, respectively. The limit of detection and the limit of quantitation were 0.03 and 0.09 µg/mL, respectively.

Reference

Taha EA, Salama NN, Fattah LEA. Stability-indicating chromatographic methods for the determination of some oxicams. *J AOAC Int.* 2004; 87: 366–73.

Method 2

Bartsch et al. reported a validated HPLC method for the determination of meloxicam in the presence of its degradation products. The chromatographic system consisted of a Shimadzu model LC-10AS pump, a model SIL-10ADVP autosampler, a model SPD-M10A diode-array detector, and a model CTO-10AC column oven. The stationary phase was a LiChrospher 100 RP-18 column (119 × 3 mm, 5-μm particle size). The column temperature was maintained at 20 °C. The mobile phase was a mixture of methanol and sodium acetate buffer (pH 4.6) (50:50, vol/vol). The flow rate was 0.8 mL/min. UV detection was performed at 272 nm.

Solutions of meloxicam in 2.5% aqueous ammonia (pH 12.6) were exposed to irradiation for up to 864 minutes. Degradation products eluted within 2 minutes, while meloxicam eluted at 4.1 minutes.

A standard curve for meloxicam was obtained from 0.25 to 50 μg/mL. The correlation coefficient was greater than 0.998. Intraday and interday coefficients of variation were 3.5 and 4.1%, respectively. The limit of detection and limit of quantitation were 0.13 and 0.25 μg/mL, respectively.

Reference

Bartsch H, Eiper A, Kopelent-Frank H, et al. A validated HPLC-assay for the determination of meloxicam in presence of its degradation products. *Sci Pharm.* 2004; 72: 213–20.

Melphalan

Chemical Names

4-[Bis(2-chloroethyl)amino]-L-phenylalanine
p-Di(2-chloroethyl)amino-L-phenylalanine

Other Names

Alkeran, L-PAM, Phenylalanine Mustard, L-Sarcolysin

Form	Molecular Formula	MW	CAS
Melphalan	$C_{13}H_{18}Cl_2N_2O_2$	305.2	148-82-3
Melphalan hydrochloride	$C_{13}H_{18}Cl_2N_2O_2$.HCl	341.7	3223-07-2

Appearance

Melphalan is an off-white to buff powder. Melphalan hydrochloride for injection is available as a sterile nonpyrogenic lyophilized powder.

Solubility

Melphalan is soluble in ethanol and propylene glycol, but it is practically insoluble in water.

pK$_a$

Melphalan has an approximate pK$_a$ of 2.5.

Method 1

Brightman et al. described an HPLC method for the determination of melphalan and related impurity content. The liquid chromatograph included a Hewlett-Packard model 1100 autosampler, a gradient pump, and a detector. The stationary phase was a Hypersil BDS C$_{18}$ column (150 × 4.6 mm, 5-μm particle size). Eluent A consisted of a buffer, acetonitrile, and water (10:100:1900, vol/vol/vol), and eluent B consisted of a buffer, acetonitrile, and water (10:1200:800, vol/vol/vol), where the buffer consisted of acetic acid, triethylamine, ammonium acetate, and water (10:2:10:88, vol/vol/vol/vol). The flow rate was 1.5 mL/min. The mobile phase was delivered from 100% A to 100% B in the first 20 minutes, held at 100% B for 5 minutes, returned to 100% A in 1 minute, and then held at 100% A for 4 minutes. UV detection was performed at 260 nm and 0.1 AUFS. The total run time was 30 minutes.

The drug substance was dissolved in methanol. The injection volume was 10–20 μL. Under these conditions, the retention time of melphalan was about 10.2 minutes (estimated from the published chromatogram).

The method was demonstrated to be stability indicating by assaying a synthetic mixture of melphalan, its known potential impurities, and its potential degradation products. Melphalan was separated from those compounds. The following is a list of compounds with their relative retention times in parentheses: dihydroxymelphalan (0.21), phthalic acid (0.28), morpholino derivative (0.34), methoxyhydroxymelphalan (0.35), chloroethylamino melphalan (0.44), monohydroxymelphalan (0.52), methoxy-melphalan (0.79), ethoxymelphalan methyl ester (0.95), melphalan (1.00), chloroethoxy-melphalan (1.05), 3-chloro analogue of melphalan (1.16), melphalan dimer (1.27), melphalan methyl ester (1.31), and melphalan ethyl ester (1.44).

A standard curve for melphalan was constructed from 0 to 3 μg. The correlation coefficient was 0.9998.

Reference

Brightman K, Finlay G, Jarvis I, et al. A stability-indicating method for the determination of melphalan and related impurity content by gradient HPLC. *J Pharm Biomed Anal.* 1999; 20: 439–47.

Method 2

Pinguet et al. studied the stability of concentrated melphalan in 50-mL polyvinyl chloride bags containing sodium chloride at various concentrations at several temperatures. The HPLC system included a Jasco model 880PU reciprocating piston pump with flow feedback control, a manual injection valve equipped with a 50-μL sample loop, a Varian model 2050 variable-wavelength UV detector, and a Shimadzu C-R5A Chromatopac integrator. The stationary phase was a Shandon Spherisorb ODS-2 C_{18} stainless steel column (150 × 4.6 mm, 5-μm particle size). The mobile phase consisted of 60% methanol and 40% 0.01 M sodium dihydrogen phosphate buffer (pH 3.0). The pH was adjusted with 2 N phosphoric acid. The flow rate was 1.3 mL/min. UV detection was performed at 254 nm. Propylparaben was used as the internal standard.

Samples were diluted 1:10. The retention time for melphalan was 3.6 minutes.

The assay was stated to be stability indicating because it separated the intact melphalan peak from its degradation product peaks. Retention times for dihydroxy-melphalan, monohydroxymelphalan, melphalan, and the internal standard were 1.4, 1.8, 3.6, and 5.9 minutes, respectively.

Standard curves for melphalan were constructed from 0.05 to 0.5 mg/mL. The correlation coefficient was 0.999. Intraday and interday coefficients of variation were 5 and 4%, respectively.

A similar stability-indicating assay was used by Tabibi and Cradock.

References
Pinguet F, Martel P, Rouanet P, et al. Effect of sodium chloride concentration and temperature on melphalan stability during storage and use. *Am J Hosp Pharm*. 1994; 51: 2701–4.

Tabibi SE, Cradock JC. Stability of melphalan in infusion fluids. *Am J Hosp Pharm*. 1984; 41: 1380–2.

Method 3
Bosanquet and Gilby developed an HPLC assay for melphalan in plasma. The liquid chromatograph consisted of an LDC Constametric IG dual reciprocating pump equipped with a Rheodyne injection valve with a 200-μL loop, an LDC UV III fixed-wavelength monitor, and a Jones Chromatography Spherisorb ODS reversed-phase column (250 × 4.6 mm, 5-μm particle size). The mobile phase consisted of a mixture of a 0.675-g/L solution of sodium dodecyl sulfate and methanol (1:4, vol/vol). The mobile phase was adjusted to a pH of approximately 3.0 with concentrated sulfuric acid. UV detection was performed at 254 nm and 0.002 AUFS. The column temperature was 40 ± 1 °C. The injection volume was 200 μL. 5-Dimethylaminonaphthalene-1-sulfonyl-L-arginine was used as an internal standard.

The method was determined to be stability indicating by spiking the melphalan solution with two known hydrolysis degradation products, monohydroxy-melphalan and dihydroxy-melphalan. Retention times for monohydroxy-melphalan, dihydroxy-melphalan, melphalan, and the internal standard were 6.8, 8.0, 8.7, and 11.5 minutes, respectively.

A standard curve for melphalan was constructed from 1 to 2000 μg per injection. The correlation coefficient of the standard curve was 1.000; the coefficient of variation was 1.07%.

Reference
Bosanquet AG, Gilby ED. Measurement of plasma melphalan at therapeutic concentrations using isocratic high-performance liquid chromatography. *J Chromatogr*. 1982; 232: 345–54.

Meperidine

Chemical Names
1-Methyl-4-phenyl-4-piperidinecarboxylic acid ethyl ester
1-Methyl-4-phenylisonipecotic acid ethyl ester

Other Names
Demerol, Pethidine

Form	Molecular Formula	MW	CAS
Meperidine	$C_{15}H_{21}NO_2$	247.3	57-42-1
Meperidine hydrochloride	$C_{15}H_{21}NO_2 \cdot HCl$	283.8	50-13-5

Appearance
Meperidine hydrochloride is a fine white crystalline powder.

Solubility
Meperidine hydrochloride is very soluble in water; it is soluble in acetone and ethyl acetate and slightly soluble in alcohol.

Method 1
Xu et al. studied the stability of meperidine hydrochloride 4 mg/mL with ondansetron hydrochloride in 0.9% sodium chloride injection. The liquid chromatograph consisted of a Waters 600E multisolvent-delivery pump, a Waters 490E programmable multiple-wavelength UV detector, and a Waters 712 WISP autosampler. The system was controlled and integrated by an NEC PowerMate SX/16 personal computer. The stationary phase was an Alltech Spherisorb cyano HPLC analytical column (300 × 4.6 mm, 5-μm particle size). The mobile phase was 65% acetonitrile in 0.02 M monobasic potassium phosphate with pH adjusted to 5.40 with 1 N sodium hydroxide and was delivered isocratically at 1.5 mL/min. UV detection was performed at 216 nm.

Samples were diluted 1:20. The injection volume was 10 μL. Under these conditions, the retention times for ondansetron and meperidine were 8.3 and 9.5 minutes, respectively.

The method was determined to be stability indicating by accelerated decomposition of meperidine hydrochloride. Heating at 95 °C for 5 hours yielded a reduction in the intact meperidine peak, with the formation of a new peak eluting at 3.4 minutes. The degradation product peak did not interfere with the intact meperidine peak.

A standard curve was generated from a linear plot of peak area versus concentration from 0.075 to 0.25 mg/mL. Its correlation coefficient was greater than 0.998. The coefficient of variation was 1.5%. The intraday and interday coefficients of variation of the assay were 1.9 and 1.6%, respectively.

Reference

Xu QA, Trissel LA, Fox JL. Compatibility of ondansetron hydrochloride with meperidine hydrochloride for combined administration. *Ann Pharmacother*. 1995; 29: 1106–9.

Method 2

Macias et al. evaluated the stability of meperidine hydrochloride in a parenteral nutrient formulation. The HPLC system consisted of a Waters model 45 solvent-delivery system, a Beckman model 160 fixed-wavelength UV detector, and a Linear model 500 strip-chart recorder. The stationary phase was a Waters μBondapak C_{18} radial-compression cartridge (10-μm particle size). The mobile phase consisted of 42% methanol and 58% water containing 0.025 M monobasic potassium phosphate and 0.75% *N,N*-octyldimethylamine. Its pH was adjusted to 7.0 with 5 N sodium hydroxide. The flow rate was 4 mL/min. UV detection was performed at 254 nm and 0.005 AUFS. Propylparaben 0.16 mg/mL in methanol was used as an internal standard. Retention times for propylparaben and meperidine were 6.3 and 8.5 minutes, respectively.

The method was determined to be stability indicating by an accelerated degradation study. Meperidine hydrochloride was exposed to strong acidic (sulfuric acid) and basic (1 N sodium hydroxide) conditions. These solutions were then boiled for 30 minutes. No degradation products interfered with the analysis of meperidine.

Standard curves were constructed for meperidine from 0.05 to 0.15 mg/mL and from 0.5 to 1.5 mg/mL. The correlation coefficient was greater than 0.999. The intraday and interday coefficients of variation for the assay were less than 1.4 and 2.2%, respectively.

Reference

Macias JM, Martin WJ, Lloyd CW. Stability of morphine sulfate and meperidine hydrochloride in a parenteral nutrient formulation. *Am J Hosp Pharm*. 1985; 42: 1087–94.

Method 3

Gupta developed an HPLC method for the quantitation of meperidine hydrochloride in pharmaceutical dosage forms. A Waters ALC 202 chromatograph was equipped with a Schoeffel SF 770 Spectroflow monitor, a Houston Omniscribe recorder, a Spectra-Physics Autolab Minigrator, and a Waters μBondapak C_{18} column (300 × 4 mm). The mobile phase was 60% acetonitrile in a 0.02 M aqueous solution of ammonium acetate and was delivered isocratically at 3.0 mL/min. UV detection was performed at 232 nm and 0.04 AUFS. Hydroxyprogesterone caproate was used as the internal standard.

Samples were diluted with water; the injection volume was 20 μL. Retention times for meperidine and the internal standard were about 4.2 and 6.0 minutes (estimated from the published chromatogram).

The analytical method was determined to be stability indicating by accelerated decomposition of meperidine hydrochloride. A solution of meperidine hydrochloride 2 mg/mL was mixed with about 1 N sodium hydroxide and was stored in an electric oven at 50 °C for 48 hours. The resulting solution was neutralized and assayed. Decomposition products eluted immediately after the solvent front and did not interfere with the intact meperidine peak.

Standard curves for meperidine hydrochloride were constructed from 5 to 13 μg/mL. The coefficient of variation was 1.2%.

Reference
Gupta VD. Quantitation of meperidine hydrochloride in pharmaceutical dosage forms by high-performance liquid chromatography. *J Pharm Sci.* 1983; 72: 695–7.

Mequitazine

Chemical Name
10-(3-Quinuclidinylmethyl)phenothiazine

Other Names
Butix, Metaplexan, Mircol, Primalan

Form	Molecular Formula	MW	CAS
Mequitazine	$C_{20}H_{22}N_2S$	322.5	29216-28-2

Appearance
Mequitazine occurs as crystals.

Method
El-Ragehy et al. described an HPLC method for the determination of mequitazine in the presence of its degradation products. The Shimadzu LC-10 AD liquid chromatograph was equipped with an SPD-10A UV-visible detector. The stationary phase was a µBondapak C_{18} column (250 × 4.6 mm, 10-µm particle size). The mobile phase consisted of acetonitrile and phosphoric acid solution (50:50, vol/vol). The flow rate was 1.5 mL/min. UV detection was performed at 256 nm. Caffeine 0.1 mg/mL in methanol was used as an internal standard.

A portion of powder from 20 tablets equivalent to 10 mg of mequitazine was accurately weighed, transferred to a beaker, extracted with 80 mL of methanol, filtered, diluted further with methanol, and mixed with internal standard. The injection volume was 20 µL. Under these conditions, retention times for caffeine and mequitazine were about 3.9 and 5.3 minutes, respectively.

The stability-indicating nature of the method was demonstrated by accelerated degradation of mequitazine. Mequitazine was refluxed with 30% hydrogen peroxide for 4 hours. The degradation product did not interfere with the determination of mequitazine.

A standard curve for mequitazine was constructed from 1 to 9 µg/mL. The correlation coefficient was 0.9999. The limit of detection and the limit of quantitation were 0.12 and 1.0 µg/mL, respectively.

Reference

El-Ragehy NA, Badawey AM, El Khateeb SZ. Stability-indicating methods for assay of mequitazine in presence of its degradate. *J Pharm Biomed Anal.* 2002; 29: 121–37.

Mercaptopurine

Chemical Name
6-Purinethiol monohydrate

Other Names
Purinethol, Puri-Nethol

Form	Molecular Formula	MW	CAS
Mercaptopurine monohydrate	$C_5H_4N_4S.H_2O$	170.2	6112-76-1

Appearance
Mercaptopurine is a yellow and odorless or almost odorless crystalline powder.

Solubility
Mercaptopurine is practically insoluble in water, acetone, and ether. It is slightly soluble in alcohol. It is soluble in hot alcohol and in dilute solutions of alkali.

pK_a
Mercaptopurine has a pK_a of 7.6.

Method
Fell et al. reported a simple procedure by reversed-phase HPLC for the rapid analysis of 6-mercaptopurine and azathioprine. The chromatograph included an Applied Chromatography Systems constant-flow pump and a Cecil Instruments CE-212 variable-wavelength UV monitor. The stationary phase was a Shandon-Southern Instruments ODS Hypersil stainless steel column (100 × 5 mm, 5-μm particle size). The mobile phase was a mixture of methanol, 25 mM monobasic potassium phosphate, and glacial acetic acid (20:79.5:0.5, vol/vol/vol) adjusted to pH 4.5. The flow rate was 1.5 mL/min. Theophylline was used as an internal standard. UV detection was performed at 240 nm and 0.02 AUFS.

Mercaptopurine tablets were crushed, mixed with 0.02 M sodium hydroxide solution, and filtered. Then theophylline 150 μg/mL was added, and the solution was diluted with 0.02 M sodium hydroxide solution to about 1 mg/mL of mercaptopurine before injection. The injection volume was 1.5 µL. Under these conditions, the retention times for mercaptopurine, theophylline, and azathioprine were about 1.7, 3.5, and 4.2 minutes, respectively (estimated from the published chromatogram).

The assay was stability indicating since mercaptopurine was well resolved from all its known potential impurities and degradation products.

The limit of detection for mercaptopurine was 0.4 ng.

Reference

Fell AF, Plag SM, Neil JM. Stability-indicating assay for azathioprine and 6-mercapto-purine by reversed-phase high-performance liquid chromatography. *J Chromatogr.* 1979; 186: 691–704.

Meropenem

Chemical Name

[4R-[3(3S*,5S*),4α,5β,6β(R*)]]-3-[[5-[(Dimethylamino)carbonyl]-3-pyrrolidinyl]thio]-6-(1-hydroxyethyl)-4-methyl-7-oxo-1-azabicyclo[3.2.0]hept-2-ene-2-carboxylic acid tri-hydrate

Other Name

Merrem

Form	Molecular Formula	MW	CAS
Meropenem	$C_{17}H_{25}N_3O_5S.3H_2O$	437.5	119478-56-7

Appearance

Meropenem occurs as colorless to white crystals.

Solubility

Meropenem is sparingly soluble in water and very slightly soluble in alcohol. It is soluble in dimethylformamide.

Method 1

Caufield and Stewart reported an HPLC assay for the simultaneous analysis of meropenem and dopamine in an intravenous fluid mixture. The instrument was a Hewlett-Packard model 1090 system, including a pump, an autosampler, and a Gilson model 117 variable-wavelength UV detector. A Waters model 996 photodiode-array detector was used for the peak purity analysis. The stationary phase was a Rainin Microsorb ODS column (150 × 2.1 mm, 5-µm particle size). The mobile phase consisted of aqueous acetic acid solution (pH 3) and acetonitrile (85:15, vol/vol). The flow rate was 0.2 mL/min. UV detection was performed at 280 nm.

Samples were diluted with a diluent of 0.9 mM ethylenediaminetetraacetic acid disodium salt aqueous solution and acetonitrile (85:15, vol/vol). The sample injection volume was 5 µL. Under these conditions, retention times of meropenem and dopamine were about 9.4 and 5.7 minutes, respectively.

The method was confirmed to be stability indicating by accelerated degradation of meropenem. Solutions of meropenem were treated with 0.1 N hydrochloric acid for 15 minutes, 0.01 N sodium hydroxide solution for 1 minute, or 0.03% hydrogen peroxide for 2 hours or were heated at 60 °C for 4 hours. None of the degradation product peaks interfered with the intact drug peak.

A standard curve for meropenem was constructed from 61.25 to 245 µg/mL. The correlation coefficient was about 0.9996. Intraday and interday coefficients of variation for the analysis of meropenem were 0.26 and 0.33%, respectively. The limit of detection was 167 ng/mL.

Reference
Caufield WV, Stewart JT. HPLC separations of meropenem and selected pharmaceuticals using a polar endcapped octadecylsilane narrow bore column. *Chromatographia*. 2000; 51: 308–14.

Method 2
Caufield and Stewart also described an HPLC separation of meropenem and aminophylline. The Hewlett-Packard model 1090 system comprised a pump, an autosampler equipped with a 25-µL loop, and a model 117 variable-wavelength detector. A Waters model 996 photodiode-array detector was used for the peak purity analysis. The stationary phase was a polar endcapped narrow bore YMC-ODS-AQ column (150 × 2.0 mm, 3-µm particle size). The mobile phase consisted of water (adjusted to pH 3.0 with glacial acetic acid) and acetonitrile (89:11, vol/vol). The flow rate was 0.2 mL/min. UV detection was performed at 270 nm.

Samples were diluted with a mixture of water and acetonitrile (89:11, vol/vol). The injection volume was 5 µL. Under these conditions, retention times for meropenem and aminophylline were about 4.4 and 5.9 minutes, respectively.

The method was demonstrated to be stability indicating by intentional degradation of the drug. Meropenem samples were treated with 0.1 N hydrochloric acid for 15 minutes, 0.01 N sodium hydroxide solution for 1 minute, or 0.03% hydrogen peroxide for 2 hours or were heated at 60 °C for 4 hours. Degradation products did not interfere with the determination of meropenem.

A calibration curve for meropenem was obtained from 126.88 to 507.50 µg/mL. The correlation coefficient was greater than 0.9996. Intraday and interday coefficients of variation were 0.71 and 0.46%, respectively. The limit of detection was 150 ng.

Reference
Caufield WV, Stewart JT. HPLC separations of meropenem and selected pharmaceuticals using a polar endcapped octadecylsilane narrow bore column. *Chromatographia*. 2000; 51: 308–14.

Method 3
Patel and Cook investigated the stability of meropenem in intravenous solutions. The liquid chromatograph consisted of a Hitachi model L600A pump, a Hitachi model AS4000 autosampler, and an Applied Biosystems model ABI 783A UV detector. The stationary phase was an Alltech Hypersil ODS C_{18} column (250 × 4.6 mm, 5-µm particle size). The mobile phase consisted of acetonitrile, methanol, and 0.02 M tetrabutylammonium hydroxide (pH 7.5) (15:10:75, vol/vol/vol). The flow rate was 1.5 mL/min. UV detection was performed at 300 nm.

Samples were diluted with the mobile phase to achieve a meropenem concentration of approximately 0.1 mg/mL. The injection volume was 20 μL. The retention time for meropenem was about 4.6 minutes (estimated from the published chromatogram).

The stability-indicating capability of the assay was demonstrated by accelerated decomposition of meropenem. Hydrolysis of meropenem at pH 3, 5, 7, and 9 produced two major degradation products, a β-lactam ring-opened hydrolysis product and a meropenem dimer product (two tautomers). These degradation products were only detected at 215 nm, with retention times of 5.9, 9.1, and 10.1 minutes.

A standard curve for meropenem was constructed from 0.05 to 0.150 mg/mL; the correlation coefficient was 0.99999. The coefficient of variation was less than 1%.

Reference
Patel PR, Cook SE. Stability of meropenem in intravenous solutions. *Am J Health Syst Pharm.* 1997; 54: 412–21.

Mesalamine

Chemical Names
5-Amino-2-hydroxybenzoic acid
5-Aminosalicylic acid

Other Names
Mesalazine, Rowasa

Form	Molecular Formula	MW	CAS
Mesalamine	$C_7H_7NO_3$	153.1	89-57-6

Appearance
Mesalamine occurs as light tan to pink needle-shaped crystals.

Solubility
Mesalamine is slightly soluble in water and very slightly soluble in dehydrated alcohol, acetone, and methanol.

pK$_a$
Mesalamine has pK$_a$ values of 3, 6, and 13.9.

Method
Henderson et al. used a stability-indicating HPLC method for the analysis of mesalamine in a rectal suspension diluted with distilled water. The liquid chromatograph included a

Waters model 501 pump, a Waters model U6K variable-volume injector, a Waters model 484 tunable UV detector, and a Hewlett-Packard model 3390 integrator-recorder. The stationary phase was an Applied Biosystems Spheri-5 ODS C_{18} column (250 × 4.6 mm, 5-μm particle size). The mobile phase consisted of an aqueous phase and methanol (70:30, vol/vol). The aqueous phase was prepared by mixing 900 mL of 0.05 M anhydrous dibasic sodium phosphate and 18.75 mL of tetrabutylammonium phosphate adjusted to pH 6.8 with 1 N phosphoric acid. The flow rate was 1.2 mL/min. UV detection was performed at 254 nm and 0.1 AUFS. Acetaminophen 300 μg/mL was used as an internal standard.

Samples were diluted with mobile phase. The injection volume was 5 μL. Retention times for acetaminophen and mesalamine were 3.8 and 6.0 minutes, respectively.

The HPLC assay was determined to be stability indicating by accelerated decomposition of mesalamine. An aqueous solution of mesalamine 1 mg/mL was decomposed by adjusting the pH to 12 with 1 N sodium hydroxide and to pH 1 with 1 N sulfuric acid and then storing the solution in natural light at room temperature for 24 hours. Degradation product peaks did not interfere with the intact mesalamine peak.

Standard curves were constructed from 200 to 300 μg/mL. The correlation coefficient was greater than 0.997. The intraday and interday coefficients of variation for the assay were 1.13 and 1.12%, respectively.

A similar method was used by Montgomery et al.

References

Henderson LM, Johnson CE, Berardi RR. Stability of mesalamine in rectal suspension diluted with distilled water. *Am J Hosp Pharm*. 1994; 51: 2955–7.

Montgomery HA, Smith FM, Scott BE, et al. Stability of 5-aminosalicylic acid suspension. *Am J Hosp Pharm*. 1986; 43: 118–20.

Mesna

Chemical Names
2-Mercaptoethanesulfonic acid monosodium salt
Sodium 2-mercaptoethanesulfonate

Other Name
Mesnex

Form	Molecular Formula	MW	CAS
Mesna	$C_2H_5NaO_3S_2$	164.2	19767-45-4

Appearance
Mesna is a hygroscopic powder.

Solubility
Mesna is freely soluble in water but sparingly soluble in organic solvents.

Method
Mayron and Gennaro used an HPLC method to evaluate the stability of mesna and granisetron hydrochloride during simulated Y-site administration. The liquid chromatographic system consisted of a piston pump with a pulse dampener, a rotary injection port with a 20-μL loop, a variable-wavelength spectrometric detector, and an integrator. The stationary phase was a cyano reversed-phase column (250 × 4.6 mm, 10-μm particle size). The mobile phase consisted of acetonitrile and 0.1 M monobasic sodium phosphate (20:80, vol/vol) adjusted to pH 4.2 with phosphoric acid. The flow rate was 2.50 mL/min. UV detection was performed at 200 nm.

Samples were diluted with mobile phase. The retention times for mesna and granisetron were 1.16 and 4.53 minutes, respectively.

The stability-indicating capability of the analytical method was demonstrated by an accelerated degradation study. Solutions of mesna and granisetron hydrochloride were adjusted to pH 2 and 11, boiled for 1 hour, and then readjusted to pH 5, diluted with the mobile phase, and analyzed. No degradation product peaks interfered with the intact mesna or granisetron peaks.

Reference
Mayron D, Gennaro AR. Stability and compatibility of granisetron hydrochloride in i.v. solutions and oral liquids and during simulated Y-site injection with selected drugs. *Am J Health Syst Pharm.* 1996; 53: 294–304.

Metadoxine

Chemical Name
Pyridoxine L-5-oxopyrrolidine-2-carboxylate

Form	Molecular Formula	MW	CAS
Metadoxine	$C_8H_{11}NO_3.C_5H_7NO_3$	298.3	74536-44-0

Method
Kaul et al. developed and validated an HPLC method for the determination of metadoxine in bulk drug and in pharmaceutical dosage formulations. The liquid chromatographic

system consisted of a Jasco model PU 1580 pump, a model AS 1555 autosampler, and a model UV 1575 UV-visible detector. The stationary phase was a Kromasil C_{18} column (250 × 4.6 mm, 5-μm particle size). The mobile phase was a mixture of methanol and water (50:50, vol/vol). The flow rate was 1.0 mL/min. UV detection was performed at 286 nm. The injection volume was 20 μL. Under these conditions, the retention time of metadoxine was 2.9 minutes.

Twenty metadoxine tablets were ground and a portion of the powder equivalent to 500 mg of metadoxine was accurately weighed, mixed with 50 mL of methanol, sonicated for 30 minutes, diluted with another 50 mL of methanol, and centrifuged at 3000 rpm for 5 minutes. The supernatant was appropriately diluted, filtered through a 0.45-μm Millipore filter, and assayed.

The method was demonstrated to be stability indicating by intentional degradation. Metadoxine in phosphate buffer was adjusted to pH 7.4 and pH 9.0 and refluxed at 70 °C for 2 hours. Metadoxine in methanol was treated separately with concentrated hydrochloric acid and 10.0 N sodium hydroxide solution and refluxed at 70 °C for 6 hours. Metadoxine methanolic solution was treated with 50% hydrogen peroxide and refluxed at 70 °C for 8 hours. Metadoxine solution was refluxed in boiling water bath for 2 hours. Metadoxine drug was incubated in an oven at 100 °C for 4 hours. Metadoxine solutions were also exposed to direct sunlight for 4 days and to UV radiation for 8 days. Peaks of degradation products were resolved from the peak of the intact metadoxine.

A standard curve for metadoxine was constructed from 10 to 60 μg/mL; the correlation coefficient was 0.9998. Interday and intraday coefficients of variation were 1.25 and 0.97%, respectively. The limit of detection and the limit of quantitation were 0.15 and 0.40 μg/mL, respectively.

Reference

Kaul N, Agrawal H, Patil B, et al. Stability-indicating HPLC method for the determination of metadoxine as bulk drug and in pharmaceutical dosage form. *Chromatographia*. 2004; 60: 501–10.

Metanephrine

Chemical Name

4-Hydroxy-3-methoxy-α-(methylaminomethyl)benzenemethanol

Form	Molecular Formula	MW	CAS
Metanephrine	$C_{10}H_{15}NO_3$	197.2	5001-33-2

Appearance

Metanephrine occurs as crystals or prisms.

Method

Chan et al. used a reversed-phase ion-pair HPLC method with electrochemical detection to evaluate the degradation of urine catecholamines and metanephrine. The Shimadzu model LC-10AT apparatus consisted of a solvent-delivery system, a low-pressure gradient flow pump, an on-line degasser, and a model L-ECD-6A electrochemical detector. The temperature for the detector was maintained at 32 °C. The stationary phase was two serially connected Waters Nova-Pak C_{18} columns (150 × 3.9 mm, 4-μm particle size). The column was maintained at 52 °C. The mobile phase was 2.2% acetonitrile in a buffer solution. The buffer solution consisted of 200 mM monobasic sodium phosphate, 4 mM sodium 1-heptanesulfonate, and 0.2 g/L disodium ethylenediamine-tetraacetic acid (EDTA), adjusted to pH 3.0 with 1 M phosphoric acid. The flow rate was 1.2 mL/min. A potential of +0.85 V was applied at a glassy carbon working electrode versus a silver–silver chloride (Ag/AgCl) reference electrode. 3,4-Dihydroxy-benzylamine (0.4 mM in 0.1 M hydrochloric acid) was used as an internal standard.

A 5-mL sample was mixed with 5 μL of the internal standard, 100 μL of 0.1 g/mL of EDTA, and 100 μL of 0.1 g/mL sodium metabisulfite; the solution was adjusted to pH 6.5 with 1 M sodium hydroxide, transferred to a preconditioned Bio-Rex 70 resin column, and allowed to drain. The column was washed with 10 mL of HPLC water, and then metanephrine was eluted with 7 mL of 4 M formic acid. The eluate was collected. The injection volume was 250 μL. Under these conditions, retention times for norepinephrine, epinephrine, the internal standard, nor-metanephrine, dopamine, and metanephrine were 4.02, 6.35, 8.22, 10.01, 13.37, and 16.67 minutes, respectively.

The stability-indicating ability of the method was demonstrated by accelerated degradation of drugs. Drugs were degraded in 1 M hydrochloric acid, 1 M sodium hydroxide solution, and 30% hydrogen peroxide at 90 °C for 1 hour. None of the degradation peaks interfered with the determination of the drugs.

A calibration curve for metanephrine was constructed from 40 to 4000 nM. The correlation coefficient was 0.9994. Intraday and interday coefficients of variation were 2.58 and 12.97%, respectively.

Reference
Chan ECY, Wee PY, Ho PC. Evaluation of degradation of urinary catecholamines and metanephrines and deconjugation of their sulfoconjugates using stability-indicating reversed-phase ion-pair HPLC with electrochemical detection. *J Pharm Biomed Anal.* 2000; 22: 515–26.

Methenolone

Chemical Name
17β-Hydroxy-1-methyl-5α-androst-1-en-3-one

Other Name
Primobolan

Form	Molecular Formula	MW	CAS
Methenolone	$C_{20}H_{30}O_2$	302.5	153-00-4
Methenolone acetate	$C_{22}H_{32}O_3$	344.5	434-05-9
Methenolone enanthate	$C_{27}H_{42}O_3$	414.6	303-42-4

Solubility
Methenolone acetate is soluble in methanol, ether, and chloroform.

Method 1
Cavrini et al. reported an assay for the determination of methenolone acetate in pharmaceutical formulations. A Varian model 5020 system was equipped with a Valco injector with a 10-μL loop, a Varian model UV-50 variable-wavelength detector, and a Bryans Southern integrator. The stationary phase was a MicroPack MCH-10 reversed-phase column (300 × 4 mm, 10-μm particle size). The mobile phase was a mixture of methanol and water (90:10, vol/vol) and was isocratically delivered at 1.5 mL/min. UV detection was performed at 240 nm and 0.05 AUFS. Prednisolone was used as an internal standard.

A sample of ground tablets equivalent to 5 mg of methenolone acetate was weighed and extracted three times with 7 mL of acetonitrile by agitation for 10 minutes. The solution was filtered, mixed with internal standard, diluted with acetonitrile, and filtered through a 0.45-μm Gelman Teflon membrane filter. The injection volume was 10 μL. Under these conditions, retention times for prednisolone, methenolone, and methenolone acetate were about 4.1, 4.8, and 7.2 minutes, respectively (estimated from the published chromatogram).

The method was reported to be stability indicating.

A standard curve for methenolone acetate was obtained from 40 to 100 μg/mL. The correlation coefficient was 0.9992.

Reference
Cavrini V, Di Pietra AM, Raggi MA. High-performance liquid chromatographic (HPLC) analysis of methenolone esters in pharmaceutical formulations. *Int J Pharm*. 1983; 13: 333–43.

Method 2
Cavrini et al. also described an HPLC assay for the determination of methenolone enanthate in pharmaceutical formulations. A Varian model 5020 system was equipped with a Valco injector with a 10-μL loop, a Varian model UV-50 variable-wavelength detector, and a Bryans Southern integrator. The stationary phase was a LiChrosorb RP8 column (250 × 4 mm, 7-μm particle size). The mobile phase was a mixture of methanol and water (90:10, vol/vol) and was isocratically delivered at 1 mL/min. UV detection was performed at 240 nm and 0.05 AUFS. Methenolone acetate was used as an internal standard.

A sample of injection equivalent to 100 mg of methenolone enanthate was mixed with internal standard, diluted with acetonitrile, and filtered through a 0.45-μm Gelman Teflon membrane filter. The injection volume was 10 μL. Under these conditions, retention times for methenolone, methenolone acetate, and methenolone enanthate were about 2.9, 3.5, and 9.1 minutes, respectively.

The method was reported to be stability indicating.

A standard curve for methenolone enanthate was obtained from 40 to 120 µg/mL. The correlation coefficient was 0.9980.

Reference

Cavrini V, Di Pietra AM, Raggi MA. High-performance liquid chromatographic (HPLC) analysis of methenolone esters in pharmaceutical formulations. *Int J Pharm.* 1983; 13: 333–43.

Methicillin

Chemical Name

Sodium (6R)-6-(2,6-dimethoxybenzamido)penicillanate monohydrate

Other Names

Staficyn, Staphcillin

Form	Molecular Formula	MW	CAS
Methicillin sodium	$C_{17}H_{19}N_2NaO_6S.H_2O$	420.4	7246-14-2

Appearance

Methicillin sodium occurs as a fine white crystalline powder that is odorless or has a slight odor.

Solubility

Methicillin sodium is very soluble in water and methanol. It is practically insoluble in ether.

Method 1

Al-Majed et al. reported an HPLC method for the determination of methicillin in bulk, injections, and spiked human plasma and urine. A Shimadzu LC-10AD VP liquid system was equipped with a model FCV-10 AL VP mixing chamber, a model SCL-10A VP system controller, a model RF-10A XL fluorescence detector, and a Rheodyne model 7725i injector. The stationary phase was a Waters µBondapak C_{18} column (300 × 3.9 mm) coupled with a guard column of the same packing material. The mobile phase consisted of acetonitrile and acetic anhydride 2% (vol/vol) (55:45, vol/vol). The flow rate was 1.8 mL/min. The column eluent was monitored by a fluoresecence detector using an excitation wavelength of 280 nm and an emission wavelength of 360 nm. The injection volume was 20 µL. Under these conditions, the retention time of methicillin was 3.1 minutes.

To show the stability-indicating property of the method, methicillin solutions were prepared in 0.005 M hydrochloric acid or 0.005 M sodium hydroxide solution and stored at room temperature for different periods of time. Retention times of degradation products were about 2.0, 3.9, and 5.7 minutes.

A standard curve for methicillin was constructed from 1.0 to 10.0 μg/mL; the correlation coefficient was 0.9996. Between-day and within-day coefficients of variation for the analysis were less than 2%. The limit of detection was 0.1 μg/mL.

Reference
Al-Majed A, Belal F, Khalil NY, et al. Stability-indicating HPLC method for the determination of methicillin in vials and biological fluids with fluorometric detection. *J Liq Chromatogr Rel Technol*. 2005; 28: 1737–50.

Method 2
Grover et al. determined by HPLC analysis the degradation rate of methicillin at 35 °C in borate buffer (pH 9.2). The Shimadzu LC-10A liquid chromatograph included model LC-10AS pumps, a model SPD-10A dual-wavelength detector, a model C-R7A data processor, and a Rheodyne model 7125 injector with a 20-μL loop. The stationary phase was a Phenomenex C_{18} column (300 × 3.9 mm, 5-μm particle size). The mobile phase was a mixture of 85% water containing 20 mM monobasic potassium phosphate and 10 mM tetramethylammonium chloride (pH 5.0) and 15% acetonitrile. The flow rate was 1.0 mL/min. UV detection was performed at 220 and 240 nm.

Samples were incubated at 35 °C and assayed. Under these conditions, the retention time for methicillin was 15.5 minutes.

The method was reported to be stability indicating.

A standard curve for methicillin was generated from 5 to 150 μg/mL. The correlation coefficient was greater than 0.99. The coefficient of variation for the analysis of methicillin was less than 1%.

Reference
Grover M, Gulati M, Singh B, et al. Correlation of penicillin structure with rate constants for basic hydrolysis. *Pharm Pharmacol Commun*. 2000; 6: 355–63.

Methocarbamol

Chemical Names
3-(2-Methoxyphenoxy)-1,2-propanediol 1-carbamate
3-(*o*-Methoxyphenoxy)-2-hydroxypropyl 1-carbamate

Other Name
Robaxin

Form	Molecular Formula	MW	CAS
Methocarbamol	$C_{11}H_{15}NO_5$	241.2	532-03-6

Appearance

Methocarbamol is a white powder.

Solubility

Methocarbamol has a solubility of 2.5 g/100 mL in water at 20 °C. It is also soluble in alcohol and propylene glycol.

Method

Pouli et al. developed a stability-indicating reversed-phase HPLC method to investigate the hydrolysis of methocarbamol and to determine the degradation products of methocarbamol. The liquid chromatograph consisted of a Waters model 590 solvent-delivery system, a Rheodyne 7125 injector fitted with a 10-μL loop, a Lambda-Max model 481 multiple-wavelength UV detector, and a BBC model 5E120 strip-chart recorder. The stationary phase was a Perkin-Elmer reversed-phase C_{18} column (3 × 3 cartridge, part no. 0258-0164). The mobile phase consisted of a water and acetonitrile (9:1, vol/vol) solution containing 1% acetic acid. The flow rate was 2.25 mL/min. UV detection was performed at 274 nm and 0.01–0.05 AUFS. Acetanilide served as an internal standard. The injection volume was 10 μL.

The method was demonstrated to be stability indicating by spiking methocarbamol solution with its known impurities and degradation products. Peaks from its impurity and degradation products did not interfere with the intact methocarbamol peak. Retention times for the internal standard and methocarbamol were 2.1 and 4.7 minutes, respectively. The retention times for 3-(2-methoxyphenoxy)propanediol guaifenesin and 2-carbamate (two degradation products) were 4 and 2.9 minutes, respectively.

A standard curve for methocarbamol was constructed from 0.05 to 1.2 mg/mL. The correlation coefficient was greater than 0.999.

Reference

Pouli N, Antoniadou-Uyzas A, Foscolos GB. Methocarbamol degradation in aqueous solutions. *J Pharm Sci.* 1994; 83: 499–501.

Methotrexate

Chemical Name

N-[4-[[(2,4-Diamino-6-pteridinyl)methyl]methylamino]benzoyl]-L-glutamic acid

Other Names
Folex, Mexate, Rheumatrex

Form	Molecular Formula	MW	CAS
Methotrexate	$C_{20}H_{22}N_8O_5$	454.5	59-05-2
Methotrexate sodium	$C_{20}H_{20}N_8Na_2O_5$	498.4	7413-34-5

Appearance
Methotrexate is an orange-brown crystalline powder. Methotrexate sodium occurs as a yellow powder.

Solubility
Methotrexate is practically insoluble in water and alcohol. Methotrexate sodium is soluble in water.

Method 1
Using HPLC, Zhang et al. evaluated the stability of methotrexate sodium 2 mg/mL (as the sodium salt) in Elliott's B solution, an artificial cerebrospinal fluid. The liquid chromatograph included a multisolvent-delivery pump, a multiple-wavelength UV detector, and an autosampler in one unit (Waters LC Module 1 Plus), which was controlled and integrated by a Digital workstation utilizing Waters Millennium 2010 chromatography manager. The stationary phase was a Vydac reversed-phase C_{18} analytical column (250×4.6 mm, 5-μm particle size). The mobile phase consisted of 15.5% acetonitrile in aqueous phosphate buffer composed of 0.2 M dibasic sodium phosphate and 0.1 M citric acid to adjust the pH to 6.0. The flow rate was 1 mL/min. UV detection was performed at 270 nm.

Samples were diluted 1:20 with mobile phase. The injection volume was 15 μL. Under these conditions, the retention time for methotrexate was 4.7 minutes.

The HPLC method was demonstrated to be stability indicating by accelerated methotrexate decomposition. Boiling methotrexate in 1 N sodium hydroxide for 2 minutes yielded a 75% reduction in the intact drug peak and the formation of a new peak at 3.7 minutes. The decomposition product peak did not interfere with the intact drug peak.

Calibration curves were constructed from a linear plot of peak area versus concentration for methotrexate from 25 to 150 μg/mL. The correlation coefficient was greater than 0.9999. The coefficient of variation was 0.5%. The intraday and interday coefficients of variation were 0.5 and 1.4%, respectively.

Reference
Zhang Y, Xu QA, Trissel LA, et al. Physical and chemical stability of methotrexate sodium, cytarabine, and hydrocortisone sodium succinate in Elliott's B solution. *Hosp Pharm.* 1996; 31: 965–70.

Method 2
Mayron and Gennaro assessed the stability of methotrexate with granisetron hydrochloride during simulated Y-site administration. The liquid chromatograph consisted of a piston pump with a pulse dampener, a rotary injection port with a 20-μL loop, a variable-wavelength spectrometric detector, and an integrator. The stationary phase was a C_{18} reversed-phase column (300×4.6 mm, 5-μm particle size). The mobile phase was a

mixture of acetonitrile and 0.1 M monobasic sodium phosphate dihydrate (20:80, vol/vol) adjusted to pH 4.2 with phosphoric acid. The flow rate was 1.2 mL/min. UV detection was performed at 300 nm.

Samples were diluted with the mobile phase. The retention times for methotrexate and granisetron were 2.73 and 5.93 minutes, respectively.

The stability-indicating capability of the analytical method was demonstrated by an accelerated degradation study. Solutions of methotrexate and granisetron hydrochloride were adjusted to pH 2 and 11, boiled for 1 hour, readjusted to pH 5, diluted with the mobile phase, and analyzed. The degradation product peaks did not interfere with the intact methotrexate or granisetron peaks.

Reference

Mayron D, Gennaro AR. Stability and compatibility of granisetron hydrochloride in i.v. solutions and oral liquids and during simulated Y-site injection with selected drugs. *Am J Health Syst Pharm*. 1996; 53: 294–304.

Method 3

Wright and Newton used an HPLC method to evaluate the stability of methotrexate injection in prefilled plastic disposable syringes. The HPLC system was composed of an Applied Chromatography Systems model 300 pump, a Rheodyne 10-µL injection valve, an Applied Chromatography Systems LC 750 monitor, and a Venture Servoscribe Is RE 541.20 potentiometric recorder. The stationary phase was a Waters Nova-Pak C_{18} steel analytical column (150 × 3.9 mm) coupled with a preinjection column packed with Whatman Precolumn Gel (37–53-µm particle size). The mobile phase was an acetate buffer at a pH of 2.8, and it was delivered isocratically at a rate of 0.5 mL/min. UV detection was performed at 280 nm and 0.1–0.5 AUFS. *p*-Aminosalicylic acid 0.035 mg/mL was used as the internal standard. The sample injection volume was 10 µL. The retention time for the internal standard was about 4.5 minutes and the retention time for methotrexate was about 11.0 minutes (estimated from the published chromatogram).

The method was determined to be stability indicating by spiking the methotrexate sample with the known degradation products. Retention times for *p*-aminobenzoylglutamic acid and 2-amino-4-hydroxyperidine-6-carboxylic acid (photolytic degradation products), methotrexate, and amethopterin (thermal degradation product) were 3.5, 11.0, and 12.0 minutes, respectively.

Reference

Wright MP, Newton JM. Stability of methotrexate injection in prefilled, plastic disposable syringes. *Int J Pharm*. 1988; 45: 237–44.

Method 4

McElnay et al. examined the stability of methotrexate in burette administration sets. The chromatograph consisted of an Altex 110A pump, a Rheodyne 20-µL loop injection valve, a Perkin-Elmer LC-75 spectrophotometer, a Hewlett-Packard 3390A integrator, and a Spherisorb ODS column (250 × 4.6 mm, 5-µm particle size). The mobile phase consisted of water and methanol (70:30, vol/vol) with tetrabutylammonium hydrogen sulfate (5 mmol/L) as an ion-pairing agent. The flow rate was 0.9 mL/min. UV detection was performed at 303 nm. The retention time for methotrexate was about 3.7 minutes (estimated from the published chromatogram).

The method was determined to be stability indicating by accelerated degradation of methotrexate. A solution of methotrexate was allowed to stand in sunlight for 12 hours over a 2-day period. The method was able to differentiate between the intact methotrexate and its degradation products.

A calibration curve was generated from 0.4 to 1.0 mg/mL. The correlation coefficient was 0.998. The coefficient of variation was 0.70% ($n = 6$).

Reference

McElnay JC, Elliott DS, Cartwright-Shamoon J, et al. Stability of methotrexate and vinblastine in burette administration sets. *Int J Pharm.* 1988; 47: 239–47.

Method 5

Cheung et al. determined the stability of methotrexate in Elliott's B solution, 0.9% sodium chloride injection, 5% dextrose injection, and lactated Ringer's injection. The liquid chromatograph was equipped with a Spectra-Physics model 3500B reciprocating piston pump with flow feedback control, a Valco rotary injector valve fitted with a 10-µL loop, a Spectra-Physics model 770 variable-wavelength UV detector, and an Omniscribe strip-chart recorder. The stationary phase was an Alltech C_{18} stainless steel column (250 × 4.6 mm, 10-µm particle size). The mobile phase consisted of 1% acetic acid, 70% distilled water, and 29% acetonitrile and was delivered at 2 mL/min. UV detection was performed at 240 nm and 0.1 AUFS. Propylparaben was used as an internal standard.

Samples were diluted with the internal standard solution, and the injection volume was 10 µL. Under these conditions, retention volumes for cytarabine, hydrocortisone, methotrexate, hydrocortisone sodium succinate, and propylparaben were 9.6, 13.0, 25.6, 32.8, and 40.4 mL, respectively.

The method was determined to be stability indicating using drug solutions that were degraded by heating on a steam bath for 6 hours. Degradation product peaks did not interfere with the intact methotrexate peak.

The standard curve for methotrexate was linear ($r > 0.999$) from 27 to 108 µg/mL. The variation determined on 2 consecutive days was about 2%.

Reference

Cheung Y-W, Vishnuvajjala BR, Flora KP. Stability of cytarabine, methotrexate sodium, and hydrocortisone sodium succinate admixture. *Am J Hosp Pharm.* 1984; 41: 1802–6.

Methyldopa

Chemical Names

3-Hydroxy-α-methyl-L-tyrosine

L-3-(3,4-Dihydroxyphenyl)-2-methylalanine

Other Name
Aldomet

Form

Form	Molecular Formula	MW	CAS
Methyldopa	$C_{10}H_{13}NO_4$	211.2	555-30-6
Methyldopate hydrochloride	$C_{12}H_{17}NO_4 \cdot HCl$	275.7	2508-79-4

Appearance
Methyldopa is a white to yellowish white fine powder. Methyldopate hydrochloride is a white or practically white and odorless or practically odorless crystalline powder.

Solubility
Methyldopa is sparingly soluble in water and slightly soluble in alcohol. Methyldopate hydrochloride is freely soluble in water and alcohol.

pK$_a$
Methyldopa has pK_a values of 2.25, 9.0, 10.35, and 12.6.

Method 1
Metwally presented an ion-pair HPLC method for the analysis of methyldopa in capsules and in the presence of its degradation products and impurities. The liquid chromatograph consisted of a Rheodyne model 7125 injector with a 20-μL loop, a Waters model 590 pump, a Schoeffel Spectroflow model 757 variable-wavelength detector, and a Hewlett-Packard model 3392A integrator. The stationary phase was a Phenomenex CN analytical column (250 × 4.6 mm, 5-μm particle size). The mobile phase consisted of water containing 2% acetic acid and 0.005 M sodium 1-heptanesulfonate and methanol (80:20, vol/vol), adjusted to pH 2.60. The flow rate was 1.6 mL/min. UV detection was performed at 280 nm.

Microcapsules equivalent to 500 mg of methyldopa were ground, transferred to a 100-mL volumetric flask, mixed with 50 mL of 0.05 M sulfuric acid, and sonicated for 15 minutes. This mixture was diluted to volume with 0.05 M sulfuric acid and filtered. The filtrate was diluted 1:10 with 0.05 M sulfuric acid. Under these conditions, retention times for methyldopa and 3-*O*-methyl-methyldopa were 4.5 and 6.0 minutes, respectively.

The stability-indicating nature of the method was shown by accelerated decomposition of methyldopa. A methyldopa solution was treated with 0.1 or 1.0 M sodium hydroxide at 25 °C. Degradation product peaks did not interfere with the intact methyldopa peak.

A standard curve for methyldopa was generated from 0.5 to 200 μg/mL. The correlation coefficient was 0.999.

Reference

Metwally M E-S. Stability-indicating high-performance liquid chromatographic assay for α-methyldopa in sustained-release capsules. *J Chromatogr*. 1991; 549: 221–8.

Method 2

Williams et al. described a stability-indicating ion-pair HPLC method for methyldopate hydrochloride in the presence of aminophylline in intravenous solutions. The chromatograph consisted of a Waters model M6000A solvent-delivery system, a Waters model U6K injector, and a Perkin-Elmer model 204A spectrofluorometer or a model LC-55 spectrophotometer. The stationary phase was a Waters C_{18} column. The mobile phase consisted of 0.005 M 1-heptanesulfonate sodium and 0.35 M acetic acid in 35% (vol/vol) methanol. The flow rate was 1.6 mL/min. The fluorometric detector was set at an excitation wavelength of 285 nm and an emission wavelength of 315 nm. The injection volume was 15 μL. The retention times for aminophylline and methyldopate hydrochloride were 2.75 and 4.5 minutes, respectively.

The stability-indicating nature of the assay was demonstrated by an accelerated degradation study. A solution of methyldopate hydrochloride was exposed to fluorescent light for 17 hours. Degradation product peaks did not interfere with the parent methyldopate hydrochloride peak.

A standard curve was constructed by plotting drug concentrations versus the peak areas from 120 to 1100 μg/mL. The correlation coefficient was greater than 0.997.

Reference

Williams DA, Fung EYY, Newton DW. Ion-pair high-performance liquid chromatography of terbutaline and catecholamines with aminophylline in intravenous solutions. *J Pharm Sci*. 1982; 71: 956–8.

Methylergonovine Maleate

Chemical Name

N-[(*S*)-2-Hydroxy-1-methylethyl]-D-lysergamide hydrogen maleate

$\cdot\ C_4H_4O_4$

Other Names

Methergin, Methergine, Methylergometrine Maleate

Form	Molecular Formula	MW	CAS
Methylergonovine maleate	$C_{20}H_{25}N_3O_2 \cdot C_4H_4O_4$	455.5	57432-61-8

Appearance

Methylergonovine maleate occurs as a white to pinkish-tan odorless microcrystalline powder.

Solubility

Methylergonovine maleate is slightly soluble in water and in alcohol.

Method

Marigny et al. used an HPLC method to study the stability of oral liquid preparations of methylergonovine. A Kontron liquid chromatograph was equipped with a model 325 pump, a model 550 photodiode-array detector, and an autoinjector. The stationary phase was a Hypersil C_{18} column (250 × 3.9 mm, 5-μm particle size). The mobile phase consisted of 0.015 M monobasic potassium phosphate buffer (pH 4.6) and acetonitrile (80:20, vol/vol). The flow rate was 1 mL/min. UV detection was performed at 312 nm. Standard solutions were prepared by dissolving methylergonovine maleate in ethanol. Under these conditions, the retention time of methylergonovine maleate was about 6.5 minutes.

In order to demonstrate the stability-indicating nature of the method, methyl-ergonovine maleate solutions were exposed to humidity, high temperature, light, and oxygen. In all these cases, degradation products of methylergonovine maleate did not interfere with the determination of the intact drug.

A five-point standard curve for methylergonovine maleate was constructed from 0.04 to 0.06 mg/mL. The correlation coefficient was 0.999. Intraday and interday coefficients of variation were less than 2 and 4%, respectively.

Reference

Marigny K, Lohezic-Ledevehat F, Aubin F, et al. Stability of oral liquid preparations of methylergometrine. *Pharmazie*. 2006; 61: 701–5.

Methyl Nicotinate

Chemical Name

Methyl pyridine-3-carboxylate

Other Name
Vitathone

Form	Molecular Formula	MW	CAS
Methyl nicotinate	$C_7H_7NO_2$	137.1	93-60-7

Appearance
Methyl nicotinate occurs as white or almost white crystals or crystalline powder.

Solubility
Methyl nicotinate is very soluble in water, alcohol, and chloroform. It is freely soluble in ether.

Method
Abounassif et al. described an HPLC analytical method for the simultaneous determination of diethylamine salicylate and methyl nicotinate in the presence of parabens. A Varian model 5000 liquid chromatograph was equipped with a model UV50 variable-wavelength detector, a Rheodyne model 7125 injector with a 20-µL loop, a Varian model 9176 recorder, and a model DSC 111L data system controller. The stationary phase was a Waters µBondapak C_{18} stainless steel column (300 × 4 mm, 10-µm particle size). Solvent A was acetonitrile and solvent B was 1% aqueous acetic acid. The mobile phase was delivered in 15% A at 1.70 mL/min for 12 minutes and stepped to 75% A at 2.5 mL/min for another 6 minutes. UV detection was performed at 254 nm. Acetaminophen (paracetamol) 8 µg/mL in water was used as an internal standard.

A sample of about 150–250 mg of the ointment was accurately weighed and transferred into a 50-mL flask, followed by 40 mL of the mobile phase and 4 mL of the internal standard. The mixture was shaken for 5 minutes, brought to volume using mobile phase, and filtered. The injection volume was 20 µL. Under these conditions, retention times for acetaminophen, methyl nicotinate, and diethylamine salicylate were about 2.5, 4.2, and 8.6 minutes, respectively (estimated from the published chromatogram).

The analytical method was reported to be stability indicating since the intact drug peaks were resolved from their degradation product peaks.

A calibration curve for methyl nicotinate was constructed from 20 to 100 µg/mL. The correlation coefficient was 0.9999.

Reference
Abounassif MA, Abdel-Moety EM, Gad-Kariem RA. HPLC-quantification of diethylamine salicylate and methyl nicotinate in ointments. *J Liq Chromatogr*. 1992; 15: 625–36.

Methylparaben

Chemical Name
Methyl 4-hydroxybenzoate

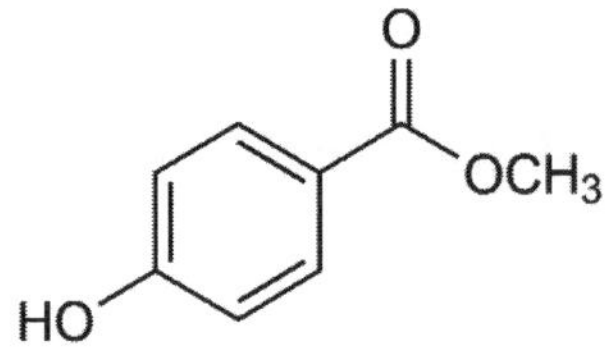

Form

Methylparaben

Molecular Formula

$C_8H_8O_3$

MW

152.1

CAS

99-76-3

Appearance

Methylparaben occurs as colorless crystals or as a white crystalline powder.

Solubility

Methylparaben is very slightly soluble in water. It is freely soluble in alcohol and methanol.

Method

Radus and Gyr reported an HPLC method for the determination of methylparaben and butylparaben in pharmaceutical formulations. The liquid chromatograph consisted of a Waters model 6000A pump or a Milton-Roy minipump, a Rheodyne model 7126 injector, a Waters model 440 or an LDC model 1203 UV detector, and an Upjohn autosampler. The stationary phase was a Waters μBondapak C_{18} column. The mobile phase was a mixture of water, acetonitrile, and glacial acetic acid (58:40:2, vol/vol/vol) and was delivered isocratically at 2 mL/min. Calusterone in water–acetonitrile was used as an internal standard.

An ointment sample (1.0 g) was mixed with 10 mL of dimethylformamide and 20 mL of internal standard, shaken at 50 °C, and centrifuged. The injection volume was 10 μL. Under these conditions, retention times for methylparaben, ethylparaben, propylparaben, and butylparaben were about 3.7, 4.6, 6.3, and 9.3 minutes, respectively (estimated from the published chromatogram).

The method was reported to be stability indicating. The primary degradation product, p-hydroxybenzoic acid, eluted with the solvent front.

A standard curve for methylparaben was constructed from 0.1 to 0.3 mg/g.

Similar methods were used by the same authors to analyze methylparaben in creams, lotions, sterile solutions, and suspensions.

Reference

Radus TP, Gyr G. Determination of antimicrobial preservatives in pharmaceutical formulations using reverse-phase liquid chromatography. *J Pharm Sci.* 1983; 72: 221–4.

Methylprednisolone

Chemical Name
(6α,11β)-11,17,21-Trihydroxy-6-methylpregna-1,4-diene-3,20-dione

Other Name
Medrol

Form	Molecular Formula	MW	CAS
Methylprednisolone	$C_{22}H_{30}O_5$	374.5	83-43-2
Methylprednisolone acetate	$C_{24}H_{32}O_6$	416.5	53-36-1
Methylprednisolone hemisuccinate	$C_{26}H_{34}O_8$	474.6	2921-57-5
Methylprednisolone sodium succinate	$C_{26}H_{33}NaO_8$	496.5	2375-03-3

Appearance
Methylprednisolone acetate occurs as a white or practically white crystalline powder. Methylprednisolone hemisuccinate is a white or almost white hygroscopic solid. Methylprednisolone sodium succinate occurs as a white or nearly white and hygroscopic, amorphous solid.

Solubility
Methylprednisolone is practically insoluble in water; it is soluble in alcohol (1 in 100), chloroform (1 in 800), and ether (1 in 800). It is slightly soluble in dehydrated alcohol and acetone but sparingly soluble in dioxane and methanol. Methylprednisolone acetate is soluble in water (1 in 1500), in alcohol (1 in 400), in chloroform (1 in 250), and in ether (1 in 1500); it is slightly soluble in dehydrated alcohol but sparingly soluble in acetone and methanol. Methylprednisolone hemisuccinate is very slightly soluble in water but freely soluble in alcohol and soluble in acetone. Methylprednisolone sodium succinate is soluble in water (1 in 1.5) and in alcohol (1 in 12); it is very slightly soluble in acetone but practically insoluble in chloroform and ether.

Method
Strom and Miller evaluated the stability and compatibility of methylprednisolone sodium succinate and cimetidine hydrochloride in 5% dextrose injection. The chromatograph was composed of a Micromeritics model 750 solvent-delivery system, a Micromeritics

model 787 variable-wavelength UV-visible detector, an Altex model 210A variable-volume injector, and a Micromeritics model 740 microprocessor. The stationary phase was an Alltech C_{18} reversed-phase column (250 × 4.6 mm, 10-μm particle size). The mobile phase consisted of 33% acetonitrile and 67% 0.05 M sodium acetate adjusted with acetic acid to pH 5.6. The flow rate was 1.0 mL/min. UV detection was performed at 248 nm. Hydrocortisone 50 μg/mL was used as the internal standard. Retention times for the 21-succinate ester of methylprednisolone and hydrocortisone were approximately 8.5 and 10.5 minutes, respectively.

The HPLC method was determined to be stability indicating by accelerated decomposition. The admixtures of methylprednisolone sodium succinate and cimetidine hydrochloride were subjected to high temperatures and extreme pH levels. In all cases, degradation product peaks did not interfere with the intact drug and internal standard peaks. Two degradation products, methylprednisolone 17-succinate ester and free alcohol, had retention times of 6 and 15 minutes, respectively.

A standard curve was constructed from 24 to 140 μg/mL for methylprednisolone 21-succinate ester; the correlation coefficient was greater than 0.999. The relative standard deviations were less than 0.58% at the methylprednisolone-equivalent concentration of 1.25 mg/mL and less than 0.9% at the methylprednisolone-equivalent concentration of 0.4 mg/mL.

Similar methods were used by the other researchers cited here.

References

Strom JG Jr, Miller SW. Stability and compatibility of methylprednisolone sodium succinate and cimetidine hydrochloride in 5% dextrose injection. *Am J Hosp Pharm.* 1991; 48: 1237–41.

Nahata MC, Morosco RS, Hipple TF. Stability of diluted methylprednisolone sodium succinate injection at two temperatures. *Am J Hosp Pharm.* 1994; 51: 2157–9.

Trissel LA, Zhang Y. Stability of methylprednisolone sodium succinate in AutoDose infusion system bags. *J Am Pharm Assoc.* 2002; 42: 868–70.

Metoclopramide

Chemical Names
4-Amino-5-chloro-*N*-[2-(diethylamino)ethyl]-2-methoxybenzamide
4-Amino-5-chloro-*N*-[2-(diethylamino)ethyl]-*o*-anisamide

Other Name
Reglan

Form	Molecular Formula	MW	CAS
Metoclopramide	$C_{14}H_{22}ClN_3O_2$	299.8	364-62-5
Metoclopramide hydrochloride	$C_{14}H_{22}ClN_3O_2.HCl.H_2O$	354.3	54143-57-6

Appearance
Metoclopramide hydrochloride is a monohydrate white odorless crystalline powder.

Solubility
Metoclopramide hydrochloride has solubilities of approximately 1.43 g/mL in water and 333 mg/mL in alcohol.

pK$_a$
Metoclopramide hydrochloride has pK$_a$ values of 0.6 and 9.3.

Method 1
Caufield and Stewart investigated an HPLC assay for the simultaneous analysis of meropenem and metoclopramide in an intravenous fluid mixture. The instrument was a Hewlett-Packard model 1090 system including a pump, an autosampler, and a Gilson model 117 variable-wavelength UV detector. A Waters model 996 photodiode-array detector was used for the peak purity analysis. The stationary phase was a YMC ODS-AQ column (150 × 2.0 mm, 3-μm particle size). The mobile phase consisted of 0.01 M triethylamine aqueous solution (which was adjusted to pH 3 with acetic acid) and acetonitrile (88:12, vol/vol). The flow rate was 0.2 mL/min. UV detection was performed at 290 nm.

Samples were diluted with the mobile phase. The sample injection volume was 5 μL. Under these conditions, retention times of meropenem and metoclopramide were about 3.6 and 11.5 minutes, respectively.

The method was demonstrated to be stability indicating by accelerated degradation of meropenem. Solutions of metoclopramide were treated with 0.1 N hydrochloric acid for 15 minutes, with 0.01 N sodium hydroxide solution for 1 minute, or with 0.03% hydrogen peroxide for 4 hours or were heated at 60 °C for 4 hours. None of the degradation product peaks interfered with the intact drug peak.

A standard curve for metoclopramide was constructed from 131.25 to 525 μg/mL. The correlation coefficient was 0.9996–1.0. Intraday and interday coefficients of variation for the analysis of metoclopramide were 2.12 and 0.46%, respectively. The limit of detection was 75 ng/mL.

Reference
Caufield WV, Stewart JT. HPLC separations of meropenem and selected pharmaceuticals using a polar endcapped octadecylsilane narrow bore column. *Chromatographia*. 2000; 51: 308–14.

Method 2
Peterson et al. investigated by HPLC analysis the compatibility and stability of fentanyl in combination with metoclopramide and midazolam. The liquid chromatographic

system consisted of a Varian model 9010 pump, a model 9050 variable-wavelength UV-visible detector, and a Rheodyne 7161 injector. The stationary phase was a Nova-Pak C_{18} column (150 × 3.9 mm, 4-μm particle size). Solvent A was acetonitrile and solvent B was 0.1 M monobasic potassium phosphate buffer (pH 7.8). The mobile phase was delivered from 25 to 60% A in 10 minutes, held at 60% A for 2 minutes, returned to 25% A in 2 minutes, and held at 25% A for 1 minute. The flow rate was 1.0 mL/min. UV detection was performed at 210 nm. Lorazepam 125 μg/mL in ethanol was used as an internal standard. The injection volume was 10 μL. Under these conditions, retention times for metoclopramide, lorazepam, midazolam, and fentanyl were about 5.6, 7.4, 10.8, and 12.8 minutes, respectively.

The method was demonstrated to be stability indicating by accelerated degradation. Solutions of metoclopramide were autoclaved at 121 °C for 5 hours or acidified to pH 1.0 with 0.1 M hydrochloric acid and autoclaved. None of the degradation products interfered with the determination of metoclopramide.

A standard curve for metoclopramide was constructed from 50 to 100 μg/mL. The correlation coefficient was 0.993. The intraday and interday coefficients of variation were less than 4%.

Reference
Peterson GM, Miller KA, Galloway JF, et al. Compatibility and stability of fentanyl admixtures in polypropylene syringes. *J Clin Pharm Ther.* 1998; 23: 67–72.

Method 3
Zhang et al. evaluated the stability of undiluted metoclopramide hydrochloride injection in 3-mL polypropylene infusion-pump syringes. A Waters LC Module 1 Plus liquid chromatograph included an autosampler and a UV detector. The stationary phase was a Vydac C_{18} reversed-phase column (250 × 4.6 mm, 5-μm particle size). The system was controlled and the peak area was integrated by a Digital Venturis personal computer with Waters Millennium 2010 chromatography manager. The mobile phase consisted of acetonitrile, 25 mM sodium acetate, and 25% tetramethylammonium hydroxide in methanol (450:550:1.6, vol/vol/vol) adjusted to pH 4 with acetic acid. The flow rate was 1.2 mL/min. UV detection was performed at 270 nm and 0.5 AUFS.

Samples were diluted 1:50 with HPLC-grade water. Under these conditions, the retention time for metoclopramide was 3.6 minutes.

The stability-indicating property of the method was shown by accelerated degradation of metoclopramide hydrochloride. Exposure to 0.3% hydrogen peroxide for 1 hour led to a 20% reduction in the peak area for metoclopramide and the formation of a new peak at 2.6 minutes. The degradation product peak did not interfere with the metoclopramide peak.

A standard curve was generated from a linear plot of peak area versus concentration of metoclopramide from 50 to 150 μg/mL. The correlation coefficient was greater than 0.9998. The intraday and interday coefficients of variation were 0.9 and 1.1%, respectively.

Reference
Zhang Y, Trissel LA, Martinez JF, et al. Stability of metoclopramide hydrochloride in plastic syringes. *Am J Health Syst Pharm.* 1996; 53: 1300–2.

Method 4

Wang et al. used an HPLC method to study the stability of a fluorouracil–metoclopramide hydrochloride admixture. The chromatograph consisted of a Shimadzu model LC-6A single piston pump, a Shimadzu model SPD-6AV dual UV-visible detector, a Waters model 745 electronic integrator, and a Waters μBondapak phenyl column (150 × 4.6 mm). The mobile phase consisted of acetonitrile, methanol, 0.01% acetic acid, and 0.005 N sulfonic acid solution (20:15:40:25, vol/vol/vol/vol) and was delivered isocratically at 1.0 mL/min. UV detection was performed at 230 nm.

Samples were diluted 1:10 with mobile phase. Retention times for fluorouracil and metoclopramide were 3.5 and 10.0 minutes, respectively.

The stability-indicating capability of the assay was demonstrated by analyzing a fluorouracil–metoclopramide hydrochloride admixture that had been stored at 50 °C for 10 days. Degradation product peaks did not interfere with the intact metoclopramide peak.

Standard curves were constructed from 0.01 to 0.25 mg/mL. Correlation coefficients were greater than 0.99. Intraday and interday coefficients of variation for the assay were 0.78 and 1.01%, respectively.

Reference

Wang D-P, Chang L-C, Lee DKT, et al. Stability of fluorouracil–metoclopramide hydrochloride admixture. *Am J Hosp Pharm*. 1995; 52: 98–9.

Method 5

Fatmi and Williams described an HPLC method for the determination of metoclopramide and its related compounds in tablet dosage forms. The liquid chromatographic system included a Waters model 510 pump, a model 481 variable-wavelength detector, a model 710B WISP autosampler, and a Fisher Scientific Recordall chart recorder. The stationary phase was a Waters μBondapak C_{18} column (300 × 3.9 mm, 10-μm particle size). The mobile phase consisted of 0.15 M ammonium acetate and acetonitrile (80:20, vol/vol) adjusted to pH 6.5 by ammonium hydroxide or acetic acid. The flow rate was 1.0 mL/min. UV detection was performed at 268 nm and 0.05 AUFS.

Metoclopramide tablets were weighed and ground. A portion of this powder equivalent to the weight of one tablet was transferred to a 100-mL volumetric flask, dissolved with 50 mL of methanol, and diluted to volume with methanol. The injection volume was 5 μL. The retention time of metoclopramide was about 5.9 minutes.

The stability-indicating nature of the method was confirmed by assaying a synthetic mixture of metoclopramide and its related compounds. Retention times for 4-acetylamino-5-chloro-*N*-[2-(diethylamino)ethyl]-2-methoxybenzamide hydrochloride and 4-amino-5-chloro-2-methoxybenzoic acid were about 4.7 and 2.0 minutes, respectively.

A standard curve for metoclopramide was obtained from 0.1785 to 1.7851 mg/mL. The correlation coefficient was 0.999.

Reference

Fatmi AA, Williams GV Sr. Analysis of metoclopramide and related compounds in tablets by liquid chromatography. *Drug Dev Ind Pharm*. 1989; 15: 1365–73.

Method 6

Suleiman et al. described an HPLC assay for the determination of metoclopramide hydrochloride in pharmaceutical dosage forms. The Beckman liquid chromatograph consisted of a model 114M single-piston pump, a model 210A 100-μL injector, and a Lloyd Instrument model JJ CR 452 recorder. The stationary phase was a Phase Separations Spherisorb RP-C$_8$ column (250 × 4.6 mm, 5-μm particle size). The mobile phase consisted of 10 mM monobasic sodium phosphate buffer, methanol, and acetonitrile (50:28:22, vol/vol/vol), adjusted to pH 4.8 with phosphoric acid. The flow rate was 1.5 mL/min. UV detection was performed at 214 nm and 0.05 AUFS. Phenobarbital was used as an internal standard.

For tablets, ground tablet powder equivalent to 10 mg of metoclopramide was weighed, mixed with 50 mL of methanol, and sonicated for 5 minutes. The mixture was filtered and then diluted. For an injection, a syrup, or drops, each sample was diluted with methanol, filtered, and diluted further. The injection volume was 50 μL. Under these conditions, the retention times of phenobarbital and metoclopramide were about 2.4 and 7.1 minutes, respectively (estimated from the published chromatogram).

The method was reported to be stability indicating since excipients in the formulation did not interfere with the analysis of metoclopramide.

A standard curve for metoclopramide was obtained from 1 to 10 μg/mL. The correlation coefficient was 0.9997. Intraday and interday coefficients of variation were less than 1.70 and 4.07%, respectively.

Reference
Suleiman MS, Najib NM, El-Sayed YM. Stability-indicating high-performance liquid chromatographic assay for the determination of metoclopramide hydrochloride in pharmaceutical dosage forms. *Analyst*. 1989; 114: 365–8.

Metolazone

Chemical Name
7-Chloro-1,2,3,4-tetrahydro-2-methyl-3-(2-methylphenyl)-4-oxo-6-quinazolinesulfon-amide

Other Name
Zaroxolyn

Form	Molecular Formula	MW	CAS
Metolazone	C$_{16}$H$_{16}$ClN$_3$O$_3$S	365.8	17560-51-9

Appearance
Metolazone is a white powder.

Solubility
Metolazone is practically insoluble in water and sparingly soluble in alcohol.

pK$_a$
Metolazone has a pK$_a$ of 9.7.

Method
Allen and Erickson used HPLC to study the stability of metolazone in extemporaneously compounded oral liquids. A Hewlett-Packard series 1050 automated high-performance liquid chromatograph included a multisolvent mixing and pumping system, an auto-injector, a diode-array detector, and a computer running Chem Station data-handling software. The stationary phase was a Hewlett-Packard Hypersil silica column (200 × 4.6 mm, 5-μm particle size). The mobile phase consisted of 700 mL of methanol, 300 mL of water, 1 mL of diisopropylamine, and 1.5 g of ammonium acetate. The flow rate used was 1.0 mL/min.

Samples were diluted 1:1000 before injection. UV detection was performed at 254 nm. The retention time for metolazone was 5.9 minutes.

The stability-indicating capacity of this method was confirmed by accelerated decomposition of metolazone. A composite chromatogram of metolazone after accelerated degradation by heat, acid, base, an oxidizing agent, and light showed that the intact metolazone peak was well separated from the degradation product peaks.

Standard curves were constructed for metolazone from 0.1 to 1.5 μg/mL. The intraday and interday coefficients of variation were 2.2 and 2.3%, respectively.

Reference
Allen LV, Erickson MA. Stability of ketoconazole, metolazone, metronidazole, procainamide hydrochloride and spironolactone in extemporaneously compounded oral liquids. *Am J Health Syst Pharm.* 1996; 53: 2073–8.

Metoprolol

Chemical Name
1-[4-(2-Methoxyethyl)phenoxy]-3-[(1-methylethyl)amino]-2-propanol

Other Names
Lopressor, Toprol XL

Form	Molecular Formula	MW	CAS
Metoprolol	$C_{15}H_{25}NO_3$	267.4	37350-58-6
Metoprolol succinate	$(C_{15}H_{25}NO_3)_2 \cdot C_4H_6O_4$	652.8	98418-47-4
Metoprolol tartrate	$(C_{15}H_{25}NO_3)_2 \cdot C_4H_6O_6$	684.8	56392-17-7

Appearance

Metoprolol, metoprolol succinate, and metoprolol tartrate occur as white crystalline powders.

Solubility

Metoprolol succinate is freely soluble in water and sparingly soluble in alcohol. Metoprolol tartrate has the following solubilities (mg/mL) at 25 °C: >1000 in water, >500 in methanol, 496 in chloroform, 1.1 in acetone, 0.89 in acetonitrile, and 0.001 in hexane.

pK_a

Metoprolol tartrate has a pK_a of 9.68.

Method 1

Allen and Erickson investigated the stability of metoprolol tartrate 10 mg/mL in extemporaneously compounded oral liquids. The method used an automated Hewlett-Packard series 1050 high-performance liquid chromatograph consisting of a multisolvent mixing and pumping system, an autoinjector, a diode-array detector, and a computer with Chem Station data-handling software. The stationary phase was a Bakerbond C_{18} analytical column (250 × 4.6 mm, 5-μm particle size). The mobile phase consisted of methanol (550 mL) and water (470 mL) with 961 mg of 1-pentanesulfonic acid sodium salt monohydrate, 82 mg of anhydrous sodium acetate, and 0.57 mL of acetic acid. The flow rate was 1.0 mL/min.

Samples were diluted 1:100. UV detection was performed at 254 nm. Under these conditions, the retention time for metoprolol tartrate was 7.3 minutes.

The HPLC analytical method was determined to be stability indicating by accelerated decomposition of metoprolol tartrate. A composite chromatogram of metoprolol tartrate after accelerated degradation by heat, acid, base, an oxidizing agent, and light showed no interference from the peaks of the degradation products with the peak of the intact metoprolol tartrate.

Calibration curves were constructed from the linear plot of metoprolol peak areas versus its concentrations from 10 to 150 μg/mL. The intraday and interday coefficients of variation were 1.3 and 2.1%, respectively.

Reference

Allen LV, Erickson MA. Stability of labetolol hydrochloride, metoprolol tartrate, verapamil hydrochloride, and spironolactone with hydrochlorothiazide in extemporaneously compounded oral liquids. *Am J Health Syst Pharm.* 1996; 53: 2304–9.

Method 2

Belliveau et al. evaluated the stability of metoprolol tartrate 0.40 mg/mL in 5% dextrose injection and 0.9% sodium chloride injection stored at room temperature in polyvinyl chloride bags for 36 hours. The liquid chromatograph consisted of a Shimadzu SPD-6A

UV detector, a Shimadzu C-R6A integrator, a Shimadzu SIL-6A autosampler, and a Fisher model 5000 chart recorder. The stationary phase was a Waters μBondapak C_{18} column (300 × 3.9 mm). The mobile phase consisted of methanol in water (75:25, vol/vol) with 0.6% ammonium hydroxide. The flow rate was 2 mL/min. UV detection was performed at 273 nm. The sample injection volume was 20 μL. Tetracaine hydrochloride was used as the internal standard. Under these conditions, retention times for tetracaine hydrochloride and metoprolol tartrate were 3.6 and 5.1 minutes, respectively.

The stability-indicating capability of the assay was demonstrated by an accelerated degradation study of metoprolol tartrate. Metoprolol tartrate was degraded in 12 N hydrochloric acid at 70 °C for 48 hours and exposed to UV radiation for 24 hours. Metoprolol tartrate degradation product peaks did not interfere with the intact metoprolol tartrate and tetracaine hydrochloride peaks.

A standard curve was constructed from 0.1 to 0.3 mg/mL. Its correlation coefficient was 0.997. Intraday and interday coefficients of variation were less than 5%.

Reference
Belliveau PP, Shea BF, Scavone JM. Stability of metoprolol tartrate in 5% dextrose injection or 0.9% sodium chloride injection. *Am J Hosp Pharm.* 1993; 50: 950–2.

Metronidazole

Chemical Names
2-Methyl-5-nitroimidazole-1-ethanol
1-(2-Hydroxyethyl)-2-methyl-5-nitroimidazole

Other Name
Flagyl

Form	Molecular Formula	MW	CAS
Metronidazole	$C_6H_9N_3O_3$	171.2	443-48-1
Metronidazole benzoate	$C_{13}H_{13}N_3O_4$	275.3	13182-89-3
Metronidazole hydrochloride	$C_6H_9N_3O_3.HCl$	207.6	69198-10-3

Appearance
Metronidazole occurs as white to pale yellow crystals or crystalline powder. Metronidazole benzoate occurs as a white or slightly yellowish crystalline powder or flakes. Metronidazole hydrochloride is an off-white lyophilized powder.

Solubility

Metronidazole has an aqueous solubility of 1.0 g in 100 mL of water at 20 °C. It is sparingly soluble in alcohol. Metronidazole hydrochloride is very soluble in water and soluble in alcohol.

pK_a

Metronidazole has a pK_a of 2.6.

Method 1

Nahata et al. investigated the stability and compatibility of intravenous metronidazole and ceftizoxime sodium in containers stored at 4 and 25 °C. The liquid chromatograph consisted of a Varian model 2010 solvent-delivery pump, a Shimadzu model SIL-10A autosampler with a sample cooler, a Varian model 2050 variable-wavelength detector, a Hewlett-Packard 3396A integrator, and a Mac-Mod Zorbax Reliance SB C_{18} analytical column (160 × 4 mm, 5-µm particle size) with a guard column (12.5 × 4 mm). The mobile phase consisted of 82% 0.05 M acetate buffer and 18% methanol with 4 mL of 40% (wt/wt) tetrabutylammonium hydroxide per liter. The final pH of the mobile phase was adjusted to 5.1 with acetic acid. The flow rate was 1.0 mL/min. UV absorption of metronidazole was monitored at 240 nm. The temperature of vials in the autosampler was maintained at 10 °C whereas that of the column was at ambient temperature. Cefaclor 1 mg/mL was used as the internal standard.

Samples were diluted 1:10 with water. The injection volume was 10 µL. Under these conditions, retention times for metronidazole, cefaclor, and ceftizoxime were 3.5, 7.8, and 9.8 minutes, respectively.

The stability-indicating capability of the assay was demonstrated by an accelerated hydrolysis of metronidazole with an acid and a base. Solutions of the drug admixture were mixed with 1.0 M hydrochloric acid or 1.0 M sodium hydroxide solution for 1 hour at 60 °C. The degradation product peaks appeared before the metronidazole peak.

Standard curves for metronidazole were constructed by linear regression analysis of drug concentration versus peak height ratio for the drug and internal standard from 0.05 to 0.75 mg/mL. The correlation coefficient was greater than 0.999. The intraday and interday coefficients of variation of the assay were less than 0.84%.

Reference

Nahata MC, Edmonds JJ, Morosco RS. Stability of metronidazole and ceftizoxime sodium in ready-to-use metronidazole bags stored at 4 and 25 °C. *Am J Health Syst Pharm.* 1996; 53: 1046–8.

Method 2

Allen and Erickson used HPLC to evaluate the stability of metronidazole in extemporaneously compounded oral liquids. The chromatograph was a Hewlett-Packard series 1050 automated high-performance liquid chromatograph including a multisolvent mixing and pumping system, an autoinjector, a diode-array detector, and a computer running Chem Station data-handling software. The stationary phase was a Bakerbond C_8 analytical column (250 × 4.6 mm, 5-µm particle size). The mobile phase was 20% methanol in water. The flow rate was 1.0 mL/min. UV detection was performed at 254 nm.

Samples were diluted 1:100 before injection. Under these conditions, metronidazole had a retention time of 3.4 minutes.

The HPLC method was determined to be stability indicating. The composite chromatogram of metronidazole after accelerated degradation by heat, acid, base, an oxidizing agent, and light did not show any interference from degradation product peaks with the intact metronidazole peak.

Calibration curves were constructed for metronidazole from 100 to 600 µg/mL. The intraday and interday coefficients of variation were 1.4 and 1.9%, respectively.

Reference

Allen LV, Erickson MA. Stability of ketoconazole, metolazone, metronidazole, procainamide hydrochloride, and spironolactone in extemporaneously compounded oral liquids. *Am J Health Syst Pharm.* 1996; 53: 2073–8.

Method 3

Belliveau et al. evaluated the stability of metronidazole and cefotaxime at refrigerated temperature (5 °C) and at room temperature (28 °C) after admixture in ready-to-use metronidazole minibags or in 0.9% sodium chloride injection in minibags. The HPLC system included a Waters model 510 pump, a Waters model 710A autosampler, a Waters model M440 detector, and a Hewlett-Packard 3396 series II integrator. The stationary phase was a Millipore Resolve analytical column (150 × 3.9 mm, 5-µm particle size). The mobile phase was a mixture of acetate buffer (2.46 g of anhydrous sodium acetate, 8 mL of acetic acid, and 0.2 g of tetrabutylammonium hydrogen sulfate per 1000 mL, pH 3.0) and acetonitrile (86:18, vol/vol). The flow rate was 1.2 mL/min. UV detection was performed at 254 nm. Cefoxitin was used as an internal standard.

Samples were diluted 1:2 with the internal standard and then further diluted with water. The injection volume was 20 µL. Under these conditions, the retention times for metronidazole, cefotaxime, and cefoxitin were 1.8, 2.3, and 3.0 minutes, respectively.

The assay was determined to be stability indicating by accelerated decomposition of metronidazole. A solution of metronidazole was heated at 80 °C for 50 hours, causing partial decomposition of metronidazole, with degradation products eluting before the intact metronidazole eluted.

Standard curves were generated from 2 to 6 µg/mL; the correlation coefficients were 0.9948. The interday coefficient of variation ($n = 5$) was 2.1%.

Reference

Belliveau PP, Nightingale CH, Quintiliani R. Stability of cefotaxime sodium and metronidazole in 0.9% sodium chloride injection or in ready-to-use metronidazole bags. *Am J Health Syst Pharm.* 1995; 52: 1561–3.

Method 4

Rivers and Webster studied the stability of three cephalosporin antibiotics with metronidazole in an admixture. A fully automated, computer-controlled HPLC system consisted of a Waters model 510 solvent-delivery pump, a Waters 712 WISP refrigerated autosampler, a Waters model 490E programmable multiple-wavelength detector, and a Waters Nova-Pak analytical column (150 × 3.9 mm). The mobile phase contained 0.02 M potassium dihydrogen phosphate and acetonitrile (93:7, vol/vol) with 0.01 M triethylamine and was adjusted to pH 4.8 with hydrochloric acid. The flow rate was 1.5 mL/min. UV detection was performed at 270 nm.

Each sample was diluted 1:50 with distilled water. The injection volume was 20 µL. Under these conditions, the retention times for ceftazidime, ceftriaxone, ceftizoxime, and metronidazole were 1.7, 2.1, 2.4, and 3.1 minutes, respectively.

The method was demonstrated to be stability indicating by accelerated decomposition of ceftizoxime sodium and metronidazole. One sample was diluted 1:5 with 0.1 M sodium hydroxide and another was diluted 1:5 with 0.05 M sulfuric acid. They were then heated under reflux conditions for 1 hour. In each case, degradation product peaks did not interfere with the intact drug peak. In acidic conditions, reflux resulted in peaks at 0.7, 0.9, 1.1, 1.3, 1.6, 2.1, 2.9, and 4.4 minutes. In alkaline conditions, reflux resulted in peaks at 0.2, 1.0, 1.1, 1.5, 1.6, 2.2, and 2.8 minutes.

A five-point standard curve was constructed for metronidazole from 0.05 to 0.2 mg/mL. The correlation coefficient was 0.994.

Reference

Rivers TE, Webster AA. Stability of ceftizoxime sodium, ceftriaxone sodium, and ceftazidime with metronidazole in ready-to-use metronidazole bags. *Am J Health Syst Pharm*. 1995; 52: 2568–70.

Method 5

Mathew et al. developed an HPLC method for the quantification of metronidazole benzoate and determined the stability of the ester in suspensions and simulated gastric and intestinal fluids. A Waters ALC 202 liquid chromatograph was equipped with a Rheodyne model 7125 universal injector, a Schoeffel SF 770 multiple-wavelength detector, a Houston Omniscribe recorder, and a Whatman micro C_{18} column (250 × 4.5 mm). The mobile phase consisted of 40% (vol/vol) acetonitrile and 0.1% acetic acid in 0.01 M monobasic potassium phosphate aqueous buffer. The flow rate was 2.0 mL/min. UV detection was performed at 271 nm and 0.1 AUFS. Dextromethorphan hydrobromide in water 2.0 mg/mL was used as the internal standard.

Samples were diluted 1:10 with methanol and then 1:10 with water. The injection volume was 20 µL. The retention times for the internal standard and metronidazole were about 4.0 and 6.6 minutes, respectively (estimated from the published chromatogram).

The method was determined to be stability indicating by accelerated decomposition of metronidazole. A stock solution of metronidazole benzoate was mixed with 1 N sodium hydroxide or 1 M sulfuric acid and boiled for 4 minutes. The product of hydrolysis and benzoic acid did not interfere with the assay.

A standard curve was constructed from 20 to 160 µg/mL with a correlation coefficient of 0.99.

A similar method was reported by Bempong et al.

References

Mathew M, Gupta VD, Bethea C. Stability of metronidazole benzoate in suspensions. *J Clin Pharm Ther*. 1994; 19: 31–4.

Bempong DK, Manning RG, Mirza T, et al. A stability-indicating HPLC assay for metronidazole benzoate. *J Pharm Biomed Anal*. 2005; 38: 776–80.

Method 6

Rivers et al. evaluated the stability of metronidazole at 8 °C in an intravenous admixture. A fully automated, computer-controlled Waters liquid chromatographic system consisted of a model 510 pump, a model 712 refrigerated autosampler, and a model 481 variable-wavelength UV detector. The stationary phase was a Waters μBondapak C_{18} analytical column (150 × 3.9 mm, 4-μm particle size). The mobile phase was a mixture of acetonitrile and 0.02 M potassium dihydrogen phosphate with 0.005 M triethylamine adjusted to pH 4.8 with sodium hydroxide (7:93, vol/vol). The flow rate was 1.75 mL/min. UV detection was performed at 270 nm. The injection volume was 20 μL. The run time was 12.5 minutes and the retention times for metronidazole and cefazolin were 4.46 and 9.25 minutes, respectively.

The HPLC analytical method was determined to be stability indicating by accelerated decomposition of the sample with acid and base. One test solution was mixed with 0.1 M sodium hydroxide and the other with 0.05 M sulfuric acid. Both solutions were refluxed for 1 hour. The chromatograms of each solution indicated no interference of the degradation products with the drug peak.

References

Rivers TE, McBride HA, Trang JM. Stability of cefazolin sodium and metronidazole at 8 °C for use as an IV admixture. *J Parenter Sci Technol.* 1993; 47: 135–7.

Rivers TE, McBride HA, Trang JM. Stability of cefotaxime sodium and metronidazole in an i.v. admixture at 8 °C. *Am J Hosp Pharm.* 1991; 48: 2638–40.

Method 7

Barnes determined the stability of metronidazole and cefuroxime sodium in an admixture for intravenous infusion. The HPLC system consisted of a Cecil CE1100 pump, a Pye LC3 variable-wavelength detector with an 8-μL cell, a Rheodyne 7125 or 7010 injection valve, and a Shimadzu C-R3A electrical integrator. The stationary phase was a reversed-phase Spherisorb ODS column (250 × 4.5 mm, 5-μm particle size). The mobile phase was a mixture of methanol and 0.005 M potassium dihydrogen phosphate buffer pH 4.0 (9:91, vol/vol) and was delivered isocratically at a rate of 2 mL/min. UV detection was performed at 320 nm.

Each sample was diluted 1:5 with water and then 1:6 with methanol and mobile phase (1:5, vol/vol). The injection volume was 20 μL. The retention times for cefuroxime and metronidazole were about 5.0 and 7.8 minutes, respectively (estimated from the published chromatogram).

The method was determined to be stability indicating by accelerated degradation of metronidazole. Boiling a metronidazole sample in 0.1 M sodium hydroxide or 0.05 M sulfuric acid for 1 hour produced chromatograms with a series of peaks, but these did not interfere with the intact drug peaks.

Reference

Barnes AR. Chemical stability of cefuroxime sodium and metronidazole in an admixture for intravenous infusion. *J Clin Pharm Ther.* 1990; 15: 187–96.

Method 8

Rovers et al. evaluated the stability and compatibility of metronidazole when admixed with gentamicin sulfate in the same dosage unit. The HPLC system consisted of a Waters M6000A high-pressure pump, a Waters U6K injector, a Perkin-Elmer LC-55 variable-wavelength spectrophotometer, and a Hewlett-Packard model 3390A integrator. The stationary phase was a Waters μBondapak C_{18} reversed-phase column (300 × 3.9 mm). The mobile phase was a mixture of methanol, acetonitrile, and 0.005 M monobasic potassium phosphate (4:3:99, vol/vol/vol), adjusted to pH 4.0 and delivered isocratically. The flow rate was 2.0 mL/min. UV detection was performed at 320 nm.

Samples were diluted 1:200 with distilled water, and the injection volume was 50 μL. Under these conditions, the retention time of metronidazole was approximately 9 minutes.

The analytical method was demonstrated to be stability indicating by accelerated decomposition of metronidazole. The pH of the commercial metronidazole solution was adjusted to approximately 12 and then boiled for about 15 minutes. The degradation product did not influence the metronidazole assay.

Reference

Rovers JP, Meneilly G, Souney PF, et al. The use of stability-indicating assays to determine the in-vitro compatibility and stability of metronidazole/gentamicin admixtures. *Can J Hosp Pharm.* 1989; 42: 143–6.

Method 9

Bell et al. used HPLC to study the stability of intravenous admixtures of metronidazole with aztreonam. The liquid chromatograph consisted of a ConstaMetric III G pump, a SpectroMonitor D detector, a Negretti model 190 injection valve fitted with a 20-μL loop, and a series 5000 Fisher Recordall dual-pen recorder. The stationary phase was a Varian MCH silica column (300 × 4 mm). The mobile phase consisted of 0.6% acetonitrile in 0.1% phosphoric acid. The flow rate was 2.0 mL/min. UV detection was performed at 238 nm.

Samples were diluted 1:100 with mobile phase. Under these conditions, the retention times for aztreonam and metronidazole were 2.4 and 4.4 minutes, respectively.

The analytical method was confirmed to be stability indicating. Solutions of metronidazole 1 mg/mL in 0.1 N hydrochloric acid or 0.1 N sodium hydroxide were heated for 6 hours at 70 °C. The peaks from the resulting decomposition products did not interfere with the intact metronidazole peak.

Reference

Bell RG, Lipford LC, Massanari MJ, et al. Stability of intravenous admixtures of aztreonam and cefoxitin, gentamicin, metronidazole, or tobramycin. *Am J Hosp Pharm.* 1986; 43: 1444–53.

Mexiletine Hydrochloride

Chemical Name

1-(2,6-Dimethylphenoxy)-2-propanamine hydrochloride

Other Names

Mexitil, Ritalmex

Form	Molecular Formula	MW	CAS
Mexiletine hydrochloride	$C_{11}H_{17}NO.HCl$	215.7	5370-01-4

Appearance

Mexiletine hydrochloride occurs as a white to off-white crystalline powder with a slightly bitter taste.

Solubility

Mexiletine hydrochloride is freely soluble in water and in alcohol.

pK_a

Mexiletine has a pK_a of 9.2

Method

Kaushik and Alexander developed a reversed-phase HPLC method for the analysis of mexiletine hydrochloride. The Waters instrument consisted of two model 501 pumps with a 20-μL loop injector, a model 486 tunable UV detector, and a model 712 WISP autosampler. The stationary phase was a Microsorb C_{18} column (150 × 4 mm, 100 Å pore size). The mobile phase consisted of 0.053 M sodium acetate buffer and methanol (50:50, vol/vol). The flow rate was 1.0 mL/min. UV detection was performed at 254 nm. Thiamine hydrochloride 1.1 mg/mL in mobile phase was used as an internal standard. The injection volume was 20 μL. Under these conditions, retention times for thiamine and mexiletine were about 3.0 and 8.2 minutes, respectively.

In order to demonstrate the stability-indicating capability of the method, mexiletine was degraded with 0.87 M hydrochloric acid or 1 M sodium hydroxide solution and heated for 30 minutes. Chromatograms showed that the intact drug peak was separated from the degradation product peaks.

A calibration curve for mexiletine was constructed from 1.502 to 10.514 mg/mL. The correlation coefficient was 0.993. Intraday and interday coefficients of variation for the analysis of mexiletine were 0.81 and 1.47%, respectively.

Reference

Kaushik S, Alexander KS. A modified reverse-phase HPLC method for the analysis of mexiletine hydrochloride. *J Liq Chromatogr Rel Technol.* 2003; 26: 1287–96.

Mezlocillin

Chemical Names

[2S-[2α,5α,6β(S^*)]]-3,3-Dimethyl-6-[[[[[3-(methylsulfonyl)-2-oxo-1-imidazolidinyl]-carbonyl]amino]phenylacetyl]amino]-7-oxo-4-thia-1-azabicyclo[3.2.0]heptane-2-carboxylic acid

6R-[2-[3-(Methylsulfonyl)-2-oxo-1-imidazolidinecarboxamido]-2-phenylacetamido]-penicillanic acid

Other Name

Mezlin

Form	Molecular Formula	MW	CAS
Mezlocillin	$C_{21}H_{25}N_5O_8S_2$	539.6	51481-65-3
Mezlocillin sodium	$C_{21}H_{24}N_5NaO_8S_2.H_2O$	579.6	42057-22-7

Appearance

Mezlocillin sodium is a pale yellow crystalline powder.

Solubility

Mezlocillin sodium is soluble in water, methanol, and dimethylformamide. It is insoluble in acetone and ethanol.

pK$_a$

Mezlocillin has a pK$_a$ of approximately 2.7.

Method 1

Grover et al. used HPLC to investigate the degradation rate of mezlocillin at 35 °C in borate buffer (pH 9.2). The Shimadzu LC-10A liquid chromatograph included model LC-10AS pumps, a model SPD-10A dual-wavelength detector, a model C-R7A data processor, and a Rheodyne model 7125 injector with a 20-μL loop. The stationary phase was a Phenomenex C_{18} column (300 × 3.9 mm, 5-μm particle size). The mobile phase was a mixture of 84% water containing 20 mM monobasic potassium phosphate and 10 mM tetramethylammonium chloride (pH 5.0) and 16% acetonitrile. The flow rate was 1.0 mL/min. UV detection was performed at 220 and 240 nm.

Samples were incubated at 35 °C and assayed. Under these conditions, the retention time for mezlocillin was 16.5 minutes.

The method was reported to be stability indicating.

A standard curve for mezlocillin was generated from 5 to 150 μg/mL. The correlation coefficient was greater than 0.99. The coefficient of variation for the analysis of mezlocillin was less than 1%.

Reference

Grover M, Gulati M, Singh B, et al. Correlation of penicillin structure with rate constants for basic hydrolysis. *Pharm Pharmacol Commun.* 2000; 6: 355–63.

Method 2

Mayron and Gennaro used an HPLC method to determine the stability of mezlocillin sodium with granisetron hydrochloride during Y-site administration. The liquid chromatographic system consisted of a piston pump with a pulse dampener, a rotary injection port with a 20-μL loop, a variable-wavelength spectrometric detector, and an integrator. The stationary phase was a cyano reversed-phase column (250 × 4.6 mm, 5-μm particle size). The mobile phase was a mixture of acetonitrile and 0.02 M sodium dihydrogen phosphate (10:90, vol/vol) adjusted to pH 4.2 with phosphoric acid. The flow rate was 2.50 mL/min. UV detection was performed at 210 nm.

Samples were diluted with the mobile phase. The retention times for mezlocillin sodium and granisetron were 4.80 and 8.32 minutes, respectively.

The stability-indicating capability of the analytical method was demonstrated by an accelerated degradation study. Solutions of mezlocillin sodium and granisetron hydrochloride were adjusted to pH 2 and 11, boiled for 1 hour, then readjusted to pH 5, diluted with the mobile phase, and analyzed. No degradation product peaks interfered with the intact mezlocillin or granisetron peaks.

Reference

Mayron D, Gennaro AR. Stability and compatibility of granisetron hydrochloride in i.v. solutions and oral liquids and during simulated Y-site injection with selected drugs. *Am J Health Syst Pharm.* 1996; 53: 294–304.

Method 3

Perry et al. evaluated the chemical stability of mezlocillin sodium in total parenteral nutrient solutions at concentrations commonly used in adults. The liquid chromatograph consisted of a Rheodyne injector 7125 valve, a Waters model 6000A solvent-delivery system, a Waters Lambda-Max 481 UV detector, and a BBC Servogen 120 chart recorder. The stationary phase was a Waters cyano-bonded reversed-phase C_{18} column (300 × 4.5 mm, 5-μm particle size). The mobile phase consisted of 35% acetonitrile, 0.2% phosphoric acid, and 64.8% water. The flow rate was 1.0 mL/min. UV detection was performed at 218 nm and 0.05 AUFS.

Samples were diluted 1:1000 with purified water. The injection volume was 20 μL. Under these conditions, the retention time for mezlocillin was 14.0 minutes.

The assay was determined to be stability indicating by accelerated decomposition of mezlocillin sodium. The first solution of mezlocillin was adjusted to pH 1.6, the second solution was evaporated on a hot plate, and the third solution was heated in a boiling water bath for 15 minutes. In all cases, no degradation product interfered with the analysis of the intact mezlocillin.

A standard curve for mezlocillin was constructed from 0 to 50 mg/mL. The intra-assay coefficient of variation was 4.3%.

Reference

Perry M, Khalidi N, Sanders CA. Stability of penicillins in total parenteral nutrient solution. *Am J Hosp Pharm.* 1987; 44: 1625–8.

Method 4

Gupta evaluated the stability of mezlocillin sodium at varying pH values in some commonly used vehicles for intravenous administration. The liquid chromatographic system was a Waters model ALC 202 pump equipped with a U6K universal injector, a Schoeffel Spectroflow SF 770 multiple-wavelength detector, a Houston Omniscribe recorder, and an Autolab minigrator. The stationary phase was a Waters μBondapak phenyl semipolar column (300 × 4 mm). The mobile phase contained 42% (vol/vol) methanol in water and 0.02 M ammonium acetate. The flow rate was 2.0 mL/min. UV detection was performed at 230 nm and 0.2 AUFS.

Samples were diluted with water. The injection volume was 20 μL. The retention time for mezlocillin was about 5 minutes (estimated from the published chromatogram).

The method was stated to be stability indicating. A chromatogram from a solution of mezlocillin sodium stored at room temperature for 4 days showed that degradation product peaks did not interfere with the intact mezlocillin peak.

Reference

Gupta VD. Stability of mezlocillin sodium as determined by high-performance liquid chromatography. *J Pharm Sci.* 1983; 72: 1479–81.

Miconazole

Chemical Names

1-[2-(2,4-Dichlorophenyl)-2-[(2,4-dichlorophenyl)methoxy]ethyl]-1*H*-imidazole

1-[2,4-Dichloro-β-[(2,4-dichlorobenzyl)oxy]phenethyl]imidazole

Other Names

Micatin, Monistat

Form	Molecular Formula	MW	CAS
Miconazole	$C_{18}H_{14}Cl_4N_2O$	416.1	22916-47-8
Miconazole nitrate	$C_{18}H_{14}Cl_4N_2O.HNO_3$	479.1	22832-87-7

Appearance

Miconazole is a white crystalline or microcrystalline powder. Miconazole nitrate occurs as a white or almost white and almost odorless crystalline powder.

Solubility

Miconazole is practically insoluble in water and freely soluble in alcohol. Miconazole nitrate is very slightly soluble in water and ether; it is slightly soluble in alcohol but freely soluble in methanol.

pK$_b$

Miconazole has a pK$_b$ of 7.35.

Method 1

Faouzi et al. studied the stability, compatibility, and plasticizer extraction of miconazole injection added to infusion solutions and stored in polyvinyl chloride containers. The liquid chromatograph was a Hewlett-Packard 1090M HPLC system equipped with a variable-volume injector, an automatic sampler, a Hewlett-Packard 79994 linear photodiode-array UV detector, a Hewlett-Packard 9000 model 300 integrator, and a Hewlett-Packard ThinkJet printer. The stationary phase was an Interchim C$_{18}$ Spherisorb ODS column (150 × 4.6 mm, 5-µm particle size). The mobile phase was 15% acetonitrile in 0.01 M monobasic sodium phosphate buffer (adjusted to pH 8 with trimethylamine). The flow rate was 1.5 mL/min. UV detection was performed at 250 nm. Metronidazole was used as the internal standard. The sample injection volume was 10 µL. Retention times for the internal standard and miconazole were about 1.8 and 3.9 minutes, respectively (estimated from the published chromatogram).

The assay was determined to be stability indicating by a degradation study. Miconazole in 0.9% sodium chloride injection and 5% dextrose injection was exposed to concentrated hydrochloric acid, 0.1 M sodium hydroxide, and 1% hydrogen peroxide. Degradation product peaks did not interfere with the intact miconazole peak.

Calibration curves were constructed from 100 to 200 µg/mL. The correlation coefficient was 0.998. Intraday and interday coefficients of variation of the assay were less than 1.02 and 1.25%, respectively.

Reference

Faouzi MEA, Dine T, Luyckx M, et al. Stability, compatibility and plasticizer extraction of miconazole injection added to infusion solutions and stored in PVC containers. *J Pharm Biomed Anal.* 1995; 13: 1363–72.

Method 2

Holmes and Aldous determined the stability of miconazole in peritoneal dialysis fluid when stored in plastic bags or glass ampules for up to 9 days. The liquid chromatograph consisted of a Waters model M45 solvent-delivery system, a Waters model 441 UV detector, and a Waters µBondapak C$_{18}$ stainless steel column (300 × 3.9 mm) preceded with a µBondapak C$_{18}$/Corasil guard column (37–50 µm). The mobile phase consisted of 15% (vol/vol) 0.05 M ammonium dihydrogen phosphate (pH 4.6) in methanol. The flow rate was 1.0 mL/min. UV detection was performed at 229 nm. *p*-Dichlorobenzene 4 mg/L in methanol was used as an internal standard.

Each sample was diluted with an equal volume of internal standard solution. The injection volume was 20 µL. Retention times for the internal standard and miconazole were about 4.6 and 8.2 minutes, respectively (estimated from the published chromatogram).

The assay was determined to be stability indicating by an accelerated degradation study of miconazole. Samples of freshly prepared solution of miconazole in peritoneal dialysis fluid at pH 2, 6, and 11 were transferred into glass ampules and autoclaved for 1 hour at 118 °C. Degradation product peaks did not interfere with the intact miconazole peak.

A standard curve was constructed from 4 to 20 mg/L of miconazole nitrate. The intraday coefficient of variation was 3.6% at 20 mg/L and 5.8% at 4 mg/L. The interday coefficient of variation was 3.9% at 20 mg/L and 11.8% at 4 mg/L.

Similar methods were used by the other researchers cited here.

References

Holmes SE, Aldous S. Stability of miconazole in peritoneal dialysis fluid. *Am J Hosp Pharm.* 1991; 48: 286–90.

McGookin AG, Millership JS, Scott EM. Miconazole sorption to intravenous infusion sets. *J Clin Pharm Ther.* 1987; 12: 433–7.

Puranajoti P, Kasina R, Tenjarla S. Microbiological and HPLC analysis of miconazole in skin, serum and phase-solubility studies. *J Clin Pharm Ther.* 1999; 24: 445–50.

Midazolam

Chemical Name

8-Chloro-6-(2-fluorophenyl)-1-methyl-4*H*-imidazo[1,5-*a*][1,4]benzodiazepine

Other Name

Versed

Form	Molecular Formula	MW	CAS
Midazolam	$C_{18}H_{13}ClFN_3$	325.8	59467-70-8
Midazolam hydrochloride	$C_{18}H_{13}ClFN_3 \cdot HCl$	362.2	59467-96-8
Midazolam maleate	$C_{18}H_{13}ClFN_3 \cdot C_4H_4O_4$	441.8	59467-94-6

Appearance

Midazolam is a white to light yellow crystalline powder.

Solubility

Midazolam has solubilities of approximately 0.24, 1.09, 3.67, 10.3, or greater than 22 mg/mL in water at pH 6.2, 5.1, 3.8, 3.4, or 2.8, respectively, at 25 °C. It is freely soluble in alcohol and acetone and is soluble in methanol.

pK$_a$

Midazolam has a pK$_a$ of 6.15.

Method 1

Peterson et al. evaluated by HPLC analysis the compatibility and stability of fentanyl in combination with scopolamine and midazolam. The liquid chromatographic system consisted of a Varian model 9010 pump, a model 9050 variable-wavelength UV-visible detector, and a Rheodyne 7161 injector. The stationary phase was a Nova-Pak C$_{18}$ column (150 × 3.9 mm, 4-μm particle size). Solvent A was acetonitrile and solvent B was 0.1 M monobasic potassium phosphate buffer (pH 7.8). The mobile phase was delivered from 25 to 60% solvent A in 10 minutes, held at 60% solvent A for 2 minutes, returned to 25% solvent A in 2 minutes, and held at 25% solvent A for 1 minute. The flow rate was 1.0 mL/min. UV detection was performed at 210 nm. Lorazepam 125 μg/mL in ethanol was used as an internal standard. The injection volume was 10 μL. Under these conditions, retention times for scopolamine, lorazepam, midazolam, and fentanyl were about 5.4, 7.4, 10.8, and 12.8 minutes, respectively.

The method was demonstrated to be stability indicating by accelerated degradation. Solutions of midazolam were autoclaved at 121 °C for 5 hours or acidified to pH 1.0 with 0.1 M hydrochloric acid and autoclaved. None of the degradation product peaks interfered with the determination of midazolam.

A standard curve for midazolam was obtained from 30 to 100 μg/mL. The correlation coefficient was 0.991. The intraday and interday coefficients of variation were less than 4%.

Reference

Peterson GM, Miller KA, Galloway JF, et al. Compatibility and stability of fentanyl admixtures in polypropylene syringes. *J Clin Pharm Ther.* 1998; 23: 67–72.

Method 2

Using an HPLC assay, Stiles et al. studied the stability of midazolam hydrochloride in polypropylene infusion-pump syringes. The liquid chromatograph consisted of a Waters model 501 pump, a Waters model 441 UV detector, a Waters model 745 data module, an Alcott model 728 autosampler, and a Rheodyne 7010 injector. The stationary phase was a Bakerbond phenyl column. The mobile phase was a mixture of acetonitrile and 0.02 M sodium dihydrogen phosphate (80:20, vol/vol) and was delivered isocratically at 2.0 mL/min. UV detection was performed at 240 nm.

Samples were diluted 1:1000 before injection. Under these conditions, the retention time for midazolam was 9.2 minutes.

The stability-indicating ability of the assay was investigated by accelerated degradation of midazolam with sulfuric acid, 1 N sodium hydroxide, or heat at 80 °C for 24 hours. Degradation product peaks did not interfere with the intact midazolam peak.

A standard curve for midazolam was constructed from 1 to 5 μg/mL. Intraday and interday coefficients of variation were 1.6 and 1.7%, respectively.

Reference

Stiles ML, Allen LV, Prince SJ. Stability of deferoxamine mesylate, floxuridine, fluorouracil, hydromorphone hydrochloride, lorazepam, and midazolam hydrochloride in polypropylene infusion-pump syringes. *Am J Health Syst Pharm.* 1996; 53: 1583–8.

Method 3

Allwood et al. used an HPLC assay to determine the stability of injections containing diamorphine hydrochloride and midazolam hydrochloride in plastic syringes. The chromatograph consisted of a Kontron model 325 pump, a Kontron model 332 UV detector, a Rheodyne model 7125 fixed-volume loop injector, and a Techsphere ODS column (250 × 4.6 mm, 5-μm particle size). The mobile phase consisted of acetonitrile and 0.1 M ammonium acetate (70:30, vol/vol), with the addition of 24 mmol/L of diethylamine. The mobile phase was delivered isocratically at 1 mL/min. UV detection was performed at 220 nm and 0.2 AUFS.

Samples were diluted with distilled water. The injection volume was 20 μL. Retention times for diamorphine and midazolam were about 5.0 and 7.5 minutes, respectively (estimated from the published chromatogram).

The stability-indicating nature of the assay was confirmed by accelerated decomposition of midazolam hydrochloride. Solutions containing midazolam 50 μg/mL were heated at 50 °C for 2 hours in either distilled water, 0.01 M hydrochloric acid, 0.01 M sodium hydroxide, or hydrogen peroxide. Chromatograms of resulting solutions showed that degradation product peaks did not interfere with the intact midazolam peak.

A standard curve for midazolam was generated from 2 to 12 μg/mL. It was linear; the correlation coefficient was 0.9958. The coefficient of variation was 1.95% ($n = 7$). The interday coefficient of variation of the analytical method was 1.95%.

Reference

Allwood MC, Brown PW, Lee M. Stability of injections containing diamorphine and midazolam in plastic syringes. *Int J Pharm Prac*. 1994; 57–9.

Method 4

Gregory et al. evaluated the stability of midazolam after its preparation in simple syrup NF and peppermint oil. The chromatograph consisted of an SSI model 300 LC pump with an SSI model 210 Guardian pressure monitoring accessory, a Nicolet IBM model LC/9505 autosampler handler, a Nicolet IBM model 9563 variable-wavelength UV detector, and a SynChrome Synchropak SCD-100-25 analytical column. The mobile phase consisted of 400 mL of acetonitrile and 600 mL of 0.1 N phosphoric acid. The phosphoric acid was adjusted to pH 5.15 with triethylamine. The flow rate was 1.0 mL/min. UV detection was performed at 230 nm. Samples were diluted 1:1000.

The stability-indicating nature of the assay was shown by accelerated degradation of midazolam. Two solutions of midazolam in oral syrup were mixed with 1 M sodium hydroxide and heated to 50 °C. Degradation product peaks did not interfere with the intact midazolam peak.

A calibration curve was generated from 0.1 to 10 μg/mL. The coefficients of variation varied from 0.7 to 7.0%.

Reference

Gregory DF, Koestner JA, Tobias JD. Stability of midazolam prepared for oral administration. *Southern Med J*. 1993; 86: 771–2.

Method 5

Bhatt-Mehta et al. studied the chemical stability of midazolam hydrochloride 0.1 and 0.5 mg/mL in three commonly used parenteral nutrient solutions. The HPLC system consisted of a Waters model 501 pump, a Waters model 481 variable-wavelength UV detector, a Waters model 712 autoinjector, and a Hewlett-Packard model 3390A integrator. The stationary phase was an Alltech Hypersil ODS C_{18} column (250 × 4.6 mm, 5-μm particle size). The mobile phase was a phosphate buffer (pH 7) with organic modifiers. The flow rate was 1.0 mL/min. UV detection was performed at 254 nm. Ethylparaben was used as an internal standard. The injection volume was 10 μL. Under these conditions, retention times for benzyl alcohol, ethylparaben, and midazolam were 4.0, 6.0, and 9.0 minutes, respectively.

The midazolam assay was determined to be stability indicating by accelerated degradation of midazolam. Aqueous solutions of midazolam were adjusted to pH 12 with 1 N sodium hydroxide or to pH 2 with 1 N sulfuric acid and then heated to 100 °C for 30 minutes. Degradation product peaks did not interfere with the midazolam peak.

Standard curves were constructed each day. The correlation coefficient was 0.99. The intraday and interday coefficients of variation were less than 5%.

References

Bhatt-Mehta V, Rosen DA, King RS, et al. Stability of midazolam hydrochloride in parenteral nutrient solutions. *Am J Hosp Pharm*. 1993; 50: 285–8.

Bhatt-Mehta V, Johnson CE, Leininger N, et al. Stability of fentanyl citrate and midazolam hydrochloride during simulated intravenous coadministration. *Am J Health Syst Pharm*. 1995; 52: 511–3.

Johnson CE, Bhatt-Mehta V, Mancari SC, et al. Stability of midazolam hydrochloride and morphine sulfate during simulated intravenous coadministration. *Am J Hosp Pharm*. 1994; 51: 2812–5.

Method 6

Steedman et al. assessed the stability of midazolam hydrochloride 2.5 mg/mL in an oral preparation over 56 days. The liquid chromatograph consisted of a Waters model 600E pump, a Waters model 715 Ultra WISP sample processor, a Waters model 994 programmable photodiode-array detector, and a Waters μBondapak C_{18} column (300 × 3.9 mm, 10-μm particle size). The mobile phase consisted of 60% acetonitrile and 40% 0.02 M monobasic sodium phosphate aqueous solution. The flow rate was 2.0 mL/min. UV detection was performed at 214 nm and 0.0027 AUFS.

Samples were diluted with mobile phase. The injection volume was 20 μL. Retention times for benzyl alcohol and midazolam were 2.26 and 4.33 minutes, respectively.

The HPLC method was determined to be stability indicating by accelerated degradation of midazolam. Solutions of midazolam 5 mg/mL were mixed with 1 N hydrochloric acid or 1 N sodium hydroxide and then were stored for several weeks at room temperature. The degradation product did not interfere with the quantification of midazolam.

A standard curve was constructed from 0.725 to 4.05 mg/mL. The correlation coefficient was 0.998.

Similar stability-indicating methods for midazolam were used by the other researchers cited here.

References

Steedman SL, Koonce JR, Wynn JE, et al. Stability of midazolam hydrochloride in a flavored, dye-free oral solution. *Am J Hosp Pharm.* 1992; 49: 615–8.

Pramar YV, Loucas VA, El-Rachidi A. Stability of midazolam hydrochloride in syringes and i.v. fluids. *Am J Health Syst Pharm.* 1997; 54: 913–5.

Hagan RL, Jacobs LF Jr, Pimsler M, et al. Stability of midazolam hydrochloride in 5% dextrose injection or 0.9% sodium chloride injection over 30 days. *Am J Hosp Pharm.* 1993; 50: 2379–81.

Peterson GM, Khoo BHC, Galloway JG, et al. A preliminary study of the stability of midazolam in polypropylene syringes. *Aust J Hosp Pharm.* 1991; 21: 115–8.

Milrinone

Chemical Names

1,6-Dihydro-2-methyl-6-oxo-[3,4′-bipyridine]-5-carbonitrile
1,2-Dihydro-6-methyl-2-oxo-5-(4-pyridinyl)nicotinonitrile

Other Name

Primacor

Form	Molecular Formula	MW	CAS
Milrinone	$C_{12}H_9N_3O$	211.2	78415-72-2
Milrinone lactate	$C_{12}H_9N_3O.C_3H_6O_3$	301.3	100286-97-3

Appearance

Milrinone is an off-white to tan crystalline compound.

Solubility

Milrinone is slightly soluble in methanol and very slightly soluble in chloroform and water.

Method

Wilson et al. used HPLC to evaluate the stability of milrinone lactate in commonly used infusion solutions. The liquid chromatograph consisted of a Varian model 5000 pump, a Rheodyne model 7126 20-μL loop injector, a Waters model 441 UV detector, a

Fisher Scientific model 5000 strip-chart recorder, and a Hewlett-Packard model 3354 laboratory automation system. The stationary phase was a Whatman PXS ODS-3 stainless steel column (250 × 4.6 mm, 10-μm particle size). The mobile phase consisted of water, methanol, and 0.5 M sodium borate (700:300:20, vol/vol/vol) with the pH adjusted to 7. The flow rate was 1.0 mL/min. UV detection was performed at 254 nm.

Each sample was diluted 1:20 with mobile phase. The retention time for milrinone was about 7.2 minutes (estimated from the published chromatogram).

The stability-indicating capability of the assay was demonstrated by accelerated decomposition of milrinone. Milrinone lactate was refluxed in 0.1 N sodium hydroxide for 24 hours. The only degradation product was 1,6-dihydro-2-methyl-6-oxo-[3,4′-bipyridine]-5-carboxamide, and it was well separated from the milrinone peak.

Standard curves for milrinone were constructed from 0 to 0.240 mg/mL and were linear.

References
Wilson TD, Forde MD, Crain AVR, et al. Stability of milrinone in 0.45% sodium chloride, 0.9% sodium chloride, or 5% dextrose injections. *Am J Hosp Pharm.* 1986; 43: 2218–20.

Wilson TD, Forde MD. Stability of milrinone and epinephrine, atropine sulfate, lidocaine hydrochloride, or morphine sulfate injection. *Am J Hosp Pharm.* 1990; 47: 2504–7.

Minocycline

Chemical Names
[4S-(4α,4aα,5aα,12aα)]-4,7-Bis(dimethylamino)-1,4,4a,5,5a,6,11,12a-octahydro-3,10,12,12a-tetrahydroxy-1,11-dioxo-2-naphthacenecarboxamide

7-Dimethylamino-6-demethyl-6-deoxytetracycline

Other Names
Dynacin, Minocin

Form	Molecular Formula	MW	CAS
Minocycline	$C_{23}H_{27}N_3O_7$	457.5	10118-90-8
Minocycline hydrochloride	$C_{23}H_{27}N_3O_7 \cdot HCl$	493.9	13614-98-7

Appearance

Minocycline hydrochloride occurs as a yellow crystalline powder.

Solubility

Minocycline hydrochloride is soluble in water and slightly soluble in alcohol. It is practically insoluble in chloroform and ether.

Method

Pearson and Trissel determined the chemical stability and physical compatibility of minocycline hydrochloride 0.1 mg/mL and rifampin 0.1 mg/mL separately in 5% dextrose injection or 0.9% sodium chloride injection at 24 and 4 °C and together at these concentrations in 0.9% sodium chloride injection at 37, 24, and 4 °C over periods of up to 7 days. The liquid chromatograph consisted of a Waters model 600E multisolvent-delivery system, a Waters model 490E programmable multiple-wavelength UV detector, a Waters model 712 WISP autosampler, and a Vydac C_8 reversed-phase column (250 × 4.6 mm, 5-μm particle size). The system was controlled and integrated by an NEC PowerMate SX/16 computer. The mobile phase consisted of 22% acetonitrile in 50 mM dipotassium hydrogen phosphate with the pH adjusted to 6.5 with phosphoric acid. The flow rate was 1 mL/min. UV detection was performed at 254 nm.

Samples were assayed without dilution. Under these conditions, the retention time for minocycline was 7.2 minutes. Rifampin did not elute within 15 minutes.

The assay was determined to be stability indicating by accelerated degradation of minocycline hydrochloride. Heating at 95 °C in a water bath for 60 minutes yielded a reduction in the peak of the intact minocycline, with the formation of new peaks at 5.9 and 12.3 minutes.

Standard curves were generated from a linear plot of peak area versus concentration from 0.025 to 0.15 mg/mL. The correlation coefficient of the standard curve was greater than 0.9999; the coefficient of variation of the assay was 0.21%.

Reference

Pearson SD, Trissel LA. Stability and compatibility of minocycline hydrochloride and rifampin in intravenous solutions at various temperatures. *Am J Hosp Pharm.* 1993; 50: 698–701.

Mitolactol

Chemical Name

1,6-Dibromo-1,6-dideoxy-D-galactitol

Other Name

Dibromodulcitol

Form	Molecular Formula	MW	CAS
Mitolactol	$C_6H_{12}Br_2O_4$	308.0	10318-26-0

Method

Clark et al. developed an HPLC method for the determination of mitolactol (dibromodulcitol). The chromatographic system included a Beckman model 112 solvent pump, a Rheodyne model 7010 injector with a 50-μL loop, a Waters model R401 differential refractometer, and a Kipp and Zonen strip-chart recorder. The stationary phase was an Alltech Spherisorb ODS column (300 mm, 5-μm particle size). The mobile phase consisted of water and methanol (98:2, vol/vol). The flow rate was 1.2 mL/min. Under these conditions, the retention time for mitolactol was about 1.1 minutes (estimated from the published chromatogram).

The stability-indicating capability of the method was determined by assaying a mixture of mitolactol and its expected degradation products. No interference from the degradation products was observed. Mitolactol was also treated with sodium hydroxide solution and the degradation product peaks were well separated from the intact drug peak.

A calibration curve for mitolactol was generated from 0 to 543 μg/mL. The correlation coefficient was greater than 0.99. Intraday and interday coefficients of variation were 4.37 and 3.93%, respectively. The limit of detection was 250 ng.

Reference

Clark BA, Cadwallader DE, Salamone MJ. Stability indicating HPLC analysis of dibromodulcitol in aqueous solutions. *Anal Lett*. 1988; 21: 411–22.

Mitomycin

Chemical Name

[1aS-(1aα,8β,8aα,8bα)]-6-amino-8-[[(aminocarbonyl)oxy]methyl]-1,1a,2,8,8a,8b-hexahydro-8a-methoxy-5-methylazirino[2′,3′:3,4]pyrrolo[1,2-a]indole-4,7-dione

Other Names

Mitomycin C, Mutamycin

Form	Molecular Formula	MW	CAS
Mitomycin	$C_{15}H_{18}N_4O_5$	334.3	50-07-7

Appearance

Mitomycin is a blue-violet crystalline powder.

Solubility

Mitomycin is soluble in water, methanol, acetone, butyl acetate, and cyclohexanone. It is slightly soluble in benzene, carbon tetrachloride, and ether. It is practically insoluble in petroleum ether.

Method 1

Quebbeman et al. studied the stability of mitomycin admixtures. The HPLC system consisted of a Waters model 6000A pump, a Waters model 440 UV detector, a Waters model WISP 710B autosampler, and a Waters model 730 data module. The stationary phase was an Alltech C_{18} column (250 × 4.6 mm, 10-μm particle size). The mobile phase consisted of 50 mM monobasic sodium phosphate buffer and methanol (70:30, vol/vol) (pH 5.5). The flow rate was 2 mL/min. UV detection was performed at 365 and 546 nm and 0.10 AUFS.

For the unbuffered mitomycin admixture, the injection volume was 10 μL. For the buffered mitomycin admixture, each sample was mixed with an equal volume of the internal standard (methylparaben 50 μg/mL); the injection volume was 25 μL. Under these conditions, the retention time for mitomycin was approximately 5 minutes.

The HPLC assay was determined to be stability indicating by a decomposition study of mitomycin. Unbuffered mitomycin 50 μg/mL in 5% dextrose injection was stored at room temperature for 24 hours. The intact mitomycin peak at 5.0 minutes decreased substantially, whereas degradation product peaks around 2.5 minutes increased. Degradation product peaks did not interfere with the intact mitomycin peak.

A standard curve was generated from 0 to 400 μg/mL; the correlation coefficient was 0.9999.

A similar analytical method was used by Gupta.

References

Quebbeman EJ, Hoffman NE, Ausman RK, et al. Stability of mitomycin admixtures. *Am J Hosp Pharm.* 1985; 42: 1750–4.

Gupta VD. Stability of mitomycin aqueous solution when stored in tuberculin syringes. *Int J Pharm Compound.* 1997; 1: 282–3.

Method 2

Beijnen et al. quantitated the degradation of mitomycin after reconstitution and dilution for intravesical instillation therapy. The liquid chromatographic system consisted of a Waters model 6000A pump, a Waters model U6K injection device, a Waters model 440 UV detector, a Spectra-Physics SP4290 integrator, and a ChromSpher C_8 analytical column (100 × 3.0 mm, 5-μm particle size). The mobile phase consisted of 450 g of water, 50 g of methanol, and 5 mL of 0.5 M phosphate buffer (pH 7.0). The flow rate was 0.7 mL/min. UV detection was performed at 254 nm. The sample injection volume was 10 μL.

The analytical method was determined to be stability indicating. The accelerated degradation of mitomycin was achieved by acid, alkali, ions (buffer), and high temperatures. The assay allowed the selective determination of mitomycin with no interference of the diastereomeric mitosene degradation products and the 7-hydroxy-mitosane.

Calibration curves were constructed from 0.1 to 1.0 mg/mL. The correlation coefficient was greater than 0.999. Coefficients of variation for the lower and upper limits of the calibration curve were 3.1 and 0.9%, respectively.

References

Beijnen JH, Hartigh JD, Underberg WJM. Quantitative aspects of the degradation of mitomycin C in alkaline solution. *J Pharm Biomed Anal.* 1985; 3: 59.

Beijnen JH, Gijn RV, Underberg WJM. Chemical stability of the antitumor drug mitomycin C in solutions for intravesical installation. *J Parenter Sci Technol.* 1990; 44: 332–5.

Mitoxantrone

Chemical Name

1,4-Dihydroxy-5,8-bis[[2-[(2-hydroxyethyl)amino]ethyl]amino]-9,10-anthracenedione

Other Names

Mitozantrone, Novantrone

Form

Form	Molecular Formula	MW	CAS
Mitoxantrone	$C_{22}H_{28}N_4O_6$	444.5	65271-80-9
Mitoxantrone dihydrochloride	$C_{22}H_{28}N_4O_6 \cdot 2HCl$	517.4	70476-82-3

Appearance

Mitoxantrone is a dark blue powder.

Solubility

Mitoxantrone dihydrochloride is sparingly soluble in water and slightly soluble in methanol. It is practically insoluble in acetonitrile, chloroform, and acetone.

Method 1

Wang et al. evaluated by HPLC analysis the stability of mitoxantrone dihydrochloride in solution. The liquid chromatograph was equipped with a Shimadzu model LC-6A dual-piston pump, a model SCI-6A recorder, and a model SPD-M6A photodiode-array detector. The stationary phase was a Waters Nova-Pak C_{18} column (150 × 3.9 mm, 5-μm particle size). The mobile phase consisted of 25% acetonitrile and 75% 0.2 M ammonium acetate at pH 4.0 including 0.005 mM 1-heptanesulfonic acid sodium salt. The flow rate was 0.8 mL/min. UV detection was performed at 242 nm. Pentobarbital sodium was used as an internal standard. Under these conditions, retention times of mitoxantrone and pentobarbital sodium were 3.5 and 11.2 minutes, respectively.

The stability-indicating ability of the assay was demonstrated by accelerated degradation. A mitoxantrone solution in hydrochloric acid at pH 2.1 was stored at 50 °C for 96 hours. No interference with the peak of mitoxantrone by degradation products was observed.

A calibration curve for mitoxantrone was constructed from 0.005 to 0.015 mg/mL. The correlation coefficient was 0.994. Intraday and interday coefficients of variation at 0.01 mg/mL were 0.10 and 1.18%, respectively.

Reference

Wang D-P, Liang G-Z, Tu Y-H. Stability of mitoxantrone hydrochloride in solution. *Drug Dev Ind Pharm.* 1994; 20: 1895–903.

Method 2

Northcott et al. used HPLC to evaluate the chemical stability of mitoxantrone dihydrochloride infusion 0.2 mg/mL in an ambulatory pump under storage and prolonged in-use conditions. The chromatographic system was composed of a ConstaMetric 300 pump, a SpectroMonitor 3100 variable-wavelength UV-visible detector, a Rheodyne 7125 loop valve fitted with a 20-μL loop, and a CI-10B integrator/printer plotter. The stationary phase was a Kontron S5C1 analytical column (250 × 4.8 mm), and the column temperature was 37 °C. The mobile phase was a mixture of 0.007 M sodium hydrogen phosphate, acetonitrile, methanol, and tetrahydrofuran (75.7:15:10:0.3, vol/vol/vol/vol) adjusted to pH 2.2 with phosphoric acid. The flow rate was 1.10 mL/min. UV detection was performed at 658 nm.

Samples were diluted 1:2000 with water, and the injection volume was 20 μL.

The HPLC assay was determined to be stability indicating by an accelerated degradation study. Four sets of mitoxantrone dihydrochloride samples were subjected to 1 M sodium hydroxide, 1 M hydrochloric acid, hydrogen peroxide, and heat. Each of the stress treatments resulted in reduction of the intact mitoxantrone peak height. In all cases, the intact drug peak was well resolved from the degradation product peaks.

Standard curves for mitoxantrone dihydrochloride were constructed from a linear plot of mitoxantrone peak height versus its concentration from 0.02 to 0.5 μg/mL. The correlation coefficient was 0.999. The coefficient of variation of the drug peak height was 0.57% ($n = 7$).

Reference

Northcott M, Allsopp MA, Powell H, et al. The stability of carboplatin, diamorphine, 5-fluorouracil and mitoxantrone infusions in an ambulatory pump under storage and prolonged in use conditions. *J Clin Pharm Ther.* 1991; 16: 123–9.

Method 3

Walker et al. investigated the stability of mitoxantrone hydrochloride solution in an opened original vial and in syringes. The HPLC system consisted of a Spectra-Physics model 8700 isocratic solvent pump, a Kratos Spectroflow 783 variable-wavelength UV detector, a Spectra-Physics model SP4270 chromatography integrator, and a Beckman

Ultrasphere ODS reversed-phase C_{18} column (250 × 4.6 mm, 5-μm particle size). The mobile phase consisted of acetonitrile, 0.2% triethylammonium phosphate buffer (pH 2.5), and 0.01 M tetrabutylammonium hydrogen sulfate. The ratio of acetonitrile to triethylammonium phosphate buffer was 40:60 (vol/vol). The flow rate was 1.5 mL/min. UV detection was performed at 243 nm. The injection volume was 1 μL. Under these conditions, the retention time for mitoxantrone was between 1.9 and 2 minutes.

The stability-indicating nature of the assay was demonstrated by accelerated degradation of mitoxantrone. One solution of mitoxantrone hydrochloride 2 mg/mL in 0.9% sodium chloride was incubated in a water bath at 90 °C for 4 hours. The other solution of mitoxantrone hydrochloride 2 mg/mL was adjusted to pH 11.5 with 0.5 N sodium hydroxide and incubated in a water bath at 90 °C for 4 hours. Degradation product peaks did not interfere with the mitoxantrone peak.

A standard curve was constructed each day from 0.25 to 1.5 mg/mL. Intraday and interday reproducibilities of samples were 1 and 3.5%, respectively.

Reference
Walker SE, Lau DWC, DeAngelis C, et al. Mitoxantrone stability in syringes and glass vials and evaluation of chemical contamination. *Can J Hosp Pharm.* 1991; 44: 143–51.

Montelukast Sodium

Chemical Name
1-[[[1-[3-[2-(7-Chloro-2-quinolinyl)ethenyl]phenyl]-3-[2-(1-hydroxy-1-methylethyl)-phenyl]propyl]thio]methyl]cyclopropaneacetic acid sodium salt

Other Name
Singulair

Form	Molecular Formula	MW	CAS
Montelukast sodium	$C_{35}H_{35}ClNNaO_3S$	608.2	151767-02-1

Appearance
Montelukast sodium is a hygroscopic white to off-white powder.

Solubility
Montelukast sodium has a solubility of 0.2 to 0.5 μg/mL in water at 25 °C. It is freely soluble in alcohol.

pK_a
Montelukast sodium has pK_a values of 2.8 and 5.7.

Method
Radhakrishna et al. reported an HPLC method for the simultaneous determination of montelukast and loratadine. The instrument consisted of a Waters model 510 pump, a model 996 photodiode-array detector, and a Rheodyne injector with a 10-μL loop. The stationary

phase was a Waters Symmetry C_{18} analytical column (250 × 4.6 mm, 5-μm particle size). The mobile phase consisted of 0.025 M monobasic sodium phosphate aqueous buffer (pH 3.7 with phosphoric acid) and acetonitrile (20:80, vol/vol). The flow rate was 1.0 mL/min. 5-Methyl 2-nitrophenol 330 μg/mL was used as an internal standard. UV detection was performed at 225 nm.

A portion of 20 ground tablets equivalent to 10 mg of montelukast and 10 mg of loratadine was accurately weighed, transferred to a 50-mL volumetric flask, mixed with 40 mL of acetonitrile and water (80:20, vol/vol), sonicated for 15 minutes, and centrifuged. The supernatant was collected and assayed. The injection volume was 10 μL. Under these conditions, retention times for the internal standard, loratadine, and montelukast were about 4.0, 5.8, and 11.3 minutes, respectively.

The analytical method was demonstrated to be stability indicating by accelerated degradation of montelukast. An initial montelukast solution was refluxed with 0.1 N hydrochloric acid at 60 °C for 12 hours; a second solution was refluxed with 0.1 N sodium hydroxide solution at 60 °C for 12 hours; a third solution was exposed to ultraviolet light (254 nm) for 24 hours; and a final solution was kept at 50 °C for 12 hours. The peak of intact montelukast was resolved from its degradation products on the chromatogram.

A calibration curve for montelukast was constructed from 100 to 600 μg/mL. The intraday and interday coefficients of variation were less than 1.4 and 0.7%, respectively.

Reference

Radhakrishna T, Narasaraju A, Ramakrishna M, et al. Simultaneous determination of montelukast and loratadine by HPLC and derivative spectrophotometric methods. *J Pharm Biomed Anal.* 2003; 31: 359–68

Moricizine

Chemical Name

[10-[3-(4-Morpholinyl)-1-oxopropyl]-10*H*-phenothiazin-2-yl]carbamic acid ethyl ester

Other Name

Ethmozine

Form	Molecular Formula	MW	CAS
Moricizine	$C_{22}H_{25}N_3O_4S$	427.5	31883-05-3
Moricizine hydrochloride	$C_{22}H_{25}N_3O_4S \cdot HCl$	464.0	29560-58-5

Appearance
Moricizine hydrochloride is a white to tan crystalline powder.

Solubility
Moricizine hydrochloride is soluble in water and in alcohol.

Method
King et al. studied degradation kinetics and mechanisms of moricizine hydrochloride in acidic medium. The liquid chromatograph consisted of a Waters model 590 solvent-delivery system, a Waters model 710B WISP autosampler, a Kratos Spectroflow 773 UV detector, and a Hewlett-Packard 3390A recording integrator. The stationary phase was a DuPont Zorbax C_8 analytical column (250 × 4.6 mm) used at 35 °C. The mobile phase consisted of 42% acetonitrile in water containing 5 µM sodium octanesulfonate, 0.2% acetic acid, and 0.1% triethylamine. The flow rate was 2.5 mL/min. UV detection was performed at 254 nm. n-Butyl p-aminobenzoate was used as an internal standard.

Samples were diluted with methanol. Under these conditions, retention times for moricizine hydrochloride and the internal standard were about 5 and 7 minutes, respectively.

The analytical method was shown to be stability indicating. Moricizine hydrochloride solutions in pH 0.6 and 4.0 buffers were incubated at 80 °C for 2 days. The degradation product peaks did not interfere with the intact moricizine hydrochloride peak.

Reference
King SP, Sigvardson KW, Dudzinski J, et al. Degradation kinetics and mechanisms of moricizine hydrochloride in acidic medium. *J Pharm Sci.* 1992; 81: 586–91.

Morphine

Chemical Name
(5α,6α)-7,8-Didehydro-4,5-epoxy-17-methylmorphinan-3,6-diol

Other Names
Duramorph, Oramorph

Form	Molecular Formula	MW	CAS
Morphine	$C_{17}H_{19}NO_3$	285.3	57-27-2
Morphine hydrochloride	$C_{17}H_{19}NO_3.HCl.3H_2O$	375.8	52-26-6
Morphine sulfate	$(C_{17}H_{19}NO_3)_2.H_2SO_4$	668.8	64-31-3

Appearance

Morphine sulfate occurs as white feathery silky crystals, as cubical masses of crystals, or as a white crystalline powder. Morphine hydrochloride occurs as colorless silky needles or as a white or almost white crystalline powder.

Solubility

Morphine sulfate has solubilities of approximately 62.5 mg/mL in water and 1.75 mg/mL in alcohol at 25 °C. Morphine hydrochloride is soluble in water, slightly soluble in alcohol, and practically insoluble in ether.

pK$_a$

Morphine has pK$_a$ values of 8.31 for the amino group and 9.51 for the phenolic group at 25 °C.

Method 1

Johnson et al. evaluated the chemical compatibility of bupivacaine hydrochloride and morphine sulfate in concentrations used clinically for epidural administration. The HPLC system included a Waters model 501 constant-flow solvent-delivery system, a Waters model U6K variable-volume injector, a Waters model 484 variable-wavelength UV detector, and a Hewlett-Packard model 3390 recording integrator. The stationary phase was a Waters Nova-Pak C$_{18}$ column (150 × 3.9 mm, 4-μm particle size). The mobile phase, a mixture of 25% acetonitrile and 75% 0.05 M phosphate buffer adjusted to pH 7.5, was delivered isocratically at 1 mL/min. UV detection was performed at 280 nm and 0.1 AUFS. The injection volume was 15 μL. The retention time for morphine was about 2.6 minutes (estimated from the published chromatogram).

The stability-indicating nature of the assay was verified by accelerated degradation of bupivacaine hydrochloride–morphine sulfate samples. Addition of acid produced degradation products eluting at 1.7 and 1.9 minutes. Addition of base generated a degradation product eluting at 9.4 minutes. Degradation product peaks did not interfere with the intact morphine peak. Bupivacaine hydrochloride was not detected.

Two sets of standard curves were generated from 80 to 120 μg/mL and from 450 to 550 μg/mL by linear regression. Correlation coefficients were greater than 0.999. The intraday and interday coefficients of variation were 1.4 and 1.9%, respectively.

Reference

Johnson CE, Christen C, Perez MM, et al. Compatibility of bupivacaine hydrochloride and morphine sulfate. *Am J Health Syst Pharm.* 1997; 54: 61–4.

Method 2

Mayron and Gennaro used an HPLC method to assess the stability of morphine sulfate with granisetron hydrochloride during Y-site administration. The liquid chromatograph consisted of a piston pump with a pulse dampener, a rotary injection port with a 20-μL loop, a variable-wavelength spectrometric detector, and an integrator. The stationary phase was a cyano reversed-phase column (250 × 4.6 mm, 5-µm particle size). The mobile phase was a mixture of acetonitrile and 0.1 M monobasic sodium phosphate dihydrate (20:80, vol/vol) adjusted to pH 4.2 with phosphoric acid. The flow rate was 1.0 mL/min. UV detection was performed at 300 nm.

Samples were diluted with mobile phase. The retention times for morphine and granisetron were 3.70 and 7.69 minutes, respectively.

The stability-indicating capability of the analytical method was demonstrated by an accelerated degradation study. Solutions of morphine sulfate and granisetron hydrochloride were adjusted to pH 2 and 11, boiled for 1 hour, readjusted to pH 5, diluted with the mobile phase, and analyzed. The degradation product peaks did not interfere with the intact morphine or granisetron peaks.

Reference
Mayron D, Gennaro AR. Stability and compatibility of granisetron hydrochloride in i.v. solutions and oral liquids and during simulated Y-site injection with selected drugs. *Am J Health Syst Pharm.* 1996; 53: 294–304.

Method 3
Trissel et al. used an HPLC method to determine the stability of morphine sulfate when admixed with ondansetron hydrochloride in 0.9% sodium chloride injection and stored at 32 °C for up to 7 days and at 4 and 22 °C for up to 31 days. The liquid chromatograph consisted of a Waters model 600E multisolvent-delivery system, a Waters model 490E programmable multiple-wavelength UV detector, a Waters model 712 WISP auto-sampler, and an Alltech Spherisorb cyano column (300 × 4.6 mm, 5-µm particle size). The system was controlled and the peak area was integrated by an NEC PowerMate SX/16 personal computer. The mobile phase was 50% acetonitrile in 0.02 M monobasic potassium phosphate with the pH adjusted to 5.40 with 1 N sodium hydroxide solution and was delivered isocratically at a rate of 1.5 mL/min. UV detection was performed at 216 nm.

Samples were diluted 1:10 before injection. The injection volume was 20 µL. Under these conditions, the retention time for morphine was 6.7 minutes. The retention time for ondansetron was 8.8–10 minutes.

The assay was demonstrated to be stability indicating by accelerated degradation of morphine sulfate. Heating morphine sulfate solution at 95 °C for 5 hours yielded a reduction in the parent morphine sulfate peak and the formation of new peaks eluting around the solvent front and at 4.9 and 5.5 minutes. Neither ondansetron nor the degradation product peaks interfered with the parent morphine sulfate peak.

Standard curves were constructed from a linear plot of peak area versus concentration from 0.01 to 0.15 mg/mL. The correlation coefficient was greater than 0.9999. The intraday and interday coefficients of variation were 1.5 and 0.7%, respectively.

References
Trissel LA, Xu Q, Martinez JF, et al. Compatibility and stability of ondansetron hydrochloride with morphine sulfate and with hydromorphone hydrochloride in 0.9% sodium chloride injection at 4, 22, and 32 °C. *Am J Hosp Pharm.* 1994; 51: 2138–42.

Trissel LA, Xu QA, Pham L. Physical and chemical stability of morphine sulfate 5 mg/mL and 50 mg/mL packaged in plastic syringes. *Int J Pharm Compound.* 2002; 6: 62–5.

Xu QA, Trissel LA, Pham L. Physical and chemical stability of low and high concentration of morphine sulfate with clonidine hydrochloride packaged in plastic syringes. *Int J Pharm Compound*. 2002; 6: 66–9.

Trissel LA, Xu QA, Pham L. Physical and chemical stability of low and high concentration of morphine sulfate with bupivacaine hydrochloride packaged in plastic syringes. *Int J Pharm Compound*. 2002; 6: 70–3.

Method 4

Wu et al. developed an HPLC assay for morphine hydrochloride. A liquid chromatographic system included Waters model 510 pumps, a Shimadzu model SIL-9A autosampler, and a Waters model 490E programmable multiwavelength detector. A Hewlett-Packard model 1040M diode-array detector was also used. The stationary phase was a Nucleosil C_{18} column (300 × 4.0 mm, 10-μm particle size). The mobile phase consisted of methanol and an aqueous solution (15:85, vol/vol). The aqueous solution contained 1% ammonium acetate, 1% acetic acid, 0.8% triethylamine, and 0.017% sodium 1-heptanesulfonate. The flow rate was 1.0 mL/min. UV detection was performed at 284 nm. The injection volume was 10 μL. Under these conditions, retention times for morphine, morphine *N*-oxide, and pseudomorphine were about 7.6, 9.6, and 11.8 minutes, respectively.

The assay was demonstrated to be stability indicating by accelerated degradation of morphine hydrochloride. Morphine hydrochloride solutions were treated with 0.1 N sodium hydroxide solution at 80 °C for 28 hours, 1% hydrogen peroxide at 80 °C for 6 hours, exposed to light for 124 days, or refluxed with sodium bisulfite in water for 4 days. None of the degradation products interfered with the determination of morphine.

A standard curve for morphine hydrochloride was constructed from 0.1 to 0.8 mg/mL. The correlation coefficient was 0.9999.

Reference

Wu S-C, Li J-H, Chen F-S, et al. A stability-indicating method for morphine assay by high-performance liquid chromatography. *Chin Pharm J*. 1993; 45: 133–44.

Method 5

Duafala et al. evaluated the stability of morphine sulfate in infusion devices and containers for intravenous administration. The HPLC system consisted of a Waters model M6000A pump, a Waters model WISP 710B autosampler, a Waters model 480 UV detector, and a Nelson analytical 4430 chromatography data system. The stationary phase was a Supelco Supelcosil LC-18DB column (250 × 4.6 mm, 5-μm particle size) preceded by a C_{18} guard column. The mobile phase was a mixture of 35% methanol and 65% aqueous buffer. The buffer contained 0.0079 M dibasic potassium phosphate and 0.0059 M monobasic potassium phosphate. The flow rate was 1.5 mL/min. UV detection was performed at 280 nm. Acetophenone 0.40 mg/mL was used as an internal standard.

Samples were diluted with methanol. The injection volume was 10 μL. Retention times for morphine and the internal standard were about 6.9 and 13.0 minutes, respectively (estimated from the published chromatogram).

The stability-indicating nature of the method was demonstrated by accelerated decomposition of morphine sulfate with heat, light, acid, base, or 0.0073 mM hydrogen peroxide. In all cases, degradation product peaks did not interfere with the morphine sulfate peak.

Standard curves for morphine sulfate were generated from 0.07 to 1.51 mg/mL; correlation coefficients were greater than 0.999. The coefficient of variation was less than 1.2%.

A similar method was used by Altman et al.

References

Duafala ME, Kleinberg ML, Nacov C, et al. Stability of morphine sulfate in infusion devices and containers for intravenous administration. *Am J Hosp Pharm.* 1990; 47: 143–6.

Altman L, Hopkins RJ, Ahmed S, et al. Stability of morphine sulfate in Cormed III (Kalex) intravenous bags. *Am J Hosp Pharm.* 1990; 47: 2040–2.

Method 6

Wilson and Forde determined the stability of morphine sulfate using an HPLC method. The liquid chromatograph consisted of a Varian model 5000 pump and autosampler, a Waters model 441 UV detector, a Fisher Scientific model 5000 strip-chart recorder, and a Hewlett-Packard model 3354 laboratory automation system. The stationary phase was a Whatman PXS ODS-3 stainless steel column (250×4.6 mm, 10-μm particle size). The mobile phase was a mixture of water, methanol, acetonitrile, and 85% phosphoric acid (800:125:75:5.1, vol/vol/vol/vol) with 1.08 g of sodium octanesulfonate. The flow rate was 1.5 mL/min. UV detection was performed at 254 nm and 0.1 AUFS.

Samples were diluted to about 0.4 mg/mL with a mixture of water, methanol, and 0.5 M sodium borate (pH 7.0) (650:350:20, vol/vol/vol). The injection volume was 20 μL.

This method was stated to be stability indicating.

A standard curve for morphine sulfate was constructed from 0 to 4.68 mg/mL.

Reference

Wilson TD, Forde MD. Stability of milrinone and epinephrine, atropine sulfate, lidocaine hydrochloride, or morphine sulfate injection. *Am J Hosp Pharm.* 1990; 47: 2504–7.

Method 7

Stiles et al. studied the stability of morphine sulfate in prefilled drug reservoirs for an infusion pump at two storage temperatures (5 and 25 °C) for 30 days, followed by a 3-day delivery period at body temperature (37 °C). The liquid chromatograph consisted of a Waters model 501 pump, a Waters model U6K injector, a Waters model 441 UV detector, a Waters 745 data module, and a Waters Nova-Pak C_{18} column (150×3.9 mm, 5-μm particle size). The mobile phase consisted of 12% methanol and 88% 0.05 M phosphate buffer at pH 7.5 and was isocratically delivered at 0.5 mL/min. UV detection was performed at 254 nm.

Each sample was diluted with distilled water. The retention time for morphine sulfate was 4.35 minutes.

The stability-indicating capability of the assay was verified by accelerated degradation of morphine sulfate 1 mg/mL at 90 °C for 11 days. The decomposition products of morphine sulfate and the excipients used in the formulations eluted separately and did not interfere with the intact morphine sulfate peak.

Standard curves were constructed each day from 10 to 100 µg/mL. The correlation coefficients were greater than 0.999. The intraday coefficient of variation was less than 2%.

Similar methods were used by the other researchers cited here.

References

Stiles ML, Tu Y-H, Allen LV Jr. Stability of morphine sulfate in portable pump reservoirs during storage and simulated administration. *Am J Hosp Pharm.* 1989; 46: 1404–7.

Macias JM, Martin WJ, Lloyd CW. Stability of morphine sulfate and meperidine hydrochloride in a parenteral nutrient formulation. *Am J Hosp Pharm.* 1985; 42: 1087–94.

Shrivastava R, Makhija SN. Stability indicating assay method for morphine sulfate. *India Drug.* 1996; 33: 525–6.

Method 8

Vecchio et al. assessed the stability of morphine in normal saline and 5% dextrose in water. The liquid chromatograph consisted of a Spectra-Physics model 8770 dual-piston pump, a Spectra-Physics model 8780 autosampler, a Spectra-Physics model 4200 integrator, and an ESA Coulochem model 5100A electrochemical detector. The stationary phase was a Brownlee Labs RP8 C_8 column (10-µm particle size). The mobile phase consisted of 8% acetonitrile and 92% 0.05 M potassium dihydrogen phosphate buffer. The flow rate was 2.0 mL/min. The electrochemical detector was operated with an applied potential of 0.60 V on the column elute. Naloxone 0.2 mg/mL was used as an internal standard.

Fifty microliters of morphine sample was mixed with 100 µL of internal standard and 1.0 mL of water. The injection volume was 50 µL. Retention times for morphine and the internal standard were about 4.2 and 7.0 minutes, respectively (estimated from the published chromatogram).

The assay was determined to be stability indicating by accelerated degradation of morphine. A solution of morphine in normal saline was adjusted to pH 11 with sodium hydroxide and heated at 95 °C for 48 hours. Degradation product peaks did not interfere with the peaks of morphine or the internal standard.

A standard curve was constructed from 0.1 to 0.4 mg/mL.

Reference

Vecchio M, Walker SE, Iazzetta J, et al. The stability of morphine intravenous infusion solution. *Can J Hosp Pharm.* 1988; 41: 5–9.

Method 9

Walker et al. used an HPLC method to investigate the stability of morphine in portable infusion pump cassettes and minibags. The liquid chromatograph included a Spectra-Physics model SP8770 isocratic solvent-delivery pump, a Schoeffel SF 770 variable-wavelength UV detector, a Spectra-Physics model SP4270 integrator, and a Hamilton PRP-1 rigid macroporous poly(styrene-divinylbenzene) copolymer column (10-µm particle size). The mobile phase consisted of 7% acetonitrile and 93% 0.01 M dibasic sodium phosphate adjusted to pH 11 with 10 M sodium hydroxide. The flow rate was 2 mL/min. UV detection was performed at 280 nm. The sample injection volume was

5 µL. The retention time for morphine was about 2.5 minutes (estimated from the published chromatogram).

The analytical method was determined to be stability indicating by an accelerated degradation study. The 0.9% sodium chloride injection with morphine was adjusted to pH 11 using sodium hydroxide and stored at 66 °C for 125 hours. The degradation product peak did not interfere with the morphine peak.

Standard curves were prepared daily.

Reference

Walker SE, Coons C, Matte D, et al. Hydromorphone and morphine stability in portable infusion pump cassettes and minibags. *Can J Hosp Pharm.* 1988; 41: 177–82.

Moxalactam

Chemical Name

7-[[Carboxy(4-hydroxyphenyl)acetyl]amino]-7-methoxy-3-[[(1-methyl-1*H*-tetrazol-5-yl)-thio]methyl]-8-oxo-5-oxa-1-azabicyclo[4.2.0]oct-2-ene-2-carboxylic acid

Other Names

Latamoxef, Moxam

Form	Molecular Formula	MW	CAS
Moxalactam	$C_{20}H_{20}N_6O_9S$	520.5	64952-97-2
Moxalactam disodium	$C_{20}H_{18}N_6Na_2O_9S$	564.4	64953-12-4

Appearance

Moxalactam is a colorless powder.

Method

Gupta and Stewart used a stability-indicating HPLC method to study the chemical stability of moxalactam disodium when mixed with metronidazole injection for intravenous infusion. A Waters ALC 202 high-performance liquid chromatograph was equipped with a Schoeffel SF 770 multiple-wavelength detector, an Omniscribe recorder, and a Waters µBondapak phenyl column. The mobile phase was 0.02 M ammonium acetate in water with the pH adjusted to 5.8 with 1% phosphoric acid. The flow rate was 2.0 mL/min. UV detection was performed at 254 nm and 0.2 AUFS. About 16 µg of moxalactam was injected.

The method was determined to be stability indicating. The standard solution of moxalactam disodium in 5% dextrose was stored at 24 °C for 7 days. Degradation product peaks did not interfere with the moxalactam peak. Retention times for degradation products were shorter than the retention time of the intact moxalactam.

References

Gupta VD, Stewart KR. Chemical stability of hydrocortisone sodium succinate and several antibiotics when mixed with metronidazole injection for intravenous infusion. *J Parenter Sci Technol.* 1985; 39: 145–8.

Gupta VD, Stewart KR, Gunter JM. Stability of cefotaxime sodium and moxalactam disodium in 5% dextrose and 0.9% sodium chloride injections. *Am J IV Ther Clin Nutr.* 1983; 10: 20–9.

Mycophenolate Mofetil

Chemical Name
2-Morpholinoethyl (*E*)-6-(4-hydroxy-6-methoxy-7-methyl-3-oxo-5-phthalanyl)-4-methyl-4-hexenoate

Other Name
CellCept

Form	Molecular Formula	MW	CAS
Mycophenolate mofetil	$C_{23}H_{31}NO_7$	433.5	115007-34-6
Mycophenolic acid	$C_{17}H_{20}O_6$	320.3	24280-93-1

Appearance
Mycophenolate mofetil occurs as a white to off-white crystalline powder.

Solubility
Mycophenolate mofetil is slightly soluble in water, freely soluble in acetone, soluble in methanol, and sparingly soluble in ethanol.

pK$_a$
Mycophenolate mofetil has pK$_a$ values of 5.6 for the morpholino group and 8.5 for the phenolic group.

Method
Anaizi et al. studied the stability of mycophenolate mofetil in an oral liquid extemporaneously prepared from commercially available capsules and a cherry-flavored vehicle. The liquid chromatograph consisted of a Perkin-Elmer model 250 binary LC pump, a Perkin-Elmer model LC 290 UV-visible detector, a Perkin-Elmer model ISS 100 auto-injector, and an Epson computer. The stationary phase was a Perkin-Elmer Pecosphere C_{18} reversed-phase column (33 × 4.6 mm, 3-μm particle size). The mobile phase consisted of 60% methanol in water and was delivered isocratically at 2.0 mL/min. UV detection was performed at 254 nm and 0.01 AUFS.

Samples were diluted 1:2500 with HPLC-grade water and filtered through a 0.2-μm filter. Spironolactone 10 μg/mL in methanol was used as an internal standard. The sample injection volume was 10 μL. Under these conditions, retention times for spironolactone and mycophenolate mofetil were approximately 0.9 and 1.3 minutes, respectively.

The stability-indicating capability of the method was demonstrated by accelerated degradation of mycophenolate mofetil. Mycophenolate mofetil 100 mg/mL aqueous mixtures were adjusted to pH 1.2 with 10 N hydrochloric acid and pH 12.4 with 10 N sodium hydroxide. These two mixtures and a third mixture at the initial pH of 6.7 were incubated in an oven at 60 °C for 6 days. Mycophenolate mofetil decomposition occurred in all three mixtures, but degradation product peaks did not interfere with the intact mycophenolate mofetil peak.

A standard curve for mycophenolate mofetil was constructed from 5 to 40 µg/mL; the correlation coefficient was 0.9999. The maximum intraday and interday coefficients of variation were 1.54 and 0.98%, respectively.

Reference

Anaizi NH, Swenson CF, Dentinger PJ. Stability of mycophenolate mofetil in an extemporaneously compounded oral liquid. *Am J Health Syst Pharm.* 1998; 55: 926–9.

Nadolol

Chemical Name

5-[3-[(1,1-Dimethylethyl)amino]-2-hydroxypropoxy]-1,2,3,4-tetrahydro-2,3-naphthalenediol

Other Name

Corgard

Form	**Molecular Formula**	**MW**	**CAS**
Nadolol	$C_{17}H_{27}NO_4$	309.4	42200-33-9

Appearance

Nadolol occurs as a white to off-white and practically odorless crystalline powder.

Solubility

Nadolol is freely soluble in alcohol. It is soluble in water at pH 2 and is slightly soluble in water at pH 7–10.

pK$_a$

Nadolol has a pK$_a$ of 9.67.

Method 1

Yang et al. used HPLC analysis to study the stability of nadolol in extemporaneously compounded powder packets from tablets. The liquid chromatograph consisted of a

Waters model 510 pump, a model 717 plus autosampler, and a model 486 tunable UV detector. A Waters model 2690 separation module and a model 996 photodiode-array detector were also used in validating the method. The stationary phase was a Waters Spherisorb S5 ODS2 column (250 × 4.6 mm). The mobile phase consisted of methanol, 5 mM sodium acetate buffer (pH 4.0), and tetrahydrofuran (40:60:2.5, vol/vol/vol). The flow rate was 1 mL/min. UV detection was performed at 254 nm.

A nadolol powder from tablets was mixed and diluted to 100 mL with a mixture of methanol and water (1:1, vol/vol), sonicated for 10 minutes, and centrifuged. The supernatant was collected. The injection volume was 10 μL. Under these conditions, the retention time for nadolol was about 7.6 minutes.

The method was shown to be stability indicating by intentional degradation of the drug. Nadolol samples were prepared in 0.1 N hydrochloric acid, in 1 N sodium hydroxide solution, or in water and then were stored at 70 °C for 1 month. Nadolol was resolved from its degradation products on the chromatogram.

A standard curve for nadolol was generated from 40 to 120 μg/mL.

References
Yang Y-HK, Lin T-R, Huang Y-F, et al. Stability of furosemide, nadolol and propranolol hydrochloride in extemporaneously compounded powder packets from tablets. *Chin Pharm J*. 2000; 52: 51–8.

Lin T-R, Yang Y-HK, Huang Y-F, et al. Content uniformity of captopril, furosemide, nadolol and propranolol hydrochloride powder packets extemporaneously compounded from tablets. *Chin Pharm J*. 2000; 52: 59–68.

Method 2
Patel et al. developed an HPLC method for nadolol and other β-adrenergic blocking drugs. The liquid chromatograph consisted of an Altex model 110A or a Waters M6000 pump, a Rheodyne model 7010 precision 20-μL loop injector, an Altex model 153 or Chromatronic model 220 fixed-wavelength detector, a Schoeffel model 770 or Varian Varichrome variable-wavelength monitor, and Linear strip-chart recorders. The stationary phase was a Merck LiChrosorb reversed-phase ethylsilane (RP2) column (250 × 4.6 mm, 10-μm particle size). The mobile phase consisted of 35% methanol and 65% 0.0005 M hydrochloric acid–0.05 M sodium chloride aqueous solution. The final pH of the mobile phase was 4.5. UV detection was performed at 254 or 220 nm. Atenolol was used as an internal standard. Retention times for atenolol and nadolol were about 4.5 and 6 minutes, respectively.

The analytical method was determined to be stability indicating. A solution of nadolol was spiked with its synthetic intermediates and possible impurities. All compounds were completely resolved. Relative retention times of β-adrenergic blocking drugs to nadolol were as follows: acebutolol (1.50), alprenolol (4.68), atenolol (0.65), metoprolol (1.75), nadolol (1.00), oxprenolol (2.7), pindolol (1.29), practolol (0.80), propranolol (2.7), sotalol (1.38), and timolol (1.45).

Linearity was obtained from 0.025 to 0.2 μg/mL of nadolol. The coefficient of variation was 0.5%.

Reference
Patel BR, Kirschbaum JJ, Poet RB. High-pressure liquid chromatography of nadolol and other β-adrenergic blocking drugs. *J Pharm Sci*. 1981; 70: 336–8.

Nafcillin Sodium

Chemical Name

[2S-(2α,5α,6β)]-6-[[(2-Ethoxy-1-naphthalenyl)carbonyl]amino]-3,3-dimethyl-7-oxo-4-thia-1-azabicyclo[3.2.0]heptane-2-carboxylic acid monosodium salt

Other Names

Nafcil, Nallpen, Unipen

Form	Molecular Formula	MW	CAS
Nafcillin sodium	$C_{21}H_{21}N_2NaO_5S$	436.5	985-16-0

Appearance

Nafcillin sodium is a white to yellowish-white powder.

Solubility

Nafcillin sodium is freely soluble in water and soluble in alcohol.

pK$_a$

Nafcillin sodium has a pK$_a$ of approximately 2.7.

Method 1

Stiles and Allen used an HPLC assay to evaluate the stability of nafcillin sodium in polyvinyl chloride drug reservoirs. The liquid chromatograph included a Waters model 501 solvent-delivery system, a Micromeritics model 728 autoinjector, a Micromeritics model 732 injection valve, a Waters Lambda-Max model 481 variable-wavelength UV detector, and a Waters model 745 data module. The stationary phase was a Bakerbond C_{18} column (250 × 2.6 mm, 5-μm particle size). The mobile phase consisted of acetonitrile and 0.05 M sodium acetate (30:70, vol/vol) and was delivered isocratically at 1.0 mL/min. UV detection was performed at 254 nm.

Samples were diluted with mobile phase.

This method was determined to be stability indicating by accelerated degradation of nafcillin sodium. Degradation product peaks did not interfere with the nafcillin peak. The retention time for nafcillin sodium was 6.1 minutes, and retention times for its degradation products were 1.8, 1.9, 3.2, 4.1, and 10.6 minutes.

A standard curve for nafcillin sodium was generated from 50 to 400 μg/mL. Intraday and interday coefficients of variation were 3.5 and 4.3%, respectively.

A similar method was used by Zhang and Trissel.

References

Stiles ML, Allen LV Jr. Stability of nafcillin sodium, oxacillin sodium, penicillin G potassium, penicillin G sodium, and tobramycin sulfate in polyvinyl chloride drug reservoirs. *Am J Health Syst Pharm*. 1997; 54: 1068–70.

Zhang Y, Trissel LA. Stability of ampicillin sodium, nafcillin sodium, and oxacillin sodium in AutoDose infusion system bags. *Int J Pharm Compound*. 2002; 6: 226–9.

Method 2

Hagan et al. investigated the stability of nafcillin sodium in the presence of lidocaine hydrochloride. A Hewlett-Packard model 1090M chromatograph was equipped with a UV diode-array detector, an autosampler, and a PASCAL-based computer system. The stationary phase was a Hewlett-Packard Hypersil C_{18} column (100×2.1 mm, 3-μm particle size). The mobile phase consisted of 0.02 M sodium acetate (pH 6.0) and acetonitrile (78:22, vol/vol) delivered isocratically at 0.5 mL/min. Extraction wavelengths for the diode-array detector were 230 and 254 nm, with 4-nm bandwidths, and the diode-array detector reference wavelength was 550 nm.

Samples were diluted 1:100 with HPLC-grade water. The injection volume was 10 μL. The column temperature was maintained at 40 °C. The retention times for nafcillin and lidocaine hydrochloride were 2.45 and 7.26 minutes, respectively.

The stability-indicating capability of the assay was demonstrated by accelerated degradation of nafcillin–lidocaine admixture samples. Samples of a nafcillin–lidocaine admixture were treated with 0.1 N hydrochloric acid or 0.1 N sodium hydroxide and then subjected to a standard autoclave cycle (121 °C for 15 minutes). The resulting samples were cooled, neutralized, and assayed. Degradation product peaks did not interfere with the nafcillin peak.

Standard curves for nafcillin were constructed from 2.5 to 20 mg/mL; correlation coefficients were greater than 0.9999. Intraday and interday coefficients of variation were 2.52% ($n = 10$) and 1.88% ($n = 5$), respectively.

Reference

Hagan RL, Carr-Lopez SM, Strickland JS. Stability of nafcillin sodium in the presence of lidocaine hydrochloride. *Am J Health Syst Pharm.* 1995; 52: 521–3.

Method 3

Riley and Lipford reported the interaction of aztreonam with nafcillin sodium in intravenous admixtures. The HPLC system consisted of a Varian model 5000 pump, a Micromeritics 728 autosampler with a Valco RC6U injection valve fitted with a 20-μL loop, a Kratos Spectroflow 757 UV-visible detector, and a Hewlett-Packard 3392 A integrator. The stationary phase was a Zorbax ODS column (150×4.6 mm). The mobile phase consisted of 43.5% methanol in 0.05 M monobasic potassium phosphate adjusted to pH 6.4 with 10 N sodium hydroxide. The flow rate was 2.0 mL/min. UV detection was performed at 238 nm.

Samples were centrifuged at 10,000 rpm for 20 minutes, and the supernatant was diluted 1:100 with 0.05 M phosphate buffer (pH 6.0). Under these conditions, aztreonam eluted at the solvent front and nafcillin eluted with a retention time of 5.2 minutes.

The stability-indicating nature of the assay was confirmed by accelerated degradation of nafcillin. Storing solutions of nafcillin 200 μg/mL in 0.1 N hydrochloric acid, 0.1 N sodium hydroxide, or 0.05 M monobasic potassium phosphate (pH 6.0) at room temperature (23–25 °C) for 2 weeks produced degradation products with peaks that did not interfere with the nafcillin peak.

Reference

Riley CM, Lipford LC. Interaction of aztreonam with nafcillin in intravenous admixtures. *Am J Hosp Pharm.* 1986; 43: 2221–4.

Method 4

Gupta and Stewart developed a stability-indicating HPLC method for a quantitation and stability study of nafcillin sodium in 0.9% sodium chloride injection and 5% dextrose injection. The liquid chromatograph was a Waters model ALC 202 system equipped with a U6K universal injector, a Schoeffel Spectroflow monitor SF 770 multiple-wavelength detector, a Houston Omniscribe recorder, and a Spectra-Physics autolab minigrator-integrator. The stationary phase was a Waters μBondapak phenyl column (300 × 4 mm). The mobile phase consisted of 50% methanol and 50% 0.01 M ammonium acetate in water. The flow rate was 2.0 mL/min. UV detection was performed at 280 nm and 0.04 AUFS.

Samples were diluted with water, and the injection volume was 20 μL.

The coefficient of variation was 1.59%.

Reference

Gupta VD, Stewart KR. Quantitation of carbenicillin disodium, cefazolin sodium, cephalothin sodium, nafcillin sodium, and ticarcillin disodium by high-pressure liquid chromatography. *J Pharm Sci.* 1980; 69: 1264–7.

Nafimidone

Chemical Name

2-(1*H*-Imidazol-1-yl)-1-(2-naphthalenyl)ethanone

Form

Form	Molecular Formula	MW	CAS
Nafimidone	$C_{15}H_{12}N_2O$	236.3	64212-22-2
Nafimidone hydrochloride	$C_{15}H_{12}N_2O.HCl$	272.7	70891-37-1

Method

Taylor et al. described an HPLC method for nafimidone. The chromatographic equipment consisted of a Waters pump, a model M441 detector, and a Rheodyne 7125 injector with a 20-μL loop. The stationary phase was a Hypersil ODS column (100 × 2 mm, 5-μm particle size). The mobile phase consisted of acetonitrile and 0.04 M phosphate buffer (pH 7) (25:75, vol/vol) containing 2 mM tetrabutylammonium bromide. The flow rate was 0.5 mL/min. Under these conditions, the retention time for nafimidone was about 6.5 minutes (estimated from the published chromatogram).

The method was reported to be stability indicating.

Reference

Taylor RB, Durham DG, Shivji ASH, et al. Development of a stability-indicating assay for nafimidone [1-(2-naphthoylmethyl)imidazole hydrochloride] by high-performance liquid chromatography. *J Chromatogr.* 1986; 353: 51–9.

Nalbuphine Hydrochloride

Chemical Name
17-Cyclobutylmethyl-7,8-dihydro-14-hydroxy-17-normorphine hydrochloride

Other Name
Nubain

Form	Molecular Formula	MW	CAS
Nalbuphine hydrochloride	$C_{21}H_{27}NO_4$.HCl	393.9	23277-43-2

Appearance
Nalbuphine hydrochloride is a white to slightly off-white powder.

Solubility
Nalbuphine hydrochloride is soluble in water and slightly soluble in alcohol.

pK$_a$
Nalbuphine has pK$_a$ values of 8.71 and 9.96.

Method
Quarry et al. developed an HPLC method for the determination of nalbuphine hydrochloride, methylparaben, and propylparaben in a parenteral product. The first chromatograph consisted of a Waters model 600 pump, a model 600E system controller, a model 717 Plus autosampler, a temperature control module, and an Applied Biosystem model 759A absorbance detector. The second chromatograph included a Waters model 510 pump, a model 680 system controller, a temperature control module, a model 712 WISP autosampler, and a Kratos Spectroflow 757 UV detector. The stationary phase was a Mac-Mod Zorbax SB-C8 or Zorbax Rx-C8 column (150 × 4.6 mm, 5-μm particle size). The column temperature was maintained at 35 °C. Solvent A consisted of 0.05% trifluoroacetic acid in water, acetonitrile, and tetrahydrofuran (92:5:3, vol/vol/vol). Solvent B was acetonitrile. The mobile phase was delivered initially at 100% A for 6 minutes and then was increased linearly from 0 to 60% B in 10 minutes, held at 60% B for 2 minutes, returned to 100% A, and equilibrated for 8 minutes. The flow rate was 2.0 mL/min. UV detection was performed at 280 nm.

Samples were diluted with 0.1 N hydrochloric acid to about 0.2 mg/mL of nalbuphine. The injection volume was 75 μL. Under these conditions, retention times for nalbuphine, β-nalbuphine, methylparaben, and propylparaben were 4.3, 5.8, 13, and 16 minutes, respectively.

The method was demonstrated to be stability indicating. All known degradation products and synthetic impurities were well separated from the intact nalbuphine. Nalbuphine solutions were also treated with 0.1 N sodium hydroxide solution, refluxed in 0.1 N hydrochloric acid, exposed to intense light (600 foot-candles), and subjected to heat (50 °C). Degradation products, impurities, and their parent compounds (with their retention times in parentheses in minutes) were: α-nor-14-hydroxydihydromorphine (1.2), noroxymorphone (1.6), nalbuphine (4.5), β-nalbuphine (5.8), nalbuphine-*N*-oxide (5.8), 6-ketonalbuphine (5.8), *p*-hydroxybenzoic acid (6.2), 2,2′-bisnalbuphine (11.2), 10-ketonalbuphine (11.7), methylparaben (13.5), and propylparaben (16.6).

Standard curves for nalbuphine were constructed from 50 to 150% of the working concentration. The correlation coefficients were greater than 0.999.

Reference

Quarry MA, Sebastian DS, Williams RC. Determination of nalbuphine hydrochloride, methylparaben, and propylparaben in nalbuphine hydrochloride injection by high performance liquid chromatography. *Chromatographia*. 1998; 47: 515–22.

Nalidixic Acid

Chemical Name

1-Ethyl-1,4-dihydro-7-methyl-4-oxo-1,8-naphthyridine-3-carboxylic acid

Other Names

NegGram, Negram, Nogram, Uriben

Form	Molecular Formula	MW	CAS
Nalidixic acid	$C_{12}H_{12}N_2O_3$	232.2	389-08-2

Appearance

Nalidixic acid occurs as a white to very pale yellow and odorless crystalline powder.

Solubility

Nalidixic acid is practically insoluble in water and slightly soluble in alcohol.

pK$_a$

Nalidixic acid has a pK$_a$ of 6.7.

Method

Argekar and Shah reported an assay for the simultaneous determination of nalidixic acid and metronidazole–metronidazole benzoate in drug formulations. The Waters liquid chromatograph included a quaternary gradient low-pressure pump, an autoinjector, and a variable-wavelength UV-visible detector. The stationary phase was a Machery-Nagel RP8 column (125 × 3.9 mm, 5-μm particle size). The mobile phase consisted of water, acetonitrile, and triethylamine (68:32:0.1, vol/vol/vol), adjusted to pH 3.1 with phosphoric acid. The flow rate was 1.0 mL/min. UV detection was 310 nm. The diluent was prepared by mixing methanol and acetonitrile (1:1, vol/vol).

For tablets, a sample of ground tablet equivalent to 100 mg of nalidixic acid was weighed into a 100-mL volumetric flask, mixed with 50 mL of diluent, sonicated for 10 minutes, brought to volume, centrifuged, and diluted 1:5 with diluent. For syrup, a 5-mL sample of syrup was weighed and transferred into a 100-mL volumetric flask, mixed with 5 mL of water by sonication for 5 minutes, sonicated with 50 mL of diluent for 10 minutes, brought to volume, centrifuged, and diluted 1:2.5 with the diluent. The injection volume was 10 μL.

The assay was reported to be stability indicating.

A standard curve for nalidixic acid was constructed from 5 to 1500 μg/mL. The limit of detection and the limit of quantification were 2 and 10 ng per injection, respectively.

Reference

Argekar AP, Shah SJ. Stability indicating HPLC method for the simultaneous estimation of nalidixic acid and metronidazole/metronidazole benzoate in their formulations. *Indian Drugs*. 1997; 34: 520–6.

Naloxone Hydrochloride

Chemical Name

17-Allyl-6-deoxy-7,8-dihydro-14-hydroxy-6-oxo-17-normorphine hydrochloride dihydrate

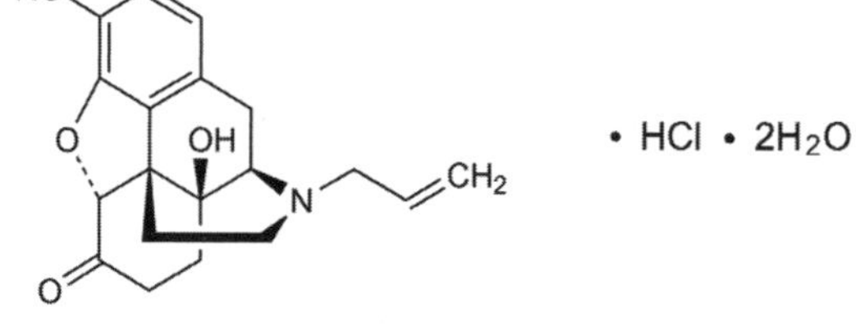

• HCl • 2H$_2$O

Other Names

Narcan, Narcanti

Form	Molecular Formula	MW	CAS
Naloxone hydrochloride	C$_{19}$H$_{21}$NO$_4$.HCl.2H$_2$O	399.9	51481-60-8

Appearance

Naloxone hydrochloride occurs as a white or almost white crystalline hygroscopic powder.

Solubility

Naloxone hydrochloride is freely soluble in water and soluble in alcohol.

pK$_a$

Naloxone hydrochloride has a pK$_a$ of 7.94.

Method

Hanna et al. described an HPLC analysis of naloxone hydrochloride in injectable solutions. A Micromeritics model 6000 system was equipped with a Schoeffel model SF 770 multiple-wavelength UV detector and a Hewlett-Packard model 3385A integrator. The stationary phase was a Waters μBondapak reversed-phase column (300 × 6.3 mm, 10-μm particle size). The mobile phase consisted of methanol and 0.1% (wt/vol) aqueous ammonium carbonate (45:55, vol/vol). The flow rate was 3.0 mL/min. UV detection was performed at 220 nm. Papaverine 0.5 mg/mL in methanol was used as an internal standard. The injection volume was 10 μL. Under these conditions, retention times of methylparaben, propylparaben, naloxone hydrochloride, and papaverine were 1.8, 4.6, 6.7, and 9.6 minutes, respectively.

The assay was proven to be stability indicating by an accelerated degradation study. Naloxone hydrochloride was degraded by heat and oxygen. No degradation product peaks interfered with the intact naloxone peak.

A standard curve was obtained from 1.60 to 2.40 ratios of naloxone hydrochloride concentration versus internal standard concentration. The correlation coefficient was 0.9999.

Reference

Hanna S, Insler M, Zapata R, et al. High-performance liquid chromatographic analysis of naloxone hydrochloride in injectable solutions. *J Chromatogr*. 1980; 200: 277–81.

Naltrexone Hydrochloride

Chemical Name

N-Cyclopropylmethyl-14-hydroxydihydromorphinone hydrochloride

Other Names

Antaxone, EN-1639A, Nalorex, ReVia, UM-792

Form	Molecular Formula	MW	CAS
Naltrexone hydrochloride	$C_{20}H_{23}NO_4.HCl$	377.9	16676-29-2

Appearance

Naltrexone hydrochloride occurs as crystals.

Method

Iyer et al. used an HPLC method to evaluate a new medium for a naltrexone implant. The Shimadzu liquid chromatograph consisted of a model SCL-10A system controller, a model SIL-10AD high-pressure pump, a model SIL-10AD autosampler, a model CTO-10AC column oven, and a model SPD-M10A photodiode-array detector. The stationary phase was a Supelco Supelcosil C_{18} column (150 × 4.6 mm, 5-µm particle size). The column temperature was maintained at 50 ℃. The mobile phase consisted of an aqueous buffer and acetonitrile (88:12, vol/vol). The buffer contained 40 mM monobasic potassium phosphate and 0.06% (vol/vol) triethylamine and was adjusted to pH 4.75 with phosphoric acid. The isocratic flow rate was 1.25 mL/min. UV detection was carried out at 204 nm.

Naltrexone hydrochloride working standard was prepared in a mixture of methanol and water (50:50, vol/vol). The injection volume was 5 µL. Under these conditions, the retention time of naltrexone hydrochloride was about 4.5 minutes (estimated from the published chromatogram).

Naltrexone hydrochloride in modified Hank's balanced salts solution was well separated from its major degradation product, demonstrating the stability-indicating nature of the analytical method.

Linear calibration curves for naltrexone hydrochloride were obtained from 0.16 to 20 µg/mL. Correlation coefficients were greater than 0.99. The interday coefficient of variation was less than 1.7%. The percent deviation in accuracy was less than 1.0%. The limit of quantitation was 0.16 µg/mL.

Reference

Iyer SS, Barr WH, Karnes HT. Characterization of a potential medium for "biorelevant" in vitro release testing of a naltrexone implant, employing a validated stability-indicating HPLC method. *J Pharm Biomed Anal.* 2007; 43: 845–53.

Naphazoline

Chemical Names

4,5-Dihydro-2-(1-naphthalenylmethyl)-1*H*-imidazole
2-(1-Naphthylmethyl)imidazoline

Other Names
Nafazair, Vasoclear

Form	Molecular Formula	MW	CAS
Naphazoline	$C_{14}H_{14}N_2$	210.3	835-31-4
Naphazoline hydrochloride	$C_{14}H_{14}N_2.HCl$	246.7	550-99-2
Naphazoline nitrate	$C_{14}H_{14}N_2.HNO_3$	273.3	5144-52-5

Appearance
Naphazoline hydrochloride and naphazoline nitrate occur as white odorless crystalline powders.

Solubility
Naphazoline hydrochloride has an aqueous solubility of 40 g in 100 mL of water; it is soluble in alcohol and slightly soluble in chloroform. It is practically insoluble in ether. Naphazoline nitrate is sparingly soluble in water; it is soluble in alcohol and very slightly soluble in chloroform. It is practically insoluble in ether.

Method 1
Ruckmick et al. identified the primary degradation product of naphazoline hydrochloride and antazoline phosphate in a commercial ophthalmic formulation. The liquid chromatograph consisted of a Waters 600E low-pressure mixing gradient solvent-delivery controller and pump, a Waters 715 Ultra WISP autosampler, a Waters 490 multichannel variable-wavelength UV detector, and a Perkin-Elmer Nelson series 900 A/D converter interface to a VAX 6210 computer. The stationary phase was a Beckman Ultrasphere C_{18} column (250 × 4.6 mm, 5-μm particle size). The mobile phase consisted of 57% methanol in water containing 22 mM heptanesulfonic acid, 0.1% dibutylamine, and 1% acetic acid. The flow rate was 1 mL/min. UV detection was performed at 280 nm.

Samples were diluted 1:10 with water. The injection volume was 20 μL. Retention times for naphazoline and antazoline were about 8.5 and 20 minutes, respectively (estimated from the published chromatogram).

The method was determined to be stability indicating by spiking the sample with the known degradation products, 1-naphthylacetylethylenediamine and 1-naphthylacetic acid. Degradation product peaks did not interfere with the naphazoline peak.

Standard curves for naphazoline were linear from 10 to 100 μg/mL, with correlation coefficients greater than 0.9999.

Reference
Ruckmick SC, Marsh DF, Duong ST. Synthesis and identification of the primary degradation product in a commercial ophthalmic formulation using NMR, MS and a stability-indicating HPLC method for antazoline and naphazoline. *J Pharm Sci.* 1995; 84: 502–7.

Method 2
Bauer and Krogh developed a stability-indicating assay for naphazoline in ophthalmic preparations. The liquid chromatograph consisted of a Waters model 6000A pump, a Rheodyne model 7120 injector with a 20-μL loop, a DuPont variable-wavelength detector, and a Hewlett-Packard 3385A recording integrator. The stationary phase was a Waters μBondapak C_{18} column. The mobile phase consisted of 700 mL of water containing 6 g

of sodium citrate dihydrate and 4 g of anhydrous citric acid with the pH adjusted to 2.2 ± 0.2 with perchloric acid and 300 mL of methanol. The flow rate was 2.0 mL/min. UV detection was performed at 265 nm. Tetrahydrozoline served as an internal standard.

Samples were diluted 1:2 with internal standard solution. Retention times for naphazoline and tetrahydrozoline were 4.37 and 5.31 minutes, respectively.

The assay was stated to be stability indicating because it could separate naphazoline from its degradation product, *N*-(2-aminoethyl)-1-(1-naphthyl)acetamide.

Calibration curves were constructed from 0.02 to 0.1 mg/mL, and the correlation coefficient was 0.9999.

Reference

Bauer J, Krogh S. High-performance liquid chromatographic stability-indicating assay for naphazoline and tetrahydrozoline in ophthalmic preparations. *J Pharm Sci.* 1983; 72: 1347–9.

Nateglinide

Chemical Name

N-[[*trans*-4-(1-methylethyl)cyclohexyl]carbonyl]-D-phenylalanine

Other Names

Fasti, Starlix

Form	Molecular Formula	MW	CAS
Nateglinide	$C_{19}H_{27}NO_3$	317.4	105816-04-4

Appearance

Nateglinide occurs as crystals.

Method

Qi et al. described an HPLC method for the determination of nateglinide in a tablet formulation. The liquid chromatographic system consisted of a Hewlett-Packard 1100 solvent-delivery system, a variable UV-visible detector, and an injector with a 20-μL loop. The stationary phase was a Hypersil BDS C_{18} analytical column (200 × 4.6 mm, 5-μm particle size). The mobile phase consisted of acetonitrile, methanol, and phosphate buffer (70:4:120, vol/vol/vol, pH 6.6) and was delivered at a flow rate of 1 mL/min. UV detection was performed at 216 nm.

A portion of ground tablets (20) equivalent to 60 mg of nateglinide was accurately weighed, transferred to a 50-mL volumetric flask, sonicated for 5 minutes, brought to volume with mobile phase, and filtered. The filtrate was diluted 50-fold with mobile phase. The injection volume was 20 μL. Under these conditions, the retention time for nateglinide was about 6.9 minutes.

The method was verified to be stability indicating by intentional degradation of nateglinide. Tablets were exposed to acid and base. No degradation products or any excipients interfered with the analysis of the drug.

A standard curve was obtained from 4.94 to 49.9 μg/mL with a correlation coefficient of 0.9999. The limit of detection and the limit of quantitation were 0.51 and 1.56 μg/mL, respectively.

Reference

Qi M, Wang P, Wang J, et al. A stability-indicating HPLC method for the determination of nateglinide in a new tablet formulation. *J Chin Pharm Sci*. 2002; 11: 101–4.

Nefazodone Hydrochloride

Chemical Name

2-[3-[4-(3-Chlorophenyl)-1-piperazinyl]propyl]-5-ethyl-2,4-dihydro-4-(2-phenoxyethyl)-3*H*-1,2,4-triazol-3-one hydrochloride

Other Names

Nefadar, Nefirel, Serzone

Form	Molecular Formula	MW	CAS
Nefazodone hydrochloride	$C_{25}H_{32}ClN_5O_2.HCl$	506.5	82752-99-6

Method

Rao et al. described a method for the quantitative determination of nefazodone in bulk materials and pharmaceutical dosage forms. The liquid chromatograph consisted of a Waters model 600 solvent-delivery system, a model 486 tunable absorbance detector, a model 996 photodiode-array detector, and a Rheodyne model 7725i injector with a 10-μL loop. The stationary phase was an Inertsil ODS-3V column (250 × 4.6 mm, 5-μm particle size). The mobile phase was a mixture of 0.05 M monobasic potassium phosphate aqueous buffer (pH 3.0), acetonitrile, and methanol (50:40:10, vol/vol/vol). The flow rate was 1.0 mL/min. UV detection was performed at 220 nm.

A standard stock solution of nefazodone was prepared by dissolving the drug substance in mobile phase. For tablets, a portion of powder from five tablets equivalent to 40 mg of nefazodone was accurately weighed, transferred into a 25-mL volumetric flask, extracted with mobile phase by vortexing and ultrasonicating, and filtered through a 0.45-μm membrane filter. The filtrate was further diluted in mobile phase. The sample injection volume was 10 μL. Under these conditions, the retention time of nefazodone was 8.6 minutes.

To show the stability-indicating nature of the method, nefazodone was refluxed with 0.5 N hydrochloric acid, 2.0 N sodium hydroxide solution, and 3% hydrogen peroxide at 60 °C for 6 hours. A nefazodone sample was also exposed to UV light at 254 nm at 70 °C for 6 hours. Degradation product peaks were well resolved from the peak of the intact nefazodone. The excipient in pharmaceutical dosage forms did not interfere with the analysis of the drug. The peak purity of nefazodone was also confirmed using a photodiode-array detector.

A linear calibration curve of nefazodone was constructed from 0.2 to 0.6 mg/mL. The correlation coefficient was 0.9997. The intraday and interday coefficients of variation were less than 0.7 and 1.2%, respectively.

Reference

Sreenivas Rao D, Geetha S, Srinivasu MK, et al. LC determination and purity evaluation of nefazodone HCl in bulk drug and pharmaceutical formulations. *J Pharm Biomed Anal.* 2001; 26: 629–36.

Nefopam Hydrochloride

Chemical Name
3,4,5,6-Tetrahydro-5-methyl-1-phenyl-1*H*-2,5-benzoxazocine hydrochloride

• HCl

Other Names
Acupam, Acupan

Form	Molecular Formula	MW	CAS
Nefopam hydrochloride	$C_{17}H_{19}NO.HCl$	289.8	23327-57-3

Method
Tu et al. used HPLC analysis to investigate the degradation kinetics of nefopam hydrochloride in solutions. The liquid chromatograph included a Waters model 6000A

dual-piston pump and a Houston Omniscribe strip-chart recorder. The stationary phase was a Waters Nova-Pak C_{18} column (150 × 3.9 mm, 5-μm particle size). The mobile phase consisted of acetonitrile and 0.2 M phosphate buffer at pH 3.0 containing 0.05 M 1-pentanesulfonic acid sodium salt (25:75, vol/vol). The flow rate was 2.3 mL/min. UV detection was performed at 201 nm.

Samples were appropriately diluted before analysis. Under these conditions, the retention time for nefopam was 5.83 minutes.

The stability-indicating nature of the method was shown by accelerated degradation of nefopam. Nefopam hydrochloride samples in pH 2.0 and 9.0 buffer solutions were stored at 90 °C for 30 days. The degradation products did not interfere with nefopam hydrochloride. Retention times for the degradation products were 2.40, 2.64, 2.96, 3.66, 3.88, 5.11, 6.71, and 7.60 minutes, respectively.

A standard curve for nefopam hydrochloride was constructed from 0.1 to 0.8 mg/mL. The correlation coefficient was 0.9997. Intraday and interday coefficients of variation were 0.86 and 1.82%, respectively.

Reference
Tu Y-H, Wang D-P, Allen LV Jr. Nefopam hydrochloride degradation kinetics in solution. *J Pharm Sci.* 1990; 79: 48–52.

Nelfinavir Mesylate

Chemical Name
[3S-[2(2S*,3S*),3α,4aβ,8aβ]]-N-(1,1-dimethyethyl)decahydro-2-[2-hydroxy-3-[(3-hydroxy-2-methylbenzoyl)amino]-4-(phenylthio)butyl]-3-isoquinolinecarboxamide monomethanesulfonate

Other Names
AG1343, Viracept

Form	Molecular Formula	MW	CAS
Nelfinavir mesylate	$C_{32}H_{45}N_3O_4S.CH_4O_3S$	663.9	159989-65-8

Solubility
Nelfinavir mesylate has a solubility of 4.5 mg/mL in water. It is very soluble in absolute alcohol, acetonitrile, and methanol.

pK$_a$
Nelfinavir mesylate has a pK$_{a1}$ of 6.0 and a pK$_{a2}$ of 11.1.

Method 1
Jing et al. described an HPLC method for the determination of nelfinavir mesylate as bulk drug and in a pharmaceutical dosage formulation. An Agilent 1100 liquid chromatograph including a model G1310A pump, a model G1314A variable UV-visible detector, a model G1328A manual injector, and an Agilent ChemStation chromatography workstation was used for the analysis. The stationary phase was a Kromasil-CN column (250 × 4.6 mm, 5-μm particle size). The mobile phase consisted of acetonitrile and 25 mM monobasic ammonium phosphate (containing 25 mM triethylamine, pH 3.4 with phosphoric acid) (40:60, vol/vol). The flow rate was 1.0 mL/min. UV detection was carried out at 210 nm. Analysis was performed at 40 °C.

Standard solutions were prepared by dissolving reference standard in mobile phase. For capsules, a portion of the powder from 20 capsules equivalent to 25 mg of nelfinavir mesylate was weighed, transferred to a 50-mL volumetric flask, mixed with 20 mL of mobile phase, shaken well, filled to volume with mobile phase, and filtered. The injection volume was 10 μL. Under these conditions, the retention time of nelfinavir mesylate was 14.8 minutes.

Nelfinavir mesylate samples were treated with 1 M hydrochloric acid and 1 M sodium hydroxide solution and heated at 80 °C for 2 hours and treated with 3.0% hydrogen peroxide; were exposed to direct sunlight for 24 hours; or were heated at 80 °C for 6 hours. The degradation products did not interfere with the determination of the intact nelfinavir mesylate, demonstrating the stability-indicating ability of the method.

A standard curve was constructed from 5.0 to 150.0 μg/mL. The correlation coefficient was 0.9999. The within-day and between-day coefficients of variation were less than 0.6 and 1.1%, respectively.

Reference
Jing Q, Shen Y, Tang Y, et al. Determination of nelfinavir mesylate as bulk drug and in pharmaceutical dosage form by stability indicating HPLC. *J Pharm Biomed Anal.* 2006; 41: 1065–9.

Method 2
Longer et al. used an HPLC method in preformulation studies of a novel HIV protease inhibitor, nelfinavir mesylate. A Hewlett-Packard model 1050 system included a photo-diode-array detector and an autosampler. The stationary phase was a Phenomenex Primesphere 5 C$_{18}$-HC reversed-phase column (150 × 4.6 mm, 5-μm particle size). Solvent A was acetonitrile and solvent B was 25 mM monobasic ammonium phosphate with 20 mM triethylamine, pH 3.4. The flow rate was 1.0 mL/min. The mobile phase was isocratically delivered at 40% A for 10 minutes and followed by a gradient to 70% A from 10 to 20 minutes, with a total run time of 25 minutes. UV detection was performed at 205 and 254 nm. Under these conditions, the retention time of the main peak for nelfinavir mesylate was between 6.5 and 7.5 minutes.

The stability-indicating nature of the method was confirmed by intentional degradation of nelfinavir mesylate. Suspensions of nelfinavir mesylate were stored in 0.1 N hydrochloric acid and 0.1 N sodium hydroxide solution for up to 1 week at 80 °C in glass ampules. The degradation product peaks did not interfere with the main peak of nelfinavir mesylate.

A calibration curve for nelfinavir mesylate was constructed from 32.5 to 520 μg/mL. The correlation coefficient was greater than 0.999.

Reference

Longer M, Shetty B, Zamansky I, et al. Preformulation studies of a novel HIV protease inhibitor, AG1343. *J Pharm Sci.* 1995; 84: 1090–3.

Neomycin

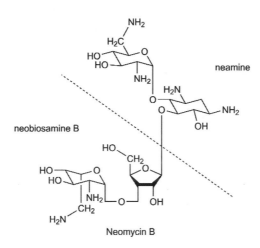

Neomycin B

Other Names

Bykomycin, Mycifradin, Neosulf

Form	Molecular Formula	MW	CAS
Neomycin A	$C_{12}H_{26}N_4O_6$	322.4	3947-65-7
Neomycin B	$C_{23}H_{46}N_6O_{13}$	614.6	119-04-0
Neomycin C	$C_{23}H_{46}N_6O_{13}$	614.6	66-86-4
Neomycin sulfate			1405-10-3

Appearance

Neomycin sulfate is a white or yellowish-white and odorless or almost odorless hygroscopic powder.

Solubility

Neomycin sulfate is very soluble in water and very slightly soluble in alcohol. It is practically insoluble in acetone, chloroform, and ether.

Method 1

Adams et al. described the determination of neomycin sulfate by HPLC coupled with pulsed electrochemical detection. The liquid chromatographic system consisted of a Merck/Hitachi model L-6200 pump, a Marathon autosampler with a 20-μL fixed loop, a

Hewlett-Packard model 3393A integrator, a laboratory-made pneumatic device, and a Dionex model PED-1 pulsed electrochemical detector equipped with a gold working electrode, an Ag/AgCl reference electrode, and a stainless-steel counterelectrode. The detector temperature was maintained at 35 °C. The stationary phase was a Polymer Laboratories (UK) PLRP-S poly(styrenedivinylbenzene) copolymer column (250 × 4.6 mm, 8-μm particle size). The column temperature was maintained at 35 °C. The mobile phase consisted of 70 g/L of sodium sulfate, 1.4 g/L of sodium 1-octanesulfonate, and 50 mL/L of 0.2 M monobasic potassium phosphate (pH 3.0) in water. The flow rate was 1 mL/min. A 0.5 M sodium hydroxide solution was delivered postcolumn at 0.3 mL/min and mixed in a Dionex reaction coil to improve the sensitivity of detection. The potential applied to the flow cell was set at +0.05 V (E_1) with a pulse duration from 0 to 0.40 second (t_1), then +0.75 V (E_2) from 0.41 to 0.60 second (t_2), followed by –0.15 V (E_3) from 0.61 to 1.00 second (t_3). The signal was integrated between 0.2 and 0.4 second. Under these conditions, retention times of neomycin C and neomycin B were about 17.1 and 21.9 minutes, respectively (estimated from the published chromatogram).

The method was reported to be stability indicating.

Standard curves were constructed from 0.025 to 0.800 mg/mL. The correlation coefficient was 0.9989 for neomycin B and 0.9975 for neomycin C. Coefficients of variation for neomycin B and C were 1.3 and 1.5%, respectively. The limit of detection for neomycin C was 5 ng. The limit of quantitation was 15 ng.

Reference

Adams E, Schepers R, Roets E, et al. Determination of neomycin sulfate by liquid chromatography with pulsed electrochemical detection. *J Chromatogr A*. 1996; 741: 233–40.

Method 2

Tsuji et al. examined the stability of neomycin by HPLC. The liquid chromatograph was composed of a Waters model 6000A pump, a Rheodyne model 7010 20-μL fixed-loop injector or a Waters model 710B WISP autosampler, and an LDC model 1201 variable-wavelength UV detector. The stationary phase was a Brownlee LiChrosorb SI-100 silica column (250 × 4.6 mm, 5-μm particle size). The mobile phase consisted of chloroform, tetrahydrofuran, and water (600:392:8, vol/vol/vol). UV detection was performed at 254 nm.

Neomycin was derivatized with 1-fluoro-2,4-dinitrobenzene (DNFB). A 5-mL sample of neomycin solution was transferred into a 250-mL volumetric flask, mixed with 15 mL of 0.15 M DNFB in methanol, heated in a silicone oil bath at 100 °C for 45 minutes, cooled, and brought to volume with the mobile phase. The organic layer was collected and assayed. Under these conditions, retention times of DNFB-neomycin C and DNFB-neomycin B were about 9.6 and 13.5 minutes, respectively (estimated from the published chromatogram).

The method was reported to be stability indicating.

References

Tsuji K, Kane MP, Rahn PD, et al. [60]Co irradiation for sterilization of veterinary mastitis products containing antibiotics and steroids. *Radiat Phys Chem*. 1981; 18: 583–93.

Tsuji K, Goetz JF, Van Meter W, et al. Normal-phase high-performance liquid chromatographic determination of neomycin sulfate derivatized with 1-fluoro-2,4-dinitrobenzene. *J Chromatogr.* 1979; 175: 141–52.

Nevirapine

Chemical Name

11-Cyclopropyl-5,11-dihydro-4-methyl-6*H*-dipyrido[3,2-*b*:2′,3′-*e*][1,4]diazepin-6-one

Other Name

Viramune

Form	Molecular Formula	MW	CAS
Nevirapine	$C_{15}H_{14}N_4O$	266.3	129618-40-2

Appearance

Nevirapine is a white to off-white crystalline powder.

Method

Li et al. described an HPLC method for the assay of nevirapine. The Waters system consisted of a model 616 four-solvent-delivery system and controller, a model 717 WISP injector, and a model 996 photodiode-array detector. The stationary phase was a Supelco Supelcosil LC-ABZ column (150 × 4.6 mm, 5-μm particle size). The column oven temperature was maintained at 35 °C. The mobile phase consisted of acetonitrile and 25 mM monobasic ammonium phosphate (pH 5.0). The flow rate was 1 mL/min. UV detection was performed at 220 nm.

Each sample was dissolved in acetonitrile and diluted with the mobile phase. The injection volume was 20 μL. Under these conditions, the retention time of nevirapine was 7.44 minutes.

The method was confirmed to be stability indicating by assaying a synthetic mixture of nevirapine and its related compounds. Nevirapine samples were also stressed by acid, base, hydrogen peroxide, UV light, and heat/humidity. Nevirapine was completely resolved from all other compounds. Photodiode-array peak analysis also confirmed the peak purity of nevirapine.

A standard curve for nevirapine was constructed from 0.11 to 0.36 mg/mL. The correlation coefficient was greater than 0.999. The coefficient of variation for the analysis of nevirapine was 0.7%.

Reference

Li QC, Tougas T, Cohen K, et al. Validation of a high-performance liquid chromatography method for the assay of and determination of related organic impurities in nevirapine drug substance. *J Chromatogr Sci.* 2000; 38: 246–54.

Nicardipine

Chemical Name

1,4-Dihydro-2,6-dimethyl-4-(3-nitrophenyl)-3,5-pyridinedicarboxylic acid methyl 2-[methyl(phenylmethyl)amino]ethyl ester

Other Name

Cardene

Form	Molecular Formula	MW	CAS
Nicardipine	$C_{26}H_{29}N_3O_6$	479.5	55985-32-5
Nicardipine hydrochloride	$C_{26}H_{29}N_3O_6 \cdot HCl$	516.0	54527-84-3

Appearance

Nicardipine hydrochloride is a greenish-yellow odorless crystalline powder.

Solubility

Nicardipine is freely soluble in chloroform, methanol, and acetic acid; sparingly soluble in anhydrous ethanol; slightly soluble in *n*-butanol, water, acetone, and dioxane; very slightly soluble in ethyl acetate; and practically insoluble in benzene, ether, and hexane.

Method

Baaske et al. determined the stability of nicardipine hydrochloride in commonly used large-volume intravenous solutions. The liquid chromatograph consisted of a Perkin-Elmer System 2/2 pump, a Waters WISP autosampler, a Perkin-Elmer model LC75 or LC90 variable-wavelength UV detector, a Scientific model 285/mm recorder, and a VG Laboratory Systems data system. The stationary phase was a Whatman Partisil ODS-3 column (250 × 4.6 mm, 5-μm particle size). The mobile phase was a mixture of a buffer, methanol, and acetonitrile (40:10:50, vol/vol/vol). The buffer contained 0.02 M monobasic potassium phosphate in water adjusted to pH 4.4–5.0 with phosphoric acid. The flow rate was 1.5 mL/min. UV detection was performed at 237 nm.

Samples were diluted with unbuffered mobile phase. The injection volume was 20 μL. The retention time for nicardipine was 7.3 minutes.

The stability-indicating nature of the assay was demonstrated by accelerated decomposition of nicardipine hydrochloride. Nicardipine hydrochloride was exposed to 0.5 M hydrochloric acid, 0.1 M sodium hydroxide, and 5% hydrogen peroxide and then heated to reflux for at least 3 hours. A solution of nicardipine hydrochloride in water was stored at 75 °C for 10 weeks. Another solution was exposed to intense fluorescent light (1000–2000 foot-candles) for 10 weeks. In all cases, degradation product peaks did not interfere with the intact nicardipine hydrochloride peak. Retention times for degradation products were about 2.5, 5.5, 5.8, and 9.6 minutes (estimated from the published chromatogram).

Standard curves for nicardipine were constructed from 0.025 to 0.075 mg/mL; the correlation coefficient was greater than 0.9999. Intraday and interday coefficients of variation were less than 0.8 and 1.2%, respectively.

Reference

Baaske DM, DeMay JF, Latona CA, et al. Stability of nicardipine hydrochloride in intravenous solutions. *Am J Health Syst Pharm.* 1996; 53: 1701–5.

Nicergoline

Chemical Name

10-Methoxy-1,6-dimethylergoline-8β-methanol 5-bromonicotinate (ester)

Other Name

Sermion

Form	Molecular Formula	MW	CAS
Nicergoline	$C_{24}H_{26}BrN_3O_3$	484.4	27848-84-6

Appearance

Nicergoline occurs as a yellowish to white crystalline powder.

Solubility

Nicergoline is practically insoluble in water but soluble in ethanol and chloroform. It is slightly soluble in ether.

Method

Ahmad et al. developed an HPLC method for the determination of nicergoline in the presence of its hydrolysis-induced degradation products. The Shimadzu chromatograph included a model LC-10AS isocratic pump, a model SPD-10A variable-wavelength detector, a model C-R7A integrator, and a Rheodyne 7161 injector with a 20-μL loop. The stationary phase was a Nucleosil C$_{18}$ column (250 × 4.6 mm, 10-μm particle size). The mobile phase consisted of methanol, water, and glacial acetic acid (80:20:0.1, vol/vol/vol). The flow rate was 1.3 mL/min. Chlorpromazine hydrochloride was used as an internal standard. UV detection was performed at 280 nm.

A portion of ground tablets equivalent to 100 mg of nicergoline was accurately weighed, dissolved, diluted to 100 mL with methanol, and filtered. The injection volume was 20 μL. Under these conditions, retention times for nicergoline and the internal standard were 6.5 and 8.5 minutes, respectively.

Nicergoline was dissolved in ethanol and then refluxed in 1 N sodium hydroxide solution or in 1 N hydrochloric acid at 90 °C for 5 hours. The retention time of the degradation product was about 5.2 minutes.

A standard curve for nicergoline was constructed from 5 to 80 μg/mL. The correlation coefficient was 0.9999. The intraday and interday coefficients of variation were 1.3 and 1.1%, respectively. The limit of detection and the limit of quantitation were 0.96 and 2.90 μg/mL, respectively.

Reference

Ahmad AK, Kawy MA, Nebsen M. First derivative ratio spectrophotometric, HPTLC-densitometric, and HPLC determination of nicergoline in presence of its hydrolysis-induced degradation product. *J Pharm Biomed Anal.* 2002; 30: 479–89.

Nicotinamide (Vitamin B$_3$)

Chemical Name

Pyridine-3-carboxamide

Other Names

Niacin, Niacinamide, Niacor, Nicobion, Nicovitol, Papulex

Form	Molecular Formula	MW	CAS
Nicotinic acid (niacin)	$C_6H_5NO_2$	123.1	59-67-6
Nicotinamide	$C_6H_6N_2O$	122.1	98-92-0

Appearance

Nicotinic acid occurs as white crystals or crystalline powder. Nicotinamide occurs as a white crystalline powder or as colorless crystals.

Solubility

Nicotinic acid has a solubility of 16.7 mg/mL in water. It is freely soluble in boiling water and in boiling alcohol. Nicotinamide is freely soluble in water.

pK_a

Nicotinic acid has a pK_a of 4.85.

Method

Van der Horst et al. described the HPLC analysis of nicotinamide in a total parenteral nutrient solution. The liquid chromatographic system consisted of a Waters model M6000A pump, a model WISP autosampler, a model 481 variable-wavelength detector, a Perkin-Elmer model 204-A fluorescence spectrophotometer, and a Shimadzu Chromatopac C-R3A integrator. The stationary phase was a LiChrosorb RP8 analytical column (250 × 4.6 mm, 10-μm particle size) with an RCSS Guard-Pak C_{18} precolumn (4 × 6 mm). The mobile phase was a mixture of 0.1% triethylamine in 0.067 M monobasic potassium phosphate buffer (pH 6.7) and acetonitrile (100:4, vol/vol). The flow rate was 2 mL/min. UV detection was performed at 260 nm. The injection volume was 20 μL. Under these conditions, the retention time for nicotinamide was about 4.9 minutes (estimated from the published chromatogram).

The method was demonstrated to be stability indicating by accelerated degradation of nicotinamide. Nicotinamide solutions were treated with 1 M sodium hydroxide solution or 1 M hydrochloric acid and heated to boiling for about 30 minutes. Degradation product peaks were separated from the peak of nicotinamide.

A standard curve for nicotinamide was generated from 3 to 30 mg/L. The correlation coefficient was 0.9970. The limit of detection was 0.8 mg/L.

Reference

Van der Horst A, Martens HJM, de Goede PNFC. Analysis of water-soluble vitamins in total parenteral nutrition solution by high pressure liquid chromatography. *Pharm Weekbl [Sci]*. 1989; 11: 169–74.

Nicotine

Chemical Name

(*S*)-3-(1-Methylpyrrolidin-2-yl)pyridine

Other Names
Habitrol, NicoDerm, Nicorette, Nicotinell

Form	Molecular Formula	MW	CAS
Nicotine	$C_{10}H_{14}N_2$	162.2	54-11-5

Appearance
Nicotine occurs as a basic colorless to pale yellow and very hygroscopic and oily volatile liquid.

Solubility
Nicotine is soluble in water and in alcohol.

pK$_a$
Nicotine has a pK_{a1} of about 7.8–7.9 and a pK_{a2} of about 3 at 15 °C.

Method
Carlisle et al. reported an assay for nicotine in transdermal patches. The chromatograph consisted of a Beckman model 126 pump, a model 507 autosampler, and a model 168 diode-array detector. The stationary phase was a Dionex PCX-500 polymer reversed-phase/ion-exchange column (250 × 4.6 mm) with a Dionex PCX-500 guard column (55 × 4.6 mm). The mobile phase was a mixture of 880 mL of acetonitrile, 75 mL of 0.6 M perchloric acid solution, 450 mL of 1.0 M potassium chloride solution, and 595 mL of water. The flow rate was 1 mL/min. UV detection was performed at 254 nm.

For raw material, a sample of nicotine free base was accurately weighed and dissolved in water to yield approximately 40 μg/mL of nicotine. For transdermal patches, a nicotine patch removed from its release liner was transferred to a 50-mL volumetric flask containing tetrahydrofuran; this mixture was sonicated for 30 minutes and slowly brought to volume with absolute alcohol. Five milliliters of this solution was transferred to a 100-mL volumetric flask containing 5 mL of 0.6 M perchloric acid, evaporated to about 3 mL by a gentle nitrogen flow at room temperature, and diluted with water to volume. The sample injection volume was 20 μL.

The analytical assay was evaluated to be stability indicating by spiking the nicotine solution with its related substances and known degradation products. Retention times in minutes were about 1.9 for cotinine, 2.0 for nicotine 1-N-oxide, 3.7 for nicotine 1,1′-di(N-oxide), 5.6 for salicylate, and 6.9 for nicotine.

A standard curve for nicotine was obtained from 0.434 to 62.0 μg/mL. The correlation coefficient was 1.0000. The limit of detection and the limit of quantitation for nicotine were 0.2 and 0.4 μg/mL, respectively.

Reference
Carlisle MR, Chicoine ML, Wygant MB. A stability-indicating high-performance liquid chromatographic assay for nicotine in transdermal patches. *Int J Pharm.* 1992; 80: 227–42.

Nifedipine

Chemical Name
Dimethyl 1,4-dihydro-2,6-dimethyl-4-(2-nitrophenyl)pyridine-3,5-dicarboxylate

Other Names
Adalate, Procardia

Form	Molecular Formula	MW	CAS
Nifedipine	$C_{17}H_{18}N_2O_6$	346.3	21829-25-4

Appearance
Nifedipine is a yellow crystalline powder.

Solubility
Nifedipine is practically insoluble in water and is soluble in alcohol.

Method 1
Using an HPLC method, Dentinger et al. determined the stability of nifedipine in an extemporaneously compounded oral solution. The liquid chromatographic system consisted of a Perkin-Elmer model 250 binary pump, a model 290 UV-visible detector, and a model ISS 100 autoinjector. The stationary phase was a MetaChem Inertsil ODS-3 column (100 × 4.6 mm, 5-μm particle size) with a MetaGuard Inertsil ODS-3 guard column (15 × 4.6 mm, 5-μm particle size). The mobile phase consisted of methanol and water (65:35, vol/vol). The flow rate was 1.1 mL/min. UV detection was performed at 265 nm and 0.01 AUFS. The injection volume was 10 μL. Methylparaben 0.5 mg/mL in water was used as an internal standard. Under these conditions, the retention times of methylparaben and nifedipine were 2.5 and 4.7 minutes, respectively.

The method was demonstrated to be stability indicating. Nifedipine samples 10 mg/mL were adjusted separately to pH 1.4 with 1 N hydrochloric acid and to pH 12.3 with 1 N sodium hydroxide solution. Nifedipine samples 10 mg/mL were also stored at 76 °C for 24 hours or placed under fluorescent light for 24 hours. No peaks that would interfere with the nifedipine peak were observed on the chromatogram.

A standard curve for nifedipine was constructed from 10 to 40 μg/mL; the correlation coefficient was 1.0000. The interday and intraday coefficients of variation were 1.0 and 1.8%, respectively.

Reference
Dentinger PJ, Swenson CF, Anaizi NH. Stability of nifedipine in an extemporaneously compounded oral solution. *Am J Health Syst Pharm.* 2003; 60: 1019–22.

Method 2

Nahata et al. used HPLC analysis to determine the stability of nifedipine in two oral suspensions stored at 4 and 25 °C in plastic prescription bottles over a 3-month period. The Hewlett-Packard 1050 liquid chromatograph consisted of a pump, an autosampler, a variable-wavelength detector, and a Hewlett-Packard 3396A integrator. The stationary phase was a Mac-Mod analytical Zorbax C_8 column (150 × 3.0 mm). The mobile phase consisted of 43% water with 0.8% diethylamine and 57% acetonitrile. The flow rate was 0.4 mL/min. UV detection was performed at 220 nm.

Each sample was diluted with mobile phase and centrifuged. The supernatant liquid was collected and analyzed. The injection volume was 10 μL. Under these conditions, the retention time for nifedipine was about 3.5 minutes.

The stability-indicating nature of the method was demonstrated by accelerated degradation. Nifedipine was treated with 2.0 M hydrochloric acid, 2.0 M sodium hydroxide, or 0.3% hydrogen peroxide or was heated at 80 °C. The degradation products eluted from the column in less than 2.3 minutes.

A standard curve for nifedipine was constructed from 0.4 to 5.0 mg/mL. The correlation coefficient was greater than 0.999. Intraday and interday coefficients of variation were less than 2.3 and 3.1%, respectively.

Reference

Nahata MC, Morosco RS, Willhite EA. Stability of nifedipine in two oral suspensions stored at two temperatures. *J Am Pharm Assoc*. 2002; 42: 865–7.

Method 3

Helin-Tanninen et al. determined nifedipine concentrations in suspension. The HPLC equipment was a Merck/Hitachi model L-6200A pump, a model AS-2000 autosampler, a model L-4500 diode-array detector, and a model D-7000 chromatography data analysis system. The stationary phase was a Beckman Ultrasphere ODS reversed-phase column (250 × 4.6 mm, 5-μm particle size). The mobile phase consisted of a mixture of deionized distilled water and methanol (32:68, vol/vol), adjusted to pH 5.8 with acetic acid. The aqueous solution contained 0.1 M ammonium acetate and 0.1% (vol/vol) triethylamine. The flow rate was 1.0 mL/min. UV detection was performed at 238 nm. Bupivacaine hydrochloride 300 μg/mL in deionized distilled water was used as an internal standard.

Samples were diluted 1:10 with methanol and centrifuged at 5000 rpm for 10 minutes. The supernatant liquid was diluted 1:10 with methanol and mixed appropriately with the internal standard. The injection volume was 50 μL. Under these conditions, the retention times for nifedipine and bupivacaine were 5.95 and 7.52 minutes, respectively.

The stability-indicating capability of the assay was demonstrated by accelerated degradation of nifedipine. Nifedipine solutions were mixed with 1 M hydrochloric acid (pH 0.25) or 1 M sodium hydroxide (pH 13.64) and heated at 60 °C for 2 days. Nifedipine powder and tablets were also heated at 120 °C for 2 hours or were degraded by excessive light. Retention times of the acid hydrolysis degradation products were 2.70, 4.30, 4.55, and 5.05 minutes. Retention times for the base decomposition products were 2.70 and 7.75 minutes. Retention times for the photodegradation products were 4.55 and 5.15 minutes.

A standard curve for nifedipine was constructed from 5 to 15 μg/mL. The correlation coefficient was 0.998.

References

Helin-Tanninen M, Naaranlahti T, Kontra K, et al. Enteral suspension of nifedipine for neonates. Part 1. Formulation of nifedipine suspension for hospital use. *J Clin Pharm Ther.* 2001; 26: 49–57.

Helin-Tanninen M, Naaranlahti T, Kontra K, et al. Enteral suspension of nifedipine for neonates. Part 2. Stability of an extemporaneously compounded nifedipine suspension. *J Clin Pharm Ther.* 2001; 26: 59–66.

Method 4

Liu and Chen reported the HPLC determination of nifedipine. The HPLC system consisted of a Shimadzu model LC-4A pump, a model SPD-2AS UV detector, and a model C-R2AX integrator. The stationary phase was a reversed-phase Zorbax ODS column (150 × 4.6 mm). The mobile phase was a mixture of methanol and water (65:35, vol/vol). The flow rate was 1.1 mL/min. UV detection was performed at 235 nm and 0.08 AUFS. The column temperature was 40 °C. Thymol 5 mg/mL in methanol and water (1:1, vol/vol) was used as an internal standard.

Samples were diluted with water. The injection volume was 10 µL. Under these conditions, retention times for nifedipine and thymol were about 5.9 and 10.1 minutes, respectively.

Nifedipine injection was exposed to fluorescent light for 10 days. The degradation product having a retention time of 6.9 minutes did not interfere with the analysis of nifedipine.

A standard curve for nifedipine was constructed from 11.35 to 23.16 µg/mL. The correlation coefficient was 0.9999.

Reference

Liu C, Chen G. Studies on the determination and stability of nifedipine injection by HPLC. *Yaowu Fenxi Zazhi.* 1993; 13: 314–7.

Method 5

Bammi et al. described an HPLC method for the analysis of nifedipine and its related compounds in soft gelatin capsules. A liquid chromatograph consisted of a Waters model 510 pump, a model WISP 712 autosampler, a model 484 tunable detector, a model 745B data module, and a Rheodyne model 7125 injector. The stationary phase was a TSK-GEL ODS column (250 × 3.9 mm, 5-µm particle size). The mobile phase consisted of methanol and water (55:45, vol/vol). UV detection was performed at 265 nm. Under these conditions, retention times for the nitrophenyl pyridine analog of nifedipine and the nitrosophenyl pyridine analog of nifedipine, propylparaben, and nifedipine were about 8.9, 10.7, 12.7, and 15.3 minutes, respectively (estimated from the published chromatogram).

The method was reported to be stability indicating.

A standard curve for nifedipine was constructed from 0.05 to 0.25 mg/mL.

Reference

Bammi RK, Nayak VG, Bhate VR, et al. Analysis of nifedipine and related compounds in soft gelatin capsules by liquid chromatography. *Drug Dev Ind Pharm.* 1991; 17: 2239–44.

Nifuroxime

Chemical Name
5-Nitro-2-furaldehyde oxime

Other Name
Ginecofuran

Form	Molecular Formula	MW	CAS
Nifuroxime	$C_5H_4N_2O_4$	156.1	6236-05-1

Appearance
Nifuroxime occurs as tasteless pale yellow or greenish crystals.

Solubility
Nifuroxime has solubilities of 1 g in 1 liter of water and 89.0 g in 1 liter of methanol.

Method
Hassan et al. reported an analytical method for nifuroxime. The chromatographic system included an LKB.2150 pump, a 50-μL loop injector, an LKB.2151 variable-wavelength detector, and an LKB.2220 integrator. The stationary phase was a LiChrosorb RP18 column (250 × 4 mm, 5-μm particle size). The mobile phase was a mixture of methanol, water, and Britton Robinson buffer (pH 3) (40:55:5, vol/vol/vol). The flow rate was 1 mL/min. UV detection was performed at 340 nm.

A portion of ground tablets containing about 50 mg of nifuroxime was accurately weighed, dissolved in dimethylformamide, and filled to a volume of 100 mL with dimethylformamide. Under these conditions, the retention time of nifuroxime was about 8.7 minutes.

The method was demonstrated to be stability indicating. Samples with different concentrations of the drug were exposed to sunlight for up to 24 hours. None of the degradation products interfered with the analysis of the intact drug.

A standard curve for nifuroxime was constructed from 3 to 18 μg/mL. The correlation coefficient was 0.9998.

Reference
Hassan SM, Ibrahim FA, El-Din MS, et al. A stability-indicating high-performance liquid chromatographic assay for the determination of some pharmaceutically important nitrocompounds. *Chromatographia*. 1990; 30: 176–80.

Nilvadipine

Chemical Name
2-Cyano-1,4-dihydro-6-methyl-4-(3-nitrophenyl)-3,5-pyridinedicarboxylic acid 3-methyl 5-(1-methylethyl)ester

Other Names
Escor, Nivadil

Form

Form	Molecular Formula	MW	CAS
Nilvadipine	$C_{19}H_{19}N_3O_6$	385.4	75530-68-6

Appearance
Nilvadipine occurs as yellow prisms.

Method
Using HPLC, Abdine et al. studied the stability of nilvadipine in bulk materials and its pharmaceutical capsules. The stationary phase was a μBondapak C_{18} column (250 × 4.6 mm). The mobile phase consisted of 45% of 0.01 M sodium acetate solution (adjusted to pH 3.5) and 55% of acetonitrile. The flow rate was 1.8 mL/min. UV detection was performed at 245 nm for the potency test and 254 nm for the degradation product detection. Procaine hydrochloride 0.4 mg/mL in methanol was used as an internal standard.

A portion of the powder from 10 capsules equivalent to 16.0 mg of nilvadipine was weighed, transferred into a 100-mL volumetric flask, mixed with 40 mL of methanol and water (8:2, vol/vol), shaken for 30 minutes, filled to the mark with the same solvent, and centrifuged for 5 minutes. Aliquots of the supernatant were transferred into a 10-mL volumetric flask, mixed with 2 mL of internal standard solution and diluted to the mark with mobile phase. The injection volume was 20 μL. Under these conditions, the retention times for procaine and nilvadipine were about 3.6 and 5.7 minutes, respectively.

The analytical method was demonstrated to be stability indicating by accelerated degradation of nilvadipine. Solutions of nilvadipine 2 mg/mL in methanol were mixed with 1 M sodium hydroxide solution or 1 M hydrochloric acid and heated at different temperatures (20 to 60 °C) for up to 150 minutes. Degradation products eluted in 1.9 and 2.9 minutes for samples treated with 1 M sodium hydroxide solution.

A standard curve for nilvadipine was constructed from 0.096 to 0.64 μg/mL. The correlation coefficient was 0.9992. The within-day and between-day coefficients of variation of the analysis of nilvadipine were less than 0.74 and 1.47%, respectively.

Reference
Abdine H, Belal F, Gadkariem EA, et al. Stability study of nilvadipine in bulk drug and pharmaceutical capsules by high performance liquid chromatography. *J Liq Chromatogr Rel Technol.* 2001; 24: 3213–25.

Nimodipine

Chemical Name
Isopropyl 2-methoxyethyl 1,4-dihydro-2,6-dimethyl-4-(3-nitrophenyl)pyridine-3,5-dicarboxylate

Other Names
Admon, Nimotop

Form	Molecular Formula	MW	CAS
Nimodipine	$C_{21}H_{26}N_2O_7$	418.4	66085-59-4

Appearance
Nimodipine is a light yellow or yellow crystalline powder.

Solubility
Nimodipine is practically insoluble in water, sparingly soluble in absolute alcohol, and freely soluble in ethyl acetate.

Method
Wang et al. reported the determination of nimodipine and its stability by HPLC. The chromatograph consisted of a Shimadzu model LC-4A system, a model SPD-2AS UV detector, and a model C-R2AX data integrator. The stationary phase was a YWG (Tianjin Chemical Company #2, China) C_{18} stainless steel column (250 × 4 mm). The mobile phase was a mixture of methanol, water, and ethyl ether (70:30:8, vol/vol/vol). The flow rate was 1 mL/min. UV detection was performed at 238 nm and 0.08 AUFS. Beclomethasone dipropionate was used as an internal standard.

Samples were appropriately diluted with the mobile phase. The injection volume was 20 μL. Under these conditions, retention times for nimodipine and the internal standard were about 6 and 10 minutes, respectively.

The assay was stability indicating since the degradation products did not interfere with the determination of nimodipine.

A standard curve was constructed from 0.08 to 0.72 μg/injection of nimodipine The correlation coefficient was 0.9999.

Reference
Wang S-l, Ma W-s, Wang X. Studies on the determination and stability of nimodipine injection by HPLC. *Yaowu Fenxi Zazhi*. 1991; 11: 81–4.

Nimustine Hydrochloride

Chemical Name

3-[(4-Amino-2-methylpyrimidin-5-yl)methyl]-1-(2-chloroethyl)-1-nitrosourea hydrochloride

Other Names

ACNU, Nidran

Form	Molecular Formula	MW	CAS
Nimustine hydrochloride	$C_9H_{13}ClN_6O_2 \cdot HCl$	309.2	55661-38-6

Appearance

Nimustine hydrochloride occurs as a white to light yellow crystalline powder.

Solubility

Nimustine hydrochloride is soluble in methanol and practically insoluble in ether.

Method

Van der Houwen et al. evaluated the stability of nimustine using an HPLC method. The chromatographic apparatus included a Waters model M6000 solvent-delivery system, a model U6K injector, a Spectroflow 773 UV detector, and an SP4270 integrator. The stationary phase was a Merck LiChrosorb RP8 stainless steel column (125 × 4 mm, 5-µm particle size). The mobile phase consisted of 250 mL of methanol and 750 mL of 0.05 M phosphate buffer, adjusted to pH 5.5 with phosphoric acid. The flow rate was 1.0 mL/min. UV detection was performed at 274 nm. The sample injection volume was 10 µL. Under these conditions, the retention time for nimustine was 5.5 minutes.

The stability-indicating nature of the method was evaluated using a photodiode-array detector. The nimustine peak was well separated from its degradation product peaks.

Reference

Van der Houwen OAGJ, Beijnen JH, Bult A, et al. Degradation kinetics of the antitumor drug nimustine (ACNU) in aqueous solution. *Int J Pharm.* 1993; 99: 73–8.

Niridazole

Chemical Name
1-(5-Nitrothiazol-2-yl)imidazolidin-2-one

Other Name
Ambilhar

Form
Niridazole

Molecular Formula
$C_6H_6N_4O_3S$

MW
214.2

CAS
61-57-4

Appearance
Niridazole occurs as yellow crystals.

Method
Hassan et al. presented an analytical method for niridazole. The system included an LKB.2150 pump, a 50-μL loop injector, an LKB.2151 variable-wavelength detector, and an LKB.2220 integrator. The stationary phase was a LiChrosorb RP18 column (250 × 4 mm, 5-μm particle size). The mobile phase was a mixture of methanol, water, and Britton Robinson buffer (pH 3) (40:55:5, vol/vol/vol). The flow rate was 1 mL/min. UV detection was performed at 368 nm.

A portion of ground tablets containing about 50 mg of niridazole was accurately weighed, dissolved in dimethylformamide, and brought to a volume of 100 mL with the same solvent. Under these conditions, the retention time of niridazole was about 10.5 minutes.

The method was demonstrated to be stability indicating. Samples with different concentrations of the drug were exposed to sunlight for up to 7 days. None of the degradation products interfered with the analysis of the intact drug.

A standard curve for niridazole was constructed from 3 to 18 μg/mL. The correlation coefficient was 0.9998.

Reference
Hassan SM, Ibrahim FA, El-Din MS, et al. A stability-indicating high-performance liquid chromatographic assay for the determination of some pharmaceutically important nitrocompounds. *Chromatographia*. 1990; 30: 176–80.

Nitrendipine

Chemical Name

Ethyl methyl 1,4-dihydro-2,6-dimethyl-4-(3-nitrophenyl)pyridine-3,5-dicarboxylate

Other Names

Baypress, Nitrendepat

Form	Molecular Formula	MW	CAS
Nitrendipine	$C_{18}H_{20}N_2O_6$	360.4	39562-70-4

Appearance

Nitrendipine is a yellow crystalline powder.

Solubility

Nitrendipine is practically insoluble in water, sparingly soluble in alcohol, and freely soluble in ethyl acetate.

Method

Tipre and Vavia used HPLC analysis to study the oxidative degradation of nitrendipine. The Jasco system was equipped with a Jasco UV-visible intelligent detector and a Rheodyne injector with a 20-μL loop. The stationary phase was a LiChroCART column (125 × 4 mm, 5-μm particle size). The mobile phase consisted of methanol, water, and acetonitrile (45:45:10, vol/vol/vol). The flow rate was 1.2 mL/min. UV detection was performed at 235 nm. Under these conditions, the retention time of nitrendipine was 5.94 minutes.

The method was verified to be stability indicating by accelerated degradation. Nitrendipine solutions were treated with 0.1 M hydrochloric acid or 0.1 M sodium hydroxide and kept at 100 °C for 72 hours. A nitrendipine solution was also exposed to sunlight for 8 hours. The degradation product peaks did not interfere with the analysis of nitrendipine.

A standard curve for nitrendipine was generated from 5 to 50 μg/mL. The limit of quantitation was 5.0 μg/mL.

Reference

Tipre DN, Vavia PR. Oxidative degradation study of nitrendipine using stability indicating, HPLC, HPTLC and spectrophotometric method. *J Pharm Biomed Anal.* 2001; 24: 705–14.

Nitrofurantoin

Chemical Name
1-(5-Nitrofurfurylideneamino)imidazolidine-2,4-dione

Other Names
Cistofuran, Cystit, Furadantin, Macrobid

Form	Molecular Formula	MW	CAS
Nitrofurantoin	$C_8H_6N_4O_5$	238.2	67-20-9
Nitrofurantoin sodium	$C_8H_5N_4NaO_5$	260.1	54-87-5

Appearance
Nitrofurantoin occurs as yellow odorless or almost odorless crystals or fine crystalline powder.

Solubility
Nitrofurantoin is very slightly soluble in water and in alcohol.

pK$_a$
Nitrofurantoin has a pK$_a$ of 7.2.

Method
Hassan et al. developed an analytical method for nitrofurantoin. The system included an LKB.2150 pump, a 50-µL loop injector, an LKB.2151 variable-wavelength detector, and an LKB.2220 integrator. The stationary phase was a LiChrosorb RP18 column (250 × 4 mm, 5-µm particle size). The mobile phase was a mixture of methanol, water, and Britton Robinson buffer (pH 3) (40:55:5, vol/vol/vol). The flow rate was 1 mL/min. UV detection was performed at 367 nm.

A portion of ground tablets containing about 50 mg of nitrofurantoin was accurately weighed, dissolved in dimethylformamide, and brought to a final volume of 100 mL with dimethylformamide. Under these conditions, the retention time of nitrofurantoin was about 4.6 minutes.

The method was demonstrated to be stability indicating. Samples with different concentrations of the drug were exposed to sunlight for up to 24 hours. None of the degradation products interfered with the analysis of the intact drug.

A standard curve for nitrofurantoin was constructed from 1 to 10 µg/mL. The correlation coefficient was 0.9994.

Reference

Hassan SM, Ibrahim FA, El-Din MS, et al. A stability-indicating high-performance liquid chromatographic assay for the determination of some pharmaceutically important nitrocompounds. *Chromatographia*. 1990; 30: 176–80.

Nitrofurazone

Chemical Name

5-Nitro-2-furaldehyde semicarbazone

Other Names

Amifur, Furacin

Form	Molecular Formula	MW	CAS
Nitrofurazone	$C_6H_6N_4O_4$	198.1	59-87-0

Appearance

Nitrofurazone is a lemon to brownish-yellow odorless crystalline powder.

Solubility

Nitrofurazone is very slightly soluble in water and slightly soluble in alcohol.

Method

Hassan et al. reported an analytical method for the determination of nitrofurazone. The system included an LKB.2150 pump, a 50-μL loop injector, an LKB.2151 variable-wavelength detector, and an LKB.2220 integrator. The stationary phase was a LiChrosorb RP18 column (250 × 4 mm, 5-μm particle size). The mobile phase was a mixture of methanol, water, and Britton Robinson buffer (pH 3) (40:55:5, vol/vol/vol). The flow rate was 1 mL/min. UV detection was performed at 375 nm.

An ointment sample containing about 4 mg of nitrofurazone was accurately weighed, mixed with 25 mL of water, dissolved, and brought to a volume of 100 mL with dimethylformamide. Under these conditions, the retention time of nitrofurazone was about 11.7 minutes.

The method was demonstrated to be stability indicating. Samples with different concentrations of the drug were exposed to sunlight for up to 24 hours. None of the degradation products interfered with the analysis of the intact drug.

A standard curve for nitrofurazone was constructed from 1 to 10 μg/mL. The correlation coefficient was 0.9992.

Reference

Hassan SM, Ibrahim FA, El-Din MS, et al. A stability-indicating high-performance liquid chromatographic assay for the determination of some pharmaceutically important nitrocompounds. *Chromatographia*. 1990; 30: 176–80.

Nitroglycerin

Chemical Names

1,2,3-Propanetriol trinitrate
Glyceryl trinitrate

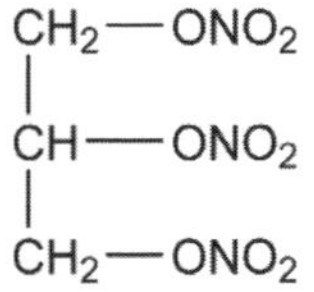

$$CH_2—ONO_2$$
$$|$$
$$CH—ONO_2$$
$$|$$
$$CH_2—ONO_2$$

Other Names

NitroBid, Nitrogard, Nitroglyn, Nitrospan

Form	Molecular Formula	MW	CAS
Nitroglycerin	$C_3H_5N_3O_9$	227.1	55-63-0

Appearance

Undiluted nitroglycerin is a volatile white to pale yellow thick and flammable explosive liquid with a sweet, burning taste. Diluted nitroglycerin is a white odorless powder when diluted with lactose and a clear colorless or pale yellow liquid when diluted with alcohol or propylene glycol.

Solubility

Undiluted nitroglycerin is slightly soluble in water and soluble in alcohol.

Method 1

Peddicord et al. studied the stability of nitroglycerin in 5% dextrose injection at ambient temperature and humidity. The liquid chromatograph included a Shimadzu model LC-10AS solvent-delivery system, a Shimadzu model SIL-10A autoinjector, a Shimadzu model CTO-10A column oven, a Shimadzu model SPD-10A variable-wavelength UV spectro-photometer, and a Phenomenex C_{18} column (300 × 3.9 mm, 5-μm particle size). The mobile phase consisted of water, methanol, and acetonitrile (66:12:24, vol/vol/vol). The flow rate was 2.8 mL/min. The UV detection wavelength was 210 nm. The oven temperature was 35 °C.

Each sample was diluted with 5% dextrose injection and the injection volume was 25 μL.

The method was determined to be stability indicating by accelerated decomposition of nitroglycerin. Solutions of nitroglycerin were adjusted to pH 2 with 1.0 N hydrochloric acid and to pH 12 with 10 N sodium hydroxide and were raised to 100 °C. Degradation product peaks did not interfere with the intact nitroglycerin peak.

A calibration curve for nitroglycerin was constructed from 0 to 200.0 µg/mL. The correlation coefficient was greater than 0.999. The intraday and interday coefficients of variation were 2.1 and 8.7%, respectively.

A similar method was used by Driver et al.

References

Peddicord TE, Olsen KM, ZumBrunnen TL, et al. Stability of high-concentration dopamine hydrochloride, norepinephrine bitartrate, epinephrine hydrochloride, and nitroglycerin in 5% dextrose injection. *Am J Health Syst Pharm.* 1997; 54: 1417–9.

Driver PS, Jarvi EJ, Gratzer PL. Stability of nitroglycerin as nitroglycerin concentrate for injection stored in plastic syringes. *Am J Hosp Pharm.* 1993; 50: 2561–3.

Method 2

Tracy et al. used an HPLC method to evaluate the adsorption and delivery of nitroglycerin through an intravenous administration set. The liquid chromatograph consisted of a Rabbit-HP isocratic solvent-delivery system, a Rheodyne injector, a Knauer variable-wavelength monitor, and a Hewlett-Packard integrator. The stationary phase was a Rainin Microsorb C_{18} reversed-phase column. The mobile phase consisted of methanol and distilled water (73:27, vol/vol). The flow rate was 0.9 mL/min. UV detection was performed at 244 nm. The injection volume was 20 µL. The retention time for nitroglycerin was 5.3 minutes.

The assay was stated to be stability indicating.

A standard curve was constructed from 10 to 200 µg/mL; the correlation coefficient was 0.9962. The coefficient of variation was 2.52%.

Reference

Tracy TS, Bowman L, Black CD. Nitroglycerin delivery through a polyethylene-lined intravenous administration set. *Am J Hosp Pharm.* 1989; 46: 2031–5.

Method 3

Wagenknecht et al. determined the stability of nitroglycerin solutions in polyolefin and glass containers. A Perkin-Elmer system 2/2 chromatograph was equipped with a Perkin-Elmer model 420 autosampler, a Perkin-Elmer model LC-55A UV photometer, a Scientific Products model 284/mm recorder, and a Waters µBondapak phenyl column (300 × 3.9 mm). The mobile phase was a mixture of 64% water, 10% tetrahydrofuran, and 26% acetonitrile and was delivered isocratically at 2.0 mL/min. UV detection was performed at 218 nm and 0.1 AUFS. Isosorbide dinitrate was used as an internal standard.

Samples were diluted with water. The injection volume was 100 µL. Retention times for isosorbide dinitrate and nitroglycerin were about 5.9 and 10.0 minutes, respectively (estimated from the published chromatogram).

The method was stated to be stability indicating.

Reference

Wagenknecht DM, Baaske DM, Alam AS, et al. Stability of nitroglycerin solutions in polyolefin and glass containers. *Am J Hosp Pharm.* 1984; 41: 1807–11.

Nizatidine

Chemical Name

N-[2-[[[2-[(Dimethylamino)methyl]-4-thiazolyl]methyl]thio]ethyl]-*N'*-methyl-2-nitro-1,1-ethenediamine

Other Name

Axid

Form	Molecular Formula	MW	CAS
Nizatidine	$C_{12}H_{21}N_5O_2S_2$	331.5	76963-41-2

Appearance

Nizatidine occurs as an off-white to buff crystalline powder.

Solubility

Nizatidine is freely soluble in chloroform, soluble in methanol, and sparingly soluble in water and buffered solution. It is slightly soluble in ethyl acetate and isopropanol.

pK_a

Nizatidine has pK_a values of 2.1 and 6.8.

Method 1

Hatton et al. studied the stability of nizatidine in total nutrient admixtures. The chromatographic system was composed of a Waters model M45 solvent-delivery system, a Rheodyne model 7010 injector, a Waters model 484 multiple-wavelength UV detector, and a Waters Nova-Pak C_{18} column (300 × 3.9 mm). The mobile phase consisted of 20% acetonitrile and 80% 0.02 M potassium dihydrogen phosphate with the pH adjusted to 6.5. The flow rate was 0.8 mL/min. UV detection was performed at 228 nm. Ranitidine hydrochloride 2 mg/mL was used as the internal standard.

Samples were diluted with HPLC-grade water and extracted in a chloroform and methanol (1:1, vol/vol) mixture. The injection volume was 20 µL. The retention time for nizatidine was 7.7 minutes.

The assay was determined to be stability indicating by an accelerated degradation study. Solutions of nizatidine in water, 1 N sodium hydroxide, and 1 N hydrochloric acid were incubated at 85 °C for 10 hours. Another solution of nizatidine in 1% hydrogen peroxide was exposed to UV radiation for 4 hours. In all cases, degradation product peaks did not interfere with the parent nizatidine peak. Retention times for degradation products were about 4.8, 6.0, and 6.9 minutes.

A calibration curve was generated from 3.00 to 250 µg/mL. The interassay coefficient of variation was 4.5%.

Reference

Hatton J, Holstad SG, Rosenbloom AD, et al. Stability of nizatidine in total nutrient admixtures. *Am J Hosp Pharm.* 1991; 48: 1507–10.

Method 2

Lantz and Wozniak evaluated the stability of nizatidine in extemporaneous oral liquid preparations. The chromatographic system consisted of a Hewlett-Packard model 1050 pump, an Applied Biosystems model 757 variable-wavelength detector, a Perkin-Elmer model ISS-100 refrigerated autosampler, and a Jones Chromatography Apex ODS reversed-phase C_{18} column (150 × 4.6 mm, 5-μm particle size). The mobile phase consisted of 75% 0.1 M ammonium acetate buffer (pH adjusted to 7.5 with 0.1% diethylamine) and 25% methanol. The flow rate was 1.0 mL/min. UV detection was performed at 230 nm and 0.1 AUFS.

Samples were filtered through Whatman No. 1 filter paper and diluted 1:25 with the mobile phase. The retention time for nizatidine was about 11.7 minutes (estimated from the published chromatogram).

The assay was determined to be stability indicating by accelerated decomposition of nizatidine. Samples of nizatidine were exposed to basic, acidic, oxidative, thermal, light, and humidity conditions. Degradation product peaks did not interfere with the nizatidine peak. Retention times for the degradation products, nizatidine sulfoxide and nizatidine amide, were about 3.5 and 7.2 minutes, respectively (estimated from the published chromatogram).

Calibration curves were constructed from 70 to 130 μg/mL. The intraday coefficient of variation was 0.82%.

Reference

Lantz MD, Wozniak TJ. Stability of nizatidine in extemporaneous oral liquid preparations. *Am J Hosp Pharm.* 1990; 47: 2716–9.

Norepinephrine

Chemical Names

4-(2-Amino-1-hydroxyethyl)-1,2-benzenediol
α-(Aminomethyl)-3,4-dihydroxybenzyl alcohol

Other Names

Levarterenol, Levophed, Noradrenaline

Form	Molecular Formula	MW	CAS
Norepinephrine	$C_8H_{11}NO_3$	169.2	51-41-2
Norepinephrine bitartrate	$C_8H_{11}NO_3.C_4H_6O_6.H_2O$	337.3	69815-49-2
Norepinephrine hydrochloride	$C_8H_{11}NO_3.HCl$	205.6	329-56-6

Appearance

Norepinephrine bitartrate is an odorless white or faintly gray crystalline powder. Norepinephrine hydrochloride is a white or brownish-white crystalline powder.

Solubility

Norepinephrine bitartrate has solubilities of approximately 400 mg/mL in water and 3.33 mg/mL in alcohol at 25 °C. Norepinephrine hydrochloride is very soluble in water and slightly soluble in alcohol.

Method

Peddicord et al. evaluated the stability of norepinephrine bitartrate in 5% dextrose injection at ambient temperature and humidity. The liquid chromatograph consisted of a Shimadzu model LC-10AS solvent-delivery system, a Shimadzu model SIL-10A auto-injector, a Shimadzu model CTO-10A column oven, a Shimadzu model SPD-10A variable-wavelength UV spectrophotometer, and a Phenomenex C_{18} column (300 × 3.9 mm, 5-μm particle size). The mobile phase consisted of water, methanol, and acetic acid (72.3:25:2.7, vol/vol/vol) and 1.34 g of 1-heptanesulfonic acid. The flow rate was 1.6 mL/min. UV detection wavelength was 280 nm. The oven temperature was 50 °C.

Each sample was diluted with 5% dextrose injection. The injection volume was 50 μL.

The method was determined to be stability indicating by accelerated decomposition of norepinephrine. Solutions of norepinephrine were adjusted to pH 2 with 1.0 N hydrochloric acid and to pH 12 with 10 N sodium hydroxide and were raised to 100 °C. Degradation product peaks did not interfere with the intact norepinephrine peak.

A calibration curve for norepinephrine was constructed from 0 to 100.0 μg/mL. The correlation coefficient was greater than 0.999. The intraday and interday coefficients of variation were all 0.8%.

Reference

Peddicord TE, Olsen KM, ZumBrunnen TL, et al. Stability of high-concentration dopamine hydrochloride, norepinephrine bitartrate, epinephrine hydrochloride, and nitroglycerin in 5% dextrose injection. *Am J Health Syst Pharm.* 1997; 54: 1417–9.

Norethindrone

Chemical Name

17β-Hydroxy-19-nor-17β-pregn-4-en-20-yn-3-one

Other Names

Aygestin, Menzol, Micronor, Norethisterone, Noristerat

Form	Molecular Formula	MW	CAS
Norethindrone	$C_{20}H_{26}O_2$	298.4	68-22-4
Norethindrone acetate	$C_{22}H_{28}O_3$	340.5	51-98-9
Norethindrone enanthate	$C_{27}H_{38}O_3$	410.6	3836-23-5

Appearance

Norethindrone is a white to creamy white odorless crystalline powder. Norethindrone acetate also is a white to creamy white odorless crystalline powder.

Solubility

Norethindrone is practically insoluble in water and sparingly soluble in alcohol. Norethindrone acetate is practically insoluble in water and soluble in alcohol.

Method

Mayberry et al. reported an HPLC method for the determination of norethindrone on Red Delicious apples. The apparatus consisted of a Waters model M6000 pump, a model 440 fixed-wavelength detector, and a Rheodyne model 7120 injector with a 10-µL loop. The stationary phase was a Waters µBondapak C_{18} reversed-phase column (300 × 3.9 mm). The mobile phase consisted of water, tetrahydrofuran, and methanol (63:26:11, vol/vol/vol). The flow rate was 1.7 mL/min. UV detection was performed at 254 nm.

Apple slices were cut and homogenized in 200 mL of a buffer solution that contained 30.9 g of boric acid, 37.3 g of potassium chloride, and 462 mL of 0.05 M sodium hydroxide in distilled water per liter (adjusted to pH 10 with acid or base). This mixture was transferred into one of the compartments of a diffusion chamber with 350 mL of chloroform in the other compartment, stirred at 500 rpm for 18 hours, filtered through a Millipore filter, evaporated to dryness under a stream of nitrogen at 40–50 °C, and reconstituted in tetrahydrofuran. The injection volume was 10 µL. Under these conditions, the retention time of norethindrone was 11.6 minutes.

The method was demonstrated to be stability indicating. Norethindrone was degraded in an oxygen atmosphere (82 psi) at 170 °C for 4.5 hours. Degradation products were separated and characterized on the chromatogram.

A standard curve for norethindrone was obtained from 15 to 350 µg/mL. The correlation coefficient was 0.9999.

Reference

Mayberry DO, Kowblansky M, Lane PA, et al. Determination of norethindrone stability on Red Delicious apples. *J Pharm Sci.* 1990; 79: 746–9.

Norfloxacin

Chemical Name
1-Ethyl-6-fluoro-1,4-dihydro-4-oxo-7-(piperazin-1-yl)quinoline-3-carboxylic acid

Other Names
Chibroxin, Chibroxol, Lexinor, Noroxin, Zoroxin

Form	Molecular Formula	MW	CAS
Norfloxacin	$C_{16}H_{18}FN_3O_3$	319.3	70458-96-7

Appearance
Norfloxacin is a white to pale yellow hygroscopic crystalline powder.

Solubility
Norfloxacin has solubilities of approximately 0.28 mg/mL in water and 1.9 mg/mL in alcohol.

pK$_a$
Norfloxacin has pK$_a$ values of 6.34 and 8.75.

Method 1
Johnson et al. used an HPLC method to study the stability of norfloxacin in an extemporaneously prepared oral liquid at room temperature and under refrigeration over 56 days. The instrument included a Waters model 501 pump, a model U6K variable-volume injector, a model 486 tunable absorbance detector, and a Hewlett-Packard model 3394 integrator-recorder. The stationary phase was a Spherisorb phenyl column (150 × 4.6 mm, 5-µm particle size). The mobile phase was a mixture of 0.094 M monobasic potassium phosphate buffer, acetonitrile, methanol, and trifluoroacetic acid (80:15:5:0.3, vol/vol/vol/vol). The flow rate was 1.0 mL/min. UV detection was performed at 278 nm.

Norfloxacin suspension (250 µL) was diluted to a final volume of 10 mL with 0.1 N hydrochloric acid and mixed well. This solution was further diluted to an expected concentration of 5 µg/mL with sample diluent (water, acetonitrile, and methanol; 80:15:5, vol/vol/vol). The injection volume was 5 µL. Under these conditions, the retention time for norfloxacin was 3.3 minutes.

The stability-indicating capability of the method was verified by subjecting a sample of norfloxacin suspension to extreme conditions of heat, light, and pH. Two separate 20-mg/mL samples were adjusted either to pH 12 with 1 N sodium hydroxide solution or to pH 2 with 1 N hydrochloric acid and exposed to direct sunlight for 56 days. The solutions were then heated to 60 ºC for 2 hours. No interfering peaks of degradation products were observed in either chromatogram.

A calibration curve for norfloxacin was constructed from 4.0 to 6.0 µg/mL. The correlation coefficient was greater than 0.999. The intraday and interday coefficients of variation for the norfloxacin analysis were 1.2 and 2.9%, respectively.

Reference
Johnson CE, Price J, Hession JM. Stability of norfloxacin in an extemporaneously prepared oral liquid. *Am J Health Syst Pharm*. 2001; 58: 577–9.

Method 2
Al-Deeb et al. presented an HPLC method for the determination of norfloxacin in bulk form and in tablets. The Shimadzu LC-10 AD liquid chromatograph included a model SPD-10A tunable UV detector, a model CTO-10A column oven controller, a model DGU-3A mechanical degasser, a model C-R4A data unit, and a Rheodyne 20-µL loop injector. The stationary phase was a Varian Micropak NH2-10 column (300 × 4 mm, 10-µm particle size). The mobile phase consisted of acetonitrile, tetrabutylammonium hydroxide, phosphoric acid, and water (10:1.5:0.167:100, vol/wt/wt/vol). The flow rate was 2 mL/min. UV detection was performed at 278 nm.

A sample of the ground tablets equivalent to 25 mg of norfloxacin was mixed with 75 mL of mobile phase in a 100-mL calibrated flask, shaken for 7 minutes, filled to volume with mobile phase, and filtered. The filtrate was appropriately diluted. Under these conditions, the retention time for norfloxacin was 3.75 minutes.

The stability-indicating nature of the method was demonstrated by accelerated drug photolysis. An aqueous solution of norfloxacin was irradiated by a 60-W UV lamp (254 nm). The intact drug was resolved from its degradation products.

Reference
Al-Deeb OA, Abdel-Moety EM, Abounassif MA, et al. Stability-indicating high performance liquid chromatographic method for determination of norfloxacin in bulk form and tablets. *Boll Chim Farmaceutico-Anno*. 1995; 134: 497–502.

Method 3
Chen and Wu reported the determination and preliminary stability of norfloxacin in bulk powder and dosage forms by HPLC analysis. A Hewlett-Packard 1050 chromatograph included a model HP 1050 photodiode-array detector and a model HP 3396A integrator. The stationary phase was a Hypersil ODS column (100 × 4.6 mm). The mobile phase was a mixture of methanol, water, and diethylamine (50:50:0.4, vol/vol/vol) (pH 5.5). The flow rate was 1.0 mL/min. UV detection was performed at 278 nm. Chloronitrodiazepine 20 µg/mL in methanol was used as an internal standard.

Samples were dissolved or diluted in the mobile phase. Under these conditions, retention times for norfloxacin and chloronitrodiazepine were 3.3 and 6.9 minutes, respectively.

The method was stability indicating since the degradation products did not interfere with the determination of norfloxacin.

A standard curve for norfloxacin was generated from 5 to 63 µg/mL. The correlation coefficient was 0.9999.

References
Chen C, Wu R. Determination and preliminary stability study of norfloxacin in bulk and dosage forms by HPLC. *J China Pharm Univ*. 1992; 23: 370–2.

Chen C, Liu X, Wu R. High-performance liquid chromatographic method for the determination of norfloxacin glutamate and glucuronate in solid and liquid dosage forms and its application to stability testing. *J Pharm Biomed Anal.* 1993; 11: 717–21.

Norgestimate

Chemical Name
13β-Ethyl-3-hydroxyimino-18,19-dinor-17α-pregn-4-en-20-yn-17β-yl acetate

Other Names
Cilest, Ortho-Cyclen

Form	Molecular Formula	MW	CAS
Norgestimate	$C_{23}H_{31}NO_3$	369.5	35189-28-7

Appearance
Norgestimate occurs as a white to pale yellow powder.

Solubility
Norgestimate is insoluble in water and sparingly soluble in acetonitrile, but it is freely to very soluble in dichloromethane.

Method
Lane et al. described an HPLC method for the simultaneous determination of norgestimate and ethinyl estradiol in oral contraceptive tablets. The liquid chromatograph included an LDC/Milton Roy model IIIG ConstaMetric pump, a model 1204A variable-wavelength detector, a Waters WISP autosampler, and a Hewlett-Packard 3357 laboratory automation system. The stationary phase was an IBM reversed-phase C_{18} column (50 × 4.5 mm, 5-μm particle size). The mobile phase was a mixture of water, tetrahydrofuran, and methanol (65:25:10, vol/vol/vol). The flow rate was 2.1 mL/min. UV detection was performed at 230 nm. Dibutylphthalate 0.05 mg/mL in methanol was used as an internal standard.

Five tablets were weighed and placed into a 50-mL centrifuge tube along with two glass beads; 25 mL of internal standard was then added and mixed. The mixture was vortexed for at least 15 minutes, sonicated for 5 minutes, and filtered through a 2-μm filter. The injection volume was 25 μL. Under these conditions, retention times for ethinyl estradiol, *syn*-norgestimate, *anti*-norgestimate, and internal standard were about 3.6, 7.6, 9.3, and 13.3 minutes, respectively (estimated from the published chromatogram).

The method was demonstrated to be stability indicating by stress degradation studies. Bulk norgestimate was heated at 50 °C for 90 days or exposed to 1000 foot-candles of light for 30 days. Norgestimate solutions were adjusted to pH 4.9, 7.0, and 10.0 with 0.1 M hydrochloric acid or 0.1 M sodium hydroxide and heated at 60 °C for 20 days. There was no interference of degradation products with the intact drug.

A standard curve for norgestimate was obtained from 0.0180 to 0.0755 mg/mL. The correlation coefficient was 0.9999.

Reference
Lane PA, Mayberry DO, Young RW. Determination of norgestimate and ethinyl estradiol in tablets by high-performance liquid chromatography. *J Pharm Sci.* 1987; 76: 44–7.

Norgestrel

Chemical Name
(17α)-(±)-13-Ethyl-17-hydroxy-18,19-dinorpregn-4-en-20-yn-3-one

and enantiomer

Other Name
Ovrette

Form	Molecular Formula	MW	CAS
Norgestrel	$C_{21}H_{28}O_2$	312.4	6533-00-2

Appearance
Norgestrel is a white or almost white and almost odorless crystalline powder.

Solubility
Norgestrel is practically insoluble in water and sparingly soluble in alcohol and methylene chloride. It is freely soluble in chloroform.

Method
Reif et al. developed a stability-indicating HPLC method for levonorgestrel in oral contraceptive tablets. The liquid chromatograph included a DuPont model 870 pump, two detectors connected in series (a Laboratory Data Control SpectroMonitor III UV detector and a Schoeffel FS 970 fluorescence detector set with excitation at 210 nm and no filter in the emission path [photomultiplier window: 80% transmission at 310 nm]),

and a Rheodyne model 7125 fixed-loop septumless injector with a 100-μL loop. The stationary phase was a DuPont Zorbax C_8 reversed-phase column (150 × 4.6 mm, 7-μm particle size) or a Hypersil ODS column (100 × 4.6 mm, 3–5-μm particle size). The precolumn was an Upchurch Scientific C-135 Uptight precolumn (40 × 4.6 mm), dry packed with silica gel 60 (230–400 mesh) from EM Reagents. The mobile phase consisted of acetonitrile, methanol, and water (35:15:45, vol/vol/vol). The flow rate was 1 mL/min. UV detection was performed at 215 nm. The sample injection volume was 100 μL. The retention time for norgestrel was 4.1 minutes.

The HPLC method was demonstrated to be stability indicating by the separation of potential impurities and degradation products from the intact levonorgestrel. Retention times for degradation products were 1.8, 2.0, 5.5, 5.8, 9.7, and 10.3 minutes.

A standard curve was generated from 0.01 to 0.3 mg/mL.

Reference

Reif VD, Eickhoff WM, Jackman JK, et al. Automated stability-indicating high-performance liquid chromatographic assay for ethinyl estradiol and (Levo)norgestrel tablets. *Pharm Res.* 1987; 4: 54–8.

Nortriptyline Hydrochloride

Chemical Name

3-(10,11-Dihydro-5*H*-dibenzo[*a,d*]cyclohepten-5-ylidene)-*N*-methyl-1-propanamine hydrochloride

Other Names

Acetexa, Allegron, Aventyl Hydrochloride, Nortrilen, Pamelor, Vividyl

Form	Molecular Formula	MW	CAS
Nortriptyline	$C_{19}H_{21}N$	263.4	72-69-5
Nortriptyline hydrochloride	$C_{19}H_{21}N \cdot HCl$	299.8	894-71-3

Appearance

Nortriptyline hydrochloride occurs as a white to off-white powder with a slight characteristic odor.

Solubility

Nortriptyline hydrochloride is soluble in water and in alcohol.

Method

El-Ragehy et al. developed a stability-indicating HPLC method for the analysis of nortriptyline hydrochloride in commercial tablets and in the presence of its degradates. The Shimadzu liquid chromatograph consisted of a model LC-10AD pump, a model SPD-10A UV-visible detector, and a model C-R6A Chromatopac recorder. The stationary phase was a Nova-Pak C_{18} column (150 × 3.9 mm, 4-μm particle size). The mobile phase consisted of 0.05 M ammonium acetate aqueous solution, methanol, and acetonitrile (4:1:5, vol/vol/vol), adjusted to pH 7.5. The flow rate was 1 mL/min. UV detection was performed at 254 nm. The injection volume was 20 μL. Under these conditions, the retention time of nortriptyline was about 4.8 minutes.

The method was shown to be stability indicating by intentional degradation of the drug. Nortriptyline solution was treated with 30% hydrogen peroxide and heated. The retention time of its degradation product was about 3.9 minutes. The intact drug peak and its degradation product peak were well resolved on the chromatogram.

A standard curve for nortriptyline was constructed from 0.60 to 3.60 μg/mL. The correlation coefficient was greater than 0.9999. The limit of detection and the limit of quantitation were 0.20 and 0.54 μg/mL, respectively.

Reference

El-Ragehy NA, Abbas SS, El-Khateeb SZ. Spectrophotometric and stability indicating high performance liquid chromatographic determination of nortriptyline hydrochloride and fluphenazine hydrochloride. *Anal Lett.* 2002; 35: 1171–91.

Novobiocin

Chemical Name

4-Hydroxy-3-[4-hydroxy-3-(3-methylbut-2-enyl)benzamido]-8-methyl-coumarin-7-yl 3-*O*-carbamoyl-5,5-di-*C*-methyl-α-L-lyxofuranoside

Other Name

Albamycin

Form	Molecular Formula	MW	CAS
Novobiocin	$C_{31}H_{36}N_2O_{11}$	612.6	303-81-1
Novobiocin calcium	$(C_{31}H_{35}N_2O_{11})_2Ca$	1263.3	4309-70-0
Novobiocin sodium	$C_{31}H_{35}N_2NaO_{11}$	634.6	1476-53-5

Appearance

Novobiocin sodium is a white or yellowish-white odorless hygroscopic crystalline powder.

Solubility

Novobiocin sodium is freely soluble in water and in alcohol.

Method

Tsuji et al. reported a normal phase HPLC method for the separation and quantification of novobiocin, its isomers, and its degradation compounds. The instrument consisted of an LDC minipump, a Rheodyne M7010 injector with a 20-μL loop or a Waters 710B WISP autosampler, and an LDC SpectroMonitor I variable-wavelength detector or an LDC UV III fixed-wavelength detector. The stationary phase was a Brownlee LiChrosorb SI-100 column (250 × 4.6 mm, 5-μm particle size). The mobile phase consisted of butyl chloride (50% water saturated), tetrahydrofuran, methanol, and acetic acid (88:5:4:3, vol/vol/vol/vol). The flow rate was 1 mL/min. UV detection was performed at 254 nm and 0.032 AUFS. Prednisone was used as an internal standard.

A sample of 12–14 mg of novobiocin sodium or calcium was weighed into a 100-mL volumetric flask, mixed with 10 mL of tetrahydrofuran, and diluted to volume with mobile phase. Under these conditions, the retention time of novobiocin was about 18.7 minutes (estimated from the published chromatogram).

The stability-indicating nature of the method was demonstrated by assaying a synthetic mixture of novobiocin, its isomers, and its degradation products. Degradation products did not influence the analysis of novobiocin. Retention times for novobiocin acid and isonovobiocin were about 8.2 and 17.0 minutes, respectively (estimated from the published chromatogram).

A standard curve for novobiocin was generated from 0.2 to 6 mg/mL. The correlation coefficient was 0.9999.

References

Tsuji K, Rahn PD, Kane MP. High-performance liquid chromatographic method for the determination of novobiocin. *J Chromatogr*. 1982; 235: 205–14.

Tsuji K, Kane MP, Rahn PD, et al. ^{60}Co irradiation for sterilization of veterinary mastitis products containing antibiotics and steroids. *Radiat Phys Chem*. 1981; 18: 583–93.

Noxythiolin

Chemical Name

1-Hydroxymethyl-3-methyl-2-thiourea

Other Names

Noxyflex S, Noxytiolin

Form	Molecular Formula	MW	CAS
Noxythiolin	$C_3H_8N_2OS$	120.2	15599-39-0

Appearance
Noxythiolin occurs as crystals.

Solubility
Noxythiolin is soluble in water (10% wt/vol) and in ethanol (4% wt/vol).

Method
Irwin et al. described an HPLC assay of noxythiolin. The liquid chromatograph consisted of an Altex model 100A dual-piston pump, a Rheodyne model 7120 injector with a 20-μL loop, and a Pye LC3 variable-wavelength UV detector. The stationary phase was a Shandon Hypersil ODS column (100 × 4.6 mm, 5-μm particle size). The column temperature was held at 33 °C. The mobile phase was double-distilled water. The flow rate was 1 mL/min. UV detection was performed at 250 nm. Thiourea 0.01% was used as an internal standard.

A sample was mixed with internal standard and diluted to ~ 0.0025% noxythiolin with water. The injection volume was 20 μL. Under these conditions, retention times for thiourea and noxythiolin were about 1.5 and 2.8 minutes, respectively (estimated from the published chromatogram).

The assay was reported to be stability indicating.

A standard curve for noxythiolin was obtained from 0 to 0.0025%. The correlation coefficient was 0.998.

Reference
Irwin WJ, Po ALW, Stephens JS. Noxythiolin—high-performance liquid chromatographic assay and stability. *J Clin & Hosp Pharm.* 1984; 9: 41–51.

NPC 1161C

Chemical Name
±8-[(4-Amino-1-methylbutyl)amino-5-(3,4-dichlorophenoxy)-6-methoxy-4-methylquinoline succinate]

Form	Molecular Formula	MW	CAS
NPC 1161C	$C_{26}H_{31}Cl_2N_3O_6$	552.44	—

Solubility
NPC 1161C is poorly soluble in water.

Method
Dutta et al. developed a simple reversed-phase HPLC method for the simultaneous determination of NPC 1161C, its impurities, and its degradation products. Two Waters HPLC systems were used: a Waters 2695 HPLC separation module equipped with a Waters 2996 photodiode-array detector, and a Waters 991 HPLC system equipped with Waters 510 pumps and a Waters 991 photodiode-array detector. The stationary phase was a Phenomenex Luna reversed-phase column (150 × 4.6 mm, 5-μm particle size). The mobile phase was a mixture of methanol, water, and trifluoroacetic acid in the ratio of

(71:29:0.065, vol/vol/vol) and was delivered isocratically at a flow rate of 1.2 mL/min. NPC 1161C and its impurities were monitored at 254 nm. Forced degradation products were detected from 210 to 400 nm.

A stock solution of NPC 1161C was prepared by dissolving the drug in a methanol and water (71:29, vol/vol) mixture. The injection volume was 50 μL. The run time was 15 minutes. Under these conditions, the retention time of NPC 1161C was 7.9 minutes and those of its two impurities were about 2.8 and 4.6 minutes.

In order to verify the stability-indicating nature of the analytical method, NPC 1161C was subjected to various forced degradation conditions. It was refluxed with 10 mL of 1 M hydrochloric acid for 7 hours, with 10 mL of 0.1 M sodium hydroxide solution for 24 hours, and with 10 mL of 10% (wt/vol) sodium bisulfite solution for 4 hours. It was also subjected to dry heat at 80 °C for 24 hours and refluxed with 10 mL of 3% (vol/vol) hydrogen peroxide solution for up to 4 hours. The peak of NPC 1161C was well separated from peaks of its degradation products and its impurities.

A standard curve was constructed for the drug from 0.3125 to 250 μg/mL with a correlation coefficient of 0.9999. Intraday and interday coefficients of variation were 4.9%. The limit of detection and the limit of quantitation were 0.035 and 0.3125 μg/mL, respectively.

Reference

Dutta AK, Avery BA, Wyandt CM. Development and validation of a stability-indicating reversed-phase high performance liquid chromatography method for NPC 1161C, a novel 8-aminoquinoline anti-malarial drug. *J Chromatogr A*. 2006; 1110: 35–45.

Nylidrin

Chemical Names

4-Hydroxy-α-[1-[(1-methyl-3-phenylpropyl)amino]ethyl]benzenemethanol

p-Hydroxy-α-[1-[(1-methyl-3-phenylpropyl)amino]ethyl]benzyl alcohol

Other Name

Buphenine

Form	Molecular Formula	MW	CAS
Nylidrin	$C_{19}H_{25}NO_2$	299.4	447-41-6
Nylidrin hydrochloride	$C_{19}H_{25}NO_2 \cdot HCl$	335.9	849-55-8

Appearance

Nylidrin hydrochloride is an odorless white crystalline powder.

Solubility
Nylidrin hydrochloride is sparingly soluble in water and slightly soluble in alcohol. It is practically insoluble in ether, chloroform, and benzene.

Method
Volpe et al. developed a stability-indicating HPLC method for the quantitative determination of nylidrin hydrochloride. The HPLC system consisted of a Perkin-Elmer model 601 pump, a Rheodyne model 7120 microliter loop injector, a Perkin-Elmer model LC-55 variable-wavelength spectrophotometric detector, and a Hewlett-Packard model 3385A electronic integrator. The stationary phase was a Waters µBondapak C_{18} column (300 × 3.9 mm). The mobile phase was a mixture of methanol and an aqueous buffer (75:25, vol/vol). The buffer contained 1.32 g of dibasic ammonium phosphate in 1000 mL of distilled water, adjusted to pH 7.5 with 85% phosphoric acid. The flow rate was 1.5 mL/min. UV detection was performed at 276 nm. The internal standard was fluorene at 0.5 mg/mL in the mobile phase. Retention times for nylidrin and the internal standard were 5.2 and 7.0 minutes, respectively.

The method was shown to be stability indicating by an accelerated degradation study. Nylidrin hydrochloride underwent base hydrolysis, acid hydrolysis, and photo-thermal oxidation. Decomposition product peaks did not interfere with the parent nylidrin hydrochloride peak.

Reference
Volpe F, Zintel J, Spiegel D. High-performance liquid chromatography of two peripheral vasodilators, nylidrin hydrochloride and isoxsuprine hydrochloride, in pharmaceutical dosage forms. *J Pharm Sci.* 1979; 68: 1264–7.

Octreotide Acetate

Chemical Name
D-Phenylalanyl-L-cysteinyl-L-phenylalanyl-D-tryptophyl-L-lysyl-L-threonyl-*N*-[2-hydroxy-1-(hydroxymethyl)propyl]-L-cysteinamide cyclic (2→7)-disulfide acetate

Other Names
Longastatina, Samilstin, Sandostatin

Form	Molecular Formula	MW	CAS
Octreotide acetate	$C_{49}H_{66}N_{10}O_{10}S_2 \cdot xC_2H_4O$	1019.2	79517-01-4

Method 1
Ripley et al. determined the stability of octreotide acetate in polypropylene syringes. The HPLC system consisted of a Shimadzu SIL9A autosampler, a Waters model 484 multiple-wavelength UV detector, and a Shimadzu CR501 Chromatopac integrator. The stationary phase was a Brownlee/Applied Biosystems Spheri-5 C_{18} reversed-phase column (250 × 4.6 mm). The mobile phase consisted of acetonitrile, water, and 1.0 M tetramethylammonium hydroxide pentahydrate (33:65:2, vol/vol/vol) adjusted to pH 4.5

with phosphoric acid. The flow rate was 0.8 mL/min. UV detection was performed at 280 nm. The injection volume was 30 μL.

The analytical method was stated to be stability indicating because it could separate octreotide and its known primary degradation compound, des-threninol. Retention times for des-threninol and octreotide were 4.3 and 6.2 minutes, respectively.

A standard curve was constructed from 50 to 225 μg/mL. The correlation coefficient was greater than 0.99. The intraassay and interassay coefficients of variation were all less than 5%.

Reference
Ripley RG, Ritchie DJ, Holstad SG. Stability of octreotide acetate in polypropylene syringes at 5 and −20 °C. *Am J Health Syst Pharm.* 1995; 52: 1910–1.

Method 2
Stiles et al. studied the stability of octreotide acetate in polypropylene syringes using HPLC. The chromatograph consisted of a Waters model 660 gradient system, a Waters model M6000 solvent-delivery system, a Micromeritics 728 autosampler, a Rheodyne 7010 injection valve, a Shimadzu model SPD-LA detector, and a Shimadzu model CR-601 integrator. The stationary phase was a Bakerbond C_{18} column (250 × 2.6 mm, 5-μm particle size). Mobile phase A consisted of tetramethylammonium hydroxide solution, water, and acetonitrile (2:88:10, vol/vol/vol) adjusted to pH 4.5 with concentrated phosphoric acid. The tetramethylammonium hydroxide solution was a 1 M solution of tetramethylammonium hydroxide pentahydrate in water. Mobile phase B consisted of tetramethylammonium hydroxide solution, water, and acetonitrile (2:38:60, vol/vol/vol) adjusted to pH 4.5 with phosphoric acid. The mobile phase was delivered gradiently from 100% mobile phase A to 100% mobile phase B over 14 minutes at 1.3 mL/min. UV detection was performed at 210 nm. The injection volume was 20 μL.

The stability-indicating capability of the method was demonstrated by spiking the octreotide acetate solution with the known degradation product, des-threninol. The degradation product peak did not interfere with the intact drug peak. Retention times for des-threninol and octreotide were 7.5 and 11.5 minutes, respectively.

Reference
Stiles ML, Allen LV Jr, Resztak KE, et al. Stability of octreotide acetate in polypropylene syringes. *Am J Hosp Pharm.* 1993; 50: 2356–8.

Ofloxacin

Chemical Name
(±)-9-Fluoro-2,3-dihydro-3-methyl-10-(4-methyl-1-piperazinyl)-7-oxo-7*H*-pyrido[1,2,3-*de*]-1,4-benzoxazine-6-carboxylic acid

Other Name
Floxin

Form	Molecular Formula	MW	CAS
Ofloxacin	$C_{18}H_{20}FN_3O_4$	361.4	82419-36-1

Appearance
Ofloxacin is an off-white to pale yellow crystalline powder.

Solubility
Ofloxacin has aqueous solubilities of 3.5–4 mg/mL at pH 7, 60 mg/mL at pH 2–5, and 303 mg/mL at pH 9.8.

pK$_a$
Ofloxacin has pK$_a$ values of 5.74 and 7.9.

Method 1
Shervington et al. described an isocratic reversed-phase HPLC method for the simultaneous separation and determination of nalidixic acid, norfloxacin, ofloxacin, ciprofloxacin, and lomefloxacin. Two liquid chromatographic systems were used: (1) a Perkin-Elmer series 410 Bio LC pump equipped with a Dionex UVD 340S photodiode-array detector and a Rheodyne model 9125 injector with a 20-μL loop and (2) a Varian Vista 5000 pump coupled to a Waters 486 tunable absorbance UV detector, a Rheodyne model 9125 injector with a 20-μL loop, and a Hewlett-Packard HP3396 integrator. The stationary phase was a Phenomenex ODS C$_{18}$(2) column (150 × 4.6 mm, 5-μm particle size) with a guard column of the same packing material (30 × 4.6 mm). The mobile phase consisted of a buffer and acetonitrile (65:35, vol/vol), pH 3.4. The aqueous buffer contained 10 mM tetrabutylammonium acetate, 10 mM sodium dodecyl sulfate, and 25 mM citric acid. The flow rate was 1.3 mL/min. UV detection was performed at 235, 254, 275, and 300 nm.

A portion of the powder from 10 tablets was weighed, transferred to a 200-mL volumetric flask, treated with 100 mL of 35% acetonitrile in water, sonicated, filled to the mark with the same diluent, centrifuged at 6000 rpm for 25 minutes, and filtered through a 0.45-μm membrane filter. An aliquot of the filtrate was diluted with the diluent and assayed. Under these conditions, retention times for nalidixic acid, ofloxacin, lomefloxacin, norfloxacin, and ciprofloxacin were 4.5, 7.9, 10.5, 11.1, and 11.9 minutes, respectively.

Ofloxacin samples were treated separately with 0.1 M sodium hydroxide solution, 0.1 M hydrochloric acid, sodium periodate, and 0.3% hydrogen peroxide and then stored at 85 °C for 96 hours. Ofloxacin powders were also exposed to natural daylight for 10 days and to UV light for 24 hours. Degradation products did not interfere with the analysis of ofloxacin.

A calibration curve for ofloxacin was constructed from 12 to 48 µg/mL. The correlation coefficient was 0.9992. The limits of detection and quantitation for ofloxacin were 0.18 and 0.36 µg/mL, respectively.

Reference

Shervington LA, Abba M, Hussain B, et al. The simultaneous separation and determination of five quinolone antibiotics using isocratic reversed-phase HPLC: Application to stability studies on an ofloxacin tablet formulation. *J Pharm Biomed Anal.* 2005; 39: 769–75.

Method 2

Bornstein et al. studied the stability of an ofloxacin injection in various infusion fluids. The HPLC system consisted of pulse-free high-pressure pumps, automatic injectors, and variable-wavelength UV detectors. The stationary phase was a Whatman Partisil-5 ODS-3 reversed-phase column (100 × 4.6 mm, 5-µm particle size). The mobile phase consisted of acetonitrile and 0.05 M potassium phosphate buffer (pH 2.4) (12:88, vol/vol). The buffer was prepared with monobasic potassium phosphate and 85% phosphoric acid. The flow rate was 1.6 mL/min. UV detection was performed at 294 nm.

Samples were diluted with the mobile phase, and the injection volume was 10 µL. A typical run time was 25 minutes.

The method was determined to be stability indicating by an accelerated degradation study. Ofloxacin was subjected to various conditions of stress such as acid, base, heat, light, and oxidation. Chromatography was also performed on samples of all known synthesis-related impurities and degradation products. In all cases, peaks from impurities and degradation products were separated from that of ofloxacin. Retention times were 3.5 minutes for descarboxy-ofloxacin, 4.3 minutes for desfluoro-ofloxacin, 6.0 minutes for the diamine derivative of ofloxacin, 7.6 minutes for ofloxacin, 12.7 minutes for ofloxacin-*N*-oxide, and 20.8 minutes for the amine derivative of ofloxacin (estimated from the published chromatogram).

A calibration curve for ofloxacin was constructed from 0.2 µg/mL to 0.5 mg/mL. The correlation coefficient of the standard curve was 0.9999.

Reference

Bornstein M, Kao S-H, Mercorelli M, et al. Stability of an ofloxacin injection in various infusion fluids. *Am J Hosp Pharm.* 1992; 49: 2756–60.

Omeprazole

Chemical Name

5-Methoxy-2-[[(4-methoxy-3,5-dimethyl-2-pyridinyl)methyl]sulfinyl]-1*H*-benzimidazole

Other Name
Prilosec

Form	Molecular Formula	MW	CAS
Omeprazole	$C_{17}H_{19}N_3O_3S$	345.4	73590-58-6

Appearance
Omeprazole is a white to off-white crystalline powder.

Solubility
Omeprazole is freely soluble in ethanol and methanol, slightly soluble in acetone and isopropanol, and very slightly soluble in water.

Method 1
DiGiacinto et al. used HPLC analysis to evaluate the stability of omeprazole prepared in suspensions and stored in amber-colored plastic oral syringes at 4 and 22 °C for up to 60 days. The apparatus included a Shimadzu model SPD-10A UV-visible detector, a model SIL-10A autoinjector, a model LC-1AS pump, a model SCL-10A system controller, and a model CTO-10A column oven. The stationary phase was a Phenomenex Maxsil C_{18} column (250 × 4.6 mm, 5-μm particle size). The column oven temperature was kept at 35 °C. The mobile phase consisted of acetonitrile and water (45:55, vol/vol) adjusted to pH 7.5 with 1 M monobasic sodium phosphate. The flow rate was 1 mL/min. UV detection was performed at 285 nm.

Samples were diluted with a mixture of acetonitrile and water (45:55, vol/vol) and vortexed for 1 minute. The injection volume was 150 μL. Under these conditions, the retention time for omeprazole was about 4.3 minutes.

The stability-indicating nature of the assay was assessed by accelerated degradation of omeprazole. Stock solutions of omeprazole were adjusted to pH 2 with 1 N hydrochloric acid and allowed to stand for 15 minutes, were adjusted to pH 13 with 1 M sodium hydroxide solution and heated to 100 °C for 20 minutes, or were stored at room temperature for 72 hours. In all cases, degradation products did not interfere with the analysis of omeprazole.

A standard curve for omeprazole was constructed from 0 to 200 μg/mL. The correlation coefficient was 0.9990. Intraday and interday coefficients of variation were 1.57 and 6.71%, respectively.

Reference

DiGiacinto JL, Olsen KM, Bergman KL, et al. Stability of suspension formulations of lansoprazole and omeprazole stored in amber-colored plastic oral syringes. *Ann Pharmacother.* 2000; 34: 600–5.

Method 2

Quercia et al. determined the stability of omeprazole in an extemporaneously prepared oral liquid. The liquid chromatograph consisted of a Waters model 6000A pump, a Waters model 717 plus autosampler, a Milton Roy SpectroMonitor 3000 variable-wavelength UV detector, and a Hewlett-Packard model 3396 Series II integrator. The stationary phase was a Waters C_{18} reversed-phase column (150×4.6 mm, 5-μm particle size). The mobile phase consisted of 0.05 M monobasic sodium phosphate buffer (pH 8.5) and acetonitrile (75:25, vol/vol). The flow rate was 1 mL/min. UV detection was performed at 302 nm. Phenacetin was used as an internal standard. The sample injection volume was 20 μL. Under these conditions, retention times for phenacetin and omeprazole were 8.5 and 9.9 minutes, respectively.

The stability-indicating capability of the method was demonstrated by accelerated degradation of omeprazole. A solution of omeprazole 2 mg/mL in water (pH 4.0) was incubated at 24 °C for 2 days. Omeprazole completely degraded and at least four degradation products appeared in the first 6 minutes. Degradation product peaks did not interfere with the intact omeprazole peak.

A standard curve for omeprazole was constructed from 0 to 250 μg/mL; the correlation coefficient was greater than 0.999. Intraday and interday coefficients of variation were 1.04–1.77% and 1.99–3.40%, respectively.

Reference

Quercia RA, Fan C, Liu X, et al. Stability of omeprazole in an extemporaneously prepared oral liquid. *Am J Health Syst Pharm.* 1997; 54: 1833–6.

Method 3

Mathew et al. evaluated by HPLC analysis the stability of omeprazole solutions at various pH values. The chromatographic apparatus was a Waters ALC 202 system equipped with a Rheodyne model 7125 injector, a Schoeffel SF 770 multiple-wavelength detector, and an Omniscribe recorder. The stationary phase was a Waters μBondapak C_{18} column (300×3.9 mm, 10-μm particle size). The mobile phase was 40% (vol/vol) acetonitrile in 0.02 M ammonium acetate in water (pH 7.1) delivered at 1.7 mL/min. UV detection was performed at 235 nm and 0.1 AUFS. Methyltestosterone 1.0 mg/mL in methanol was used as an internal standard.

A fine powder of ground capsule contents equivalent to 25 mg of omeprazole was weighed and mixed with 20 mL of methanol and 0.2 mL of 2.5 N sodium hydroxide. This mixture was stirred occasionally for 10 minutes, brought to 25 mL with methanol, and filtered. The middle portion of filtrate was collected. One milliliter of the filtrate was mixed with 2.0 mL of the internal standard and 0.5 mL of 1 N sodium hydroxide and diluted to 25 mL with water. The injection volume was 20 μL. Under these conditions, retention times for omeprazole and methyltestosterone were about 4.2 and 11.3 minutes, respectively (estimated from the published chromatogram).

The method was established to be stability indicating by accelerated degradation of omeprazole. Solutions of omeprazole were treated with either 1 N sulfuric acid or 1 N sodium hydroxide solution and heated to boiling for 5 minutes. The drug peak was resolved from the degradation product peaks.

A standard curve for omeprazole was obtained from 20 to 60 μg/mL, with a correlation coefficient of 0.999.

Reference
Mathew M, Gupta VD, Bailey RE. Stability of omeprazole solutions at various pH values as determined by high-performance liquid chromatography. *Drug Dev Ind Pharm.* 1995; 21: 965–71.

Ondansetron

Chemical Name
(±)-1,2,3,9-Tetrahydro-9-methyl-3-[(2-methyl-1*H*-imidazol-1-yl)methyl]-4*H*-carbazol-4-one

Other Name
Zofran

Form	Molecular Formula	MW	CAS
Ondansetron	$C_{18}H_{19}N_3O$	293.4	99614-02-5
Ondansetron hydrochloride	$C_{18}H_{19}N_3O.HCl.2H_2O$	365.9	103639-04-9

Appearance
Ondansetron hydrochloride is a white to off-white powder.

Solubility
Ondansetron hydrochloride is soluble in water and in normal saline.

pK$_a$
Ondansetron hydrochloride has a pK$_a$ of 7.4.

Method 1
Evrard et al. used an HPLC method to evaluate the stability and compatibility of ondansetron hydrochloride and dexamethasone sodium phosphate combinations in 0.9% sodium chloride injection and 5% dextrose injection. The liquid chromatograph consisted of a Merck-Hitachi L-6000 pump, a Merck-Hitachi L-4000 UV detector, and a Merck-Hitachi D-2500 chromato-integrator. The stationary phase was a Merck Lichrocart C$_8$ column with LiChrospher 60 RP-Select B (125 × 4 mm, 5-μm particle

size). The mobile phase was a mixture of 25% acetonitrile and 75% aqueous buffer. The buffer contained 0.02 M monobasic potassium phosphate adjusted to pH 6.0 with sodium hydroxide. UV detection was performed at 241 nm.

Samples were diluted with the mobile phase, and the injection volume was 20 μL.

The stability-indicating capability of the method was confirmed by separation of the intact ondansetron, ondansetron-related substances, dexamethasone, and 4-hydroxybenzoic acid. The mean capacity factors (k') for ondansetron, related substances (GR37896A and GR44122X), and dexamethasone were 11.2, 3.3, 14.9, and 1.3, respectively.

Five-point calibration curves for the ondansetron reference standard from 9 to 27 μg/mL were generated by linear regression analysis of peak area versus concentration. The correlation coefficient was 0.9998. Intraday and interday coefficients of variation for ondansetron hydrochloride were 1.2 and 3.1%, respectively.

Reference

Evrard B, Ceccato A, Gaspard O, et al. Stability of ondansetron hydrochloride and dexamethasone sodium phosphate in 0.9% sodium chloride injection and in 5% dextrose injection. *Am J Health Syst Pharm*. 1997; 54: 1065–8.

Method 2

Kirkham et al. studied the stability of ondansetron hydrochloride at therapeutic concentrations in a total parenteral nutrient admixture. The liquid chromatograph consisted of a Varian 9010 solvent-delivery system with a UV-visible detector and an Applied Biosystems Spheri-10 column (220 × 4.6 mm, 10-μm particle size). The mobile phase was a mixture of 85% sodium acetate (pH 6.5) in 4% dimethylformamide and 15% acetonitrile and was delivered isocratically. The flow rate was 1.0 mL/min. UV detection was performed at 305 nm.

Samples were diluted with sterile deionized water.

The stability-indicating nature of the method was demonstrated by accelerated decomposition of ondansetron. A solution of ondansetron hydrochloride 0.5 mg/mL was mixed with hydrochloric acid at 90 °C to bring it to pH 1. A second solution was mixed with 10% hydrogen peroxide at room temperature. In all cases, the degradation product peaks were distinct from the intact ondansetron peak.

A standard curve was generated, with a correlation coefficient of 0.9955.

Reference

Kirkham JC, Rutherford ET, Cunningham GN, et al. Stability of ondansetron hydrochloride in a total parenteral nutrient admixture. *Am J Health Syst Pharm*. 1995; 52: 1557–8.

Method 3

Trissel et al. determined the stability of ondansetron hydrochloride 0.1 and 1.0 mg/mL with morphine sulfate 1 mg/mL and with hydromorphone hydrochloride 0.5 mg/mL when admixed in 0.9% sodium chloride injection. The HPLC analysis used a Waters model 600E multisolvent-delivery system, a Waters model 490E programmable multiple-wavelength UV detector, a Waters model 712 WISP autosampler, and an Alltech Spherisorb cyano column (300 × 4.6 mm, 5-μm particle size). The system was controlled and the peak area was integrated by an NEC PowerMate SX/16 personal computer. The mobile phase was 50% acetonitrile in 0.02 M monobasic potassium phosphate with the pH adjusted to 5.40 with 1 N sodium hydroxide and was delivered isocratically at 1.5 mL/min. The UV detector was operated at 216 nm.

Samples were diluted 1:10, and the injection volume was 10 μL. Under these conditions, the retention time for ondansetron was 8.8–10 minutes. Morphine eluted at 6.7 minutes and hydromorphone at 7.5 minutes.

The assay was demonstrated to be stability indicating by accelerated degradation of ondansetron hydrochloride. Heating ondansetron hydrochloride solution at 95 °C for 4 hours yielded a reduction in the intact ondansetron peak and the formation of new peaks at 3.2 and 5.6 minutes. Neither the hydromorphone hydrochloride, the morphine sulfate, or the degradation product peaks interfered with the intact ondansetron peak.

Standard curves were constructed from a linear plot of peak area versus concentration from 0.01 to 0.15 mg/mL. The correlation coefficient was greater than 0.9998. The intraday and interday coefficients of variation were 0.6 and 1.0%, respectively.

Similar stability-indicating methods were used by the other researchers cited here.

References

Trissel LA, Xu Q, Martinez JF, et al. Compatibility and stability of ondansetron hydrochloride with morphine sulfate and with hydromorphone hydrochloride in 0.9% sodium chloride injection at 4, 22, and 32 °C. *Am J Hosp Pharm.* 1994; 51: 2138–42.

Stewart JT, Warren FW, King DT, et al. Stability of ondansetron hydrochloride, doxorubicin hydrochloride, and dacarbazine or vincristine sulfate in elastomeric portable infusion devices and polyvinyl chloride bags. *Am J Health Syst Pharm.* 1997; 54: 915–20.

Stewart JT, Warren FW, King DT, et al. Stability of ondansetron hydrochloride and five antineoplastic medications. *Am J Health Syst Pharm.* 1996; 53: 1297–300.

Fleming RA, Olsen DJ, Savage PD, et al. Stability of ondansetron hydrochloride and cyclophosphamide in injectable solutions. *Am J Health Syst Pharm.* 1995; 52: 514–6.

Chung KC, Moon YSK, Chin A, et al. Compatibility of ondansetron hydrochloride and piperacillin sodium–tazobactam sodium during simulated Y-site administration. *Am J Health Syst Pharm.* 1995; 52: 1554–6.

Bosso JA, Prince RA, Fox JL. Compatibility of ondansetron hydrochloride with fluconazole, ceftazidime, aztreonam, and cefazolin sodium under simulated Y-site conditions. *Am J Hosp Pharm.* 1994; 51: 389–90.

Xu QA, Trissel LA, Fox JL. Compatibility of ondansetron hydrochloride with meperidine hydrochloride for combined administration. *Ann Pharmacother.* 1993; 29: 1106–9.

McGuire TR, Narducci WA, Fox JL. Compatibility and stability of ondansetron hydrochloride, dexamethasone, and lorazepam in injectable solutions. *Am J Hosp Pharm.* 1993; 50: 1410–4.

Stiles ML, Allen LV Jr, Fox JL. Stability of ondansetron hydrochloride in portable infusion-pump reservoirs. *Am J Hosp Pharm.* 1992; 49: 1471–3.

Bosso JA, Prince RA, Fox JL. Stability of ondansetron hydrochloride in injectable solutions at −20, 5, and 25 °C. *Am J Hosp Pharm*. 1992; 49: 2223–5.

Graham CL, Dukes GE, Kao C-F, et al. Stability of ondansetron in large-volume parenteral solutions. *Ann Pharmacother*. 1992; 26: 768–71.

Orlistat

Chemical Name
(*S*)-1-[(2*S*,3*S*)-3-Hexyl-4-oxo-oxetan-2-ylmethyl]dodecyl *N*-formyl-L-leucinate

Other Name
Xenical

Form	Molecular Formula	MW	CAS
Orlistat	$C_{29}H_{53}NO_5$	495.7	96829-58-2

Appearance
Orlistat occurs as a white to off-white crystalline powder.

Solubility
Orlistat is practically insoluble in water, freely soluble in chloroform, and very soluble in methanol and ethanol.

Method
Mohammadi et al. described an HPLC method for the determination of orlistat in capsules and bulk form. The Waters liquid apparatus comprised a model 600 solvent-delivery system, a model 717 plus autosampler, and a model 2487 dual λ absorbance detector. The stationary phase was a Perfectsil target ODS-3 column (250 × 4.6 mm, 5-μm particle size). The mobile phase consisted of methanol, acetonitrile, and trifluoroacetic acid (82.5:17.5:0.01, vol/vol/vol). The flow rate was 0.7 mL/min. The detector was set at 210 nm.

A portion of the powder obtained from 20 capsules equivalent to the weight of one capsule was accurately weighed, transferred into a 100-mL volumetric flask, made up to the mark with methanol, sonicated for 30 minutes, filtered through a 0.45-μm nylon filter, diluted with mobile phase, and assayed. The injection volume was 20 μL. Under these conditions, the retention time of orlistat was about 9.0 minutes (estimated from the published chromatogram).

This method was confirmed to be stability indicating by accelerated degradation. Orlistat solutions were prepared in methanol and 3% hydrogen peroxide (80:20, vol/vol), in methanol and 2 M hydrochloric acid (80:20, vol/vol), or in methanol and 2 M sodium hydroxide solution (80:20, vol/vol) and refluxed for 30 minutes. Orlistat solution was also prepared in methanol and water (80:20, vol/vol) and refluxed for 2 hours. Orlistat capsules and powder were exposed to dry heat at 60 °C in an oven for 3 days or exposed to UV radiation at 320–400 nm at 25 °C for 40 hours. The degradation product peaks did not interfere with the peak of the intact drug.

A standard curve for orlistat was constructed from 0.02 to 0.75 mg/mL. The correlation coefficient was 0.9998. The intraday and interday coefficients of variation were less than 2.30 and 3.59%, respectively. The limit of detection and the limit of quantitation were 0.006 and 0.02 mg/mL, respectively.

Reference
Mohammadi A, Haririan I, Rezanour N, et al. A stability-indicating high performance liquid chromatographic assay for the determination of orlistat in capsules. *J Chromatogr A.* 2006; 1116: 153–7.

Ornidazole

Chemical Name
1-Chloro-3-(2-methyl-5-nitroimidazol-1-yl)propan-2-ol

Other Names
Invigan, Mebaxol, Tiberal, Tinerol

Form	Molecular Formula	MW	CAS
Ornidazole	$C_7H_{10}ClN_3O_3$	219.6	16773-42-5

Appearance
Ornidazole occurs as crystals.

pK$_a$
Ornidazole has a pK$_a$ value of 2.4.

Method
Bakshi et al. reported an HPLC assay for degradation studies on ornidazole. The Waters chromatograph consisted of a model 600E pump and a model 717 autosampler coupled with a Waters model 996 photodiode-array detector and Millennium software v2.1. The

stationary phase was a Supelcosil LC-18-DB column (250 × 4.6 mm, 10-μm particle size). The mobile phase consisted of water and acetonitrile (86:14, vol/vol) and was delivered isocratically at 1 mL/min. UV detection was performed at 310 nm. The sample injection volume was 20 μL. Under these conditions, the retention time for ornidazole was about 17.3 minutes.

For degradation studies, ornidazole 1 mg/mL in 0.1 M hydrochloric acid was heated at 80 °C for 72 hours, ornidazole 1 mg/mL in 0.1 M sodium hydroxide solution was heated at 80 °C for 8 hours, ornidazole 1 mg/mL in water was heated at 80 °C for 120 hours, and ornidazole 1 mg/mL was treated with 3% hydrogen peroxide. The peak of the intact ornidazole was well resolved from all degradation product peaks.

A calibration curve for ornidazole was generated from 5 to 500 μg/mL. The correlation coefficient was 0.9996. The intraday and interday coefficients of variation for the analysis of ornidazole were less than 1 and 2.5%, respectively.

Reference
Bakshi M, Singh B, Singh A, et al. The ICH guidance in practice: Stress degradation studies on ornidazole and development of a validated stability-indicating assay. *J Pharm Biomed Anal.* 2001; 26: 891–7.

Otilonium Bromide

Chemical Name
Diethylmethyl[2-[4-(2-octyloxybenzamido)benzoyloxy]ethyl]ammonium bromide

Other Names
Spasen, Spasmomen

Form	Molecular Formula	MW	CAS
Otilonium bromide	$C_{29}H_{43}BrN_2O_4$	563.6	26095-59-0

Appearance
Otilonium bromide occurs as crystals.

Method
Mannucci et al. reported an HPLC method for the simultaneous determination of otilonium bromide, diazepam, and related compounds in tablets. A Hewlett-Packard 1090L system was equipped with an autosampler and a model 1040M diode-array

detector. The stationary phase was a Merck octyl derivative silica column (250 × 4.0 mm, 5-μm particle size) with a precolumn (4 × 4 mm) of the same packing. The column temperature was maintained at 50 °C. The mobile phase consisted of 0.5 M sodium acetate trihydrate buffer containing 5 mM 1-heptanesulfonic acid monohydrate sodium salt and methanol (30:70, vol/vol), adjusted to pH 6.0 with glacial acetic acid. The flow rate was 1 mL/min. n-Butyl p-hydroxybenzoate 1 mg/mL in acetonitrile was used as an internal standard. Diazepam, n-butyl p-hydroxybenzoate, and otilonium bromide were monitored at 230, 254, and 290 nm, respectively.

A tablet was accurately weighed and placed into a 50-mL volumetric flask, dissolved with 5 mL of water with stirring, mixed with 5 mL of internal standard, diluted to volume with acetonitrile, and centrifuged at 3500 rpm (2472 × g) for 5 minutes. The injection volume of the supernatant was 10 μL. The total run time was 15 minutes. Under these conditions, retention times for the internal standard, diazepam, and otilonium bromide were 3.98, 4.72, and 10.48 minutes, respectively.

The method was verified to be stability indicating by assaying a related-compounds standard solution. The intact drug peaks were well separated from their related-compounds peaks. Furthermore, the excipients did not interfere with the analysis of drugs.

Standard curves for otilonium bromide were constructed from 0.099 to 0.790 mg/mL and from 0.198 to 1.586 mg/mL. The correlation coefficients were both 1.0.

Reference
Mannucci C, Bertini J, Cocchini A, et al. High-performance liquid chromatographic method for assay of otilonium bromide, diazepam, and related compounds in finished pharmaceutical forms. *J Pharm Sci*. 1993; 82: 367–70.

Oxacillin

Chemical Name
[2S-(2α,5α,6β)]-3,3-Dimethyl-6-[[(5-methyl-3-phenyl-4-isoxazolyl)carbonyl]amino]-7-oxo-4-thia-1-azabicyclo[3.2.0]heptane-2-carboxylic acid

Other Names
Bactocill, Prostaphlin

Form	Molecular Formula	MW	CAS
Oxacillin	$C_{19}H_{19}N_3O_5S$	401.4	66-79-5
Oxacillin sodium	$C_{19}H_{18}N_3NaO_5S.H_2O$	441.4	7240-38-2

Appearance
Oxacillin sodium is a fine white crystalline powder.

Solubility
Oxacillin sodium is freely soluble in water and methanol, slightly soluble in dehydrated alcohol and chloroform, and practically insoluble in ether.

pK$_a$
Oxacillin sodium has a pK$_a$ of approximately 2.8.

Method 1
Zhang and Trissel evaluated by HPLC the stability of oxacillin sodium in AutoDose infusion system bags. The Waters model LC Module 1 Plus chromatograph included a multisolvent-delivery pump, an autosampler, and a multiple-wavelength UV detector. The stationary phase was a Phenomenex Kromasil C$_{18}$ column (250 × 4.6 mm, 5-μm particle size). The mobile phase consisted of 76% 0.04 M sodium acetate (pH 7) and 24% acetonitrile. The flow rate was 1.25 mL/min. UV detection was performed at 225 nm and 0.5 AUFS.

Each sample was diluted 1:100 with water. The injection volume was 10 μL. Under these conditions, the retention time of oxacillin was 10.8 minutes.

The stability-indicating nature of the method was demonstrated by accelerated degradation. Oxacillin solutions were mixed with 1 N sodium hydroxide, 1 N hydrochloric acid, or 3% hydrogen peroxide and heated. There was no interference from the degradation products with the peak of the intact oxacillin.

A standard curve for oxacillin was generated from 25 to 150 μg/mL. The correlation coefficient was greater than 0.9999. The coefficient of variation for the analysis of oxacillin was 0.2%. Intraday and interday coefficients of variation were 1.6 and 0.4%, respectively.

Reference
Zhang Y, Trissel LA. Stability of ampicillin sodium, nafcillin sodium, and oxacillin sodium in AutoDose infusion system bags. *Int J Pharm Compound*. 2002; 6: 226–9.

Method 2
Grover et al. developed an HPLC method for the analysis of oxacillin in the presence of its degradation products. The Shimadzu model LC-10A liquid chromatograph included model LC-10AS pumps, a model SPD-10A dual-wavelength UV-visible detector, a model C-R7A data processor, and a Rheodyne model 7125 injector with a 20-μL loop. The stationary phase was a Phenomenex Resolve C$_{18}$ column (300 × 3.9 mm, 5-μm particle size). The mobile phase was 15% acetonitrile in 20 mM monobasic potassium phosphate and 10 mM tetramethylammonium chloride in water (pH 5.0). The flow rate was 1 mL/min. UV detection was performed at 220 and 240 nm. Under these conditions, the retention time for oxacillin was 35.5 minutes.

The stability-indicating nature of the assay was demonstrated by degrading the drug. Oxacillin was degraded at pH 2.0. The intact oxacillin peak was separated from its degradation product peaks.

A standard curve for oxacillin was obtained from 5 to 150 μg/mL. The correlation coefficient was greater than 0.99.

References
Grover M, Gulati M, Singh S. Stability-indicating analysis of isoxazolyl penicillins using dual wavelength high-performance liquid chromatography. *J Chromatogr B*. 1998; 708: 153–9.

Grover M, Gulati M, Singh B, et al. Correlation of penicillin structure with rate constants for basic hydrolysis. *Pharm Pharmacol Commun*. 2000; 6: 355–63.

Method 3

Stiles and Allen evaluated the stability of oxacillin sodium in polyvinyl chloride drug reservoirs. The liquid chromatograph consisted of a Waters model 501 solvent-delivery system, a Micromeritics model 728 autoinjector, a Micromeritics model 732 injection valve, a Waters Lambda-Max model 481 variable-wavelength UV detector, and a Waters model 745 data module. The stationary phase was a Bakerbond C_{18} column (250×2.6 mm, 5-μm particle size). The mobile phase consisted of 700 mL of water containing 14 mM monobasic sodium phosphate, 300 mL of acetonitrile, and 100 mL of methanol. The flow rate was 1.5 mL/min. UV detection was performed at 225 nm.

Samples were diluted 1:250 with the mobile phase.

The method was determined to be stability indicating by accelerated degradation of oxacillin sodium. The retention time for oxacillin was 5.1 minutes and retention times for its degradation products were 2.1, 2.8, and 3.5 minutes.

A standard curve for oxacillin was constructed from 50 to 500 μg/mL. Intraday and interday coefficients of variation were 1.0 and 2.9%, respectively.

Reference

Stiles ML, Allen LV Jr. Stability of nafcillin sodium, oxacillin sodium, penicillin G potassium, penicillin G sodium, and tobramycin sulfate in polyvinyl chloride drug reservoirs. *Am J Health Syst Pharm.* 1997; 54: 1068–70.

Oxaliplatin

Chemical Name

cis-[(1*R*,2*R*)-1,2-Cyclohexanediamine-*N*,*N′*][oxalato(2-)-*O*,*O′*]platinum

Other Names

Eloxatin, Uxalun, Xaliplat

Form	Molecular Formula	MW	CAS
Oxaliplatin	$C_8H_{14}N_2O_4Pt$	397.3	61825-94-3

Appearance

Oxaliplatin occurs as a white or almost white crystalline powder.

Solubility

Oxaliplatin is slightly soluble in water, practically insoluble in dehydrated alcohol, and very slightly soluble in methanol.

Method

Using an HPLC method, Trissel and Zhang evaluated the chemical stability of oxaliplatin with palonosetron hydrochloride in 5% dextrose. The Waters liquid chromatograph consisted of an Alliance 2690 separation module and a model 2480 dual-wavelength detector. The stationary phase was a Phenomenex Hypersil ODS C_{18} analytical column (250 × 4.6 mm, 5-μm particle size). The mobile phase was a mixture of 172 mL of acetonitrile and 1000 mL of water containing 1 g of heptanesulfonic acid sodium salt. The flow rate was 1.0 mL/min. The detector was set at 254 nm and 0.5 AUFS. The sample injection volume was 15 μL. Under these conditions, the retention time of oxaliplatin was 7.1 minutes.

The analytical method was demonstrated to be stability indicating by accelerated degradation. Oxaliplatin solutions were mixed with 1 N sodium hydroxide solution, 1 N hydrochloric acid, or 3% hydrogen peroxide and then subjected to heating. Loss of the intact drug was observed, and there was no interference of the degradation product peaks with the peak of the intact drug.

A standard curve for oxaliplatin was constructed from 100 to 400 μg/mL. The correlation coefficient was greater than 0.9999. The coefficient of variation of the analysis was 0.2%.

Reference

Trissel LA, Zhang Y. Physical and chemical stability of palonosetron HCl with cisplatin, carboplatin, and oxaliplatin during simulated Y-site administration. *J Oncol Pharm Pract.* 2004; 10: 191–5.

Oxaprozin

Chemical Name
3-(4,5-Diphenyloxazol-2-yl)propionic acid

Other Names
Daypro, Prozina

Form	Molecular Formula	MW	CAS
Oxaprozin	$C_{18}H_{15}NO_3$	293.3	21256-18-8

Appearance
Oxaprozin occurs as crystals.

Solubility
Oxaprozin is slightly soluble in alcohol and insoluble in water.

pK$_a$
Oxaprozin has a pK$_a$ of 4.3.

Method

Ibrahim determined by HPLC analysis oxaprozin and its related compounds in pharmaceutical products. A Varian 5000 liquid chromatograph was equipped with a Varian model 8055 autosampler and an Applied Biosystems model 783A programmable absorbance detector. Another liquid chromatograph was used for peak purity analysis and consisted of a Varian model 9010 pump, a model 9096 autosampler, and a model 9065 photodiode-array detector. The stationary phase was a Waters µBondapak C$_{18}$ column (300 × 3.9 mm). The mobile phase was a mixture of an aqueous solution containing 0.01 M monobasic potassium phosphate and 0.005 M 1-decanesulfonic acid sodium salt, methanol, and acetonitrile (2:1:1, vol/vol/vol) adjusted to pH 4.2 with phosphoric acid. The flow rate was 1 mL/min. UV detection was performed at 254 nm.

Samples were prepared in the mobile phase. The injection volume was 10 µL. Under these conditions, the retention times for benzoin and oxaprozin were 6.8 and 19.8 minutes, respectively.

The stability-indicating ability of the assay was shown by an accelerated degradation study. Oxaprozin solutions were exposed to UV light (254 and 366 nm) for 40 hours or stored at 50 ºC for 2 days. No interference due to the degradation products was observed. Photodiode-array detection also confirmed the peak purity of oxaprozin.

A calibration curve for oxaprozin was constructed from 0.4 to 1.1 mg/mL. The correlation coefficient was greater than 0.999. The limit of detection was about 0.05 µg/mL.

Reference

Ibrahim FB. Quantitative determination of oxaprozin and several of its related compounds by high-performance reversed-phase liquid chromatography. *J Liq Chromatogr.* 1995; 18: 2621–33.

Oxathiin Carboxanilide

Chemical Name

2-Chloro-5-[[(5,6-Dihydro-2-methyl-1,4-oxathiin-3-yl)carbonyl]amino]benzoic acid 1-methylethyl ester

Other Name

NSC 615985

Form	Molecular Formula	MW	CAS
Oxathiin carboxanilide	C$_{16}$H$_{18}$ClNO$_4$S	355.8	135812-0403

Method

Oh et al. investigated the stability and solubilization of oxathiin carboxanilide by HPLC analysis. The liquid chromatograph consisted of a Waters model M-45 pump, a model 441 detector, and a model 740 data module. The stationary phase was a Beckman

Ultrasphere ODS column (150 × 4.6 mm, 5-µm particle size) or a Supelco LC-18 column (250 × 4.6 mm, 5-µm particle size). The mobile phase consisted of acetonitrile and water at pH 4. UV detection was performed at 254 nm.

An emulsion sample was filtered through a 5-µm Millipore SM type membrane filter and diluted 1:25 with absolute alcohol. An aliquot (250 µL) of this solution was applied to a prewetted Waters Sep-Pak C_{18} cartridge, eluted with 10 mL of 60% acetonitrile in phosphate buffer (pH 4), and assayed. Under these conditions, the retention time of oxathiin carboxanilide was about 9.4 minutes (estimated from the published chromatogram).

The method was stated to be stability indicating.

Reference
Oh I, Chi S-C, Vishnuvajjala BR, et al. Stability and solubilization of oxathiin carbox-anilide, a novel anti-HIV agent. *Int J Pharm*. 1991; 73: 23–31.

Oxazepam

Chemical Names
7-Chloro-1,3-dihydro-3-hydroxy-5-phenyl-2*H*-1,4-benzodiazepin-2-one
7-Chloro-3-hydroxy-5-phenyl-1,3-dihydro-2*H*-1,4-benzodiazepin-2-one

Other Name
Serax

Form	Molecular Formula	MW	CAS
Oxazepam	$C_{15}H_{11}ClN_2O_2$	286.7	604-75-1

Appearance
Oxazepam is a creamy white or pale yellow powder.

Solubility
Oxazepam is soluble in alcohol, chloroform, and dioxane. It is practically insoluble in water.

pK$_a$
Oxazepam has pK$_a$ values of 1.7 and 11.6.

Method

Reif and DeAngelis developed a stability-indicating HPLC assay for oxazepam tablets and capsules. The method used a Laboratory Data Control Constametric II chromatograph equipped with a Schoeffel Spectroflow Monitor SF 770 UV detector, a Rheodyne 7010 injector, and a Spectra-Physics SP4100 integrator. The stationary phase was various 10-μm microparticulate reversed-phase columns [ES Industries Chromegabond C_{18}, Waters μBondapak C_{18}, DuPont Zorbax ODS (150×4.6 mm), and Whatman Partisil ODS-3]. The mobile phase was a mixture of methanol, water, and acetic acid (60:40:1, vol/vol/vol) and was delivered isocratically at 2 mL/min. UV detection was performed at 254 nm.

Sample powders were dissolved with water and methanol and centrifuged. The injection volume was 10 μL. The retention time for oxazepam was 5.6 minutes.

The stability-indicating capability of the assay was demonstrated by accelerated decomposition of oxazepam. Oxazepam was degraded in hydrochloric acid, pH 5.5 phosphate buffer, pH 3.2 acetate buffer, and sodium hydroxide. Degradation products follow, with their retention times in minutes given in parentheses: 7-chloro-5-phenyl-4,5-dihydro-2H-benzodiazepine-2,3-(1H)-dione (3.6), 6-chloro-3,4-dihydro-4-phenyl-2-quinazolinecarboxylic acid (4.5), 6-chloro-4-phenyl-2-(1H)-quinazolinone (8.2), 2′-benzoyl-4′-chloroglyoxanilide (11.2), 6-chloro-4-phenyl-2-quinazolinecarboxaldehyde (11.8), quinazoline alcohol (12.4), and 2-amino-5-chlorobenzophenone (18.4).

A standard curve for oxazepam was constructed from 0.06 to 0.12 mg/mL.

Reference

Reif VD, DeAngelis NJ. Stability-indicating high-performance liquid chromatographic assay for oxazepam tablets and capsules. *J Pharm Sci.* 1983; 72: 1330–2.

Oxcarbazepine

Chemical Name

10,11-Dihydro-10-oxo-5H-dibenz[b,f]azepine-5-carboxamide

Other Names

Atoxecar, Aurene, Timox, Trileptal

Form	Molecular Formula	MW	CAS
Oxcarbazepine	$C_{15}H_{12}N_2O_2$	252.3	28721-07-5

Method

Pathare et al. reported a validation of a method for oxcarbazepine. The Shimadzu liquid chromatograph consisted of model LC-10AT VP pumps, a model SCL-10A autosampler, and a model SPD-10M VP photodiode-array detector. The stationary phase was an Inertsil C_{18} column (250 × 4.6 mm, 5-µm particle size). The mobile phase was a mixture of 0.02 M monobasic potassium phosphate buffer, acetonitrile, and methanol (45:35:20, vol/vol/vol). The flow rate was 1 mL/min. UV detection was performed at 256 nm.

Oxcarbazepine solution was prepared in mobile phase. The injection volume was 20 µL. Under these conditions, the retention time of oxcarbazepine was about 4.9 minutes.

To demonstrate the stability-indicating property of the method, oxcarbazepine was subjected to stress conditions of UV light (254 nm, 10 days), heat (60 °C, 10 days), acid (0.5 N hydrochloric acid, 2 days), base (0.5 N sodium hydroxide solution, 2 days), and oxidation (3.0% hydrogen peroxide, 2 days). The analysis of the drug was unaffected in the presence of its degradation products and impurities.

A standard curve of oxcarbazepine was constructed from 50 to 150 µg/mL. The correlation coefficient was greater than 0.9999. The intraday and interday coefficients of variation were 0.8 and 0.9%, respectively. The accuracy expressed as the percentage recovery of the drug in samples ranged from 98.6 to 100.7%.

Reference

Pathare DB, Jadhav AS, Shingare MS. A validated stability indicating LC method for oxcarbazepine. *J Pharm Biomed Anal.* 2007: 43: 1825–30

Oxomemazine

Chemical Name

10-(3-Dimethylamino-2-methylpropyl)phenothiazine 5,5-dioxide

Other Names

Doxergan, Rectoplexil, Toplexil

Form	Molecular Formula	MW	CAS
Oxomemazine	$C_{18}H_{22}N_2O_2S$	330.4	3689-50-7
Oxomemazine hydrochloride	$C_{18}H_{22}N_2O_2S.HCl$	366.9	4784-40-1

Appearance

Oxomemazine and oxomemazine hydrochloride occur as crystals.

Method

Hewala described an HPLC method for paracetamol (acetaminophen), guaifenesin, sodium benzoate, and oxomemazine in the presence of degradation products. The Beckman Gold system consisted of a model 125 programmable pump, a model 166 programmable UV detector, and a Rheodyne 20-µL loop injector. The stationary phase was a stainless steel C_{18} ODS column (250 × 4.6 mm, 5-µm particle size) with a guard column (50 × 4.6 mm) of the same packing material. The mobile phase was a mixture of methanol and water (40:60, vol/vol) containing 0.05% heptanesulfonic acid and adjusted to pH 3.0 with phosphoric acid. The flow rate was 1.5 mL/min. UV detection was performed at 235 nm and 0.05 AUFS. Salbutamol sulfate was used as an internal standard.

A 10-mL portion of the cough syrup was diluted to 100 mL with methanol, mixed with internal standard solution, and further diluted with methanol to 2.5 µg/mL of salbutamol (albuterol) sulfate in the final solution. Under these conditions, retention times for salbutamol sulfate and oxomemazine were about 6.4 and 8.6 minutes, respectively (estimated from the published chromatogram).

The method was evaluated to be stability indicating by assaying a mixture of active compounds and possible degradation products.

A standard curve for oxomemazine was generated from 0.2 to 5.0 µg/mL. The correlation coefficient was 0.9998. Intraday and interday coefficients of variation were 1.67 and 2.66%, respectively.

Reference

Hewala II. Stability-indicating HPLC assay for paracetamol, guaiphenesin, sodium benzoate and oxomemazine in cough syrup. *Anal Lett*. 1994; 27: 71–93.

Oxybutynin Chloride

Chemical Name

4-Diethylaminobut-2-ynyl α-cyclohexylmandelate hydrochloride

Other Names

Cystrin, Ditropan

Form	Molecular Formula	MW	CAS
Oxybutynin chloride	$C_{22}H_{31}NO_3$.HCl	394.0	1508-65-2

Appearance
Oxybutynin chloride is a white or almost white and practically odorless crystalline powder.

Solubility
Oxybutynin chloride is freely soluble in water and in alcohol.

pK_a
Oxybutynin chloride has a pK_a value of 6.96.

Method
Schutter and Moerloose reported the determination by HPLC analysis of oxybutynin chloride in pharmaceuticals. The Spectra-Physics SP8000 liquid chromatograph was equipped with a Spectra-Physics model 770 variable-wavelength detector and a BD8 single channel recorder. The stationary phase was an Alltech RSil C_{18} column (150 × 4.1 mm, 5-μm particle size). The mobile phase was a mixture of methanol and water (65:35, vol/vol) containing 30 mM sodium octanesulfonate and 7.5 mM N,N-dimethyl-octylamine. Its pH was adjusted to 5.0 with phosphoric acid. The flow rate was 1 mL/min. UV detection was performed at 220 nm. The diluent consisted of methanol and water (65:35, vol/vol). Oxyphencyclimine hydrochloride in diluent was used as an internal standard.

For tablets, ground tablet powder equivalent to 2.5 tablets was weighed and placed into a 50-mL volumetric flask, mixed with 25 mL of diluent and 10 mL of internal standard, sonicated for 2 minutes, diluted to volume with diluent, and centrifuged at 3000 × g for 5 minutes. For syrup, a 12.5-mL sample of syrup was weighed and placed into a 50-mL volumetric flask, mixed with 25 mL of diluent and 10 mL of internal standard, sonicated for 2 minutes, and diluted to volume with diluent. The injection volume was 10 μL. Under these conditions, retention times for oxybutynin and oxyphencyclimine were about 7.5 and 8.8 minutes, respectively (estimated from the published chromatogram).

The method was evaluated to be stability indicating by accelerated degradation. Oxybutynin was degraded by acid, base, hydrogen peroxide, light, and boiling. Degradation products did not influence the determination of oxybutynin.

A standard curve for oxybutynin was constructed from 0.81 to 1.16 mg/mL. The correlation coefficient was 0.9998.

Reference
Schutter JAD, Moerloose PD. Determination of oxybutinin chloride in pharmaceuticals by reversed-phase ion-pair liquid chromatography with two counter-ions in eluent. *J Chromatogr*. 1988; 450: 337–42.

Oxycodone

Chemical Name
6-Deoxy-7,8-dihydro-14-hydroxy-3-O-methyl-6-oxomorphine

Other Names
Endocodone, OxyContin, Percolone, Roxicodone

Form	Molecular Formula	MW	CAS
Oxycodone	$C_{18}H_{21}NO_4$	315.3	76-42-6
Oxycodone hydrochloride	$C_{18}H_{21}NO_4.HCl$	351.8	124-90-3
Oxycodone terephthalate	$(C_{18}H_{21}NO_4)_2.C_8H_6O_4$	796.9	64336-55-6

Appearance
Oxycodone occurs as long rods or as tautomeric and strongly refringent scales. Oxycodone hydrochloride occurs as white to off-white odorless hygroscopic crystals or as a powder.

Solubility
Oxycodone is insoluble in water and soluble in alcohol. Oxycodone hydrochloride and oxycodone terephthalate are freely soluble in water and slightly soluble in alcohol.

Method
Gebauer et al. developed a method for the stability study of oxycodone and lidocaine in rectal gel. The HPLC instrument consisted of a Waters model 510 pump, a model W717 autoinjector, and a model 996 photodiode-array detector and was managed by Waters Millenium 32 software. The stationary phase was a Hewlett-Packard Zorbax SB-C_8 reversed-phase column (250 × 4.6 mm, 5-µm particle size). The mobile phase was a mixture of 350 mL of methanol, 150 mL of water, 10 mL of acetic acid, and 1.60 g of sodium dodecyl sulfate. The flow rate was 1.5 mL/min. The eluant was monitored by a photodiode-array detector from 250 to 300 nm. The chromatogram for oxycodone was extracted from the contour plot at 285 nm.

A sample (0.5 mL) of the gel was placed into a disposable filtration tube and centrifuged at 2600 × g for 30 minutes at 10 °C. The filtrate was collected and assayed. Under these conditions, retention times for the preservative methyl 4-hydroxybenzoate, oxycodone, and lidocaine were about 2.8, 5.1, and 8.0 minutes, respectively.

The method was demonstrated to be stability indicating by intentional degradation of oxycodone. Aqueous solutions of oxycodone hydrochloride were heated to dryness or mixed with hydrogen peroxide. The peak of the intact oxycodone hydrochloride was resolved from its degradation product peaks on the chromatogram.

A calibration curve for oxycodone was constructed from 0.05 to 1.5% (wt/wt). The correlation coefficient was greater than 0.998. The interday coefficient of variation for the analysis was 2.7%. The recovery of oxycodone was 98.2%.

Reference
Gebauer MG, McClure AF, Vlahakis TL. Stability-indicating HPLC method for the estimation of oxycodone and lidocaine in rectal gel. *Int J Pharm.* 2001; 223: 49–54.

Oxyphenbutazone

Chemical Name
4-Butyl-1-(4-hydroxyphenyl)-2-phenylpyrazolidine-3,5-dione monohydrate

Other Names
Californit, Tanderil

Form	Molecular Formula	MW	CAS
Oxyphenbutazone	$C_{19}H_{20}N_2O_3.H_2O$	342.4	7081-38-1

Appearance
Oxyphenbutazone is a white to yellowish-white odorless crystalline powder.

Solubility
Oxyphenbutazone is practically insoluble in water and freely soluble in alcohol.

Method
Fabre et al. described a reversed-phase HPLC procedure for separating and determining oxyphenbutazone and six potential decomposition products. The Merck model LMC chromatograph was equipped with a model LC 313 variable-wavelength detector and a 10-μL loop injector. The stationary phase was a Merck LiChrosorb RP18 stainless steel column (250 × 4.0 mm, 7-μm particle size). The mobile phase was a mixture of 0.1 M tromethamine citrate buffer (pH 5.25), acetonitrile, and tetrahydrofuran (65:29:6, vol/vol/vol). The flow rate was 1.3 mL/min. UV detection was performed at 239 nm and 0.02 AUFS.

A sample of oxyphenbutazone (about 250 mg of ointment, 20 mg of tablet cores, or 34 mg of capsule powder) was accurately weighed, sonicated for 5 min in 25 mL of methanol, and centrifuged at 4000 rpm for 10 minutes. The supernatant from the ointment, tablets, or capsules was injected without dilution. The injection volume was 10 μL. Under these conditions, the retention time for oxyphenbutazone was about 9.9 minutes (estimated from the published chromatogram).

The assay was evaluated to be stability indicating by assaying a mixture of the drug and its degradation products. The intact drug peak was well resolved from the degradation product peaks.

A standard curve for oxyphenbutazone was constructed from 0 to 100 μg/mL. The correlation coefficient was greater than 0.999.

Reference
Fabre H, Ramiaramana A, Blanchin M-D, et al. Stability-indicating assay for oxyphenbutazone. Part II. High-performance liquid chromatographic determination of oxyphenbutazone and its degradation products. *Analyst.* 1986; 111: 133–7.

Oxytocin

Chemical Name
Cys-Tyr-Ile-Gln-Asn-Cys-Pro-Leu-Gly-NH$_2$ cyclic(1→6) disulfide

Cys-Tyr—Ile—Gln—Asn—Cys—Pro—Leu—Gly-NH$_2$

Other Names
Pitocin, Syntocinon

Form	Molecular Formula	MW	CAS
Oxytocin	$C_{43}H_{66}N_{12}O_{12}S_2$	1007.2	50-56-6

Appearance
Oxytocin is a white or almost white powder.

Solubility
Oxytocin is soluble in water.

Method
Wang et al. reported a reversed-phase HPLC method for the determination of oxytocin acetate in oxytocin injection USP, synthetic. The instrument comprised a Waters model 680 automated gradient controller, two model 510 pumps, a model 717 WISP auto-sampler, and a model 486 variable-wavelength UV detector. The stationary phase was an Alltech Hypersil ODS column (120 × 4.6 mm, 5-μm particle size) or a Beckman Ultra-

sphere ODS column (150 × 4.6 mm, 5-μm particle size). Mobile phase A was 100 mM monobasic sodium phosphate buffer (adjusted to pH 3.1–4.5 with phosphoric acid). Mobile phase B was a mixture of acetonitrile and water (1:1, vol/vol). The flow rate was 1.5 mL/min. The mobile phase was delivered from 70% A to 50% A from 0 to 30 minutes, held at 50% A for 5 minutes, and then increased from 50% A to 70% A from 35 to 45 minutes and held at 70% A for another 15 minutes. UV detection was performed at 220 nm. Chlorobutanol was used as an internal standard. The diluent contained 5 g/L of chlorobutanol, 5 mL/L of glacial acetic acid, 5 g/L of ethanol, and 1.82 g/L of sodium acetate trihydrate in water.

Samples were assayed without dilution. The injection volume was 100 μL. Under these conditions, retention times for oxytocin acetate and chlorobutanol were 10.2 and 21.1 minutes, respectively.

The stability-indicating nature of the method was demonstrated by an accelerated degradation study. Oxytocin was degraded by heat, acid, base, oxidizing agent, and fluorescent radiation for up to 48 hours. Degradation products did not interfere with the analysis of oxytocin. Photodiode-array detection also confirmed the peak purity of oxytocin.

A standard curve for oxytocin in the diluent was generated from 5 to 15 units/mL. The correlation coefficient was 1.000.

Reference

Wang G, Miller RB, Melendez L, et al. A stability-indicating HPLC method for the determination of oxytocin acetate in Oxytocin Injection, USP, Synthetic. *J Liq Chromatogr Rel Technol.* 1997; 20: 567–81.

Paclitaxel

Chemical Name

[2aR-[2aα,4β,4aβ,6β,9α(αR*, βS*),11α,12α,12aα,12bα]]-β-(Benzoylamino)-α-hydroxy-benzenepropanoic acid 6,12b-bis(acetyloxy)-12-(benzoyloxy)-2a,3,4,4a,5,6,9,10,11,-12,12a,12b-dodecahydro-4,11-dihydroxy-4a,8,13,13-tetramethyl-5-oxo-7,11-methano-1H-cyclodeca[3,4]benz[1,2-b]oxet-9-yl ester

Other Name
Taxol

Form	Molecular Formula	MW	CAS
Paclitaxel	$C_{47}H_{51}NO_{14}$	853.9	33069-62-4

Appearance
Paclitaxel occurs as a white to off-white crystalline powder.

Solubility
Paclitaxel is insoluble in water but highly lipophilic.

Method 1
Burm et al. used an HPLC method to evaluate the stability of paclitaxel with ondansetron hydrochloride or ranitidine hydrochloride during simulated Y-site administration. The liquid chromatograph consisted of a Hitachi model L-6200 pump, a model L-4200 UV-visible detector, a model AS2000 autosampler, and a model D-2500 integrator. The stationary phase was an Alltech Adsorbosphere C_{18} column (250 × 4.6 mm, 5-μm particle size). The mobile phase consisted of acetonitrile and a buffer (60:40, vol/vol). The buffer contained 12.5 mM ammonium phosphate adjusted to pH 4.5 with 0.1 N hydrochloric acid. The flow rate was 1.0 mL/min. UV detection was performed at 227 nm.

Samples were diluted with mobile phase. The injection volume was 20 μL. Retention time for paclitaxel was 10.0 minutes.

The assay was determined to be stability indicating by accelerated degradation of paclitaxel. Paclitaxel samples 240 μg/mL were exposed to 1 N hydrochloric acid or 1 N sodium hydroxide for 5 hours at 58 °C, to 3% hydrogen peroxide for 17 hours at room temperature, or to UV radiation and 1 N hydrochloric acid for 22 hours at room temperature. No degradation products of paclitaxel interfered with the intact paclitaxel peak.

A standard curve for paclitaxel was constructed from 40 to 200 μg/mL. The correlation coefficient was greater than 0.999. The intraday and interday coefficients of variation were less than 2%.

References
Burm J-P, Jhee SS, Chin A, et al. Stability of paclitaxel with ondansetron hydro-chloride or ranitidine hydrochloride during simulated Y-site administration. *Am J Hosp Pharm*. 1994; 51: 1201–4.

Burm J-P, Choi J-S, Jhee SS, et al. Stability of paclitaxel and fluconazole during simulated Y-site administration. *Am J Hosp Pharm*. 1994; 51: 2704–6.

Method 2
Xu et al. determined the chemical stability of paclitaxel 0.1 and 1 mg/mL in 5% dextrose injection and in 0.9% sodium chloride injection. The HPLC system consisted of a Waters 600E multisolvent-delivery system, a Waters 490E programmable multiple-wavelength UV detector, and a Waters 710B WISP autosampler. The system was controlled and the peak area was integrated by an NEC PowerMate personal computer with Waters Maxima 820 chromatography manager. The stationary phase was a Vydac C_{18} reversed-phase

column (250 × 4.6 mm, 5-μm particle size). The mobile phase was 53% acetonitrile in water and was delivered isocratically at 1.5 mL/min. UV detection was performed at 254 nm and 0.5 AUFS.

Samples were diluted with the respective infusion solution before analysis. Under these conditions, the retention time for paclitaxel was 6.09 minutes.

The assay was determined to be stability indicating by accelerated degradation of paclitaxel. Paclitaxel solution 0.1 mg/mL was adjusted to pH 11.1 with 0.1 N sodium hydroxide. After 1 hour at room temperature, only 78% of the intact paclitaxel remained. Retention times for degradation products were 2.00, 4.18, and 9.02 minutes. These peaks did not interfere with the parent peak.

A calibration curve for paclitaxel was constructed from a linear plot of peak area versus concentration of paclitaxel reference standard from 0.025 to 0.150 mg/mL. The correlation coefficient was greater than 0.9999. The intraday and interday coefficients of variation were 1.4 and 2.0%, respectively.

Similar methods were used by the other researchers cited here.

References

Xu Q, Trissel LA, Martinez JF. Stability of paclitaxel in 5% dextrose injection or 0.9% sodium chloride injection at 4, 22, or 32 °C. *Am J Hosp Pharm.* 1994; 51: 3058–60.

Xu QA, Trissel LA, Davis M. Compatibility of paclitaxel in 5% glucose and 0.9% sodium chloride injections with EVA minibags. *Aust J Hosp Pharm.* 1998; 28: 156–9.

Trissel LA, Xu Q, Martinez JF. Compounding an extended-stability admixture of paclitaxel for long-term infusion. *Int J Pharm Compound.* 1997; 1: 49–52.

Xu QA, Trissel LA, Gilbert DL. Paclitaxel compatibility with a TOTM-plasticized PVC administration set. *Hosp Pharm.* 1997; 32: 1635–8.

Chin A, Ramakrishnan RR, Yoshimura NN, et al. Paclitaxel stability and compatibility in polyolefin containers. *Ann Pharmacother.* 1994; 28: 35–6.

Method 3

Using an HPLC method, Waugh et al. investigated the stability of paclitaxel in infusion solutions. The chromatograph consisted of an Altex model 110A constant-flow pump, a Waters model 440 UV detector, a Rheodyne model 7125 injector, and a Shimadzu C-R3A Chromatopac integrator. The stationary phase was a Shandon ODS Hypersil column (150 × 4.6 mm, 5-μm particle size). The mobile phase was 40% acetonitrile in water. The flow rate was 4 mL/min. UV detection was performed at 254 nm.

Samples were diluted with the mobile phase before injection. The retention volume for paclitaxel was 24 mL.

The analytical method was determined to be stability indicating by accelerated decomposition of paclitaxel. A solution of paclitaxel was adjusted to pH 11.5 and allowed to stand for 1.5 hours. Numerous degradation product peaks were observed. These peaks did not interfere with the paclitaxel peak.

Standard curves for paclitaxel were constructed from a linear plot of peak area versus concentration from 0.04 to 0.3 mg/mL. The coefficient of variation was 0.59%.

A similar method was used by Mayron and Gennaro.

References

Waugh WN, Trissel LA, Stella VJ. Stability, compatibility, and plasticizer extraction of taxol (NSC-125973) injection diluted in infusion solutions and stored in various containers. *Am J Hosp Pharm.* 1991; 48: 1520–4.

Mayron D, Gennaro AR. Stability and compatibility of granisetron hydrochloride in i.v. solutions and oral liquids and during simulated Y-site injection with selected drugs. *Am J Health Syst Pharm.* 1996; 53: 294–304.

Palonosetron Hydrochloride

Chemical Name

(3a*S*)-2-[(*S*)-1-Azabicyclo[2.2.2]oct-3-yl]-2,3,3a,4,5,6-hexahydro-1-oxo-1*H*-benz[*de*]iso-quinoline monohydrochloride

Other Name

Aloxi

Form	Molecular Formula	MW	CAS
Palonosetron hydrochloride	$C_{19}H_{24}N_2O \cdot HCl$	332.9	135729-55-4

Appearance

Palonosetron hydrochloride occurs as a white to off-white crystalline powder.

Solubility

Palonosetron hydrochloride is freely soluble in water, soluble in propylene glycol, and slightly soluble in ethanol and 2-propanol.

Method

Using an HPLC method, Xu and Trissel evaluated the stability of palonosetron hydrochloride with paclitaxel and with docetaxel in 5% dextrose injection during simulated Y-site administration. The Waters Alliance separation module model 2690 system was coupled with a Waters model 996 photodiode-array detector. The stationary phase was an Agilent Zorbax SB-C$_8$ column (250 × 4.6 mm, 5-μm particle size). The mobile phase consisted of 720 mL of water, 280 mL of acetonitrile, and 0.67 mL of trifluoroacetic acid. The flow rate was 1.2 mL/min. UV detection was performed at 260 nm and 0.5 AUFS. The injection volume was 10 μL. Under these conditions, the retention time of palonosetron was 8.9 minutes and no peak for paclitaxel or docetaxel was detected.

The method was demonstrated to be stability indicating by accelerated degradation. Sample solutions were mixed with 1 N hydrochloric acid, 1 N sodium hydroxide solution, and 3% hydrogen peroxide and heated. Loss of the intact drugs was observed, and there was no interference with the peak of palonosetron from the degradation product peaks or other drug peaks.

A standard curve for palonosetron was constructed from 2 to 50 µg/mL; the correlation coefficient was greater than 0.9999. The coefficient of variation for the analysis was 1.6%.

Similar methods were reported.

References

Xu QA, Trissel LA. Stability of palonosetron hydrochloride with paclitaxel and docetaxel during simulated Y-site administration. *Am J Health Syst Pharm.* 2004; 61: 1596–8.

Trissel LA, Zhang Y, Xu QA. Physical and chemical stability of palonosetron hydrochloride with dacarbazine and with methylprednisolone sodium succinate during simulated Y-site administration. *Int J Pharm Compound.* 2006; 10: 234–6.

Trissel LA, Xu QA. Physical and chemical stability of palonosetron HCl in 4 infusion solutions. *Ann Pharmacother.* 2004; 38: 1608–11.

Trissel LA, Zhang Y. Compatibility and stability of Aloxi (palonosetron hydrochloride) admixed with dexamethasone sodium phosphate. *Int J Pharm Compound.* 2004; 8: 398–403.

Trissel LA, Zhang Y. Physical and chemical stability of palonosetron HCl with cisplatin, carboplatin, and oxaliplatin during simulated Y-site administration. *J Oncol Pharm Pract.* 2004; 10: 191–5.

Pamidronate Disodium

Chemical Names

Disodium 3-amino-1-hydroxypropylidene diphosphonate
Disodium 3-amino-1-hydroxypropane-1,1-diphosphonate

Other Name

Aredia

Form	Molecular Formula	MW	CAS
Pamidronate disodium	$C_3H_9NNa_2O_7P_2$	279.0	57248-88-1
Pamidronic acid	$C_3H_{11}NO_7P_2$	235.1	40391-99-9

Appearance

Pamidronate disodium is a white to practically white powder.

Solubility

Pamidronate disodium is soluble in water and practically insoluble in organic solvents.

Method

Hartigh et al. developed a method for the quantitative determination of bisphosphonates in the quality control of pharmaceutical preparations. The high-performance liquid chromatograph consisted of an LKB 2150 HPLC pump equipped with a Waters high-sensitivity noise filter and restrictor tube, a Promis autosampler, a Waters temperature control system, a Waters 430 conductivity detector, and a Spectra-Physics model SP4270 integrator. The stationary phase was a column (50 × 4.6 mm, 10-μm particle size) packed with Waters IC-pac anion and used in combination with a Waters Guard-Pak precolumn module. The mobile phase was 2 mM nitric acid filtered through a 0.5-μm filter and deaerated ultrasonically before use. The flow rate was 1.0 mL/min. All chromatographic analyses were performed at a column temperature of 35 ± 0.5 °C. 3-Phenyl-3-amino-1-hydroxypropylidene-1,1-bisphosphonic acid was used as an internal standard.

Samples were dissolved in purified water or mobile phase and diluted to an appropriate concentration before analysis. The injection volume was 20 μL. The retention time for pamidronate was 4 minutes.

The HPLC method was determined to be stability indicating by the separation between pamidronate and phosphate, a potential impurity and decomposition product of the bisphosphonates. Retention times for pamidronate, the phosphate, and the internal standard were about 4.0, 5.6, and 6.4 minutes, respectively (estimated from the published chromatogram).

A calibration curve for pamidronate was constructed from 200 ng to 10 μg per injection. The correlation coefficient was 0.9992. The intraday and interday coefficients of variation were 1.7 and 2.7%, respectively.

The limit of detection of pamidronate was 100 ng per injection.

Reference

Hartigh JD, Langebroek R, Vermeij P. Ion-exchange liquid chromatographic analysis of bisphosphonates in pharmaceutical preparations. *J Pharm Biomed Anal.* 1993; 11: 977–83.

Pantoprazole

Chemical Name

5-Difluoromethoxybenzimidazol-2-yl-3,4-dimethoxy-2-pyridylmethyl sulfoxide

Other Names

Controloc, Pantoloc, Pantozol, Protonix

Form	Molecular Formula	MW	CAS
Pantoprazole	$C_{16}H_{15}F_2N_3O_4S$	383.4	102625-70-7
Pantoprazole sodium	$C_{16}H_{14}F_2N_3NaO_4S$	405.4	138786-67-1

Appearance
Pantoprazole sodium occurs as a white to off-white crystalline powder.

Solubility
Pantoprazole sodium is freely soluble in water and practically insoluble in *n*-hexane.

Method
Nesseem and Bebawy developed an HPLC method for the analysis of pantoprazole. The Waters liquid chromatograph included an autosampler and a UV-visible detector. The stationary phase was a Nova-Pak R CN column (150 × 3.9 mm, 10-μm particle size). The mobile phase consisted of phosphate buffer (pH 3.0) and methanol (40:60, vol/vol). The flow was 1.5 mL/min. The sample injection volume was 20 μL. Under these conditions, the retention time of pantoprazole was 2.6 minutes.

The assay method was reported to quantitatively determine pantoprazole without any interference from its degradation products.

A calibration curve for pantoprazole was constructed from 10 to 70 μg/mL. The correlation coefficient was 0.9998. The limit of detection and the limit of quantitation were 1.78 and 5.87 μg/mL, respectively.

Reference
Nesseem DI, Bebawy LI. Coacervation method and stability study of pantoprazole as an antiulcerative drug. *Bull Fac Pharm Cairo Univ.* 2003; 41: 223–34.

Papaverine

Chemical Names
1-[(3,4-Dimethoxyphenyl)methyl]-6,7-dimethoxyisoquinoline
6,7-Dimethoxy-1-veratrylisoquinoline

Other Name
Pavabid

Form	Molecular Formula	MW	CAS
Papaverine	$C_{20}H_{21}NO_4$	339.4	58-74-2
Papaverine hydrochloride	$C_{20}H_{21}NO_4 \cdot HCl$	375.9	61-25-6

Appearance
Papaverine hydrochloride occurs as white crystals or as a white crystalline powder.

Solubility
Papaverine hydrochloride is soluble in water and slightly soluble in alcohol.

Method
Tu et al. determined the stability of papaverine hydrochloride and phentolamine mesylate mixed in a single vial. The liquid chromatograph consisted of a Waters model 6000A dual-piston pump, a Waters model 440 dual UV detector, a Houston Omniscribe recorder, and a Waters Nova-Pak C_{18} column (150 × 3.9 mm, 5-μm particle size). The mobile phase consisted of 30% acetonitrile and 70% 0.05 M monobasic potassium phosphate buffer at pH 7.85. The flow rate was 1.3 mL/min. UV detection was performed at 254 nm. The retention times for phentolamine and papaverine were 3.6 and 8.5 minutes, respectively.

This assay was determined to be stability indicating by accelerated decomposition of papaverine hydrochloride. A solution of papaverine hydrochloride and phentolamine mesylate was stored at 90 °C for 5 days. The degradation product peaks did not interfere with the peaks of the intact papaverine and phentolamine.

Standard curves were constructed each day and their correlation coefficients were greater than 0.99.

A similar method was reported by other researchers.

References
Tu Y-H, Allen LV Jr, Wang D-P. Stability of papaverine hydrochloride and phentolamine mesylate in injectable mixtures. *Am J Hosp Pharm.* 1987; 44: 2524–7.

Trissel LA, Zhang Y. Long-term stability of Trimix: A three-drug injection used to treat erectile dysfunction. *Int J Pharm Compound.* 2004; 8: 231–4.

Paroxetine

Chemical Name
3-[(1,3-Benzodioxol-5-yloxy)methyl]-4-(4-fluorophenyl)piperidine

Other Names
Paxil

Form	Molecular Formula	MW	CAS
Paroxetine	$C_{19}H_{20}FNO_3$	329.4	61869-08-7
Paroxetine hydrochloride	$C_{19}H_{20}FNO_3.HCl$	365.8	78246-49-8
Paroxetine mesylate	$C_{19}H_{20}FNO_3.CH_4O_3S$	425.5	217797-14-3

Appearance
Paroxetine hydrochloride occurs as a white to off-white solid.

Solubility
Paroxetine hydrochloride is slightly soluble in water and soluble in alcohol and in methyl alcohol.

Method
Lambropoulos et al. reported the development and validation of an assay for paroxetine hydrochloride tablets. The liquid chromatographic system consisted of a Hitachi model L-6200A Intellient pump, a Micromeritics model 728 autosampler, an Applied Biosystems model 785A programmable absorbance detector, and a Hewlett-Packard 1050 photodiode-array detector. The stationary phase was an Inertsil C_{18} column (150 × 4.6 mm, 5-µm particle size). The mobile phase consisted of acetonitrile and a buffer containing 10 mM decanesulfonic acid and 10 mM monobasic sodium phosphate (pH 3.0) (60:40, vol/vol). The flow rate was 1.2 mL/min. UV detection was performed at 235 nm.

A working standard solution of paroxetine was prepared in mobile phase. For tablets, a portion of powder from 20 tablets equivalent to two tablets was accurately weighed, transferred into a 100-mL volumetric flask, mixed with 70 mL of mobile phase, shaken for 10 minutes, sonicated for 10 minutes, reshaken for another 10 minutes, filled to volume with mobile phase, mixed well, and filtered through a Gelman Acrodisc polytetrafluoroethylene (PTFE) 0.45-µm filter. The injection volume was 25 µL. Under these conditions, the retention time of paroxetine was about 12 minutes.

Paroxetine tablets were degraded at 80 °C in 0.1 N hydrochloric acid, 0.1 N sodium hydroxide solution, and 3% hydrogen peroxide. Samples were also heated at 80 °C dry and wet. Samples were exposed to light equivalent to 1000 foot candles in dry and in wet conditions. Degradation product peaks were resolved from the intact paroxetine peak, demonstrating the stability-indicating nature of the method.

A linear standard curve was constructed from 0.20 to 0.67 mg/mL. The correlation coefficient was greater than 0.999. The limit of quantitation was 0.8 µg/mL.

Reference
Lambropoulos J, Spanos GA, Lazaridis NV. Method development and validation for the HPLC assay (potency and related substances) for 20 mg paroxetine tablets. *J Pharm Biomed Anal.* 1999; 19: 793–802.

Penicillin G

Chemical Name
[2S-(2α,5α,6β)]-3,3-Dimethyl-7-oxo-6-[(phenylacetyl)amino]-4-thia-1-azabicyclo-[3.2.0]heptane-2-carboxylic acid

Other Names
Benzylpenicillin, Pfizerpen

Form	Molecular Formula	MW	CAS
Penicillin G	$C_{16}H_{18}N_2O_4S$	334.4	61-33-6
Penicillin G potassium	$C_{16}H_{17}KN_2O_4S$	372.5	113-98-4
Penicillin G sodium	$C_{16}H_{17}N_2NaO_4S$	356.4	69-57-8

Appearance
Penicillin G potassium and penicillin G sodium occur as white or almost white crystalline powders.

Solubility
Penicillin G potassium and penicillin G sodium are very soluble in water. They are practically insoluble in ether, chloroform, fixed oil, and liquid paraffin.

pK_a
Penicillin G has a pK_a of 2.76.

Method 1
Stiles and Allen evaluated the stability of penicillin G potassium and penicillin G sodium in polyvinyl chloride drug reservoirs. The analytical method was an HPLC method and the liquid chromatograph included a Waters model 501 solvent-delivery system, a Micromeritics model 728 autoinjector, a Micromeritics model 732 injection valve, a Waters Lambda-Max model 481 variable-wavelength UV detector, and a Waters model 745 data module. The stationary phase was a Bakerbond C_{18} column (250 × 2.6 mm, 5-μm particle size). The mobile phase was a mixture of methanol and 0.074 M monobasic potassium phosphate aqueous buffer (pH 4.15) (47.5:52.5, vol/vol). The flow rate was 1.5 mL/min. UV detection was performed at 225 nm.

Samples were diluted with the mobile phase.

The analytical method was determined to be stability indicating by accelerated degradation of penicillin G. The retention time for penicillin G was 6.6 minutes, and retention times for its degradation products were 3.3, 3.8, and 4.7 minutes.

A standard curve for penicillin G was constructed from 50 to 200 μg/mL. Intraday and interday coefficients of variation were 1.6 and 1.5% for penicillin G potassium and 2.4 and 2.6% for penicillin G sodium.

Reference
Stiles ML, Allen LV Jr. Stability of nafcillin sodium, oxacillin sodium, penicillin G potassium, penicillin G sodium, and tobramycin sulfate in polyvinyl chloride drug reservoirs. *Am J Health Syst Pharm.* 1997; 54: 1068–70.

Method 2

Allwood and Brown utilized a stability-indicating HPLC method to investigate the stability of penicillin G sodium injection after reconstitution. The injector was a Rheodyne model 7125. The stationary phase was a 10-cm column packed with Spherisorb 5-μm ODS. The mobile phase consisted of acetonitrile and 0.031 M potassium acid phosphate (20:80, vol/vol). The flow rate was 1.5 mL/min. UV detection was performed at 227 nm and 0.2 AUFS. The sample injection volume was 20 μL. Under these conditions, the retention time for penicillin G was approximately 3.5 minutes.

This method was determined to be stability indicating by comparing chromatograms for unstressed and heat-stressed (60 °C for more than 5 hours) samples in citrate buffers (pH 6.0 and 8.0). The degradation product peaks did not interfere with the intact penicillin G peak.

Calibration curves were constructed by plotting the peak area of penicillin G against concentration from 0.6 to 6.0 mg/mL. The correlation coefficient was 0.9992, and the coefficient of variation of repeated injections ($n = 5$) was 1.25%.

A similar method was used by Gupta and Stewart.

References

Allwood MC, Brown PW. The effect of buffering on the stability of reconstituted benzylpenicillin injection. *Int J Pharm Pract.* 1992; 1: 242–4.

Gupta VD, Stewart KR. Chemical stability of hydrocortisone sodium succinate and several antibiotics when mixed with metronidazole injection for intravenous infusion. *J Parenter Sci Technol.* 1985; 39: 145–8.

Method 3

Using HPLC, Stiles et al. studied the stability of penicillin G sodium in portable pump reservoirs. The liquid chromatograph consisted of a Waters model 501 pump, a U6K universal injector, a model 441 UV detector, and a model 745 data module. The stationary phase was a Waters Nova-Pak C_{18} analytical column (150×3.9 mm, 5-μm particle size). The mobile phase consisted of 50% methanol and 50% 0.005 M phosphate buffer at pH 7.5 with 0.0013 M tetrabutylammonium hydroxide and was delivered isocratically at 0.5 mL/min. UV detection was performed at 254 nm.

Samples were diluted with distilled water. The retention time for penicillin G sodium was 2.4 minutes.

The assay was confirmed to be stability indicating by degrading the sample solution at 25 °C for 24 hours. Penicillin G sodium was well separated from its degradation products and excipients.

Standard curves were constructed from 100 to 400 units/mL; the correlation coefficient was greater than 0.99. The coefficient of variation of the assay was less than 5%.

Reference

Stiles M, Tu Y-H, Allen LV. Stability of cefazolin sodium, cefoxitin sodium, ceftazidime, and penicillin G sodium in portable pump reservoirs. *Am J Hosp Pharm.* 1989; 46: 1408–12.

Pentamidine

Chemical Names

4,4'-[1,5-Pentanediylbis(oxy)]bis-benzenecarboximidamide

4,4'-(Pentamethylenedioxy)dibenzamidine

Other Names

NebuPent, Pentacarinat, Pentam 300

Form	Molecular Formula	MW	CAS
Pentamidine	$C_{19}H_{24}N_4O_2$	340.4	100-33-4
Pentamidine isethionate	$C_{19}H_{24}N_4O_2.2C_2H_6O_4S$	592.7	140-64-7
Pentamidine mesylate	$C_{19}H_{24}N_4O_2.2CH_3SO_3H$	532.6	6823-79-6

Appearance

Pentamidine isethionate occurs as white or almost white crystals or powder. Pentamidine mesylate is a white or very faintly pink granular powder.

Solubility

Pentamidine isethionate is soluble in water (~100 mg/mL at 25 °C) and is slightly soluble in alcohol. Pentamidine mesylate is slightly soluble in water and alcohol and practically insoluble in acetone, chloroform, and ether.

pK_a

Pentamidine has a pK_{a1} and pK_{a2} of 11.4.

Method

De et al. evaluated the stability of pentamidine isethionate in 5% dextrose injection and 0.9% sodium chloride injection when stored in polyvinyl chloride bags and when infused through a polyvinyl chloride infusion set. The liquid chromatograph consisted of a Waters M730 data module, a Waters model 720 system controller, a Waters 710B WISP automatic injector, a Waters model 6000A pump, a Waters model 480 variable-wavelength UV detector, and a Micropak CN-10 stainless steel column (300 × 4.0 mm). The mobile phase consisted of 20% 0.005 M methanesulfonic acid and 80% methyl-*tert*-butyl ether and was delivered gradiently at 1 mL/min. After 4 minutes, methanesulfonic acid was increased from 20 to 60% in 16 minutes. The final composition was held constant for another 12 minutes. Therefore, the total run time was approximately 40 minutes. UV detection was performed at 254 nm and 0.5 AUFS. The sample injection volume was 20 µL. Under these conditions, the retention time of pentamidine was 23 minutes.

The method was stated to be stability indicating because the assay separated the contaminants and degradation products of pentamidine from the intact drug. Retention times of impurities such as 1,5-di(*p*-amidophenoxy)pentane, 1-(*p*-amidinophenoxy)-5-(*p*-carboxyphenoxy)pentane, and 1-(*p*-amidinophenoxy)-5-(*p*-amidophenoxy)pentane were 5, 16, and 18 minutes, respectively.

The intrarun and interrun coefficients of variation were less than 2 and 5%, respectively.

Reference

De NC, Alam AS, Kapoor JN. Stability of pentamidine isethionate in 5% dextrose and 0.9% sodium chloride injections. *Am J Hosp Pharm.* 1986; 43: 1486–8.

Pentobarbital

Chemical Names

5-Ethyl-5-(1-methylbutyl)-2,4,6(1*H*,3*H*,5*H*)-pyrimidinetrione
5-Ethyl-5-(1-methylbutyl)barbituric acid

Other Names

Nembutal, Pentobarbitone

Form	Molecular Formula	MW	CAS
Pentobarbital	$C_{11}H_{18}N_2O_3$	226.3	76-74-4
Pentobarbital sodium	$C_{11}H_{17}N_2NaO_3$	248.3	57-33-0

Appearance

Pentobarbital sodium occurs as white crystalline granules or as a white powder.

Solubility

Pentobarbital sodium is freely soluble in water and alcohol but practically insoluble in ether.

pK$_a$

Pentobarbital has a pK$_a$ of 7.85–8.03.

Method 1

Morley and Elrod reported the HPLC determination of pentobarbital and pentobarbital sodium in bulk drug substance and dosage forms. The apparatus consisted of a Shimadzu model LC-10AD pump, a model SIL-10A autosampler, a model SPD-10A UV detector, and a model C-R7A data-handling system. The stationary phase was a Nucleosil C_{18}

column (150 × 4.6 mm, 5-µm particle size). The mobile phase consisted of 0.01 M monobasic potassium phosphate buffer (adjusted to pH 3.5 with phosphoric acid) and acetonitrile (72:28, vol/vol). The flow rate was 1 mL/min. UV detection was performed at 214 nm and 0.10 AUFS. 4'-Ethoxyacetophenone 0.70 mg/mL in acetonitrile was used as an internal standard.

Samples of elixir were diluted in mobile phase. Samples of capsule contents were extracted with mobile phase, diluted, and filtered. For suppositories, samples were dissolved, diluted with tetrahydrofuran–water (9:1, vol/vol), and filtered. The injection volume was 50 µL. Under these conditions, the retention time for pentobarbital was about 14.9 minutes.

The stability-indicating ability of the method was shown by accelerated degradation of pentobarbital. Pentobarbital sodium was degraded with high-intensity UV light, acid reflux, and base reflux. A solid sample was also stored at 150 °C for 1 hour. No degradation products interfered with the retention time of pentobarbital.

A calibration curve was constructed from 0.0892 to 89.2 µg/mL. The correlation coefficient was greater than 0.9999.

Reference

Morley JA, Elrod L Jr. Determination of pentobarbital and pentobarbital sodium in bulk drug substance and dosage forms by high-performance liquid chromatography. *J Pharm Biomed Anal.* 1997; 16: 119–29.

Method 2

Reif et al. developed a stability-indicating HPLC method for pentobarbital sodium injection. The system consisted of a Waters 6000A pump, an LDC SpectroMonitor III variable-wavelength UV detector, and a Rheodyne 7010 injector equipped with a 20-µL loop. The stationary phase was a µBondapak C_{18} column (250–300 × 4 mm, 10-µm particle size). The mobile phase was a mixture of acetate buffer, methanol, and propylene glycol (65:35:4, vol/vol/vol). Acetate buffer was prepared by diluting 4.1 g of anhydrous sodium acetate and 3 mL of acetic acid to a total volume of 1 L with distilled water. The flow rate was 2.0 mL/min. UV detection was performed at 230 nm and 0.4 AUFS.

Samples were diluted with the mobile phase. The injection volume was 20 µL.

The HPLC assay was demonstrated to be stability indicating by spiking the pentobarbital sample with its known degradation products. Pentobarbital was separated from its degradation products on the chromatogram. Retention times for 2-ethyl-2-(1-methylbutyl)propanediamide, pentobarbital, and *N*-(aminocarbonyl)-2-ethyl-3-methyl-hexanamide were about 1.6, 3.0, and 5.0 minutes, respectively (estimated from the published chromatogram).

A standard curve was constructed from 0.14 to 1.30 mg/mL. The correlation coefficient was greater than 0.9999.

Reference

Reif VD, Kaufmann KL, DeAngelis NJ, et al. Liquid chromatographic assays for barbiturate injections. *J Pharm Sci.* 1986; 75: 714–6.

Pentostatin

Chemical Name
(*R*)-3-(2-Deoxy-β-D-*erythro*-pentofuranosyl)-3,6,7,8-tetrahydroimidazo[4,5-*d*]-[1,3]diazepin-8-ol

Other Name
Nipent

Form	Molecular Formula	MW	CAS
Pentostatin	$C_{11}H_{16}N_4O_4$	268.3	53910-25-1

Appearance
Pentostatin occurs as a white to off-white solid.

Solubility
Pentostatin is freely soluble in water (>100 mg/mL).

pK_a
Pentostatin has a pK_a of 5.4.

Method
Al-Razzak et al. studied the chemical stability of pentostatin at various pH values, buffer concentrations, and temperatures. The liquid chromatograph consisted of a Kratos Spectroflow 400 solvent-delivery system, a Kratos Spectroflow 480 injector fitted with a 20-μL loop, a Kratos spectroflow 783 variable-wavelength detector, and a Nelson analytical integrator. The stationary phase was a Shandon Hypersil ODS C_{18} reversed-phase column (150 × 4.6 mm, 5-μm particle size). The mobile phase consisted of methanol and 0.01 M phosphate buffer (pH 7.0) (2.5:97.5, vol/vol). The flow rate was 2 mL/min. UV detection was performed at 280 nm. The retention volume for pentostatin was 15.4 mL.

The assay was stated to be stability indicating because the major acidic degradation product and the major alkaline degradation product had retention volumes of 2.96 and 9.60 mL, respectively. They did not interfere with the intact pentostatin peak.

Calibration curves were constructed from 0.003 to 0.04 mg/mL.

Reference
Al-Razzak LA, Benedetti AE, Waugh WN, et al. Chemical stability of pentostatin (NSC-218321), a cytotoxic and immunosuppressant agent. *Pharm Res.* 1990; 7: 452–60.

Perphenazine

Chemical Name

2-[4-[3-(2-Chlorophenothiazin-10-yl)propyl]piperazin-1-yl]ethanol

Other Names

Fentazin, Trilafon

Form	Molecular Formula	MW	CAS
Perphenazine	$C_{21}H_{26}ClN_3OS$	404.0	58-39-9
Perphenazine decanoate	$C_{31}H_{44}ClN_3O_2S$	558.2	
Perphenazine enanthate	$C_{28}H_{38}ClN_3O_2S$	516.1	17528-28-8

Appearance

Perphenazine is a white to creamy white odorless powder.

Solubility

Perphenazine is practically insoluble in water and freely soluble in alcohol.

Method 1

Using an HPLC method, Li et al. investigated the stability of perphenazine aerosols generated using the capillary aerosol generator. The Waters 2695 separation module was equipped with a model 2995 photodiode-array detector. The stationary phase was a Waters Nova-Pak C_8 column (150 × 3.9 mm, 4-μm particle size). The mobile phase consisted of 0.01 M ammonium acetate/acetic acid buffer (pH 3.0) and acetonitrile (52:48, vol/vol). The flow rate was 1.0 mL/min. UV detection was performed at 256 nm. The run time was 16 minutes.

Perphenazine was dissolved in methanol and then diluted in mobile phase. The injection volume was 60 μL. Under these conditions, the retention time of perphenazine was about 3.8 minutes (estimated from the published chromatogram).

The stability-indicating ability of the method was demonstrated by accelerated degradation studies. Methanolic perphenazine solutions were treated with 0.1 N hydrochloric acid or 0.1 N sodium hydroxide solution at 24 or 50 °C for 3 days, with 0.5% hydrogen peroxide aqueous solution at 24 °C for up to 5 hours, or stored at 24 °C in the dark for 2 months. There was good baseline separation of the intact drug from its major degradation products formed under these stress conditions.

A calibration curve was constructed from 0.03 to 26.8 μg/mL. The correlation coefficient of the linear regression line was 0.9999. The coefficient of variation of the assay was 0.25%. The limit of detection and the limit of quantitation were 0.01 and 0.03 μg/mL, respectively.

Reference
Li X, Blondino FE, Hindle M, et al. Stability and characterization of perphenazine aerosols generated using the capillary aerosol generator. *Int J Pharm.* 2005; 303: 113–24.

Method 2
Al-Obaid et al. presented the simultaneous quantitation of perphenazine and its sulfoxide. The Waters 600E liquid chromatograph was equipped with a Waters model U6K injector and a model 486 tunable absorbance detector. The stationary phase was a Waters Nova-Pak phenyl column (150 × 3.9 mm, 4-μm particle size). The mobile phase consisted of methanol and 0.015 M sodium acetate buffer (pH 6.5) (81:19, vol/vol). The flow rate was 1.0 mL/min. UV detection was performed at 254 nm and 0.1 AUFS. Pindolol 3 μg/mL was used as an internal standard.

A portion of ground tablet powder equivalent to 4 mg of perphenazine was weighed, mixed with 5 mL of water, heated for 3 minutes in a water bath, cooled, mixed with 50 mL of water, shaken for 15 minutes, filled to 100 mL with methanol, filtered, added with internal standard, and diluted 1:10 with the mobile phase. The injection volume was 10 μL. Under these conditions, retention times for perphenazine sulfoxide, pindolol, and perphenazine were 1.7, 2.5, and 4.4 minutes, respectively.

The method was reported to be stability indicating since it separated perphenazine from perphenazine sulfoxide.

A standard curve for perphenazine was constructed from 0.5 to 6 μg/mL. The correlation coefficient was 0.9999. The limit of detection was 0.1 ng.

Reference
Al-Obaid AM, Hagga MEM, El-Khawad IE, et al. Simultaneous quantitation of some phenothiazine drug substances and their monosulphoxide degrades by high performance liquid chromatography (HPLC). *J Liq Chromatogr Rel Technol.* 1996; 19: 1369–89.

Phenelzine Sulfate

Chemical Name
Phenethylhydrazine sulfate

Other Name
Nardil

Form	Molecular Formula	MW	CAS
Phenelzine sulfate	$C_8H_{12}N_2 \cdot H_2SO_4$	234.3	156-51-4

Appearance
Phenelzine sulfate is a white to yellowish-white powder or pearly platelets with a characteristic odor.

Solubility
Phenelzine sulfate is freely soluble in water. It is practically insoluble in alcohol, chloroform, and ether.

Method
George and Stewart reported an assay for phenelzine sulfate. The liquid chromatographic system comprised a Beckman model 110B pump, a Rheodyne model 7125 injector, a Kratos Spectroflow model 757 variable-wavelength detector, and a Shimadzu model C-R3A integrator or a Houston Instrument 4500 recorder. The stationary phase was a Brownlee ODS column (100 × 4.6 mm, 5-μm particle size). The mobile phase consisted of 800 mL of double-distilled water containing 4.4 g of 1-heptanesulfonic acid sodium salt and 6.9 g of monobasic sodium phosphate (adjusted to pH 2.5 with phosphoric acid) and 200 mL of acetonitrile. The flow rate was 1.0 mL/min. UV detection was performed at 209 nm. Under these conditions, the retention time for phenelzine was approximately 6.8 minutes.

The method was reported to be stability indicating since it separated the degradants from the parent compound.

A standard curve for phenelzine was constructed from 0.075 to 0.225 mg/mL. The correlation coefficient was 0.9991.

Reference
George GD, Stewart JT. HPLC assay for phenelzine sulfate drug substance and the decomposition product phenethyl alcohol using short wavelength UV detection. *J Liq Chromatogr*. 1988; 11: 2399–407.

Phenobarbital

Chemical Names
5-Ethyl-5-phenyl-2,4,6($1H,3H,5H$)-pyrimidinetrione
5-Ethyl-5-phenylbarbituric acid

Other Names
Luminal, Phenobarbitone

Form	Molecular Formula	MW	CAS
Phenobarbital	$C_{12}H_{12}N_2O_3$	232.2	50-06-6
Phenobarbital sodium	$C_{12}H_{11}N_2NaO_3$	254.2	57-30-7

Appearance

Phenobarbital occurs as white crystals or as a white crystalline powder. Phenobarbital sodium occurs as flaky crystals, as white crystalline granules, or as a white powder.

Solubility

Phenobarbital is very slightly soluble in water and soluble in alcohol. One gram of phenobarbital sodium dissolves in about 1 mL of water and about 10 mL of alcohol. It is insoluble in ether and chloroform.

pK$_a$

Phenobarbital has a pK$_a$ of 7.3.

Method 1

Walker et al. evaluated the chemical stability of phenobarbital using an HPLC method. The chromatographic system consisted of a Spectra-Physics SP4200 ternary gradient solvent-delivery pump, a Schoeffel SF 770 variable-wavelength UV detector, and a Beckman Ultrasphere ODS C_{18} reversed-phase column (250 × 4.2 mm, 5-μm particle size). The mobile phase consisted of acetonitrile and phosphate buffer, which also contained heptanesulfonic acid 1 mg/mL as a counterion. It was delivered gradiently from 18 to 35% acetonitrile in 16 minutes at 2 mL/min. The UV detection wavelength was 230 nm.

The method was demonstrated to be stability indicating by accelerated decomposition of phenobarbital. A solution of phenobarbital 0.4 mg/mL in 80% methanol and 20% water was incubated at 90 °C for 363 minutes and then assayed. No degradation products interfered with the determination of the intact phenobarbital.

The intraday coefficient of variation for the assay was 2.27%.

Reference

Walker SE, DeAngelis C, Iazzetta J. Stability and compatibility of combinations of hydromorphone and a second drug. *Can J Hosp Pharm.* 1991; 44: 289–95.

Method 2

Reif et al. developed a stability-indicating HPLC assay for phenobarbital sodium injection. The liquid chromatograph consisted of a Waters 6000A pump, an LDC SpectroMonitor III variable-wavelength detector, and a Rheodyne 7010 injector with a 20-μL loop. The stationary phase was a Partisil ODS-3 (250–300 × 4 mm, 10-μm particle size). The mobile phase consisted of acetate buffer, methanol, and propylene glycol (45:55:4, vol/vol/vol). Acetate buffer contained 4.1 g of anhydrous sodium acetate and 3 mL of acetic acid in a total volume of 1 L of distilled water. The flow rate was 2 mL/min. UV detection was performed at 230 nm and 0.4 AUFS.

Samples were diluted in mobile phase. The injection volume was 20 μL.

The assay was demonstrated to be stability indicating by spiking phenobarbital samples with known degradation products. Degradation product peaks did not interfere with the intact drug peak. Retention times for ethylphenylpropanediamide, phenobarbital, α-phenylbutyric acid, and *N*-(aminocarbonyl)-2-phenylbutanamide were about 2.0, 4.6, 7.3, and 11.5 minutes, respectively (estimated from the published chromatogram).

A standard curve for phenobarbital was constructed from 0.154 to 1.43 mg/mL. The correlation coefficient was greater than 0.999.

Reference

Reif VD, Kaufmann KL, DeAngelis NJ, et al. Liquid chromatographic assays for barbiturate injections. *J Pharm Sci.* 1986; 75: 714–6.

Method 3

Nahata et al. studied the stability of phenobarbital sodium in 0.9% sodium chloride injection at an unadjusted and adjusted pH at 4 °C. The analysis used a Laboratory Data Control high-performance liquid chromatograph. The stationary phase was a reversed-phase Spherisorb ODS column (220 × 4 mm). The mobile phase was 50% methanol in water and was delivered isocratically at 1 mL/min. The column eluent was monitored at 235 nm.

Samples were diluted to a phenobarbital concentration of 1 mg/mL before injection.

The stability-indicating capability of the method was demonstrated by accelerated decomposition of phenobarbital. A sample of phenobarbital sodium was boiled at 100 °C for 14 hours. Degradation product peaks did not interfere with the phenobarbital peak. Retention times for phenobarbital and the degradation product were about 0.6 and 3.2 minutes, respectively (estimated from the published chromatogram).

The coefficient of variation of the assay of phenobarbital 1 mg/mL was less than 4%.

Reference

Nahata MC, Hipple TF, Strausbauch SD. Stability of phenobarbital sodium diluted in 0.9% sodium chloride injection. *Am J Hosp Pharm.* 1986; 43: 384–5.

Phenoxybenzamine

Chemical Names

N-(2-Chloroethyl)-*N*-(1-methyl-2-phenoxyethyl)benzenemethanamine

N-(2-Chloroethyl)-*N*-(1-methyl-2-phenoxyethyl)benzylamine

Other Name

Dibenzyline

Form	Molecular Formula	MW	CAS
Phenoxybenzamine	$C_{18}H_{22}ClNO$	303.8	59-96-1
Phenoxybenzamine hydrochloride	$C_{18}H_{22}ClNO.HCl$	340.2	63-92-3

Appearance

Phenoxybenzamine hydrochloride occurs as a white crystalline powder.

Solubility

Phenoxybenzamine hydrochloride has solubilities of approximately 40 mg/mL in water and 167 mg/mL in alcohol at 25 °C.

pK$_a$

Phenoxybenzamine has a pK$_a$ of 4.4.

Method

Lim et al. determined the stability of phenoxybenzamine hydrochloride in various vehicles. The liquid chromatograph was a Hewlett-Packard model 1050 Chem Station equipped with a Hewlett-Packard model 1050 diode-array detector. The stationary phase was a Phenomenex C_{18} column (250 × 4.6 mm, 10-μm particle size). The mobile phase consisted of monobasic potassium phosphate buffer (pH 7.5) and methanol (9:1, vol/vol). The flow rate was 1 mL/min. UV detection was performed at 254 nm. The column temperature was maintained at 30 °C.

Each sample was diluted with methanol to an expected drug concentration of 0.8 mg/mL. The injection volume was 25 μL.

The stability-indicating capability of the method was demonstrated by accelerated decomposition of phenoxybenzamine hydrochloride. Aqueous solutions of phenoxybenzamine hydrochloride 2 mg/mL were heated in 1 M sulfuric acid (pH 2) or 1 M sodium hydroxide (pH 12) at 100 °C for 30 minutes. Degradation product peaks were obtained at 2.9–3.6 minutes and at 4.5 minutes. In each case, degradation product peaks did not interfere with the intact drug peak.

A standard curve for phenoxybenzamine hydrochloride was constructed from 0.60 to 2.40 mg/mL in methanol; the correlation coefficient was greater than 0.99. The intraday and interday coefficients of variation were 0.15–3.51 and 1.43–3.86%, respectively.

Reference

Lim L-Y, Tan LL, Chan EWY, et al. Stability of phenoxybenzamine hydrochloride in various vehicles. *Am J Health Syst Pharm.* 1997; 54: 2073–8.

Phentolamine

Chemical Names

3-[[(4,5-Dihydro-1*H*-imidazol-2-yl)methyl](4-methylphenyl)amino]phenol

2-[*N*-(*m*-Hydroxyphenyl)-*p*-toluidinomethyl]imidazoline

Other Name
Regitine

Form

Form	Molecular Formula	MW	CAS
Phentolamine	$C_{17}H_{19}N_3O$	281.4	50-60-2
Phentolamine mesylate	$C_{17}H_{19}N_3O.CH_4O_3S$	377.5	65-28-1

Appearance
Phentolamine mesylate is a white or off-white crystalline powder.

Solubility
One gram of phentolamine mesylate dissolves in 50 mL of water, 23 mL of alcohol, and 660 mL of chloroform.

pK$_a$
Phentolamine mesylate has a pK$_a$ of 8.01.

Method
Tu et al. determined the stability of phentolamine mesylate and papaverine hydrochloride mixed in a single vial. The liquid chromatograph consisted of a Waters model 6000A dual-piston pump, a Waters model 440 dual UV detector, a Houston Omniscribe recorder, and a Waters Nova-Pak C_{18} column (150 × 3.9 mm, 5-µm particle size). The mobile phase consisted of 30% acetonitrile and 70% 0.05 M monobasic potassium phosphate buffer at pH 7.85. The flow rate was 1.3 mL/min. UV detection was performed at 254 nm. The retention times for phentolamine and papaverine were 3.6 and 8.5 minutes, respectively.

This assay was determined to be stability indicating by accelerated decomposition of phentolamine. A solution of papaverine hydrochloride and phentolamine mesylate was stored at 90 °C for 5 days. The degradation product peaks did not interfere with peaks of the intact papaverine and phentolamine.

Standard curves were constructed each day and their correlation coefficients were greater than 0.99.

Reference
Tu Y-H, Allen LV Jr, Wang D-P. Stability of papaverine hydrochloride and phentolamine mesylate in injectable mixtures. *Am J Hosp Pharm.* 1987; 44: 2524–7.

Phenylbutazone

Chemical Name
4-Butyl-1,2-diphenylpyrazolidine-3,5-dione

Other Names
Ambene, Butazolidin, Butazolidine, Butazone

Form	Molecular Formula	MW	CAS
Phenylbutazone	$C_{19}H_{20}N_2O_2$	308.4	50-33-9

Appearance
Phenylbutazone is a white or off-white odorless crystalline powder.

Solubility
Phenylbutazone is very slightly soluble in water, soluble in alcohol, and freely soluble in acetone and ether.

Method
Fabre et al. reported an assay for the determination of phenylbutazone. The Spectra-Physics model SP8000 liquid chromatograph was equipped with a Schoeffel model SF 770 variable-wavelength UV detector. The stationary phase was a laboratory-made LiChrosorb RP18 column (150 × 4.0 mm, 5-μm particle size). The mobile phase was a mixture of 0.1 M Tris-citrate buffer (pH 5.25) and acetonitrile (52:48, vol/vol). The flow rate was 2 mL/min. UV detection was performed at 237 nm and 0.1 AUFS. Diphenylamine 500 μg/mL in mobile phase was used as an internal standard.

Samples were diluted with the mobile phase. The injection volume was 10 μL. Under these conditions, retention times for phenylbutazone and diphenylamine were about 3.0 and 6.2 minutes, respectively (estimated from the published chromatogram).

The stability-indicating capability of the assay was shown by assaying a mixture of phenylbutazone and its degradation products. The intact phenylbutazone peak was resolved from its degradation product peaks.

A standard curve for phenylbutazone was generated from 5 to 160 μg/mL. The correlation coefficient was greater than 0.998.

Reference
Fabre H, Hussam-Eddine N, Mandrou B. Stability-indicating assay for phenylbutazone: High-performance liquid chromatographic determination of hydrazobenzene and azobenzene in degraded aqueous phenylbutazone solutions. *J Pharm Sci.* 1984; 73: 1706–9.

Phenylephrine

Chemical Names
(*R*)-3-Hydroxy-α-[(methylamino)methyl]benzenemethanol
1-*m*-Hydroxy-α-[(methylamino)methyl]benzyl alcohol

Other Names
Neo-Synephrine, Prefrin

Form	Molecular Formula	MW	CAS
Phenylephrine	$C_9H_{13}NO_2$	167.2	59-42-7
Phenylephrine bitartrate	$C_9H_{13}NO_2.C_4H_6O_6$	317.3	13998-27-1
Phenylephrine hydrochloride	$C_9H_{13}NO_2.HCl$	203.7	61-76-7
Phenylephrine oxazolidine	$C_{15}H_{23}NO_2$	249.2	126766-90-3

Appearance
Phenylephrine bitartrate is a white crystalline powder. Phenylephrine hydrochloride occurs as odorless white or practically white crystals.

Solubility
Phenylephrine bitartrate is soluble in water and insoluble in alcohol. Phenylephrine hydrochloride is freely soluble in water and alcohol.

Method 1
Qiu et al. developed an HPLC method for the quantitation of phenylephrine oxazolidine. The liquid chromatograph consisted of a Shimadzu model LC-6A solvent-delivery pump, a Rheodyne model 7164 20-μL loop injection valve, a Shimadzu model SPD-6 AV variable-wavelength detector, a model RF-535 fluorescence detector, and a model CR-601 integrator. The stationary phase was a Waters μBondapak CN normal-phase column (300 × 4.0 mm, 10-μm particle size). The mobile phase was a mixture of hexane and isopropyl alcohol (95:5, vol/vol). The flow rate was 1.0 mL/min. UV detection was performed at 280 nm and 0.01 AUFS. The injection volume was 20 μL. Under these conditions, retention times for pivalaldehyde, phenylephrine oxazolidine, propylparaben, and methylparaben were about 3.4, 6.0, 7.0, and 8.0 minutes, respectively.

The method was reported to be stability indicating since it resolved the degradation product and preservatives from phenylephrine oxazolidine.

A standard curve for phenylephrine oxazolidine was constructed from 4 to 200 μg/mL.

Reference

Qiu Y, Guillory JK, Schoenwald RD. Formulation, *in vitro* dissolution, and ocular bioavailability of high- and low-melting phenylephrine oxazolidines. *Pharm Res.* 1993; 10: 1627–31.

Method 2

Gupta and Stewart investigated the chemical stabilities of phenylephrine and lidocaine (lignocaine) in combination in the most commonly used concentrations. A Waters ALC 202 high-pressure liquid chromatograph was equipped with an Omniscribe recorder, a Kratos SF 770 multiple-wavelength detector, and a Waters μBondapak C_{18} nonpolar column (300 × 3.9 mm). The mobile phase was a 2% (vol/vol) aqueous solution of acetonitrile containing 0.02 M monobasic potassium phosphate, adjusted to pH 6.0 with phosphoric acid. The flow rate was 2.0 mL/min. UV detection was performed at 271 nm and 0.04 AUFS. A solution of acetaminophen 0.1% in methanol was used as an internal standard.

Samples were diluted with water, and the injection volume was 20 μL. Retention times for phenylephrine and acetaminophen were about 4.5 and 9.0 minutes, respectively (estimated from the published chromatogram).

The assay was stated to be stability indicating.

Reference

Gupta VD, Stewart KR. Chemical stabilities of lignocaine hydrochloride and phenylephrine hydrochloride in aqueous solution. *J Clin Hosp Pharm.* 1986; 11: 449–52.

Method 3

Schieffer et al. reported a simultaneous determination of phenylephrine hydrochloride, phenylpropanolamine hydrochloride, guaifenesin, and sodium benzoate in dosage forms. The Waters model ALC 204 system was equipped with a Waters model 6000A reciprocating pump, a model 440 absorbance detector, a Rheodyne model 725 injector, and a Hewlett-Packard model 3352B integrator. The stationary phase was a Whatman Partisil-10 C_8 column. The mobile phase consisted of 50 mL of methanol, 170 mL of acetonitrile, 755 mL of water, and 25 mL of pentanesulfonic acid sodium salt in glacial acetic acid. The flow rate was 2.0 mL/min. UV detection was performed at 254 nm.

For liquid forms, samples were diluted with water. For solid forms, samples were leached with water or mobile phase and then diluted. Under these conditions, retention times for phenylephrine hydrochloride, phenylpropanolamine hydrochloride, guaifenesin, and benzoic acid were about 2.8, 4.4, 7.4, and 14.4 minutes, respectively (estimated from the published chromatogram).

The stability-indicating nature of the method was demonstrated by spiking the sample solution with possible impurities and degradation products. Phenylephrine hydrochloride, phenylpropanolamine hydrochloride, guaifenesin, and benzoic acid were separated from α-aminopropiophenone, 2-(2-methoxyphenoxy)-1,3-propanediol, *m*-hydroxybenzaldehyde, guaicol, and benzaldehyde.

A standard curve for phenylephrine was constructed from 2 to 5.6 μg/mL. The correlation coefficient was 0.9999.

References

Schieffer GW, Smith WO, Lubey GS, et al. Determination of the structure of a synthetic impurity in guaifenesin: Modification of a high-performance liquid chromatographic method for phenylephrine hydrochloride, phenylpropanolamine hydrochloride, guaifenesin, and sodium benzoate in dosage forms. *J Pharm Sci.* 1984; 73: 1856–8.

Schieffer GW, Hughes DE. Simultaneous stability-indicating determination of phenylephrine hydrochloride, phenylpropanolamine hydrochloride, and guaifenesin in dosage forms by reversed-phase paired-ion high-performance liquid chromatography. *J Pharm Sci.* 1983; 72: 55–9.

Phenylmercuric Nitrate

Chemical Name
Nitratophenylmercury

Form	Molecular Formula	MW	CAS
Phenylmercuric nitrate	$C_6H_5HgOH.C_6H_5HgNO_3$	634.4	55-68-5

Appearance
Phenylmercuric nitrate is a mixture of phenylmercuric hydroxide and phenylmercuric nitrate. It occurs as a white or pale yellow crystalline powder.

Solubility
Phenylmercuric nitrate is very slightly soluble in water and in alcohol. It is more soluble in the presence of nitric acid or alkali hydroxide.

Method
Parkin used HPLC analysis to investigate the interaction of phenylmercuric nitrate and sodium metabisulfite in eye drop formulations. The Waters liquid chromatograph included a Waters model 501 pump, a model 712 WISP autosampler, a model 490 variable-wavelength detector, and a Hewlett-Packard model 3396A integrator. The stationary phase was a Waters ODS column (300 × 3.9 mm, 10-μm particle size). The mobile phase was a mixture of acetonitrile and water (75:25, vol/vol) containing 0.1 mM disodium ethylenediaminetetraacetate. The flow rate was 1.8 mL/min. UV detection was performed at 258 nm.

Phenylmercuric nitrate was determined through derivatization. An aliquot of 1 mL of phenylmercuric nitrate sample was reacted with 1 mL of diethylamine salt of diethylaminedithiocarbamate (DEADTC). The injection volume was 20 μL. Under these conditions, retention times for DEADTC, DEADTC oxidation product, DEADTC complex of phenylmercuric nitrate, and DEADTC complex of mercury were 1.3, 3.6, 4.5, and 6.2 minutes, respectively.

The method was reported to be stability indicating.

A standard curve for phenylmercuric nitrate was generated from 0 to 0.25 mM. The correlation coefficient was 0.9998.

Reference

Parkin JE. High-performance liquid chromatographic investigation of the interaction of phenylmercuric nitrate and sodium metabisulphite in eye drop formulations. *J Chromatogr.* 1990; 511: 233–42.

Phenylpropanolamine

Chemical Name

(1*RS*,2*SR*)-2-Amino-1-phenylpropan-1-ol

(+)-Form

Other Names

Acutrim, Dexatrim, Fasupond, Procol

Form	Molecular Formula	MW	CAS
Phenylpropanolamine	$C_9H_{13}NO$	151.2	14838-15-4
Phenylpropanolamine hydrochloride	$C_9H_{13}NO.HCl$	187.7	154-41-6

Appearance

Phenylpropanolamine hydrochloride is a white or almost white crystalline powder.

Solubility

Phenylpropanolamine hydrochloride is freely soluble in water and in alcohol.

Method 1

Heidemann et al. reported an HPLC method for the analysis of phenylpropanolamine hydrochloride, pseudoephedrine hydrochloride, dextromethorphan hydrobromide, and chlorpheniramine maleate in cough-cold products. The apparatus consisted of a Waters model 510 dual-piston pump, a model 481 variable-wavelength detector, and a model 710B WISP autosampler. The stationary phase was a DuPont Zorbax 300-SCX column (150 × 4.6 mm, 7–8-µm particle size). The mobile phase consisted of acetonitrile and ethylenediamine sulfate buffer (pH 4.52) (35:65, vol/vol). Ethylenediamine sulfate buffer was prepared by dissolving 225 mg of the salt in 1 L of water and adjusting the pH to 4.52 with 0.2 N sulfuric acid. The flow rate was 2 mL/min. UV detection was performed at 216 nm.

For immediate-release tablets, 10 tablets were placed in a 1000-mL volumetric flask containing 250 mL of 0.1 N hydrochloric acid in 50% alcohol, shaken for 2 hours, diluted to volume with water, and filtered through glass-fiber paper. The injection volume was 10 µL. For sustained-release tablets, 10 tablets were powdered, transferred into a 1000-mL volumetric flask containing 600 mL of 0.01 N hydrochloric acid, boiled vigorously with magnetic stirring for 10–13 minutes, cooled to room temperature, diluted to volume with 0.01 N hydrochloric acid, filtered through glass-fiber paper, and diluted 1:10 with water. The injection volume was 20 µL. Under these conditions, retention times of phenylpropanolamine hydrochloride and chlorpheniramine maleate were about 2.6 and 9.5 minutes, respectively (estimated from the published chromatogram).

The method was evaluated to be stability indicating by the analysis of degraded samples of the different formulations. Degradation products did not interfere with the determination of the drugs.

A standard curve for phenylpropanolamine hydrochloride was constructed from 80 to 120% of the labeled concentrations. The correlation coefficient was 1.000.

Reference

Heidemann DR, Groon KS, Smith JM. HPLC determination of cough-cold products. *LC-GC.* 1987; 5: 422–6.

Method 2

Schieffer et al. reported a simultaneous determination of phenylephrine hydrochloride, phenylpropanolamine hydrochloride, guaifenesin, and sodium benzoate in dosage forms. The Waters model ALC 204 system was equipped with a Waters model 6000A reciprocating pump, a model 440 absorbance detector, a Rheodyne model 725 injector, and a Hewlett-Packard model 3352B integrator. The stationary phase was a Whatman Partisil-10 C_8 column. The mobile phase consisted of 50 mL of methanol, 170 mL of acetonitrile, 755 mL of water, and 25 mL of pentanesulfonic acid sodium salt in glacial acetic acid. The flow rate was 2.0 mL/min. UV detection was performed at 254 nm.

For liquid forms, samples were diluted with water. For solid forms, samples were leached with water or mobile phase and then diluted. Under these conditions, retention times for phenylephrine hydrochloride, phenylpropanolamine hydrochloride, guaifenesin, and benzoic acid were about 2.8, 4.4, 7.4, and 14.4 minutes, respectively (estimated from the published chromatogram).

The stability-indicating nature of the method was demonstrated by spiking the sample solution with possible impurities and degradation products. Phenylephrine hydrochloride, phenylpropanolamine hydrochloride, guaifenesin, and benzoic acid were separated from α-aminopropiophenone, 2-(2-methoxyphenoxy)-1,3-propanediol, *m*-hydroxybenzaldehyde, guaicol, and benzaldehyde.

A standard curve for phenylpropanolamine was constructed from 0.24 to 0.45 µg/mL. The correlation coefficient was 0.9999.

References

Schieffer GW, Smith WO, Lubey GS, et al. Determination of the structure of a synthetic impurity in guaifenesin: Modification of a high-performance liquid chromatographic method for phenylephrine hydrochloride, phenylpropanolamine hydrochloride, guaifenesin, and sodium benzoate in dosage forms. *J Pharm Sci.* 1984; 73: 1856–8.

Schieffer GW, Hughes DE. Simultaneous stability-indicating determination of phenyl-ephrine hydrochloride, phenylpropanolamine hydrochloride, and guaifenesin in dosage forms by reversed-phase paired-ion high-performance liquid chromatography. *J Pharm Sci.* 1983; 72: 55–9.

Phenyltoloxamine Citrate

Chemical Name

2-(2-Benzylphenoxy)-*N,N*-dimethylmethylamine dihydrogen citrate

Other Names

Biocidan, Codipront, Flextra, Lobac

Form	Molecular Formula	MW	CAS
Phenyltoloxamine citrate	$C_{17}H_{21}NO.C_6H_8O_7$	447.5	1176-08-5

Appearance

Phenyltoloxamine citrate occurs as crystals.

Solubility

Phenyltoloxamine citrate is soluble in water.

Method

Aukunuru et al. presented the simultaneous determination of acetaminophen, salicyl-amide, phenyltoloxamine, and related products by HPLC. A Varian system consisted of a model 9010 pump, a model 9095 autosampler, a model 9050 UV absorbance detector, and a Rainin Dinamax MacIntegrator. The stationary phase was a Phenomenex Prodigy C_8 column (150 × 4.6 mm, 5-μm particle size). Mobile phase A was 0.1 M phosphate buffer (adjusted to pH 2.7 with phosphoric acid). Mobile phase B was acetonitrile. The mobile phase was linearly delivered from 5% B to 45% B in 17 minutes followed by 10 minutes for equilibration. The flow rate was 1 mL/min. UV detection was performed at 220 nm. The injection volume was 50 μL. Under these conditions, retention times of acetaminophen, salicylamide, and phenyltoloxamine were 5.7, 10.9, and 15.9 minutes, respectively.

The stability-indicating nature of the assay was verified by accelerated degra-dation of the drugs. Drug solutions were prepared in 1 N hydrochloric acid, 1 N sodium hydroxide, or 10% hydrogen peroxide and heated at 60 °C. No degradation products interfered with the analysis of the drugs.

A standard curve for phenyltoloxamine was generated from 0.04 to 200 µg/mL. The correlation coefficient was 0.999.

Reference
Aukunuru JV, Kompella UB, Betageri GV. Simultaneous high performance liquid chromatographic analysis of acetaminophen, salicylamide, phenyltoloxamine, and related products. *J Liq Chromatogr Rel Technol.* 2000; 23: 565–78.

Phenytoin

Chemical Name
5,5-Diphenyl-2,4-imidazolidinedione

Other Name
Dilantin

Form	Molecular Formula	MW	CAS
Phenytoin	$C_{15}H_{12}N_2O_2$	252.3	57-41-0
Phenytoin sodium	$C_{15}H_{11}N_2NaO_2$	274.3	630-93-3

Appearance
Phenytoin is a white powder. Phenytoin sodium is a white hygroscopic powder.

Solubility
Phenytoin is practically insoluble in water, soluble in hot alcohol, and slightly soluble in cold alcohol. Phenytoin sodium is freely soluble in water, soluble in alcohol, and freely soluble in warm propylene glycol.

pK$_a$
Phenytoin has a pK$_a$ of 8.06–8.33.

Method 1
Splinter et al. used an HPLC method to determine the recovery of phenytoin from a suspension after in vitro administration through percutaneous endoscopic gastrostomy Pezzer catheters. The liquid chromatograph was composed of a Beckman model 1108 pump, a Beckman model 163 variable-wavelength UV detector, a Beckman model 427 integrator, and a Waters µBondapak C$_{18}$ reversed-phase column with a Brownlee Labs OD-GU precolumn. The mobile phase was a mixture of acetonitrile and water (2:3, vol/vol) and was delivered isocratically at 1.5 mL/min. The column eluent was monitored at 214 nm and 0.2 AUFS.

Samples were filtered through a 0.22-µm Millipore polyvinylidene difluoride hydrophobic filter before injection.

The assay was determined to be stability indicating. Four ampules containing phenytoin were placed in a mechanical convection oven at 100 ± 0.25 °C and were chromatographed after 17 and 23 days of incubation. The degradation products did not interfere with the intact phenytoin peak.

A standard curve for phenytoin was generated from 10 to 150 µg/mL. Intraday and interday coefficients of variation were 1.47 and 1.67%, respectively.

Reference

Splinter MY, Seifert CF, Bradberry JC, et al. Recovery of phenytoin suspension after in vitro administration through percutaneous endoscopic gastrostomy Pezzer catheters. *Am J Hosp Pharm.* 1990; 47: 373–6.

Method 2

Walker et al. evaluated the chemical stability of phenytoin sodium using an HPLC method. The chromatographic system consisted of a Spectra-Physics SP4200 ternary gradient solvent-delivery pump, a Schoeffel SF 770 variable-wavelength UV detector, and a Beckman Ultrasphere ODS C_{18} reversed-phase column (250 × 4.2 mm, 5-µm particle size). The mobile phase consisted of acetonitrile and phosphate buffer, which also contained heptanesulfonic acid 1 mg/mL as a counterion. It was gradiently delivered from 20 to 50% acetonitrile in 12 minutes at 2 mL/min. UV detection wavelength was set at 230 nm.

The assay was determined to be stability indicating. A solution of phenytoin 5 mg/mL in methanol was incubated at 90 °C for 2 hours and then assayed. Degradation product peaks did not interfere with the intact phenytoin peak.

The intraday coefficient of variation for the assay was 1.63%.

Reference

Walker SE, DeAngelis C, Iazzetta J. Stability and compatibility of combinations of hydromorphone and a second drug. *Can J Hosp Pharm.* 1991; 44: 289–95.

Method 3

Miller and Strom studied the stability of phenytoin sodium in three enteral nutrient formulas. The HPLC system consisted of a Micromeritics model 750 solvent-delivery system, a Micromeritics model 787 variable-wavelength UV-visible detector, and a Micromeritics model 740 microprocessor. The stationary phase was an Alltech C_{18} column (250 × 4.6 mm, 10-µm particle size). The mobile phase consisted of methanol and water (60:40, vol/vol) and was run isocratically at 1.25 mL/min. The detector was operated at 245 nm. Carbamazepine 0.4 mg/mL in methanol was used as the internal standard.

A 2.0-mL aliquot of the sample was mixed with 1 mL of the internal standard, diluted to 10.0 mL with methanol, and then centrifuged for 2 minutes. The supernatant was filtered through a 0.45-µm Gelman Alpha-450 filter and assayed. The injection volume was 25 µL. Under these conditions, retention times for phenytoin and carbamazepine were 6.7 and 8.7 minutes, respectively.

The method was stated to be stability indicating because it separated precursors, intermediates, and degradation products from the intact phenytoin peak. These compounds follow, with their relative retention times given in parentheses: diphenylhydantoic acid (0.26), benzilic acid (0.31), α,α-diphenylglycine (0.52), benzoin (1.36), benzophenone (3.81), and benzil (3.84).

A similar method was used by Shah and Ogger.

References

Miller SW, Strom JG Jr. Stability of phenytoin in three enteral nutrient formulas. *Am J Hosp Pharm.* 1988; 45: 2529–32.

Shah VP, Ogger KE. Comparison of ultraviolet and liquid chromatographic methods for dissolution testing of sodium phenytoin capsules. *J Pharm Sci*. 1986; 75: 1113–5.

Phosphanilic Acid

Chemical Name
(4-Aminophenyl)phosphonic acid

Other Name
4-Phosphonoaniline

Form	Molecular Formula	MW	CAS
Phosphanilic acid	$C_6H_8NO_3P$	173.1	5337-17-7

Method
Gadde et al. described an HPLC analysis of phosphanilic acid. The Waters liquid chromatograph consisted of a model 6000A pump, a model 710 WISP autosampler, and a model 440 absorbance detector. A Hewlett-Packard model 1040A photodiode-array detector was used for the peak purity analysis. The stationary phase was a Waters μBondapak NH_2 column. The mobile phase consisted of acetonitrile and 0.012 M aqueous ammonium dihydrogen phosphate (10:90, vol/vol). The flow rate was 2.0 mL/min. UV detection was performed at 254 nm. Salicylic acid 14 mg/mL was used as an internal standard.

Phosphanilic acid and salicylic acid solutions were prepared in 0.5 N sodium hydroxide solution and diluted with a mixture of acetonitrile/water (45:55, vol/vol). The injection volume was 10 μL. Under these conditions, the retention times of salicylic acid and phosphanilic acid were about 4.1 and 7.0 minutes, respectively (estimated from the published chromatogram).

The stability-indicating capability of the assay was shown by accelerated degradation. Chlorhexidine phosphanilate samples were heated at 60 ℃ or exposed to 1000-foot candles of light for 8 days. The intact phosphanilic acid peak was separated from degradation product peaks of chlorhexidine phosphanilate.

A standard curve for phosphanilic acid was obtained from 0.02 to 0.06 mg/mL. The correlation coefficient was greater than 0.9999.

Reference
Gadde RR, McNiff EF, Peer MM. High-performance liquid chromatographic analysis of chlorhexidine phosphanilate, a new antimicrobial agent. *J Pharm Biomed Anal*. 1991; 9: 1031–6.

Physostigmine

Chemical Name

(3aS,8aR)-1,2,3,3a,8,8a-Hexahydro-1,3a,8-trimethylpyrrolo[2,3-b]indol-5-ol methylcarbamate

Other Names

Anticholium, Antilirium

Form	Molecular Formula	MW	CAS
Physostigmine	$C_{15}H_{21}N_3O_2$	275.3	57-47-6
Physostigmine salicylate	$C_{15}H_{21}N_3O_2.C_7H_6O_3$	413.5	57-64-7
Physostigmine sulfate	$(C_{15}H_{21}N_3O_2)_2.H_2SO_4$	648.8	64-47-1

Appearance

Physostigmine is a white odorless microcrystalline powder. Physostigmine salicylate occurs as colorless or white odorless crystals or as a white powder. Physostigmine sulfate is a white or almost white odorless and hygroscopic crystalline powder.

Solubility

Physostigmine is slightly soluble in water and freely soluble in alcohol. Physostigmine salicylate has solubilities of approximately 13.3 mg/mL in water and 62.5 mg/mL in alcohol at 25 °C. Physostigmine sulfate is very soluble in water and freely soluble in alcohol.

Method 1

Rubnov et al. developed an HPLC method for the determination of physostigmine in the presence of its degradation products. The Hewlett-Packard 1090 liquid chromatograph was equipped with a model HP3396A integrator, a 20-µL loop injector, and a diode-array detector. The stationary phase was an HPLC Technologies (UK) Bondapak C_{18} column (250 × 5 mm, 10-µm particle size). The mobile phase was a mixture of acetonitrile and 0.1 M ammonium acetate (pH 6.0) (50:50, vol/vol). The flow rate was 1.2 mL/min. UV detection was performed at 248 nm.

Samples were diluted with water and then with acetonitrile to approximately 200 µg/mL of physostigmine. Under these conditions, retention times for salicylate and physostigmine were 2.5 and 8.3 minutes, respectively.

The stability-indicating nature of the method was demonstrated by accelerated degradation of physostigmine. An aqueous sample of physostigmine salicylate was heated in boiling water for 2 hours. The degradation products did not interfere with the physostigmine peak.

Reference
Rubnov S, Levy D, Schneider H. Liquid chromatographic analysis of physostigmine salicylate and its degradation products. *J Pharm Biomed Anal*. 1999; 18: 939–45.

Method 2
Lau et al. described the simultaneous determination of tacrine and physostigmine in skin samples. An LDC ConstaMetric I chromatograph was equipped with an LDC UV-III fixed-wavelength detector, a Valco injection valve with a 50-µL loop, and a Linear Instrument recorder. The stationary phase was a Custom Spherisorb ODS-1 column (150 × 4.6 mm, 5-µm particle size) with an ODS guard column (20 × 4 mm, 40-µm particle size). The mobile phase consisted of acetonitrile and water (52:48, vol/vol) (pH 3.5) containing 0.01 M octanesulfonic acid and 1% (vol/vol) acetic acid. The flow rate was 1 mL/min. UV detection was performed at 254 nm and 0.008 AUFS. Diazepam was used as an internal standard.

A 150–200-mg skin sample was homogenized with two 4-mL portions of chloroform; the mixture was filtered through phase separation paper, alkalinized with 2 mL of 10% (wt/vol) sodium hydroxide solution, washed with two 2-mL portions of water, filtered, and dried under a stream of air; the residue was reconstituted with 1 mL of the mobile phase and filtered through a microfilter. The injection volume was 50 µL. Under these conditions, retention times of physostigmine, diazepam, and tacrine were 4.5, 6.0, and 7.7 minutes, respectively.

The method was demonstrated to be stability indicating by accelerated degradation of the drugs. Sample solutions were boiled for 0.5 hour or adjusted to pH less than 1 or pH greater than 13 and boiled for 1.25 hours. The intact drugs were separated from their degradation products.

A standard curve for physostigmine was generated from 1 to 50 µg/200 mg of skin sample. The correlation coefficient was greater than 0.998. Intraday and interday coefficients of variation were 5.2 and 10.2%, respectively. The limit of detection was 1 µg/200 mg of skin sample.

Reference
Lau SWJ, Chow D, Feldman S. Simultaneous determination of physostigmine and tetrahydroaminoacridine in a transdermal permeation study by high-performance liquid chromatography. *J Chromatogr*. 1990; 526: 87–95.

Pilocarpine

Chemical Name
(3*S*,4*R*)-3-Ethyldihydro-4-[(1-methyl-1*H*-imidazol-5-yl)methyl]furan-2(3*H*)-one

Other Names

Akarpine, Pilocar, Pilocarpol, Pilogel

Form	Molecular Formula	MW	CAS
Pilocarpine	$C_{11}H_{16}N_2O_2$	208.3	92-13-7
Pilocarpine borate	$C_{11}H_{16}N_2O_2 \cdot xBH_3O_3$		16509-56-1
Pilocarpine hydrochloride	$C_{11}H_{16}N_2O_2 \cdot HCl$	244.7	54-71-1
Pilocarpine nitrate	$C_{11}H_{16}N_2O_2 \cdot HNO_3$	271.3	148-72-1

Appearance

Pilocarpine occurs as a viscous oily liquid or as crystals. Pilocarpine hydrochloride occurs as hygroscopic colorless and translucent odorless crystals. Pilocarpine nitrate occurs as colorless or white crystals or as a white or almost white crystalline powder.

Solubility

Pilocarpine is soluble in water and alcohol. Pilocarpine hydrochloride has solubilities of approximately 3.33 g/mL in water and 333 mg/mL in alcohol at 25 °C. Pilocarpine nitrate has solubilities of approximately 0.25 g/mL in water and 13.3 mg/mL in alcohol at 25 °C.

pKa

Pilocarpine has pK_a values of 7.15 and 12.57.

Method 1

Aromdee et al. investigated the enzymatic hydrolysis of pilocarpine in human serum in vitro using a stability-indicating HPLC method. The HPLC apparatus consisted of a Jasco model 880-PU isocratic pump, a model 875 UV detector, and a model 851-AS autosampler. The stationary phase was a Nova-Pak C_{18} column (150 × 3.9 mm) with a Hypersil C_{18} guard column (20 × 4.6 mm, 5-μm particle size). The mobile phase was prepared by dissolving 50.0 g of monobasic potassium phosphate in 900 mL of water and then mixing with 8.0 mL of triethylamine, 3.5 mL of phosphoric acid, 9.0 mL of acetonitrile, and finally an additional 90 mL of water. The flow rate was 1 mL/min. UV detection was performed at 216 nm.

The diluent was prepared by adding 50 mg of pseudoephedrine hydrochloride as internal standard and 4.1 g of sodium fluoride as enzyme inhibitor to McIlvaine's citrate buffer (pH 5.5) and diluting to 1 liter with buffer. Samples of serum (50 μL) were diluted appropriately with the diluent before assay. The injection volume was 20 μL. Under these conditions, retention times for pilocarpic acid, pilocarpine, and pseudoephedrine were about 4.1, 8.0, and 15.7 minutes, respectively (estimated from the published chromatogram).

The HPLC method was reported to be stability indicating. The intraday and interday coefficients of variation for the analysis at 100 µmol/L were 3.5 and 7.7%, respectively.

References

Aromdee C, Fawcett JP, Ferguson MM, et al. Serum pilocarpine esterase activity and response to oral pilocarpine. *Biochem Molecular Med.* 1996; 59: 57–61.

Aromdee C, Ledger R. Hydrolysis of pilocarpine and isopilocarpine in plasma. *Proc Univ Otago Med Sch.* 1994; 72: 15–6.

Fawcett JP, Tucker IG, Davies MN, et al. Formulation and stability of pilocarpine solution. *Int J Pharm Pract.* 1994; 3: 14–8.

Method 2

Wong et al. reported an assay for pilocarpine nitrate in reservoirs. The Hewlett-Packard 1090M liquid chromatograph was equipped with a Hewlett-Packard 1040M diode-array detector and a Hewlett-Packard Chem Station data acquisition system. The stationary phase was a Supelcosil LC-18 reversed-phase column (250 × 4.6 mm, 5-µm particle size) with a Supelcosil LC-18 guard column. The mobile phase consisted of 80% 0.073 M monobasic potassium phosphate buffer (pH 7.7) and 20% methanol. The flow rate was 1.5 mL/min. The column temperature was 60 °C. UV detection was performed at 215 nm. The sample injection volume was 50 µL. Under these conditions, retention times were about 2.7 minutes for pilocarpic acid, 3.4 minutes for isopilocarpic acid, 8.4 minutes for isopilocarpine, and 9.2 minutes for pilocarpine.

The assay was proven to be stability indicating by accelerated degradation of pilocarpine. Pilocarpine nitrate and isopilocarpine nitrate were hydrolyzed in 0.1 M potassium hydroxide solution for 2 hours. Degradation products did not interfere with the determination of pilocarpine.

A calibration curve for pilocarpine was constructed from 8.0 to 970.0 µg/mL. The correlation coefficient was 1.00. The limit of detection was 1.62 µg/mL.

Reference

Wong O, Anderson C, Allaben L, et al. Stability-indicating assay method for pilocarpine nitrate in reservoirs used in the cystic fibrosis indicator system. *Int J Pharm.* 1991; 76: 171–5.

Pioglitazone Hydrochloride

Chemical Name

5-[[4-[2-(5-Ethyl-2-pyridinyl)ethoxy]phenyl]methyl]-2,4-thiazolidinedione monohydrochloride

Other Name
Actos

Form	Molecular Formula	MW	CAS
Pioglitazone hydrochloride	$C_{19}H_{20}N_2O_3S.HCl$	392.9	112529-15-4

Appearance
Pioglitazone hydrochloride occurs as an odorless white crystalline powder.

Solubility
Pioglitazone hydrochloride is practically insoluble in water, soluble in dimethylformamide, and slightly soluble in anhydrous ethanol.

Method
Wanjari and Gaikwad reported an HPLC method for the determination of pioglitazone in a tablet dosage form. The Shimadzu LC-10AT VP system was equipped with a Rheodyne injector with a 20-μL loop. The stationary phase was a Hypersil C_8 column (250 × 4.6 mm). The mobile phase consisted of acetonitrile and 0.15% (vol/vol) triethylamine solution (40:60, vol/vol) adjusted to pH 4.6 with phosphoric acid. The flow rate was 1.5 mL/min. UV detection was performed at 210 nm and 0.010 AUFS. The sample injection volume was 20 μL. Under these conditions, the retention time of pioglitazone was 7.6 minutes.

A portion of ground pioglitazone tablets equivalent to 25 mg of pioglitazone was weighed, extracted in methanol, filtered through Whatman No. 1 filter paper, diluted with mobile phase, and injected.

To demonstrate the stability-indicating capability of the method, pioglitazone samples were exposed to 0.1 N hydrochloric acid at 40 °C, 0.1 N sodium hydroxide solution at 40 °C, 3% hydrogen peroxide at 40 °C, heat at 60 °C, UV light at 254 nm, or 75% relative humidity at 40 °C for 24 hours. No interference with the analysis of pioglitazone from excipients, impurities, or degradation products was observed.

A calibration curve for pioglitazone was constructed from 10 to 80 μg/mL; the correlation coefficient was 0.9999. The limit of detection and the limit of quantitation were 0.6 and 2.0 μg/mL, respectively.

Reference
Wanjari DB, Gaikwad NJ. Stability indicating RP-HPLC method for determination of pioglitazone from tablets. *Indian J Pharm Sci.* 2005; 67: 256–8.

Pipamperone Hydrochloride

Chemical Name
1'-[4-(4-Fluorophenyl)-4-oxobutyl]-[1,4'-bipiperidine]-4'-carboxamide dihydrochloride

Other Names
Dipiperon, Piperonil

Form	Molecular Formula	MW	CAS
Pipamperone	$C_{21}H_{30}FN_3O_2$	375.5	1893-33-0
Pipamperone hydrochloride	$C_{21}H_{30}FN_3O_2 \cdot 2HCl$	448.4	2448-68-2

Appearance
Pipamperone hydrochloride occurs as crystals.

Method
Trabelsi et al. reported a method for the determination of pipamperone in the presence of impurities and its degradation products. The Shimadzu system consisted of a model LC-10AT VP pump, a model SPD-10A VP variable UV-visible detector, a model C-R8A integrator, and a Rheodyne 7725i injector. The stationary phase was a Supelco LC-ABZ$^+$ column (250 × 4.6 mm). The mobile phase was a mixture of acetonitrile, tetrahydrofuran, and 0.05 M monobasic sodium phosphate aqueous buffer (pH 6.5) (16:11:73, vol/vol/vol). The flow rate was 1.0 mL/min. The detector was set at 246 nm.

A portion of the powder obtained from 20 tablets equivalent to 40 mg of pipamperone dihydrochloride was accurately weighed, transferred into a 100-mL volumetric flask, sonicated for 5 minutes with 100 mL of methanol, filtered through a 0.22-μm membrane filter, diluted with methanol, and assayed. The injection volume was 20 μL. Under these conditions, the retention time of pipamperone was about 6.8 minutes (estimated from the published chromatogram).

Pipamperone dihydrochloride substance was treated with 1 N hydrochloric acid or 1 N sodium hydroxide solution and refluxed for 5 hours. The drug substance was also treated with hydrogen peroxide and heated at 80 ºC for 2 hours. The degradation product peaks did not interfere with the peak of the intact drug, demonstrating the stability-indicating nature of this assay method.

A calibration curve for pipamperone dihydrochloride was obtained from 10 to 30 μg/mL with a correlation coefficient of 0.9994. The intraday and interday coefficients of variation were less than 1.54 and 1.70%, respectively. The limit of detection and the limit of quantitation were 0.1 and 0.3 μg/mL, respectively.

Reference
Trabelsi H, Hassen IE, Bouabdallah S, et al. Stability indicating LC method for the determination of pipamperone. *J Pharm Biomed Anal.* 2005; 39: 914–9.

Piperacillin

Chemical Names

[2S-[2α,5α,6β(S*)]]-6-[[[[(4-Ethyl-2,3-dioxo-1-piperazinyl)carbonyl]amino]-phenylacetyl]-amino]-3,3-dimethyl-7-oxo-4-thia-1-azabicyclo[3.2.0]heptane-2-carboxylic acid

(2S,5R,6R)-6-[(R)-2-(4-Ethyl-2,3-dioxo-1-piperazinecarboxamido)-2-phenylacetamido]-3,3-dimethyl-7-oxo-4-thia-1-azabicyclo[3.2.0]heptane-2-carboxylic acid

Other Name

Pipracil

Form	Molecular Formula	MW	CAS
Piperacillin	$C_{23}H_{27}N_5O_7S$	517.6	61477-96-1
Piperacillin sodium	$C_{23}H_{26}N_5NaO_7S$	539.5	59703-84-3

Appearance

Piperacillin sodium is a white to off-white hygroscopic lyophilized crystalline powder.

Solubility

Piperacillin sodium is freely soluble in water and sparingly soluble in alcohol.

Method 1

Zhang et al. investigated the chemical stability of linezolid injection at 2 mg/mL admixed with piperacillin sodium. The Waters liquid chromatograph consisted of a model 600E multisolvent-delivery pump, a model 996 photodiode-array detector, and a WISP model 712 autosampler. The stationary phase was a Waters Symmetry C_8 column (250 × 4.6 mm, 5-µm particle size). The mobile phase consisted of 0.02 M monobasic sodium phosphate buffer, acetonitrile, and 1 N tetrabutylammonium hydroxide (70:30:12, vol/vol/vol), adjusted to pH 5.4 with phosphoric acid. The flow rate was 1.0 mL/min. UV detection was performed at 220 nm and 0.5 AUFS.

Samples were diluted 1:200 with water. The injection volume was 10 µL. Under these conditions, the retention times for linezolid and piperacillin were 6.0 and 10.4 minutes, respectively.

The method was shown to be stability indicating by accelerated degradation. Sample solutions were treated with 1 N sodium hydroxide, 1 N hydrochloric acid, or 3% hydrogen peroxide or were subjected to heating. There was no interference of the degradation product peaks with the intact drug peak.

A standard curve for piperacillin was generated from 50 to 250 µg/mL. The correlation coefficient was more than 0.9999. Intraday and interday coefficients of variation were 1.4 and 0.8%, respectively.

Reference

Zhang Y, Xu QA, Trissel LA, et al. Compatibility and stability of linezolid injection admixed with aztreonam or piperacillin sodium. *J Am Pharm Assoc.* 2000; 40: 520–4.

Method 2

Moon et al. studied the stability of piperacillin sodium–tazobactam sodium in polypropylene syringes and polyvinyl chloride minibags. The liquid chromatograph consisted of a Hitachi model L-6200 pump, a model L-4200 UV-visible detector, a model AS2000 autosampler, and a model D-2500 integrator. The stationary phase was an Alltech Adsorbosphere C_{18} column (250 × 4.6 mm, 5-μm particle size). The mobile phase consisted of 0.01 M sodium phosphate and acetonitrile (70:30, vol/vol) adjusted with 85% phosphoric acid to pH 3.9. The flow rate was 1.2 mL/min. UV detection was performed at 220 nm.

Samples were diluted 1:100 with 0.9% sodium chloride injection. The injection volume was 20 μL. Under these conditions, the retention time for piperacillin was 7.34 minutes.

The assay was demonstrated to be stability indicating by accelerated decomposition of piperacillin. Samples of piperacillin 80 mg/mL were exposed to 1 N sodium hydroxide at 58 °C for 4 hours, 1 N hydrochloric acid at 58 °C for 4 hours, and 1% hydrogen peroxide at room temperature for 4 hours. The degradation product peak (at about 4 minutes) did not interfere with the piperacillin peak.

A standard curve for piperacillin was constructed from 200 to 1050 μg/mL; the correlation coefficient was greater than 0.998. The intraday and interday coefficients of variation were less than 5%.

References

Moon YSK, Chung KC, Chin A, et al. Stability of piperacillin sodium–tazobactam sodium in polypropylene syringes and polyvinyl chloride minibags. *Am J Health Syst Pharm.* 1995; 52: 999–1000.

Chung KC, Moon YSK, Chin A, et al. Compatibility of ondansetron hydrochloride and piperacillin sodium–tazobactam sodium during simulated Y-site administration. *Am J Health Syst Pharm.* 1995; 52: 1554–6.

Inagaki K, Gill MA, Okamoto MP, et al. Stability of ranitidine hydrochloride with aztreonam, ceftazidime, or piperacillin sodium during simulated Y-site administration. *Am J Hosp Pharm.* 1992; 49: 2769–72.

Method 3

Sarkar et al. studied the stability of piperacillin, clindamycin, and ranitidine in polyolefin bags. The HPLC system included two Gilson model 302 pumps, a Gilson model 811 dynamic mixer, a Gilson model 802b manometric module, a Gilson model 231 autoinjector, and a Shimadzu model SPD-6A UV detector. The stationary phase was an Alltech phenyl column (100 × 4.6 mm). The mobile phase consisted of 0.02 M potassium phosphate (pH 3.0) and acetonitrile and was delivered gradiently at 1.5 mL/min. The UV detector was set at 214 nm and 0.64 AUFS.

Samples were diluted 1:100 in 5% dextrose injection. The injection volume was 20 μL and the retention time for piperacillin was 5.2 minutes.

The stability-indicating capability of the method was demonstrated by accelerated decomposition of piperacillin. Piperacillin solution was mixed with concentrated acid and allowed to degrade for 24 hours. The resulting solution was neutralized and assayed. The intact piperacillin peak was well separated from its degradation product peaks. Retention times for the degradation products were 1.4, 2.2, 3.1, and 4.0 minutes.

A calibration curve was constructed from 200 to 700 µg/mL. The coefficient of variation was 3.0%.

Reference

Sarkar MA, Rogers E, Reinhard M, et al. Stability of clindamycin phosphate, ranitidine hydrochloride, and piperacillin sodium in polyolefin containers. *Am J Hosp Pharm*. 1991; 48: 2184–6.

Method 4

Perry et al. investigated the chemical stability of piperacillin in total parenteral nutrient solutions at concentrations commonly used in adults. The analytical system consisted of a Rheodyne injector 125 valve, a Waters model 6000A solvent-delivery system, a Waters Lambda-Max 481 UV detector, and a BBC Servogen 120 chart recorder. The stationary phase was a Waters cyano-bonded reversed-phase C_{18} column (300 × 4.5 mm, 5-µm particle size). The mobile phase consisted of 32% acetonitrile and 68% water adjusted to pH 2.5 with phosphoric acid. The flow rate was 2.0 mL/min. UV detection was performed at 254 nm and 0.05 AUFS.

Samples were diluted 1:1000 with purified water. The injection volume was 20 µL. Under these conditions, the retention time for piperacillin was 6.0 minutes.

The assay was determined to be stability indicating by accelerated decomposition of piperacillin. A solution of piperacillin was adjusted to pH 1.6, a second solution was evaporated on a hot plate, and a third solution was heated in a boiling water bath for 15 minutes. In all cases, degradation product peaks did not interfere with the intact piperacillin peak.

A standard curve was constructed from 0 to 50 mg/mL. The intraassay coefficient of variation was 1.2%.

Reference

Perry M, Khalidi N, Sanders CA. Stability of penicillins in total parenteral nutrient solution. *Am J Hosp Pharm*. 1987; 44: 1625–8.

Method 5

Marble et al. evaluated the stability of piperacillin sodium with clindamycin phosphate in 5% dextrose injection and 0.9% sodium chloride injection. The HPLC system consisted of an Altex model 110A solvent-metering pump, an Altex model 210 injector (or a Waters WISP 710B autoinjector), a Waters Lambda-Max model 481 LC spectrophotometer, an Altex model 153 analytical UV detector, and a Linear Instruments model 555 recorder (or a Hewlett-Packard model 3392A integrator). The stationary phase was an Alltech C_8 reversed-phase analytical column (250 × 4.6 mm, 10-µm particle size). The mobile phase was 40% acetonitrile in a 0.02 M potassium phosphate buffer. The pH was adjusted to 3.0 with phosphoric acid. The mobile phase was delivered isocratically at 2 mL/min. The UV detector was set at 254 nm and 0.64 AUFS. *p*-Nitrobenzenesulfonamide 0.25 mg/mL in 50% methanol and 50% water was used as an internal standard.

Samples were diluted by adding 100 µL of test solutions to 1 mL of internal standard solution. The injection volume was 10 µL. Under these conditions, retention times for the internal standard and for piperacillin were 3 and 9.5 minutes, respectively.

The stability-indicating capability of the assay was verified by accelerated degradation of piperacillin. Acidification (pH 2.0) or heating of piperacillin solutions resulted in degradation products that did not interfere with the determination of the intact drug.

Intrarun and interrun coefficients of variation were 1.3 and 4.5%, respectively.

Reference

Marble DA, Bosso JA, Townsend RJ. Stability of clindamycin phosphate with aztreonam, ceftazidime sodium, ceftriaxone sodium, or piperacillin sodium in two intravenous solutions. *Am J Hosp Pharm.* 1986; 43: 1732–6.

Method 6

Gupta et al. determined the stability of piperacillin sodium in 5% dextrose and 0.9% sodium chloride injections. The liquid chromatograph was a Waters ALC 202 system. The stationary phase was a Waters microphenyl column. The mobile phase was 38% methanol in 0.02 M aqueous ammonium acetate buffer. The flow rate was 2.5 mL/min. UV detection was performed at 230 nm and 0.04 AUFS.

Samples were diluted to 0.1 mg/mL of piperacillin sodium with water. The injection volume was 20 μL. The retention time for piperacillin was about 6 minutes (estimated from the published chromatogram).

The assay was demonstrated to be stability indicating by a degradation study. A piperacillin solution (1% in water) was stored at 25 °C for 16 days. Degradation product peaks did not interfere with the piperacillin peak.

A standard curve for piperacillin was constructed from 0.05 to 0.12 mg/mL.

Reference

Gupta VD, Davis DD, Stewart KR. Stability of piperacillin sodium in dextrose 5% and sodium chloride 0.9% injections. *Am J IV Ther Clin Nutr.* 1984; 11: 14–9.

Pirenzepine Hydrochloride

Chemical Name

5,11-Dihydro-11-(4-methylpiperazin-1-ylacetyl)pyrido[2,3-*b*][1,4]benzodiazepine-6-one dihydrochloride monohydrate

Other Names

Abrinac, Gastricur, Gastrozepin

Form	Molecular Formula	MW	CAS
Pirenzepine hydrochloride	$C_{19}H_{21}N_5O_2.2HCl.H_2O$	442.3	29868-97-1

Appearance

Pirenzepine hydrochloride occurs as a white or yellowish crystalline powder.

Solubility

Pirenzepine hydrochloride is freely soluble in water, very slightly soluble in dehydrated alcohol, and practically insoluble in dichloromethane.

Method

Abounassif developed a normal-phase HPLC method for the analysis of pirenzepine hydrochloride in a pharmaceutical formulation. The Varian model 5000 liquid chromatograph included a model UV50 variable-wavelength detector, a model 9176 recorder, a model DCS 111L data system, and a Rheodyne 7125 10-μL loop injector. The stationary phase was a Micro-Pack Si-10 column (300 × 3.9 mm). The mobile phase consisted of concentrated ammonium hydroxide (28–30% NH_3) and methanol (0.75:99.25, vol/vol). The flow rate was 2 mL/min. UV detection was performed at 254 nm and 0.05 AUFS. Clobazam 1.5 μg/mL was used as an internal standard.

A sample of ground tablets equivalent to about 10 mg of pirenzepine hydrochloride was transferred into a 100-mL volumetric flask, mixed with 5 mL of water, allowed to stand for 15 minutes, diluted with 90 mL of methanol, shaken vigorously for 15 minutes, brought to volume with methanol, and filtered. The filtrate was diluted with mobile phase containing the internal standard to give a concentration of approximately 15 μg/mL of pirenzepine hydrochloride. The injection volume was 10 μL. Under these conditions, retention times for clobazam and pirenzepine hydrochloride were 1.9 and 2.8 minutes, respectively.

The stability-indicating nature of the method was established. A 1-mL sample of pirenzepine hydrochloride in 6 N hydrochloric acid was kept at room temperature in the dark for 7 days. The degradation products did not influence the determination of the intact pirenzepine hydrochloride.

A standard curve for pirenzepine hydrochloride was obtained from 5 to 25 μg/mL. The correlation coefficient was 0.999.

Reference

Abounassif MA. High performance liquid chromatographic determination of pirenzepine dihydrochloride in its pharmaceutical formulation. *Anal Lett.* 1985; 18: 2083–9.

Piretanide

Chemical Name

4-Phenoxy-3-(pyrrolidin-1-yl)-5-sulfamoylbenzoic acid

Other Names
Arlix, Prilace, Trialix

Form

Form	Molecular Formula	MW	CAS
Piretanide	$C_{17}H_{18}N_2O_5S$	362.4	55837-27-9

Appearance
Piretanide occurs as a yellowish-white to yellowish powder.

Solubility
Piretanide is very slightly soluble in water and sparingly soluble in dehydrated alcohol.

Method
Youssef developed an HPLC method for the determination of piretanide in the presence of its degradation products. The Waters system consisted of a model 600 controller, a model 486 tunable absorbance detector, a model 600 pump, and a model 746 integrator. The stationary phase was a µBondapak C_{18} column (250 × 4.6 mm, 10-µm particle size). The mobile phase consisted of methanol, water, and glacial acetic acid (70:30:1, vol/vol/vol). The flow rate was 1.0 mL/min. UV detection was performed at 275 nm and 0.05 AUFS.

A portion of the powder obtained from 10 tablets equivalent to 50 mg of piretanide was accurately weighed, dissolved in 40 mL of methanol, shaken for 1 hour, filtered, transferred to a 50-mL volumetric flask, brought to the mark with methanol, and assayed. The injection volume was 20 µL. Under these conditions, the retention time of piretanide was about 5.0 minutes.

This method was demonstrated to be stability indicating by accelerated degradation. Piretanide was dissolved in aqueous sodium hydroxide (pH 9) and exposed to sunlight for 10 days. The degradation product peak did not interfere with the determination of piretanide.

A calibration curve for piretanide was obtained from 0.02 to 0.2 µg/20 µL. The correlation coefficient was 0.9992. The intraday and interday coefficients of variation were 0.82 and 1.64%, respectively.

Reference
Youssef NF. Stability-indicating methods for the determination of piretanide in presence of the alkaline induced degradates. *J Pharm Biomed Anal.* 2005; 39: 871–6.

Piroxicam

Chemical Name
4-Hydroxy-2-methyl-*N*-2-pyridyl-2*H*-1,2-benzothiazine-3-carboxamide 1,1-dioxide

Other Names
Candyl, Cycladol, Feldene, Novo-Pirocam, Piroflam

Form	Molecular Formula	MW	CAS
Piroxicam	$C_{15}H_{13}N_3O_4S$	331.3	36322-90-4

Appearance
Piroxicam is a white crystalline powder.

Solubility
Piroxicam is very slightly soluble in water and slightly soluble in alcohol.

pK$_a$
Piroxicam has a pK$_a$ of 5.1.

Method
Bartsch et al. described the determination of piroxicam by HPLC. The Shimadzu system consisted of a model LC-10AS pump, a model SPD-M10A diode-array detector, a model CTO-10AC column oven, and a Rheodyne injector with a 20-µL loop. The stationary phase was a Merck LiChrospher 100 RP18 endcapped column (119 × 3 mm, 5-µm particle size). The mobile phase consisted of methanol and 0.4 M acetate buffer (pH 4.3) (45:55, vol/vol). Under these conditions, the retention time of piroxicam was 4.7 minutes.

The assay was reported to be stability indicating. No degradation product peaks interfered with the peak of the intact piroxicam.

Standard curves were constructed from 8.8 to 44.0 µg/mL, from 60 to 300 µg/mL, and from 0.44 to 2.20 mg/mL. Correlation coefficients were greater than 0.999.

References
Bartsch H, Eiper A, Kopelent-Frank H. Stability indicating assays for the determination of piroxicam—comparison of methods. *J Pharm Biomed Anal.* 1999; 20: 531–41.

Bartsch H, Eiper A, Kopelent-Frank H, et al. Untersuchungen zur photostabilitat von piroxicam mit hilfe der densitometrie. *Sci Pharm.* 1996; 64: 263–70.

Pivmecillinam Hydrochloride

Chemical Name

(2S, 5R, 6R)-6-[[(Hexahydro-1H-azepin-1-yl)methylene]amino]-3,3-dimethyl-7-oxo-4-thia-1-azabicyclo[3.2.0]heptane-2-carboxylic acid (2,2-dimethyl-1-oxopropoxy)methyl ester monohydrochloride

Other Names

Amdinocillin Pivoxil, Selexid

Form	Molecular Formula	MW	CAS
Pivmecillinam hydrochloride	$C_{21}H_{33}N_3O_5S.HCl$	476.0	32887-03-9

Appearance

Pivmecillinam hydrochloride is a white or almost white crystalline powder.

Solubility

Pivmecillinam hydrochloride is freely soluble in water, dehydrated alcohol, and methanol.

Method

Hagel and Waysek developed an HPLC assay for a stability study of pivmecillinam hydrochloride both as the pure compound and in its capsule dosage form. The chromatograph consisted of an LDC model 396 minipump or a Waters model 6000A pump, a Waters model U6K injector or a Rheodyne model 7120 injector, and a Tracor model 970A variable-wavelength detector or an LDC model 1202 variable-wavelength detector. The stationary phase was an ES Industries Chromegabond C_{18} column (300 × 4.6 mm, 10-μm particle size) or a Chromegabond C_8 column (150 × 4.6 mm, 5-μm particle size). The mobile phase consisted of 0.01 M sodium phosphate buffer (adjusted to pH 3.0 with phosphoric acid) and acetonitrile (40:60, vol/vol). The flow rate was 2.0 mL/min. UV detection was performed at 220 nm and 0.16 AUFS.

Capsules of pivmecillinam hydrochloride were accurately weighed and dissolved in acetonitrile, and the mixture was then filtered. The filtrate was diluted with acetonitrile to a pivmecillinam hydrochloride concentration of about 0.1 mg/mL. The injection volume was 20 μL. Under these conditions, the retention time of pivmecillinam hydrochloride was about 12.5 minutes (estimated from the published chromatogram).

The stability-indicating capability of the method was demonstrated by accelerated solution degradation and thermal degradation. The degradation products did not interfere with the analysis of pivmecillinam hydrochloride.

A standard curve was constructed from 28.5 to 456 ng of pivmecillinam hydrochloride per injection. The correlation coefficient was 0.9916.

Reference

Hagel RB, Waysek EH. Stability-indicating assay for pivmecillinam hydrochloride and pivmecillinam hydrochloride capsules. *J Chromatogr*. 1979; 178: 97–104.

Pizotyline (Pizotifen)

Chemical Name

4-(9,10-Dihydro-4*H*-benzo[4,5]cyclohepta[1,2-*b*]thien-4-ylidene)-1-methylpiperidine

Other Name

Sandomigran

Form

Form	Molecular Formula	MW	CAS
Pizotyline	$C_{19}H_{21}NS$	295.4	15574-96-6
Pizotyline malate	$C_{19}H_{21}NS.C_4H_6O_5$	429.5	5189-11-7

Appearance

Pizotyline malate occurs as a white or slightly yellowish-white and odorless or almost odorless crystalline powder.

Solubility

Pizotyline malate is very slightly soluble in water, slightly soluble in alcohol and in chloroform, and sparingly soluble in methanol.

Method

Abounassif et al. developed a method for the analysis of pizotyline in the presence of its photodegradation products. The Waters system consisted of a model 600E system controller, a model 486 tunable absorbance detector, a model 746 data module, and a Rheodyne model 7161 injector. The stationary phase was a Varian Bondesil CN column (250 × 4.6 mm, 5-μm particle size). The mobile phase consisted of acetonitrile and 0.01 M sodium acetate (pH 3.5) (75:25, vol/vol). The flow rate was 2 mL/min. UV detection was carried out at 254 nm.

Pizotyline stock solution was prepared in methanol. For tablets, a portion of powder from six tablets equivalent to one tablet was accurately weighed into a 20-mL volumetric flask, shaken with mobile phase for 25 minutes, filtered, and diluted by a factor of 6 with mobile phase. For syrup, the syrup sample was diluted by a factor of 6 with water. The injection volume was 20 μL. Under these conditions, the retention time of pizotyline was about 4.5 minutes (estimated from the published chromatogram).

The stability-indicating power of the method was confirmed by injecting a UV-degraded solution of pizotyline onto the column. No interference of degradation products with the analysis of the intact drug was observed.

A linear calibration curve for pizotyline was constructed from 5 to 25 μg/mL. The correlation coefficient was greater than 0.9995. The interday coefficient of variation was less than 2%.

Reference
Abounassif MA, El-Obeid HA, Gadkariem EA. Stability studies on some benzocycloheptane antihistaminic agents. *J Pharm Biomed Anal.* 2005; 36: 1011–8.

Polymyxin B

Chemical Name
Mixture of polymyxins B_1 and B_2

Other Names
Component of Bacisporin, Maxitrol, Neosporin, Neotopic, Otobiotic

Form	Molecular Formula	MW	CAS
Polymyxin B			1404-26-8
Polymyxin B sulfate			1405-20-5
Polymyxin B_1	$C_{56}H_{98}N_{16}O_{13}$	1203.5	4135-11-9
Polymyxin B_2	$C_{55}H_{96}N_{16}O_{13}$	1189.5	34503-87-2
Polymyxin B_3	$C_{55}H_{96}N_{16}O_{13}$	1189.5	71140-58-4

Appearance
Polymyxin B sulfate occurs as a white to buff-colored powder. It is odorless or has a faint odor.

Solubility
Polymyxin B sulfate is freely soluble in water and slightly soluble in alcohol.

Method
Taylor et al. used HPLC analysis to study the chemical stability of polymyxin B in aqueous solution. The liquid chromatographic system included a Jasco model PU 980 pump, a model UV975 detector, and a Rheodyne model 7125 injector with a 20-μL loop. The stationary phase was a laboratory-made Hypersil ODS column (100 × 4.6 mm, 3-μm particle size). The mobile phase consisted of 34.5% acetonitrile and 65.5% 10 mM aqueous phosphate buffer (pH 3.0) containing 0.33% (wt/vol) perchloric acid. The flow rate was 1.5 mL/min. UV detection was performed at 200 nm and 0.16 AUFS. The injection volume was 20 μL. Retention times of polymyxin B_2, B_3, and B_1 were about 7.6, 10.3, and 12.4 minutes, respectively (estimated from the published chromatogram).

The stability-indicating nature of the method was demonstrated by assaying an extensively decomposed polymyxin B solution. No degradation product peaks interfered with the retention times of polymyxin B_1 or polymyxin B_2.

A standard curve for polymyxin B was constructed from 125 to 500 μg/mL. The correlation coefficient was 0.9987.

Reference
Taylor RB, Richards RME, Low AS, et al. Chemical stability of polymyxin B in aqueous solution. *Int J Pharm.* 1994; 102: 201–6.

Polythiazide

Chemical Name
6-Chloro-3,4-dihydro-2-methyl-3-(2,2,2-trifluoroethylthiomethyl)-2*H*-1,2,4-benzo-thiadiazine-7-sulfonamide 1,1-dioxide

Other Names
Drenusil, Nephril, Renese

Form	Molecular Formula	MW	CAS
Polythiazide	$C_{11}H_{13}ClF_3N_3O_4S_3$	439.9	346-18-9

Appearance
Polythiazide occurs as a white or almost white crystalline powder with an odor.

Solubility
Polythiazide is practically insoluble in water and slightly to sparingly soluble in alcohol.

Method
Bachman described an HPLC method for the determination of prazosin and polythiazide. The liquid chromatograph consisted of an LDC ConstaMetric I pump, a SpectroMonitor II variable-wavelength detector, a Rheodyne model 7125 loop injector, and a Hewlett-Packard 3380A integrator. The stationary phase was an IBM cyano column (250 × 4.5 mm, 5-μm particle size). The mobile phase consisted of 650 mL of 0.05 M monobasic sodium phosphate buffer and 350 mL of acetonitrile, adjusted to pH 3.0 with phosphoric acid. The flow rate was 1.7 mL/min. UV detection was performed at 268 nm. Benzophenone 0.03 mg/mL in the mobile phase was used as an internal standard.

A sample of ground capsules was accurately weighed, mixed with internal standard solution, sonicated for 5 minutes, diluted in mobile phase, and filtered. The injection volume was 10 μL. Under these conditions, retention times for prazosin, benzophenone, and polythiazide were about 4.2, 5.5, and 6.5 minutes, respectively.

The method was reported to be stability indicating since the method resolved the intact drug from its degradation products and impurities.

A standard curve for polythiazide was generated from 25 to 625 ng. The correlation coefficient was 0.9999. The limit of detection was 1.0 ng.

Reference

Bachman WJ. High performance liquid chromatographic determination of diuretic-antihypertensive combination products. I. Prazosin and polythiazide. *J Liq Chromatogr.* 1986; 9: 1033–49.

Pralidoxime Chloride

Chemical Names

2-[(Hydroxyimino)methyl]-1-methylpyridinium chloride
2-Formyl-1-methylpyridinium chloride oxime

Other Name

Protopam Chloride

Form	Molecular Formula	MW	CAS
Pralidoxime chloride	$C_7H_9ClN_2O$	172.6	51-15-0
Pralidoxime iodide	$C_7H_9IN_2O$	264.1	94-63-3
Pralidoxime mesylate	$C_7H_9N_2O.CH_3O_3S$	232.3	154-97-2
Pralidoxime methylsulfate	$C_7H_9N_2O.CH_3O_4S$	248.3	1200-55-1

Appearance

Pralidoxime chloride is a white to pale yellow crystalline powder.

Solubility

Pralidoxime chloride has the following solubilities in grams per 100 mL at 25 °C: 65.5 in water, 8.5 in methanol, 0.89 in ethanol, and 0.09 in isopropanol.

pK_a

Pralidoxime has a pK_a of 7.8–8.

Method

Schroeder et al. developed a stability-indicating method for the quantitation of pralidoxime chloride and the major UV-absorbing decomposition products and studied the stability of the drug in water and stored in autoinjectors at room temperature for about 10 years. The chromatographic system consisted of a Waters 710A WISP autosampler, a Waters 6000A pump, a Perkin-Elmer LC-85 variable-wavelength spectrophotometric detector, a Spectrum Scientific model 1021A electronic filter, and a Kipp & Zonen model BD41 strip-chart recorder. The stationary phase was a Waters μPorasil column (300 × 3.9 mm, 10-μm particle size). The mobile phase consisted of acetonitrile and water (86:14, vol/vol), in which the aqueous component (pH 2.9) contained 8.36 mM tetraethylammonium chloride and 52.5 mM acetic acid. The flow rate was 1.0 mL/min. UV detection was performed at 295 nm and 1.28 AUFS.

Samples were diluted 1:4000 with mobile phase, and the injection volume was 5 μL.

The method was determined to be stability indicating by spiking a pralidoxime chloride solution with its known decomposition products. Degradation product peaks did not interfere with the pralidoxime peak. Degradation product and pralidoxime chloride peaks follow with retention times in minutes given in parentheses: 1-methyl-2(1H)-pyridinone (4.2), pralidoxime chloride (8.3), 2-formyl-1-methylpyridinium chloride (9.7), 2-cyano-1-methylpyridinium chloride (10.5), 2-(hydroxymethyl)-1-methylpyridinium chloride (11.4), 2-(aminocarbonyl)-1-methylpyridinium chloride (12.3), and 2-carboxy-1-methylpyridinium chloride (16.5) (estimated from the published chromatogram).

The method is sensitive to at least 5 ng of the drug.

Reference
Schroeder AC, DiGiovanni JH, Bredow JV, et al. Pralidoxime chloride stability-indicating assay and analysis of solution samples stored at room temperature for ten years. *J Pharm Sci.* 1989; 78: 132–6.

Prazosin Hydrochloride

Chemical Name
2-[4-(2-Furoyl)piperazin-1-yl]-6,7-dimethoxyquinazolin-4-ylamine hydrochloride

Other Names
Alphavase, Alpress, Eurex, Minipress

Form	Molecular Formula	MW	CAS
Prazosin hydrochloride	$C_{19}H_{21}N_5O_4 \cdot HCl$	419.9	19237-84-4

Appearance
Prazosin hydrochloride is a white to tan powder.

Solubility
Prazosin hydrochloride is slightly soluble in water and very slightly soluble in alcohol.

pK$_a$

Prazosin hydrochloride has a pK$_a$ of 6.5 in a mixture of water and ethanol (1:1, vol/vol).

Method 1

Bakshi et al. developed a stability-indicating HPLC method for the determination of prazosin in the presence of its degradation products. The liquid chromatograph consisted of a Waters model 600E pump, a model 996 photodiode-array detector, and a model 717 autosampler. The stationary phase was a Waters Spherisorb C$_{18}$ column (250 × 4.6 mm, 5-μm particle size). The mobile phase was composed of acetonitrile, water, glacial acetic acid, and diethylamine (65:35:1:0.02, vol/vol/vol/vol). The flow rate was 1 mL/min. UV detection was performed at 254 nm. The injection volume was 5 μL. Under these conditions, the retention time for prazosin was about 13.0 minutes.

The prazosin solutions were heated at 80 °C for 10 days, treated with 0.1 N hydrochloric acid and heated at 80 °C for 90 hours, treated with 0.1 N sodium hydroxide solution and heated at 80 °C for 4 hours, treated with 3% hydrogen peroxide at room temperature for 6 hours, treated with 30% hydrogen peroxide for 24 hours, exposed to the light, or subjected to dry heat at 50 °C for 3 months. The peak of the intact drug was well resolved from peaks of its degradation products. The absence of any coeluting peak was also confirmed by photodiode-array peak purity analysis.

A standard curve for prazosin was constructed from 50 to 500 μg/mL. The correlation coefficient was greater than 0.9991. The intraday and interday coefficients of variation were less than 0.7 and 1.6%, respectively.

Reference

Bakshi M, Ojha T, Singh S. Validated specific HPLC methods for determination of prazosin, terazosin and doxazosin in the presence of degradation products formed under ICH-recommended stress conditions. *J Pharm Biomed Anal.* 2004; 34: 19–26.

Method 2

Tenjarla and Tseggai described an assay for the quantification of prazosin in the diffusate and in extracts of human skin. An LDC liquid chromatograph was equipped with a ConstaMetric I pump, a SpectroMonitor 3 UV detector, and a Hewlett-Packard 3394A integrator. The stationary phase was an Alltech cyano column (150 × 4.6 mm). The mobile phase consisted of acetonitrile, methanol, and water (45:5:50, vol/vol/vol) containing 3 mM sodium heptanesulfonate. The flow rate was 1.5 mL/min. UV detection was performed at 230 nm and 0.01 AUFS. Verapamil hydrochloride 100 μg/mL was used as an internal standard.

The skin sample was cut into small pieces, homogenized with 5 mL of methanol and filtered (repeated three times), evaporated to dryness, and reconstituted with 1 mL of the mobile phase. The injection volume was 10 μL. Under these conditions, retention times of prazosin and verapamil were 4.4 and 5.8 minutes, respectively.

The stability-indicating ability of the assay was shown by accelerated degradation of prazosin. The drug solutions were boiled for 3 minutes with either 0.1 N hydrochloric acid or 0.1 N sodium hydroxide solution. There was no interference from degradation products at the retention times of prazosin and verapamil.

Standard curves for prazosin were obtained from 0.5 to 7 μg/mL for the diffusate, with a correlation coefficient greater than 0.99, and from 1 to 10 μg/mL for the skin extracts, with a correlation coefficient greater than 0.97. Intraday coefficients of variation were 3.2 and 6.5% for the diffusate and the skin extract, respectively. Interday coefficients of variation were 4.2 and 11.7% for the diffusate and the skin extract, respectively. The limit of detection was 0.05 µg/mL.

Reference
Tenjarla SN, Tseggai A. High-performance liquid chromatographic assay of prazosin for transdermal screening studies. *J Clin Pharm Ther*. 1992; 17: 37–42.

Method 3
Bachman presented an HPLC method for the determination of prazosin and polythiazide. The liquid chromatograph consisted of an LDC ConstaMetric I pump, a SpectroMonitor II variable-wavelength detector, a Rheodyne model 7125 loop injector, and a Hewlett-Packard 3380A integrator. The stationary phase was an IBM cyano column (250 × 4.5 mm, 5-µm particle size). The mobile phase consisted of 650 mL of 0.05 M monobasic sodium phosphate buffer and 350 mL of acetonitrile, adjusted to pH 3.0 with phosphoric acid. The flow rate was 1.7 mL/min. UV detection was performed at 268 nm. Benzophenone 0.03 mg/mL in the mobile phase was used as an internal standard.

A sample of ground capsules was accurately weighed, mixed with internal standard solution, sonicated for 5 minutes, diluted in mobile phase, and filtered. The injection volume was 10 µL. Under these conditions, retention times for prazosin, benzophenone, and polythiazide were about 4.2, 5.5, and 6.5 minutes, respectively.

The method was reported to be stability indicating since the method resolved the intact drug from its degradation products and impurities.

A standard curve for prazosin hydrochloride was generated from 50 to 1250 ng. The correlation coefficient was 0.9999. The limit of detection was 0.6 ng.

Reference
Bachman WJ. High performance liquid chromatographic determination of diuretic–antihypertensive combination products. I. Prazosin and polythiazide. *J Liq Chromatogr*. 1986; 9: 1033–49.

Prednisolone Sodium Phosphate

Chemical Name
11β,17α,21-Trihydroxypregna-1,4-diene-3,20-dione 21-(disodium phosphate)

Other Names
Econopred, Vasocidin

Form	Molecular Formula	MW	CAS
Prednisolone	$C_{21}H_{28}O_5$	360.4	50-24-8
Prednisolone acetate	$C_{23}H_{30}O_6$	402.5	52-21-1
Prednisolone hemisuccinate	$C_{25}H_{32}O_8$	460.5	2920-86-7
Prednisolone hexanoate	$C_{27}H_{38}O_6$	458.6	69164-69-8
Prednisolone pivalate	$C_{26}H_{36}O_6$	444.6	1107-99-9

Prednisolone sodium metasulfobenzoate	$C_{28}H_{31}NaO_9S$	566.6	630-67-1
Prednisolone sodium phosphate	$C_{21}H_{27}Na_2O_8P$	484.4	125-02-0
Prednisolone sodium succinate	$C_{25}H_{31}NaO_8$	482.5	1715-33-9
Prednisolone steaglate	$C_{41}H_{64}O_8$	684.9	5060-55-9
Prednisolone tebutate	$C_{27}H_{38}O_6 \cdot H_2O$	476.6	7681-14-3

Appearance

Prednisolone is a white or almost white odorless hygroscopic crystalline powder. Prednisolone acetate is a white or almost white odorless crystalline powder. Prednisolone hemisuccinate is a fine and creamy white almost odorless powder. Prednisolone pivalate is a white or almost white crystalline powder. Prednisolone sodium phosphate occurs as a white or slightly yellow hygroscopic crystalline powder or friable granules. Prednisolone tebutate is a white to slightly yellow hygroscopic powder.

Solubility

Prednisolone is very slightly soluble in water and soluble in alcohol. Prednisolone acetate is practically insoluble in water and slightly soluble in alcohol. Prednisolone hemisuccinate is very slightly soluble in water and freely soluble in alcohol. Prednisolone pivalate is practically insoluble in water and slightly soluble in alcohol. Prednisolone sodium phosphate is freely soluble in water and soluble in methanol. Prednisolone tebutate is very slightly soluble in water and sparingly soluble in alcohol.

Method 1

Bachman and Gambertoglio developed an HPLC assay for the quantitation of prednisolone sodium phosphate in the presence of formulation excipients, impurities, and degradation products. The chromatographic instrument included a Waters model 510 pump, a model WISP 710B autosampler, a model 490 programmable multiwavelength detector, and a model 840 data module. The stationary phase was a Waters µBondapak phenyl analytical column (300 × 3.9 mm, 10-µm particle size). The mobile phase was prepared by mixing 750 mL of deionized water containing 1.38 g of monobasic sodium phosphate and 250 mL of acetonitrile. The flow rate was 1.6 mL/min. UV detection was performed at 243 nm and 1.0 AUFS. Phenacetin 80 µg/mL in acetonitrile–deionized water (1:100, vol/vol) was used as an internal standard.

Samples were diluted with deionized water before injection. The injection volume was 20 µL. Under these conditions, retention times in minutes were 2.17 for a degradation product, 2.23 for niacinamide, 2.78 for a degradation product, 4.03 for hydrocortisone sodium phosphate, 4.13 for prednisolone sodium phosphate, 4.55 for benzyl alcohol, 5.60 for phenol, 6.58 for dexamethasone sodium phosphate, 7.26 for phenacetin, 11.23 for hydrocortisone, 11.60 for prednisolone, and 22.23 for dexamethasone.

The assay was stability indicating because it was capable of separating prednisolone sodium phosphate from its formulation excipients, impurities, and degradation products.

A calibration curve for prednisolone sodium phosphate was constructed from 10 to 90 µg/mL. The correlation coefficient was 0.9998.

Reference
Bachman WJ, Gambertoglio JG. A stability indicating HPLC assay for prednisolone sodium phosphate in implantable infusion pumps. *Anal Lett.* 1990; 23: 893–900.

Method 2
Dekker and Beijnen evaluated the stability of prednisolone by HPLC. A Waters liquid chromatograph was equipped with a Pye Unicam LC3 variable-wavelength detector and a Spectra-Physics SP4000 integrator. The stationary phase was a Waters µBondapak C_{18} column (300 × 3.9 mm). The mobile phase consisted of water and methanol (50:50, wt/wt) containing 1% (vol/wt) of 0.4 M sodium phosphate solution (pH 7.0). The flow rate was 1.0 mL/min. UV detection was performed at 240 nm. The injection volume was 25 µL.

The method was reported to be stability indicating since the decomposition products did not influence the determination of prednisolone.

Reference
Dekker D, Beijnen JH. Stability of corticosteroids under anaerobic conditions. *Pharm Weekbl Sci Ed.* 1980; 2: 1116–20.

Prifinium Bromide

Chemical Name
3-Diphenylmethylene-1,1-diethyl-2-methylpyrrolidinium bromide

Other Names
Padrin, Riabal

Form	Molecular Formula	MW	CAS
Prifinium bromide	$C_{22}H_{28}BrN$	386.4	4630-95-9

Appearance
Prifinium bromide occurs as crystals.

Method
Sasa et al. described the reversed-phase HPLC determination of prifinium bromide in pharmaceutical formulations. The chromatographic system included a Varian 2010

pump, a Rheodyne 10-µL loop injector, a Varian 2050 variable-wavelength UV detector, and a Varian 4290 integrator. The stationary phase was a Supelco Supelcosil LC-8-DB reversed-phase column (250 × 4.6 mm, 5-µm particle size). The mobile phase was a mixture of 0.03 M ammonium acetate in acetonitrile–water (65:35, vol/vol), adjusted to pH 4.0 with glacial acetic acid. The flow rate was 1.5 mL/min. UV detection was performed at 254 nm and 0.2 AUFS. Benzophenone 10 µg/mL in methanol was used as an internal standard.

Samples were dissolved in the internal standard solution, sonicated if necessary, diluted to a prifinium bromide concentration of approximately 60 µg/mL with the internal standard solution, and filtered through a 0.45-µm membrane filter. The injection volume was 10 µL. Under these conditions, the retention times of prifinium bromide and benzophenone were about 3.3 and 4.0 minutes, respectively.

The method was reported to be stability indicating since the excipients in the formulations did not interfere with the determination of prifinium bromide.

A linear relationship between the peak area ratios and the prifinium bromide concentration was established from 0.15 to 0.90 µg. The correlation coefficient was 0.9995. The limit of detection was less than 1.0 ng.

Reference

Sasa SI, Jalal IM, Khalil HS. Determination of prifinium bromide in six pharmaceutical formulations by reverse-phase HPLC. *J Liq Chromatogr*. 1988; 11: 447–62.

Prilocaine

Chemical Name
2-(Propylamino)propiono-*o*-toluidide

Other Names
Citanest, Xylonest

Form	Molecular Formula	MW	CAS
Prilocaine	$C_{13}H_{20}N_2O$	220.3	721-50-6
Prilocaine hydrochloride	$C_{13}H_{20}N_2O.HCl$	256.8	1786-81-8

Appearance
Prilocaine is a white or almost white crystalline powder. Prilocaine hydrochloride occurs as a white crystalline powder or as colorless crystals.

Solubility

Prilocaine is slightly soluble in water and very soluble in alcohol and acetone. Prilocaine hydrochloride is freely soluble in water and alcohol. It is very slightly soluble in acetone.

Method

Storms and Stewart developed an HPLC method for the determination of a prilocaine hydrochloride and epinephrine drug combination. The HPLC system consisted of a Beckman model 110-B pump, a Rheodyne model 7125 injector with a 20-μL loop, a Waters model 486 UV-visible detector, and a Shimadzu model C-R3A Chromatopac integrator. The stationary phase was a Waters μBondapak phenyl column (300 × 3.9 mm). The mobile phase consisted of an ion-pair aqueous buffer and acetonitrile (80:20, vol/vol), where the ion-pair aqueous buffer was 25 mM monobasic potassium phosphate in water (pH 3.0) with 50 mM heptanesulfonic acid sodium salt. The flow rate was 1 mL/min. UV detection was performed at 254 nm. The injection volume was 20 μL. Under these conditions, retention times of epinephrine and prilocaine hydrochloride were 4.2 and 13.4 minutes, respectively.

The method was shown to be stability indicating by accelerated degradation of prilocaine and epinephrine. At ambient temperature and 90 °C, each drug solution was separately mixed with 0.1 N hydrochloric acid, 0.1 N sodium hydroxide solution, and 3% hydrogen peroxide. When present, degradation product peaks did not interfere with the intact drug of interest.

A calibration curve for prilocaine hydrochloride was constructed from 8 to 200 μg/mL; the correlation coefficient was 0.9998. The intraday and interday coefficients of variation were between 0.04 and 0.60%.

Reference

Storms ML, Stewart JT. Stability-indicating HPLC assays for the determination of prilocaine and procaine drug combinations. *J Pharm Biomed Anal*. 2002; 30: 49–58.

Procainamide Hydrochloride

Chemical Name

4-Amino-*N*-[2-(diethylamino)ethyl]benzamide monohydrochloride

Other Names

Amisalin, Procamide, Procanbid, Procapan, Pronestyl

Form	Molecular Formula	MW	CAS
Procainamide hydrochloride	$C_{13}H_{21}N_3O \cdot HCl$	271.8	614-39-1

Appearance

Procainamide hydrochloride is a white to tan hygroscopic crystalline powder.

Solubility

Procainamide hydrochloride is freely soluble in water and soluble in alcohol. It is also slightly soluble in chloroform.

pK$_a$

Procainamide hydrochloride has a pK$_a$ of 9.23.

Method 1

Allen and Erickson studied the stability of procainamide hydrochloride in extemporaneously compounded oral liquids. The Hewlett-Packard series 1050 automated high-performance liquid chromatograph with Chem Station software consisted of a multi-solvent mixing and pumping system, an autoinjector, a diode-array detector, and a computer. The stationary phase was a Bakerbond C$_{18}$ analytical column (250 × 4.6 mm, 5-μm particle size). The mobile phase consisted of water, methanol, and triethylamine (140:60:1, vol/vol/vol), with the pH adjusted to 7.5 with phosphoric acid. The flow rate was 1.0 mL/min. UV detection was performed at 280 nm.

Samples were diluted 1:100. Under these conditions, procainamide hydrochloride eluted with a retention time of 3.8 minutes.

This assay was determined to be stability indicating. After accelerated degradation by heat, acid, base, oxidizing agent, and light, the composite chromatogram of procainamide hydrochloride showed no interference from the degradation product peaks with the intact procainamide peak.

Standard curves were constructed for procainamide hydrochloride from 100 to 600 μg/mL. The intraday and interday coefficients of variation were 1.7 and 2.1%, respectively.

Reference

Allen LV, Erickson MA. Stability of ketoconazole, metolazone, metronidazole, procainamide hydrochloride, and spironolactone in extemporaneously compounded oral liquids. *Am J Health Syst Pharm.* 1996; 53: 2073–8.

Method 2

Metras et al. investigated the stability of procainamide hydrochloride in an extemporaneously compounded oral liquid. The chromatographic system consisted of a Perkin-Elmer model 250 binary pump, a Perkin-Elmer model LC 290 variable-wavelength UV detector, a Perkin-Elmer model ISS100 autoinjector, and an Epson +1 computer workstation. The stationary phase was a Perkin-Elmer Pecosphere C$_8$ reduced-activity column (33 × 4.6 mm, 3-μm particle size). The mobile phase consisted of 14% acetonitrile, 86% 0.01 M ammonium phosphate buffer, and 0.005% triethylamine adjusted to pH 6.0. The flow rate was 2 mL/min. UV detection was performed at 280 nm and 0.05 AUFS. Procaine hydrochloride 20 μg/mL served as an internal standard.

Procainamide hydrochloride samples were diluted with organic-free deionized water. The injection volume was 10 μL. Under these conditions, the retention times for procainamide and procaine were 0.48 and 1.18 minutes, respectively. The assay run time was approximately 1.2 minutes.

The analytical method was determined to be stability indicating by spiking the procainamide hydrochloride solution with its known degradation product, p-amino-benzoic acid. The degradation product peak did not interfere with the intact procain-amide hydrochloride peak. The retention times for p-aminobenzoic acid, procainamide, and procaine were 0.28, 0.48, and 1.18 minutes, respectively.

Standard curves for procainamide hydrochloride were constructed from 12.5 to 25 µg/mL; the correlation coefficients were 0.9943–0.9995.

A similar method was used by Riley.

References
Metras JI, Swenson CF, McDermott MP. Stability of procainamide hydrochloride in an extemporaneously compounded oral liquid. *Am J Hosp Pharm.* 1992; 49: 1720–4.

Riley CM. Stability of milrinone and digoxin, furosemide, procainamide hydrochloride, propranolol hydrochloride, quinidine gluconate, or verapamil hydrochloride in 5% dextrose injection. *Am J Hosp Pharm.* 1988; 45: 2079–91.

Method 3
Riley and Junkin determined the stability of procainamide hydrochloride in intravenous admixtures. The HPLC system consisted of a Kratos Spectroflow 400 solvent-delivery system, a Waters model 450 variable-wavelength UV detector, a Micromeritics model 728 autosampler, a Valco VICI six-port electronically actuated injection valve fitted with a 20-µL loop, a Fisher Scientific series 5000 Recordall strip-chart recorder, and a Shimazdu model C-R3A electronic integrator. The stationary phase was a Phase Separations Spherisorb phenyl analytical column (150 × 4.6 mm, 5-µm particle size). The mobile phase consisted of acetonitrile, water, and potassium dihydrogen phosphate buffer (0.5 M) (12:78:10, vol/vol/vol) with a pH of 6.8. The flow rate was 2 mL/min. UV detection was performed at 268 nm.

Samples were diluted 1:10 with the mobile phase.

The HPLC method was determined to be stability indicating. It was free from interference by excipients and degradation products.

The peak areas were linearly related to the concentration of the samples from 1 to 15 mg/mL. The coefficient of variation was less than 1%.

A similar method was used by Henry et al.

References
Riley CM, Junkin P. Stability of amrinone and digoxin, procainamide hydrochloride, propranolol hydrochloride, sodium bicarbonate, potassium chloride, or verapamil hydrochloride in intravenous admixtures. *Am J Hosp Pharm.* 1991; 48: 1248–52.

Henry DW, Lacerte JA, Klutman NE, et al. Irreversibility of procainamide–dextrose complex in plasma in vitro. *Am J Hosp Pharm.* 1991; 48: 2426–9.

Method 4
Schaaf et al. determined the stability of procainamide hydrochloride with esmolol hydrochloride in 5% dextrose injection using an HPLC method. The chromatographic system consisted of a Shimadzu model LC-6A pump, an LDC/Milton Roy Spectrometer III variable-wavelength UV detector, and an LDC/Milton Roy Spherisorb ODS-2

reversed-phase column (250 × 4.6 mm, 5-μm particle size). The mobile phase consisted of 34 mM aqueous monobasic potassium phosphate, methanol, and acetonitrile (65:10:25, vol/vol/vol) and was delivered isocratically at a rate of 1 mL/min. UV detection was performed at 280 nm. Propyl p-hydroxybenzoate was used as an internal standard.

Samples were diluted 1:100 with the mobile phase. The injection volume was 20 μL.

The stability-indicating nature of this assay was verified by spiking the procainamide hydrochloride solution with the known decomposition products and impurities. The degradation product peaks did not interfere with the intact procainamide peak. Retention times for the procainamide–dextrose complex, p-aminobenzoic acid, the esmolol degradation product, procainamide, esmolol, and the internal standard were 3.0, 3.8, 4.8, 7.0, 17.6, and 33.8 minutes, respectively.

A standard curve for procainamide was constructed from 0 to 50 μg/mL; the correlation coefficient was 0.999.

Reference

Schaaf LJ, Robinson DH, Vogel GJ, et al. Stability of esmolol hydrochloride in the presence of aminophylline, bretylium tosylate, heparin sodium, and procainamide hydrochloride. *Am J Hosp Pharm*. 1990; 47: 1567–71.

Method 5

Raymond et al. evaluated the stability of procainamide hydrochloride in neutralized 5% dextrose injection. The liquid chromatograph consisted of an Isco model 2300 pump, an Isco model V4 variable-wavelength UV detector, and Zenith Z-150 and Chromadapt-PC computers. The stationary phase was an Altex Ultrasphere C_8 column (150 × 4.6 mm, 5-μm particle size). The mobile phase consisted of 40% acetonitrile and 60% 0.01 M phosphate buffer adjusted to pH 6.0. The flow rate was 1.25 mL. UV detection was performed at 280 nm and 0.1 AUFS. Procaine hydrochloride 400 μg/L in distilled water was used as an internal standard.

Each sample was diluted with distilled water. The injection volume was 10 μL. The assay run time was approximately 12 minutes. The retention times for procainamide and procaine were approximately 5 and 8 minutes, respectively.

The stability-indicating capability of the assay was demonstrated by spiking the procainamide hydrochloride solution with the known major degradation product, p-aminobenzoic acid. The degradation product peak did not interfere with the procainamide peak. The retention time for p-aminobenzoic acid was 1.5 minutes.

A standard curve for procainamide hydrochloride was constructed from 133 to 283 μg/mL; the correlation coefficient was 0.9948.

Reference

Raymond GG, Reed MT, Teagarden JR, et al. Stability of procainamide hydrochloride in neutralized 5% dextrose injection. *Am J Hosp Pharm*. 1988: 45: 2513–7.

Procaine Hydrochloride

Chemical Name
2-Diethylaminoethyl 4-aminobenzoate hydrochloride

Other Names
Gero, Novanaest, Novocain

Form	Molecular Formula	MW	CAS
Procaine hydrochloride	$C_{13}H_{20}N_2O_2.HCl$	272.8	51-05-8

Appearance
Procaine hydrochloride occurs as small colorless or white odorless crystals or as a white crystalline powder.

Solubility
Procaine hydrochloride is freely soluble in water and soluble in alcohol.

pK$_a$
Procaine hydrochloride has a pK$_a$ of 9.1.

Method
Wang described a method for the study of procaine stability in aqueous systems. The chromatograph was equipped with a Waters model 440 254-nm UV detector and a Mini-Lab model CSI 38 integrator. The stationary phase was a Waters μBondapak C_{18} stainless steel column (300 × 4 mm). The mobile phase consisted of methanol and 1% acetic acid (40:60, vol/vol) at pH 4.7. The flow rate was 1.1 mL/min. UV detection was performed at 254 nm and 0.5 AUFS. Benzoic acid was used as an internal standard.

The sample was diluted with water. The injection volume was 5 μL. Under these conditions, retention times of p-aminobenzoic acid, procaine hydrochloride, and benzoic acid were 3.5, 4.5, and 5.5 minutes, respectively.

The method was demonstrated to be stability indicating because p-amino-benzoic acid, diethylaminoethanol, methylparaben, and propylparaben did not interfere with the analysis of procaine hydrochloride.

A calibration curve for procaine hydrochloride was generated from 0.200 to 0.800 mg/mL. The correlation coefficient was 0.9999.

Reference
Wang D-P. Stability of procaine in aqueous systems. *Analyst*. 1983; 108: 851–6.

Procarbazine Hydrochloride

Chemical Name
N-Isopropyl-α-(2-methylhydrazino)-p-toluamide monohydrochloride

Other Names
Matulane, Natulan

Form	Molecular Formula	MW	CAS
Procarbazine hydrochloride	$C_{12}H_{19}N_3O.HCl$	257.8	366-70-1

Appearance
Procarbazine hydrochloride is a white to pale yellow crystalline powder with a slight odor.

Solubility
Procarbazine hydrochloride is freely soluble in water and sparingly soluble in alcohol.

pK$_a$
Procarbazine hydrochloride has a pK$_a$ of 6.8.

Method
Burce and Boehlert presented a method for the separation and quantitation of possible degradation products of procarbazine hydrochloride in its dosage form. A DuPont model 848 chromatograph was equipped with a 254-nm UV detector, a septum injector, and a Linear Instruments model 282 dual-pen recorder with integrator. The stationary phase was a Whatman Partisil PXS 10/25 ODS-2 C_{18} reversed-phase column (250 mm, 10-μm particle size). The mobile phase was 44% methanol in 0.05 M ammonium phosphate (pH 5.5). Cinnamyl alcohol 0.1 mg/mL in mobile phase was used as an internal standard.

Five capsules of procarbazine hydrochloride were accurately weighed into a centrifuge tube, mixed with 10 mL of the mobile phase, and shaken for 15 seconds. The mixture was immediately filtered through a 0.45-μm Millipore filter, and the filtrate was immediately injected. The injection volume was 4 μL. Under these conditions, the retention times for procarbazine and cinnamyl alcohol were 3.31 and 16.85 minutes, respectively.

The analytical method was reported to be stability indicating since it could separate the drug from its degradation products. Retention times for procarbazine, N-isopropyl-4-formylbenzamide, N-(1-methylethyl)-4-[(2-methylamino)hydrazo]benzamide, and *N*-(1-methylethyl)-4-[(2-methylazo)methyl]benzamide were 3.31, 7.61, 10.61, and 14.21 minutes, respectively.

A standard curve for procarbazine was constructed from 50 to 350 µg/mL. The correlation coefficient was 0.9999. The limit of detection was 4 µg/mL and the limit of quantitation was 40 µg/mL.

Reference

Burce GL, Boehlert JP. Separation and quantitation of possible degradation products of procarbazine hydrochloride in its dosage form. *J Pharm Sci.* 1978; 67: 424–6.

Prochlorperazine

Chemical Names

2-Chloro-10-[3-(4-methyl-1-piperazinyl)propyl]-10*H*-phenothiazine
3-Chloro-10-[3-(4-methyl-1-piperazinyl)propyl]phenothiazine

Other Name

Compazine

Form	Molecular Formula	MW	CAS
Prochlorperazine	$C_{20}H_{24}ClN_3S$	374.0	58-38-8
Prochlorperazine edisylate	$C_{20}H_{24}ClN_3S.C_2H_6O_6S_2$	564.1	1257-78-9
Prochlorperazine maleate	$C_{20}H_{24}ClN_3S.2C_4H_4O_4$	606.1	84-02-6
Prochlorperazine mesylate	$C_{20}H_{24}ClN_3S.2CH_4O_3S$	566.2	5132-55-8

Appearance

Prochlorperazine occurs as a clear pale yellow viscous liquid. Prochlorperazine edisylate is a white to very light yellow odorless crystalline powder. Prochlorperazine maleate is a white to pale yellow and practically odorless crystalline powder. Prochlorperazine mesylate is a white or almost white odorless or almost odorless powder.

Solubility

Prochlorperazine is very slightly soluble in water (<0.1% at 20 °C). It is slightly soluble in methanol and ethanol and practically insoluble in ether, benzene, and chloroform. Prochlorperazine edisylate has approximate solubilities of 500 mg/mL in water and 0.67 mg/mL in alcohol at 25 °C. Prochlorperazine maleate is practically insoluble in water and has a solubility of approximately 0.83 mg/mL in alcohol at 25 °C. Prochlorperazine mesylate is very soluble in water, sparingly soluble in alcohol, slightly soluble in chloroform, and practically insoluble in ether.

Method

Walker et al. studied the stability of prochlorperazine using an HPLC assay. The analysis was performed with a Schoeffel SF 770 UV detector, a Spectra-Physics SP4200 chromatographic integrator, and a Beckman Ultrasphere C_{18} reversed-phase column (250 × 4.2 mm, 5-μm particle size). The mobile phase consisted of acetonitrile and phosphate buffer containing 1 mg/mL of heptanesulfonic acid; it was delivered gradiently at 2.0 mL/min from 16 to 46% acetonitrile in 14 minutes. UV detection was performed at 230 nm.

The analytical method was determined to be stability indicating by accelerated degradation of prochlorperazine using acid, base, or heat and by inspection of chromatograms for the appearance of additional peaks and changes in retention time and peak shape. Degradation product peaks did not interfere with the intact prochlorperazine peak.

Intraday and interday coefficients of variation for the assay were 0.76 and 4.22%, respectively.

Reference

Walker SE, Iazzetta J, DeAngelis C, et al. Stability and compatibility of combination of hydromorphone and dimenhydrinate, lorazepam or prochlorperazine. *Can J Hosp Pharm*. 1993; 46: 61–5.

Procinonide

Chemical Name

(6α,11β,16α)-6,9-Difluoro-11-hydroxy-16,17-[(1-methylethylidene)bis(oxy)]-21-(1-oxo-propoxy)-pregna-1,4-diene-3,20-dione

Other Name

Trisyn

Form	Molecular Formula	MW	CAS
Procinonide	$C_{27}H_{34}F_2O_7$	508.6	58497-00-0

Method

Shek et al. described a reversed-phase HPLC method for a study of procinonide pro-pionate stability in a cream. The instrument was a Spectra-Physics model 3500 system equipped with an oven, a Valco loop injector, and a SpectraMonitor III UV detector or a Spectra-Physics model 8300 UV detector. The stationary phase was a Spectra-Physics Spherisorb ODS stainless steel column (250 × 4.6 mm) with a Whatman Co:Pell ODS guard column (70 × 2.1 mm). The mobile phase was a mixture of tetrahydrofuran, acetonitrile, and water (1:3:6, vol/vol/vol); it was delivered isocratically at 1.2 mL/min. UV detection was performed at 254 nm and 0.02 AUFS. The column temperature was 45 °C. Flucloronide in acetonitrile 190 μg/mL was used as an internal standard.

Two grams of cream was transferred to a mixture of 75 mL isooctane and 25 mL acetonitrile containing 1 mL of the internal standard and vigorously shaken for 1 minute. The lower acetonitrile layer was collected. This extraction was repeated three or four times. The combined acetonitrile solution was evaporated to dryness using a steam bath and a stream of nitrogen. The residue was redissolved in 10 mL of acetoni-trile, diluted with 10 mL of water, and centrifuged at 3000 rpm for 10 minutes. The clear solution was collected and assayed. The injection volume was 50 μL. Under these conditions, retention times for the major degradation product (fluocinolone acetonide), the internal standard, and procinonide were about 7.2, 13.3, and 27.2 minutes, respec-tively (estimated from the published chromatogram).

The stability-indicating nature of the method was demonstrated by assaying a mixture of the drug and its possible degradation products. The chromatogram showed the excellent separation of the intact drug from all degradation products.

Reference

Shek E, Bragonje J, Benjamin EJ, et al. A stability indicating high-performance liquid chromatography determination of a triple corticoid integrated system in a cream. *Int J Pharm.* 1982; 11: 257–9.

Proguanil Hydrochloride

Chemical Name

1-(4-Chlorophenyl)-5-isopropylbiguanide hydrochloride

Other Names

Chloroguanide, Malarone, Paludrine

Form	Molecular Formula	MW	CAS
Proguanil hydrochloride	$C_{11}H_{16}ClN_5.HCl$	290.2	637-32-1

Appearance

Proguanil hydrochloride is a white and odorless or almost odorless crystalline powder.

Solubility

Proguanil hydrochloride is slightly soluble in water, more soluble in hot water, and sparingly soluble in alcohol.

Method

Taylor et al. reported a chemical stability study of proguanil hydrochloride using an HPLC method. The instrument consisted of a Waters model 6000A pump, model 440 and model 441 fixed-wavelength detectors, and a Varian Polychrom 9060 diode-array detector or a Shimadzu SPDM6A diode-array detector. The stationary phase was a laboratory-made Hypersil ODS column (100 × 2 mm, 5- or 3-µm particle size). The mobile phase consisted of acetonitrile and 20 mM phosphate buffer containing 200 mM sodium lauryl sulfate (40:60, vol/vol). UV detection was performed at 254 nm. The injection volume was 20 µL. Under these conditions, the retention time of proguanil hydrochloride was about 9.8 minutes (estimated from the published chromatogram).

This analytical method was reported to be stability indicating because the intact drug was separated from its degradation products, which were generated after partial decomposition at 75 °C for 26 hours.

References

Taylor RB, Moody RR, Ochekpe NA, et al. A chemical stability study of proguanil hydrochloride. *Int J Pharm.* 1990; 60: 185–90.

Taylor RB, Moody RR, Ochekpe NA. Determination of proguanil and its metabolites cycloguanil and 4-chlorophenylbiguanide in plasma, whole blood and urine by high-performance liquid chromatography. *J Chromatogr.* 1987; 416: 394–9.

Promethazine

Chemical Name

Dimethyl (1-methyl-2-phenothiazin-10-ylethyl)amine

Other Names
Phenergan, Prorex

Form	Molecular Formula	MW	CAS
Promethazine	$C_{17}H_{20}N_2S$	284.4	60-87-7
Promethazine hydrochloride	$C_{17}H_{20}N_2S.HCl$	320.9	58-33-3
Promethazine theoclate	$C_{17}H_{20}N_2S.C_7H_7ClN_4O_2$	499.0	17693-51-5

Appearance
Promethazine hydrochloride is a white or faintly yellow crystalline powder. Promethazine theoclate is a white or almost white and odorless or almost odorless powder.

Solubility
Promethazine hydrochloride is very soluble in water, hot dehydrated alcohol, and chloroform. Promethazine theoclate is very slightly soluble in water and sparingly soluble in alcohol.

pK$_a$
Promethazine hydrochloride has a pK$_a$ of 9.1.

Method 1
Mathew et al. described the HPLC determination of promethazine hydrochloride in pharmaceutical dosage forms. The system was a Waters ALC 202 chromatograph equipped with a Rheodyne model 7125 universal injector, a Schoeffel SF 770 multiple-wavelength detector, and a Houston Omniscribe recorder. The stationary phase was a Whatman micro C_{18} column (250 × 3.9 mm). The mobile phase consisted of 40% acetonitrile and 0.1% glacial acetic acid in 0.01 M monobasic potassium phosphate buffer. The flow rate was 2.0 mL/min. UV detection was performed at 271 nm and 0.1 AUFS. Dextromethorphan hydrobromide 200 µg/mL in water was used as an internal standard.

For promethazine hydrochloride injection, for syrup, and for syrup with phenylephrine hydrochloride, the sample was diluted with water to contain about 300 µg/mL of promethazine hydrochloride and 200 µg/mL of dextromethorphan hydrobromide. For promethazine hydrochloride suppositories, one or two suppositories were crushed and mixed with 10 mL of 1 N hydrochloric acid and 35 mL of water for 30 minutes in the dark. The supernatant was collected, washed with water, and diluted with water to contain about 300 µg/mL of promethazine hydrochloride and 200 µg/mL of internal standard. For promethazine hydrochloride tablets, the ground powder equivalent to one tablet was mixed with 5 mL of 1 N sodium hydroxide solution for 30 minutes and filtered. The middle portion of the filtrate was diluted with water to contain about 300 µg/mL of promethazine hydrochloride and 200 µg/mL of dextromethorphan hydrobromide. The injection volume was 20 µL. Under these conditions, retention times for dextromethorphan and promethazine were about 4.0 and 5.3 minutes, respectively (estimated from the published chromatogram).

The method was demonstrated to be stability indicating by intentional degradation of promethazine hydrochloride. A solution of promethazine hydrochloride was boiled for 2–3 minutes; the clear solution turned pink. Degradation products and excipients present in the dosage formulations did not interfere with the analysis of promethazine hydrochloride.

Reference
Mathew M, Das Gupta V, Bethea C. Quantitation of promethazine hydrochloride in pharmaceutical dosage forms using high performance liquid chromatography. *Drug Dev Ind Pharm.* 1994; 20: 1693–8.

Method 2
Stavchansky et al. developed an HPLC method for the determination of promethazine hydrochloride in the presence of its thermal and photolytic degradation products. The liquid chromatograph was equipped with a Beckman model 100A pump and a Hitachi model 200-40 spectrophotometer. The stationary phase was an Altex Ultrasphere ODS stainless steel column (250 × 4.6 mm, 5-µm particle size). The column temperature was 45 °C. The mobile phase consisted of 15 mM monobasic potassium phosphate (pH 5) and acetonitrile containing 0.009 mM triethanolamine (15:85, vol/vol). The flow rate was 1.6 mL/min. UV detection was performed at 249 nm. Promazine hydrochloride 0.25 mg/mL was used as an internal standard. The injection volume was 10 µL. Under these conditions, retention times for promethazine and promazine were 5.9 and 9.0 minutes, respectively.

The stability-indicating ability of the method was demonstrated by accelerated degradation of promethazine. Promethazine samples were exposed to UV light (350 nm) or kept at 70 °C for up to 60 hours. Degradation products did not interfere with the determination of promethazine hydrochloride.

A standard curve for promethazine in water was constructed from 30 to 600 µg/mL. The correlation coefficient was 0.999.

Reference
Stavchansky S, Wallace J, Chue M, et al. High pressure liquid chromatographic determination of promethazine hydrochloride in the presence of its thermal and photolytic degradation products: A stability indicating assay. *J Liq Chromatogr.* 1983; 6: 1333–44.

Propantheline Bromide

Chemical Name
Diisopropylmethyl[2-(xanthen-9-ylcarbonyloxy)ethyl]ammonium bromide

Other Names
Probamide, Pro-Banthine

Form	Molecular Formula	MW	CAS
Propantheline bromide	$C_{23}H_{30}BrNO_3$	448.4	50-34-0

Appearance
Propantheline bromide occurs as a white or yellowish-white odorless and slightly hygroscopic powder or crystals.

Solubility
Propantheline bromide is very soluble in water and alcohol.

Method
Jalal et al. reported the determination of propantheline bromide in tablet formulations by HPLC analysis. The Varian 5000 liquid chromatograph was equipped with a Valco 10-μL manual loop injector, a Varian UV-50 spectrophotometric detector, and a Spectra-Physics 4100 digital integrator. The stationary phase was a Supelco LC-8 column (250 × 4.6 mm, 5-μm particle size). The mobile phase contained 2.32 g of ammonium acetate, 600 mL of acetonitrile, 20 mL of tetrahydrofuran, and water to make 1 liter. The pH of the mobile phase was adjusted to 4.5 with glacial acetic acid. The flow rate was 1.5 mL/min. UV detection was performed at 254 nm and 0.10 AUFS. Methyl xanthanoate 24.8 μg/mL in methanol was used as an internal standard.

A sample of powdered tablets was accurately weighed, dissolved in 10 mL of the internal standard solution, and sonicated for 5 minutes. The mixture was filtered through a 0.45-μm membrane filter and diluted 1:10 with the internal standard solution. The injection volume was 10 μL. Under these conditions, retention times for xanthanoic acid, xanthone, methyl xanthanoate, and propantheline bromide were about 2.8, 3.8, 4.9, and 5.3 minutes, respectively.

The method was demonstrated to be stability indicating since it separated the intact drug from its degradation products and excipients.

A linear relationship of peak area ratio versus propantheline bromide concentration from 0.05 to 0.2 mg/mL was obtained. The correlation coefficient was greater than 0.999. The limit of detection was 150 μg.

Reference
Jalal IM, Sasa SS, Rjoob AW, et al. Determination of propantheline bromide in tablet formulations by reverse-phase HPLC. *J Liq Chromatogr*. 1987; 10: 2525–45.

Propofol

Chemical Names
2,6-Bis(1-methylethyl)phenol
2,6-Diisopropylphenol

Other Name
Diprivan

Form	Molecular Formula	MW	CAS
Propofol	$C_{12}H_{18}O$	178.3	2078-54-8

Appearance
Propofol occurs as a clear and colorless to slightly yellowish liquid.

Solubility
Propofol is very slightly soluble in water, very soluble in dehydrated alcohol and in methanol, and slightly soluble in cyclohexane and in isopropanol.

pK_a
Propofol has a pK_a of 11.

Method 1
Zhang et al. reported an HPLC method for the determination of cisatracurium besylate and propofol in a mixture. The apparatus consisted of a Beckman model 110B pump, an Alcott model 738 autosampler, a Waters Lambda-Max model 481 LC spectrophotometer, and a Hewlett-Packard model 3394A integrator. The stationary phase was an Alltech Spherisorb ODS-2 column (250 × 4.6 mm, 5-µm particle size). The mobile phase was a mixture of acetonitrile and 0.3 M ammonium formate (adjusted to pH 5.2 with formic acid) (50:50, vol/vol). The flow rate was 1.0 mL/min. UV detection was performed at 280 nm. The injection volume was 20 µL. Under these conditions, retention times for cisatracurium besylate and propofol were about 8.8 and 13.1 minutes, respectively.

The method was demonstrated to be stability indicating. Mixtures of cisatracurium besylate and propofol were treated with 0.01 N hydrochloric acid, 0.01 N sodium hydroxide solution, or 3–30% hydrogen peroxide. The degradation product peaks did not interfere with the analysis of cisatracurium besylate and propofol.

A standard curve for propofol was generated from 37 to 592 µg/mL. The correlation coefficient was 0.9999.

Reference
Zhang H, Wang P, Bartlett MG, et al. HPLC determination of cisatracurium besylate and propofol mixtures with LC-MS identification of degradation products. *J Pharm Biomed Anal.* 1998; 16: 1241–9.

Method 2
Chernin et al. determined the stability and compatibility of thiopental sodium 12.5 mg/mL and propofol 5 mg/mL stored together in polypropylene syringes at room temperature

and at 4 °C. The HPLC system consisted of an Alcott model 760 pump, a Hewlett-Packard model 3395 integrator, an Alcott model 728 autosampler, an Applied Biosystems model 749A UV-visible detector, and a Waters model 990 photodiode-array detector. The stationary phase was a Mac-Mod Analytical Zorbax SB phenyl column (150 × 4.6 mm, 5-µm particle size). The mobile phase was a mixture of 45% acetonitrile and 55% aqueous 0.01 M monobasic potassium phosphate buffer with the pH adjusted to 4.0 with 10% phosphoric acid. The flow rate was 1.0 mL/min. UV detection was performed at 235 nm.

Samples were diluted 1:200. The injection volume was 20 µL. Under these conditions, the retention times for thiopental sodium and propofol were 4.9 and 12.0 minutes, respectively.

The stability-indicating capability of the assay was shown by accelerated decomposition of propofol. A propofol sample was incubated for 7 hours at 23 °C in 0.1 N hydrochloric acid, for 7 hours at 23 °C in 0.1 N sodium hydroxide, and for 6 hours at 50 °C in 3% hydrogen peroxide. No degradation product peaks interfered with the parent propofol peak.

A standard curve for propofol was generated from 12.7 to 38.0 µg/mL by linear regression of peak areas against drug concentrations. The correlation coefficient was greater than 0.9998. Intraday and interday variabilities were 1.13 and 0.82%, respectively.

Reference
Chernin EL, Stewart JT, Smiler B. Stability of thiopental sodium and propofol in polypropylene syringes at 23 and 4 °C. *Am J Health Syst Pharm.* 1996; 53: 1576–9.

Method 3
Prankerd and Jones evaluated the physicochemical compatibility of propofol with thiopental sodium. The liquid chromatograph included a Waters model M501 pump, a Waters model 712 WISP automatic injector, a Waters model M470 scanning fluorescence detector, a Waters model M490E UV detector, and a Max 810 workstation. The stationary phase was a Waters Nova-Pak C_{18} reversed-phase radial-compression column. The mobile phase consisted of 10 mM phosphate buffer (pH 6.0) and acetonitrile (45:55, vol/vol). The flow rate was 1.6 mL/min. The fluorescence detector was set at an excitation wavelength of 276 nm and an emission wavelength of 314 nm. UV detection was performed at 290 nm.

Each sample was diluted 1:5000 with the mobile phase. The injection volume was 40 µL. Under these conditions, the retention times were 13.4 minutes for propofol (fluorescence detection) and 4.0 minutes for thiopental (UV detection).

The assay was determined to be stability indicating by examining three propofol–thiopental sodium admixtures stored at 4 °C for 6 months. At 6 months, less than 5% of the intact propofol remained. The degradation product peaks did not interfere with the intact propofol peak.

A calibration curve for propofol was constructed from 20 to 1000 µg/mL. The correlation coefficient of the standard curve was greater than 0.9988. The interassay coefficients of variation were 2.0% at the upper end of the range and 4.3% at the lower end of the range.

Reference
Prankerd RJ, Jones RD. Physicochemical compatibility of propofol with thiopental sodium. *Am J Health Syst Pharm.* 1996; 53: 2606–10.

Method 4

Bhatt-Mehta et al. studied the stability of propofol with parenteral nutrient solutions during simulated Y-site injection. The liquid chromatographic system consisted of a Waters model 501 pump, a Waters model 712 autosampler, a Waters model 481 variable-wavelength UV detector, and a Hewlett-Packard model 3390A integrator. The stationary phase was an Alltech Hypersil ODS C_{18} column (250 × 4.6 mm, 5-μm particle size). The mobile phase was a mixture of acetonitrile, methanol, and water (55:10:35, vol/vol/vol); it was delivered isocratically at 2 mL/min. UV detection was performed at 270 nm. The internal standard was thymol 1 mg/mL.

All samples containing parenteral nutrients were filtered with Micron Separation nylon filters (0.45-μm pore size, 3-mm diameter membrane). The injection volume was 50 μL. Under these conditions, the retention times for thymol and propofol were 4.0 and 7.5 minutes, respectively.

The propofol assay was shown to be stability indicating by accelerated degradation of propofol. The solutions of propofol 10 mg/mL were adjusted to pH 12 with 4 N sodium hydroxide or to pH 2 with 85% phosphoric acid. Another propofol solution was exposed to sunlight over 14 days at room temperature and then subjected to heat at 100 °C for 30 days. In all cases, degradation product peaks did not interfere with the intact propofol peak.

A calibration curve for propofol was generated from 2.5 to 90 μg/mL. The correlation coefficient of the calibration curve was 0.99. The intraday and interday coefficients of variation were less than 5%.

Reference

Bhatt-Mehta V, Paglia RE, Rosen DA. Stability of propofol with parenteral nutrient solutions during simulated Y-site injection. *Am J Health Syst Pharm*. 1995; 52: 192–6.

Propoxycaine Hydrochloride

Chemical Name

2-(Diethylamino)ethyl-4-amino-2-propoxybenzoate monohydrochloride

Other Name

Ravocaine and Novocain

Form	Molecular Formula	MW	CAS
Propoxycaine hydrochloride	$C_{16}H_{26}N_2O_3$.HCl	330.9	550-83-4

Appearance

Propoxycaine hydrochloride occurs as a white odorless crystalline solid.

Solubility

Propoxycaine hydrochloride is soluble 1 in 2 of water, 1 in 10 of alcohol, and 1 in 80 of ether. It is practically insoluble in acetone and chloroform.

Method

Storms and Stewart developed an HPLC assay for the determination of a propoxycaine hydrochloride, norepinephrine hydrochloride, and procaine hydrochloride drug combination. The HPLC system consisted of a Beckman model 110-B pump, a Rheodyne model 7125 injector with a 20-µL loop, a Waters model 486 UV-visible detector, and a Shimadzu model C-R3A Chromatopac integrator. The stationary phase was a Waters µBondapak phenyl column (300 × 3.9 mm). The mobile phase consisted of an ion-pair aqueous buffer and acetonitrile (70:30, vol/vol), where the ion-pair aqueous buffer was 25 mM monobasic potassium phosphate in water (pH 3.0) with 50 mM heptanesulfonic acid sodium salt. The flow rate was 1 mL/min. UV detection was performed at 254 nm. The injection volume was 20 µL. Under these conditions, retention times of norepinephrine, procaine hydrochloride, and propoxycaine hydrochloride were 3.9, 9.8, and 18.9 minutes, respectively.

The assay was demonstrated to be stability-indicating by intentional degradation of norepinephrine, procaine hydrochloride, and propoxycaine hydrochloride. At ambient temperature and 90 ºC, each drug solution was mixed with 0.1 N hydrochloric acid, 0.1 N sodium hydroxide solution, and 3% hydrogen peroxide, separately. When present, degradation product peaks did not interfere with the analysis of the intact drug.

Reference

Storms ML, Stewart JT. Stability-indicating HPLC assays for the determination of prilocaine and procaine drug combinations. *J Pharm Biomed Anal.* 2002; 30: 49–58.

Propoxyphene

Chemical Name

(+)-(1*S*,2*R*)-1-Benzyl-3-dimethylamino-2-methyl-1-phenylpropyl propionate

Other Names

Darvon-N, Darvon Pulvules, Doloxene

Form	Molecular Formula	MW	CAS
Propoxyphene	$C_{22}H_{29}NO_2$	339.5	469-62-5
Propoxyphene hydrochloride	$C_{22}H_{29}NO_2.HCl$	375.9	1639-60-7
Propoxyphene napsylate	$C_{22}H_{29}NO_2.C_{10}H_8O_3S.H_2O$	565.7	26570-10-5

Appearance

Propoxyphene hydrochloride is a white crystalline powder. Propoxyphene napsylate is an odorless or almost odorless white crystalline powder.

Solubility

Propoxyphene hydrochloride is freely soluble in water and soluble in alcohol. Propoxyphene napsylate is slightly soluble in water and soluble in alcohol.

Method

Ibrahim described an HPLC method for the determination of propoxyphene. The Varian 5000 liquid chromatograph was equipped with a Varian model 9010 pump, a model 9095 auto-sampler, and a model 9065 photodiode-array detector. The stationary phase was a Waters µBondapak C_{18} column (300 × 3.9 mm). The mobile phase consisted of 500 mL of 0.01 M monobasic potassium phosphate buffer containing 0.005 M 1-decanesulfonic acid sodium salt, 500 mL of acetonitrile, and 1 mL of heptylamine, adjusted to pH 7.9 with phosphoric acid. The flow rate was 2 mL/min. UV detection was performed at 220 nm. Papaverine hydrochloride 0.2 mg/mL in the mobile phase was used as an internal standard. The injection volume was 10 µL. Under these conditions, the retention times for papaverine hydrochloride, acetoxy, carbinol hydrochloride, and propoxyphene were about 2.7, 10.1, 11.9, and 13.6 minutes, respectively.

The stability-indicating nature of the method was demonstrated by accelerated decomposition of the drug. Solutions of propoxyphene were exposed to UV light (254 and 366 nm) for 30 hours or were stored at 50 ℃ for 2 days. The intact propoxyphene was well resolved from its degradation products.

A linear relationship between the peak area and the concentration of propoxyphene was obtained. The correlation coefficient was 0.9998. The limit of detection was 0.05 µg/mL.

Reference

Ibrahim FB. Simultaneous determination and separation of several barbiturates and analgesic products by ion-pair high-performance liquid chromatography. *J Liq Chromatogr*. 1993; 16: 2835–51.

Propranolol

Chemical Names

1-[(1-Methylethyl)amino]-3-(1-naphthalenyloxy)-2-propanol
1-(Isopropylamino)-3-(1-naphthyloxy)-2-propanol

Other Name
Inderal

Form

Form	Molecular Formula	MW	CAS
Propranolol	$C_{16}H_{21}NO_2$	259.4	525-66-6
Propranolol hydrochloride	$C_{16}H_{21}NO_2.HCl$	295.8	318-98-9

Appearance
Propranolol hydrochloride is a white or off-white crystalline powder.

Solubility
Propranolol hydrochloride is soluble in water and alcohol but practically insoluble in ether, benzene, and ethyl acetate.

Method 1
Riley and Junkin used HPLC to study the stability of propranolol hydrochloride in intravenous admixtures. The HPLC system consisted of a Kratos Spectroflow 400 solvent-delivery system, a Micromeritics model 728 autosampler, a Waters model 450 variable-wavelength UV detector, a Valco VICI six-port electronically actuated injection valve fitted with a 20-μL loop, a Fisher Scientific series 5000 Recordall strip-chart recorder, and a Shimadzu model C-R3A electronic integrator. The stationary phase was a Phase Separations Spherisorb phenyl analytical column (150 × 4.6 mm, 5-μm particle size). The mobile phase was a mixture of acetonitrile, water, tetrabutylammonium hydrogen sulfate (10 mM), and potassium dihydrogen phosphate (0.5 M) (15:58:17:10, vol/vol/vol/vol) with a pH of 7.0. The flow rate was 2 mL/min. UV detection was performed at 268 nm.

Samples were diluted 1:10 with the mobile phase.

The HPLC method was determined to be stability indicating. It was free from interference by formulatory excipients and degradation products.

The peak areas were linearly related to the drug concentration from 50 to 750 μg/mL. The coefficient of variation was less than 1%.

Reference
Riley CM, Junkin P. Stability of amrinone and digoxin, procainamide hydrochloride, propranolol hydrochloride, sodium bicarbonate, potassium chloride, or verapamil hydrochloride in intravenous admixture. *Am J Hosp Pharm.* 1991; 48: 1245–52.

Method 2
Strom and Miller determined the in vitro compatibility, chemical stability, and interaction of cephalexin, cimetidine, diazepam, and propranolol with three selected enteral

nutrient products. A Micromeritics liquid chromatograph included a solvent-delivery system, a variable-wavelength UV-visible detector, and a variable-volume injector. Chromatograms were recorded and peak heights calculated with a Micromeritics microprocessor. The stationary phase was a cyano column. The mobile phase consisted of acetonitrile, water, acetic acid, and triethylamine (42.4:56.6:0.9:0.1, vol/vol/vol/vol); it was delivered isocratically at 1.25 mL/min. UV detection was performed at 290 nm. Tyramine 300 μg/mL was used as an internal standard.

Each sample was diluted with methanol and centrifuged for 2 minutes. The supernatant was filtered through a Gelman Alpha-450 0.45-μm filter. The injection volume was 25 μL.

The analytical method was demonstrated to be stability indicating by accelerated decomposition of propranolol. Propranolol sample solutions were prepared in 1 N hydrochloric acid and 1 N sodium hydroxide and were heated at 80 °C for 24 hours. The intact propranolol peak was well separated from its degradation product and from other component peaks in the enteral nutrient formulas.

Reference
Strom JG, Miller SW. Stability of drugs with enteral nutrient formulas. *Ann Pharmacother.* 1990; 24: 130–4.

Method 3
Hitscherich et al. presented a simple analytical method for the simultaneous determination of hydrochlorothiazide and propranolol hydrochloride in tablets. The liquid chromatograph included a Waters model 501 pump, a Perkin-Elmer model LC-420B autoinjector, a Perkin-Elmer model LC-95 variable-wavelength detector, and a Spectra-Physics model SP-4270 integrator. The stationary phase was an Altex Ultrasphere cyano column (250 × 4.6 mm, 5-μm particle size). The mobile phase was prepared by mixing 15 mL of acetonitrile with 85 mL of 0.05 M monobasic ammonium phosphate, adjusted to pH 3.0 with phosphoric acid. The flow rate was 2 mL/min. UV detection was carried out at 290 nm.

A portion of powder from 20 tablets equivalent to 25 mg of hydrochlorothiazide was accurately weighed, transferred to a 100-mL volumetric flask, mixed with 5 mL of 0.1 N hydrochloric acid, diluted with 50 mL of methanol, sonicated for 10 minutes, cooled down, filled to volume with methanol, mixed well, filtered through a 0.45-μm solvent-resistant filter, diluted with mobile phase, and assayed. The injection volume was 20 μL. Under these conditions, retention times for hydrochlorothiazide and propranolol hydrochloride were 4.5 and 14.8 minutes, respectively.

The method was confirmed to be stability indicating by an accelerated degradation study. Propranolol hydrochloride was refluxed for 24 hours in 1.0 N hydrochloric acid. The intact drug was separated from degradation products at retention times of 9.0 and 15.9 minutes.

A standard curve was constructed from 0 to 160.5 μg/mL with a correlation coefficient of 0.9986. The average recovery for hydrochlorothiazide was 100.2%.

Reference
Hitscherich ME, Rydberg EM, Tsilifonis DC, et al. Simultaneous determination of hydrochlorothiazide and propranolol hydrochloride in tablets by high performance liquid chromatography. *J Liq Chromatogr.* 1987; 10: 1011–21.

Method 4

Henry et al. investigated the stability of propranolol hydrochloride in an extemporaneous suspension compounded from tablets (Inderal). The system consisted of a Waters model 720 system controller, a Waters model 6000A solvent-delivery system, a Waters model 710B WISP autosampler, a Waters model 440 UV detector, and a Waters data module. The stationary phase was a Shandon ODS Hypersil reversed-phase C_{18} column (150 × 5 mm, 5-μm particle size) with a precolumn of the same packing material. The mobile phase consisted of 30% acetonitrile, 4 mM sodium octanesulfonate, 3.5 mM tetrabutylammonium hydrogen sulfate, and 0.1% sulfuric acid in sterile water for irrigation. The flow rate was 1.0 mL/min. UV detection was performed at 280 nm and 0.1 AUFS.

Samples were diluted with the mobile phase and filtered through a 0.22-μm nylon filter. The injection volume was 10 μL. Under these conditions, the retention time for propranolol hydrochloride was 6.3 minutes.

To demonstrate that the assay was stability indicating, solutions of propranolol hydrochloride were acidified with sulfuric acid (pH 2), alkalinized with sodium hydroxide (pH 8.5), or heated in an oven (100 °C). The new degradation product peaks did not interfere with the intact propranolol hydrochloride peak.

A calibration curve for propranolol hydrochloride was constructed from 30 to 160 μg/mL; the correlation coefficient of the curve was 0.99. The largest intraday variation was 3.8%.

Reference

Henry DW, Repta AJ, Smith FM, et al. Stability of propranolol hydrochloride suspension compounded from tablets. *Am J Hosp Pharm.* 1986; 43: 1492–5.

Propylene Carbonate

Chemical Name
4-Methyl-1,3-dioxolan-2-one

Form	Molecular Formula	MW	CAS
Propylene carbonate	$C_4H_6O_3$	102.1	108-32-7

Appearance
Propylene carbonate is a clear colorless mobile liquid.

Solubility
Propylene carbonate is freely soluble in water and miscible with alcohol and chloroform.

Method

Cheng and Gadde reported an HPLC method for analysis of propylene carbonate. The Waters model 204 liquid chromatograph consisted of a model M6000A pump, a model U6K injector, a model R401 refractive index detector, and a Hewlett-Packard model 3352B integrator. The stationary phase was a DuPont Zorbax ODS column (250 × 4.6 mm). The mobile phase consisted of water and methanol (9:1, vol/vol). The flow rate was 1 mL/min. Resorcinol 2.5 mg/mL in mobile phase was used as an internal standard.

A propylene carbonate sample (~ 100 mg) was mixed with 10 mL of internal standard and 15 mL of the mobile phase, heated in a boiling water bath for 40 seconds, cooled, and filtered. The injection volume was 20 µL. Under these conditions, the retention times of propylene carbonate and resorcinol were about 9.2 and 15.2 minutes, respectively (estimated from the published chromatogram).

The method was reported to be stability indicating. Propylene carbonate was separated from its degradation products. Retention times for propylene glycol, allyl alcohol, and propionaldehyde were about 4.1, 7.3, and 11.2 minutes, respectively (estimated from the published chromatogram).

A standard curve for propylene carbonate was generated from 2.4 to 12.3 mg/mL. The correlation coefficient was 0.9999. The coefficient of variation for the analysis of propylene carbonate was 0.6%.

Reference

Cheng H, Gadde RR. Determination of propylene carbonate in pharmaceutical formulations using liquid chromatography. *J Pharm Sci*. 1985; 74: 695–6.

Propylthiouracil

Chemical Name

2,3-Dihydro-6-propyl-2-thioxo-4(1*H*)-pyrimidinone

Other Names

Propycil, Prothiucil

Form	Molecular Formula	MW	CAS
Propylthiouracil	$C_7H_{10}N_2OS$	170.2	51-52-5

Appearance

Propylthiouracil occurs as white or practically white crystals or crystalline powder.

Solubility

Propylthiouracil is slightly or very slightly soluble in water and in ether. It is sparingly soluble in alcohol.

Method 1

Mitra et al. reported a simple assay method for the determination of propylthiouracil. The Waters liquid chromatograph consisted of a model 501 pump, a model 712 WISP autosampler with a 100-μL loop, and a model 486 absorbance detector. The stationary phase was a Nova-Pak C_8 analytical column (150 × 3.9 mm, 4-μm particle size). The mobile phase was a mixture of water and methanol (70:30, vol/vol). The flow rate was 0.5 mL/min. UV detection was performed at 276 nm.

Propylthiouracil was dissolved in alcohol and diluted with mobile phase. Methylthiouracil was used as an internal standard. The injection volume was 15 μL. The total run time was 7 minutes. Under these conditions, the retention times of methylthiouracil and propylthiouracil were about 3.1 and 6.6 minutes, respectively.

Propylthiouracil solutions were mixed with 1 N hydrochloric acid or 1 N sodium hydroxide solution at 70–80 °C for 1.5 hours, cooled down, neutralized, and assayed. The degradation products did not interfere with the determination of the intact propylthiouracil and internal standard, confirming the stability-indicating nature of the analytical method.

A calibration curve for propylthiouracil was constructed linearly from 1 to 100 μg/mL. The correlation coefficient was 0.9958. The intraday and interday coefficients of variation were less than 1.28 and 1.65%, respectively. The limit of detection and the limit of quantitation were 4.5 and 15 ng, respectively.

Reference

Mitra P, Riga A, Alexander KS. A rapid high-performance liquid chromatographic determination of propylthiouracil. *J Liq Chromatogr Rel Technol.* 2004; 27: 549–58.

Method 2

Abdul-Fattah and Bhargava developed an HPLC method for the determination of propylthiouracil in tablets and topicals. A Hewlett-Packard model 1100 system was equipped with a multiple-wavelength UV detector and a model 3395 integrator. The stationary phase was a Mac-Mod Zorbax SB-CN column (150 × 4.6 mm, 5-μm particle size). The column temperature was 35 °C. The mobile phase was a mixture of 5% (vol/vol) acetonitrile in 0.025 M monobasic potassium phosphate aqueous buffer (adjusted to pH 4.6 with 0.5 N sodium hydroxide). The flow rate was 1 mL/min. UV detection was performed at 282 nm.

For tablets, a portion of ground tablet powder equivalent to 50 mg of propylthiouracil was weighed, dissolved in 100 mL of a diluent of water–methanol (1:1, vol/vol), filtered through a 0.45-μm syringe filter, and diluted with the diluent. For creams, a 0.5-g cream sample was weighed and placed into a 100-mL volumetric flask, dissolved in 100 mL of the diluent at 70 °C in a water bath, sonicated, diluted, and filtered through a 0.45-μm syringe filter. The injection volume was 20 μL. Under these conditions, the retention time of propylthiouracil was about 7.7 minutes.

The method was demonstrated to be stability indicating by accelerated degradation of the drug. Propylthiouracil was degraded by 3% hydrogen peroxide, 1 N hydrochloric acid, or 1 N sodium hydroxide and heated at 80 °C for 1 hour. Degradation products did not interfere with the analysis of propylthiouracil.

A standard curve for propylthiouracil was generated from 0.25 to 25 µg/mL. The correlation coefficient was 0.999. The limit of detection was 0.05 µg/mL. The limit of quantitation was 0.06 µg/mL.

Reference

Abdul-Fattah AM, Bhargava HN. Development and validation of a high-performance liquid chromatographic method for the analysis of propylthiouracil in pharmaceuticals. *Drug Dev Ind Pharm.* 2001; 27: 831–5.

Method 3

Nahata et al. determined the stability of propylthiouracil in extemporaneously prepared oral suspensions at 4 and 25 °C. The Hewlett-Packard model 1050 liquid chromatograph consisted of a pump, an autosampler, a variable-wavelength detector, and a model 3396A integrator. The stationary phase was a Mac-Mod Zorbax C_{18} column (150 × 3.0 mm). The column temperature was 40 °C. The mobile phase consisted of 5 mM 1-heptane-sulfonic acid, 1% (vol/vol) acetic acid, and methanol (40:45:15, vol/vol/vol). The flow rate was 0.4 mL/min. UV detection was performed at 276 nm. 6-Methyl-2-thiouracil 25 µg/mL in the mobile phase was used as an internal standard.

Samples were diluted with the internal standard and centrifuged. The injection volume was 10 µL. Under these conditions, retention times of the internal standard and propylthiouracil were about 2.1 and 5.3 minutes, respectively.

The method was demonstrated to be stability indicating by accelerated degradation of the drug. Propylthiouracil solutions were degraded with 2 M hydrochloric acid, 2 M sodium hydroxide, or 0.03% hydrogen peroxide at 60 °C. Degradation products did not influence the quantification of propylthiouracil.

A standard curve for propylthiouracil was constructed from 4 to 130 µg/mL. The correlation coefficient was greater than 0.999. Intraday and interday coefficients of variation were less than 2.3 and 4.2%, respectively.

Reference

Nahata MC, Morosco RS, Trowbridge JM. Stability of propylthiouracil in extemporaneously prepared oral suspensions at 4 and 25 °C. *Am J Health Syst Pharm.* 2000; 57: 1141–3.

Pseudoephedrine Hydrochloride

Chemical Name

(+)-(1*S*,2*S*)-2-Methylamino-1-phenyl-1-propanol hydrochloride

Other Names

PediaCare, Sudafed

Form	Molecular Formula	MW	CAS
Pseudoephedrine	$C_{10}H_{15}NO$	165.2	90-82-4
Pseudoephedrine hydrochloride	$C_{10}H_{15}NO.HCl$	201.7	345-78-8
Pseudoephedrine sulfate	$(C_{10}H_{15}NO)_2.H_2SO_4$	428.5	7460-12-0

Appearance

Pseudoephedrine hydrochloride occurs as fine white to off-white crystals or powder. Pseudoephedrine sulfate occurs as white odorless crystals or as a crystalline powder.

Solubility

Pseudoephedrine hydrochloride has solubilities of approximately 2 g/mL in water and 278 mg/mL in alcohol at 25 °C; it is sparingly soluble in chloroform. Pseudoephedrine sulfate is freely soluble in alcohol.

Method 1

Fong and Eickhoff reported an HPLC method for the determination of pseudoephedrine hydrochloride, doxylamine succinate, and dextromethorphan hydrobromide in a commercial cough-cold liquid formulation. The instrument consisted of two Beckman model 112 pumps, a model 165 variable-wavelength UV-visible detector, a Waters WISP 710B autosampler, and a Beckman model 450 data system. The stationary phase was an Analytichem Sepralyte CN column (150 × 4.6 mm, 5-μm particle size). The mobile phase consisted of acetonitrile, methanol, and an aqueous buffer (75:10:15, vol/vol/vol). The buffer solution contained 5 mM dibasic potassium phosphate and 59 mM triethylamine, adjusted to pH 5.3 with glacial acetic acid. The flow rate was 2.0 mL/min. UV detection was performed at 262 nm and 0.03 AUFS.

The sample was diluted 1:10 with water and centrifuged. The injection volume was 20 μL. Under these conditions, the retention times for pseudoephedrine hydrochloride, doxylamine succinate, and dextromethorphan hydrobromide were about 4.0, 4.9, and 5.7 minutes, respectively.

The method was reported to be stability indicating since it separated chemically similar compounds.

Standard curves were constructed from 10 to 200% of the normal working concentration levels. Intraday and interday coefficients of variation were less than 1 and 0.6%, respectively. The limit of detection was 4 μg/mL.

Reference

Fong GW, Eickhoff WM. Liquid chromatographic determination of amines in complex cough-cold formulations. *Int J Pharm*. 1989; 53: 91–7.

Method 2

George and Contario described an HPLC method for the determination of terfenadine, pseudoephedrine hydrochloride, and ibuprofen in a liquid formulation. The liquid chromatographic system comprised a Waters model M6000A pump, a model 710B WISP autosampler, a Spectroflow 757 variable-wavelength detector, and a Beckman CALS data system. The stationary phase was a Whatman Partisil SCX column (250 × 4.6 mm, 10-μm particle size). The mobile phase consisted of 500 mL of water, 2.0 g of monobasic sodium phosphate monohydrate, 1.0 g of phosphoric acid (85%), 3.0 g of sodium perchlorate monohydrate, and 500 mL of acetonitrile. The flow rate was 1.5 mL/min. UV detection was performed at 210 nm and 0.32 AUFS.

Samples were diluted with methanol and then with the mobile phase. The injection volume was 20 μL. Under these conditions, retention times for pseudoephedrine, ibuprofen, and terfenadine were about 1.1, 2.4, and 4.4 minutes, respectively (estimated from the published chromatogram).

The stability-indicating capability of the method was established by intentional degradation of pseudoephedrine. Samples of pseudoephedrine were treated with 0.1 M hydrochloric acid at 90 °C for up to 72 hours. The intact drug was well resolved from its degradation products.

Reference

George RC, Contario JJ. Quantitation of terfenadine, pseudoephedrine hydrochloride, and ibuprofen in a liquid animal dosing formulation using high performance liquid chromatography. *J Liq Chromatogr*. 1988; 11: 475–88.

Method 3

Yacobi et al. described a stability-indicating HPLC method for the simultaneous determination of pseudoephedrine and chlorpheniramine in pharmaceutical dosage forms. A Waters model ALC/GPC 204 liquid chromatograph was equipped with a UV detector, a Houston Omniscribe recorder, and a Waters μBondapak C_{18} nonpolar column (300 × 4 mm). The mobile phase was a mixture of acetonitrile, methanol, and sodium nitrate (35:40:25, vol/vol/vol) and 1-heptanesulfonic acid (0.001 M) with a pH of 5. The flow rate was 2 mL/min. UV detection was performed at 254 nm. Chlorpromazine was used as an internal standard. The injection volume was 5 μL. Retention times for pseudoephedrine and chlorpromazine were 3.1 and 7.3 minutes, respectively.

To demonstrate that the method was stability indicating, solutions containing pseudoephedrine hydrochloride 600 μg/mL and chlorpheniramine maleate 40 μg/mL in water, 3 N sodium hydroxide, and buffers of various pH values (2.5, 4.5, 6.5, and 8) were heated either at 90 °C for 1 hour or at 100 °C for 24 hours. Degradation product peaks did not interfere with the pseudoephedrine peak.

Reference

Yacobi A, Look ZM, Lai C-M. Simultaneous determination of pseudoephedrine and chlorpheniramine in pharmaceutical dosage forms. *J Pharm Sci*. 1978; 67: 1668–70.

Pyrantel Pamoate

Chemical Name

(*E*)-1,4,5,6-Tetrahydro-1-methyl-2-[2-(2-thienyl)vinyl]pyrimidine 4,4′-methylenebis(3-hydroxy-2-naphthoate)

Other Names
Antiminth, Pyrantel Embonate

Form	Molecular Formula	MW	CAS
Pyrantel pamoate	$C_{11}H_{14}N_2S.C_{23}H_{16}O_6$	594.7	22204-24-6

Appearance
Pyrantel pamoate is a yellow to tan solid.

Solubility
Pyrantel pamoate is practically insoluble in water and methanol. It is soluble in dimethyl sulfoxide.

Method
Argekar et al. developed an HPLC method for the simultaneous determination of pyrantel pamoate and mebendazole from tablets. The chromatograph consisted of a Tosho CCPE pump equipped with a universal injector and an Oracle 2 integrator. The stationary phase was a reversed-phase Shodex C_8 column (250 × 3.9 mm, 5-µm particle size). The mobile phase was a mixture of 0.05 M monobasic sodium phosphate, acetonitrile, and triethylamine (60:40:1.5, vol/vol/vol). The pH was adjusted to 6.8 with dilute phosphoric acid. UV detection was performed at 290 nm.

Tablets were weighed and finely ground. An amount of powder equivalent to 50 mg of pyrantel pamoate and 75 mg of mebendazole was mixed with 25 mL of glacial acetic acid and ultrasonicated for 30 minutes and then kept in a water bath for 20 minutes, cooled, and diluted with 65 mL of a mixture of acetonitrile, diethylamine, and acetic acid (90:10:10, vol/vol/vol) (diluent A). The mixture was sonicated for another 20 minutes and the final volume was made to 100 mL with diluent A. The resulting solution was filtered through Whatman No. 42 filter paper. Six milliliters of the filtrate was diluted to 50 mL with the mobile phase and used for the assay. Under these conditions, retention times for pamoic acid, pyrantel, and mebendazole were about 3.6, 6.4, and 12.8 minutes, respectively.

The HPLC method was demonstrated to be stability indicating by accelerated decomposition of pyrantel pamoate. Tablets were kept in an oven at 105 °C for 1 month and the solution of the tablets was kept in sunlight for 1 hour. The test samples were analyzed by the HPLC method. Degradation products of pyrantel pamoate and mebendazole did not interfere with the intact pyrantel pamoate.

A standard curve was constructed for pyrantel pamoate from 20 to 160 µg. The correlation coefficient was 0.999. The limit of detection and the limit of quantification for pyrantel pamoate were 10 and 40 µg, respectively.

Reference
Argekar AP, Raj SV, Kapadia SU. Simultaneous determination of mebendazole and pyrantel pamoate from tablets by high performance liquid chromatography–reverse phase (RP-HPLC). *Talanta*. 1997; 44: 1959–65.

Pyrazinamide

Chemical Names
Pyrazinecarboxamide
Pyrazinoic acid amide

Other Name
Rifater

Form	Molecular Formula	MW	CAS
Pyrazinamide	$C_5H_5N_3O$	123.1	98-96-4

Appearance
Pyrazinamide is a white to practically white and odorless or practically odorless crystalline powder.

Solubility
Pyrazinamide is sparingly soluble in water and slightly soluble in alcohol.

pK$_a$
Pyrazinamide has a pK$_a$ of 0.5.

Method 1
Glass et al. optimized an assay for the simultaneous determination of isoniazid, rifampicin, and pyrazinamide in a fixed dose combination. The Waters liquid chromatograph included a WISP model 6000A solvent-delivery system, a WISP model 710B autosampler, a model 481 Lambda Max UV detector, and a data module. The stationary phase was a Waters µBondapak analytical column (250 × 4.6 mm, 10-µm particle size). The mobile phase consisted of acetonitrile and 0.0002 M tetrabutylammonium hydroxide solution (42.5:57.5, vol/vol) (pH 3.1). The flow rate was 1.0 mL/min. The eluent was monitored at 260 nm.

Isoniazid, rifampicin, and pyrazinamide standard solutions were prepared in water. The injection volume was 20 µL. Under these conditions, the retention times of rifampicin, isoniazid, and pyrazinamide were 2.85, 3.54, and 10.97 minutes, respectively.

The assay was stated to be stability indicating.

A standard curve was obtained with a correlation coefficient of greater than 0.9998. The limit of detection was 0.150 µg/mL.

Reference
Glass BD, Agatonovic-Kustrin S, Chen Y-J, et al. Optimization of a stability indicating HPLC method for the simultaneous determination of rifampicin, isoniazid, and pyrazinamide in fixed-dose combination using artificial neural networks. *J Chromatogr Sci.* 2007; 45: 38–44.

Method 2

Nahata et al. determined the stability of pyrazinamide in two suspensions stored in plastic bottles and glass bottles at two temperatures for up to 2 months. The HPLC system consisted of a Hewlett-Packard 1050 series pump, an autosampler, a variable-wavelength detector, and a 3396A integrator. The stationary phase was a Mac-Mod Analytical Zorbax Rx-C$_8$ column (160 × 4.0 mm, 5-μm particle size). The mobile phase consisted of 30% 0.005 M tetrabutylammonium hydroxide adjusted to pH 3.0 with phosphoric acid and 70% methanol. The flow rate was 0.7 mL/min. UV detection was performed at 265 nm. Rifampin 1.5 mg/mL was used as an internal standard.

Pyrazinamide samples were diluted 10-fold with methanol. Twenty microliters of this solution was mixed with 20 μL of the internal standard and 1 mL of the mobile phase. The injection volume was 10 μL. Retention times for pyrazinamide and rifampin were about 2.1 and 3.8 minutes, respectively.

The HPLC method was determined to be stability indicating by accelerated decomposition of pyrazinamide. Pyrazinamide suspension 100 mg/mL was mixed with 1 mL of 1.0 M hydrochloric acid or 1 mL of 1.0 M sodium hydroxide; these solutions were then heated to 60 °C for 90 minutes. Another pyrazinamide suspension was mixed with 1 mL of 0.3% hydrogen peroxide. In all cases, degradation product peaks did not interfere with the pyrazinamide peak.

Standard curves for pyrazinamide were constructed from 0.5 to 15.0 mg/mL. The correlation coefficient of the standard curve was 0.999. The intraday and interday coefficients of variation were less than 4%.

Reference

Nahata MC, Morosco RS, Peritore SP. Stability of pyrazinamide in two suspensions. *Am J Health Syst Pharm.* 1995; 52: 1558–60.

Pyridoxine (Vitamin B$_6$) Hydrochloride

Chemical Name

3-Hydroxy-4,5-bis(hydroxymethyl)-2-picoline hydrochloride

Other Names

Benadon, Hexavibex, Rodex

Form	Molecular Formula	MW	CAS
Pyridoxine hydrochloride	C$_8$H$_{11}$NO$_3$.HCl	205.6	58-56-0

Appearance

Pyridoxine hydrochloride occurs as a white or almost white crystalline powder or crystals.

Solubility

Pyridoxine hydrochloride is freely soluble in water and slightly soluble in alcohol.

Method 1

Al-Jallad et al. developed an HPLC method for the simultaneous determination of pyridoxine hydrochloride and meclizine hydrochloride in tablets. The Beckman Gold system comprised a model 116 pump, a model 116 UV spectrophotometric detector, and a Rheodyne model 7725 20-μL loop injector. The stationary phase was a Waters Symmetry C$_{18}$ silica column (250 × 4.6 mm, 5-μm particle size). The mobile phase was a mixture of 200 mL of water containing 1.5 g of dodecyl sulfate sodium salt, 500 mL of acetonitrile, 250 mL of methanol, 15 mL of glacial acetic acid, and 15 mL of tetrahydrofuran. The flow rate was 1.5 mL/min. UV detection was performed at 254 nm.

A sample equivalent to 50 mg of pyridoxine hydrochloride and 75 mg of meclizine hydrochloride was accurately weighed, transferred to a 100-mL volumetric flask, mixed with the mobile phase, shaken for 15 minutes, brought to volume with the mobile phase, and centrifuged. The injection volume was 20 μL. Under these conditions, retention times for pyridoxine hydrochloride and meclizine hydrochloride were about 2.2 and 7.4 minutes, respectively (estimated from the published chromatogram).

The stability-indicating capability of the assay was demonstrated by accelerated decomposition of pyridoxine hydrochloride. The drug was treated with 1 M sodium hydroxide solution, 1 M sulfuric acid, or 3.3% hydrogen peroxide. No interference from degradation products with the determination of pyridoxine hydrochloride was observed.

A calibration curve for pyridoxine hydrochloride was obtained from 0.25 to 0.76 mg/mL. The correlation coefficient was 0.9999. The limit of detection was 0.12 μg/mL.

Reference

Al-Jallad T, Al-Kurdi Z, Badwan A, et al. Simultaneous determination of pyridoxine hydrochloride and meclizine hydrochloride in tablet formulations by HPLC. *Pharm Pharmacol Commun.* 1999; 5: 479–83.

Method 2

Van der Horst et al. reported the HPLC determination of pyridoxine in total parenteral nutrient solution. The liquid chromatographic system consisted of a Waters model M6000A pump, a model WISP autosampler, a model 481 variable-wavelength detector, a Perkin-Elmer model 204-A fluorescence spectrophotometer, and a Shimadzu Chromatopac C-R3A integrator. The stationary phase was a LiChrosorb RP8 analytical column (250 × 4.6 mm, 10-μm particle size) with an RCSS Guard-Pak C$_{18}$ precolumn (4 × 6 mm). The mobile phase was a mixture of 0.1% triethylamine in 0.067 M monobasic potassium phosphate buffer (pH 6.7) and acetonitrile (100:4, vol/vol). The flow rate was 2 mL/min. Fluorescence detection was performed at an excitation wavelength of 335 nm and an emission wavelength of 390 nm. The injection volume was 20 μL. Under these conditions, the retention time for pyridoxine was about 3.3 minutes (estimated from the published chromatogram).

The method was demonstrated to be stability indicating by accelerated degradation of pyridoxine. Pyridoxine solutions were treated with 1 M sodium hydroxide solution or 1 M hydrochloric acid and then were heated to boiling for about 30 minutes. Degradation products were separated from the peak of pyridoxine.

A standard curve for pyridoxine was generated from 0.4 to 4.0 mg/L. The correlation coefficient was 0.9995. The limit of detection was 0.06 mg/L.

Reference
Van der Horst A, Martens HJM, de Goede PNFC. Analysis of water-soluble vitamins in total parenteral nutrition solution by high pressure liquid chromatography. *Pharm Weekbl [Sci]*. 1989; 11: 169–74.

Quinidine

Chemical Names
(9S)-6′-Methoxycinchonan-9-ol
α-(6-Methoxy-4-quinolyl)-5-vinyl-2-quinuclidinemethanol

Other Names
Cardioquin, Quinora

Form	Molecular Formula	MW	CAS
Quinidine	$C_{20}H_{24}N_2O_2$	324.4	56-54-2
Quinidine bisulfate	$C_{20}H_{24}N_2O_2.H_2SO_4$	422.5	747-45-5
Quinidine gluconate	$C_{20}H_{24}N_2O_2.C_6H_{12}O_7$	520.6	7054-25-3
Quinidine polygalacturonate	$(C_{20}H_{24}N_2O_2.C_6H_{10}O_7.H_2O)_x$		65484-56-2
Quinidine sulfate	$(C_{20}H_{24}N_2O_2)_2.H_2SO_4.2H_2O$	783.0	6591-63-5

Appearance
Quinidine bisulfate occurs as colorless and odorless or almost odorless crystals. Quinidine gluconate is a white odorless powder. Quinidine polygalacturonate is a creamy white amorphous powder. Quinidine sulfate occurs as fine needlelike white crystals or as a fine crystalline powder.

Solubility

Quinidine bisulfate is freely soluble in water and alcohol and is practically insoluble in ether. Quinidine gluconate is freely soluble in water and slightly soluble in alcohol. Quinidine polygalacturonate is sparingly soluble in water and freely soluble in hot 40% alcohol. Quinidine sulfate is slightly soluble in water and soluble in alcohol.

pK$_a$

Quinidine has pK$_a$ values of 4 and 8.6.

Method

Allen and Erickson investigated the stability of quinidine sulfate in extemporaneously compounded oral liquids. A Hewlett-Packard series 1050 automated high-performance liquid chromatograph included a multisolvent mixing and pumping system, an auto-injector, a diode-array detector, and a computer with Chem Station software. The stationary phase was a Bakerbond C$_{18}$ column (250 × 4.6 mm, 5-μm particle size). The mobile phase consisted of water, acetonitrile, methanesulfonic acid solution, and diethylamine solution (80:20:2:2, vol/vol/vol/vol). The methanesulfonic acid solution consisted of 35 mL of methanesulfonic acid and 20 mL of acetic acid diluted to 500 mL with water. The diethylamine solution consisted of 10 mL of diethylamine diluted to 100 mL with water. The mobile phase was delivered isocratically at 1.0 mL/min. UV detection was performed at 235 nm.

Samples were diluted 1:100. Under these conditions, the retention time for quinidine was 8.5 minutes.

The HPLC method was determined to be stability indicating. A composite chromatogram of quinidine sulfate after accelerated degradation showed that degradation product peaks did not interfere with the intact quinidine peak.

A standard curve for quinidine sulfate was constructed from 25 to 125 μg/mL. The intraday and interday coefficients of variation were 0.81 and 1.4%, respectively.

Reference

Allen LV Jr, Erickson MA III. Stability of bethanechol chloride, pyrazinamide, quinidine sulfate, rifampin, and tetracycline hydrochloride in extemporaneously compounded oral liquids. *Am J Health Syst Pharm.* 1998; 55:1804–9.

Quinine

Chemical Name

(αR)-α-(6-Methoxy-4-quinolyl)-α-[(2S,4S,5R)-(5-vinylquinuclidin-2-yl)]methanol

Other Names
Coco-Quinine, Pholcones, Quinate, Quindan

Form	Molecular Formula	MW	CAS
Quinine	$C_{20}H_{24}N_2O_2$	324.4	130-95-0
Quinine bisulfate	$C_{20}H_{24}N_2O_2.H_2SO_4.7H_2O$	548.6	549-56-4
Quinine dihydrochloride	$C_{20}H_{24}N_2O_2.2HCl$	397.3	60-93-5
Quinine ethylcarbonate	$C_{23}H_{28}N_2O_4$	396.5	83-75-0
Quinine hydrochloride	$C_{20}H_{24}N_2O_2.HCl.2H_2O$	396.9	6119-47-7
Quinine sulfate	$(C_{20}H_{24}N_2O_2)_2.H_2SO_4.2H_2O$	782.9	6119-70-6

Appearance
Quinine bisulfate occurs as colorless crystals or as a white crystalline powder. Quinine dihydrochloride is a white or almost white powder. Quinine hydrochloride occurs as colorless fine silky needles. Quinine sulfate occurs as colorless or white odorless fine needlelike crystals or as a white or almost white crystalline powder.

Solubility
Quinine bisulfate is freely soluble in water and sparingly soluble in alcohol. Quinine dihydrochloride is very soluble in water and freely soluble in alcohol. Quinine hydrochloride is soluble in water and freely soluble in alcohol. Quinine sulfate is slightly soluble in water and in alcohol.

Method
Faouzi et al. developed an assay for the analysis of quinine. The Hewlett-Packard 1090M system comprised a variable-volume injector, an automatic sampling system, a model 79994 linear photodiode-array detector, and a model 300 integrator. The stationary phase was an Interchim Hypersil ODS C_{18} column (150 × 4.6 mm, 5-μm particle size). The mobile phase consisted of acetonitrile, methanol, and 0.2% triethylamine buffer (adjusted to pH 3.5 with phosphoric acid) (40:20:40, vol/vol/vol). The flow rate was 0.4 mL/min. UV detection was performed at 330 nm.

Samples were diluted with mobile phase. The injection volume was 20 μL. Under these conditions, the retention time for quinine was about 4.5 minutes.

The method was reported to be stability indicating. The peak purity of quinine was confirmed by photodiode-array detection.

Standard curves were generated from 2.5 to 20 and 5 to 20 μg/mL. Correlation coefficients were greater than 0.999. The intraday coefficient of variation was 2.96%.

Reference
Faouzi MA, Khalfi F, Dine T, et al. Stability, compatibility and plasticizer extraction of quinine injection added to infusion solutions and stored in polyvinyl chloride (PVC) containers. *J Pharm Biomed Anal.* 1999; 21: 923–30.

Quinupristin–Dalfopristin

Chemical Names

Quinupristin: [(6R, 9S, 10R, 13S, 15aS, 18R, 22S, 24aS)-22-[p-(Dimethylamino)benzyl]-6-ethyldocosahydro-10,23-dimethyl-5,8,12,15,17,21,24-heptaoxo-13-phenyl-18-[[(3S)-3-quinuclidinylthio]methyl]-12H-pyrido[2,1-f]pyrrolo[2,1-l][1,4,7,10,13,16]-oxapentaazacyclononadecin-9-yl]-3-hydroxypicolinamide

Dalfopristin: (3R,4R,5E,10E,12E,14S,26R,26aS)-26-[[2-(Diethylamino)ethyl]sulfonyl]-8,9,14,15,24,25,26,26a-octahydro-14-hydroxy-3-isopropyl-4,12-dimethyl-3H-21,18-nitrilo-1H,22H-pyrrolo[2,1-c][1,8,4,19]dioxadiazacyclotetracosine-1,7,16,22-(4H,17H)-tetrone

Other Name

Synercid

Form	Molecular Formula	MW	CAS
Quinupristin	$C_{53}H_{67}N_9O_{10}S$	1022.2	120138-50-3
Dalfopristin	$C_{34}H_{50}N_4O_9S$	690.9	112362-50-2

Appearance

Quinupristin occurs as a white to very slightly yellow hygroscopic powder. Dalfopristin occurs as a slightly yellow to yellow hygroscopic powder.

Method

Vasselle et al. reported a gradient HPLC method for the determination of Synercid freeze-dried powders. The system was a Varian model 9012 pump, a Waters model 717 plus autosampler, a Milton/Roy model 3100 UV detector, and a Prolabo oven. A Hewlett-Packard model 1050 photodiode-array detector was also used for the peak purity analysis. The stationary phase was a Merck LiChrospher 100 RP18 column (125 × 4 mm, 5-μm particle size). The column temperature was 40 °C. Mobile phase A was a mixture of 200 mL of acetonitrile and 800 mL of 0.03 M monobasic potassium phosphate buffer (adjusted to pH 2.9 with phosphoric acid). Mobile phase B was a mixture of 650 mL of acetonitrile and 350 mL of phosphate buffer (pH 2.9). The mobile phase was linearly decreased from 100% A to 30% A from 0 to 42.5 minutes, returned to 100% A from 42.5 to 44 minutes, and held at 100% A for another 5 minutes. The flow rate was 1.1 mL/min. UV detection was performed at 254 nm.

Samples were diluted 1:100 with the mobile phase A. The sample temperature in the autosampler was maintained at 4–10 °C. The injection volume was 10 μL. Under these conditions, dalfopristin eluted in 7.8–9.2 minutes and the major component of quinupristin eluted in 25.8–28.2 minutes. The other two components of quinupristin eluted in 22–25 minutes.

The method was reported to be stability indicating.

Standard curves for dalfopristin and quinupristin were obtained from 5 to 15 μg per injection. The limit of quantitation was 0.05% for dalfopristin and 0.12% for quinupristin.

Reference

Vasselle B, Gousset G, Bounine J-P. Development and validation of a high-performance liquid chromatographic stability-indicating method for the analysis of Synercid in quality control, stability and compatibility. *J Pharm Biomed Anal.* 1999; 19: 641–57.

Ramipril

Chemical Name

[2S-[1[R*(R*)],2α,3aβ,6aβ]]-1-[2-[[1-(Ethoxycarbonyl)-3-phenylpropyl]amino]-1-oxo-propyl]octahydrocyclopenta[b]pyrrole-2-carboxylic acid

Other Name

Altace

Form

Form	Molecular Formula	MW	CAS
Ramipril	$C_{23}H_{32}N_2O_5$	416.5	87333-19-5

Appearance

Ramipril is a white crystalline substance.

Solubility

Ramipril is soluble in polar organic solvents and in buffered aqueous solutions.

Method 1

Belal et al. reported a method for the simultaneous determination of ramipril and hydrochlorothiazide in dosage forms. The Waters liquid chromatograph consisted of a model 600E multisolvent-delivery system, a model U6K injector, a model 486 tunable absorbance detector, and a model 746 data module. The stationary phase was a Supelco Supelcosil LC-8 stainless steel column (150 × 4.6 mm, 5-µm particle size). The mobile phase consisted of acetonitrile and 0.1 M sodium perchlorate solution (46:54, vol/vol), adjusted to pH 2.5 with phosphoric acid. The flow rate was 1.5 mL/min. UV detection was performed at 210 nm. Clobazam 0.30 mg/mL in the mobile phase was used as an internal standard.

Ten tablets were ground. A portion of this powder equivalent to 2.5 mg of ramipril and 12.5 mg of hydrochlorothiazide was weighed, transferred into a 50-mL beaker, extracted with three 30-mL portions of acetonitrile, filtered into a 100-mL volumetric flask, brought to volume with acetonitrile, mixed with internal standard, and diluted with the mobile phase. The injection volume was 10 µL. Under these conditions, the retention times of hydrochlorothiazide, clobazam, and ramipril were 1.5, 3.4, and 5.1 minutes, respectively.

The stability-indicating nature of the method was confirmed by accelerated degradation. Ramipril solutions were treated with 2 M sodium hydroxide or 2 M hydrochloric acid and heated in a boiling water bath for 1 hour. Degradation products of ramipril did not influence its determination.

A standard curve for ramipril was obtained from 4.5 to 45 µg/mL. The correlation coefficient was 0.9999. The limit of detection was 180 ng/mL.

Reference

Belal F, Al-Zaagi IA, Gadkariem EA, et al. A stability-indicating LC method for the simultaneous determination of ramipril and hydrochlorothiazide in dosage forms. *J Pharm Biomed Anal.* 2001; 24: 335–42.

Method 2

Allen et al. studied the stability of ramipril when mixed with water, apple juice, and applesauce. A Hewlett-Packard model 1050 series automated liquid chromatograph was equipped with a UV detector, an injector, a computer, a printer, and a J.T. Baker C_{18} reversed-phase column. The mobile phase was a mixture of perchlorate buffer and acetonitrile (61.5:38.5, vol/vol), adjusted to pH 2.1 ± 0.1 with concentrated phosphoric acid. The perchlorate buffer contained 14 g of sodium perchlorate monohydrate and

100 mL of 0.5 M phosphoric acid in deionized and filtered water in a final volume of 1000 mL. Its pH was adjusted to 2.5 ± 0.1 with triethylamine. The flow rate was 1.0 mL/min. The UV detection wavelength was 210 nm.

Samples of ramipril in water and apple juice were centrifuged at $350 \times g$ for 10 minutes. Samples of ramipril in applesauce were mixed in equal volume with 0.1 N hydrochloric acid, vortexed for 2 minutes, and then centrifuged at $350 \times g$ for 10 minutes. The supernatant was then filtered through a 0.45-μm Millipore Millex-HA filter and diluted 1:2 or 1:3 with deionized water. The injection volume was 50 μL. Under these conditions, the retention time for ramipril was 12.1 minutes.

The assay was determined to be stability indicating by spiking ramipril solution with two resolution standards: 2,3-dibenzoyl-L-tartaric acid (a precursor of ramipril), and ramipril diketopiperazine (a degradation product). Retention times for its precursor, for ramipril, and for a degradation product were 6.6, 12.1, and 23.1 minutes, respectively.

Calibration curves for ramipril were constructed from 0 to 0.015 mg/mL; the correlation coefficient was 0.999. The intraday and interday coefficients of variation were 0.84 and 0.75%, respectively.

Reference

Allen LV Jr, Stiles ML, Prince SJ, et al. Stability of ramipril in water, apple juice, and applesauce. *Am J Health Syst Pharm.* 1995; 52: 2433–6.

Ranitidine

Chemical Name

N-[2-[[[5-[(Dimethylamino)methyl]-2-furanyl]methyl]thio]ethyl]-*N′*-methyl-2-nitro-1,1-ethenediamine

Other Name

Zantac

Form	Molecular Formula	MW	CAS
Ranitidine	$C_{13}H_{22}N_4O_3S$	314.4	66357-35-5
Ranitidine hydrochloride	$C_{13}H_{22}N_4O_3S.HCl$	350.9	66357-59-3

Appearance

Ranitidine hydrochloride is a white to pale yellow granular substance.

Solubility
Ranitidine hydrochloride has solubilities of 660 mg/mL in water and 190 mg/mL in alcohol.

pK$_a$
Ranitidine has pK$_a$ values of 2.7 and 8.2.

Method 1
Munro and Walker developed an isocratic HPLC method for ranitidine analysis. The Thermo Separations liquid chromatograph consisted of a model SCM1000 degasser, a model P4000 quaternary pump, a model AS3000 variable-loop autosampler, and a model UV6000 photodiode-array detector. The stationary phase was a YMC-Pack ODS-AM column (150 × 4.6 mm, 5-μm particle size) with a YMC guard column (20 × 4.0 mm, 5-μm particle size). The column temperature was 35 °C. The mobile phase consisted of 57% buffer solution and 43% methanol, where the buffer contained 10 mM sodium dodecyl sulfate and 50 mM phosphoric acid, adjusted to pH 6.8 with triethylamine. The flow rate was 1.0 mL/min. UV detection was 228 nm. The injection volume was 25 μL. Under these conditions, the retention time of ranitidine was 5.5 minutes.

The method was demonstrated to be stability indicating by accelerated degradation. Ranitidine samples were degraded in 1.0 M hydrochloric acid or 1.0 M sodium hydroxide solution for 1 day, were degraded in 1.0% hydrogen peroxide for 4 hours, were heated in a 105 °C oven for 7 days, or were exposed to 500-W UV light for 30 days. Ranitidine was separated from all its degradation products.

A calibration curve for ranitidine was generated from 0.056 to 44.4 μg/mL. The correlation coefficient was greater than 0.9996. The limit of detection was 0.028 μg/g. The limit of quantitation was 0.056 μg/g.

Reference
Munro JS, Walker TA. Ranitidine hydrochloride: Development of an isocratic stability-indicating high-performance liquid chromatographic separation. *J Chromatogr A*. 2001; 914: 13–21.

Method 2
Caufield and Stewart described an HPLC assay for the simultaneous analysis of meropenem and ranitidine in intravenous fluid mixture. The instrument was a Hewlett-Packard model 1090 system including a pump, an autosampler, and a Gilson model 117 variable-wavelength UV detector. A Waters model 996 photodiode-array detector was used for the peak purity analysis. The stationary phase was a YMC ODS-AQ column (150 × 2.0 mm, 3-μm particle size). The mobile phase consisted of aqueous acetic acid solution (pH 3) and acetonitrile (92:8, vol/vol). The flow rate was 0.2 mL/min. UV detection was performed at 317 nm.

Samples were diluted with the mobile phase. The sample injection volume was 5 μL. Under these conditions, retention times of ranitidine and meropenem were about 4.7 and 7.7 minutes, respectively.

The method was evaluated to be stability indicating by accelerated degradation of ranitidine. Solutions of ranitidine were treated with 0.1 N hydrochloric acid for 15 minutes, with 0.01 N sodium hydroxide solution for 2 minutes, or with 0.03% hydrogen peroxide for 1.75 hours or were heated at 60 °C for 4 hours. None of the degradation product peaks interfered with the intact drug peak.

A standard curve for ranitidine was constructed from 131.25 to 525 µg/mL. The correlation coefficient was 0.9996–1.0. Intraday and interday coefficients of variation were 0.63 and 0.93%, respectively. The limit of detection was 131 ng/mL.

Reference

Caufield WV, Stewart JT. HPLC separations of meropenem and selected pharmaceuticals using a polar endcapped octadecylsilane narrow bore column. *Chromatographia*. 2000; 51: 308–14.

Method 3

Using an HPLC method, Nahata et al. investigated the stability of ranitidine hydrochloride in water for injection in glass vials and plastic syringes. A Hewlett-Packard series 1050 liquid chromatograph included a solvent-delivery pump, an autosampler, a variable-wavelength UV detector, and a Hewlett-Packard 3396A integrator. The stationary phase was a Waters µBondapak C_{18} column (300 × 3.9 mm, 10-µm particle size). The mobile phase was a mixture of acetonitrile and 0.1 M ammonium phosphate buffer. The pH of the mobile phase was adjusted to 3.0 with phosphoric acid. The flow rate was 1.5 mL/min. UV detection was performed at 228 nm. The autosampler vials were at 10 °C whereas the analytical column remained at room temperature. Caffeine 1 mg/mL was used as an internal standard.

Samples were diluted 1:10 with water. One hundred microliter portions of this dilution were mixed with 100 µL of the internal standard before analysis. The injection volume was 10 µL. Under these conditions, retention times for ranitidine and caffeine were 3.8 and 7.8 minutes, respectively.

To determine the stability-indicating nature of the assay, 1 mL of ranitidine 1 mg/mL was mixed with 1 mL of 1.0 M sodium hydroxide and 1.0 M hydrochloric acid for 1 hour. Chromatograms of these solutions showed that the quantification of the peak area of the intact ranitidine was not affected by the peaks of degradation products.

A standard curve was generated for ranitidine from 0.25 to 3.50 mg/mL. Its linearity was determined by linear regression analysis of the ranitidine concentration versus peak area ratios of ranitidine and caffeine. The correlation coefficient was greater than 0.999. The intraday and interday coefficients of variation of the assay were less than 2%.

Reference

Nahata MC, Morosco RS, Fox J. Stability of ranitidine hydrochloride in water for injection in glass vials and plastic syringes. *Am J Health Syst Pharm*. 1996; 53: 1588–90.

Method 4

Inagaki et al. studied the stability of ranitidine hydrochloride with cefmetazole sodium during simulated Y-site administration. The liquid chromatographic system consisted of a Hitachi model 6200 intelligent pump, a Hitachi model L-4200 UV-visible detector, a Hitachi model AS-2000 autosampler, and a Hitachi model D-2500 chromato-integrator. The stationary phase was an Alltech Adsorbosphere C_{18} analytical column (250 × 4.6 mm, 5-µm particle size) with a guard column of the same packing material. The mobile phase was a mixture of methanol and 0.1 M ammonium acetate (65:35, vol/vol) and was delivered isocratically at 1.0 mL/min. UV detection was performed at 322 nm and 0.128 AUFS.

Samples were diluted 1:20 with 0.9% sodium chloride injection. The injection volume was 50 µL. The retention time for ranitidine was 6.5 minutes.

The assay was shown to be stability indicating by accelerated decomposition of ranitidine hydrochloride. Solutions of the drug were degraded in 1 N hydrochloric acid, 1 N sodium hydroxide, and water for 10 hours at 80 °C and in 1% hydrogen peroxide for 2 hours at room temperature. Solutions were also exposed to UV radiation for 20 hours at room temperature. The resultant solutions were assayed by HPLC. Degradation product peaks did not interfere with the intact ranitidine and cefmetazole peaks.

Standard curves were constructed from a linear plot of the peak area against the concentration from 0.05 to 1.0 mg/mL. The correlation coefficient was greater than 0.999.

A similar method was used by Crowther et al.

References

Inagaki K, Gill MA, Okamoto MP, et al. Chemical compatibility of cefmetazole sodium with ranitidine hydrochloride during simulated Y-site administration. *J Parenter Sci Technol.* 1993; 47: 35–9.

Crowther RS, Bellanger R, Szauter KEM. In vitro stability of ranitidine hydrochloride in enteral nutrient formulas. *Ann Pharmacother.* 1995; 29: 859–62.

Inagaki K, Gill MA, Okamoto MP, et al. Stability of ranitidine hydrochloride with aztreonam, ceftazidime, or piperacillin sodium during simulated Y-site administration. *Am J Hosp Pharm.* 1992; 49: 2769–72.

Method 5

Williams et al. evaluated the stability of ranitidine hydrochloride in total parenteral nutrient mixtures. The liquid chromatograph consisted of a Waters 6000A solvent-delivery system, a Waters Lambda-Max model 480 variable-wavelength UV detector, a Shimadzu SIL-6A autoinjector, a Shimadzu C-R3A Chromatopac integrator, and a Chromanetics Scientific Spherisorb ODS-1 reversed-phase analytical column (250 × 4.6 mm, 5-µm particle size) with an Applied Biosystems Brownlee NewGuard RP18 ODS guard column (1.5-cm cartridge, 7-µm particle size). The mobile phase consisted of 24% 0.05 M sodium dibasic phosphate buffer at pH 6 and 76% methanol. The flow rate was 1.0 mL/min. UV detection was performed at 228 nm and 0.005 AUFS. *N*-[3-[[3-(Dimethylamino)-methyl]phenoxy]propyl]-*N'*-methyl-2-nitro-1,1-ethenediamine hydrochloride 1 mg/mL was used as an internal standard.

Ranitidine was extracted from the admixture with chloroform–methanol (1:1, vol/vol). The injection volume was 25 µL. Under these conditions, retention times for ranitidine and the internal standard were 5.8 and 8.9 minutes, respectively.

The assay was determined to be stability indicating by accelerated degradation of ranitidine hydrochloride. Ranitidine hydrochloride solutions in deionized water, 1 N hydrochloric acid, and 1 N sodium hydroxide were incubated at 70 °C for 18 hours. In all cases, degradation product peaks did not interfere with the ranitidine peak.

The interday coefficient of variation was 3.3%.

Reference

Williams MF, Hak LJ, Dukes G. In vitro evaluation of the stability of ranitidine hydrochloride in total parenteral nutrient mixtures. *Am J Hosp Pharm.* 1990; 47: 1574–9.

Method 6

Stewart et al. used an HPLC method to evaluate the stability of ranitidine hydrochloride in intravenous admixtures stored frozen, refrigerated, and at room temperature. The chromatograph consisted of a Beckman model 110A pump, a Waters WISP autosampler, a Kratos model 757 UV detector, a Waters model 990 diode-array UV detector, and a Hewlett-Packard model 3392A integrator. The stationary phase was a Keystone Spherisorb ODS 1 column (200 × 4.6 mm, 10-μm particle size). The mobile phase consisted of methanol and 0.1 M aqueous ammonium acetate (85:15, vol/vol). The flow rate was 1.5 mL/min. UV detection was performed at 322 nm. Samples were diluted to a ranitidine concentration of 0.1 mg/mL with the mobile phase. The injection volume was 10 μL.

The stability-indicating capability of the method was demonstrated by accelerated decomposition of ranitidine hydrochloride. Solutions of ranitidine hydrochloride 0.5 mg/mL were degraded in 1 N hydrochloric acid and 1 N sodium hydroxide for 6 hours at 80 °C. The purity of the ranitidine peak was examined with a diode-array UV detector. Degradation product peaks did not interfere with the intact ranitidine peak.

References

Stewart JT, Warren FW, Johnson SM, et al. Stability of ranitidine in intravenous admixtures stored frozen, refrigerated, and at room temperature. *Am J Hosp Pharm.* 1990; 47: 2043–6.

Galante LJ, Stewart JT, Warren FW, et al. Stability of ranitidine hydrochloride at dilute concentration in intravenous infusion fluids at room temperature. *Am J Hosp Pharm.* 1990; 47: 1580–4.

Galante LJ, Stewart JT, Warren FW, et al. Stability of ranitidine hydrochloride with eight medications in intravenous admixtures. *Am J Hosp Pharm.* 1990; 47: 1606–10.

Method 7

Cano et al. assessed the stability of ranitidine hydrochloride in total nutrient admixtures. A Perkin-Elmer series 4 set high-performance liquid chromatograph was equipped with a 6-μL Rheodyne type injector, an LC-85B UV detector, a Perkin-Elmer Sigma 15 integrator, and a 3-cm-long Perkin-Elmer HS-3 C_{18} column. The mobile phase was a mixture of water, acetonitrile, and 5% ammonium hydroxide in isopropyl alcohol (83:10:7, vol/vol/vol) that was delivered isocratically at 1 mL/min. UV detection was performed at 228 nm. Cimetidine 200 μg/mL was used as an internal standard.

One milliliter of ranitidine hydrochloride sample was mixed with 1 mL of chloroform–methanol (1:1, vol/vol) and 200 μL of the internal standard, vortexed for 5 minutes, and then centrifuged at 2500 × *g* for 5 minutes. The aqueous phase was filtered through a 0.45-μm Millipore HU filter. The injection volume was 6 μL.

To demonstrate that the method was stability indicating, ranitidine hydrochloride solutions 1 mg/mL in water, 1 N hydrochloric acid, and 1 N sodium hydroxide were heated at 85 °C for 10 hours. Another ranitidine hydrochloride solution 1 mg/mL was exposed to UV radiation for 4 hours at room temperature. In all cases, the degradation product peaks did not interfere with the intact ranitidine peak.

A standard curve for ranitidine hydrochloride was constructed from 10 to 200 μg/mL.

Reference

Cano SM, Montoro JB, Pastor C, et al. Stability of ranitidine hydrochloride in total nutrient admixtures. *Am J Hosp Pharm.* 1988; 45: 1100–2.

Method 8

Gupta et al. determined the chemical stabilities of famotidine and ranitidine hydrochloride in intravenous admixtures using HPLC methods. The chromatographic system consisted of a Waters ALC 202 system, a Schoeffel SF 770 multiple-wavelength detector, and a Houston Omniscribe recorder. The stationary phase was a Waters μBondapak C_{18} nonpolar column (300 × 3.9 mm). The mobile phase consisted of 10% methanol, 7% acetonitrile in 0.01 M aqueous phosphate buffer with the pH adjusted to 5.85 with either 0.1 N sodium hydroxide or 0.1 N hydrochloric acid. The flow rate was 2.0 mL/min. UV detection was performed at 262 nm and 0.04 AUFS. Caffeine was used as an internal standard.

Samples were diluted 1:10 with water. The injection volume was 20 μL. Under these conditions, retention times for ranitidine and the internal standard were about 4.5 and 6.5 minutes, respectively (estimated from the published chromatogram).

The analytical method was demonstrated to be stability indicating by accelerated decomposition of ranitidine hydrochloride. Stock solutions of ranitidine hydrochloride were mixed with either 1 N sulfuric acid or 1 N sodium hydroxide and were then boiled for about 15 minutes. Degradation product peaks did not interfere with the intact ranitidine peak.

Reference

Gupta VD, Parasrampuria J, Bethea C. Chemical stabilities of famotidine and ranitidine hydrochloride in intravenous admixtures. *J Clin Pharm Ther.* 1988; 13: 329–34.

Method 9

Lampasona et al. studied the stability of ranitidine hydrochloride admixtures frozen and refrigerated in minibags. The chromatograph consisted of a Perkin-Elmer 4B HPLC pump, a Valco six-port injector, a Perkin-Elmer LC 75 variable-wavelength UV detector, and a Waters C_{18} Nova-Pak radial compression cartridge. The mobile phase contained 184 mL of 0.1 M monobasic potassium phosphate, 5 mM pentanesulfonic acid, and 240 mL of acetonitrile in a final volume of 2 L in deionized water. The mobile phase was adjusted to pH 6.3 with 1 N potassium hydroxide. The flow rate was 1.5 mL/min. UV detection was performed at 228 nm and 0.16 AUFS. Caffeine was used as the internal standard.

Samples were diluted 1:10 with deionized water. The injection volume was 60 μL. Under these conditions, retention times for caffeine and ranitidine were about 5.8 and 7.3 minutes, respectively (estimated from the published chromatogram).

The stability-indicating nature of the assay was shown by accelerated decomposition of ranitidine. Solutions of ranitidine hydrochloride 1 mg/mL in water, 1 N hydrochloric acid, and 1 N sodium hydroxide were heated at 90 °C for 10 hours. Another solution of ranitidine hydrochloride 1 mg/mL in 1% hydrogen peroxide was exposed to UV radiation at 21 °C for 4 hours. In all cases, the intact ranitidine peak was well separated from its degradation product peaks.

A standard curve for ranitidine was generated from 0.25 to 2.50 mg/mL. The interday coefficient of variation was 3.3%.

References

Lampasona V, Mullins RE, Parks RB. Stability of ranitidine admixtures frozen and refrigerated in minibags. *Am J Hosp Pharm.* 1986; 43: 921–5.

Bullick L, Parks RB, Lampasona V, et al. Stability of ranitidine hydrochloride and amino acids in parenteral nutrient solutions. *Am J Hosp Pharm.* 1985; 42: 2683–7.

Method 10

Walker et al. determined the stability of ranitidine hydrochloride in a total parenteral nutrient solution. The liquid chromatograph consisted of a Spectra-Physics model SP8700 solvent-delivery system, a Waters model 710B WISP autosampler, a Schoeffel SF 770 variable-wavelength UV detector, and a Spectra-Physics model SP4270 integrator. The stationary phase was a Brownlee C_2 reversed-phase column (10-µm particle size). The mobile phase consisted of acetonitrile and 0.5 M potassium phosphate buffer (34:66, vol/vol) at pH 6.8. The flow rate was 2.0 mL/min. UV detection was performed at 235 nm.

Two hundred microliters of ranitidine hydrochloride samples were mixed with 100 µL of the internal standard solution (*n*-propyl *p*-hydroxybenzoate 275 µg/mL) and then diluted with 4.0 mL of water. The injection volume was 200 µL.

The method was determined to be stability indicating by accelerated decomposition of ranitidine hydrochoride. Ranitidine hydrochloride was exposed to 6 M hydrochloric acid and 6 M sodium hydroxide at 60 °C for 18 hours. Degradation product peaks did not interfere with the intact ranitidine peak.

A similar method was used by Sarkar et al.

References

Walker SE, Bayliff CD. Stability of ranitidine hydrochloride in total parenteral nutrient solution. *Am J Hosp Pharm.* 1985; 42: 590–2.

Sarkar MA, Rogers E, Reinhard M, et al. Stability of clindamycin phosphate, ranitidine hydrochloride, and piperacillin sodium in polyolefin containers. *Am J Hosp Pharm.* 1991; 48: 2184–6.

Walker SE, Kirby K. Stability of ranitidine hydrochloride admixtures refrigerated in polyvinyl chloride minibags. *Can J Hosp Pharm.* 1988; 41: 105–8.

Reserpine

Chemical Name

(3β,16β,17α,18β,20α)-11,17-Dimethoxy-18-[(3,4,5-trimethoxybenzoyl)oxy]yohimban-16-carboxylic acid methyl ester

Other Name
Serpasil

Form	Molecular Formula	MW	CAS
Reserpine	$C_{33}H_{40}N_2O_9$	608.7	50-55-5

Appearance
Reserpine is a white or pale buff to slightly yellow crystalline powder.

Solubility
Reserpine is insoluble in water and very slightly soluble in alcohol.

pKa
Reserpine has a pK_a of 6.6.

Method
Vincent and Awang reported an HPLC method for the determination of reserpine in pharmaceutical formulations. An LDC chromatograph was equipped with a Valco valve and a 20-μL injection loop, an LDC variable-wavelength detector, a Schoeffel FS 970 fluorescence detector, a Honeywell Electronik 196 recorder, a Pharmacia Fine Chemicals model 410 recorder, and a Spectra-Physics minigrator-integrator. The stationary phase was a Brownlee Laboratories LiChrosorb RP8 stainless steel column (250 × 4.6 mm, 10-μm particle size). The mobile phase consisted of 0.05 M monobasic sodium phosphate buffer (pH 4.5) and methanol (50:50, vol/vol) and was delivered isocratically at 2.0 mL/min. UV detection was performed at 254 nm and 0.01 AUFS. The fluorescence detector was set at an excitation wavelength of 330 nm and an emission wavelength of 470 nm. Propiophenone was used as an internal standard.

Reserpine was extracted using water saturated by ethyl acetate. The injection volume was 10 μL.

The analytical method was shown to be stability indicating because it could separate reserpine from its degradation products (3,4,5,6-tetradehydroreserpine, 3,4-dehydroreserpine, and 3-isoreserpine). The retention times for propiophenone, reserpine, 3,4,5,6-tetradehydroreserpine, 3,4-dehydroreserpine, and 3-isoreserpine were 6.7, 15.2, 18, 20, and 29.6 minutes, respectively.

A standard curve for reserpine was constructed from 0.250 to 4.0 μg/mL. The correlation coefficient was 0.9998.

The same method was used by Coffman et al.

References
Vincent A, Awang DVC. Determination of reserpine in pharmaceutical formulations by high performance liquid chromatography. *J Liq Chromatogr.* 1981; 4: 1651–61.

Coffman HD, Crabbs WC, Kolinski RE, et al. Stability of reserpine injections and tablets submitted by US hospitals. *Am J Hosp Pharm.* 1986; 43: 103–9.

Retinol (Vitamin A)

Chemical Name
3,7-Dimethyl-9-(2,6,6-trimethyl-1-cyclohexen-1-yl)-2,4,6,8-nonatetraen-1-ol

Other Names
Arovit, Ro-A-Vit

Form	Molecular Formula	MW	CAS
Retinol	$C_{20}H_{30}O$	286.4	68-26-8

Appearance
Vitamin A occurs as a light yellow to red oil.

Solubility
Vitamin A is insoluble in water and soluble in dehydrated alcohol.

Method
Allwood and Martin used HPLC to investigate the photodegradation of vitamins A and E in parenteral nutrient mixtures during infusion. The liquid chromatograph consisted of a Kontron model 422 pump, a Gynkotek model UVD 340S diode-array detector, a Gynkotek model Gina 50 autosampler, and a Gynkotek PC integrator. The stationary phase was a Hypersil ODS column (250 × 4.6 mm, 5-μm particle size). The mobile phase was methanol. The flow rate was 1.5 mL/min. UV detection was performed at 292 nm. The injection volume was 20 μL.

The method was reported to be stability indicating.

A standard curve for retinol was constructed from 0.6 to 3 IU/mL. The correlation coefficient was 0.9933.

Reference
Allwood MC, Martin HJ. The photodegradation of vitamins A and E in parenteral nutrition mixtures during infusion. *Clin Nutr.* 2000; 19: 339–42.

Ribavirin

Chemical Name
1-β-D-Ribofuranosyl-1*H*-1,2,4-triazole-3-carboxamide

Other Names
Rebetol, Rebetron, Tribavirin, Virazile, Virazole

Form	Molecular Formula	MW	CAS
Ribavirin	$C_8H_{12}N_4O_5$	244.2	36791-04-5

Appearance
Ribavirin is a white crystalline powder.

Solubility
Ribavirin is freely soluble in water and slightly soluble in alcohol.

Method
Shah et al. described a specific HPLC method for the determination of ribavirin in its pharmaceutical dosage forms. The liquid chromatograph consisted of a Waters model 510 dual-piston reciprocating pump, a model 486 UV detector, a model 746 integrator, and a Rheodyne model 7125 injector. The stationary phase was a μBondapak C_{18} reversed-phase column (300 × 3.9 mm). The mobile phase consisted of methanol and 0.01 M monobasic potassium phosphate (5:95, vol/vol), adjusted to pH 4.6. The flow rate was 1.0 mL/min. UV detection was performed at 207 nm and 1.0 AUFS.

For capsules, a powder sample was mixed with mobile phase, stirred well, and filtered. The filtrate was diluted to approximately 25 μg/mL of ribavirin in the mobile phase. For syrup, samples were diluted with the mobile phase to approximately 25 μg/mL of ribavirin. Under these conditions, the retention time for ribavirin was about 4.6 minutes (estimated from the published chromatogram).

The method was reported to be stability indicating since it separated ribavirin from its impurities.

A calibration curve for ribavirin was obtained from 0 to 100 μg/mL.

Reference
Shah Y, Joshi S, Jindal KC, et al. Stability indicating HPLC method for ribavirin and its pharmaceutical dosage forms. *Drug Dev Ind Pharm.* 1994; 20: 85–91.

Riboflavin (Vitamin B$_2$)

Chemical Name
3,10-Dihydro-7,8-dimethyl-10-(D-*ribo*-2,3,4,5-tetrahydroxypentyl)benzopteridine-2,4-dione

Other Names
Beflavina, Flavaxin

Form	Molecular Formula	MW	CAS
Riboflavin	C$_{17}$H$_{20}$N$_4$O$_6$	376.4	83-88-5
Riboflavin 5′-phosphate sodium	C$_{17}$H$_{20}$N$_4$NaO$_9$P	478.3	130-40-5

Appearance
Riboflavin is a yellow to orange-yellow crystalline powder. Riboflavin sodium phosphate is a fine yellow to orange-yellow and odorless or almost odorless hygroscopic crystalline powder.

Solubility
Riboflavin is very slightly soluble in water and practically insoluble in alcohol. Riboflavin sodium phosphate is sparingly soluble in water and very slightly soluble in alcohol.

pK$_a$
Riboflavin has a pK$_a$ of 10.2.

Method
Van der Horst et al. reported the HPLC analysis of riboflavin 5′-phosphate sodium in total parenteral nutrient solution. The liquid chromatographic system consisted of a Waters model M6000A pump, a model WISP autosampler, a model 481 variable-wavelength detector, a Perkin-Elmer model 204-A fluorescence spectrophotometer, and a Shimadzu Chromatopac C-R3A integrator. The stationary phase was a LiChrosorb RP8 analytical column (250 × 4.6 mm, 10-μm particle size) with an RCSS Guard-Pak C$_{18}$ precolumn (4 × 6 mm). The mobile phase was a mixture of 0.1% triethylamine in 0.067 M monobasic potassium phosphate buffer (pH 6.4) and acetonitrile (85:15, vol/vol). The flow rate was 2 mL/min. Fluorescence detection was performed at an excitation wavelength of 465 nm and an emission wavelength of 525 nm. The injection volume was

20 μL. Under these conditions, the retention time for riboflavin 5'-phosphate sodium was about 5.6 minutes (estimated from the published chromatogram).

The method was demonstrated to be stability indicating by accelerated degradation of riboflavin 5'-phosphate sodium. Riboflavin 5'-phosphate sodium solutions were treated with 1 M sodium hydroxide solution or 1 M hydrochloric acid and then heated to boiling for about 30 minutes. Degradation products were separated from the intact drug.

A standard curve for riboflavin 5'-phosphate sodium was generated from 0.1 to 1.7 mg/L. The correlation coefficient was 0.9995. The limit of detection was 0.03 mg/L.

Reference
Van der Horst A, Martens HJM, de Goede PNFC. Analysis of water-soluble vitamins in total parenteral nutrition solution by high pressure liquid chromatography. *Pharm Weekbl [Sci]*. 1989; 11: 169–74.

Rifampin

Chemical Names
5,6,9,17,19,21-Hexahydroxy-23-methoxy-2,4,12,16,18,20,22-heptamethyl-8-[*N*-(4-methyl-1-piperazinyl)formimidoyl]-2,7-(epoxypentadeca[1,11,13]trienimino)naphtho-[2,1-*b*]furan-1,11(2*H*)-dione 21-acetate

3-[[(4-Methyl-1-piperazinyl)imino]methyl]rifamycin

Other Names
Rifadin, Rifampicin

Form	Molecular Formula	MW	CAS
Rifampin	$C_{43}H_{58}N_4O_{12}$	822.9	13292-46-1

Appearance
Rifampin is a red-brown crystalline powder.

Solubility
Rifampin is freely soluble in chloromethane and dimethyl sulfoxide; soluble in ethyl acetate, methanol, and tetrahydrofuran; and slightly soluble in water (pH <6), acetone, and carbon tetrachloride.

pK$_a$
Rifampin has a pK$_a$ of 7.9.

Method 1

Glass et al. optimized an assay for the simultaneous determination of rifampicin, isoniazid, and pyrazinamide in a fixed-dose combination. The Waters liquid chromatograph included a WISP model 6000A solvent-delivery system, a WISP model 710B autosampler, a model 481 Lambda Max UV detector, and a data module. The stationary phase was a Waters µBondapak analytical column (250 × 4.6 mm, 10-µm particle size). The mobile phase consisted of acetonitrile and 0.0002 M tetrabutylammonium hydroxide solution (42.5:57.5, vol/vol) (pH 3.1). The flow rate was 1.0 mL/min. The eluent was monitored at 260 nm.

Rifampicin, isoniazid, and pyrazinamide standard solutions were prepared in water. The injection volume was 20 µL. Under these conditions, the retention times of rifampicin, isoniazid, and pyrazinamide were 2.85, 3.54, and 10.97 minutes, respectively.

The method was stated to be stability indicating.

A standard curve was obtained with a correlation coefficient of greater than 0.9998. The limit of detection was 0.200 µg/mL.

Reference

Glass BD, Agatonovic-Kustrin S, Chen Y-J, et al. Optimization of a stability-indicating HPLC method for the simultaneous determination of rifampicin, isoniazid, and pyrazinamide in fixed-dose combination using artificial neural networks. *J Chromatogr Sci.* 2007; 45: 38–44.

Method 2

Nahata et al. determined the effect of four preparation methods and extended storage on rifampin concentration in extemporaneously prepared suspensions. The liquid chromatographic system was composed of a Varian 2010 series pump, a 9090 autosampler, a 2050 variable-wavelength detector, and a D2000 integrator. The stationary phase was a Beckman Ultrasphere ODS column (250 × 4.6 mm, 5-µm particle size). The mobile phase consisted of 39% 0.01 M dibasic potassium phosphate with pH adjusted to 6.0 with phosphoric acid and 61% methanol. The flow rate was 1.5 mL/min. UV detection was performed at 280 nm.

Five hundred microliters of rifampin formulation was diluted with internal standard solution containing 330 µg of *n*-butyl *p*-hydroxybenzoate and filtered through a 0.45-µm nylon 66 filter. The injection volume was 10 µL. Retention times for the internal standard and rifampin were 5.6 and 7.1 minutes, respectively.

The stability-indicating nature of the method was demonstrated by accelerated decomposition of rifampin. Rifampin solutions 1 mg/mL were mixed with 1.0 M sodium hydroxide and 1.0 M sulfuric acid and were heated to 60 °C for 30 minutes. Degradation product peaks did not interfere with the rifampin peak.

A calibration curve for rifampin was generated from 5.0 to 15.0 mg/mL. The correlation coefficient was greater than 0.999.

Reference

Nahata MC, Morosco RS, Hipple TF. Effect of preparation method and storage on rifampin concentration in suspensions. *Ann Pharmacother.* 1994; 28: 182–5.

Method 3

Pearson and Trissel determined the chemical stability and physical compatibility of rifampin 0.1 mg/mL and minocycline hydrochloride 0.1 mg/mL separately in 5% dextrose injection or 0.9% sodium chloride injection at 24 and 4 °C and together at these concentrations in 0.9% sodium chloride injection at 37, 24, and 4 °C over periods of up to 7 days. The system consisted of a Waters model 600E multisolvent-delivery system, a Waters model 490E programmable multiple-wavelength UV detector, a Waters model 712 WISP autosampler, and a Vydac C_8 reversed-phase column (250 × 4.6 mm, 5-μm particle size). The system was controlled and integrated by an NEC PowerMate SX/16 computer. The mobile phase consisted of 32.8% acetonitrile in 50 mM dipotassium hydrogen phosphate with the pH adjusted to 6.5 with phosphoric acid. The flow rate was 1 mL/min. UV detection was performed at 254 nm.

Samples were analyzed without further dilution. Under these conditions, the retention time for rifampin was 6.8 minutes.

The assay was demonstrated to be stability indicating by accelerated degradation of rifampin. Heating at 95 °C in a water bath for 30 minutes yielded a reduction in the peak of the intact minocycline, with the formation of new peaks at 4.7, 5.1, 7.7, 11.3, and 13.0 minutes. Neither minocycline nor the degradation products interfered with the parent rifampin peak.

Standard curves were generated from a linear plot of peak area versus concentration from 0.025 to 0.15 mg/mL. The correlation coefficient of the standard curve was greater than 0.9997.

Reference

Pearson SD, Trissel LA. Stability and compatibility of minocycline hydrochloride and rifampin in intravenous solutions at various temperatures. *Am J Hosp Pharm.* 1993; 50: 698–701.

Method 4

Krukenberg et al. evaluated the stability of 1% rifampin suspensions prepared in five syrups. The HPLC system consisted of an Altex 110A pump, a Waters model 710B WISP autosampler, a Waters 441 fixed-wavelength UV detector, and a Hewlett-Packard 3357 data system. The stationary phase was a Merck LiChrosorb RP2 column (250 × 4.6 mm, 5-μm particle size). The mobile phase was a mixture of methanol, 0.1 M ammonium formate, and tetrahydrofuran (10:11:1, vol/vol/vol) and was delivered isocratically at 1.5 mL/min. UV detection was performed at 254 nm.

Samples were diluted 1:20 with water and methanol. The injection volume was 10 μL. Under these conditions, the retention time for rifampin was 11.3 minutes.

This method was stability indicating because it could separate rifampin, 3-formyl rifampin, 25-desacetyl rifampin (the solution degradation product), and rifampin quinone (the major oxidation product). The retention times for 25-desacetyl rifampin, rifampin, 3-formyl rifampin, and rifampin quinone were 5.8, 11.3, 18.5, and 36.2 minutes, respectively.

A standard curve for rifampin was constructed from 3 to 9 μg/mL.

Reference

Krukenberg CC, Mischler PG, Massad EN, et al. Stability of 1% rifampin suspensions prepared in five syrups. *Am J Hosp Pharm.* 1986; 43: 2225–8.

Risedronate Sodium

Chemical Name
Sodium trihydrogen [1-hydroxy-2-(3-pyridyl)ethylidene]diphosphonate

Other Name
Actonel

Form	Molecular Formula	MW	CAS
Risedronate sodium	$C_7H_{10}NNaO_7P_2$	305.1	115436-72-1

Appearance
Risedronate sodium occurs as a fine white to off-white odorless crystalline powder.

Solubility
Risedronate sodium is soluble in water and essentially insoluble in common organic solvents.

Method
Aluoch et al. described the development and validation of an ion-pair HPLC method for the determination of risedronate in its pharmaceutical formulation. The liquid chromatographic system consisted of a Dionex 40 gradient pump, an LC 20 chromatography enclosure, an AD 20 absorbance detector, and a Rheodyne injector with a 25-µL loop. The stationary phase was a Zorbax Eclipse XDB C_{18} column (150 × 4.6 mm, 3.5-µm particle size). The mobile phase consisted of acetonitrile and 0.01 M sodium phosphate buffer (pH 7.5) (10:90, vol/vol) containing 5 mM tetrabutylammonium phosphate and 1 mM ethylenediaminetetraacetic acid (EDTA). The sodium phosphate buffer was prepared by dissolving 2.68 g of dibasic sodium phosphate in 1000 mL of water and adjusting the pH to 7.5 with dilute phosphoric acid. The flow rate was 1.0 mL/min. UV detection was performed at 262 nm. The injection volume was 25 µL. Under these conditions, the retention time of risedronate was 11.02 minutes. Powdered risedronate sodium tablets were treated with mobile phase.

In order to demonstrate the stability-indicating property of the method, risedronate sodium samples were refluxed in water or with 1 N hydrochloric acid or 1 N sodium hydroxide solution at 70 °C for 16 hours. Risedronate sodium solutions were also exposed to ultraviolet light (254 nm) or stored at 70 °C for 24 hours. The peak of the intact risedronate was separated from its degradation product peaks.

A calibration curve for risedronate was obtained from 50 to 150 µg/mL; the correlation coefficient was 0.999. The coefficient of variation for the assay was less than 2.0%. The limit of detection and the limit of quantitation were 30 and 100 ng, respectively.

Reference
Aluoch A, Tatini R, Parsons DM, et al. Stability indicating ion-pair HPLC method for the determination of risedronate in a commercial formulation. *J Liq Chromatogr Rel Technol.* 2004; 27: 2799–813.

Risperidone

Chemical Name
3-[2-[4-(6-Fluoro-1,2-benzisoxazol-3-yl)piperidino]ethyl]-6,7,8,9-tetrahydro-2-methyl-4H-pyrido[1,2-a]pyrimidin-4-one

Other Names
Belivon, Risperdal

Form
Risperidone

Molecular Formula
$C_{23}H_{27}FN_4O_2$

MW
410.5

CAS
106266-06-2

Appearance
Risperidone occurs as a white or almost white powder.

Solubility
Risperidone is practically insoluble in water, sparingly soluble in alcohol, and freely soluble in dichloromethane.

Method
El-Sherif et al. described an assay for the determination of risperidone in the presence of its degradation products in bulk powder and in tablets. The Hewlett-Packard series 1100 liquid chromatograph was equipped with a quaternary pump, a photodiode-array detector, and a manual injector with a 20-μL loop. The stationary phase was a LiChrosorb RP-18 column (250 × 4.6 mm, 10-μm particle size). The mobile phase consisted of methanol and 0.05 M potassium dihydrogen phosphate buffer, pH 7 (65:35, vol/vol). The flow rate was 1 mL/min. UV detection was performed at 280 nm.

A portion of powder from 20 tablets equivalent to 10 mg of risperidone was accurately weighed, transferred into a 25-mL volumetric flask, dissolved in 20 mL of methanol, filled to the mark with methanol, shaken for a few minutes, and filtered. The injection volume was 20 μL. Under these conditions, the retention time of risperidone was 17.0 minutes.

To demonstrate the stability-indicating property of the method, a standard mixture of risperidone and its degradation products was assayed. The degradation products had retention times between 2.0 and 8.7 minutes, showing that risperidone was well separated from its degradation products.

A calibration curve for risperidone was constructed from 25 to 500 μg/mL. The correlation coefficient was 0.9999. The intraday and interday coefficients of variation were 1.12 and 0.36%, respectively. The limit of detection and the limit of quantitation were 3.00 and 12.4 μg/mL, respectively.

Reference

El-Sherif ZA, El-Zeany B, El-Houssini OM. High performance liquid chromatographic and thin layer densitometric methods for the determination of risperidone in the presence of its degradation products in bulk powder and in tablets. *J Pharm Biomed Anal.* 2005; 36: 975–81.

Rizatriptan Benzoate

Chemical Name

N,N-Dimethyl-5-(1*H*-1,2,4-triazol-1-ylmethyl)-1*H*-indole-3-ethanamine monobenzoate

Other Names

Maxalt, Rizalief

Form	Molecular Formula	MW	CAS
Rizatriptan benzoate	$C_{15}H_{19}N_5.C_7H_6O_2$	391.5	145202-66-0

Appearance

Rizatriptan occurs as a white to off-white crystalline solid.

Solubility

Rizatriptan is soluble in water at about 42 mg/mL at 25 °C.

Method

Rao et al. developed a reversed-phase analytical method for rizatriptan benzoate. The liquid chromatograph was an Agilent 1100 series system equipped with a diode-array detector. The stationary phase was an Agilent Zorbax SB-CN column (250 × 4.6 mm, 5-μm particle size). The mobile phase consisted of 10 mM monobasic potassium phosphate adjusted to pH 3.4 with phosphoric acid (solvent A), acetonitrile (solvent B), and methanol (solvent C). The flow rate was 1.0 mL/min. The mobile phase was gradiently delivered as shown on the following page.

Time, minutes	Solvent A, %	Solvent B, %	Solvent C, %
0	92	8	0
10	88	9	3
16	88	9	3
19	82	15	3
23	65	32	3
28	47	50	3
34	47	50	3
35	92	8	0
40	92	8	0

UV detection was performed at 225 nm. Samples were prepared in solvent A. The injection volume was 5 μL. Under these conditions, retention times of rizatriptan and benzoic acid were about 12.7 and 14.8 minutes, respectively.

To demonstrate the stability-indicating property of the assay method, bulk drug was subjected to stress conditions of photolytic degradation, thermal degradation (60 °C), acid hydrolysis (0.5 N hydrochloric acid), base hydrolysis (0.1 N sodium hydroxide solution), water hydrolysis, and oxidative hydrolysis (3.0% hydrogen peroxide). Impurities and degradation products did not interfere with the determination of rizatriptan benzoate.

A standard curve of rizatriptan benzoate was constructed from 125 to 750 μg/mL. The correlation coefficient was greater than 0.999. The coefficient of variation of the assay was 0.8%.

Reference

Rao BM, Sangaraju S, Srinivasu MK, et al. Development and validation of a specific stability indicating high performance liquid chromatographic method for rizatriptan benzoate. *J Pharm Biomed Anal.* 2006; 41: 1146–51.

Rofecoxib

Chemical Name

4-[*p*-(Methylsulfonyl)phenyl]-3-phenyl-2(*5H*)-furanone

Other Name
Vioxx

Form	Molecular Formula	MW	CAS
Rofecoxib	$C_{17}H_{14}O_4S$	314.4	162011-90-7

Appearance
Rofecoxib occurs as a white to off-white to light yellow powder.

Solubility
Rofecoxib is insoluble in water and is very slightly soluble in ethanol.

Method 1
Krishnaiah et al. developed an HPLC method for the determination of rofecoxib in pharmaceutical dosage forms. The Shimadzu liquid chromatograph included two LC-10 AT VP pumps, an SPD-10A VP variable-wavelength UV-visible detector, a CTO-10 AS VP column oven, and an SCL-10A VP system controller. The stationary phase was a YMC RP C_{18} column (250 × 4.6 mm, 5-μm particle size). The mobile phase consisted of 62% water containing 0.04% triethylamine, 0.15% glacial acetic acid, and 38% acetonitrile. The flow rate was 0.9 mL/min. The column temperature was maintained at 40 °C. Tinidazole 1 μg/mL was used as an internal standard.

A portion of ground tablets (20) equivalent to 25 mg of rofecoxib was accurately weighed, transferred into a 100-mL volumetric flask containing about 30 mL of acetonitrile, sonicated, filled to the mark, and filtered through a 0.45-μm filter. Two milliliters of the filtrate was transferred into a 10-mL volumetric flask containing 10 μg of internal standard and diluted to the mark with mobile phase. The injection volume was 20 μL. Under these conditions, the retention times for tinidazole and rofecoxib were about 5.2 and 14.8 minutes, respectively.

The assay method was stated to be stability indicating. The excipients did not interfere with the analysis of the drug.

A standard curve for rofecoxib was obtained from 0.1 to 10 μg/mL. The correlation coefficient was 0.9999.

Reference
Krishnaiah YSR, Raghumurthy V, Satyanarayana V, et al. High performance liquid chromatographic determination of rofecoxib in pharmaceutical dosage forms. *Acta Ciencia Indica.* 2001; 27: 71–4.

Method 2
Using HPLC, Mao et al. studied the decomposition of rofecoxib solution under alkaline and photolytic stress. The Hewlett-Packard model 1100 liquid chromatograph consisted of an autosampler, a binary pump, and a photodiode-array detector. The stationary phase was a Waters Symmetry C_8 column (250 × 4.6 mm, 5-μm particle size). The mobile phase consisted of 0.1% phosphoric acid aqueous solution and acetonitrile and was delivered linearly from 30 to 60% acetonitrile over 15 minutes and then to 85% acetonitrile over the next 10 minutes. The flow rate was 1.0 mL/min. UV detection was performed at 220 nm.

Rofecoxib solution was prepared in 50% (vol/vol) acetonitrile in water. The injection volume was 10 μL. Under these conditions, the retention time of rofecoxib was about 13.4 minutes.

Rofecoxib solutions (0.2 mg/mL) were treated with 1.0 N sodium hydroxide solution for 30 minutes or exposed to room light for 1 hour. Three major decomposition products from base degradation observed at 8.3, 11.6, and 17.1 minutes did not interfere with the analysis of rofecoxib.

Reference
Mao B, Abrahim A, Ge Z, et al. Examination of rofecoxib solution decomposition under alkaline and photolytic stress conditions. *J Pharm Biomed Anal*. 2002; 28: 1101–13.

Rosiglitazone Maleate

Chemical Name
(±)-5-[*p*-[2-(Methyl-2-pyridylamino)ethoxy]benzyl]-2,4-thiazolidinedione maleate

Other Names
Component of Avandamet, Avandia

Form	Molecular Formula	MW	CAS
Rosiglitazone	$C_{18}H_{19}N_3O_3S$	357.4	122320-73-4
Rosiglitazone maleate	$C_{18}H_{19}N_3O_3S.C_4H_4O_4$	473.5	155141-29-0

Appearance
Rosiglitazone maleate occurs as a white to off-white solid.

Solubility
Rosiglitazone maleate is readily soluble in ethanol and in a buffered aqueous solution with a pH of 2.3.

pK$_a$
Rosiglitazone maleate has pK$_a$ values of 6.8 and 6.1.

Method

Radhakrishna et al. reported an HPLC determination of rosiglitazone in bulk and pharmaceutical formulations. The instrument consisted of a Waters 510 pump, a Rheodyne injector with a 10-µL loop, and a Waters 996 photodiode-array detector. The stationary phase was a Waters Symmetry C_{18} analytical column (250 × 4.6 mm, 5-µm particle size). The mobile phase consisted of 0.025 M monobasic sodium phosphate buffer (adjusted to pH 6.2 with sodium hydroxide solution) and acetonitrile (50:50, vol/vol). The flow rate was 1.0 mL/min. UV detection was performed at 245 nm. Indole 1.2 mg/mL was used as an internal standard.

A portion of ground tablets (20) equivalent to 40 mg of rosiglitazone maleate was extracted with mobile phase and centrifuged. The supernatant was diluted with mobile phase before injection. The injection volume was 10 µL. Under these conditions, retention times of maleic acid, rosiglitazone, and internal standard were about 2.0, 7.4, and 9.1 minutes, respectively (estimated from the published chromatogram).

The analytical method was demonstrated to be stability indicating by accelerated degradation of rosiglitazone. Rosiglitazone samples were refluxed with 0.1 N hydrochloric acid at 61 °C for 12 hours, refluxed with 0.1 N sodium hydroxide solution at 61 °C for 12 hours, refluxed with 3% hydrogen peroxide for 30 minutes, exposed to ultraviolet light (254 nm), or incubated at 70 °C for 12 hours. Degradation product peaks were resolved from the peak of intact rosiglitazone.

A standard curve for rosiglitazone was constructed from 180 to 910 µg/mL. The correlation coefficient was more than 0.999. The intraday and interday coefficients of variation for the analysis of rosiglitazone were all less than 0.98%.

Reference

Radhakrishna T, Satyanarayana J, Satyanarayana A. LC determination of rosiglitazone in bulk and pharmaceutical formulations. *J Pharm Biomed Anal.* 2002; 29: 873–80.

Salicylamide

Chemical Name
2-Hydroxybenzamide

Other Names
Isosal, Sinedol, Urtosal

Form	Molecular Formula	MW	CAS
Salicylamide	$C_7H_7NO_2$	137.1	65-45-2

Appearance
Salicylamide is a white practically odorless crystalline powder.

Solubility
Salicylamide is slightly soluble in water and soluble in alcohol.

pK$_a$
Salicylamide has a pK$_a$ of 8.2.

Method
Aukunuru et al. reported the simultaneous HPLC determination of acetaminophen, salicylamide, phenyltoloxamine, and related products. The Varian system consisted of a model 9010 pump, a model 9095 autosampler, a model 9050 UV absorbance detector, and a Rainin Dinamax MacIntegrator. The stationary phase was a Phenomenex Prodigy C$_8$ column (150 × 4.6 mm, 5-μm particle size). Mobile phase A was 0.1 M phosphate buffer (adjusted to pH 2.7 with phosphoric acid). Mobile phase B was acetonitrile. The mobile phase was linearly delivered from 5% B to 45% B in 17 minutes followed by 10 minutes for equilibration. The flow rate was 1 mL/min. UV detection was performed at 220 nm. The injection volume was 50 μL. Under these conditions, retention times of acetaminophen, salicylamide, and phenyltoloxamine were 5.7, 10.9, and 15.9 minutes, respectively.

The stability-indicating nature of the assay was shown by accelerated degradation of drugs. Drug solutions were prepared in 1 N hydrochloric acid, 1 N sodium hydroxide, or 10% hydrogen peroxide and heated at 60 °C. No degradation products interfered with the analysis of drugs.

A standard curve for salicylamide was generated from 0.04 to 200 μg/mL. The correlation coefficient was 0.999.

Reference
Aukunuru JV, Kompella UB, Betageri GV. Simultaneous high performance liquid chromatographic analysis of acetaminophen, salicylamide, phenyltoloxamine, and related products. *J Liq Chromatogr Rel Technol.* 2000; 23: 565–78.

Saquinavir

Chemical Name
(*S*)-*N*-[(α*S*)-α-[(1*R*)-2-[(3*S*,4a*S*,8a*S*)-3-(*tert*-Butylcarbamoyl)octahydro-2(*1H*)-isoquinolyl]-1-hydroxyethyl]phenethyl]-2-quinaldamidosuccinamide

Other Names

Fortovase, Invirase

Form	Molecular Formula	MW	CAS
Saquinavir	$C_{38}H_{50}N_6O_5$	670.8	127779-20-8
Saquinavir mesylate	$C_{38}H_{50}N_6O_5 \cdot CH_4O_3S$	767.0	149845-06-7

Appearance

Saquinavir and saquinavir mesylate are white to off-white powders.

Solubility

Saquinavir is insoluble in water at 25 °C. Saquinavir mesylate has a solubility of 2.22 mg/mL in water at 25 °C.

Method

Tan et al. used HPLC to study the stability of extemporaneously prepared saquinavir formulations. The Hewlett-Packard 1050 system consisted of a quaternary pump, an autosampler, and a photodiode-array detector. The stationary phase was a reversed-phase Spherisorb C_8 column (150 × 4.5 mm, 5-μm particle size). The mobile phase consisted of acetonitrile and 5 mM dibasic potassium phosphate buffer (55:45, vol/vol), adjusted to pH 8 with phosphoric acid. The flow rate was 1.3 mL/min. UV detection was performed at 240 nm.

Each sample was appropriately diluted with a diluent of methanol and water (90:10, vol/vol) before injection. The sample injection volume was 10 μL. Under these conditions, the retention time for saquinavir was about 3.8 minutes.

To demonstrate the stability-indicating nature of the method, saquinavir solutions were mixed with either 1 N phosphoric acid or 1 N sodium hydroxide solution and heated to 100 °C for 1 hour. The saquinavir peak was resolved from other degradation product peaks.

A calibration curve for saquinavir was obtained from 0.05 to 1 mg/mL. The correlation coefficient was better than 0.999. The intraday and interday coefficients of variation for the analysis of saquinavir were less than 3 and 8%, respectively.

Reference

Tan LK, Thenmozhiyal JC, Ho PC. Stability of extemporaneously prepared saquinavir formulations. *J Clin Pharm Ther.* 2003; 28: 457–63.

Scopolamine

Chemical Name

(−)-(1S,3s,5R,6R,7S,8s)-6,7-Epoxy-3[(S)-tropoyloxy]tropane

Other Names
Buscopan, Hyoscine, Scopace

Form	Molecular Formula	MW	CAS
Methscopolamine bromide (Scopolamine methylbromide)	$C_{18}H_{24}BrNO_4$	398.3	155-41-9
Scopolamine	$C_{17}H_{21}NO_4$	303.4	51-34-3
Scopolamine butylbromide	$C_{21}H_{30}BrNO_4$	440.4	149-64-4
Scopolamine hydrobromide	$C_{17}H_{21}NO_4 \cdot HBr \cdot 3H_2O$	438.3	6533-68-2
Scopolamine methylnitrate (Methscopolamine nitrate)	$C_{18}H_{24}N_2O_7$	380.4	6106-46-3

Appearance
Scopolamine butylbromide is a white or almost white crystalline powder. Scopolamine hydrobromide occurs as efflorescent odorless and colorless crystals or as a white crystalline powder.

Solubility
Scopolamine butylbromide is freely soluble in water and in dichloromethane. It is sparingly soluble in dehydrated alcohol. Scopolamine hydrobromide is freely soluble in water, soluble in alcohol, slightly soluble in chloroform, and practically insoluble in ether.

pK_a
Scopolamine has a pK_a of 7.55 (at 23 °C) to 7.81 (at 25 °C).

Method
Peterson et al. studied by HPLC analysis the compatibility and stability of fentanyl in combination with scopolamine and midazolam. The liquid chromatographic system consisted of a Varian model 9010 pump, a model 9050 variable-wavelength UV-visible detector, and a Rheodyne 7161 injector. The stationary phase was a Nova-Pak C_{18} column (150 × 3.9 mm, 4-µm particle size). Solvent A was acetonitrile and solvent B was 0.1 M monobasic potassium phosphate buffer (pH 7.8). The mobile phase was delivered from 25 to 60% A in 10 minutes, kept at 60% A for 2 minutes, returned to 25% A in 2 minutes, and held at 25% A for 1 minute. The flow rate was 1.0 mL/min. UV detection was performed at 210 nm. Lorazepam 125 µg/mL in ethanol was used as an internal standard. The injection volume was 10 µL. Under these conditions, retention times for scopolamine, lorazepam, midazolam, and fentanyl were about 5.4, 7.4, 10.8, and 12.8 minutes, respectively.

The method was demonstrated to be stability indicating by accelerated degradation. Solutions of scopolamine were autoclaved at 121 °C for 5 hours or acidified to pH 1.0 with 0.1 M hydrochloric acid and autoclaved. None of the degradation product peaks interfered with the determination of scopolamine.

A standard curve for scopolamine was constructed from 40 to 100 μg/mL. The correlation coefficient was 0.995. The intraday and interday coefficients of variation were less than 4%.

Reference

Peterson GM, Miller KA, Galloway JF, et al. Compatibility and stability of fentanyl admixtures in polypropylene syringes. *J Clin Pharm Ther*. 1998; 23: 67–72.

Secnidazole

Chemical Name

1-(2-Methyl-5-nitroimidazol-1-yl)propan-2-ol

Other Names

Flagentyl, Secnidal

Form	Molecular Formula	MW	CAS
Secnidazole	$C_7H_{11}N_3O_3$	185.2	3366-95-8

Appearance

Secnidazole occurs as crystals.

Method 1

Bakshi and Singh used an HPLC method to study the stability of secnidazole. The Waters apparatus consisted of a model 600E pump, a model 717 autoinjector, a degasser module, and a model 996 photodiode-array detector. The stationary phase was a Waters Spherisorb ODS2 C_{18} column (250 × 4.6 mm, 5-μm particle size). The mobile phase was a mixture of water and methanol (85:15, vol/vol). The flow rate was 1 mL/min. The detector was set at 310 nm. The injection volume was 10 μL. Under these conditions, the retention time of secnidazole was 12.7 minutes.

To demonstrate the stability-indicating ability of the assay method, secnidazole was studied by accelerated degradation. Solutions of the drug were heated in water at 80 °C for 5 days, in 0.1 M hydrochloric acid at 80 °C for 12 hours, in 0.1 M sodium hydroxide solutions at 80 °C for 8 hours, treated with 3% hydrogen peroxide at room temperature for 6 hours, or stored in a photostability chamber for 8 days. The drug was also heated in dry conditions at 50 °C for 3 months. In all cases, secnidazole was resolved from its degradation products.

A standard curve for secnidazole was obtained from 50 to 500 μg/mL. The correlation coefficient was 0.9997. The intraday and interday coefficients of variation were less than 0.5 and 1.5%, respectively.

Reference
Bakshi M, Singh S. ICH guidance in practice: establishment of inherent stability of secnidazole and development of a validated stability-indicating high-performance liquid chromatographic assay method. *J Pharm Biomed Anal.* 2004; 36: 769–75.

Method 2
Moustafa and Bibawy reported an assay of secnidazole in the presence of its degradation products. The chromatographic instrument consisted of a Perkin-Elmer isocratic LC250 pump, a Rheodyne 7125 injector with a 20-μL loop, and a Perkin-Elmer diode-array detector. The stationary phase was a Shimadzu μBondapak C_{18} column (150 × 4.6 mm, 5-μm particle size). The mobile phase was 30% methanol in water. The flow rate was 1.0 mL/min. UV detection was performed at 319 nm.

A sample of ground powder equivalent to 50 mg of secnidazole was transferred into a 50-mL volumetric flask, dissolved in the mobile phase, filtered, and diluted again with the mobile phase. The injection volume was 20 μL. Under these conditions, the retention time of secnidazole was 3.91 minutes.

The stability-indicating nature of the assay was demonstrated by assaying a mixture of secnidazole and its degradation products. The degradation products did not influence the analysis of secnidazole.

A standard curve for secnidazole was obtained from 2 to 20 μg/mL. The correlation coefficient was 0.9999.

Reference
Moustafa AA, Bibawy LI. Stability-indicating assay of secnidazole in the presence of its degradation products. *Spectroscopy Lett.* 1999; 32: 1073–98.

Secobarbital

Chemical Name
5-(1-Methylbutyl)-5-(2-propenyl)-2,4,6(1*H*,3*H*,5*H*)-pyrimidinetrione

Other Names
Quinalbarbitone, Seconal

Form	Molecular Formula	MW	CAS
Secobarbital	$C_{12}H_{18}N_2O_3$	238.3	76-73-3
Secobarbital sodium	$C_{12}H_{17}N_2NaO_3$	260.3	309-43-3

Appearance
Secobarbital is a white odorless amorphous or crystalline powder. Secobarbital sodium is a white odorless powder.

Solubility
Secobarbital is very slightly soluble in water, freely soluble in alcohol and ether, and soluble in chloroform. Secobarbital sodium is very soluble in water and soluble in alcohol, but it is practically insoluble in ether.

pK_a
Secobarbital has a pK_a of 7.74–7.9.

Method
Reif et al. described a stability-indicating HPLC method for secobarbital sodium injection. The liquid chromatographic system consisted of a Waters 6000A chromatographic pump, an LDC SpectroMonitor III variable-wavelength UV detector, and a Rheodyne 7010 injector with a 20-μL loop. The stationary phase was a Zorbax ODS column (75–150 × 4.6 mm, 7-μm particle size). The mobile phase consisted of a mixture of acetate buffer, methanol, and polyethylene glycol 300 (400:600:4, vol/vol/vol). Acetate buffer was prepared by adding 4.1 g of anhydrous sodium acetate and 3 mL of acetic acid to a total volume of 1 L of distilled water. The flow rate was 1 mL/min. UV detection was performed at 230 nm and 0.4 AUFS.

Samples were diluted with the mobile phase to a secobarbital sodium concentration of 1.0 mg/mL. The injection volume was 20 μL.

The HPLC method was determined to be stability indicating by spiking the secobarbital sodium sample with its known degradation products. Degradation product peaks did not interfere with the intact secobarbital peak. Retention times for 2-allyl-2-(1-methylbutyl)propanediamide, *N*-(aminocarbonyl)-2-allyl-2-carboxy-3-methylhexanamide, secobarbital, and *N*-(aminocarbonyl)-2-allyl-3-methylhexanamide were about 2.5, 3.2, 5.7, and 10.2 minutes, respectively (estimated from the published chromatogram).

A standard curve for secobarbital was constructed from 0.127 to 1.183 mg/mL. The correlation coefficient was greater than 0.9999.

Reference
Reif VD, Kaufmann KL, DeAngelis NJ, et al. Liquid chromatographic assays for barbiturate injections. *J Pharm Sci*. 1986; 75: 714–6.

Selegiline

Chemical Names
(*R*)-*N*,α-Dimethyl-*N*-2-propynylbenzeneethanamine
(−)-*R*-*N*,α-dimethyl-*N*-2-propynylphenethylamine

Other Names
L-Deprenyl, Eldepryl

Form	Molecular Formula	MW	CAS
Selegiline	$C_{13}H_{17}N$	187.3	14611-51-9
Selegiline hydrochloride	$C_{13}H_{17}N.HCl$	223.7	14611-52-0

Appearance
Selegiline hydrochloride is a white to nearly white crystalline powder.

Solubility
Selegiline hydrochloride is freely soluble in water, chloroform, and methanol.

Method
Chafetz et al. reported a trace decomposition of selegiline hydrochloride using a stability-indicating HPLC assay. The liquid chromatograph consisted of a Varian LC Star system with a Star 9095 autosampler, a Rheodyne 7126S injector, a Polychrom 9065 diode-array detector, an Epson LQ 2550 printer, and a Compaq 386e computer running Varian Star 9020 workstation software. The stationary phase was a Rainin Microsorb RP18 column (100 × 4.6 mm, 3-μm particle size). The mobile phase was a mixture of acetonitrile and a buffer (20:80, vol/vol). The buffer was a solution of 0.1 M ammonium dihydrogen phosphate and 0.08% triethylamine adjusted to pH 3.1 with phosphoric acid. The flow rate was 1 mL/min. UV detection was performed at 205 nm.

The stability-indicating nature of the method was demonstrated by accelerated decomposition of selegiline hydrochloride. Ampules containing selegiline hydrochloride 5 mg/mL were placed in an oven at 105 °C for an appropriate time period of up to 72 hours. The degradation product peak did not interfere with the selegiline peak. The retention times for the degradation product, methamphetamine, and selegiline hydrochloride were 2.3 and 3.8 minutes, respectively.

Reference
Chafetz L, Desai MF, Sukonik L. Trace decomposition of selegiline. Use of worst-case kinetics for a stable drug. *J Pharm Sci.* 1994; 83: 1250–2.

Sibutramine Hydrochloride

Chemical Name

1-(4-Chlorophenyl)-*N*,*N*-dimethyl-α-(2-methylpropyl)cyclobutanemethanamine hydrochloride monohydrate

• HCl • H$_2$O

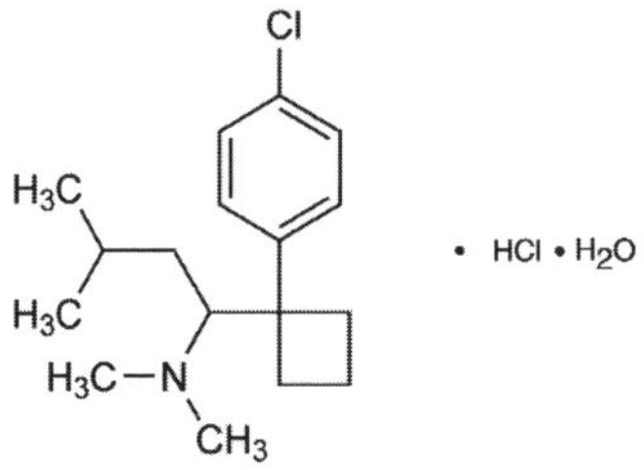

Other Names

Meridia, Reductil

Form	Molecular Formula	MW	CAS
Sibutramine	C$_{17}$H$_{26}$ClN	279.9	106650-56-0
Sibutramine hydrochloride monohydrate	C$_{17}$H$_{26}$ClN.HCl.H$_2$O	334.3	125494-59-9

Appearance

Sibutramine hydrochloride monohydrate occurs as a white to cream crystalline powder.

Solubility

Sibutramine hydrochloride monohydrate has a solubility of 2.9 mg/mL in water at pH 5.2.

Method

Segall et al. reported a reversed-phase HPLC method for the determination of sibutramine hydrochloride in tablets. The liquid chromatograph consisted of a Spectra-Physics model ISO Chrom dual-piston reciprocating pump, a Hewlett-Packard model 1050 UV-visible detector, a Hewlett-Packard model 3395 integrator, and a Rheodyne model 7125 injector. The stationary phase was a Varian Microsorb-MV analytical column (250 × 4.6 mm). The mobile phase was a mixture of methanol, water, triethylamine (80:20:0.3, vol/vol/vol), adjusted to pH 4.0 with phosphoric acid. The flow rate was 1.1 mL/min. UV detection was performed at 225 nm and 1 AUFS.

The sample injection volume was 20 μL. Under these conditions, the retention time of sibutramine was about 4 minutes.

The method was shown to be stability indicating by accelerated degradation of sibutramine. Sibutramine samples were treated with 1 N hydrochloric acid, 1 N sodium hydroxide solution, or 30% hydrogen peroxide and refluxed for 15 minutes. Sibutramine samples were also heated at 110 °C for 24 hours or exposed to daylight for 24 hours. No peaks from degradation products interfered with the peak of intact sibutramine.

A standard curve for sibutramine was constructed from 30 to 96 μg/mL. The correlation coefficient was 0.9995.

Reference

Segall AI, Collado EA, Ricci RA, et al. Reversed-phase HPLC determination of sibutramine hydrochloride in the presence of its oxidatively-induced degradation products. *J Liq Chromatogr Rel Technol.* 2003; 26: 977–86.

Sildenafil Citrate

Chemical Name

5-[2-Ethoxy-5-(4-methylpiperazin-1-ylsulfonyl)phenyl]-1,6-dihydro-1-methyl-3-propyl-pyrazolo[4,3-d]pyrimidin-7-one citrate

Other Name

Viagra

Form	Molecular Formula	MW	CAS
Sildenafil citrate	$C_{22}H_{30}N_6O_4S.C_6H_8O_7$	666.7	171599-83-0

Appearance

Sildenafil citrate is a white to off-white crystalline powder.

Solubility

Sildenafil citrate has a solubility of 3.5 mg/mL in water.

pK_a

Sildenafil citrate has pK_a values of 6.5 and 9.2.

Method 1

Abd-Elbary et al. described an HPLC assay for the determination of sildenafil citrate in bulk and in formulations. The liquid chromatographic system consisted of a Shimadzu model LC-10 AS pump, a model SPD-10A variable-wavelength detector, a model C-R6A integrator, and a Rheodyne model 7161 injector. The stationary phase was a Waters μBondapak C_{18} column (300 × 3.9 mm, 10-μm particle size). The mobile phase consisted of methanol, water, acetonitrile (60:20:20, vol/vol/vol) adjusted to pH 6.1 with 0.1% glacial acetic acid. The flow rate was 0.5 mL/min. UV detection was performed at 290 nm and 0.0001 AUFS. The injection volume was 10 μL. Cinnarizine in methanol 0.1 mg/mL was used as the internal standard. Under these conditions, retention times of sildenafil citrate and cinnarizine were 4.2 and 6.6 minutes, respectively.

Sildenafil citrate tablets were powdered, shaken in a mixture of water and methanol (1:4, vol/vol) for 5 minutes, and centrifuged at 3000 rpm for 5 minutes. A portion of the supernatant was collected, mixed with the internal standard, diluted with mobile phase, and injected.

In order to demonstrate the stability-indicating nature of the method, sildenafil solutions at pH 2.5, 4, and 8 were stored at 60 and 80 °C. The degradation product had a retention time of 10.3 minutes.

A calibration curve for sildenafil was obtained from 0.01 to 0.2 μg/mL; the correlation coefficient was 0.9999. The intraday and interday coefficients of variation were less than 0.48%.

Reference

Abd-Elbary A, Foda NH, El-Gazayerly ON. Stability-indicating high performance liquid chromatographic assay for the determination of sildenafil citrate in bulk and in formulations. *Chromatographia.* 2004; 59: 561–6.

Method 2

Using HPLC analysis, Daraghmeh et al. described the determination of sildenafil citrate and its impurities. The Beckman Gold system included a model 168 diode-array detector and a model 125 programmable pump. The stationary phase was a Waters μBondapak C_{18} stainless steel column (300 × 3.9 mm, 10-μm particle size). The mobile phase consisted of 0.2 M ammonium acetate (pH 7.0) and acetonitrile (1:1, vol/vol). The flow rate was 1 mL/min. UV detection was performed at 240 nm.

A sample equivalent to 100 mg of sildenafil citrate was weighed, dissolved and diluted to 100 mL with mobile phase, sonicated for 15 minutes, centrifuged at 4000 rpm for 10 minutes, and diluted with mobile phase. The injection volume was 20 μL. Under these conditions, the retention time for sildenafil was about 9.3 minutes (estimated from the published chromatogram).

The stability-indicating nature of the method was demonstrated by assaying a mixture of sildenafil citrate and its related compounds. Sildenafil citrate samples were also treated with 1.5% hydrogen peroxide at room temperature for 60 minutes, with 0.1 M hydrochloric acid at 65 °C for 12 days, or with 0.1 M sodium hydroxide at 65 °C for 12 days or were heated at the melting point for 5 minutes. Sildenafil citrate was well separated from its degradation products and impurities.

Standard curves for sildenafil citrate were constructed from 2.39 to 11.95 μg/mL and from 64.3 to 257.0 μg/mL. Correlation coefficients were greater than 0.999. The limit of detection was 0.451 μg/mL. The limit of quantitation was 1.68 μg/mL.

Reference

Daraghmeh N, Al-Omari M, Badwan AA, et al. Determination of sildenafil citrate and related substances in the commercial products and tablet dosage form using HPLC. *J Pharm Biomed Anal.* 2001; 25: 483–92.

Method 3

Segall et al. reported an HPLC determination of sildenafil citrate in the presence of its degradation products. The liquid chromatograph consisted of a Spectra-Physics model ISO Chrom dual-piston reciprocating pump, a Hewlett-Packard model 1050 UV-visible detector, a model 3395 integrator, and a Rheodyne model 7125 injector. The stationary phase was a Merck LiChrospher 100RP18 column (250 × 4 mm, 5-μm particle size). The mobile phase consisted of 700 mL of a buffer solution and 300 mL of acetonitrile. The buffer solution was prepared by dissolving 9.7 g of monobasic potassium phosphate and 10.4 g of triethylamine in 1 L of water and adjusting to pH 3.0 with phosphoric acid. The flow rate was 1.0 mL/min. UV detection was performed at 225 nm and 2 AUFS. The injection volume was 20 μL. Under these conditions, the retention time for sildenafil citrate was about 7 minutes.

The method was demonstrated to be stability indicating by accelerated degradation of sildenafil citrate. Sildenafil citrate samples were stored in an oven at 110 ℃ for 24 hours, exposed to daylight for 24 hours, treated with 1 N hydrochloric acid, with 1 N sodium hydroxide solution, or with hydrogen peroxide and refluxed for 15 minutes. The sildenafil citrate peak was well resolved from its degradation product peaks.

A standard curve for sildenafil citrate was generated from 50 to 800 μg/mL. The correlation coefficient was 0.9999.

Reference
Segall AI, Vitale MF, Perez VL, et al. Reversed-phase HPLC determination of sildenafil citrate in the presence of its oxidative-induced degradation products. *J Liq Chromatogr Rel Technol*. 2000; 23: 1377–86.

Sodium Nitroprusside

Chemical Names
Pentakis(cyano-*C*)nitrosylferrate(2–) disodium
Sodium nitrosylpentacyanoferrate(III)

Other Names
Nitropress, Sodium Nitroferricyanide

Form	Molecular Formula	MW	CAS
Sodium nitroprusside	$Na_2[Fe(CN)_5NO]$	261.9	14402-89-2

Appearance
Sodium nitroprusside occurs as reddish-brown practically odorless crystals or powder.

Solubility
Sodium nitroprusside is freely soluble in water and slightly soluble in alcohol.

Method

Karnatz et al. evaluated the stability of sodium nitroprusside and esmolol hydrochloride in intravenous admixtures. The liquid chromatograph consisted of a Perkin-Elmer series 2 solvent-delivery system, a Perkin-Elmer model 420 automatic sampler, a Perkin-Elmer LC-55 variable-wavelength UV detector, a Scientific Products model 285/mm recorder, and a Hewlett-Packard 3357 online data system. The stationary phase was a Waters μBondapak phenyl stainless steel column (300 × 3.9 mm, 10-μm particle size). The mobile phase consisted of 300 mL of acetonitrile and 700 mL of buffer. The buffer contained 0.01 M monobasic potassium phosphate and 0.005 M tetrabutylammonium hydroxide adjusted to a final pH of 7.1 with phosphoric acid. The mobile phase was delivered at a constant flow rate of 2 mL/min. UV detection was performed at 200 nm.

Samples were diluted 1:10 with water. The injection volume was 100 μL. Retention times for sodium nitroprusside and esmolol were 2.09 and 6.92 minutes, respectively.

The stability-indicating nature of this method was demonstrated by accelerated degradation of sodium nitroprusside. The solution of sodium nitroprusside was placed in an intensive-light cabinet (approximately 1200 foot-candles) for 1 week. Degradation product peaks did not interfere with the sodium nitroprusside peak.

Reference

Karnatz NN, Wong J, Baaske DM, et al. Stability of esmolol hydrochloride and sodium nitroprusside in intravenous admixtures. *Am J Hosp Pharm.* 1989; 46: 101–4.

Sorbic Acid

Chemical Name
(*E,E*)-Hexa-2,4-dienoate

Other Name
Kalii Sorbas

Form	Molecular Formula	MW	CAS
Potassium sorbate	$C_6H_7KO_2$	150.2	24634-61-5
Sorbic acid	$C_6H_8O_2$	112.1	22500-92-1

Appearance

Potassium sorbate occurs as white or almost white crystals, granules, or powder with a characteristic odor. Sorbic acid is a free-flowing, white or almost white crystalline powder with a characteristic odor.

Solubility

Potassium sorbate is very soluble in water and slightly soluble in alcohol. Sorbic acid is slightly soluble in water. It is freely soluble in alcohol and in ether.

Method

Barnes developed an HPLC method for the determination of caffeine and potassium sorbate in a pharmaceutical formulation for oral use. The system was composed of a Cecil model CE1100 pump, a Pye model LC3 detector, a Shimadzu model C-R3A integrator, and a Talbot model ASI-4 autosampler. The stationary phase was a Hichrom Spherisorb hexyl reversed-phase column (100 × 4.5 mm, 5-μm particle size). The mobile phase was 12% acetonitrile in 0.1 M sodium acetate buffer (pH 4.5). The flow rate was 2 mL/min. UV detection was performed at 258 nm. Orcinol aqueous solution 3 mg/mL was used as an internal standard.

A 1-g sample was diluted to 25 mL with water, mixed with internal standard, and diluted again with water. The injection volume was 20 μL. Under these conditions, the retention times of caffeine, orcinol, and sorbate were 1.7, 2.6, and 3.5 minutes, respectively.

The stability-indicating nature of the method was confirmed by accelerated degradation of the sample. Oral solutions were refluxed with 0.1 M sodium hydroxide or 0.05 M sulfuric acid for 1 hour. No degradation product peaks interfered with the analysis of potassium sorbate.

A calibration curve for potassium sorbate was constructed from 0.02 to 0.16 μg per injection. The correlation coefficient was 0.9999.

Reference

Barnes AR. Determination of caffeine and potassium sorbate in a neonatal oral solution by HPLC. *Int J Pharm.* 1992; 80: 267–70.

Sorivudine

Chemical Name

(+)-1-β-D-Arabinofuranosyl-5-[(*E*)-2-bromovinyl]uracil

Other Name

Bravavir

Form	Molecular Formula	MW	CAS
Sorivudine	$C_{11}H_{13}BrN_2O_6$	349.1	77181-69-2

Appearance
Sorivudine occurs as white crystals.

Method
Desai et al. studied the photoisomerization of sorivudine using a stability-indicating HPLC method. The chromatograph consisted of a Perkin-Elmer 410 pump, a Waters 712 WISP injector, and an Applied Biosystems ABI 783A UV detector. The stationary phase was a LiChrosorb reversed-phase C_{18} column (150 × 4.6 mm, 5-μm particle size). The mobile phase was a mixture of 15 parts of acetonitrile and 85 parts of 0.01 M triethylammonium acetate buffer, adjusted to pH 7.0. The flow rate was 1.0 mL/min. UV detection was performed at 254 nm. Under these conditions, retention times for Z-isomer of sorivudine and E-isomer of sorivudine were 3.8 and 6.0 minutes, respectively.

The HPLC method was reported to be stability indicating.

Reference
Desai D, Li D, Janjikhel R, et al. Effects of light intensity, n-alcohols, water-soluble colorants, and solution viscosity on photoisomerization of sorivudine. *Pharm Develop Technol.* 2001; 6: 99–106.

Sotalol Hydrochloride

Chemical Name
N-[4-[1-Hydroxy-2-[(1-methylethyl)amino]ethyl]phenyl]methanesulfonamide monohydrochloride

Other Names
Betapace, Sotacor, Sotalex

Form	Molecular Formula	MW	CAS
Sotalol hydrochloride	$C_{12}H_{20}N_2O_3S.HCl$	308.8	959-24-0

Appearance
Sotalol hydrochloride occurs as a white or almost white powder.

Solubility

Sotalol hydrochloride is freely soluble in water. It is slightly soluble in alcohol and practically insoluble in chloroform.

pK$_a$

Sotalol has a pK$_1$ of 8.2 and a pK$_2$ of 9.8.

Method

Using HPLC, Nahata and Morosco determined the chemical stability of sotalol hydrochloride in extemporaneously prepared suspensions. The Hewlett-Packard 1050 liquid chromatograph included a pump, an autosampler, a variable-wavelength detector, and an Hewlett-Packard 3396A integrator. The stationary phase was a MAC-MOD Zorbax C$_{18}$ column (150 × 3.0 mm). The mobile phase consisted of 75% 5 mM octanesulfonic acid aqueous solution and 25% acetonitrile, adjusted to pH 3.2. The flow rate was 0.4 mL/min. UV detection was performed at 235 nm.

One hundred microliters of the suspension sample was mixed with 25 mL of mobile phase and centrifuged. The supernatant was collected and analyzed. The sample injection volume was 10 μL. Under these conditions, the retention time for sotalol was about 5.1 minutes.

The analytical method was established to be stability indicating by forced degradation of sotalol. The drug was treated with 2 M hydrochloric acid, 2 M sodium hydroxide solution, or 3% hydrogen peroxide and heated at 80 °C. The degradation products did not interfere with the analysis of sotalol.

A calibration curve for sotalol was constructed from 2 to 26 μg/mL. The correlation coefficient was greater than 0.999. The intraday and interday coefficients of variation of the analysis were less than 1.9 and 2.6%, respectively.

Reference

Nahata MC, Morosco RS. Stability of sotalol in two liquid formulations at two temperatures. *Ann Pharmacother.* 2003; 37: 506–9.

Sparfloxacin

Chemical Name

5-Amino-1-cyclopropyl-7-(*cis*-3,5-dimethylpiperazin-1-yl)-6,8-difluoro-1,4-dihydro-4-oxoquinoline-3-carboxylic acid

Other Name
Zagam

Form	Molecular Formula	MW	CAS
Sparfloxacin	$C_{19}H_{22}F_2N_4O_3$	392.4	110871-86-8

Appearance
Sparfloxacin is a yellow crystalline powder.

Solubility
Sparfloxacin is practically insoluble in water and very slightly soluble in alcohol.

Method
Argekar and Shah reported an HPLC method for the analysis of sparfloxacin in bulk drug and tablet formulations. The Waters liquid chromatograph consisted of a low-pressure quaternary pump, an autoinjector, and a variable-wavelength UV detector. The stationary phase was a Merck RP-Select B column (125 × 4.0 mm, 5-μm particle size). The mobile phase consisted of water, acetonitrile, and triethylamine (80:20:0.2, vol/vol/vol), adjusted to pH 2.6 with phosphoric acid. The flow rate was 1.0 mL/min. UV detection was performed at 304 nm.

Powder from ground tablets equivalent to 200 mg of sparfloxacin was weighed, dissolved in 10 mL of acetonitrile by sonication for 10 minutes, diluted to 100 mL with methanol, centrifuged, diluted 1:10 (vol/vol) with methanol, and centrifuged again. The injection volume was 10 μL. The run time was 10 minutes. Under these conditions, the retention time for sparfloxacin was 7.2 minutes.

The stability-indicating capability of the method was demonstrated by induced degradation of sparfloxacin. Samples were treated with 1 M hydrochloric acid, 1 M sodium hydroxide, or 30% hydrogen peroxide for 1 week. No degradation products influenced the determination of sparfloxacin.

A standard curve for sparfloxacin was generated from 8 to 1000 μg/mL. The correlation coefficient was 0.999.

Reference
Argekar AP, Shah SJ. Stability indicating HPLC method for the determination of sparfloxacin (SPAR). *Anal Lett*. 1999; 32: 1363–70.

Spironolactone

Chemical Name
(7α,17α)-7-(Acetylthio)-17-hydroxy-3-oxopregn-4-ene-21-carboxylic acid γ-lactone

Other Name
Aldactone

Form	Molecular Formula	MW	CAS
Spironolactone	$C_{24}H_{32}O_4S$	416.6	52-01-7

Appearance
Spironolactone is a light cream to light tan crystalline powder.

Solubility
Spironolactone is practically insoluble in water, but it is soluble in alcohol.

Method 1
Alexander et al. developed an HPLC method for the quantitative analysis of spironolactone in a liquid dosage form. The system consisted of a Beckman 110B solvent-delivery pump, a Beckman system organizer injector fitted with a 20-μL loop, a model 420 system controller programmer, a Knauer spectrophotometric UV detector, and a Shimadzu C-R3A integrator. The stationary phase was a Supelcosil LC-18 column (250 × 4.6 mm, 5-μm particle size). The mobile phase was prepared by diluting 660 mL of methanol with distilled water to 1 L. The flow rate was 1.3 mL/min. UV detection was performed at 254 nm and 0.04 AUFS. Hydrocortisone 21-acetate 0.5 mg/mL in methanol was used as an internal standard. The injection volume was 50 μL. Under these conditions, the retention times for the internal standard and the drug were 8.0 and 9.5 minutes, respectively.

The method was demonstrated to be stability indicating by accelerated degradation of spironolactone. A spironolactone sample (1.5 mL) was mixed with 1 mL of 1 M hydrochloric acid or 1 mL of 1 M sodium hydroxide solution, diluted to 15 mL, and boiled for 30 minutes. The intact drug peak was well separated from its degradation product peaks.

A standard curve for spironolactone was constructed from 4 to 16 μg/mL. The correlation coefficient was 1.000. Intraday and interday coefficients of variation were 0.996 and 1.45%, respectively.

Reference
Alexander KS, Vangala SSKS, Dollimore D. An improved high-performance liquid chromatography assay for spironolactone analysis. *Drug Dev Ind Pharm.* 1998; 24: 101–7.

Method 2

Using HPLC, Allen and Erickson evaluated the stability of spironolactone 5 mg/mL in extemporaneously compounded oral liquids. The system was a Hewlett-Packard series 1050 liquid chromatograph, including a multisolvent mixing and pumping system, an autoinjector, a diode-array detector, and a computer with Chem Station software. The stationary phase was a Bakerbond C_{18} analytical column (250 × 4.6 mm, 5-µm particle size). The mobile phase was 79% methanol in water; it was delivered isocratically at 1.0 mL/min. UV detection was performed at 254 nm.

Samples were diluted 1:50. Under these conditions, spironolactone had a retention time of 7.4 minutes.

This method was demonstrated to be stability indicating by accelerated degradation of spironolactone. A composite chromatogram of spironolactone after accelerated decomposition by acid, base, heat, oxidizing agent, and light showed that the intact spironolactone peak was well resolved from degradation product peaks.

Standard curves were constructed from a linear plot of spironolactone peak areas versus concentration from 10 to 150 µg/mL. The intraday and interday coefficients of variation were 1.6 and 1.9%, respectively.

Reference

Allen LV, Erickson MA. Stability of labetalol hydrochloride, metoprolol tartrate, verapamil hydrochloride, and spironolactone with hydrochlorothiazide in extemporaneously compounded oral liquids. *Am J Health Syst Pharm.* 1996; 53: 2304–9.

Method 3

Allen and Erickson also used HPLC to study the stability of spironolactone in extemporaneously compounded oral liquids. The liquid chromatograph was the same as described in Method 1. The mobile phase was 55% acetonitrile in 0.02 M dibasic ammonium phosphate aqueous buffer. The flow rate was 1.0 mL/min. UV detection was performed at 254 nm.

Samples were diluted 1:100. Under these conditions, the retention time for spironolactone was 5.6 minutes.

This HPLC analytical method was determined to be stability indicating. The composite chromatogram of the drug spironolactone after accelerated degradation by heat, acid, base, oxidizing agent, and light showed that the spironolactone peak was well separated from its degradation product peaks.

Standard curves were generated for spironolactone from 50 to 300 µg/mL. The intraday and interday coefficients of variation were 1.5 and 1.6%, respectively.

Reference

Allen LV, Erickson MA. Stability of ketoconazole, metolazone, metronidazole, procainamide hydrochloride, and spironolactone in extemporaneously compounded oral liquids. *Am J Health Syst Pharm.* 1996; 53: 2073–8.

Method 4

Nahata et al. determined the extended stability of spironolactone in an extemporaneously prepared suspension containing simple syrup and carboxymethylcellulose that was stored at 4 and 22 °C for 3 months. The system consisted of a Hewlett-Packard 1050 series pump, an autosampler, a variable-wavelength detector, and a Hewlett-Packard 3396A integrator.

The stationary phase was a Phenomenex Zorbax ODS column (150 × 4.6 mm, 7-μm particle size). The mobile phase was a mixture of 50% acetonitrile and 50% distilled water containing 0.1% acetic acid. The flow rate was 1.0 mL/min. UV detection was performed at 238 nm. Bumetanide in methanol and water (50:50, vol/vol) 50 μg/mL was used as the internal standard.

Samples were diluted 1:20 with the mobile phase. The injection volume was 10 μL. The retention times for bumetanide and spironolactone were 8.7 and 10.1 minutes, respectively.

The stability-indicating capability of the method was demonstrated by accelerated base and acid hydrolysis of spironolactone. Spironolactone suspensions 1 mg/mL were mixed with 1.0 M sodium hydroxide or with 1.0 M sulfuric acid and heated to 60 °C for 90 minutes. Degradation product peaks did not interfere with the intact spironolactone peak.

A standard curve for spironolactone was constructed by linear regression analysis of the spironolactone concentration versus the peak height ratios of spironolactone and internal standard from 0.10 to 2.00 mg/mL. The intraday and interday coefficients of variation ranged from 0.6 to 2.7%.

A similar method was used by McKnight et al.

References
Nahata MC, Morosco RS, Hipple TF. Stability of spironolactone in an extemporaneously prepared suspension at two temperatures. *Ann Pharmacother*. 1993; 27: 1198–9.

McKnight DL, Emmott N, Sunderland VB, et al. Preliminary formulation and stability of a spironolactone oral solution for pediatric use. *Aust J Hosp Pharm*. 1997; 27: 390–4.

Method 5
Using a stability-indicating HPLC assay, Pramar and Gupta studied the effect of pH, phosphate and citrate buffers, ionic strength, and temperature on the stability of spironolactone. The liquid chromatograph consisted of a Waters ALC 202 system, a Schoeffel SF 770 multiple-wavelength detector, and a Houston Omniscribe recorder. The stationary phase was a Waters μBondapak phenyl column (300 × 3.9 mm). The mobile phase contained 39% acetonitrile in 0.01 M monobasic potassium phosphate aqueous buffer. The mobile phase was delivered isocratically at 2.2 mL/min. UV detection was performed at 254 nm and 0.04 AUFS. Methyltestosterone 25 μg/mL in methanol was used as an internal standard.

Samples were diluted with a mixture of 0.01 M monobasic potassium phosphate and 25% methanol (vol/vol) to a spironolactone concentration of 80 μg/mL.

The assay was stated to be stability indicating because the degradation product peak of spironolactone (canrenone) did not interfere with the spironolactone peak.

Reference
Pramar Y, Gupta VD. Preformulation studies of spironolactone: Effect of pH, two buffer species, ionic strength, and temperature on stability. *J Pharm Sci*. 1991; 80: 551–3.

Method 6
Mathur and Wickman used an HPLC assay to investigate the stability of extemporaneously compounded spironolactone suspensions. The liquid chromatograph was

equipped with a Waters model 710B WISP autosampler, a Waters model 440 fixed-wavelength detector, and a Linear Instruments model 500 linear strip-chart recorder. The stationary phase was a DuPont Pharmaceutics Zorbax ODS column (150×4.6 mm). The mobile phase consisted of 55% acetonitrile and 45% water. The flow rate was 1.5 mL/min. UV detection was performed at 254 nm.

Spironolactone was extracted from the suspension with methanol; the methanolic solutions were acidified with 0.15 M acetic acid. The injection volume was 25 µL.

The stability-indicating capability of the method was demonstrated by accelerated degradation of spironolactone. Spironolactone and its known degradation product, aldediene, were stored at 70 and 55 °C for 5 days. Degradation product peaks did not interfere with the spironolactone peak.

The intraday coefficient of variation was 0.8% ($n = 24$).

Reference
Mathur LK, Wickman A. Stability of extemporaneously compounded spironolactone suspensions. *Am J Hosp Pharm.* 1989; 46: 2040–2.

Stavudine

Chemical Name
1-(2,3-Dideoxy-β-D-*glycero*-pent-2-enofuranosyl)thymine

Other Name
Zerit

Form	Molecular Formula	MW	CAS
Stavudine	$C_{10}H_{12}N_2O_4$	224.2	3056-17-5

Appearance
Stavudine is a white to off-white crystalline solid.

Solubility
Stavudine has a solubility of 83 mg/mL in water at 23 °C.

Method 1
Santoro et al. reported an assay method for the quantitative determination of stavudine in capsules. The HPLC system consisted of a Varian model 5000 solvent-delivery pump, a model 4000 variable-wavelength UV detector, a model 4400 integrator, and a Rheodyne

model 7125 injector. The stationary phase was a LiChrospher 100 RP-18 column (125 × 4.0 mm, 5-µm particle size). The mobile phase was a mixture of water and methanol (85:15, vol/vol). The flow rate was 1.0 mL/min. UV detection was carried out at 265 nm.

A sample equivalent to 50.0 mg of stavudine was accurately weighed, transferred to a 100-mL volumetric flask, mixed with 50 mL of water, sonicated for 30 minutes, and diluted to the mark with water. The solution was further diluted with mobile phase and filtered before analysis. The injection volume was 20 µL. Under these conditions, the retention time of stavudine was about 3.6 minutes.

The drug products were incubated in different temperature and humidity conditions (25 °C/70%, 40 °C/75%, and 50 °C/90%) for 90 days. Under these conditions, the peak of the intact drug was resolved from that of its degradation product, thymine (1.7 minutes).

A standard curve of stavudine was obtained from 30.0 to 100.0 µg/mL. The correlation coefficient was 0.9999. The limit of detection and the limit of quantitation were 0.80 and 2.67 µg/mL, respectively. The coefficient of variation for the analysis of stavudine was 1.05%.

Reference
Santoro MI, Taborianski AM, Kedor-Hackmann AKS. Stability-indicating methods for quantitative determination of zidovudine and stavudine in capsules. *Quim Nova.* 2006; 29: 240–4.

Method 2
Dunge et al. described an HPLC method for the determination of stavudine. The Waters instrument consisted of a model 600E quaternary gradient pump, a model 996 photodiode-array detector, a model 717 autoinjector, and a degasser module. The stationary phase was a Merck reversed-phase C_{18} column (250 × 4.6 mm, 5-µm particle size). The mobile phase was a mixture of water and methanol (90:10, vol/vol). The flow rate was 1 mL/min. UV detection was carried out at 265 nm. The injection volume was 10 µL. Under these conditions, the retention time of stavudine was 12.5 minutes.

To confirm the stability-indicating property of the method, accelerated decomposition studies of stavudine were performed following ICH guidelines. Briefly, solutions of the drug in water, 0.1 M hydrochloric acid, and 0.1 M sodium hydroxide solution were incubated at 80 °C for up to 3 days. The drug was also treated in 3% hydrogen peroxide for 24 hours and then in 30% hydrogen peroxide for 48 hours at room temperature. Solutions of stavudine in water, 0.01 M hydrochloric acid, and 0.1 M sodium hydroxide were exposed in a photostability chamber for 15 days. The bulk drug material was heated dry at 50 °C for 3 months. The bulk drug substance was also treated at 40 °C and 75% relative humidity for 3 months. In all cases, the peak of the intact stavudine was well separated from the peaks of its degradation products.

A calibration curve of stavudine was constructed from 25 to 500 µg/mL. The correlation coefficient was 0.9994. The intraday and interday coefficients of variation were 0.20 and 0.95%, respectively.

Reference
Dunge A, Sharda N, Singh B, et al. Establishment of inherent stability of stavudine and development of a validated stability-indicating HPLC assay method. *J Pharm Biomed Anal.* 2005; 37: 1115–9.

Streptozocin

Chemical Names
2-Deoxy-2-[[(methylnitrosoamino)carbonyl]amino]-D-glucopyranose
2-Deoxy-2-(3-methyl-3-nitrosoureido)-D-glucopyranose

Other Name
Zanosar

Form	Molecular Formula	MW	CAS
Streptozocin	$C_8H_{15}N_3O_7$	265.2	18883-66-4

Appearance
Streptozocin is an ivory-colored crystalline powder.

Solubility
Streptozocin is very soluble in water and soluble in alcohols.

pK$_a$
Streptozocin has a pK$_a$ of 1.35.

Method
Mayron and Gennaro used an HPLC method to study the stability of streptozocin with granisetron hydrochloride during simulated Y-site administration. The liquid chromatographic system consisted of a piston pump with a pulse dampener, a rotary injection port with a 20-μL loop, a variable-wavelength spectrometric detector, and an integrator. The stationary phase was a cyano reversed-phase column (250 × 4.6 mm, 5-μm particle size). The mobile phase was a mixture of acetonitrile and 0.02 M sodium acetate trihydrate (26:74, vol/vol) adjusted to pH 4.0 with acetic acid. The flow rate was 1.50 mL/min. UV detection was performed at 300 nm.

Samples were diluted with the mobile phase. The retention times for streptozocin and granisetron were 2.01 and 5.59 minutes, respectively.

The stability-indicating capability of the analytical method was demonstrated by an accelerated degradation study. Solutions of streptozocin and granisetron hydrochloride were adjusted to pH 2 and 11, boiled for 1 hour, readjusted to pH 5, diluted with the mobile phase, and analyzed. Degradation product peaks did not interfere with the intact streptozocin or granisetron peaks.

Reference

Mayron D, Gennaro AR. Stability and compatibility of granisetron hydrochloride in i.v. solutions and oral liquids and during simulated Y-site injection with selected drugs. *Am J Health Syst Pharm.* 1996; 53: 294–304.

Succinic Acid

Chemical Name
Butanedioic acid

Other Name
Amber Acid

Form	Molecular Formula	MW	CAS
Succinic acid	$C_4H_6O_4$	118.1	110-15-6

Appearance
Succinic acid occurs as odorless monoclinic prisms.

Solubility
Succinic acid is soluble in water and in alcohol.

Method

Thoma and Ziegler determined both succinic acid and fenoldopam in dissolution media by HPLC analysis. The liquid chromatograph consisted of a Waters model 6000A pump, a TPS autosampler, and a TPS fast-scanning UV detector. The stationary phase was a Nucleosil ODS C_{18} column (250 × 5.4 mm, 5-μm particle size). The mobile phase consisted of methanol and 0.025 M monobasic sodium phosphate buffer (pH 2.5) (20:80, vol/vol). The flow rate was 1 mL/min. Succinic acid was monitored at 205 nm.

Pellet samples were ground, dispersed in 0.1 N hydrochloric acid, and filtered. The injection volume was 20 μL. Under these conditions, retention times of succinic acid and fenoldopam were 3.2 and 8.5 minutes, respectively.

The method was reported to be stability indicating.

A calibration curve for succinic acid was generated from 0.05 to 1.35 mg/mL. The correlation coefficient was 0.9993.

Reference

Thoma K, Ziegler I. Simultaneous quantification of released succinic acid and a weakly basic drug compound in dissolution media. *Eur J Pharm Biopharm.* 1998; 46: 183–90.

Succinylcholine Chloride

Chemical Names
2,2′-[(1,4-Dioxo-1,4-butanediyl)bis(oxy)]bis[N,N,N-trimethylethanaminium] dichloride

Bis[2-dimethylaminoethyl]succinate bis[methochloride]

Other Names
Anectine, Suxamethonium Chloride

Form	Molecular Formula	MW	CAS
Succinylcholine chloride	$C_{14}H_{30}Cl_2N_2O_4$	361.3	71-27-2

Appearance
Succinylcholine chloride is a white odorless crystalline powder.

Solubility
Succinylcholine chloride has solubilities of approximately 1 g/mL in water and 2.9 mg/mL in alcohol at 25 °C.

Method
Schmutz and Muhlebach determined the stability of succinylcholine chloride injection. The analysis used a Perkin-Elmer series 3B system with a Perkin-Elmer model LC-75 UV detector. The stationary phase was a Perkin-Elmer silica A column (250 × 2.6 mm). The mobile phase was a mixture of 0.5 M tetramethylammonium sulfate, water, and methanol (2:5:13, vol/vol/vol) that was delivered isocratically at 1 mL/min. UV detection was performed at 214 nm. The injection volume was 10 μL.

To ensure the stability-indicating nature of the assay, succinylcholine chloride was incubated with 0.1 M sodium hydroxide for 5 and for 10 minutes. The degradation product peaks did not interfere with the intact succinylcholine chloride peak. Retention times for methyl 4-hydroxybenzoate, succinic acid, succinylmonocholine, choline, and succinylcholine were 1.05, 1.35, 2.1, 2.9, and 5.1 minutes, respectively.

A calibration curve for succinylcholine chloride was constructed from 5.0 to 12.5 mg/mL. The correlation coefficient was 0.997.

Reference
Schmutz CW, Muhlebach SF. Stability of succinylcholine chloride injection. *Am J Hosp Pharm.* 1991; 48: 501–6.

Sufentanil

Chemical Name
N-[4-(Methoxymethyl)-1-[2-(thienyl)ethyl]-4-piperidinyl]-N-phenylpropanamide

Other Name
Sufenta

Form

Form	Molecular Formula	MW	CAS
Sufentanil	$C_{22}H_{30}N_2O_2S$	386.6	56030-54-7
Sufentanil citrate	$C_{22}H_{30}N_2O_2S.C_6H_8O_7$	578.7	60561-17-3

Appearance
Sufentanil citrate is a white crystalline powder.

Solubility
Sufentanil citrate has solubilities of 46 and 21 mg/mL in water and in alcohol, respectively.

pK$_a$
Sufentanil has a pK$_a$ of 8.01.

Method
Roos et al. investigated the stability of sufentanil citrate in a portable pump reservoir, a glass container, and a polyethylene container. The HPLC system consisted of a Pye Unicam PU 4010 pump with a fixed loop of 50 µL, a Pye Unicam PU 4020 variable-wavelength UV detector, and a Hitachi D-2000 chromato-integrator. The stationary phase was a Chrompack Chromspher C_{18} column (100 × 3.0 mm). The mobile phase consisted of 36.4% methanol, 36.4% acetonitrile, and 27.2% demineralized water containing 0.5% of ammonium acetate. The flow rate was 0.6 mL/min. UV detection was performed at 225 nm. The injection volume was 300 µL. The retention time for sufentanil citrate was 3.0 ± 0.2 minutes.

The HPLC method was demonstrated to be stability indicating by accelerated degradation of sufentanil citrate. Sufentanil citrate (1 mL of a 5-µg/mL solution) was stored for 14 days at 65 °C in a portable pump reservoir. The degradation product peaks did not interfere with the intact sufentanil citrate peak.

Calibration curves were constructed for sufentanil citrate each day from 1 to 5 µg/mL. The correlation coefficient for the curves was greater than 0.99. The interday coefficient of variation was 1.8 ± 0.2%.

References

Roos PJ, Glerum JH, Meilink JW. Stability of sufentanil citrate in a portable pump reservoir, a glass container and a polyethylene container. *Pharm Weekbl [Sci]*. 1992; 14: 196–200.

Roos PJ, Glerum JH, Schroeders MJH. Effect of glucose 5% solution and bupivacaine hydrochloride on absorption of sufentanil citrate in a portable pump reservoir during storage and simulated infusion by an epidural catheter. *Pharm World Sci*. 1993; 15: 269–74.

Sulbactam

Chemical Names

(2S-*cis*)-3,3-Dimethyl-7-oxo-4-thia-1-azabicyclo[3.2.0]heptane-2-carboxylic acid 4,4-dioxide

(2S,5R)-3,3-Dimethyl-7-oxo-4-thia-1-azabicyclo[3.2.0]heptane-2-carboxylic acid 4,4-dioxide

Other Name

Component of Unasyn

Form	Molecular Formula	MW	CAS
Sulbactam	$C_8H_{11}NO_5S$	233.2	68373-14-8
Sulbactam sodium	$C_8H_{10}NNaO_5S$	255.2	69388-84-7

Appearance

Sulbactam is a white to off-white powder.

Solubility

Sulbactam is freely soluble in aqueous diluents; it is sparingly soluble in acetone, chloroform, and ethyl acetate.

Method

Using HPLC, Belliveau et al. evaluated the stability of sulbactam sodium with aztreonam and ampicillin sodium in 0.9% sodium chloride injection. The system consisted of a Waters model 510 pump, a Waters model 710A WISP autosampler, an LDC analytical model SM 4000 variable-wavelength UV detector, and a Hewlett-Packard model 3396 series II integrator. The stationary phase was a Waters µBondapak C_{18} column (300 × 3.9 mm). The mobile phase consisted of 0.005 M tetrabutylammonium hydrogen sulfate, 7% acetonitrile, and 14% methanol in a 0.010 M phosphate buffer (pH 2.6–2.7); it was delivered isocratically at 1 mL/min. UV detection was performed at 225 nm. Cimetidine was used as an internal standard.

Samples were diluted 1:8 with water. One hundred microliters of this solution was mixed with 100 µL of internal standard solution and 200 µL of water. The injection volume was 20 µL. Under these conditions, the retention times for cimetidine, ampicillin, sulbactam, and aztreonam were 3.27, 4.35, 5.79, and 7.05 minutes, respectively.

The stability-indicating nature of the method was demonstrated by accelerated decomposition of sulbactam. A solution of sulbactam 2 mg/mL was stored in a water bath (70 °C) for 8 days. Peaks associated with the degradation products increased in height and area over 8 days. However, these peaks did not interfere with the intact sulbactam peak.

Standard curves for sulbactam were constructed from 0.5 to 2 mg/mL. The correlation coefficients were greater than 0.9977. Intraday and interday coefficients of variation were less than 7%.

Reference
Belliveau PP, Nightingale CH, Quintiliani R. Stability of aztreonam and ampicillin sodium–sulbactam sodium in 0.9% sodium chloride injection. *Am J Hosp Pharm.* 1994; 51: 901–4.

Sulfamethoxazole

Chemical Names
4-Amino-*N*-(5-methyl-3-isoxazolyl)benzenesulfonamide
N[1]-(5-Methyl-3-isoxazolyl)sulfanilamide

Other Names
Component of Bactrim, Sulphamethoxazole

Form	Molecular Formula	MW	CAS
Sulfamethoxazole	$C_{10}H_{11}N_3O_3S$	253.3	723-46-6

Appearance
Sulfamethoxazole is a white or off-white crystalline powder.

Solubility

Sulfamethoxazole is soluble 1 in 3400 of water, 1 in 50 of alcohol, 1 in 3 of acetone, and 1 in 1000 of chloroform and ether. It is soluble in dilute solutions of alkali hydroxide.

Method 1

Kaufman et al. determined the stability of undiluted trimethoprim–sulfamethoxazole in single-use plastic syringes using an HPLC analysis. A Waters chromatograph consisted of a solvent-delivery system, a model 440 UV detector, a data processor, and a model 710 B WISP automatic sampling system. The stationary phase was a Waters µBondapak column in a Z-module radial-compression separation system with a Supelco guard column with an LC-18-DB 5-µm cartridge. The mobile phase consisted of 15% acetonitrile and 1% acetic acid in water. The flow rate was 1.7 mL/min. UV detection was performed at 240 nm and 0.1 AUFS. Phenacetin 0.4 mg/mL was used as the internal standard. Under these conditions, retention times for trimethoprim, sulfamethoxazole, and phenacetin were 4.20, 9.74, and 12.45 minutes, respectively.

The assay was verified to be stability indicating by an accelerated degradation study. Concentrated hydrochloric acid was added to a sample of sulfamethoxazole and the solution was heated for 4 hours at 90 °C. After the addition of 3% hydrogen peroxide, the solution was heated for an additional 1 hour at 100 °C. The peak of sulfamethoxazole at 9.74 minutes disappeared and was replaced by peaks at 11.1 and 11.7 minutes.

A standard curve for sulfamethoxazole was constructed from 16 to 160 mg/mL. Intraday and interday coefficients of variation for the assay were less than 3%.

Reference

Kaufman MB, Scavone JM, Foley JJ. Stability of undiluted trimethoprim–sulfamethoxazole for injection in plastic syringes. *Am J Hosp Pharm*. 1992; 49: 2782–3.

Method 2

Holmes and Aldous reported the stability of sulfamethoxazole and trimethoprim in peritoneal dialysis fluid using an HPLC method. The liquid chromatograph consisted of a Waters model M45 solvent-delivery system, a Waters model 441 UV detector, and a Waters µBondapak C_{18} stainless steel column (300 × 3.9 mm) with a µBondapak C_{18}/Corasil 37–50-µm guard column. The mobile phase consisted of acetonitrile and 0.067 M phosphate buffer (pH 6.2) (1:5, vol/vol) delivered isocratically at 1 mL/min. UV detection was performed at 229 nm. Sulfamethazine (sulfadimidine) 52 mg/L in methanol was used as an internal standard.

Samples were diluted with an equal volume of internal standard solution. The injection volume was 20 µL.

To determine the stability-indicating capability of the assay, cotrimoxazole solutions (20 mg/L of trimethoprim and 100 mg/L of sulfamethoxazole) were autoclaved for 1 hour at 118 °C at pH 2, 6, and 11. Decomposition product peaks did not interfere with trimethoprim and sulfamethoxazole peaks. Retention times for the decomposition product, internal standard, trimethoprim, and sulfamethoxazole were about 7.0, 14.0, 16.5, and 19.0 minutes, respectively (estimated from the published chromatogram).

Intraday and interday coefficients of variation for the assay were 2.5 and 1.1%, respectively.

Reference

Holmes SE, Aldous S. Stability of cotrimoxazole in peritoneal dialysis fluid. *Peritoneal Dialysis Int.* 1990; 10: 157–60.

Method 3

Jarosinski et al. evaluated the stability of trimethoprim–sulfamethoxazole admixtures at various concentrations in 5% dextrose injection or in 0.9% sodium chloride injection. The liquid chromatograph consisted of a Perkin-Elmer series 10 pump, a Perkin-Elmer LC-90 detector, a Perkin-Elmer LCI-100 integrator, and a Rheodyne 7125 injector. The stationary phase was an Altex ODS column (250 × 4.6 mm, 5-μm particle size). The mobile phase consisted of 70 parts of 5 mM phosphate and 0.1% triethylamine with pH adjusted to 5 with phosphoric acid and 30 parts of acetonitrile. The flow rate was 1 mL/min. UV detection was performed at 286 nm.

Samples were diluted 1:100 with distilled-deionized water.

The analytical method was determined to be stability indicating by accelerated degradation of sulfamethoxazole. The formulation was heated with acid or base. Degradation product peaks did not interfere with the sulfamethoxazole peak.

A standard curve was generated for sulfamethoxazole from 0.80 to 3.20 mg/mL. The correlation coefficient of the standard curve was greater than 0.999. Coefficients of variation were 0.15% at 3.20 mg/mL and 4.7% at 0.80 mg/mL.

Reference

Jarosinski PF, Kennedy PE, Gallelli JF. Stability of concentrated trimethoprim–sulfamethoxazole admixtures. *Am J Hosp Pharm.* 1989; 46: 732–7.

Sulindac

Chemical Name

(Z)-[5-Fluoro-2-methyl-1-(4-methylsulfinylbenzylidene)inden-3-yl]acetic acid

Other Names

Clinoril, Saldac

Form	Molecular Formula	MW	CAS
Sulindac	$C_{20}H_{17}FO_3S$	356.4	38194-50-2

Appearance
Sulindac occurs as a yellow odorless or almost odorless, polymorphic crystalline powder.

Solubility
Sulindac is practically insoluble in water and slightly soluble in alcohol.

pK$_a$
Sulindac has an apparent pK$_a$ of 4.7.

Method
Jalal et al. developed an assay for the determination of sulindac in tablet formulations. The Beckman system included a model 112 solvent-delivery module, a model 165 variable-wavelength detector, a 10-µL fixed loop injector, and a model C-RIB integrator. The stationary phase was a Beckman Ultrasphere XL ODS column (70 × 4.6 mm, 3-µm particle size). The mobile phase consisted of 0.05 M ammonium acetate (adjusted to pH 6.0 using glacial acetic acid) and methanol (50:50, vol/vol). The flow rate was 1.0 mL/min. UV detection was performed at 280 nm and 0.1 AUFS. Propylparaben 0.2 mg/mL in methanol was used as an internal standard.

A portion of powdered tablets was accurately weighed, transferred to a 100-mL volumetric flask, brought to volume with methanol, stirred for 5 minutes, and diluted by a factor of 25 with methanol after the internal standard solution had been added. The sample injection volume was 10 µL. Under these conditions, retention times for sulindac and propylparaben were about 3.5 and 7.8 minutes, respectively.

The assay was determined to be stability indicating by a decomposition study of sulindac. The drug was refluxed in 0.1 N sodium hydroxide solution for 7 hours. Degradation products were well separated from the intact sulindac on its chromatogram.

A standard curve for sulindac was constructed from 0.04 to 0.12 mg/mL. The correlation coefficient was 0.9998.

Reference
Jalal IM, Khalil HS, Jawhari D. Stability-indicating assay for sulindac in tablet formulations by reverse-phase HPLC. *J Liq Chromatogr*. 1989; 12: 3087–101.

Sumatriptan

Chemical Name
3-[2-(Dimethylamino)ethyl]-*N*-methyl-1*H*-indole-5-methanesulfonamide

Other Name
Imitrex

Form	Molecular Formula	MW	CAS
Sumatriptan	$C_{14}H_{21}N_3O_2S$	295.4	103628-46-2
Sumatriptan succinate	$C_{14}H_{21}N_3O_2S.C_4H_6O_4$	413.5	103628-48-4

Appearance
Sumatriptan is a white to off-white powder.

Solubility
Sumatriptan is freely soluble in water and in 0.9% sodium chloride.

pK_a
Sumatriptan has pK_a values of 4.21 and 5.67 (succinic acid), 9.63 (tertiary amine group), and 12 or greater (sulfonamide group).

Method 1
Fish et al. evaluated the stability of sumatriptan succinate in extemporaneously prepared oral liquids. The liquid chromatograph consisted of a Waters model 501 pump, a Spectra-Physics SP8880 autosampler, an Applied Biosystems 759A variable-wavelength UV detector, a Waters model 746 data module, and an Alltech 2000 solvent recycler. The stationary phase was a Phase Separations Spherisorb ODS-1 stainless steel analytical column (250 mm, 5-μm particle size) with an Applied Biosystems guard column. The mobile phase consisted of 0.01 M dibutylamine in 0.025 M aqueous dihydrogen phosphate and acetonitrile (75:25, vol/vol). The final pH of the mobile phase was adjusted to 8.0 with 1 N sodium hydroxide. The flow rate was 1.5 mL/min. UV detection was performed at 282 nm. N-Hydroxymethyl-sumatriptan succinate was used as an internal standard.

Samples were diluted 1:23.8 with 0.1 M hydrochloric acid and filtered through a 0.22-μm nylon syringe filter. The injection volume was 10 μL.

The method was shown to be stability indicating by accelerated decomposition of sumatriptan succinate. Sumatriptan succinate solutions 5 mg/mL were exposed to 0.1 N sodium hydroxide and 0.1 N hydrochloric acid at 100 °C for 8 hours or exposed to 1% hydrogen peroxide at 100 °C for 2 hours. In all cases, the degradation product peaks were well separated from peaks for both intact sumatriptan and the internal standard. Retention times for the degradation products, for N-hydroxymethyl-sumatriptan, and sumatriptan were 5–8, 11, and 14 minutes, respectively.

A standard curve for sumatriptan was constructed from 0 to 8.33 mg/mL. The correlation coefficient was 0.999. The intraday and interday coefficients of variation were 0.6 and 2.2%, respectively.

Reference
Fish DN, Beall HD, Goodwin SD, et al. Stability of sumatriptan succinate in extemporaneously prepared oral liquids. *Am J Health Syst Pharm.* 1997; 54: 1619–22.

Method 2
Singh and Jain developed an HPLC method for the determination of sumatriptan succinate in pharmaceutical preparations. The system consisted of a Shimadzu LC10AD pump, an SCL10AD system controller, an SPD-6MA photodiode-array detector, and an SCL-10A autoinjector. The stationary phase was a Nova-Pak C_{18} column (150 × 3.9 mm,

4-μm particle size). The mobile phase consisted of 0.0025 M phosphoric acid (adjusted to pH 3.0 with triethylamine) and acetonitrile (95:5, vol/vol). The flow rate was 1 mL/min. UV detection was performed at 223 nm. Paracetamol (acetaminophen) 1 mg/mL in the mobile phase was used as an internal standard.

For raw material and injections, samples were diluted with the mobile phase. For tablets, a sample of ground sumatriptan tablets was accurately weighed, mixed into to the mobile phase, and filtered. The filtrate was diluted with the mobile phase. The injection volume was 10 μL. Under these conditions, retention times for paracetamol and sumatriptan were about 3.5 and 6.2 minutes, respectively (estimated from the published chromatogram).

The stability-indicating capacity of the method was demonstrated by intentional degradation of sumatriptan. A sample of pure drug was heated at 105 °C for 12 hours. The intact drug was separated from its degradation products.

A calibration curve was obtained from 200 to 1600 ng. The correlation coefficient was 0.999. The limit of detection was 10 ng.

Reference
Singh S, Jain R. Stability indicating HPLC method for the determination of sumatriptan succinate in pharmaceutical preparations and its application in dissolution rate studies. *Indian Drugs*. 1997; 34: 527–31.

Suprofen

Chemical Name
2-[4-(2-Thenoyl)phenyl]propionic acid

Other Name
Profenal

Form	Molecular Formula	MW	CAS
Suprofen	$C_{14}H_{12}O_3S$	260.3	40828-46-4

Appearance
Suprofen occurs as a white to off-white and odorless or almost odorless powder.

Solubility
Suprofen is sparingly soluble in water.

Method

Lagu et al. described an HPLC method for suprofen in the drug substance and in capsules. The Waters model ALC 202 system consisted of a model M6000 pump, a model 440 UV detector, a model WISP 710A autosampler, a model U6K septumless injector or a Rheodyne model 7125 fixed-loop injector, and an Omniscribe model B5117-IX recorder. The stationary phase was a Waters μBondapak C_{18} column (300 × 3.9 mm). The mobile phase was a mixture of 0.01 M citric acid monohydrate buffer solution (adjusted to pH 3.0 with 0.02 M dibasic sodium phosphate heptahydrate) and acetonitrile (2:3, vol/vol). The flow rate was 0.5 mL/min. UV detection was performed at 254 nm. 4-Nitrobenzoic acid 1.2 mg/mL in methanol was used as an internal standard.

A sample equivalent to 220 mg of suprofen was weighed and placed into a 50-mL volumetric flask, mixed with 25 ml of methanol by shaking for 15 minutes, brought to volume with methanol, filtered, and diluted 1:2 with the internal standard. The injection volume was 2 μL. Under these conditions, retention times of 4-nitrobenzoic acid and suprofen were about 7.8 and 9.9 minutes (estimated from the published chromatogram).

The stability-indicating nature of the method was demonstrated by assaying a synthetic mixture of intermediates, impurities, and suprofen degradation products. Suprofen substance was also degraded by heating at its melting point of 124 °C for 1 hour in an oil bath, dispersing in 1 N hydrochloric acid at 50 °C for 72 hours, dispersing in 1 N sodium hydroxide solution at 50 °C for 72 hours, or dispersing in 3% hydrogen peroxide at 50 °C for 72 hours. Suprofen was separated from the following compounds (relative retention times in parentheses): α-hydroxy-α-methyl-4-(2-thienylcarbonyl)-benzeneacetic acid (0.78), 4-carboxy-α-methylbenzeneacetic acid (0.71), methyl-α-methyl-4-(2-thienylcarbonyl)benzeneacetic acid (1.54), ethyl-4-(2-thienylcarbonyl)-benzene (2.12), 4-acetyl-α-methylbenzeneacetic acid (0.81), α-methyl-4-(2-thienylcarbonyl)benzene acetonitrile (1.34), (1-bromoethyl)-4-(2-thienylcarbonyl)benzene (2.11), diethyl-2-methyl-2-[4-(2-thienylcarbonyl)phenyl]-1,3-propanedioate (2.10), 4-fluorophenyl-2-thienylmethanone (1.49), α-methyl-4-(2-thienylcarbonyl)benzene acetamide (0.86), 2-fluorophenyl-2-thienyl-methanone (1.39), 3-fluorophenyl-2-thienylmethanone (1.53), α-methyl-4-(5-chloro-2-thienylcarbonyl)benzeneacetic acid (1.36), 2-[4-(2-thienylcarbonyl)phenoxy]propanoic acid (0.91), methyl-4-(2-thienylcarbonyl)-benzene (1.71), and 4,4-carbonylbis(α-methyl-benzeneacetic acid) (0.84).

Standard curves for suprofen were generated from 0.0997 to 2.00 μg/mL with the correlation coefficient of 0.9998 by using peak height ratios and from 0.200 to 1.50 μg/mL with the correlation coefficient of 0.9999 by using peak area ratios.

Reference

Lagu AL, Young R, McGonigle E, et al. High-performance liquid chromatographic determination of suprofen in drug substance and capsules. *J Pharm Sci.* 1982; 71: 85–7.

Suramin Sodium

Chemical Name
8,8'-[Carbonylbis[imino-3,1-phenylenecarbonylimino(4-methyl-3,1-phenylene)carbonyl-imino]]bis-1,3,5-naphthalenetrisulfonic acid hexasodium salt

Other Name
Bayer 205

Form	Molecular Formula	MW	CAS
Suramin sodium	$C_{51}H_{34}N_6Na_6O_{23}S_6$	1429.2	129-46-4

Appearance
Suramin sodium occurs as a white, pinkish-white, or slightly cream-colored and odorless or almost odorless hygroscopic powder.

Solubility
Suramin sodium is freely soluble in water. It is very slightly soluble in alcohol and is practically insoluble in chloroform and ether.

Method 1
Kettenes-van den Bosch et al. investigated the stability of suramin in aqueous solutions by HPLC analysis. The Waters liquid chromatograph consisted of a model 6000A pump, a model 680 gradient controller, a model 441 UV detector or a Hewlett-Packard model 1040A photodiode-array detector, and a Spectra-Physics model SP4270 integrator. The stationary phase was a Merck LiChrospher 100 RP8 column (125 × 4 mm, 5-µm particle size). The mobile phase A was a mixture of methanol, acetonitrile, and aqueous buffer (10:2:88, wt/wt/wt). The mobile phase B was a mixture of methanol, acetonitrile, and aqueous buffer (46:5:49, wt/wt/wt). The aqueous buffer consisted of 0.01 M triethylamine (adjusted to pH 6.5 with trifluoroacetic acid and 0.01 M ammonium acetate). The mobile phase was delivered at 100% A from 0 to 2 minutes, linearly changed from 100% A to 100% B from 2 to 8 minutes, held at 100% B from 8 to 12 minutes, and returned to 100% A from 12 to 13 minutes. UV detection was performed at 254 nm.

Samples were diluted 1:10 with phosphate buffer (pH 7). Under these conditions, the retention time for suramin was about 9.2 minutes (estimated from the published chromatogram).

The method was reported to be stability indicating. Suramin solutions at pH 1, 7, and 13 were placed in a water bath at 80 °C for 20 hours. Degradation products did not coelute with suramin.

A standard curve for suramin was obtained from 1×10^{-7} to 1×10^{-4} M. The correlation coefficient was 0.9999. The limit of detection was 0.6 ng.

Reference
Kettenes-van den Bosch JJ, Overbeek EHW, Underberg WJM, et al. Degradation of suramin in aqueous solution. *Int J Pharm*. 1997; 155: 27–34.

Method 2

Beijnen et al. studied the chemical stability of suramin sodium in commonly used infusion fluids at a concentration of approximately 1 mg/mL. The liquid chromatograph included a Waters model 510 solvent-delivery system, a Waters model U6K injector, a Waters model 441 UV detector, and a Spectra-Physics SP4290 integrator. The stationary phase was a Chromapak C_{18} cartridge column (100 × 3.0 mm, 5-μm particle size). The mobile phase consisted of methanol (460 mL), acetonitrile (50 mL), and 0.05 M phosphate buffer (pH 6.5) containing 0.005 M tetrabutylammonium phosphate reagent (490 mL). The flow rate was 0.8 mL/min. UV detection was performed at 313 nm and 0.2 AUFS. The sample injection volume was 5 μL.

The stability-indicating nature of the analytical method was demonstrated by accelerated degradation of suramin sodium. A suramin sodium solution 1 mg/mL in 0.9% sodium chloride injection was incubated for 4 hours at 90 °C. The retention times for the degradation product and suramin were 2.00 and 6.36 minutes, respectively.

Calibration curves of suramin sodium in water were constructed from 0.1 to 2.0 mg/mL. The correlation coefficients of the calibration curves were greater than 0.9999.

Reference

Beijnen JH, Gijn RV, Horenblas S, et al. Chemical stability of suramin in commonly used infusion fluids. *Ann Pharmacother*. 1990; 24: 1056–8.

Tacrine Hydrochloride

Chemical Name

1,2,3,4-Tetrahydroacridin-9-ylamine hydrochloride

Other Name

Cognex

Form	Molecular Formula	MW	CAS
Tacrine hydrochloride	$C_{13}H_{14}N_2 \cdot HCl$	234.7	1684-40-8

Appearance

Tacrine hydrochloride occurs as a white solid.

Solubility

Tacrine hydrochloride is freely soluble in distilled water, ethanol, methanol, and 0.1 N hydrochloric acid.

Method

Lau et al. reported the simultaneous determination of tacrine and physostigmine in skin samples. An LDC ConstaMetric I chromatograph was equipped with an LDC UV-III fixed-wavelength detector, a Valco injection valve with a 50-μL loop, and a Linear Instrument recorder. The stationary phase was a Custom Spherisorb ODS-1 column (150 × 4.6 mm, 5-μm particle size) with an ODS guard column (20 × 4 mm, 40-μm particle size). The mobile phase consisted of acetonitrile and water (52:48, vol/vol) (pH 3.5) containing 0.01 M octanesulfonic acid and 1% (vol/vol) acetic acid. The flow rate was 1 mL/min. UV detection was performed at 254 nm and 0.008 AUFS. Diazepam was used as an internal standard.

A 150–200-mg skin sample was homogenized with two 4-mL portions of chloroform, filtered through phase separation paper, alkalinized with 2 mL of 10% (wt/vol) sodium hydroxide solution, washed with two 2-mL portions of water, filtered, dried under a stream of air, reconstituted with 1 mL of mobile phase, and filtered through a microfilter. The injection volume was 50 μL. Under these conditions, retention times of physostigmine, diazepam, and tacrine were 4.5, 6.0, and 7.7 minutes, respectively.

The method was demonstrated to be stability indicating by accelerated degradation of drugs. Sample solutions were boiled for 0.5 hour or were adjusted to pH less than 1 or pH greater than 13 and boiled for 1.25 hours. The intact drugs were separated from their degradation products.

A standard curve for tacrine was generated from 1 to 50 μg/200 mg of skin sample. The correlation coefficient was greater than 0.998. Intraday and interday coefficients of variation were 5.6 and 4.2%, respectively. The limit of detection was 1 μg/200 mg of skin sample.

Reference

Lau SWJ, Chow D, Feldman S. Simultaneous determination of physostigmine and tetrahydroaminoacridine in a transdermal permeation study by high-performance liquid chromatography. *J Chromatogr*. 1990; 526: 87–95.

Tacrolimus

Chemical Name

[3*S*–[3*R**,[*E*(1*S**,3*S**,4*S**)],4*S**,5*R**,8*S**,9*E*,12*R**,14*R**,15*S**,16*R**,18*S**,19*S**,26a*R**]]-5,6,8,11,12,13,14,15,16,17,18,19,24,25,26,26a-Hexadecahydro-5,19-dihydroxy-3-[2-(4-hydroxy-3-methoxycyclohexyl)-1-methylethenyl]-14,16-dimethoxy-4,10,12,18-tetramethyl-8-(2-propenyl)-15,19-epoxy-3*H*-pyrido[2,1-*c*][1,4]oxaazacyclotricosine-1,7,20,21(4*H*,23*H*)-tetrone

Other Name
Prograf

Form	Molecular Formula	MW	CAS
Tacrolimus	$C_{44}H_{69}NO_{12}$	804.0	104987-11-3

Appearance
Tacrolimus occurs as colorless prisms.

Solubility
Tacrolimus is soluble in methanol, ethanol, acetone, ethyl acetate, chloroform, and diethyl ether and is sparingly soluble in hexane. It is insoluble in water.

Method
Jacobson et al. determined the short-term stability of tacrolimus in an extemporaneously prepared oral liquid dosage formulation from commercially available capsules. The system consisted of a Waters model 501 pump, a Waters model U6K variable-volume injector, a Waters model 486 tunable UV detector, a Hewlett-Packard model 3394 integrator-recorder, an Eppendorf CH-30 column heater, and an Applied Biosystems Spheri-5 ODS column (250 × 4.6 mm). The mobile phase was 65% acetonitrile in deionized distilled water and was delivered isocratically at 1.7 mL/min. UV detection was performed at 214 nm.

Samples were diluted with the mobile phase to a tacrolimus concentration of 50 μg/mL. The injection volume was 10 μL. The retention time for tacrolimus was about 6.7 minutes (estimated from the published chromatogram).

The method was demonstrated to be stability indicating. Two tacrolimus sample solutions 500 μg/mL were allowed to stand in direct sunlight for 45 days after adjustment to either pH 12 with 1 N sodium hydroxide or pH 2 with 1 N sulfuric acid. On day 45 these two samples were heated to 75 °C for 3 hours. The resulting solutions were assayed by HPLC after appropriate treatment. Degradation product peaks did not interfere with the tacrolimus peak. Retention times for degradation products were about 1.2, 2.0, and 12.5 minutes (estimated from the published chromatogram).

A standard curve was constructed on each day by linear regression of the peak heights against standard concentrations. The standard curve was linear ($r > 0.999$)

from 40 to 60 µg/mL. The intraday and interday coefficients of variation were less than 2 and 2.5%, respectively.

Reference
Jacobson PJ, Johnson CE, West NJ, et al. Stability of tacrolimus in an extemporaneously compounded oral liquid. *Am J Health Syst Pharm.* 1997; 54: 178–80.

Tamoxifen

Chemical Names
(Z)-2-[4-(1,2-Diphenyl-1-butenyl)phenoxy]-*N,N*-dimethylethanamine
1-*p*-β-Dimethylaminoethoxyphenyl-*trans*-1,2-diphenylbut-1-ene

Other Name
Nolvadex

Form	Molecular Formula	MW	CAS
Tamoxifen	$C_{26}H_{29}NO$	371.5	10540-29-1
Tamoxifen citrate	$C_{26}H_{29}NO.C_6H_8O_7$	563.6	54965-24-1

Appearance
Tamoxifen citrate is a fine white crystalline powder.

Solubility
Tamoxifen citrate has a solubility of 0.5 mg/mL in water at 37 °C; it is very slightly soluble in alcohol.

pK$_a$
Tamoxifen has a pK$_a$ of 8.85.

Method
Jalonen developed a stability-indicating HPLC method to simultaneously assay tamoxifen citrate and its *E*-isomer citrate impurity in both bulk drug and in tablets. The chromatograph consisted of a Kontron model 414 constant-flow pump, a Kontron model MSI 660 fixed-wavelength UV detector equipped with a filter of 280 nm, a Kontron model 660 automatic sampling system, and a Kontron model Anacomp 220 data system for the recording of the chromatogram and for the calculation. The stationary

phase was a Shandon Hypersil ODS stainless steel column (250 × 4.6 mm, 5-μm particle size). The mobile phase was 40% acetonitrile in an aqueous phosphate buffer (0.01 M monobasic sodium phosphate and 0.05 M *N,N*-dimethyloctylamine) adjusted to pH 2.0 with phosphoric acid and then filtered through a Millipore 0.45-μm Durapore filter. The flow rate was 1.2 mL/min.

The HPLC assay was demonstrated to be stability indicating. Holding the drug for 24 hours in a 3% hydrogen peroxide aqueous methanolic solution generated a degradation product with the same retention time as tamoxifen *N*-oxide. UV irradiation (254 nm) of an aqueous methanolic solution of tamoxifen citrate for 5 days decreased the intact drug to 90% and increased the *E*-isomer to 7.5%. Degradation product peaks did not interfere with the intact drug peak. Relative retention times for tamoxifen alcohol, *E*-isomer citrate, desmethyl-tamoxifen, the unknown decomposition product, tamoxifen citrate, and tamoxifen *N*-oxide were 0.19, 0.79, 0.82, 0.91, 1.00, and 1.49, respectively.

A calibration curve was constructed from 0 to 2.28 mg/mL. The correlation coefficient was 0.9996.

Reference

Jalonen HG. Simultaneous determination of tamoxifen citrate and its *E* isomer impurity in bulk drug and tablets by high-performance liquid chromatography. *J Pharm Sci.* 1988; 77: 810–3.

Tazobactam

Chemical Name

[2*S*-(2α,3β,5α)]-3-Methyl-7-oxo-3-(1*H*-1,2,3-triazol-1-ylmethyl)-4-thia-1-azabicyclo-[3.2.0]heptane-2-carboxylic acid 4,4-dioxide

Other Name

Component of Tazocin and Zosyn

Form	Molecular Formula	MW	CAS
Tazobactam	$C_{10}H_{12}N_4O_5S$	300.3	89786-04-9
Tazobactam sodium	$C_{10}H_{11}N_4NaO_5S$	322.3	89785-84-2

Method

Moon et al. studied the stability of piperacillin sodium–tazobactam sodium in polypropylene syringes and polyvinyl chloride minibags. The liquid chromatograph consisted of a Hitachi model L-6200 pump, a model L-4200 UV-visible detector, a model AS2000 autosampler, and a model D-2500 integrator. The stationary phase was an Alltech Adsorbosphere C_{18} column (250 × 4.6 mm, 5-μm particle size). The mobile phase consisted of 0.01 M sodium phosphate and acetonitrile (93:7, vol/vol) adjusted with 85% phosphoric acid to pH 2.7. The flow rate was 1.2 mL/min. UV detection was performed at 215 nm.

Samples were diluted 1:40 with the mobile phase. The injection volume was 20 μL. Under these conditions, the retention time for tazobactam was 6.40 minutes.

The assay was demonstrated to be stability indicating by accelerated decomposition of tazobactam. Samples of tazobactam 5 mg/mL were exposed to 1 N sodium hydroxide at 58 °C for 4 hours, to 1 N hydrochloric acid at 58 °C for 4 hours, and to 1% hydrogen peroxide at room temperature for 4 hours. The degradation product peak (about 2 minutes) did not interfere with the tazobactam peak.

A standard curve for tazobactam was constructed from 200 to 600 μg/mL; the correlation coefficient was greater than 0.998. The intraday and interday coefficients of variation were less than 5%.

References
Moon YSK, Chung KC, Chin A, et al. Stability of piperacillin sodium–tazobactam sodium in polypropylene syringes and polyvinyl chloride minibags. *Am J Health Syst Pharm.* 1995; 52: 999–1000.

Chung KC, Moon YSK, Chin A, et al. Compatibility of ondansetron hydrochloride and piperacillin sodium–tazobactam sodium during simulated Y-site administration. *Am J Health Syst Pharm.* 1995; 52: 1554–6.

Tedisamil

Chemical Name
3′,7′-Bis(cyclopropylmethyl)spiro[cyclopentane-1,9′-[3,7]diazabicyclo[3.3.1]nonane]

Form	Molecular Formula	MW	CAS
Tedisamil	$C_{19}H_{32}N_2$	288.5	90961-53-8

Appearance
Tedisamil occurs as white crystals.

Solubility
Tedisamil is soluble in water.

Method
Korner described an HPLC method with amperometric detection for tedisamil. The Merck-Hitachi HPLC system consisted of a model L6200A pump, a model AS 2000A autosampler with a 10-μL loop, and a BAS LC-4C amperometric detector. The stationary phase was an Agilent Zorbax 80-5 SB-CN column (150 × 4.6 mm, 5-μm particle size). The mobile phase

consisted of water, acetic acid, acetonitrile, and phosphoric acid (580:50:370:1, vol/vol/vol/vol), adjusted to pH 5.0. The flow rate was 1.0 mL/min. Electrochemical detection was performed using a glassy carbon electrode set at a potential of +1150 mV. The sample injection volume was 10 μL. Under these conditions, the retention time of tedisamil was about 14.6 minutes (estimated from the published chromatogram).

The stability-indicating nature of the method was demonstrated by accelerated studies of tedisamil. Tedisamil was treated in 0.1 N sodium hydroxide solution at 50 °C for 24 hours or in 1% hydrogen peroxide solution at 50 °C for 22 hours. The degradation products did not interfere with the determination of tedisamil.

Reference
Korner A. Uncovering deficiencies in mass balance using HPLC with chemiluminescence nitrogen-specific detection. *LCGC North America*. 2002; 20: 364–73.

Teicoplanin

	R
Teicoplanin A$_2$-1	(Z)-4-decenoic acid
A$_2$-2	8-methylnonanoic acid
A$_2$-3	n-decanoic acid
A$_2$-4	8-methyldecanoic acid
A$_2$-5	9-methyldecanoic acid

Other Names
Targocid, Teicoplanin A$_2$, Teicox, Teiklonal

Form	Molecular Formula	MW	CAS
Teicoplanin A$_2$-1	$C_{88}H_{95}Cl_2N_9O_{33}$	1877.6	91032-34-7
Teicoplanin A$_2$-2	$C_{88}H_{97}Cl_2N_9O_{33}$	1879.6	91032-26-7
Teicoplanin A$_2$-3	$C_{88}H_{97}Cl_2N_9O_{33}$	1879.6	91032-36-9
Teicoplanin A$_2$-4	$C_{89}H_{99}Cl_2N_9O_{33}$	1893.6	91032-37-0
Teicoplanin A$_2$-5	$C_{89}H_{99}Cl_2N_9O_{33}$	1893.6	91032-38-1

Appearance
Teicoplanin occurs as a white amorphous powder.

Method
Manduru et al. determined the stability of teicoplanin sodium and ceftazidime separately and in combination in a peritoneal dialysis solution. The HPLC system consisted of a Waters model 510 high-pressure pump, Waters Ultra WISP 715 and WISP 410B sampler processors, a Waters model 481 variable-wavelength UV detector, a Shimadzu Chromatopac C-R3A integrator, and a Perkin-Elmer model 1020 personal integrator. The stationary phase was a Waters µBondapak reversed-phase C_{18} analytical column (300×3.9 mm). The mobile phase consisted of 25% acetonitrile and 75% 0.1 M monobasic sodium phosphate in deionized water, adjusted to pH 4.0 with phosphoric acid (vol/vol). The flow rate was 2.0 mL/min. UV detection was performed at 210 nm. Under these conditions, retention times for three major teicoplanin components were 4.2, 5.3, and 9.5 minutes.

The stability-indicating ability of the assay was confirmed by accelerated decomposition of teicoplanin. Samples of teicoplanin were heated to 55 °C in a water bath for 30 minutes, cooled in an ice bath for another 30 minutes, and adjusted to pH 12.43 with 1 N sodium hydroxide and to pH 2.06 with 1 N hydrochloric acid. Peaks for teicoplanin decomposition products were well separated from the intact teicoplanin peaks.

Standard curves were constructed from 10 to 40 µg/mL. The intrarun and interrun coefficients of variation were less than 6%.

Reference
Manduru M, Fariello A, White RL, et al. Stability of ceftazidime sodium and teicoplanin sodium in a peritoneal dialysis solution. *Am J Health Syst Pharm.* 1996; 53: 2731–4.

Telmisartan

Chemical Name
4'-[(1,4'-Dimethyl-2'-propyl[2, 6'-bi-1*H*-benzimidazol]-1'-yl)methyl][1,1'-biphenyl]-2-carbox-ylic acid

Other Names
Micardis, Pritor

Form	Molecular Formula	MW	CAS
Telmisartan	$C_{33}H_{30}N_4O_2$	514.6	144701-48-4

Appearance
Telmisartan occurs as a white to slightly yellowish solid.

Solubility
Telmisartan is practically insoluble in water but soluble in strong base.

Method
Kumar and Muley reported a reversed-phase method for the determination of telmisartan in solid dosage forms. The Merck Hitachi liquid chromatograph was equipped with a model L-7100 isocratic solvent-delivery pump, a model L-7200 autosampler, and a model L-7400 variable-wavelength UV-visible detector. The stationary phase was a Supelco Discovery RP Amide C_{16} column (250 × 4.6 mm, 5-µm particle size). The mobile phase consisted of 0.02 M potassium phosphate buffer and acetonitrile (55:45, vol/vol), adjusted to pH 3.5 with orthophosphoric acid. The flow rate was 1 mL/min. UV detection was performed at 254 nm.

Standard stock solutions of the drug and its internal standard, methylparaben, were prepared in methanol. Powder equivalent to 100 mg of telmisartan from 20 tablets was accurately weighed, transferred to a 100-mL volumetric flask, mixed with 50 mL of methanol, sonicated for 10 minutes, and filled to the mark with additional methanol. An aliquot of 1 mL of this sample solution was mixed with 1 mL of internal standard, then diluted to 20 mL with mobile phase, and filtered through a 0.2-µm filter. The injection volume was 20 µL. Under these conditions, retention times of methylparaben and telmisartan were 6.10 and 8.62 minutes, respectively.

To verify the stability-indicating property of the assay, the dosage forms and placebo were treated with 6 M hydrochloric acid, with 6 M sodium hydroxide solution, with 30% hydrogen peroxide, heated at 50 °C, or exposed to ultraviolet light and sunlight. The peak of the intact drug was well resolved from peaks of its degradation products.

A standard curve for telmisartan was obtained from 25 to 75 µg/mL. The correlation coefficient was 0.9999. The limit of detection and the limit of quantitation for the drug were 5.1 and 15.8 µg/mL, respectively.

Reference
Kumar MV, Muley PR. Stability indicating RP-HPLC method for determination of telmisartan in solid dosage forms. *Indian Pharmacist.* June 2005; 69–72.

Temazepam

Chemical Name
7-Chloro-1,3-dihydro-3-hydroxy-1-methyl-5-phenyl-2*H*-1,4-benzodiazepin-2-one

Other Names
Levanxol, Normison, Restoril

Form	Molecular Formula	MW	CAS
Temazepam	$C_{16}H_{13}ClN_2O_2$	300.7	846-50-4

Appearance
Temazepam is a white or almost white crystalline powder.

Solubility
Temazepam is very slightly soluble in water and sparingly soluble in alcohol.

Method
Fatmi and Hickson reported a reversed-phase HPLC method for the determination of temazepam in capsules in the presence of its synthetic precursor and possible degradation products. The liquid chromatograph was composed of a Waters model 510 pump, a model 481 variable-wavelength detector, a model 710B WISP autosampler, and a Fisher Scientific Recordall chart recorder. The stationary phase was a Supelco Supelcosil LC8DB C_8 column (50 × 4.6 mm, 5-μm particle size). The mobile phase consisted of methanol and 1% acetic acid (40:60, vol/vol). The flow rate was 1.5 mL/min. UV detection was performed at 254 nm and 0.05 AUFS. Sulfanilamide 0.75 mg/mL in methanol was used as an internal standard.

A portion of the capsule contents equivalent to one capsule was weighed, transferred into a 100-mL volumetric flask, dissolved in 50 mL of methanol, mixed with 5 mL of internal standard, diluted to volume with methanol, and filtered through a 0.45-μm membrane filter. The injection volume was 5 μL. Under these conditions, retention times of the internal standard and temazepam were about 0.80 and 2.80 minutes, respectively.

The stability-indicating ability of the method was shown by assaying a synthetic mixture of the precursor and potential degradation products. Temazepam was well resolved from these compounds.

A standard curve for temazepam was obtained from 0.075 to 0.60 mg/mL. The correlation coefficient was 0.9999.

Reference
Fatmi AA, Hickson EA. Determination of temazepam and related compounds in capsules by high-performance liquid chromatography. *J Pharm Sci.* 1988; 77: 87–9.

Teniposide

Chemical Name

[5R-[5α,5aβ,8aα,9β(R*)]]-5,8,8a,9-Tetrahydro-5-(4-hydroxy-3,5-dimethoxyphenyl)-9-[[4,6-O-(2-thienylmethylene)-β-D-glucopyranosyl]oxy]furo[3′,4′:6,7]naphtho[2,3-d]-1,3-dioxol-6(5aH)-one

Other Name

Vumon

Form	Molecular Formula	MW	CAS
Teniposide	$C_{32}H_{32}O_{13}S$	656.7	29767-20-2

Appearance

Teniposide is a white to off-white crystalline powder.

Solubility

Teniposide is slightly soluble in methanol and very soluble in acetone and in dimethylformamide.

Method

Beijnen et al. used HPLC to study the stability of teniposide in commonly used infusion fluids. The liquid chromatographic system consisted of a Waters model M45 solvent-delivery system, a Waters U6K injector, a Spectra-Physics 200 programmable wavelength detector, and a Spectra-Physics 4270 integrator. The stationary phase was a μBondapak phenyl column (300 × 3.9 mm, 10-μm particle size). The mobile phase consisted of methanol and water (50:50, wt/wt), to which 0.5% (vol/wt) 0.5 M sodium phosphate buffer (pH 6) was added. The flow rate was 1.0 mL/min. UV detection was performed at 280 nm.

The method was stability indicating because it separated the parent drug from its *cis*-fused lactone analog.

Calibration curves for teniposide were constructed from 0.1 to 0.5 mg/mL; the correlation coefficient was greater than 0.999. The coefficient of variation was 1.0%.

Reference

Beijnen JH, Beijnen-Bandhoe AU, Dubbelman AC, et al. Chemical and physical stability of etoposide and teniposide in commonly used infusion fluids. *J Parenter Sci Technol.* 1991; 45: 108–12.

Tenoxicam

Chemical Name

4-Hydroxy-2-methyl-*N*-2-pyridinyl-2*H*-thieno[2,3-*e*]-1,2-thiazine-3-carboxamide 1,1-dioxide

Other Names

Alganex, Dolmen, Mobiflex, Rexalgan, Ro12-0068/000, Tilatil, Tilcotil

Form	Molecular Formula	MW	CAS
Tenoxicam	$C_{13}H_{11}N_3O_4S_2$	337.4	59804-37-4

Appearance

Tenoxicam occurs as a yellow polymorphic crystalline powder.

Solubility

Tenoxicam is practically insoluble in water and very slightly soluble in alcohol. It dissolves in solutions of acids and alkalis.

Method 1

Taha et al. developed an HPLC method for the determination of tenoxicam in the presence of its alkaline degradation products. The Waters chromatograph consisted of a model 600 LC series pump, a model 600 controller unit, a model 486 tunable absorbance detector, an injector with a 20-μL loop, and a model 746 data module. The stationary phase was a Nova-Pak C_{18} column (150 × 3.9 mm, 4-μm particle size). The mobile phase was a mixture of methanol, acetonitrile, and acetate buffer (pH 4.6) (1.9:0.1:3.0, vol/vol/vol). The flow rate was 0.8 mL/min. The detector was set at 280 nm.

A portion of powder equivalent to 25 mg of tenoxicam from 10 tablets was accurately weighed, mixed in 0.3 mL of 1 N sodium hydroxide solution and 15 mL of methanol, stirred for 30 minutes, further diluted with methanol, and filtered. The sample injection volume was 20 μL. Under these conditions, the retention time for tenoxicam was about 3.0 minutes.

Tenoxicam was heated in 1 N sodium hydroxide solution for 2 hours. Its degradation products did not interfere with the determination of the intact tenoxicam, demonstrating the stability-indicating nature of the method.

A calibration curve for tenoxicam was constructed from 0.5 to 20 µg/mL. The correlation coefficient was 0.9997. The intraday and interday coefficients of variation of the analysis were 1.58 and 0.68%, respectively. The limit of detection and the limit of quantitation were 0.03 and 0.09 µg/mL, respectively.

Reference
Taha EA, Salama NN, Fattah LEA. Stability-indicating chromatographic methods for the determination of some oxicams. *J AOAC International.* 2004; 87: 366–73.

Method 2
Bartsch et al. developed an assay for the quantitation of tenoxicam in the presence of its photodegradation products. The chromatograph consisted of a Shimadzu model LC-10AS pump, a model SPD-M10A diode-array detector, a model CTO-10AC column oven, and a Rheodyne 20-µL loop injector. The stationary phase was a Merck LiChrospher 100 RP18 endcapped column (119 × 3 mm, 5-µm particle size). The mobile phase consisted of 0.4 M acetate buffer (pH 4.6) containing 0.0075 M tetrabutylammonium hydrogen sulfate and methanol (60:40, vol/vol). UV detection was performed at 254 nm. Samples were prepared in 2.5% ammonium hydroxide solution (pH ~11.8). Under these conditions, the retention time of tenoxicam was about 4.4 minutes (estimated from the published chromatogram).

The method was demonstrated to be stability indicating by an intentional degradation study. Tenoxicam (40 µg/mL) solution was irradiated with a xenon source in a Suntest light cabinet for 48 minutes. The degradation products did not interfere with the quantitation of tenoxicam.

Standard curves for tenoxicam were generated from 8.8 to 44.0 µg/mL, 60 to 300 µg/mL, and 0.44 to 2.20 mg/mL. The correlation coefficients were greater than 0.9986. The intraday and interday coefficients of variation were less than 1.47 and 2.74%, respectively. The limit of detection and the limit of quantitation were 0.20 and 0.75 µg/mL, respectively.

Reference
Bartsch H, Eiper A, Kopelent-Frank H, et al. Selective assays for quantitation of tenoxicam in presence of its photodegradation products. *J Liq Chromatogr Rel Technol.* 2002; 25: 2821–31.

Terazosin Hydrochloride

Chemical Name
1-(4-Amino-6,7-dimethoxy-2-quinazolinyl)-4-[(tetrahydro-2-furanyl)carbonyl]piperazine monohydrochloride dihydrate

Other Names

Heitrin, Hytracin, Hytrin, Hytrinex, Vasocard, Vasomet, Vicard

Form	Molecular Formula	MW	CAS
Terazosin	$C_{19}H_{25}N_5O_4$	387.4	63590-64-7
Terazosin hydrochloride	$C_{19}H_{25}N_5O_4.HCl$	423.9	63074-08-8
Terazosin hydrochloride dihydrate	$C_{19}H_{25}N_5O_4.HCl.2H_2O$	459.9	70024-40-7

Appearance

Terazosin hydrochloride occurs as a white crystalline substance.

Solubility

Terazosin hydrochloride is freely soluble in water having a solubility of 761 mg/mL and also in isotonic saline. Terazosin hydrochloride dihydrate has an aqueous solubility of 24.2 mg/mL.

Method

Bakshi et al. developed a stability-indicating HPLC method for the determination of terazosin in the presence of its degradation products. The liquid chromatograph consisted of a Waters model 600E pump, a model 996 photodiode-array detector, and a model 717 autosampler. The stationary phase was a Waters Spherisorb C_{18} column (250 × 4.6 mm, 5-μm particle size). The mobile phase consisted of acetonitrile, water, glacial acetic acid, and diethylamine (65:35:1:0.02, vol/vol/vol/vol). The flow rate was 1 mL/min. UV detection was performed at 254 nm. The injection volume was 5 μL. Under these conditions, the retention time for terazosin was about 10.5 minutes.

The terazosin solutions were heated at 80 °C for 10 days, treated with 0.1 N hydrochloric acid and heated at 80 °C for 90 hours, treated with 0.1 N sodium hydroxide solution and heated at 80 °C for 4 hours, treated with 3% hydrogen peroxide at room temperature for 6 hours, treated with 30% hydrogen peroxide for 24 hours, exposed to the light, or subjected to dry heat at 50 °C for 3 months. The peak of the intact drug was well resolved from peaks of its degradation products. The absence of any coeluting peak was also confirmed by photodiode-array peak purity analysis.

A standard curve for terazosin was constructed from 50 to 500 μg/mL. The correlation coefficient was greater than 0.9997. The intraday and interday coefficients of variation were less than 0.5 and 1.8%, respectively.

Reference

Bakshi M, Ojha T, Singh S. Validated specific HPLC methods for determination of prazosin, terazosin and doxazosin in the presence of degradation products formed under ICH-recommended stress conditions. *J Pharm Biomed Anal.* 2004; 34: 19–26.

Terbutaline

Chemical Names

5-[2-[(1,1-Dimethylethyl)amino]-1-hydroxyethyl]-1,3-benzenediol

α-[(*tert*-Butylamino)methyl]-3,5-dihydroxybenzyl alcohol

Other Name

Brethine

Form	Molecular Formula	MW	CAS
Terbutaline	$C_{12}H_{19}NO_3$	225.3	23031-25-6
Terbutaline sulfate	$(C_{12}H_{19}NO_3)_2.H_2SO_4$	548.7	23031-32-5

Appearance

Terbutaline sulfate is a white to grayish-white crystalline powder.

Solubility

Terbutaline sulfate has a solubility of 1 in 4 of water. It is slightly soluble in alcohol and methanol but practically insoluble in chloroform and ether.

Method 1

Tenjarla et al. developed an assay for terbutaline. An LDC ConstaMetric I liquid chromatograph was equipped with an LDC SpectraMonitor II variable-wavelength UV detector and a Rheodyne 100-μL loop injector. The stationary phase was an Alltech Spherisorb CN column (250 × 4.6 mm, 5-μm particle size). The mobile phase was 30% acetonitrile in a phosphate buffer (pH 5.6). The flow rate was 1.4 mL/min. UV detection was performed at 225 nm. Propranolol hydrochloride 2.5 μg/mL was used as an internal standard.

The drug-exposed skin was washed twice with 3 mL of water, homogenized with 5 mL of methanol, and filtered. This process was repeated twice with the residue. The filtrate was combined, evaporated to dryness, reconstituted with the mobile phase, mixed with the internal standard, and assayed. The injection volume was 5 μL. Under these conditions, retention times of terbutaline and propranolol were 3.3 and 7.9 minutes, respectively.

The method was stated to be stability indicating. A terbutaline solution was boiled for 3 minutes in 0.1 N hydrochloric acid or 0.1 N sodium hydroxide.

A standard curve for terbutaline was generated from 5 to 15 μg/mL. The correlation coefficient was more than 0.98.

Reference

Tenjarla SN, Allen R, Mitchell B. High performance liquid chromatographic assay of terbutaline for preformulation studies. *J Liq Chromatogr*. 1995; 18: 1603–15.

Method 2

Jacobson and Peterson developed an HPLC assay for the simultaneous determination of ipratropium bromide, fenoterol, albuterol (salbutamol), and terbutaline in nebulizer solution. The liquid chromatograph consisted of a Varian model 9010 solvent-delivery system, a Rheodyne model 7161 injector with a 10-μL external loop, a Varian model 9050 variable-wavelength UV-visible detector, and a Varian GC Star workstation. The stationary phase was a Waters Nova-Pak C_{18} Radial-Pak cartridge (100 × 8 mm, 4-μm particle size) inside a Waters RCM 8 × 10 compression module. Mobile phase A consisted of tetrahydrofuran and distilled water (40:60, vol/vol) containing 0.0025 M Waters Pic B-8 Reagent Low UV. Mobile phase B was distilled water, and mobile phase C consisted of methanol and distilled water (50:50, vol/vol). The flow rate was 2.0 mL/min. A mixture of 50% mobile phase A and 50% mobile phase B was delivered up to 7.7 minutes, and then the composition of the mixture was changed linearly to 60% mobile phase A, 15% mobile phase B, and 25% mobile phase C at 13.0 minutes. The run time was 13.0 minutes with a 5.0-minute equilibration time. Mepivacaine hydrochloride 1% was used as an internal standard. The injection volume was 20 μL. Retention times for albuterol, terbutaline, ipratropium, mepivacaine, and fenoterol were 3.2, 4.3, 5.9, 8.2, and 12.7 minutes, respectively.

The assay was determined to be stability indicating by accelerated decomposition of terbutaline with heat, hydrogen peroxide, acid, and base. In all cases, degradation product peaks did not interfere with the intact terbutaline peak.

Reference

Jacobson GA, Peterson GM. High-performance liquid chromatographic assay for the simultaneous determination of ipratropium bromide, fenoterol, salbutamol, and terbutaline in nebulizer solution. *J Pharm Biomed Anal*. 1994; 12: 825–32.

Method 3

Gupta developed a stability-indicating assay for the quantitation of terbutaline sulfate in pharmaceutical dosage forms. The chromatograph consisted of a Waters ALC 202 system equipped with a U6K universal injector, a Kratos Schoeffel SF 770 multiple-wavelength detector, a Houston Omniscribe recorder, and a Waters μBondapak phenyl column (300 × 3.9 mm, 10-μm particle size). The mobile phase consisted of 8% methanol in 0.02 M monobasic potassium phosphate in water. The solution was adjusted to a pH of about 3.6 with 1.7% phosphoric acid. The flow rate was 2.0 mL/min. UV detection was performed at 278 nm and 0.4 AUFS. Salicylic acid 0.1% in methanol was used as an internal standard.

Samples were diluted with water. The injection volume was 20 μL. Under these conditions, retention times for terbutaline and the internal standard were about 4.9 and 7.0 minutes, respectively (estimated from the published chromatogram).

The analytical method was demonstrated to be stability indicating by an accelerated degradation study. A sodium hydroxide pellet (about 200 mg) was added to a standard solution of terbutaline (200 μg/mL) and heated on a hot plate. The resulting solution was cooled, neutralized with about 1 N sulfuric acid, and assayed. Degradation product peaks did not interfere with the intact terbutaline peak.

Standard curves for terbutaline were constructed and were linear from 2 to 7 µg/mL and from 100 to 350 µg/mL.

Reference
Gupta DV. Quantitation of terbutaline sulfate in pharmaceutical dosage forms using high performance liquid chromatography. *J Liq Chromatogr*. 1986; 9: 1065–74.

Method 4

Mehta et al. determined the stability of terbutaline sulfate in admixtures of terbutaline 30 µg/mL in 0.45% sodium chloride injection and in more commonly used fluids. The liquid chromatograph consisted of a Waters model M6000A pump, a Rheodyne model 7125 injector, and a Coulochem model 5100A electrochemical detector. The stationary phase was an IBM reversed-phase C_{18} analytical column (250 × 4.5 mm, 5-µm particle size) with a Whatman:Pell ODS guard column. The mobile phase was a mixture of 18 parts of acetonitrile and 82 parts of a mixture of 5 mM octanesulfonic acid and 25 mM phosphate (pH 7.6). The flow rate was 1 mL/min. The guard cell potential was set at 900 mV, and the analytical cell potentials for electrodes 1 and 2 were set at 330 and 690 mV, respectively. Metaproterenol 2.5 µg/mL was used as an internal standard.

Samples were diluted with water. The injection volume was 20 µL. Under these conditions, the retention times of metaproterenol and terbutaline were 7 and 10 minutes, respectively.

The method was stated to be stability indicating.

Standard curves were generated each day by plotting the drug concentrations versus the peak height ratios of terbutaline to metaproterenol. The interday coefficient of variation of this assay was 3.2%.

Reference
Mehta J, Searcy CJ, Jung DT. Stability of terbutaline sulfate admixtures stored in polyvinyl chloride bags. *Am J Hosp Pharm*. 1986; 43: 1760–2.

Method 5

Williams et al. described a stability-indicating ion-pair HPLC method for terbutaline in the presence of aminophylline in intravenous solutions. The liquid chromatograph consisted of a Waters model M6000A solvent-delivery system, a Waters model U6K injector, and a Perkin-Elmer model 204A spectrofluorometer or a model LC-55 spectrophotometer. The stationary phase was a Waters C_{18} column. The mobile phase was 0.005 M 1-heptanesulfonate sodium and 0.35 M acetic acid in 35% (vol/vol) methanol. The flow rate was 1.6 mL/min. The fluorometric detector was set at an excitation wavelength of 285 nm and an emission wavelength of 315 nm. The injection volume was 40 µL. Retention times for aminophylline and terbutaline were 2.75 and 5.33 minutes, respectively.

The stability-indicating nature of the assay was demonstrated by an accelerated degradation study. A solution of terbutaline was exposed to fluorescent light for 17 hours. Degradation product peaks did not interfere with the parent terbutaline peak.

A standard curve was constructed by plotting drug concentrations versus the peak areas from 0.4 to 4.2 µg/mL. The correlation coefficient was greater than 0.997.

Reference
Williams DA, Fung EYY, Newton DW. Ion-pair high-performance liquid chromatography of terbutaline and catecholamines with aminophylline in intravenous solutions. *J Pharm Sci.* 1982; 71: 956–8.

Terfenadine

Chemical Name
1-(4-*tert*-Butylphenyl)-4-[4-(α-hydroxybenzhydryl)piperidino]butan-1-ol

Other Names
Allegra, Seldane, Terlane

Form	Molecular Formula	MW	CAS
Terfenadine	$C_{32}H_{41}NO_2$	471.7	50679-08-8
Terfenadine carboxylate hydrochloride (Fexofenadine hydrochloride)	$C_{32}H_{39}NO_4.HCl$	538.1	153439-40-8

Appearance
Terfenadine and terfenadine carboxylate hydrochloride (fexofenadine hydrochloride) occur as white to off-white crystalline powders.

Solubility
Terfenadine is slightly soluble in water and soluble in alcohol. Terfenadine carboxylate hydrochloride has a solubility of 2.2 mg/mL in water at 25 °C. It is freely soluble in alcohol.

pK$_a$
Terfenadine carboxylate hydrochloride has pK$_a$ values of 4.25 and 9.53 at 25 °C.

Method
George and Contario described an HPLC method for the determination of terfenadine, pseudoephedrine hydrochloride, and ibuprofen in a liquid formulation. The liquid chromatographic system included a Waters model M6000A pump, a model 710B WISP autosampler, a Spectroflow 757 variable-wavelength detector, and a Beckman CALS data system. The stationary phase was a Phase Separations Spherisorb ODS-2 column

(100 × 4.6 mm, 3-μm particle size). The mobile phase consisted of 400 mL of water, 1.0 g of monobasic sodium phosphate monohydrate, 0.5 g of phosphoric acid (85%), 3.0 g of sodium perchlorate monohydrate, and 600 mL of acetonitrile. The flow rate was 1.5 mL/min. UV detection was performed at 210 nm and 0.32 AUFS.

Samples were diluted with methanol and then with the mobile phase. The injection volume was 20 μL. Under these conditions, retention times for pseudo-ephedrine, ibuprofen, and terfenadine were about 1.1, 2.4, and 4.4 minutes, respectively (estimated from the published chromatogram).

The stability-indicating capability of the method was established by intentional degradation of terfenadine. Samples of terfenadine were treated with 0.1 M hydrochloric acid at 90 °C for up to 72 hours. The intact terfenadine was well resolved from its degradation products.

Reference

George RC, Contario JJ. Quantitation of terfenadine, pseudoephedrine hydrochloride, and ibuprofen in a liquid animal dosing formulation using high performance liquid chromatography. *J Liq Chromatogr*. 1988; 11: 475–88.

Tetracaine

Chemical Name

2-Dimethylaminoethyl 4-butylaminobenzoate

Other Names

Amethocaine, Pontocaine, Styptocaine, Tetcaine

Form	Molecular Formula	MW	CAS
Tetracaine	$C_{15}H_{24}N_2O_2$	264.4	94-24-6
Tetracaine hydrochloride	$C_{15}H_{24}N_2O_2 \cdot HCl$	300.8	136-47-0

Appearance

Tetracaine is a white or light yellow waxy solid. Tetracaine hydrochloride is a white odorless and slightly hygroscopic polymorphic crystalline powder.

Solubility

Tetracaine is very slightly soluble in water and freely soluble in alcohol. Tetracaine hydrochloride is very soluble in water and soluble in alcohol.

pK$_a$

Tetracaine hydrochloride has a pK$_a$ of 8.39.

Method

Wang investigated by HPLC the stability of tetracaine in aqueous systems. A liquid chromatograph was equipped with a Waters model 440 UV detector and a Mini-Lab model CSI38 integrator. The stationary phase was a Waters µBondapak C$_{18}$ column (300 × 3.9 mm). The mobile phase consisted of methanol and potassium phosphate buffer (pH 7.8) (70:30, vol/vol). The flow rate was 1.5 mL/min. UV detection was performed at 254 nm and 0.1 AUFS. Methyltestosterone 1 mg/mL in water was used as an internal standard.

Samples were mixed with the internal standard solution and diluted with water. The injection volume was 5 µL. Under these conditions, retention times for methyltestosterone and tetracaine were 4.5 and 6.0 minutes, respectively.

The method was reported to be stability indicating since it separated tetracaine from its degradation products.

A calibration curve for tetracaine was obtained from 0.300 to 1.200 mg/mL. The correlation coefficient was 0.9938.

Reference

Wang D-P. Stability of tetracaine in aqueous systems. *J Taiwan Pharm Assoc*. 1983; 35: 132–41.

Tetracycline

Chemical Name

[4S-(4α,4aα,5aα,6β,12aα)]-4-(Dimethylamino)-1,4,4a,5,5a,6,11,12a-octahydro-3,6,10,12,12a-pentahydroxy-6-methyl-1,11-dioxo-2-naphthacenecarboxamide

Other Names

Panmycin, Sumycin

Form	Molecular Formula	MW	CAS
Tetracycline	C$_{22}$H$_{24}$N$_2$O$_8$	444.4	60-54-8
Tetracycline hydrochloride	C$_{22}$H$_{24}$N$_2$O$_8$.HCl	480.9	64-75-5

Appearance

Tetracycline is a yellow odorless crystalline powder. Tetracycline hydrochloride occurs as a yellow odorless hygroscopic crystalline powder.

Solubility

Tetracycline has solubilities of approximately 0.4 mg/mL in water and 20 mg/mL in alcohol at 25 °C. Tetracycline hydrochloride has solubilities of approximately 100 mg/mL in water and 10 mg/mL in alcohol at 25 °C.

Method

Allen and Erickson investigated the stability of tetracycline hydrochloride in extemporaneously compounded oral liquids. A Hewlett-Packard series 1050 automated high-performance liquid chromatograph included a multisolvent mixing and pumping system, an autoinjector, a diode-array detector, and a computer with Chem Station software. The stationary phase was a Bakerbond C_8 column (250 × 4.6 mm, 5-μm particle size). The mobile phase consisted of 0.1 M ammonium oxalate, dimethylformamide, 0.2 M dibasic ammonium phosphate (adjusted with 3 N ammonium hydroxide to pH 7.6) (68:5:27, vol/vol/vol). The mobile phase was delivered isocratically at 1.2 mL/min. UV detection was performed at 280 nm.

Samples were diluted 1:100. Under these conditions, the retention time for tetracycline was 5.9 minutes.

The HPLC method was determined to be stability indicating. A composite chromatogram of tetracycline hydrochloride after accelerated degradation showed that degradation product peaks did not interfere with the intact tetracycline peak.

A standard curve for tetracycline hydrochloride was constructed from 50 to 300 μg/mL. The intraday and interday coefficients of variation were 0.9 and 1.4%, respectively.

Reference

Allen LV Jr, Erickson MA III. Stability of bethanechol chloride, pyrazinamide, quinidine sulfate, rifampin, and tetracycline hydrochloride in extemporaneously compounded oral liquids. *Am J Health Syst Pharm.* 1998; 55: 1804–9.

Tetrahydrouridine

Chemical Name

1-(β-D-Ribofuranosyl)-4-hydroxytetrahydro-1(1*H*)-pyrimidinone

Method

Using an HPLC method, Xiang et al. investigated the epimer interconversion, isomerization, and hydrolysis of tetrahydrouridine. The Waters 2690 separation module was equipped with a model 996 photodiode-array detector. The stationary phase was an Advanced Separation Technologies Cyclobond I 2000 RSP column (250 × 4.6 mm) with a guard column (20 × 4.0 mm) of the same packing material. The mobile phase consisted of 10 mM phosphoric acid in water and acetonitrile (5:95, vol/vol), adjusted to pH 7.5 with triethylamine. The flow rate was 1.5 mL/min. UV detection was carried out at 195 nm. Under these conditions, retention times of two different epimers of tetrahydrouridine were 11.8 and 12.4 minutes.

Tetrahydrouridine solution was prepared in an acetate buffer (pH 4.0) and incubated at 70 °C for 24 hours. Three major degradation products were observed at retention times of 8.9, 13.4, and 14.4 minutes, respectively.

A standard curve was constructed from 0.1 to 2 mg/mL. The correlation coefficients were 1.0000 and 0.9999 for each of the two epimer peaks.

Reference
Xiang T-X, Niemi R, Bummer P, et al. Epimer interconversion, isomerization, and hydrolysis of tetrahydrouridine: implications for cytidine deaminase inhibition. *J Pharm Sci.* 2003; 92: 2027–39.

Tetrahydrozoline

Chemical Names
4,5-Dihydro-2-(1,2,3,4-tetrahydro-1-naphthalenyl)-1*H*-imidazole
2-(1,2,3,4-Tetrahydro-1-naphthyl)-2-imidazoline

Other Names
Optigene, Tetryzoline, Tyzine, Visine

Form	Molecular Formula	MW	CAS
Tetrahydrozoline	$C_{13}H_{16}N_2$	200.3	84-22-0
Tetrahydrozoline hydrochloride	$C_{13}H_{16}N_2.HCl$	236.7	522-48-5

Appearance
Tetrahydrozoline hydrochloride occurs as a white odorless solid.

Solubility
Tetrahydrozoline hydrochloride has solubilities of approximately 286 mg/mL in water and 133 mg/mL in alcohol at 25 °C. It is very slightly soluble in chloroform and practically insoluble in ether.

Method 1
Schutter et al. described the determination of tetrahydrozoline hydrochloride in pharmaceutical formulations by HPLC analysis. The Spectra-Physics SP8000 liquid chromatograph was equipped with a Spectra-Physics model 770 variable-wavelength detector, a Valco six-port injection valve, and a Kipp & Zonen BD8 single-channel recorder. The stationary phase was an RSIL C_{18} column (150 × 4.1 mm, 5-μm particle size). The mobile phase consisted of methanol and water (40:60, vol/vol) containing 20 mM sodium 1-octanesulfonate and 10 mM *N,N*-dimethyloctylamine and was adjusted to pH 3.0 with phosphoric acid. The flow rate was 1.0 mL/min. UV detection was performed at 220 nm. Totazoline hydrochloride 0.5 mg/mL in methanol–water (40:60, vol/vol) was used as an internal standard.

A 5-mL sample solution was mixed with 10 mL of internal standard in a 50-mL volumetric flask and diluted to volume with methanol–water (40:60, vol/vol). The injection volume was 10 μL. Under these conditions, the retention time for tetrahydrozoline was about 2.2 minutes (estimated from the published chromatogram).

The stability-indicating nature of the assay was demonstrated by accelerated degradation of tetrahydrozoline. A tetrahydrozoline hydrochloride sample was partially degraded with 0.1 M sodium hydroxide solution. The intact tetrahydrozoline was separated from its degradation products.

A standard curve for tetrahydrozoline was obtained from 7.5 to 125 μg/mL of the label claim. The correlation coefficient was 0.9996.

Reference

Schutter JAD, Bossche WVD, Moerloose PD. Stability-indicating analysis of tetryzoline hydrochloride in pharmaceutical formulations by reversed-phase ion-pair liquid chromatography. *J Chromatogr*. 1987; 391: 303–8.

Method 2

Andermann and Richard presented an HPLC assay for the determination of tetrahydrozoline hydrochloride in ophthalmic solutions. The system consisted of a Waters model 6000A pump, a model 710B WISP autosampler with a 20-μL loop, a model M730 data module, and a Schoeffel model 750 detector. The stationary phase was a Waters μBondapak C_{18} column (300 × 3.9 mm, 10-μm particle size). The mobile phase consisted of 600 mL of 0.05 M sodium tetraborate (adjusted to pH 7.0 with 0.1 M monobasic potassium phosphate) and 400 mL of acetonitrile. UV detection was performed at 254 nm.

Samples were dissolved in purified water. The injection volume was 20 μL. Under these conditions, the retention time for tetrahydrozoline hydrochloride was 5.14 minutes.

The method was demonstrated to be stability indicating by accelerated degradation of tetrahydrozoline. A sample solution was mixed with 10 M sodium hydroxide and stored in an oven at 50 °C for 24 hours. Tetrahydrozoline was resolved from its degradation product at a retention time of 3.13 minutes.

A standard curve for tetrahydrozoline was constructed from 25 to 500 μg/mL. The correlation coefficient was 0.9996. The limit of detection was 300 ng.

Reference

Andermann G, Richard A. Stability-indicating determination of tetrahydrozoline hydrochloride in ophthalmic solutions by high-performance liquid chromatography. *J Chromatogr*. 1984; 298: 189–92.

Method 3

Bauer and Krogh developed a stability-indicating assay for tetrahydrozoline hydrochloride in ophthalmic preparations. The liquid chromatograph consisted of a Waters model 6000A pump, a Rheodyne model 7120 injector with a 20-μL loop, a DuPont variable-wavelength detector, and a Hewlett-Packard 3385A recording integrator. The stationary phase was a Waters μBondapak C_{18} column. The mobile phase consisted of 700 mL of water containing 6 g of sodium citrate dihydrate and 4 g of anhydrous citric acid with the pH adjusted to 2.2 ± 0.2 with perchloric acid and 300 mL of methanol. The flow rate was 2.0 mL/min. UV detection was performed at 265 nm. Naphazoline was used as an internal standard.

Samples were diluted 1:2 with the internal standard solution. The injection volume was 20 μL. Retention times for naphazoline and tetrahydrozoline were 4.37 and 5.31 minutes, respectively.

The assay was stated to be stability indicating because it could separate tetrahydrozoline from its degradation product, *N*-(2-aminoethyl)-1,2,3,4-tetrahydro-1-naphthylamine.

Calibration curves were constructed from 0.1 to 0.5 mg/mL; the correlation coefficient was 0.9999.

Reference
Bauer J, Krogh S. High-performance liquid chromatographic stability-indicating assay for naphazoline and tetrahydrozoline in ophthalmic preparations. *J Pharm Sci.* 1983; 72: 1347–9.

Theophylline

Chemical Name
3,7-Dihydro-1,3-dimethylpurine-2,6(1*H*)-dione

Other Names
Accurbron, Lasma

Form	Molecular Formula	MW	CAS
Theophylline	$C_7H_8N_4O_2$	180.2	58-55-9

Appearance
Theophylline occurs as a white odorless crystalline powder.

Solubility
Theophylline is slightly soluble in water and sparingly soluble in alcohol.

Method
Heidemann reported an HPLC method for the determination of theophylline, guaifenesin, and benzoic acid in liquid and solid pharmaceutical dosage forms. A DuPont model 830 liquid chromatograph was equipped with a DuPont model 837 variable-wavelength detector, a Rheodyne model 7105 injector, and a Spectra-Physics autolab system 1V integrator. The stationary phase was a Whatman Partisil-10-ODS reversed-

phase column (250 × 4.6 mm, 10-µm particle size). The mobile phase was a mixture of 0.001 M sodium citrate–citric acid buffer (adjusted to pH 4.15) and acetonitrile (9:1, vol/vol). The flow rate was 2 mL/min. UV detection was performed at 230 nm. Methylparaben was used as an internal standard.

Liquid samples were diluted with water and the internal standard. Tablets were ground, wetted with alcohol, diluted with water, stirred for 1 hour, and filtered. The filtrate was diluted with water and the internal standard. The injection volume was 20 µL. Under these conditions, retention times for benzoic acid, theophylline sodium glycinate, guaifenesin, and methylparaben were about 6.8, 9.4, 11.6, and 16.0 minutes, respectively.

The method was reported to be stability indicating.

Reference
Heidemann DR. Rapid, stability-indicating, high-pressure liquid chromatographic determination of theophylline, guaifenesin, and benzoic acid in liquid and solid pharmaceutical dosage forms. *J Pharm Sci.* 1979; 68: 530–2.

Thiamine Hydrochloride (Vitamin B₁)

Chemical Name
3-(4-Amino-2-methylpyrimidin-5-ylmethyl)-5-(2-hydroxyethyl)-4-methylthiazolium chloride hydrochloride

Other Names
Benerva, Betaxin, Thiamilate

Form	Molecular Formula	MW	CAS
Thiamine hydrochloride	$C_{12}H_{17}ClN_4OS.HCl$	337.3	67-03-8
Thiamine mononitrate	$C_{12}H_{17}N_5O_4S$	327.4	532-43-4

Appearance
Thiamine hydrochloride occurs as colorless or white crystals or as a white or almost white crystalline powder. Thiamine nitrate occurs as small white or colorless crystals or as a white or almost white crystalline powder.

Solubility
Thiamine hydrochloride is freely soluble in water and slightly soluble in alcohol. Thiamine nitrate is sparingly soluble in water and slightly soluble in alcohol.

Method

Van der Horst et al. described the HPLC analysis of thiamine in total parenteral nutrient solution. The liquid chromatographic system consisted of a Waters model M6000A pump, a model WISP autosampler, a model 481 variable-wavelength detector, a Perkin-Elmer model 204-A fluorescence spectrophotometer, and a Shimadzu Chromatopac C-R3A integrator. The stationary phase was a LiChrosorb RP8 analytical column (250 × 4.6 mm, 10-μm particle size) with an RCSS Guard-Pak C_{18} precolumn (4 × 6 mm). The mobile phase was a mixture of 0.1% triethylamine in 0.2 M monobasic potassium phosphate buffer (pH 5.75) and acetonitrile (100:4, vol/vol). The flow rate was 3 mL/min. UV detection was performed at 246 nm.

An aliquot (200 μL) of the sample was injected onto the precolumn, washed with 20 mL of water, and switched to the analytical column. Under these conditions, the retention time for thiamine was about 3.2 minutes (estimated from the published chromatogram).

The method was demonstrated to be stability indicating by accelerated degradation of thiamine. Thiamine solutions were treated with 1 M sodium hydroxide solution or 1 M hydrochloric acid and heated to boiling for about 30 minutes. Degradation product peaks were separated from the peak of thiamine.

A standard curve for thiamine was generated from 0.1 to 2.4 mg/L. The correlation coefficient was 0.9995. The limit of detection was 0.036 mg/L.

Reference

Van der Horst A, Martens HJM, de Goede PNFC. Analysis of water-soluble vitamins in total parenteral nutrition solution by high pressure liquid chromatography. *Pharm Weekbl [Sci]*. 1989; 11: 169–74.

Thimerosal

Chemical Name

Sodium (2-carboxyphenylthio)ethylmercury

Other Names

Aeroaid, Merthiolate, Thiomersal

Form	Molecular Formula	MW	CAS
Thimerosal	$C_9H_9HgNaO_2S$	404.8	54-64-8

Appearance

Thimerosal occurs as a light cream-colored crystalline powder.

Solubility
Thimerosal is freely soluble in water and soluble in alcohol.

Method
Hu et al. described an assay for the simultaneous determination of thimerosal and chlorhexidine gluconate in solutions. The liquid chromatographic system consisted of a Jasco model 880-PU pump, a model 870 variable-wavelength UV detector, an SIC Autosampler 23 automatic sampler, and an SIC Chromatocorder 12 integrator. The stationary phase was a Nucleosil C_{18} column (7-μm particle size) with a Nucleosil C_{18} precolumn (7-μm particle size). The mobile phase consisted of 0.1 M monobasic potassium phosphate buffer (adjusted to pH 3.5 with phosphoric acid) and methanol (40:60, vol/vol). The flow rate was 1.0 mL/min. UV detection was performed at 254 nm and 0.04 AUFS. Methylparaben 3 mg/mL was used as an internal standard.

A 9.5-mL sample was mixed with 0.5 mL of the internal standard solution in a 10-mL volumetric flask. The injection volume was 20-μL. Under these conditions, retention times for methylparaben, thimerosal, and chlorhexidine gluconate were 7.8, 11.4, and 14.9 minutes, respectively.

The stability-indicating capability of the assay was shown by assaying a synthetic mixture of thimerosal, chlorhexidine gluconate, and internal standard and their degradation products. Thimerosal and chlorhexidine gluconate were separated from other compounds.

A standard curve for thimerosal was constructed from 2 to 150 μg/mL. The correlation coefficient was 0.999.

Reference
Hu OY-P, Wang S-Y, Fang Y-J, et al. Simultaneous determination of thimerosal and chlorhexidine in solutions for soft contact lenses and its applications in stability studies. *J Chromatogr*. 1990; 523: 321–6.

Thiopental Sodium

Chemical Names
5-Ethyldihydro-5-(1-methylbutyl)-2-thioxo-4,6(1*H*,5*H*)-pyrimidinedione monosodium salt
5-Ethyl-5-(1-methylbutyl)-2-thiobarbituric acid sodium salt

Other Name
Thiopentone Sodium

Form | Molecular Formula | MW | CAS

Form	Molecular Formula	MW	CAS
Thiopental sodium	$C_{11}H_{17}N_2NaO_2S$	264.3	71-73-8

Appearance
Thiopental sodium is a white to yellowish-white powder or a pale greenish hygroscopic powder.

Solubility

Thiopental sodium is freely soluble in water; it is partly soluble in alcohol. It is insoluble in ether, benzene, and petroleum ether.

Method 1

Chernin et al. determined the stability and compatibility of thiopental sodium 12.5 mg/mL and propofol 5 mg/mL stored together in polypropylene syringes at room temperature and 4 °C. The HPLC system consisted of an Alcott model 760 pump, a Hewlett-Packard model 3395 integrator, an Alcott model 728 autosampler, an Applied Biosystems model 749A UV-visible detector, and a Waters model 990 photodiode-array detector. The stationary phase was a Mac-Mod Analytical Zorbax SB phenyl column (150 × 4.6 mm, 5-μm particle size). The mobile phase was a mixture of 45% acetonitrile and 55% 0.01 M monobasic potassium phosphate aqueous buffer with pH adjusted to 4.0 with 10% phosphoric acid. The flow rate was 1.0 mL/min. UV detection was performed at 235 nm.

Samples were diluted 1:200. The injection volume was 20 μL. Under these conditions, the retention times for thiopental and propofol were 4.9 and 12.0 minutes, respectively.

The stability-indicating capability of the assay was shown by accelerated decomposition of thiopental sodium. Thiopental sodium samples were incubated for 2 hours at 23 °C in 0.1 N hydrochloric acid, for 30 minutes at 50 °C in 0.1 N sodium hydroxide, and for 10 minutes at 23 °C in 3% hydrogen peroxide. Degradation product peaks did not interfere with the parent thiopental peak.

A standard curve for thiopental sodium was generated from 31.4 to 94 μg/mL by linear regression of peak areas against drug concentrations. The correlation coefficient was greater than 0.9999. Intraday and interday variabilities were 1.43 and 0.66%, respectively.

Reference

Chernin EL, Stewart JT, Smiler B. Stability of thiopental sodium and propofol in polypropylene syringes at 23 and 4 °C. *Am J Health Syst Pharm.* 1996; 53: 1576–9.

Method 2

Prankerd and Jones evaluated the physicochemical compatibility of propofol with thiopental sodium. The liquid chromatograph included a Waters model M501 pump, a Waters model 712 WISP automatic injector, a Waters model M470 scanning fluorescence detector, a Waters model M490E UV detector, and a Waters Max 810 workstation. The stationary phase was a Waters Nova-Pak C_{18} reversed-phase radial-compression column. The mobile phase consisted of 10 mM phosphate buffer (pH 6.0) and acetonitrile (45:55, vol/vol). The flow rate was 1.6 mL/min. UV detection was performed at 290 nm. The fluorescence detector was set at an excitation wavelength of 276 nm and an emission wavelength of 314 nm.

Each sample was diluted 1:5000 with the mobile phase. The injection volume was 40 μL. Under these conditions, the retention times were 4.4 minutes for thiopental (UV detection) and 13.4 minutes for propofol (fluorescence detection).

The assay was determined to be stability indicating by examining three propofol–thiopental sodium admixtures stored at 4 °C for 6 months. At 6 months, less than 5% of the intact thiopental sodium remained. The degradation product peaks did not interfere with the intact thiopental peak.

A calibration curve for thiopental sodium was constructed from 20 to 1000 µg/mL. The correlation coefficient of the standard curve was greater than 0.9988. The interassay coefficients of variation were 1.9% at the upper end of the range and 8.8% at the lower end of the range.

A similar method was used by Gupta et al.

References
Prankerd RJ, Jones RD. Physicochemical compatibility of propofol with thiopental sodium. *Am J Health Syst Pharm*. 1996; 53: 2606–10.

Gupta VD, Gardner SN, Jolowsky CM, et al. Chemical stability of thiopental sodium injection in disposable plastic syringes. *J Clin Pharm Ther*. 1987; 12: 339–42.

Thiorphan

Chemical Name
N-[2-(Mercaptomethyl)-1-oxo-3-phenylpropyl]glycine

Form	Molecular Formula	MW	CAS
Thiorphan	$C_{12}H_{15}NO_3S$	253.3	76721-89-6

Method
Kuijpers et al. determined the stability of thiorphan in 0.9% sodium chloride injection containing 1% human serum albumin. The liquid chromatograph consisted of an LKB Bromma model 2150 pump, a model 2152 controller, an Applied Biosystems model 1000s diode-array detector, a Rheodyne 100-µL loop injector, and a Spectra-Physics model SP4270 integrator. The stationary phase was a Microsphere C_{18} reversed-phase column (100 × 4.6 mm, 3-µm particle size). The mobile phase was a mixture of 10 mM triethylamine, 2 mM sodium dodecyl sulfate, 0.1 M phosphate buffer adjusted to pH 2.1 with phosphoric acid, and 40% acetonitrile. The flow rate was 0.4 mL/min. UV detection was performed at 259 nm.

A 100-µL aliquot of sample was diluted with 900 µL of ice-cold distilled water. The injection volume was 20 µL. Under these conditions, the retention time of thiorphan was 4.4 minutes.

The stability-indicating ability of the method was demonstrated by accelerated degradation of thiorphan. A solution of thiorphan in 0.9% sodium chloride injection containing 1% human serum albumin was degraded by heating at 70 °C for 24 hours. The retention times of two degradation products were 5.8 and 6.5 minutes.

A standard curve for thiorphan was constructed from 0.00625 to 0.1 mg/mL.

Reference
Kuijpers EAP, Hartigh JD, Vermeij P. A stability study involving HPLC analysis of aqueous thiorphan solutions in the presence of human serum albumin. *Pharm Dev Technol*. 1998; 3: 185–92.

Thiotepa

Chemical Name
1,1',1''-Phosphinothioylidynetrisaziridine

Other Name
Thioplex

Form	Molecular Formula	MW	CAS
Thiotepa	$C_6H_{12}N_3PS$	189.2	52-24-4

Appearance
Thiotepa occurs as fine white crystalline flakes.

Solubility
Thiotepa has an aqueous solubility of 19 g in 100 mL at 25 °C and is very soluble in alcohols.

Method 1
Murray et al. investigated the stability of thiotepa (lyophilized) in 0.9% sodium chloride injection. The liquid chromatograph was a Waters HPLC system with a UV detector and a refrigerated autosampler. The stationary phase was an ES Industries Chromegabond Gamma C_{18} silica reversed-phase column (150 × 4.0 mm, 5-μm particle size). The mobile phase consisted of acetonitrile (13%) and water (87%). The flow rate was 1 mL/min. The injection volume was 10 μL. UV detection was performed at 215 nm. The retention time for thiotepa was approximately 6.7 minutes.

The method was demonstrated to be stability indicating by accelerated decomposition of thiotepa in the presence of sodium chloride. A solution containing thiotepa 1.5 mg/mL and sodium chloride 100 mg/mL was heated for 1 hour at 100 °C. The thiotepa peak decreased in area and a new peak of the chloro-adduct appeared at approximately 19 minutes.

A standard curve for thiotepa was constructed from 0 to 6.0 mg/mL; the correlation coefficient was 0.998. The intraday and interday coefficients of variation were 0.6 and 1.9%, respectively.

Reference
Murray KM, Erkkila D, Gombotz WR, et al. Stability of thiotepa (lyophilized) in 0.9% sodium chloride injection. *Am J Health Syst Pharm.* 1997; 54: 2588–91.

Method 2
Xu et al. determined the stability of thiotepa 0.5 and 5 mg/mL in 5% dextrose injection. The liquid chromatograph consisted of a Waters 600E multisolvent-delivery pump, a Waters 490E programmable multiple-wavelength UV detector, and a Waters 712 WISP

autosampler. The system was controlled and integrated by an NEC PowerMate SX/16 personal computer. The stationary phase was an Alltech Spherisorb C_{18} reversed-phase HPLC analytical column (300 × 4.6 mm, 5-μm particle size). The mobile phase consisted of 18% methanol in water and was delivered isocratically at 1.2 mL/min. UV detection was performed at 216 nm and 0.5 AUFS.

Samples were diluted 1:10 with water. The injection volume was 10 μL. The retention time for thiotepa was about 6.2 minutes.

The method was determined to be stability indicating by accelerated decomposition of thiotepa. Boiling a thiotepa solution for 20 minutes yielded a reduction in peak area for intact thiotepa and the appearance of a decomposition peak at about 2 minutes. Each run was completed within 10 minutes.

A standard curve for thiotepa was generated from 0.125 to 0.750 mg/mL from a linear plot of peak area versus thiotepa concentration. The correlation coefficient of the standard curve was greater than 0.9998. The intraday and interday coefficients of variation were 2.3 and 1.5%, respectively.

Reference
Xu QA, Trissel LA, Zhang Y, et al. Stability of thiotepa (lyophilized) admixtures in 5% dextrose injection. *Am J Health Syst Pharm*. 1996; 53: 2728–30.

Thymopentin

Chemical Name
L-Arginyl-L-lysyl-L-α-aspartyl-L-valyl-L-tyrosine

Other Names
Mepentil, Thymopoietin, Timosin

Form	Molecular Formula	MW	CAS
Thymopentin	$C_{30}H_{49}N_9O_9$	679.8	69558-55-0

Method
Helm and Muller investigated by HPLC the effect of pH and buffer on the stability of thymopentin in aqueous solution. The instrument consisted of a Pye Unicam model PU 4011 dual-piston pump, a model PU 4020 variable-wavelength UV detector, a model PU 4810 integrator, and a Rheodyne 20-μL injector. The stationary phase was a Waters μBondapak C_{18} column (250 × 3.9 mm, 10-μm particle size). The mobile phase consisted of 0.02 M potassium dihydrogen phosphate buffer (pH 3.0) and methanol (90:10, vol/vol). The flow rate was 3 mL/min. UV detection was performed at 210 nm. Under these conditions, the retention time for thymopentin was about 4.5 minutes.

The method was reported to be stability indicating since it separated the drug from its major degradation products.

Standard curves for thymopentin were constructed from 50 to 200 μg/mL. The correlation coefficients were greater than 0.999.

Reference

Helm VJ, Muller BW. Stability of the synthetic pentapeptide thymopentin in aqueous solution: Effect of pH and buffer on degradation. *Int J Pharm*. 1991; 70: 29–34.

Tiagabine Hydrochloride

Chemical Name

1-[4,4-Bis(3-methyl-2-thienyl)-3-butenyl]-3-piperidinecarboxylic acid hydrochloride

Other Name

Gabitril

Form	Molecular Formula	MW	CAS
Tiagabine hydrochloride	$C_{20}H_{25}NO_2S_2.HCl$	412.0	145821-59-6

Appearance

Tiagabine hydrochloride occurs as a white to off-white odorless crystalline powder.

Solubility

Tiagabine hydrochloride is sparingly soluble in water to about 3% and soluble in aqueous base.

pK$_a$

Tiagabine hydrochloride has a pK_{a1} of 3.3 and a pK_{a2} of 9.4.

Method

Nahata and Morosco examined by HPLC analysis the chemical stability of tiagabine hydrochloride in two extemporaneously prepared suspensions. The Hewlett-Packard 1050 chromatograph consisted of a pump, an autosampler, and a variable-wavelength detector and was equipped with a model 3396A integrator. The stationary phase was a MAC-MOD Zorbax CN column (150 × 3.0 mm). The mobile phase consisted of 5 mM octanesulfonic acid aqueous solution and acetonitrile (50:50, vol/vol). The flow rate was 0.4 mL/min. UV detection was performed at 240 nm.

One hundred microliters of suspension sample was mixed with 5 mL of mobile phase and centrifuged. The supernatant was collected and analyzed. The injection volume was 10 µL. The run time was 10 minutes. Under these conditions, the retention time of tiagabine was about 3.2 minutes.

The assay method was verified to be stability indicating by accelerated degradation of tiagabine hydrochloride. The drug was treated with 2 M hydrochloric acid, 2 M sodium hydroxide solution, or 0.03% hydrogen peroxide at 60 °C. The degradation products did not interfere with the determination of tiagabine.

A calibration curve for tiagabine was constructed from 0.10 to 1.50 mg/mL. The correlation coefficient was greater than 0.999. The intraday and interday coefficients of variation of the analysis were less than 1.7 and 2.3%, respectively.

Reference

Nahata MC, Morosco RS. Stability of tiagabine in two oral liquid vehicles. *Am J Health Syst Pharm.* 2003; 60: 75–7.

Ticarcillin

Chemical Names

[2S-[2α, 5α, 6β(S*)]]-6-[(Carboxy-3-thienylacetyl)amino]-3,3-dimethyl-7-oxo-4-thia-1-azabicyclo[3.2.0]heptane-2-carboxylic acid

N-(2-Carboxy-3,3-dimethyl-7-oxo-4-thia-1-azabicyclo[3.2.0]hept-6-yl)-3-thiophene malonamic acid

Other Name

Ticar

Form	Molecular Formula	MW	CAS
Ticarcillin	$C_{15}H_{16}N_2O_6S_2$	384.4	34787-01-4
Ticarcillin disodium	$C_{15}H_{14}N_2Na_2O_6S_2$	428.4	4697-14-7

Appearance

Ticarcillin disodium is a hygroscopic white to pale yellow powder or lyophilized cake.

Solubility

Ticarcillin disodium has a solubility of more than 100 g in 100 mL of water.

Method 1

Mayron and Gennaro used an HPLC method to assess the stability of ticarcillin disodium with granisetron hydrochloride during simulated Y-site administration. The liquid chromatograph consisted of a piston pump with a pulse dampener, a rotary injection port with a 20-μL loop, a variable-wavelength spectrometric detector, and an integrator. The stationary phase was a cyano reversed-phase column (250 × 4.6 mm, 5-μm particle size). The mobile phase was a mixture of acetonitrile and 0.1 M monobasic sodium phosphate dihydrate (20:80, vol/vol) adjusted to pH 4.2 with phosphoric acid. The flow rate was 0.8 mL/min. UV detection was performed at 195 nm.

Samples were diluted with the mobile phase. The retention times for ticarcillin and granisetron were 3.07 and 7.22 minutes, respectively.

The stability-indicating capability of the analytical method was demonstrated by an accelerated degradation study. Solutions of ticarcillin disodium and granisetron hydrochloride were adjusted to pH 2 and 11, boiled for 1 hour, readjusted to pH 5, diluted with the mobile phase, and analyzed. The degradation product peaks did not interfere with the intact ticarcillin or granisetron peaks.

Reference
Mayron D, Gennaro AR. Stability and compatibility of granisetron hydrochloride in i.v. solutions and oral liquids and during simulated Y-site injection with selected drugs. *Am J Health Syst Pharm.* 1996; 53: 294–304.

Method 2
Perry et al. investigated the chemical stability of ticarcillin disodium in total parenteral nutrient solutions at concentrations commonly used in adults. The HPLC system consisted of a Rheodyne injector 7125 valve, a Waters model 6000A solvent-delivery system, a Waters Lambda-Max 481 UV detector, and a BBC Servogen 120 chart recorder. The stationary phase was a Waters cyano-bonded reversed-phase C_{18} column (300 × 4.5 mm, 5-μm particle size). The mobile phase consisted of 30% acetonitrile, 0.1% phosphoric acid, and 0.3% tetramethylammonium chloride. The flow rate was 1.0 mL/min. UV detection was performed at 204 nm and 0.05 AUFS.

Samples were diluted 1:1000 with purified water. The injection volume was 20 μL. Under these conditions, the retention time for ticarcillin was 8.5 minutes.

The assay was demonstrated to be stability indicating by accelerated decomposition of ticarcillin disodium. A solution of ticarcillin disodium was adjusted to pH 1.6, a second solution was evaporated on a hot plate, and a third solution was heated in a boiling water bath for 15 minutes. In all cases, degradation product peaks did not interfere with the intact ticarcillin peak.

A standard curve was constructed from 0 to 50 mg/mL. The intraassay coefficient of variation was 7.8%.

Reference
Perry M, Khalidi N, Sanders CA. Stability of penicillins in total parenteral nutrient solution. *Am J Hosp Pharm.* 1987; 44: 1625–8.

Method 3
Gupta and Stewart reported a stability-indicating HPLC assay for use in a stability study of ticarcillin disodium in 0.9% sodium chloride injection and 5% dextrose injection. The liquid chromatograph consisted of a Waters model ALC 202 system equipped with a U6K universal injector, a Schoeffel Spectroflow monitor SF 770 multiple-wavelength detector, a Houston Omniscribe recorder, and a Spectra-Physics autolab minigrator-integrator. The stationary phase was a Waters μBondapak phenyl column (300 × 4 mm). The mobile phase was 0.01 M ammonium acetate. The flow rate was 1.6 mL/min. UV detection was performed at 245 nm and 0.1 AUFS. Samples were diluted with water. The injection volume was 20 μL. The coefficient of variation was 1.6%.

Reference
Gupta VD, Stewart KR. Quantitation of carbenicillin disodium, cefazolin sodium, cephalothin sodium, nafcillin sodium, and ticarcillin disodium by high-pressure liquid chromatography. *J Pharm Sci.* 1980; 69: 1264–7.

Tinidazole

Chemical Name
1-[2-(Ethylsulfonyl)ethyl]-2-methyl-5-nitroimidazole

Other Name
Fasigyn

Form	Molecular Formula	MW	CAS
Tinidazole	$C_8H_{13}N_3O_4S$	247.3	19387-91-8

Appearance
Tinidazole occurs as an almost white or pale yellow crystalline powder.

Solubility
Tinidazole is practically insoluble in water and sparingly soluble in methanol.

Method 1
Bakshi and Singh developed a stability-indicating HPLC method for tinidazole. The Waters chromatograph consisted of a model 600E pump, a model 996 photodiode-array detector, a model 717 autoinjector, and a degasser. The stationary phase was a Waters Spherisorb C_{18} column (250 × 4.6 mm, 5-μm particle size). The mobile phase consisted of water and acetonitrile (88:12, vol/vol). The flow rate was 0.8 mL/min. UV detection was performed at 310 nm. The injection volume was 10 μL. Under these conditions, the retention time of tinidazole was about 14.4 minutes.

The accelerated degradation study of tinidazole was carried out by dissolving the drug in 0.1 M hydrochloric acid (1 mg/mL) and heating it at 80 °C for 12 hours in 0.1 M sodium hydroxide solution at 80 °C, in 0.01 M phosphate buffer (pH 10) at 80 °C for 6 hours, in water at 80 °C for 5 days, in 3% hydrogen peroxide at room temperature for 6 hours, and in 30% hydrogen peroxide at room temperature for 48 hours or exposing tinidazole in 0.1 M hydrochloric acid, water, and phosphate buffer (pH 10) in a photo-stability chamber for 12 days. Photodiode-array purity analysis indicated that the degradation products did not coelute with the drug.

A calibration curve for tinidazole was constructed from 50 to 500 μg/mL; the correlation coefficient was 0.999.

Reference
Bakshi M, Singh S. HPLC and LC-MS studies on stress degradation behavior of tinidazole and development of a validated specific stability-indicating HPLC assay method. *J Pharm Biomed Anal.* 2004; 34: 11–8.

Method 2

Argekar and Powar reported the simultaneous determination of tinidazole and ciprofloxacin in tablet dosage forms by HPLC. The instrument included a Tosho CCPE pump, a universal injector, a UV detector, and an Oracle-2 integrator. The stationary phase was an Inertsil C_{18} column (250 × 4.0 mm, 5-μm particle size). The mobile phase was a mixture of 0.1% triethanolamine in water and acetonitrile (78:22, vol/vol), adjusted to pH 2.6 with 1% phosphoric acid. The flow rate was 1.0 mL/min. UV detection was performed at 310 nm.

A sample of ground tablets equivalent to 600 mg of tinidazole and 50 mg of ciprofloxacin was accurately weighed and sonicated for 30 minutes in a 100-mL volumetric flask containing distilled water; the flask was then filled to the mark with water and the mixture was filtered through a Whatman No. 42 filter. The filtrate was diluted 1:10 with the mobile phase. The injection volume was 20 μL. Under these conditions, retention times for ciprofloxacin and tinidazole were 4.2 and 8.2 minutes, respectively.

The assay was demonstrated to be stability indicating. Samples were mixed with 0.1 N hydrochloric acid, 0.1 N sodium hydroxide solution, or 30% hydrogen peroxide; were exposed to sunlight for a week; or were stored at 45 °C for a month. No interference of degradation products with the analysis of tinidazole and ciprofloxacin was observed.

A calibration curve for tinidazole was constructed from 10 to 500 μg/mL. The limits of detection and quantitation were 3 and 4 μg, respectively.

Reference

Argekar AP, Powar SG. Simultaneous determination of ciprofloxacin and tinidazole in pharmaceutical preparations by RP-HPLC. *Indian Drugs.* 1999; 36: 399–402.

Tizanidine Hydrochloride

Chemical Name

5-Chloro-*N*-(4,5-dihydro-1*H*-imidazol-2-yl)-2,1,3-benzothiadiazol-4-amine monohydrochloride

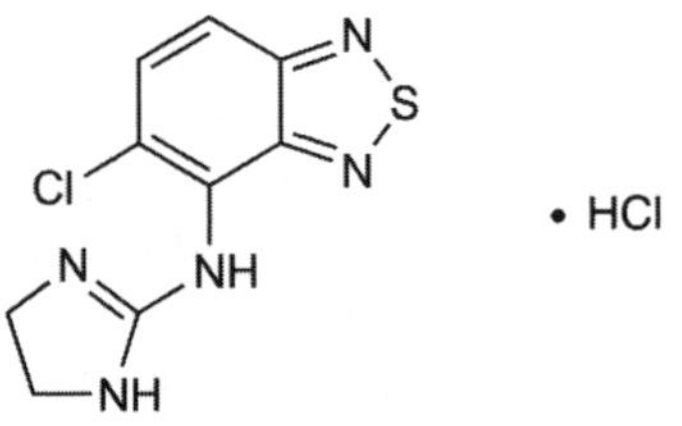

· HCl

Other Names

Sirdalud, Ternelin, Zanaflex

Form	Molecular Formula	MW	CAS
Tizanidine hydrochloride	$C_9H_8ClN_5S \cdot HCl$	290.2	64461-82-1

Appearance
Tizanidine occurs as crystals.

Method
Qi et al. developed a method for the determination of tizanidine and related substances in dosage formulations. An Agilent HP1100 series chromatographic system was used. The stationary phase was a Hypersil CN column (150 × 5.0 mm, 5-μm particle size). The mobile phase consisted of an ion-pairing solution, methanol, and acetonitrile (50:57:18, vol/vol/vol). The ion-pairing solution was prepared by mixing 2.5 mL of 0.05 M heptane-sulfonic acid sodium salt and 0.8 mL of triethylamine in 800 mL of water, adjusting to pH 3.3 with glacial acetic acid, and completing to 1000 mL with water. The flow rate was 1.0 mL/min. UV detection was performed at 227 nm.

A portion of powder from two tablets equivalent to 2 mg of tizanidine was accurately weighed, transferred into a 50-mL volumetric flask, mixed with 20 mL of mobile phase, sonicated for 5 minutes, brought to volume with mobile phase, and filtered. The filtrate was diluted by a factor of 5 with mobile phase. The injection volume was 20 μL. Under these conditions, the retention time of tizanidine was about 7.0 minutes.

The method was verified to be stability indicating by assaying a mixture of tizanidine and its five precursors, intermediates, and intentional degradation products. Drug tablets were exposed to heat, acid (0.1 M hydrochloric acid), base (0.1 M sodium hydroxide solution), and oxidizing agent (3% hydrogen peroxide). The intact drug peak was separated from its precursors, intermediates, excipients, and degradation products.

A calibration curve for tizanidine was generated from 2 to 20 μg/mL; the correlation coefficient was 0.9998. The coefficient of variation of the analysis was 1.0%. The limit of quantitation was 0.051 μg/mL.

Reference
Qi M-L, Wang P, Wang L. Validated liquid chromatography method for assay of tizanidine in drug substance and formulated products. *Anal Chim Acta.* 2003; 478: 171–7.

Tobramycin

Chemical Name
O-3-Amino-3-deoxy-α-D-glucopyranosyl-(1→6)-*O*-[2,6-diamino-2,3,6-trideoxy-α-D-*ribo*-hexopyranosyl-(1→4)]-2-deoxy-D-streptamine

Other Name

Nebcin

Form	Molecular Formula	MW	CAS
Tobramycin	$C_{18}H_{37}N_5O_9$	467.5	32986-56-4
Tobramycin sulfate	$(C_{18}H_{37}N_5O_9)_2 \cdot 5H_2SO_4$	1425.4	79645-27-5

Appearance

Tobramycin is a white to off-white hygroscopic powder.

Solubility

Tobramycin is freely soluble in water and very slightly soluble in alcohol.

Method

Stiles and Allen studied the stability of tobramycin sulfate 1 and 10 mg/mL in polyvinyl chloride drug reservoirs. The liquid chromatographic system included a Waters model 501 solvent-delivery system, a Micromeritics model 728 autoinjector, a Micromeritics model 732 injection valve, a Waters Lambda-Max model 481 variable-wavelength UV detector, and a Waters model 745 data module. The stationary phase was a Bakerbond C_{18} column (250 × 2.6 mm, 5-μm particle size). The mobile phase contained 2.0 g of tris(hydroxymethyl)aminomethane, 800 mL of water, and 20 mL of 1 N sulfuric acid and was diluted to 2000 mL. The flow rate was 1.2 mL/min. UV detection was performed at 365 nm.

Samples were diluted 1:5 (for 1-mg/mL samples) and 1:10 (for 10-mg/mL samples) with the mobile phase. Derivatizing reagent 1 was 2,4-dinitrofluorobenzene 10 mg/mL in ethanol. Derivatizing reagent 2 was prepared by dissolving 300 mg of tris(hydroxymethyl)aminomethane in 20 mL of water and slowly adding that solution to 80 mL of dimethyl sulfoxide with mixing. To 4 mL of the sample solution were added 10 mL of derivatizing reagent 1 and 10 mL of derivatizing reagent 2 with mixing. The mixture was heated on a water bath at 60 °C for 60 minutes. It was brought to a volume of 48 mL with acetonitrile and allowed to cool to room temperature. It was then brought to 50 mL with additional acetonitrile.

The method was demonstrated to be stability indicating by accelerated degradation of tobramycin. The retention time for the tobramycin derivative was 11.2 minutes and retention times for its degradation products were 6.3 and 8.5 minutes.

A standard curve was generated for tobramycin sulfate from 50 to 250 μg/mL. Intraday and interday coefficients of variation of the assay were 11.2 and 0.9%, respectively.

Similar methods were used by the other researchers cited here.

References

Stiles ML, Allen LV Jr. Stability of nafcillin sodium, oxacillin sodium, penicillin G potassium, penicillin G sodium, and tobramycin sulfate in polyvinyl chloride drug reservoirs. *Am J Health Syst Pharm*. 1997; 54: 1068–70.

Xu QA, Trissel LA, Saenz CA, et al. Stability of gentamicin sulfate and tobramycin sulfate in AutoDose infusion system bags. *Int J Pharm Compound*. 2002; 6: 152–4.

Xu QA, Trissel LA, Zhang Y, et al. Compatibility and stability of linezolid injection admixed with gentamicin sulfate and tobramycin sulfate. *Int J Pharm Compound*. 2000; 4: 476–9.

Russ H, McCleary D, Katimy R, et al. Development and validation of a stability-indicating HPLC method for the determination of tobramycin and its related substances in an ophthalmic suspension. *J Liq Chromatgr Rel Technol*. 1998; 21: 2165–81.

Tocopherol (Vitamin E)

Chemical Name
(+)-2,5,7,8-Tetramethyl-2-(4,8,12-trimethyltridecyl)chroman-6-ol

Form	Molecular Formula	MW	CAS
d-α-Tocopherol	$C_{29}H_{50}O_2$	430.7	59-02-9
dl-α-Tocopherol	$C_{29}H_{50}O_2$	430.7	10191-41-0

Appearance
Tocopherols occur as practically odorless and clear colorless or yellowish-brown viscous oils.

Solubility
Tocopherols are practically insoluble in water. They are freely soluble in dehydrated alcohol.

Method
Allwood and Martin used HPLC to study the photodegradation of vitamins A and E in parenteral nutrient mixtures during infusion. The liquid chromatograph consisted of a Kontron model 422 pump, a Gynkotek model UVD 340S diode-array detector, a Gynkotek model Gina 50 autosampler, and a Gynkotek PC integrator. The stationary phase was a Hypersil ODS column (250 × 4.6 mm, 5-μm particle size). The mobile phase was methanol. The flow rate was 1.5 mL/min. UV detection was performed at 292 nm. The injection volume was 20 μL.

The method was reported to be stability indicating.

A standard curve for tocopherol was constructed from 1 to 6 μg/mL. The correlation coefficient was 0.9943.

Reference
Allwood MC, Martin HJ. The photodegradation of vitamins A and E in parenteral nutrition mixtures during infusion. *Clin Nutr*. 2000; 19: 339–42.

Tolbutamide

Chemical Names
N-[(Butylamino)carbonyl]-4-methylbenzenesulfonamide
1-Butyl-3-(p-tolylsulfonyl)urea

Other Name
Orinase

Form	Molecular Formula	MW	CAS
Tolbutamide	$C_{12}H_{18}N_2O_3S$	270.3	64-77-7
Tolbutamide sodium	$C_{12}H_{17}N_2NaO_3S$	292.3	473-41-6

Appearance
Tolbutamide sodium is a white to off-white and practically odorless crystalline powder.

Solubility
Tolbutamide sodium is freely soluble in water and soluble in alcohol.

Method 1
Gupta described the quantitation of tolbutamide in tablets by HPLC analysis. The Waters model ALC 202 liquid chromatograph was equipped with a model U6K universal injector, a Schoeffel model SF 770 UV detector, a Spectra-Physics Autolab minigrator, and an Omniscribe recorder. The stationary phase was a Waters μBondapak phenyl column (300 × 4 mm). The mobile phase consisted of methanol, water, and glacial acetic acid (45:54.5:0.5, vol/vol/vol) containing 0.02 M ammonium acetate. The flow rate was 2.0 mL/min. UV detection was performed at 232 nm and 0.1 AUFS. Chlorpropamide 1.0 mg/mL in methanol was used as an internal standard.

A portion of ground tablet powder equivalent to 50 mg of tolbutamide was weighed, mixed with 40 mL of methanol, stirred for 4–5 minutes, brought to 50 mL with methanol, filtered, and diluted 20-fold with the internal standard and water. The injection volume was 20 μL. Under these conditions, the retention times of chlorpropamide and tolbutamide were about 5.3 and 6.9 minutes, respectively (estimated from the published chromatogram).

The method was stated to be stability indicating since the known degradation product of tolbutamide, p-toluenesulfonamide, did not influence the analysis of the drug.

A standard curve for tolbutamide was constructed from 0.8 to 2.4 μg.

Reference
Gupta VD. Quantitation of chlorpropamide and tolbutamide in tablets by stability-indicating reverse-phase high-performance liquid chromatography. *Anal Lett.* 1984; 17: 2119–28.

Method 2

Robertson et al. reported the stability-indicating HPLC determination of tolbutamide sodium and its sulfonamide degradates. A Varian model 4100 liquid chromatograph included a fixed-wavelength UV detector, a septumless injector port, and a Spectra-Physics computing integrator. The stationary phase was a Brinkmann LiChrosorb Si-60 column (250×3.2 mm, 10-μm particle size). The mobile phase consisted of 4% absolute ethanol, 8% tetrahydrofuran, and 0.06% acetic acid in n-hexane. UV detection was performed at 254 nm and 0.02 AUFS. Micronized prednisone 0.70 mg/mL in ethyl acetate was used as an internal standard.

Samples were extracted with a mixture of 10% (vol/vol) aqueous hydrochloric acid and the internal standard solution. The injection volume was 5 μL.

This assay was confirmed to be stability indicating by accelerated degradation of tolbutamide sodium samples. Tolbutamide sodium solution was exposed to 60 °C and 70% relative humidity for 6 weeks. Tolbutamide sodium partially degraded but the degradation product, p-toluenesulfonamide, did not interfere with the quantitation of the intact tolbutamide. Retention times for tolbutamide, p-toluenesulfonamide, and prednisone were about 3.1, 5.9, and 16.0 minutes, respectively (estimated from the published chromatogram).

Calibration curves were constructed from 0.3 to 3.0 mg/mL. The correlation coefficient was 0.9997.

Reference

Robertson DL, Butterfield AG, Kolasinski H, et al. Stability-indicating high-performance liquid chromatographic determination of chlorpropamide, tolbutamide, and their respective sulfonamide degradates. *J Pharm Sci.* 1979; 68: 577–80.

Topiramate

Chemical Name

2,3:4,5-Di-*O*-isopropylidene-β-D-fructopyranose sulfamate

Other Name

Topamax

Form	Molecular Formula	MW	CAS
Topiramate	$C_{12}H_{21}NO_8S$	339.4	97240-79-4

Appearance
Topiramate occurs as a white crystalline powder.

Solubility
Topiramate has a solubility of 9.8 mg/mL in water and it is freely soluble in alcohol.

Method
Micheel et al. reported an HPLC method to analyze sulfamate and sulfate ions in degraded topiramate. The Waters liquid chromatograph was composed of a model 600E pump, a model 715 autosampler, and a Dionex pulsed electrochemical detector equipped with an anion self-regenerating suppressor (4 mm). The stationary phase was a Dionex Ion Pac AS5A analytical column (150 × 4.0 mm, 5-μm particle size) with a Dionex Ion Pac AG5A guard column (50 × 4.0 mm, 5-μm particle size). Mobile phase A was deionized or distilled water (18 mΩ or better). Mobile phase B was 50 mM sodium hydroxide solution (18 mΩ or better). The mobile phase was gradiently delivered at 1.0 mL/min as follows:

Time, minutes	A, %	B, %
0	96	4
2	50	50
15	50	50
16	96	4
30	96	4

The total run time was 30 minutes. The detector was set in the conductivity mode with the anion self-regenerating suppressor in the autosuppression recycle mode.

The sample solvent was a mixture of acetonitrile and water (20:80, vol/vol). Tablet samples were shaken in sample solvent for 1 hour and filtered through a 0.2-μm Whatman Nylon-66 filter device. The injection volume was 20 μL. Under these conditions, the retention times for sulfamate and sulfate were 6 and 17 minutes, respectively.

The method was reported to be stability indicating.

A standard curve for sulfamic acid was constructed from 12.5 to 9375 ng per injection with a correlation coefficient of 0.9997. A standard curve for sodium sulfate was also generated from 49.98 to 14,992.5 ng per injection with a correlation coefficient of 0.9995. The limit of detection for sulfamate was 0.02 mol%. The limits of quantitation were 0.05 mol% for sulfamate and 0.1 mol% for sulfate.

Reference
Micheel AP, Ko CY, Guh HY. Ion chromatography method and validation for the determination of sulfate and sulfamate ions in topiramate drug substance and finished product. *J Chromatogr B*. 1998; 709: 166–72.

Topotecan

Chemical Name

(*S*)-10-[(Dimethylamino)methyl]-4-ethyl-4,9-dihydroxy-1*H*-pyrano[3′,4′:6,7]indolizino-
[1,2-*b*]quinoline-3,14(4*H*,12*H*)-dione

Other Names

Asotecan, Hycamtin

Form	Molecular Formula	MW	CAS
Topotecan	$C_{23}H_{23}N_3O_5$	421.5	123948-87-8
Topotecan hydrochloride	$C_{23}H_{23}N_3O_5 \cdot HCl$	458.0	119413-54-6

Appearance

Topotecan hydrochloride occurs as a light yellow to greenish powder.

Solubility

Topotecan hydrochloride is soluble in water up to 1 mg/mL.

Method 1

Kramer and Thiesen evaluated the stability of topotecan infusion solutions in polyvinyl chloride bags and in elastomeric portable infusion devices. The liquid chromatograph consisted of a Waters model 510 pump, a model 717 plus autosampler, and a model 996 photodiode-array detector. The stationary phase was a Waters Symmetry C_{18} reversed-phase column (150 × 3.9 mm, 5-μm particle size). The mobile phase consisted of water, acetonitrile, and 0.1% trifluoroacetic acid (85:15:0.1, vol/vol/vol). The flow rate was 1.2 mL/min. UV detection was performed at 228 nm.

Samples were diluted 1:20 with water for injection. The injection volume was 10 μL. Under these conditions, the retention time for topotecan was about 6.5 minutes.

The stability-indicating nature of the method was established by accelerated degradation of topotecan. Topotecan samples were heated at 100 °C for 45 minutes or diluted with 4 N sodium hydroxide solution (pH 14). The intact topotecan was clearly separated from its degradation products on its chromatogram.

A standard curve for topotecan was constructed from 12 to 75 μg/mL. The correlation coefficient was 0.9997.

Reference

Kramer I, Thiesen J. Stability of topotecan infusion solutions in polyvinylchloride bags and elastomeric portable infusion devices. *J Oncol Pharm Practice*. 1999; 5: 75–82.

Method 2

Patel et al. evaluated the stability of solutions of topotecan hydrochloride injection in sterile water stored for up to 28 days at 5, 25, and 30 °C. The HPLC system consisted of two Shimadzu model LC-6A solvent-delivery pumps, a Shimadzu model SPD-6AV UV detector, a Shimadzu model SIL-6A autoinjector, and a Shimadzu model SCL-6A system controller with Perkin-Elmer PE-Nelson AccessChrom software. The stationary phase was a Nacalai Tesque Cosmosil 5C18-AR C_{18} column (250 × 4.6 mm, 5-μm particle size). Mobile phase A consisted of water, acetonitrile, and trifluoroacetic acid (85:15:0.1, vol/vol/vol), and mobile phase B was a mixture of water, acetonitrile, and trifluoroacetic acid (60:40:0.1, vol/vol/vol). The flow rate was 1.0 mL/min. The mobile phase was delivered gradiently at 100% A from 0 to 16 minutes, 100% A to 100% B from 16 to 40 minutes, and then returned to 100% A and run from 42 to 60 minutes. UV detection was performed at 228 nm.

Samples were diluted with 0.001 N hydrochloric acid, yielding a topotecan concentration of 0.08 mg/mL. The injection volume was 20 μL. Under these conditions, the retention time for topotecan was 10.7–12.0 minutes.

The stability-indicating capability of the method was demonstrated by accelerated decomposition of topotecan and by spiking the topotecan hydrochloride solution with the known impurities and degradation products. The method was accepted by the Food and Drug Administration (FDA) for measuring the potency and stability of topotecan. Topotecan samples were placed under diffused light, at 40 °C for several weeks, or under acidic and basic conditions. Degradation product peaks did not interfere with the intact topotecan peak.

A calibration curve for topotecan was constructed from a linear plot of topotecan peak area versus concentration from 0.010 to 0.124 mg/mL; the correlation coefficient was 0.9999. The precisions, expressed as relative standard deviation, were 1.2, 0.8, and 0.6% for concentrations of 0.074, 0.099, and 0.124 mg/mL, respectively. The intraday and interday variations were 1.0 and 0.9%, respectively.

References

Patel K, Craig SB, McBride MG, et al. Microbial inhibitory properties and stability of topotecan hydrochloride injection. *Am J Health Syst Pharm*. 1998; 55: 1584–7.

Graig SB, Bhatt UH, Patel K. Stability and compatibility of topotecan hydrochloride for injection with common infusion solutions and containers. *J Pharm Biomed Anal*. 1997; 16: 199–205.

Torsemide

Chemical Name

N-[[(1-Methylethyl)amino]carbonyl]-4-[(3-methylphenyl)amino]-3-pyridinesulfonamide

Other Names
Demadex, Torasemid, Torasemide

Form	Molecular Formula	MW	CAS
Torsemide	$C_{16}H_{20}N_4O_3S$	348.4	56211-40-6

Appearance
Torsemide is a white to off-white crystalline powder.

pK$_a$
Torsemide has a pK$_a$ of 7.1

Method
Montgomery et al. determined the stability of torsemide in 5% dextrose injection through 72 hours. The HPLC system consisted of a Waters model 501 pump, a Waters model 712 WISP autosampler, and a Waters model 991 PDA detector and was controlled by Waters Millennium 2010 chromatography manager. The stationary phase was a Waters μBondapak C_{18} column (300 × 3.9 mm, 10-μm particle size). The mobile phase consisted of 0.01 M monobasic potassium phosphate buffer (pH 4.5) (Component A) and acetonitrile (Component B) and was delivered gradiently at 1.5 mL/min. The step-gradient was used as follows: 15% B from 0 to 6 minutes, 34% B from 6.1 to 8 minutes, 25% B from 8.1 to 13 minutes, 30% B from 13.1 to 23 minutes, and 15% B from 23.1 to 30 minutes.

Samples were diluted with deionized water. The injection volume was 15 μL. Under these conditions, the retention time for torsemide was 4.8 minutes.

The stability-indicating nature of the method was shown by accelerated decomposition of torsemide. Torsemide injection 10 mg/mL was adjusted to pH 1 with concentrated sulfuric acid and to pH 12 with sodium hydroxide and allowed to stand in normal fluorescent light for 12 days. These solutions were then heated to 100 °C for 1 hour, cooled down, and adjusted to pH 7 before being analyzed. Degradation product peaks did not interfere with the intact torsemide peak.

Standard curves for torsemide were constructed from 0.10 to 0.30 mg/mL and from 0.4 to 0.6 mg/mL. The correlation coefficient of the standard curve was 0.996. The intraday and interday coefficients of variation were 2.3 and 5.1%, respectively.

Reference
Montgomery PA, Cornish LA, Johnson CE, et al. Stability of torsemide in 5% dextrose injection. *Am J Health Syst Pharm.* 1998; 55: 1042–3.

Treprostinil Sodium

Chemical Name
[[(1*R*,2*R*,3a*S*,9a*S*)-2,3,3a,4,9,9a-Hexahydro-2-hydroxy-1-[(3*S*)-3-hydroxyoctyl]-1*H*-benz-
[*f*]inden-5-yl]oxy]acetic acid monosodium salt

Other Name
Remodulin

Form	Molecular Formula	MW	CAS
Treprostinil sodium	$C_{23}H_{33}NaO_5$	412.5	289480-64-4

Appearance
Treprostinil occurs as crystals.

Method 1
Using an HPLC method, Xu et al. evaluated the chemical stability of treprostinil sodium injection packaged in plastic syringe pump reservoirs. The Waters LC module 1 chromatograph consisted of a multisolvent-delivery pump and an ultraviolet detector. The stationary phase was an Alltech Spherisorb CN analytical column (250 × 4.6 mm, 5-µm particle size). The mobile phase consisted of acetonitrile (1200 mL) and 0.05 M monobasic potassium phosphate (300 mL), adjusted to pH 2.4 with phosphoric acid. The flow rate was 1.2 mL/min. UV detection was performed at 215 nm and 0.5 AUFS.

Samples were diluted to a nominal treprostinil concentration of 0.05 mg/mL with mobile phase. The injection volume was 5 µL. Under these conditions, retention times for the preservative metacresol and for treprostinil were about 4.8 and 8.0 minutes, respectively.

The method was validated to be stability indicating by accelerated degradation. Sample solutions were heated and exposed to 1 N hydrochloric acid, 1 N sodium hydroxide solution, and 3% hydrogen peroxide. The degradation product peaks did not interfere with the intact drug or metacresol peaks.

A standard curve was constructed from 10 to 125 µg/mL. The correlation coefficient was greater than 0.9997. The intraday and interday coefficients of variation were 1.4 and 1.2%, respectively.

Reference
Xu QA, Trissel LA, Pham L. Physical and chemical stability of treprostinil sodium injection packaged in plastic syringe pump reservoirs. *Int J Pharm Compound.* 2004; 8: 228–30.

Method 2
Using an HPLC method, Phares et al. studied the stability of treprostinil sodium diluted in sterile water for injection, 0.9% sodium chloride injection, and 5% dextrose injection. The liquid chromatograph consisted of a Hewlett-Packard 1050 quaternary gradient pump, a Thermo-Separation AS100 fixed loop injector, and a Thermo-Separation UV100 UV-visible detector. The stationary phase was a Hypersil ODS column (250 × 4.6 mm, 5-µm particle size). Mobile phase A was a mixture of acetonitrile, water, and

trifluoroacetic acid (40:60:0.1, vol/vol/vol) and mobile phase B was a mixture of acetonitrile, water, and trifluoroacetic acid (78:22:0.1, vol/vol/vol). The mobile phase was gradiently delivered at 2.0 mL/min at 100% A from 0 to 17 minutes, 100% B from 17 to 35 minutes, and 100% A from 35 to 40 minutes. UV detection was performed at 217 nm.

Treprostinil sodium injection was appropriately diluted before assay in sterile water for injection, 0.9% sodium chloride injection, or 5% dextrose injection. The sample injection volume was 20 μL. Under these conditions, retention times for metacresol and treprostinil were 3.4 and 15.0 minutes, respectively.

The analytical method was validated to be stability indicating by forced degradation of treprostinil. Solutions of treprostinil were irradiated in 1000 foot-candles of light for 7 days, incubated at 60 °C for 35 days, oxidized with 30% hydrogen peroxide solution at 60 °C for 18 hours, treated with 6 N hydrochloric acid at 60 °C for 6 days, or treated with 6 N sodium hydroxide solution at 60 °C for 35 days. Retention times for benzindenetriol and methoxybenzindenediol were 13.6 and 24.4 minutes, respectively. Degradation products did not interfere with the analysis of treprostinil.

A linear relationship was obtained from 0.5 to 5.0% and from 60 to 140% of treprostinil 1.0 mg/mL. The limit of detection and the limit of quantitation were 0.025 and 0.05% at 1.0 mg/mL of treprostinil, respectively.

Reference

Phares KR, Weiser WE, Miller SP, et al. Stability and preservative effectiveness of treprostinil sodium after dilution in common intravenous diluents. *Am J Health Syst Pharm.* 2003; 60: 916–22.

Tretinoin (Retinoic Acid)

Chemical Name

3,7-Dimethyl-9-(2,6,6-trimethylcyclohex-1-enyl)nona-2,4,6,8-all-*trans*-tetraenoic acid

Other Names

Avita, Retin-A, Vitamin A Acid

Form	Molecular Formula	MW	CAS
Tretinoin	$C_{20}H_{28}O_2$	300.4	302-79-4

Appearance

Tretinoin is a yellow to light-orange crystalline powder.

Solubility

Tretinoin is practically insoluble in water and slightly soluble in alcohol.

Method 1

Caviglioli et al. developed an assay for the determination of retinoic acid in hard-gelatin capsules containing lactose and as the bulk drug substance. The Hewlett-Packard 1050 system included a manual injector fitted with a 10-μL loop and a model 1040M diode-array detector. The stationary phase was a Merck LiChrosphere 100 RP18 column (250 × 4 mm, 5-μm particle size) with a LiChrosorb RP18 guard column (4 × 4 mm, 5-μm particle size). The mobile phase consisted of acetonitrile and 1% ammonium acetate aqueous buffer (90:10, vol/vol). The flow rate was 1.25 mL/min. UV detection was performed at 340 nm.

Samples were weighed, transferred into a 50-mL volumetric flask, mixed with methanol, sonicated for 15 minutes, and filtered through a cellulose acetate membrane. Under these conditions, the retention time for retinoic acid was about 9 minutes.

The stability-indicating nature of the method was demonstrated by accelerated degradation of retinoic acid. Retinoic acid was degraded by refluxing at 75 °C for 3.5 hours, by storing at 100 °C for 24 hours and then at 120 °C for another 24 hours, by refluxing in 0.1 N hydrochloric acid–methanol for 5 minutes, by refluxing in 0.1 N sodium hydroxide–methanol for 30 minutes, or by exposing to light at 254 nm for 2 hours. Degradation products did not interfere with the analysis of retinoic acid.

A standard curve for retinoic acid was constructed from 0.12 to 0.38 mg/mL. The correlation coefficient was 1.000.

References

Caviglioli G, Parodi B, Cafaggi S, et al. Stability indicating HPLC assay for retinoic acid in hard gelatin capsules containing lactose and as bulk drug substance. *Drug Dev Ind Pharm.* 1994; 20: 2395–408.

Caviglioli G, Parodi B, Posocco V, et al. Stability study of hard gelatin capsules containing retinoic acid. *Drug Dev Ind Pharm.* 2000; 26: 995–1001.

Method 2

Kril et al. investigated tretinoin in creams by HPLC analysis. The Hewlett-Packard 1090 system with a 1040 photodiode-array detector or a Waters system equipped with a model 6000A solvent-delivery system, a model 712 WISP autosampler, and a model 441 absorbance detector was used. The stationary phase was a Waters Nova-Pak C_{18} column (150 × 3.9 mm, 4-μm particle size) with a Waters Nova-Pak C_{18} guard column. The mobile phase consisted of 420 mL of tetrahydrofuran and 580 mL of phosphate buffer. The buffer was prepared by dissolving 1.38 g of monobasic sodium phosphate hydrate in 1 L of water and adjusting to pH 3.0 with dilute phosphoric acid. The flow rate was 1.0 mL/min. UV detection was performed at 365 nm and 0.02 AUFS. Anthracene 1 mg/mL in tetrahydrofuran was used as an internal standard.

A portion of cream containing 1.0 mg of tretinoin was weighed into a 50-mL volumetric flask, mixed with 5.0 mL of internal standard and 20 mL of tetrahydrofuran, shaken to disperse the cream, and diluted to volume with tetrahydrofuran. A 5-mL aliquot of this solution in a 25-mL volumetric flask was mixed with 10 mL of tetrahydrofuran and 0.1 mL of dilute phosphoric acid, diluted to volume with water, and filtered through a 0.45-μm disposable filter. The injection volume was 25 μL. Under these conditions, the retention times for isotretinoin and tretinoin were 22.7 and 26.1 minutes, respectively.

The stability-indicating nature of the method was shown by accelerated degradation of tretinoin. Tretinoin samples were oxidized with 0.025 M potassium persulfate for 12 hours, irradiated with light (approximately 1000 foot-candles) for 10 minutes, stored in an oven at 50 °C for 47 weeks, or treated with 0.025 M sodium hydroxide for 12 hours or 0.1 M hydrochloric acid for 47 weeks. No degradation products interfered with the analysis of tretinoin.

A standard curve for tretinoin was obtained from 0.4 to 12 μg/mL. The correlation coefficient was 0.9999. The coefficient of variation for the analysis of tretinoin was 1.0%.

Reference

Kril MB, Burke KA, DiNunzio JE, et al. Determination of tretinoin in creams by high-performance liquid chromatography. *J Chromatogr*. 1990; 522: 227–34.

Triamcinolone Acetonide

Chemical Name

9α-Fluoro-11β,21-dihydroxy-16α,17α-isopropylidenedioxypregna-1,4-diene-3,20-dione

Other Names

Albicort, Alcortin, Kenacort, Triamcinoloni Acetonidum, Triderm, Trilone

Form	Molecular Formula	MW	CAS
Triamcinolone	$C_{21}H_{27}FO_6$	394.4	124-94-7
Triamcinolone acetonide	$C_{24}H_{31}FO_6$	434.5	76-25-5
Triamcinolone diacetate	$C_{25}H_{31}FO_8$	478.5	67-78-7
Triamcinolone hexacetonide	$C_{30}H_{41}FO_7$	532.6	5611-51-8

Appearance

Triamcinolone occurs as a white or almost white odorless and slightly hygroscopic crystalline powder. Triamcinolone acetonide occurs as a white or cream-colored almost odorless crystalline powder. Triamcinolone diacetate occurs as a white to off-white almost odorless fine crystalline powder. Triamcinolone hexacetonide occurs as a white to cream-colored crystalline powder.

Solubility

Triamcinolone is very slightly soluble in water, chloroform, and ether. It is slightly soluble in ethanol and methanol. Triamcinolone acetonide is practically insoluble in water and sparingly soluble in ethanol, chloroform, and methanol. It is very slightly soluble in ether. Triamcinolone diacetate is practically insoluble in water. It is soluble in 1 in 13 of ethanol, 1 in 80 of chloroform, and 1 in 40 of methanol. Triamcinolone hexacetonide is practically insoluble in water and sparingly soluble in dehydrated alcohol. It is slightly soluble in methanol and ether.

Method

Gupta used an HPLC method to study the stability of triamcinolone acetonide solutions. The Waters liquid chromatograph was equipped with a model ALC 202 pump, a U6K universal injector, a Schoeffel model SF770 Spectroflow monitor, a Houston Omniscribe recorder, and a Spectra-Physics Autolab minigrator. The stationary phase was a Waters µBondapak C_{18} column (300×4 mm). The mobile phase consisted of 0.02 M monobasic potassium phosphate aqueous buffer and acetonitrile (68:32, vol/vol), pH 4.2. The flow rate was 3.0 mL/min. UV detection was performed at 254 nm and 0.04 AUFS.

For solutions, each sample (8 mL) was mixed with 0.8 mL of hydrocortisone solution (internal standard) and 5 mL of methanol, and filled to the final volume of 25 mL with water. For the ointment, each sample (1 g) was mixed with 0.8 mL of hydrocortisone solution, 5 mL of methanol, and 5 mL of water, stirred until dissolved, and then filled to the final volume of 25 mL with water. The sample injection volume was 20 µL. Under these conditions, retention times of hydrocortisone and triamcinolone acetonide were about 4.0 and 8.2 minutes, respectively (estimated from the published chromatogram).

This method was reported to be stability indicating since it separated the intact triamcinolone acetonide from its degradation products and internal standard. The coefficient of variation was less than 1.9%.

Reference

Gupta VD. Stability of triamcinolone acetonide solutions as determined by high-performance liquid chromatography. *J Pharm Sci.* 1983; 72: 1453–6.

Triamterene

Chemical Name

2,4,7-Triamino-6-phenylpteridine

Other Names

Dyrenium, Dytac, Urocaudal

Form	Molecular Formula	MW	CAS
Triamterene	$C_{12}H_{11}N_7$	253.3	396-01-0

Appearance

Triamterene is a yellow odorless crystalline powder.

Solubility

Triamterene is practically insoluble in water and very slightly soluble in alcohol.

pK_a

Triamterene has a pK_a of 6.2.

Method

Menon and White described a simultaneous determination of hydrochlorothiazide and triamterene in capsules by HPLC analysis. The apparatus consisted of a Waters model 6000A dual-piston reciprocating pump, a model U6K universal injector, a Schoeffel model SF 770 variable-wavelength detector, and a Hewlett-Packard model 3385A automation system. The stationary phase was a Waters μBondapak C_{18} column (300 × 4 mm, 10-μm particle size). The mobile phase was a mixture of 20 mL of 0.2 M sodium acetate (adjusted to pH 5.0 with acetic acid), 780 mL of water, 150 mL of acetonitrile, and 50 mL of methanol. The flow rate was 2 mL/min. UV detection was performed at 273 nm and 0.1 AUFS. *m*-Hydroxyacetophenone 3 mg/mL in acetonitrile–water (1:1, vol/vol) was used as an internal standard.

A portion of powder from 10 capsules equivalent to 50 mg of hydrochlorothiazide and 100 mg of triamterene was prepared as a slurry with 25 mL of acetonitrile, followed by 4 mL of acetic acid. The mixture was shaken well for 5 minutes, dissolved with 10 mL of internal standard solution, diluted to 100 mL with water, and filtered through a Nuclepore polycarbonate membrane. The injection volume was 40 μL. Under these conditions, retention times for hydrochlorothiazide, *m*-hydroxyacetophenone, and triamterene were about 4.0, 7.1, and 10.0 minutes, respectively (estimated from the published chromatogram).

The method was verified to be stability indicating by accelerated degradation of drugs. Samples were refluxed in 1 N hydrochloric acid, 1 N sodium hydroxide solution, or water. The intact drugs were resolved from their degradation products.

A standard curve for triamterene was obtained from 30 to 70 mg per capsule. The correlation coefficient was 1.000. The coefficient of variation for the analysis of triamterene was 0.9%.

Reference

Menon GN, White LB. Simultaneous determination of hydrochlorothiazide and triamterene in capsule formulations by high-performance liquid chromatography. *J Pharm Sci*. 1981; 70: 1083–5.

Trifluoperazine Hydrochloride

Chemical Name

10-[3-(4-Methylpiperazin-1-yl)propyl]-2-(trifluoromethyl)phenothiazine dihydrochloride

Other Names

Stelazine, Terfluzine

Form	Molecular Formula	MW	CAS
Trifluoperazine hydrochloride	$C_{21}H_{24}F_3N_3S.2HCl$	480.4	440-17-5

Appearance

Trifluoperazine hydrochloride is a white to pale yellow and odorless or almost odorless slightly hygroscopic crystalline powder.

Solubility

Trifluoperazine hydrochloride has solubilities of 286 mg/mL in water and 90.9 mg/mL in alcohol at 25 °C.

Method 1

Abdel-Moety and Al-Deeb reported a chromatographic determination of trifluoperazine hydrochloride in bulk form and tablets. A Shimadzu LC-10 AD liquid chromatograph included an SPD-10A tunable UV-detector, a Rheodyne 20-μL injector, a CTO-10A column oven controller, a DGU-3A mechanical degasser, and a C-R4A Chromatopac data unit. The stationary phase was a Varian MicroPak NH_2-10 column (300 × 4 mm, 10-μm particle size). The mobile phase consisted of 505 mL of acetonitrile, 8.3 g of tetra-butylammonium hydroxide, and 1 g of phosphoric acid, made up to the final volume of 1 L with water. The flow rate was 1 mL/min. UV detection was performed at 254 nm.

A sample of the powdered tablets containing about 25 mg of trifluoperazine was added to about 75 mL of the mobile phase, extracted mechanically for about 10 minutes, and brought to the final volume of 100 mL with the mobile phase. It was further diluted to a concentration of 25 μg/mL with the mobile phase. The injection volume was 20 μL. Under these conditions, the retention time of trifluoperazine was 2.5 minutes.

The assay was demonstrated to be stability indicating by accelerated drug photolysis of trifluoperazine. An aqueous solution of 10 mg/mL of the drug was irradiated by a 60-W UV-lamp at 254 nm in a glass open container. The peak of trifluoperazine was resolved from its degradation product peaks on the chromatogram.

A calibration curve for trifluoperazine was generated from 10 to 50 μg/mL. The correlation coefficient was 0.9997.

Reference

Abdel-Moety EM, Al-Deeb OA. Stability-indicating liquid chromatographic determination of trifluoperazine hydrochloride in bulk form and tablets. *Eur J Pharm Sci.* 1997; 5: 1–5.

Method 2

Al-Obaid et al. presented the simultaneous quantitation of trifluoperazine and its sulfoxide. A Waters 600E liquid chromatograph was equipped with a Waters model U6K injector and a model 486 tunable absorbance detector. The stationary phase was a Waters Nova-Pak phenyl column (150 × 3.9 mm, 4-µm particle size). The mobile phase consisted of methanol and 0.015 M sodium acetate buffer (pH 6.5) (81:19, vol/vol). The flow rate was 1.2 mL/min. UV detection was performed at 254 nm and 0.1 AUFS. Alprenolol was used as an internal standard.

Trifluoperazine tablets were ground and a portion of powder equivalent to 4 mg of trifluoperazine was weighed, mixed with 5 mL of water, heated for 3 minutes in a water bath, cooled, mixed with 50 mL of water, shaken for 15 minutes, filled to 100 mL with methanol, and filtered. Internal standard was added and the solution was diluted 1:10 with the mobile phase. The injection volume was 10 µL. Under these conditions, retention times for trifluoperazine sulfoxide, alprenolol, and trifluoperazine were 1.8, 3.4, and 5.5 minutes, respectively.

The method was reported to be stability indicating since it separated trifluoperazine from trifluoperazine sulfoxide.

A standard curve for trifluoperazine was constructed from 1 to 12 µg/mL. The correlation coefficient was 1.0000. The limit of detection was 0.15 ng.

Reference

Al-Obaid AM, Hagga MEM, El-Khawad IE, et al. Simultaneous quantitation of some phenothiazine drug substances and their monosulphoxide degrades by high performance liquid chromatography (HPLC). *J Liq Chrom & Rel Technol.* 1996; 19: 1369–89.

Method 3

Abdel-Moety et al. reported a similar HPLC method for the analysis of trifluoperazine in bulk form and pharmaceutical formulations. The HPLC system was the same as described in Method 1. The stationary phase was a Waters reversed-phase µBondapak C_{18} column (300 × 3.9 mm, 10-µm particle size). The mobile phase consisted of acetonitrile and 0.1% aqueous phosphoric acid (57.5:42.5, vol/vol). Other experimental parameters were the same as described in Method 1. Under these conditions, the retention time for trifluoperazine was about 10.4 minutes (estimated from the published chromatogram).

The method was confirmed to be stability indicating by assaying a photodegraded solution of trifluoperazine with a photodiode-array detector.

Reference

Abdel-Moety EM, Al-Rashood KA, Rauf A, et al. Photostability-indicating HPLC method for determination of trifluoperazine in bulk form and pharmaceutical formulations. *J Pharm Biomed Anal.* 1996; 14: 1639–44.

Triflupromazine

Chemical Name
N,N-Dimethyl-2-(trifluoromethyl)-10*H*-phenothiazine-10-propanamine

Other Names
Fluopromaxine, Psyquil, Vesprin

Form	Molecular Formula	MW	CAS
Triflupromazine hydrochloride	$C_{18}H_{19}F_3N_2S.HCl$	388.9	1098-60-8

Appearance
Triflupromazine hydrochloride is a white to pale tan crystalline powder with a characteristic odor.

Solubility
Triflupromazine hydrochloride is very soluble in water and alcohol and freely soluble in chloroform. It is practically insoluble in ether.

Method
Al-Obaid et al. described the simultaneous quantitation of triflupromazine and its sulfoxide. A Waters 600E liquid chromatograph was equipped with a Waters model U6K injector and a model 486 tunable absorbance detector. The stationary phase was a Waters Nova-Pak phenyl column (150 × 3.9 mm, 4-μm particle size). The mobile phase consisted of methanol and 0.015 M sodium acetate buffer (pH 6.5) (81:19, vol/vol). The flow rate was 1.2 mL/min. UV detection was performed at 254 nm and 0.1 AUFS. Alprenolol was used as an internal standard.

Triflupromazine tablets were ground and a portion of powder equivalent to 4 mg of triflupromazine was weighed, mixed with 5 mL of water, heated for 3 minutes in a water bath, cooled, mixed with 50 mL of water, shaken for 15 minutes, filled to 100 mL with methanol, and filtered. Internal standard was added and the solution was diluted 1:10 with the mobile phase. The injection volume was 10 μL. Under these conditions, retention times for triflupromazine sulfoxide, alprenolol, and triflupromazine were 2.0, 3.2, and 6.1 minutes, respectively.

The method was reported to be stability indicating since it separated triflupromazine from triflupromazine sulfoxide.

A standard curve for triflupromazine was constructed from 1 to 10 μg/mL. The correlation coefficient was 0.9999. The limit of detection was 0.1 ng.

Reference

Al-Obaid AM, Hagga MEM, El-Khawad IE, et al. Simultaneous quantitation of some phenothiazine drug substances and their monosulphoxide degrades by high performance liquid chromatography (HPLC). *J Liq Chromatogr Rel Technol*, 1996; 19: 1369–89.

Trimetazidine Hydrochloride

Chemical Name

1-(2,3,4-Trimethoxybenzyl)piperazine dihydrochloride

Other Names

Idaptan, Imovexil, Vastarel, Vastinol

Form	Molecular Formula	MW	CAS
Trimetazidine hydrochloride	$C_{14}H_{22}N_2O_3 \cdot 2HCl$	339.3	13171-25-0

Appearance

Trimetazidine hydrochloride occurs as a slightly hygroscopic white or almost white crystalline powder.

Solubility

Trimetazidine hydrochloride is freely soluble in water and sparingly soluble in alcohol.

Method

Thoppil and Amin reported a reversed-phase liquid chromatographic method for the determination of trimetazidine hydrochloride. The chromatographic instrument included a Jasco-PV 980 pump, a U-975 UV-visible intelligent detector, and a Rheodyne 7725 injector with a 20-μL loop. The stationary phase was a LiChrospher 100 RP18 stainless steel column (250 × 4 mm, 5-μm particle size). The mobile phase consisted of a mixture of water, acetonitrile, and triethylamine (90:10:1, vol/vol/vol), adjusted to pH 3.3 with phosphoric acid. The flow rate was 1.0 mL/min. UV detection was performed at 270 nm.

Crushed tablet powder equivalent to one tablet was weighed accurately, dispersed in water, and sonicated for 15 minutes. The final volume was brought to 100 mL with purified water. This solution was filtered and the filtrate was collected and appropriately diluted with the mobile phase. The injection volume was 20 μL. Under these conditions, the retention time for trimetazidine was about 5.6 minutes.

The assay was demonstrated to be stability indicating by accelerated degradation of trimetazidine. The drug was refluxed with 1 M hydrochloric acid and 1 M sodium hydroxide solution at 70 °C for 10 hours. A drug solution of 1% (wt/vol) was mixed with 30% (vol/vol) hydrogen peroxide solution for 2 hours. Another solution of 1% (wt/vol) was exposed to direct sunlight for 4 hours. None of the peaks of the degraded products interfered with the intact trimetazidine peak.

A calibration curve of trimetazidine hydrochloride in methanol was constructed from 500 to 3000 ng/mL. The correlation coefficient was 0.9986. The limits of detection and quantitation were 5 and 20 ng/mL, respectively.

Reference

Thoppil SO, Amin PD. Trimetazidine: Stability indicating RPLC assay method. *J Pharm Biomed Anal.* 2001; 25: 191–5.

Trimethoprim

Chemical Names

5-[(3,4,5-Trimethoxyphenyl)methyl]-2,4-pyrimidinediamine
2,4-Diamino-5-(3,4,5-trimethoxybenzyl)pyrimidine

Other Names

Component of Bactrim and Septra

Form	Molecular Formula	MW	CAS
Trimethoprim	$C_{14}H_{18}N_4O_3$	290.3	738-70-5

Appearance

Trimethoprim occurs as white to cream-colored bitter-tasting odorless crystals or crystalline powder.

Solubility

Trimethoprim has the following solubilities in grams per 100 mL at 25 °C: dimethyl-acetamide, 13.86; benzyl alcohol, 7.29; propylene glycol, 2.57; chloroform, 1.82; methanol, 1.21; water, 0.04; and ether, 0.003.

pK$_a$

Trimethoprim has a pK$_a$ of 6.6.

Method 1

Using an HPLC method, Kaufman et al. determined the stability of undiluted trimethoprim–sulfamethoxazole in single-use plastic syringes. A Waters chromatograph consisted of a solvent-delivery system, a model 440 UV detector, a data processor, and a model 710 B WISP automatic sampling system. The stationary phase was a Waters μBondapak column in a Z-module radial-compression separation system with a Supelco guard column with an LC-18-DB 5-μm cartridge. The mobile phase consisted of 15% acetonitrile and 1% acetic acid in water. The flow rate was 1.7 mL/min. UV detection was performed at 240 nm and 0.1 AUFS. Phenacetin 0.4 mg/mL was used as an internal standard. Under these conditions, retention times for trimethoprim, sulfamethoxazole, and phenacetin were 4.20, 9.74, and 12.45 minutes, respectively.

The assay was verified to be stability indicating by an accelerated degradation study. Concentrated hydrochloric acid was added to a sample of trimethoprim, and the solution was heated for 4 hours at 90 °C. After the addition of 3% hydrogen peroxide, the solution was heated for another 1 hour at 100 °C. Degradation product peaks did not interfere with the peak of the intact trimethoprim.

A standard curve for trimethoprim was constructed from 3.2 to 32 mg/mL. Intraday and interday coefficients of variation for the assay were less than 3%.

Reference

Kaufman MB, Scavone JM, Foley JJ. Stability of undiluted trimethoprim–sulfamethoxazole for injection in plastic syringes. *Am J Hosp Pharm*. 1992; 49: 2782–3.

Method 2

Holmes and Aldous reported the stability of trimethoprim and sulfamethoxazole in peritoneal dialysis fluid determined using an HPLC method. The liquid chromatograph consisted of a Waters model M45 solvent-delivery system, a Waters model 441 UV detector, and a Waters μBondapak C_{18} stainless steel column (300 × 3.9 mm) with a Bondapak C_{18}/Corasil 37–50-μm guard column. The mobile phase consisted of acetonitrile and 0.067 M phosphate buffer (pH 6.2) (1:5, vol/vol) and was delivered isocratically at 1 mL/min. UV detection was performed at 229 nm. Sulfamethazine (sulfadimidine) 52 mg/L in methanol was used as an internal standard.

Samples were diluted with an equal volume of internal standard. The injection volume was 20 μL.

To determine the stability-indicating capability of the assay, cotrimoxazole solutions (trimethoprim 20 mg/mL and sulfamethoxazole 100 mg/mL) were autoclaved for 1 hour at 118 °C at pH 2, 6, and 11. Decomposition product peaks did not interfere with the trimethoprim and sulfamethoxazole peaks. Retention times for the decomposition product, internal standard, trimethoprim, and sulfamethoxazole were about 7.0, 14.0, 16.5, and 19.0 minutes, respectively (estimated from the published chromatogram).

Intraday and interday coefficients of variation for the method were 3.9 and 3.7%, respectively.

Reference

Holmes SE, Aldous S. Stability of cotrimoxazole in peritoneal dialysis fluid. *Peritoneal Dialysis Int*. 1990; 10: 157–60.

Method 3

Tu et al. evaluated the stability of a nonaqueous trimethoprim preparation. The liquid chromatograph consisted of a Waters model 6000A dual-piston pump, a Waters model 450 variable-wavelength detector, and a Hewlett-Packard model 3390 A integrator. The stationary phase was a Waters Nova-Pak C_{18} column (150 × 3.9 mm, 5-μm particle size). The mobile phase was composed of 40% methanol and 60% 0.005 M heptanesulfonic acid sodium salt solution at pH 3.06. The flow rate was 1.3 mL/min. UV detection was performed at 201 nm.

Samples were diluted with 0.1 M phosphoric acid.

The method was demonstrated to be stability indicating by accelerated degradation of trimethoprim. After storage at 140 °C for 28 days, the degradation products eluted separately. The retention time for trimethoprim was 4.73 minutes and the times for the four degradation products were 1.78, 2.06, 2.51, and 3.12 minutes.

Standard curves were constructed for trimethoprim from 1 to 40 μg/mL. The correlation coefficient of the standard curve was greater than 0.99.

Reference

Tu Y-H, Wang D-P, Allen LV Jr. Stability of a nonaqueous trimethoprim preparation. *Am J Hosp Pharm.* 1989; 46: 301–4.

Method 4

Jarosinski et al. studied the stability of trimethoprim–sulfamethoxazole admixtures at various concentrations in 5% dextrose injection or 0.9% sodium chloride injection. The liquid chromatograph consisted of a Perkin-Elmer series 10 pump, a Perkin-Elmer LC-90 detector, a Perkin-Elmer LCI-100 integrator, and a Rheodyne 7125 injector. The stationary phase was an Altex ODS column (250 × 4.6 mm, 5-μm particle size). The mobile phase consisted of 70 parts of 5 mM phosphate and 0.1% triethylamine, with the pH adjusted to 5 with phosphoric acid, and 30 parts of acetonitrile. The flow rate was 1 mL/min. UV detection was performed at 286 nm.

Samples were diluted 1:100 with distilled-deionized water.

The analytical method was determined to be stability indicating by accelerated degradation of trimethoprim. The formulation was heated with acid or base. Degradation product peaks did not interfere with the trimethoprim peak.

A standard curve was generated for trimethoprim from 0.16 to 0.64 mg/mL. The correlation coefficient of the standard curve was greater than 0.999. The coefficients of variation were 1.68% at 3.20 mg/mL and 4.7% at 0.80 mg/mL.

Reference

Jarosinski PF, Kennedy PE, Gallelli JF. Stability of concentrated trimethoprim–sulfamethoxazole admixtures. *Am J Hosp Pharm.* 1989; 46: 732–7.

Method 5

Bergh and Breytenbach described an HPLC analysis of trimethoprim in pharmaceuticals. The liquid chromatograph included a Knauer model FR-30 dual-piston pump, a model 87 variable-wavelength monitor, a Rheodyne model 7010 20-μL loop injector, and a Columbia Scientific Supergrater-3A integrator. The stationary phase was a DuPont Zorbax TMS stainless steel column (250 × 4.6 mm, 7-μm particle size). The mobile phase consisted of acetonitrile, 1-propanol, methanol, tetrahydrofuran, acetic acid, and water (5:20:15:25:1:34, vol/vol/vol/vol/vol/vol). The flow rate was 2 mL/min. UV detection was performed at 271 nm.

Twenty tablets were weighed and ground. A portion of powder equivalent to 40 mg of trimethoprim was weighed, dissolved, filtered, and diluted to 0.20 mg/mL of trimethoprim in ethanol. The suspension was diluted and filtered. Under these conditions, the retention time of trimethoprim was 1.09 minutes.

The stability-indicating properties of the method were shown by assaying a mixture of trimethoprim and its degradation products. Degradation products did not interfere with the analysis of trimethoprim.

A standard curve for trimethoprim was constructed from 0.16 to 0.24 mg/mL. The correlation coefficient was 0.998.

Reference
Bergh JJ, Breytenbach JC. Stability-indicating high-performance liquid chromatographic analysis of trimethoprim in pharmaceuticals. *J Chromatogr.* 1987; 387: 528–31.

Triptorelin

Chemical Name
5-Oxo-L-prolyl-L-histidyl-L-tryptophyl-L-seryl-L-tyrosyl-D-tryptophyl-L-leucyl-L-arginyl-
    L-prolylglycinamide

Other Names
Decapeptyl, Trelstar

Form	Molecular Formula	MW	CAS
Triptorelin	$C_{64}H_{82}N_{18}O_{13}$	1311.4	57773-63-4
Triptorelin acetate	$C_{64}H_{82}N_{18}O_{13}.C_2H_4O_2$	1371.5	140194-24-7
Triptorelin diacetate	$C_{64}H_{82}N_{18}O_{13}.2C_2H_4O_2$	1431.6	105581-02-0
Triptorelin pamoate	$C_{64}H_{82}N_{18}O_{13}.C_{23}H_{16}O_6$	1699.8	124508-66-3

Appearance

Triptorelin occurs as a fluffy white solid.

Method

Helm and Muller investigated the stability of triptorelin in aqueous solution by using an HPLC method. The system consisted of a Pye Unicam model PU 4011 dual-piston pump, a model PU 4020 variable-wavelength UV absorbance detector, a model PU 4810 integrator, and a Rheodyne 20-µL loop injector. The stationary phase was a Shandon Hypersil ODS column (60 × 4.6 mm, 3-µm particle size). The mobile phase was a mixture of 0.2 M triethylammonium phosphate buffer (pH 2.0) and tetrahydrofuran (78:22, vol/vol). The flow rate was 1 mL/min. UV detection was performed at 210 nm. Under these conditions, the retention time of triptorelin was about 5.3 minutes (estimated from the published chromatogram).

The method was stability indicating because the degradation products generated by storing the drug at pH 8.0 at 50 °C for 28 days were well resolved from the peak of the intact drug.

A calibration curve for triptorelin was generated from 100 to 300 µg/mL. The correlation coefficient was 0.9998.

Reference

Helm VJ, Muller BW. Stability of gonadorelin and triptorelin in aqueous solution. *Pharm Res.* 1990; 7: 1253–6.

Trisodium Citrate (Sodium Citrate)

Chemical Name

Trisodium 2-hydroxypropane-1,2,3-tricarboxylate dihydrate

Form	Molecular Formula	MW	CAS
Trisodium citrate	$C_6H_5Na_3O_7.2H_2O$	294.1	6132-04-3

Appearance

Trisodium citrate occurs as white granular crystals or as a crystalline powder.

Solubility

Trisodium citrate has a solubility of 1 in 1.5 of water and 1 in 0.6 of boiling water. It is practically insoluble in alcohol.

Method

Using an HPLC method, Levesque et al. evaluated the stability of trisodium citrate 4.0% and 46.7% in polyvinyl chloride syringes. The Beckman chromatograph consisted of a model 126 programmable solvent module, a model 168 diode-array detector, and an Altex model 210A manual injector. The stationary phase was a Beckman Ultrasphere ODS C_{18} column (250 × 4.6 mm, 5-μm particle size). The mobile phase contained 13.6 g of monobasic potassium phosphate in 950 mL of distilled water, adjusted to pH 2.5 with phosphoric acid. The flow rate was 1.2 mL/min. UV detection was performed at 215 nm.

Each sample was diluted with distilled deionized water to a nominal concentration of 0.1 mg/mL. The injection volume was 100 μL. Under these conditions, the retention time of trisodium citrate was about 7.1 minutes (estimated from the published chromatogram).

The method was demonstrated to be stability indicating by an accelerated degradation study of trisodium citrate. Sample solutions were adjusted to pH 1 with concentrated hydrochloric acid or to pH 12 with 6 N sodium hydroxide solution or were treated with 30% hydrogen peroxide and refluxed for 24 hours. The degradation products did not interfere with the analysis of trisodium citrate.

A standard curve for the drug was constructed from 0.005 to 0.200 mg/mL. The correlation coefficient was 0.9999. The intraday and interday coefficients of variation of the analysis were 0.99 and 2.64%, respectively.

Reference

Levesque N, Girard L, Leger J, et al. Stability of trisodium citrate 4.0% and 46.7% in polyvinyl chloride syringes. *Can J Hosp Pharm.* 2001; 54: 264–8.

Tropicamide

Chemical Name
N-Ethyl-α-(hydroxymethyl)-*N*-(4-pyridinylmethyl)benzeneacetamide

Other Names
Mydriacyl, Mydriaticum, Opticyl

Form	Molecular Formula	MW	CAS
Tropicamide	$C_{17}H_{20}N_2O_2$	284.4	1508-75-4

Appearance
Tropicamide occurs as a white or almost white odorless or almost odorless crystalline powder.

Solubility

Tropicamide is slightly soluble in water. It is freely soluble in alcohol, chloroform, and solutions of strong acids.

Method

Amanlou et al. described a method for the determination of tropicamide in pharmaceutical formulations. The Waters apparatus included a model 510 pump, a model 717 plus autosampler, a model 480 variable-wavelength UV detector, and a model 746 data module. The stationary phase was a Waters μBondapak ODS (C_{18}) column (300 × 3.9 mm, 10-μm particle size). The mobile phase consisted of 50 mM monobasic potassium phosphate aqueous buffer (pH 4) and methanol (70:30, vol/vol). The flow rate was 2 mL/min. UV detection was carried out at 257 nm and 0.01AUFS. Atropine sulfate was used as an internal standard.

An aliquot of commercial eye drops equivalent to 5 mg of tropicamide was transferred to a 25-mL volumetric flask, diluted with deionized water, and assayed. The injection volume was 20 μL. Under these conditions, retention times of atropine and tropicamide were 4.10 and 11.89 minutes, respectively.

The degradation products, tropic acid and ethyl (γ-picolyl)amine, of tropicamide in diluted sulfuric acid did not interfere with the determination of the intact drug, confirming that the analytical method was stability indicating.

A standard curve for tropicamide was constructed from 2 to 300 μg/mL. The correlation coefficient was 0.9996. The intraday and interday coefficients of variation were less than 3.33 and 5.30%, respectively. The limit of detection and the limit of quantitation were 0.1 and 1 μg/mL, respectively.

Reference

Amanlou M, Asmardi G, Andalibi P, et al. Determination of tropicamide in pharmaceutical formulations using high-performance liquid chromatography. *J Chromatogr A.* 2005; 1088: 136–9.

Ursodiol

Chemical Names

(3α,5β,7β)-3,7-Dihydroxycholan-24-oic acid
17β-(1-Methyl-3-carboxypropyl)etiocholane-3α,7β-diol

Other Name

Ursodeoxycholic Acid

Form	Molecular Formula	MW	CAS
Ursodiol	$C_{24}H_{40}O_4$	392.6	128-13-2

Appearance

Ursodiol is a white or almost white crystalline powder.

Solubility

Ursodiol is freely soluble in ethanol and acetic acid. It is slightly soluble in chloroform and sparingly soluble in ether. It is practically insoluble in water.

Method 1

Nahata et al. investigated the stability of ursodiol in two extemporaneously prepared oral suspensions. The liquid chromatograph consisted of a Hewlett-Packard 1050 pump, a Hitachi model AS-400 autosampler, a model L-7400 variable-wavelength detector, and a model D-2000 integrator. The stationary phase was a Zorbax RX C_{18} column (150 × 3.0 mm). The mobile phase consisted of 10 mM monobasic potassium phosphate and methanol (25:75, vol/vol) adjusted to pH 5.25 with phosphoric acid. The flow rate was 0.4 mL/min. UV detection was performed at 201 nm.

A 100-μL ursodiol sample was mixed with 10 mL of methanol and centrifuged. The supernatant was collected and analyzed. The injection volume was 10 μL. Under these conditions, the retention time for ursodiol was about 5.4 minutes.

The stability-indicating nature of the method was established by accelerated degradation of ursodiol. Ursodiol samples were treated with 2.0 M hydrochloric acid, 2.0 M sodium hydroxide solution, or 0.03% hydrogen peroxide at 60 °C. Degradation products did not interfere with the analysis of ursodiol.

A standard curve for ursodiol was constructed from 2.5 to 25 μg/mL. The correlation coefficient was greater than 0.999. Intraday and interday coefficients of variation were less than 0.3 and 1.3%, respectively.

Reference

Nahata MC, Morosco RS, Hipple TF. Stability of ursodiol in two extemporaneously prepared oral suspensions. *J Appl Ther Res.* 1999; 2: 221–4.

Method 2

Mallett et al. evaluated the stability of ursodiol 25 mg/mL in an extemporaneously prepared oral liquid. The HPLC system was a 1090M liquid chromatograph controlled by Hewlett-Packard HPLC 3D Chem Station software V.A.03.02 and equipped with an autosampler, a UV-visible diode-array detector, and a heated column compartment. The stationary phase was an Alltech Altima C_{18} column (250 × 4.6 mm, 5-μm particle size). The mobile phase was a mixture of 45% 10 mM monobasic potassium phosphate buffer (adjusted to pH 3.0 with phosphoric acid) and 55% acetonitrile. The column temperature was maintained at 40 °C. The flow rate was 1.0 mL/min. UV detection was performed at 201 nm with a reference wavelength of 450 nm and a 4-nm bandwidth.

Ursodiol was extracted with methanol. The injection volume was 20 μL.

The analytical method was shown to be stability indicating by an accelerated degradation study. Two samples of ursodiol were treated with 0.1 M hydrochloric acid and 0.1 M sodium hydroxide and autoclaved at 121 °C for 21 minutes. Degradation product peaks did not interfere with the ursodiol peak.

Standard curves were generated from 1.0 to 8.0 mg/mL. The correlation coefficient was greater than 0.999. The intraday and interday coefficients of variation of the assay were 2.30 and 4.46%, respectively.

Reference

Mallett MS, Hagan RL, Peters DA. Stability of ursodiol 25 mg/mL in an extemporaneously prepared oral liquid. *Am J Health Syst Pharm.* 1997; 54: 1401–4.

Method 3

Johnson and Nesbitt determined the short-term stability of ursodiol in an extemporaneously compounded oral liquid. The liquid chromatograph consisted of a Waters model 501 constant-flow solvent-delivery system, a Waters model U6K pump, a Waters model 486 tunable UV detector, and a Hewlett-Packard model 3394 integrator-recorder. The stationary phase was an Applied Biosystems Spheri-5 ODS C_{18} column (250 × 4.6 mm, 5-μm particle size). The mobile phase consisted of 25% 0.01 M dihydrogen potassium phosphate buffer and 75% methanol. The final pH of the mobile phase was adjusted to 5.25 with dilute phosphoric acid. The flow rate was 1.2 mL/min. UV detection was performed at 201 nm.

Each sample was diluted to an ursodiol concentration of 1500 μg/mL with methanol and filtered through a 0.22-μm Millipore Millex-GV filter.

The stability-indicating capability of the assay was demonstrated by accelerated decomposition of ursodiol. Two separate samples of ursodiol 60 mg/mL in simple syrup were exposed to direct sunlight for 35 days after adjustment to either pH 12 with 1 N sodium hydroxide or pH 2 with 1 N sulfuric acid. These two solutions were then heated to 60 °C for 2 hours. Decomposition of ursodiol was achieved and the degradation product peaks did not interfere with the ursodiol peak. Retention time for ursodiol was 7.8 minutes; for the two degradation products, retention times were 9.2 and 14.5 minutes.

A standard curve for ursodiol was produced each day by linear regression of the peak height versus concentration. It was linear from 1200 to 1800 μg/mL; the correlation coefficient was greater than 0.998. The intraday and interday coefficients of variation were 0.6 and 1.8%, respectively.

Reference

Johnson CE, Nesbitt J. Stability of ursodiol in an extemporaneously compounded oral liquid. *Am J Health Syst Pharm.* 1995; 52: 1798–800.

Valacyclovir Hydrochloride

Chemical Name

L-Valine, 2-[(2-amino-1,6-dihydro-6-oxo-9H-purin-9-yl)methoxy]ethyl ester monohydrochloride

Other Name
Valtrex

Form	Molecular Formula	MW	CAS
Valacyclovir hydrochloride	$C_{13}H_{20}N_6O_4 \cdot HCl$	360.8	124832-27-5

Appearance
Valacyclovir hydrochloride occurs as a white to off-white powder.

Solubility
Valacyclovir hydrochloride has a solubility of 174 mg/mL in water at 25 °C.

pK_a
Valacyclovir hydrochloride has pK_a values of 1.90, 7.47, and 9.43.

Method
Fish et al. evaluated the stability of valacyclovir hydrochloride in extemporaneously prepared oral liquids using an HPLC assay. The liquid chromatograph consisted of a Waters 501 pump, a Spectra-Physics SP8880 autosampler, an Applied Biosystems 759A UV detector, and a Waters 746 data module. The stationary phase was a Waters μBondapak C_{18} stainless steel analytical column (300 × 3.9 mm) with an Applied Biosystems guard column. The mobile phase consisted of 5.0 M aqueous sodium acetate and acetonitrile (99:1, vol/vol), adjusted to pH 3.0 with 6 N hydrochloric acid. The flow rate was 3.0 mL/min. UV detection was performed at 254 nm. The internal standard was guanosine.

A 1-mL sample of suspension was diluted 1:250 with 5.0 mM sodium acetate (adjusted to pH 3.0), filtered through a 0.22-μm nylon syringe filter and immediately assayed. The injection volume was 10 μL. Under these conditions, retention times for the internal standard and valacyclovir were about 6 and 14 minutes, respectively.

The stability-indicating capability of the assay was verified by intentionally degrading valacyclovir solution (50 mg/mL) with 0.1 N sodium hydroxide at 100 °C for 8 hours, with 0.1 N hydrochloric acid at 100 °C for 8 hours, or with 1% hydrogen peroxide at 100 °C for 2 hours. The peaks for decomposition products were resolved from peaks for both the internal standard and valacyclovir.

A standard curve for valacyclovir hydrochloride was constructed from 0 to 0.4 mg/mL. The correlation coefficient was 0.998. The intraday and interday coefficients of variation for the assay of valacyclovir suspensions were 0.4 and 2.0%, respectively.

Reference
Fish DN, Vidaurri VA, Deeter RG. Stability of valacyclovir hydrochloride in extemporaneously prepared oral liquids. *Am J Health Syst Pharm*. 1999; 56: 1957–60.

Valganciclovir

Chemical Name
9-[2-Hydroxy-1-(hydroxymethyl)ethoxymethyl]guanine L-valine ester

Other Name
Valcyte

Form	Molecular Formula	MW	CAS
Valganciclovir	$C_{14}H_{22}N_6O_5$	354.4	175865-60-8
Valganciclovir hydrochloride	$C_{14}H_{22}N_6O_5$.HCl	390.8	175865-59-5

Appearance
Valganciclovir hydrochloride occurs as a white to off-white crystalline powder.

Solubility
Valganciclovir hydrochloride has a solubility of 70 mg/mL in water at 25 °C at pH 7.0.

Method
Anaizi et al. used HPLC analysis to investigate the stability of valganciclovir in an extemporaneously compounded oral liquid. The stationary phase was a MetaChem Inertsil ODS-3 column (100 × 4.6 mm, 5-µm particle size) with a MetaGuard Inertsil ODS-3 column (15 × 4.6 mm, 5-µm particle size). The mobile phase was 2.5% (vol/vol) acetonitrile in 25 mM phosphate aqueous buffer (adjusted to pH 2.5 with phosphoric acid). The flow rate was 1.6 mL/min. UV detection was performed at 254 nm and 0.01 AUFS. The injection volume was 10 µL. Hypoxanthine 5 µg/mL was used as an internal standard. Under these conditions, retention times for hypoxanthine and two valganciclovir diastereomers were 1.8, 5.9, and 6.9 minutes, respectively.

The stability-indicating nature of the method was demonstrated by accelerated degradation of valganciclovir. Valganciclovir samples were adjusted to pH 1.4 with 1 N hydrochloric acid or to pH 9.5 with 1 N sodium hydroxide solution, or samples were kept at the initial pH 3.8 and placed in an oven at 75 °C for 24 hours. Degradation products did not interfere with the determination of valganciclovir.

A standard curve for valganciclovir was obtained from 8 to 32 µg/mL. The correlation coefficient was 0.9999. Interday and intraday coefficients of variation were 0.5 and 1.0%, respectively.

Reference
Anaizi NH, Dentinger PJ, Swenson CF. Stability of valganciclovir in an extemporaneously compounded oral liquid. *Am J Health Syst Pharm.* 2002; 59: 1267–70.

Vancomycin

Other Name
Vancocin

Form

Form	Molecular Formula	MW	CAS
Vancomycin	$C_{66}H_{75}Cl_2N_9O_{24}$	1449.2	1404-90-6
Vancomycin hydrochloride	$C_{66}H_{75}Cl_2N_9O_{24}.HCl$	1485.7	1404-93-9

Appearance
Vancomycin hydrochloride is an amphoteric tan to brown free-flowing powder.

Solubility
Vancomycin is freely soluble in water and insoluble in alcohol.

Method 1
Khalfi et al. evaluated by HPLC analysis the compatibility and stability of vancomycin hydrochloride with polyvinyl chloride infusion material in various conditions. The Hewlett-Packard 1090M system was equipped with a Hewlett-Packard 79994 linear photodiode-array detector. The stationary phase was a Hypersil ODS C_{18} column (150 × 4.6 mm, 5-µm particle size). The mobile phase consisted of acetonitrile and an aqueous buffer (8:92, vol/vol); the buffer was 0.2% triethylamine in water and was adjusted to pH 3 with phosphoric acid. The flow rate was 2 mL/min. UV detection was performed at 280 nm. Ceftazidime 100 µg/mL in water was used as an internal standard.

Each sample was diluted with the mobile phase. The injection volume was 10 µL. Under these conditions, the retention times of vancomycin and ceftazidime were 5.8 and 8.4 minutes, respectively.

The method was established to be stability indicating by accelerated degradation. Vancomycin solutions were treated with 1 N sodium hydroxide, 1 N hydrochloric acid, or 1% hydrogen peroxide and boiled for 2 hours. The degradation products did not influence the analysis of vancomycin.

A standard curve for vancomycin was generated from 50 to 100 µg/mL. The correlation coefficient was 0.999.

Reference

Khalfi F, Dine T, Gressier B, et al. Compatibility and stability of vancomycin hydrochloride with PVC infusion material in various conditions using stability-indicating high-performance liquid chromatographic assay. *Int J Pharm.* 1996; 139: 243–7.

Method 2

Trissel et al. studied the stability of vancomycin hydrochloride and aztreonam in 5% dextrose and 0.9% sodium chloride injections using HPLC. The chromatographic system consisted of a Waters model 600E multisolvent-delivery pump, a Waters 490E programmable multiple-wavelength UV detector, a Waters model 712 WISP autosampler, and an Alltech Hypersil ODS C_{18} analytical column (300 × 4.6 mm, 5-µm particle size). The system was controlled and integrated by an NEC PowerMate SX/16 personal computer running Waters Maxima 820 chromatography manager. The mobile phase was a mixture of 20 mL of 1 M triethylamine buffer in 900 mL of HPLC-grade water (pH adjusted to 3.2 with 1 N sodium hydroxide), 14 mL of tetrahydrofuran, and 83 mL of acetonitrile. The mobile phase was filtered through a 0.2-µm membrane filter, degassed for 5 minutes, and delivered isocratically at 1.5 mL/min. UV detection was performed at 216 nm.

Samples were diluted with the mobile phase. The injection volume was 20 µL. Under these conditions, retention times for aztreonam and vancomycin were 5.4 and 6.4 minutes, respectively.

The HPLC assay was demonstrated to be stability indicating by accelerated degradation of vancomycin. Adding 1 N sodium hydroxide, followed by boiling for 2 hours, led to a reduction in the peak for the intact vancomycin and the formation of multiple new peaks. Addition of 1 N hydrochloric acid to the vancomycin solution also resulted in a decrease in the peak for the intact vancomycin and a new peak at 10.4 minutes. Neither aztreonam nor the degradation products of either drug interfered with the intact vancomycin peak.

Six-point calibration curves were constructed from a linear plot of peak area versus vancomycin concentration from 0.025 to 0.150 mg/mL. The correlation coefficient of the standard curve was greater than 0.996. The intraday and interday coefficients of variation were 2.27 and 1.03%, respectively.

References

Trissel LA, Xu QA, Martinez JF. Compatibility and stability of aztreonam and vancomycin hydrochloride. *Am J Hosp Pharm.* 1995; 52: 2560–4.

Trissel LA, Xu QA, Zhang Y, et al. Stability of ciprofloxacin and vancomycin hydrochloride in AutoDose Infusion System bags. *Hosp Pharm.* 2001; 36: 1170–3.

Method 3

Allen et al. investigated the stability of vancomycin hydrochloride in the presence of cefpirome sulfate during simulated Y-site injection. The HPLC method utilized either one of two systems. One unit included an Alcott model 728 Micromeritics autosampler, a Rhodyne 7010 injector with an Alcott 732 electrically actuated valve, a Waters model 501 solvent-delivery pump, a Waters model 441 UV detector, a Waters model 401 refractive index detector, and a Waters model 745 data module. The other unit consisted of an Alcott model 728 Micromeritics autosampler, a Rheodyne 7010 injector with an Alcott 732 electrically actuated valve, a Shimadzu model LA-6A solvent-delivery pump, a Shimadzu model SPD-6A UV detector, and an Orion model 901 microprocessor ion analyzer. The stationary phase was a C_{18} analytical column. The mobile phase consisted of 29% acetonitrile, 1% tetrahydrofuran, and 70% triethylamine buffer (triethylamine 4 mL and water to 2000 mL, pH adjusted to 3.2 with phosphoric acid). The mobile phase was delivered isocratically at 2.0 mL/min. UV detection was performed at 280 nm.

Samples were diluted before injection. The injection volume was 20 μL. Under these conditions, the retention times for vancomycin and cefpirome were 4.1 and 6.5 minutes, respectively.

The analytical method was demonstrated to be stability indicating by accelerated decomposition of vancomycin hydrochloride. Standard solutions of vancomycin hydrochloride were mixed with sodium hydroxide, hydrochloric acid, or potassium chlorate or were subjected to heat at 80 °C or light (150 foot-candles from a tungsten filament source). In each case, degradation product peaks did not interfere with the intact vancomycin peak.

A calibration curve was generated from 10 to 100 μg/mL; the correlation coefficient was greater than 0.99. The intraday and interday coefficients of variation were 3.4 and 3.3%, respectively.

A similar method was used by Walker and Birkhaus.

References

Allen LV, Stiles ML, Prince SJ, et al. Stability of cefpirome sulfate in the presence of commonly used intensive care drugs during simulated Y-site injection. *Am J Health Syst Pharm.* 1995; 52: 2427–33.

Walker SE, Birkhaus B. Stability of intravenous vancomycin. *Can J Hosp Pharm.* 1988; 41: 233–42.

Method 4

Wood et al. investigated the shelf life of vancomycin hydrochloride when reconstituted with water for injection, 5% dextrose, and 0.9% sodium chloride. The HPLC system included an Altex 110A pump, a Rheodyne model 7125 sample injector equipped with a 20-μL injection loop, a Pye Unicam variable-wavelength detector, and a Tekman TE 200 chart recorder. The stationary phase was a Spherisorb ODS stainless steel column (100 × 4.6 mm, 5-μm particle size). The mobile phase consisted of acetonitrile and water (15:85, vol/vol). Forty drops of diethylamine were added to each liter of the mobile phase, with the pH adjusted to 3.15 with 10% phosphoric acid. The flow rate was 1.5 mL/min. UV detection was performed at 281 nm and 0.08 AUFS.

Samples were diluted to a vancomycin concentration of 200 μg/mL.

The stability-indicating nature of the assay was demonstrated by comparing chromatograms of standard solutions and partially degraded and fully degraded solutions. Solutions were subjected to extremes of temperature and pH.

Reference

Wood MJ, Lund R, Beavan M. Stability of vancomycin in plastic syringes measured by high-performance liquid chromatography. *J Clin Pharm Ther.* 1995; 20: 319–25.

Method 5

Vaughan and Poon evaluated the stability of vancomycin hydrochloride and ceftazidime alone and in combination in heparinized and nonheparinized peritoneal dialysis solution. The system consisted of a Waters model 710B WISP autosampler, a Waters Lambda-Max model 481 UV-visible detector, a Waters model 510 pump, and a Perkin-Elmer Nelson model 1020X integrator. The stationary phase was a Waters μBondapak C_{18} column. The mobile phase consisted of 9.4% acetonitrile, 10% 0.2 M ammonium acetate, and 80.6% deionized water adjusted to pH 5.15 with sodium hydroxide. The flow rate was 2.0 mL/min. UV detection was performed at 214 nm.

The assay was demonstrated to be stability indicating by subjecting vancomycin, ceftazidime, and heparin samples to a temperature of 37 °C at pH 1.5 and 12.5 for 6 days. The intact vancomycin eluted at 8 minutes, with three degradation product peaks at 2.5, 4.3, and 6.0 minutes. Neither heparin nor any degradation products interfered with vancomycin separation.

Standard curves were constructed from 25 to 75 μg/mL. Intrarun and interrun coefficients of variation for the measurement of vancomycin were all less than 3%.

Similar analytical methods were used by the other researchers cited here.

References

Vaughan LM, Poon CY. Stability of ceftazidime and vancomycin alone and in combination in heparinized and nonheparinized peritoneal dialysis solution. *Ann Pharmacother*. 1994; 28: 572–6.

Wang D-P, Wang M-T, Wong C-Y, et al. Compatibility of vancomycin hydrochloride and famotidine in 5% dextrose injection. *Int J Pharm Compound*. 1997; 1: 354–5.

Mawhinney WM, Adaic CG, Gorman SP, et al. Stability of vancomycin hydrochloride in peritoneal dialysis solution. *Am J Hosp Pharm*. 1992; 49: 137–9.

Nahata MC. Stability of vancomycin hydrochloride in total parenteral nutrient solutions. *Am J Hosp Pharm*. 1989; 46: 2055–7.

Nahata MC, Miller MA, Durrell DE. Stability of vancomycin hydrochloride in various concentrations of dextrose injection. *Am J Hosp Pharm*. 1987; 44: 802–4.

Venlafaxine Hydrochloride

Chemical Name

1-[2-(Dimethylamino)-1-(4-methoxyphenyl)ethyl]cyclohexanol hydrochloride

Other Name
Effexor

Form	Molecular Formula	MW	CAS
Venlafaxine hydrochloride	$C_{17}H_{27}NO_2 \cdot HCl$	313.9	99300-78-4

Appearance
Venlafaxine hydrochloride occurs as a white to off-white crystalline solid.

Solubility
Venlafaxine hydrochloride has a solubility of 572 mg/mL in water (adjusted to ionic strength of 0.2 M with sodium chloride).

Method 1
Sankar et al. developed an analytical method for the determination of venlafaxine in bulk materials and pharmaceutical formulations. The Waters apparatus included a model 600E pump, a model 717 autosampler, a model 996 photodiode-array detector, and a degasser module. The stationary phase was a LiChrospher C_{18} column (250 × 4.6 mm, 5-µm particle size). The column temperature was maintained at 35 °C. The mobile phase consisted of acetonitrile and phosphate buffer (50:50, vol/vol). The buffer was prepared by dissolving 5.52 g of monobasic sodium phosphate in water; the pH was adjusted to 6.8 with 10% sodium hydroxide solution. The flow rate was 1 mL/min. UV detection was performed at 224 nm.

A portion of the powder from 20 tablets equivalent to 100 mg of venlafaxine hydrochloride was weighed, transferred into a 100-mL volumetric flask, mixed with 50 mL of the mobile phase, sonicated for 15 minutes, filled to volume with mobile phase, and filtered through a 0.45-µm nylon syringe filter. The injection volume was 20 µL. Under these conditions, the retention time of venlafaxine hydrochloride was about 10.0 minutes.

The method was demonstrated to be stability indicating by accelerated decomposition of venlafaxine hydrochloride. The drug was (1) dissolved in 0.1 M hydrochloric acid and heated at 80 °C for 24 hours, (2) dissolved in 1 M hydrochloric acid and heated at 80 °C for 12 hours, (3) dissolved in 0.1 M sodium hydroxide solution and heated at 80 °C for 24 hours, (4) dissolved in 1 M sodium hydroxide solution and heated at 80 °C for 12 hours, (5) dissolved in 3% hydrogen peroxide and stored at room temperature for 24 hours, and (6) dissolved in 30% hydrogen peroxide and stored at room temperature for 12 hours. The drug was also stored in a photostability chamber for up to 30 days, and stored at 60 °C dry for up to 14 days. In all these cases, degradation products did not interfere with the determination of venlafaxine hydrochloride.

A calibration curve for venlafaxine was obtained from 1.0 to 8.0 mg/mL with a correlation coefficient of 0.9999. The intraday and interday coefficients of variation were less than 1.3 and 1.4%, respectively.

Reference
Sankar DG, Krishna MV, Latha PVM. ICH guidance in practice: stability indicating HPLC assay method of venlafaxine in bulk form and solid dosage forms. *Orient J Chem.* 2005; 21: 591–6.

Method 2

Makhija and Vavia described an HPLC method for the determination of venlafaxine in a pharmaceutical formulation. The liquid chromatograph consisted of a Jasco PU-980 pump and a Jasco UV-975 detector. The stationary phase was a Spherisorb C_8 column (250 × 4.6 mm, 5-μm particle size). The mobile phase consisted of acetonitrile and 0.04 M monobasic sodium phosphate buffer (75:25, vol/vol, pH 6.8). The flow rate was 1.5 mL/min. UV detection was performed at 224 nm. The injection volume was 20 μL. Under these conditions, the retention time of venlafaxine was 5.3 minutes.

Twenty venlafaxine tablets were ground and a portion of the powder equivalent to 18 mg of venlafaxine was accurately weighed, transferred to a 25-mL volumetric flask, extracted in methanol, sonicated for 20 minutes, filled to the 25-mL mark, and centrifuged. The supernatant was appropriately diluted with mobile phase and assayed.

Venlafaxine was subjected to accelerated degradation by heating at 70 °C for 1 hour under acidic condition (1 M hydrochloric acid), basic condition (1 M sodium hydroxide solution), and oxidation (hydrogen peroxide). The degradation product did not interfere with the analysis of venlafaxine. The degradation product has a retention time of 4.3 minutes.

A calibration curve was constructed from 1 to 10 μg/mL. The correlation coefficient was 0.9999. The coefficient of variation of the analysis was 0.99%. The limit of detection and the limit of quantitation were 150 and 600 ng/mL, respectively.

Reference

Makhija SN, Vavia PR. Stability indicating LC method for the estimation of venlafaxine in pharmaceutical formulations. *J Pharm Biomed Anal.* 2002; 28: 1055–9.

Verapamil

Chemical Names

α-[3-[[2-(3,4-Dimethoxyphenyl)ethyl]methylamino]propyl]-3,4-dimethoxy-α-(1-methylethyl)benzeneacetonitrile

5-[(3,4-Dimethoxyphenethyl)methylamino]-2-(3,4-dimethoxyphenyl)-2-isopropyl valeronitrile

Other Names

Calan, Iproveratril, Isoptin, Verelan

Form	Molecular Formula	MW	CAS
Verapamil	$C_{27}H_{38}N_2O_4$	454.6	52-53-9
Verapamil hydrochloride	$C_{27}H_{38}N_2O_4 \cdot HCl$	491.1	152-11-4

Appearance

Verapamil hydrochloride is a white or practically white crystalline powder.

Solubility

Verapamil is practically insoluble in water, but it is freely soluble in the lower alcohols, acetone, ethyl acetate, and chloroform.

Method 1

Chen et al. developed an HPLC method for verapamil. The Shimadzu liquid chromatograph included a model LC-6A pump, a model SCL-6A system controller, a model SIL-6A autosampler, a model SPD-6AV variable-wavelength UV detector, and a model C-R3A integrator. A Hewlett-Packard 1040M photodiode-array detector was used for the peak purity analysis. The stationary phase was a Nucleosil 3 C_{18} stainless steel column (150 × 4 mm). The mobile phase consisted of acetonitrile and 7.3 mM triethylamine (pH 3.5) (33:67, vol/vol). The flow rate was 1.0 mL/min. UV detection was performed at 278 nm and 0.02 AUFS. Under these conditions, the retention time for verapamil was about 15 minutes (estimated from the published chromatogram).

The stability-indicating nature of the method was established by accelerated degradation of verapamil. Verapamil samples were treated with 3% hydrogen peroxide, 0.1 N hydrochloric acid, or 0.1 N sodium hydroxide solution at room temperature or 80 °C. The verapamil peak was well resolved from its degradation product peaks.

A standard curve for verapamil was constructed from 12.73 to 100.8 μg/mL. The correlation coefficient was 0.9999. The intraday and interday coefficients of variation were 1.65 and 1.14%, respectively. The limit of detection and the limit of quantitation were 0.833 and 2.50 μg/mL, respectively.

Reference

Chen G-L, Li Y-F, Lee F-Y. A stability indicating method for verapamil by high-performance liquid chromatography/diode-array detector. *Chin Pharm J.* 2000; 52: 113–22.

Method 2

Nahata evaluated by HPLC the stability of verapamil in an extemporaneous liquid dosage form. The system included a Varian model 2010 pump, a model SIL-10A autosampler with a cooler, a model 2050 variable-wavelength detector, and a Hewlett-Packard 3396A integrator. The stationary phase was a μBondapak C_{18} column (300 × 3.9 mm). The mobile phase consisted of 0.01 M dibasic potassium phosphate and methanol (39.9:61.0, vol/vol). The flow rate was 1.0 mL/min. UV detection was performed at 250 nm. The injection volume was 10 μL.

The method was established to be stability indicating by accelerated degradation. Verapamil was treated with acid, base, or oxidizing agent at 60 °C. Degradation products did not influence the determination of verapamil.

A standard curve for verapamil was constructed and its correlation coefficient was greater than 0.999.

Reference

Nahata MC. Stability of verapamil in an extemporaneous liquid dosage form. *J Appl Ther*. 1997; 1: 271–3.

Method 3

Allen and Erickson used HPLC to study the chemical stability of verapamil hydrochloride 50 mg/mL in extemporaneously compounded oral liquids. The system was an automated Hewlett-Packard series 1050 high-performance liquid chromatograph including a multisolvent mixing and pumping system, an autoinjector, a diode-array detector, and a computer with Chem Station data-handling software. The stationary phase was a Bakerbond C_{18} analytical column (250 × 4.6 mm, 5-µm particle size). The mobile phase consisted of 50% acetonitrile, 50% aqueous buffer solution (0.01 N sodium acetate with 33 mL of acetic acid per liter), and 0.5% 2-aminoheptane. The flow rate was 0.5 mL/min. UV detection was performed at 278 nm.

Samples were diluted 1:100. Under these conditions, verapamil eluted in 4.8 minutes.

The stability-indicating nature of this assay was demonstrated by accelerated degradation of verapamil hydrochloride. A composite chromatogram of verapamil hydrochloride after intentional decomposition by acid, base, heat, oxidizing agent, and light showed that the intact verapamil peak was well resolved from degradation product peaks.

Standard curves were constructed from a linear plot of verapamil peak areas versus its concentrations from 100 to 750 µg/mL. The intraday and interday coefficients of variation were 0.7 and 1.6%, respectively.

Reference

Allen LV, Erickson MA. Stability of labetalol hydrochloride, metoprolol tartrate, verapamil hydrochloride, and spironolactone with hydrochlorothiazide in extemporaneously compounded oral liquids. *Am J Health Syst Pharm*. 1996; 53: 2304–9.

Method 4

Riley and Junkin evaluated the stability of verapamil hydrochloride in intravenous admixtures using an HPLC assay. The liquid chromatograph consisted of a Kratos Spectroflow 400 solvent-delivery system, a Waters model 450 variable-wavelength UV detector, a Micromeritics model 728 autosampler, a Valco VICI six-port electronically actuated injection valve fitted with a 20-µL loop, a Fisher Scientific series 5000 Recordall strip-chart recorder, and a Shimadzu model C-R3A electronic integrator. The stationary phase was a Phase Separations Spherisorb phenyl column (150 × 4.6 mm, 5-µm particle size). The mobile phase was a mixture of acetonitrile, water, tetrabutylammonium hydrogen sulfate (100 mM), and potassium dihydrogen phosphate (0.5 M) (15:50:25:10, vol/vol/vol/vol) with a pH of 5.1. The flow rate was 2 mL/min. UV detection was performed at 268 nm.

Samples were diluted 1:10 with the mobile phase.

The HPLC method was demonstrated to be stability indicating and free from interference by excipients and degradation products.

The peak areas were linearly related to the concentration from 0.125 to 3.00 mg/mL.

Reference

Riley CM, Junkin P. Stability of amrinone and digoxin, procainamide hydrochloride, propranolol hydrochloride, sodium bicarbonate, potassium chloride, or verapamil hydrochloride in intravenous admixtures. *Am J Hosp Pharm.* 1991; 48: 1245–52.

Method 5

Gupta and Stewart used an HPLC method to investigate the chemical stability of dobutamine hydrochloride and verapamil hydrochloride when mixed together in 0.9% sodium chloride and 5% dextrose injections. A Waters model ALC 202 liquid chromatograph was equipped with a Waters model U6K universal injector, a Spectroflow SF 770 multiple-wavelength detector, an Omniscribe recorder, and a Waters μBondapak phenyl semipolar column (300 × 4.0 mm). The mobile phase consisted of 40% (vol/vol) acetonitrile in water with 0.02 M monobasic potassium phosphate and 0.3% (vol/vol) acetic acid (pH 3.9 ± 0.1) and was delivered isocratically at 2.4 mL/min. UV detection was performed at 278 nm and 0.04 AUFS. Dextromethorphan hydrobromide 130 μg/mL was used as an internal standard.

Samples were assayed without further dilution. The injection volume was 20 μL. Retention times for dobutamine and verapamil were about 1.8 and 4.5 minutes, respectively (estimated from the published chromatogram).

To determine if the HPLC method was stability indicating, a solution of verapamil hydrochloride 1.6 mg/mL was mixed with 1 N sodium hydroxide and heated to boiling on a hot plate for about 30 minutes. Degradation product peaks did not interfere with the verapamil peak.

A standard curve for verapamil was constructed from 1.6 to 4.2 μg per 20-μL injection.

Reference

Gupta VD, Stewart KR. Stability of dobutamine hydrochloride and verapamil hydrochloride in 0.9% sodium chloride and 5% dextrose injections. *Am J Hosp Pharm.* 1984; 41: 686–9.

Vertilmicin Sulfate

Form	Molecular Formula	MW	CAS
Vertilmicin sulfate	$(C_{22}H_{43}N_5O_7)_2 \cdot 5H_2SO_4$	1469.6	459427-94-2

Method

Liu and Duan described the determination of vertilmicin sulfate and its related compounds in bulk drug and its pharmaceutical formulations. The Agilent HP 1100 series system consisted of an Agilent G1311A QuatPump fitted with a G1322A degasser, a manual injector with a 100-μL loop, and an Agilent G1314A variable-wavelength detector. The stationary phase was an Agilent XDB-C_8 column (150 × 4.6 mm, 5-μm particle size). The mobile phase consisted of an aqueous buffer and acetonitrile (75:25, vol/vol). The aqueous buffer contained 20 mM sodium heptanesulfonate and 30 mM triethylamine and was adjusted to pH 2.5 with phosphoric acid. The flow rate was 1 mL/min. The column temperature was maintained at 30 °C. UV detection was performed at 201 nm. The sample injection volume was 10 μL. Under these conditions, the retention time of vertilmicin sulfate was 10.3 minutes.

To demonstrate the stability-indicating nature of the analytical method, vertilmicin samples were stored under stress conditions (light, heat, acid, base, and oxidation). The intact vertilmicin sulfate and its related compounds were well separated from each other.

A calibration curve was obtained from 0.25 to 5 mg/mL; the correlation coefficient was 0.9998. The within-day and between-day coefficients of variation of the analysis were less than 2%. At an injection volume of 50 μL, the limit of detection and the limit of quantitation of vertilmicin were 1 and 3 μg/mL, respectively.

Reference

Liu Z, Duan G. Stability indicating reversed-phase ion-pairing liquid chromatographic determination of vertilmicin sulfate as bulk drug and in injection. *J Pharm Biomed Anal.* 2005; 37: 577–83.

Vinblastine

Other Names
Velban, Vincaleukoblastine

Form

Form	Molecular Formula	MW	CAS
Vinblastine	$C_{46}H_{58}N_4O_9$	811.0	865-21-4
Vinblastine sulfate	$C_{46}H_{58}N_4O_9.H_2SO_4$	909.1	143-67-9

Appearance
Vinblastine sulfate is a yellowish-white solid.

Solubility
Vinblastine is practically insoluble in water and petroleum ether. It is soluble in alcohols, acetone, ethyl acetate, and chloroform. Vinblastine sulfate is very slightly soluble in ethanol. It is practically insoluble in ether. One part of vinblastine sulfate is soluble in 10 parts of water and in 50 parts of chloroform.

pK$_a$

Vinblastine has pK$_a$ values of 5.4 and 7.4.

Method 1

Beijnen et al. evaluated the stability of vinblastine in three commonly used infusion fluids. The HPLC system consisted of a Waters model M45 solvent-delivery pump, a model 440 dual-wavelength detector, a model 710 WISP autosampler, and a Spectra-Physics model SP4270 integrator. The stationary phase was a Hypersil ODS stainless steel analytical column (150 × 3.9 mm, 5-μm particle size). The mobile phase was a mixture of methanol and 10 mM phosphate buffer (pH 7.0) (60:40, wt/wt). The flow rate was 1.0 mL/min. UV detection was performed at 254 and 280 nm. The sample injection volume was 20 μL. Under these conditions, the retention time for vinblastine was 7.4 minutes.

The stability-indicating nature of the assay was determined by accelerated degradation of a sample of vinblastine solution at pH 1 and 10 and 80 °C. The degradation products were well separated from the intact drug on its chromatogram.

Standard curves for vinblastine were constructed from 5 to 20 μg/mL. The correlation coefficients were greater than 0.999.

Reference

Beijnen JH, Vendrig DEMM, Underberg WJM. Stability of vinca alkaloid anticancer drugs in three commonly used infusion fluids. *J Parenter Sci & Technol*. 1989; 43: 84–7.

Method 2

McElnay et al. examined the stability of vinblastine sulfate in burette administration sets. The HPLC system consisted of an Altex 110A pump, a Rheodyne 20-μL loop injection valve, an Altex 330 fixed-wavelength UV detector, a Hewlett-Packard 3390A integrator, and a Spherisorb ODS column (250 × 4.6 mm, 5-μm particle size). The mobile phase consisted of water and methanol (50:50, vol/vol) with the pH adjusted to 3.0 with acetic acid. The flow rate was 1.5 mL/min. UV detection was performed at 254 nm.

The method was demonstrated to be stability indicating by accelerated decomposition of vinblastine sulfate. A solution of vinblastine sulfate was autoclaved for 1 hour. The degradation product peak did not interfere with the vinblastine peak. Retention times for vinblastine and the degradation product were 2.6 and 4 minutes, respectively.

A calibration curve was generated from 1 to 3 μg/mL; the correlation coefficient was 0.998.

Reference

McElnay JC, Elliott DS, Cartwright-Shamoon J, et al. Stability of methotrexate and vinblastine in burette administration sets. *Int J Pharm*. 1988; 47: 239–47.

Vincamine

Chemical Name

Methyl (3α,16α)-14,15-dihydro-14β-hydroxyeburnamenine-14-carboxylate

Other Names

Aethroma, Cetal, Oxygeron

Form	Molecular Formula	MW	CAS
Vincamine	$C_{21}H_{26}N_2O_3$	354.4	1617-90-9

Appearance

Vincamine occurs as yellow crystals.

Method

Shehata et al. reported an HPLC analytical method for the determination of vincamine in the presence of its degradation product. The Hitachi-Merck instrument consisted of a model L-7150 pump, a model L-7455 photodiode-array detector, a Rheodyne model 7725i injector with a 20-μL loop, and a model D-7000 chromatography data station. The stationary phase was a Merck Lichrocart RP-18 column (250 × 4.6 mm, 5-μm particle size). The mobile phase consisted of acetonitrile and 0.01 M ammonium carbonate aqueous solution (7:3, vol/vol). The flow rate was 1.6 mL/min. UV detection was performed at 280 nm.

A portion of the powder from 20 capsules equivalent to 25 mg of vincamine was weighed into a 250-mL beaker, mixed with 50 mL of 0.1 N hydrochloric acid, shaken well, filtered into a 100-mL volumetric flask, filled to the mark with 0.1 N hydrochloric acid, and assayed. The injection volume was 20 μL. Under these conditions, the retention time of vincamine was 3.67 minutes.

The method was verified to be stability indicating by assaying a mixture containing the drug and its degradation product, vincaminic acid. Vincaminic acid had a retention time of 6.61 minutes.

A standard curve for vincamine was constructed from 2 to 20 μg/mL. The correlation coefficient was 0.9994. The intraday and interday coefficients of variation were less than 0.38 and 0.57%, respectively.

Reference

Shehata MAM, El Sayed MA, El Tarras MF, et al. Stability-indicating methods for determination of vincamine in presence of its degradation product. *J Pharm Biomed Anal.* 2005; 38: 72–8.

Vincristine

Chemical Name
22-Oxovincaleukoblastine

Other Names
Leurocristine, Oncovin

Form	Molecular Formula	MW	CAS
Vincristine	$C_{46}H_{56}N_4O_{10}$	825.0	57-22-7
Vincristine sulfate	$C_{46}H_{56}N_4O_{10}.H_2SO_4$	923.1	2068-78-2

Appearance
Vincristine sulfate is a white to slightly yellow odorless and very hygroscopic amorphous or crystalline powder.

Solubility
Vincristine sulfate is freely soluble in water and soluble in methanol, but it is practically insoluble in ether.

pK$_a$
Vincristine has pK$_a$ values of 5 and 7.4.

Method 1
Mayron and Gennaro assessed the stability of vincristine sulfate with granisetron hydrochloride during simulated Y-site administration. The liquid chromatograph consisted of a piston pump with a pulse dampener, a rotary injection port with a 20-μL loop, a variable-wavelength spectrometric detector, and an integrator. The stationary phase was a LiChrosorb RP8 column (250 × 4.6 mm, 5-μm particle size). The mobile phase was 5 mL of diethylamine and 295 mL of water adjusted to pH 7.5 with phosphoric acid and the final volume was adjusted to 1 L with methanol. The flow rate was 1.75 mL/min. UV detection was performed at 300 nm.

Samples were diluted with the mobile phase. The retention times for vincristine and granisetron were 4.01 and 8.10 minutes, respectively.

The stability-indicating capability of the analytical method was demonstrated by an accelerated degradation study. Solutions of vincristine sulfate and granisetron hydrochloride were adjusted to pH 2 and 11, boiled for 1 hour, readjusted to pH 5, diluted with the mobile phase, and analyzed. The degradation product peaks did not interfere with the intact vincristine or granisetron peaks.

Reference

Mayron D, Gennaro AR. Stability and compatibility of granisetron hydrochloride in i.v. solutions and oral liquids and during simulated Y-site injection with selected drugs. *Am J Health Syst Pharm.* 1996; 53: 294–304.

Method 2

Nyhammar et al. used an HPLC method to determine the stability of vincristine sulfate and doxorubicin hydrochloride in two portable infusion-pump reservoirs. The liquid chromatographic system consisted of an LDC/Milton ConstaMetric III pump, a Waters 717 autosampler, a Shimadzu SPD-6AV UV-visible variable-wavelength detector, and a Hichrom Nucleosil 100-5CN column (150 × 4.6 mm, 5-μm particle size). The mobile phase consisted of acetonitrile, methanol, and 20 mM ammonium dihydrogen phosphate aqueous solution (pH 4.5) (20:20:60, vol/vol/vol) containing sodium heptanesulfonate at a final concentration of 10 mM as an ion-pairing agent. The flow rate was 1.0 mL/min. UV detection was performed at 297 nm.

Samples were assayed without further dilution. The injection volume was 25 μL. Under these conditions, retention times for doxorubicin and vincristine were 4.1 and 11.8 minutes, respectively.

To demonstrate that the assay was stability indicating, solutions of vincristine sulfate were mixed with 0.1 M sodium hydroxide (final pH 13) and 0.1 M hydrochloric acid (final pH 1). Degradation product peaks did not interfere with the intact vincristine sulfate peak.

Calibration curves for vincristine sulfate were constructed and were linear for drug concentrations from 5 to 100 μg/mL. The correlation coefficient was 1.000. The intraday and interday coefficients of variation were 0.37 and 1.3%, respectively.

Reference

Nyhammar EK, Johansson SV, Seiving BE. Stability of doxorubicin hydrochloride and vincristine sulfate in two portable infusion-pump reservoirs. *Am J Health Syst Pharm.* 1996; 53: 1171–3.

Method 3

Beijnen et al. evaluated by HPLC analysis the stability of vincristine in three commonly used infusion fluids. The system consisted of a Waters model M45 solvent-delivery pump, a model 440 dual-wavelength detector, a model 710 WISP autosampler, and a Spectra-Physics model SP4270 integrator. The stationary phase was a Hypersil ODS stainless steel analytical column (150 × 3.9 mm, 5-μm particle size). The mobile phase was a mixture of methanol and 10 mM phosphate buffer (pH 7.0) (60:40, wt/wt). The flow rate was 1.0 mL/min. UV detection was performed at 254 and 280 nm. The sample injection volume was 20 μL. Under these conditions, the retention time for vincristine was 5.3 minutes.

The stability-indicating nature of the assay was determined by accelerated degradation of samples of vincristine solution at pH 1 and 10 and 80 °C. The degradation products were well separated from the intact drug on its chromatogram.

Standard curves for vincristine were constructed from 5 to 20 µg/mL. The correlation coefficients were greater than 0.999.

Reference

Beijnen JH, Vendrig DEMM, Underberg WJM. Stability of vinca alkaloid anticancer drugs in three commonly used infusion fluids. *J Parenter Sci & Technol.* 1989; 43: 84–7.

Method 4

Beijnen et al. evaluated the stability of intravenous admixtures of vincristine sulfate and doxorubicin hydrochloride using HPLC. The chromatograph consisted of a Waters model M45 solvent-delivery pump, a Waters model 440 dual-wavelength detector, a Waters model 710B WISP autosampler, and a Spectra-Physics SP4270 integrator. The stationary phase was a home-packed Merck LiChrosorb RP8 analytical column (300 × 3.9 mm, 10-µm particle size). The mobile phase consisted of 600 mL of acetonitrile and 400 mL of 0.1% ammonia solution. The pH of the aqueous solution was adjusted to 4.0 with formic acid. The flow rate was 1.5 mL/min. UV detection was performed at 280 and 313 nm.

Samples were assayed without further dilution. The injection volume was 15 µL.

To demonstrate the stability-indicating capacity of the assay, solutions of vincristine sulfate were adjusted to pH 1 and 13. Degradation product peaks did not interfere with the intact vincristine peak.

Calibration curves of vincristine were constructed and were linear (r >0.999) from 0.002 to 0.060 mg/mL. Coefficients of variation were 9.8% at 0.002 mg/mL and 0.9% at 0.060 mg/mL.

Reference

Beijnen JH, Neef C, Meuwissen OJAT, et al. Stability of intravenous admixtures of doxorubicin and vincristine. *Am J Hosp Pharm.* 1986; 43: 3022–7.

Vindesine

Chemical Name

3-Carbamoyl-4-*O*-deacetyl-3-de(methoxycarbonyl)vincaleukoblastine

Other Names
Eldisin, Eldisine

Form
Vindesine sulfate

Molecular Formula
$C_{43}H_{55}N_5O_7.H_2SO_4$

MW
852.0

CAS
59917-39-4

Appearance
Vindesine sulfate occurs as a white or almost white hygroscopic amorphous substance.

Solubility
Vindesine sulfate is freely soluble in water and in methanol.

Method
Beijnen et al. evaluated by HPLC analysis the stability of vindesine in three commonly used infusion fluids. The system consisted of a Waters model M45 solvent-delivery pump, a model 440 dual-wavelength detector, a model 710 WISP autosampler, and a Spectra-Physics model SP4270 integrator. The stationary phase was a Hypersil ODS stainless steel analytical column (150 × 3.9 mm, 5-μm particle size). The mobile phase was a mixture of methanol and 10 mM phosphate buffer (pH 7.0) (60:40, wt/wt). The flow rate was 1.0 mL/min. UV detection was performed at 254 and 280 nm. The sample injection volume was 20 μL. Under these conditions, the retention time for vindesine was 6.7 minutes.

The stability-indicating nature of the assay was determined by accelerated degradation of samples of vindesine solutions at pH 1 and 10 and 80 °C. The degradation products were well separated from the intact drug on its chromatogram.

Standard curves for vindesine were constructed from 5 to 20 μg/mL. The correlation coefficients were greater than 0.999.

Reference
Beijnen JH, Vendrig DEMM, Underberg WJM. Stability of vinca alkaloid anticancer drugs in three commonly used infusion fluids. *J Parenter Sci & Technol.* 1989; 43: 84–7.

Warfarin

Chemical Names
4-Hydroxy-3-(3-oxo-1-phenylbutyl)-2*H*-1-benzopyran-2-one
3-(α-Acetonylbenzyl)-4-hydroxycoumarin

Other Name
Coumadin

Form	Molecular Formula	MW	CAS
Warfarin	$C_{19}H_{16}O_4$	308.3	81-81-2
Warfarin sodium	$C_{19}H_{15}NaO_4$	330.3	129-06-6

Appearance
Warfarin sodium is a white crystalline powder.

Solubility
Warfarin sodium is very soluble in water, freely soluble in alcohol, and very slightly soluble in chloroform or ether.

pK_a
Warfarin has a pK_a of 5.

Method 1
Montgomery et al. described the development and validation of an HPLC method for the simultaneous determination of warfarin and aspirin in tablets. The chromatographic system consisted of a Hewlett-Packard 1050 series pump, an autosampler, a column oven, and a detector. The stationary phase was a Zorbax C_8 column (250 × 4.0 mm, 5-µm particle size). The column temperature was maintained at 40 °C. Mobile phase A was water, adjusted to pH 2.6 with formic acid. Mobile phase B was methanol. Mobile phase C was acetonitrile. The flow rate was 1.0 mL/min. Mobile phases (A:B:C) were initially delivered at a ratio of 68:17:15 (vol/vol/vol) for 11 minutes, increased linearly to 56:17:27 (vol/vol/vol) at 15 minutes, held at this ratio for 38 minutes and then returned in 0.5 minute to the initial ratio and equilibrated for 6 minutes. Butylparaben was used as an internal standard. UV detection was performed at 280 nm.

The sample solvent mixture was made of acetonitrile, chloroform, formic acid, citric acid, and butylparaben (49.95:49.95:0.1:0.1:0.05, vol/vol/vol/vol/vol). The injection volume was 20 µL. Under these conditions, retention times for aspirin and warfarin were about 13 and 42 minutes, respectively (estimated from the published chromatogram).

The assay was demonstrated to be stability indicating. A sample of aspirin and warfarin was spiked with known potential degradation products and the formulation ingredients, but these did not interfere with the analysis of aspirin and warfarin.

A calibration curve for warfarin was generated from 0.01 to 0.18 mg/mL. The correlation coefficient was 0.9997.

Reference
Montgomery ER, Taylor S, Segretario J, et al. Development and validation of a reversed-phase liquid chromatographic method for analysis of aspirin and warfarin in a combination tablet formulation. *J Pharm Biomed Anal.* 1996; 15: 73–82.

Method 2
Using an HPLC method, Kuhn et al. determined the in vitro recovery of warfarin from an enteral nutrient formula. The liquid chromatograph consisted of a Rheodyne model 7125

injector fitted with a 20-μL fixed-volume stainless steel loop, an Eldex Laboratories model 1007/B-94 isocratic pump, a Gilson model HM UV detector, and a Fisher model B5117-11 recorder. The stationary phase was a Phase Separations Spherisorb S10 ODS-1 column (200 × 4.6 mm, 10-μm particle size). The mobile phase consisted of 65% methanol and 35% acetic acid (pH 3.4, 0.1% vol/vol). The flow rate was 1.7 mL/min. UV detection was performed at 280 nm. Naphthoic acid 20 μg/mL was used as an internal standard.

The assay was demonstrated to be stability indicating by accelerated decomposition of warfarin sodium. A solution of warfarin sodium was exposed to alkaline conditions (2.0 M sodium hydroxide) and was subjected to 120 °C and 20 psi for 150 minutes. Degradation product peaks did not interfere with the warfarin peak.

A calibration curve for warfarin was constructed each day from 3.3 to 133.3 μg/mL; the correlation coefficient of the calibration curve was 0.9994. The coefficients of variation for the assay were 0.8 and 1.2% for 5 and 10 μg/mL, respectively.

Reference
Kuhn TA, Garnett WR, Wells BK, et al. Recovery of warfarin from an enteral nutrient formula. *Am J Hosp Pharm.* 1989; 46: 1395–9.

Xanthinol Niacinate (Xanthinol Nicotinate)

Chemical Name
7-[2-Hydroxy-3-[(2-hydroxyethyl)methylamino]propyl]theophylline niacinate

Other Name
Complamin

Form
Xanthinol niacinate

Molecular Formula
$C_{13}H_{21}N_5O_4.C_6H_5NO_2$

MW
434.4

CAS
437-74-1

Appearance
Xanthinol niacinate occurs as crystals.

Solubility
Xanthinol niacinate is freely soluble in water.

Method
Han developed an HPLC method for the determination of xanthinol niacinate (xanthinol nicotinate). The liquid chromatograph was equipped with a Waters model 440 UV

detector. The stationary phase was a Waters µBondapak C_{18} column. The mobile phase consisted of methanol and water (1:1, vol/vol). The flow rate was 1.0 mL/min. UV detection was performed at 254 nm and 0.2 AUFS. Caffeine was used as an internal standard. The injection volume was 10 µL. Under these conditions, retention times for xanthinol niacinate and caffeine were about 2.0 and 3.6 minutes, respectively.

The method was reported to be stability indicating.

A standard curve for xanthinol niacinate was obtained from 0.200 to 0.600 mg/mL. The coefficient of variation for the analysis of xanthinol niacinate was 1.42%.

Reference

Han CD. High performance liquid chromatographic determination of xanthinol nicotinate. *Yakhak Hoeji*. 1984; 28: 321–5.

Xilobam

Chemical Name

(2,6-Dimethylphenyl)(1-methyl-2-pyrrolidinylidene)urea

Form	Molecular Formula	MW	CAS
Xilobam	$C_{14}H_{19}N_3O$	245.3	50528-97-7

Method

Janicki et al. reported the analysis by HPLC of decomposition products and stability assessment of xilobam. A Waters liquid chromatograph consisted of a UV detector, a pump, and a loop injector. The stationary phase was a LiChrosorb RP18 C_{18} column (250 × 2 mm). The mobile phase was a mixture of 0.1% ammonium carbonate and acetonitrile (60:40, vol/vol). The flow rate was 1.0 mL/min. UV detection was performed at 254 nm. *N*-(3-Chlorophenyl)-*N'*-(1-methyl-2-pyrrolidylidene)urea 2 mg/mL was used as an internal standard.

An aliquot (5 mL) of sample was mixed with 4 mL of internal standard in a 10-mL volumetric flask and filled to volume with methanol. The injection volume was 10 µL. Under these conditions, the retention time for xilobam was 2.1 minutes.

The stability-indicating nature of the method was demonstrated by forced degradation of xilobam. Xilobam was dissolved in 1 N hydrochloric acid, 1 N sodium hydroxide, or water and heated at 80 °C for 24 hours. Xilobam was separated from its degradation products.

Reference

Janicki CA, Walkling WD, Erlich RH, et al. Xilobam: Analysis, determination of decomposition products, and assessment of stability. *J Pharm Sci*. 1981; 70: 778–80.

Yohimbine Hydrochloride

Chemical Name
Methyl 17α-hydroxyyohimban-16α-carboxylate hydrochloride

Other Name
Aphrodyne

Form	Molecular Formula	MW	CAS
Yohimbine hydrochloride	$C_{21}H_{26}N_2O_3$.HCl	390.9	65-19-0

Appearance
Yohimbine hydrochloride occurs as an odorless crystalline powder.

pK$_a$
Yohimbine has a pK$_a$ of 6–7.5.

Method
Mittal et al. developed an HPLC method for the quantitative analysis of yohimbine. The apparatus consisted of Waters model 501 pumps, a model 486 tunable detector, and a model 712 WISP autosampler. The stationary phase was a Nova-Pak C$_{18}$ column (150 × 4.6 mm, 5-μm particle size). The mobile phase was a mixture of 700 mL of methanol and 300 mL of water. The flow rate was 0.6 mL/min. UV detection was performed at 270 nm and 1 AUFS. Caffeine 100 μg/mL in distilled water was used as an internal standard. The injection volume was 20 μL. Under these conditions, retention times for caffeine and yohimbine were 2.3 and 4.6 minutes, respectively.

The method was reported to be stability indicating.

A standard curve for yohimbine was generated from 7.92 to 59.40 μg/mL. The correlation coefficient was 0.9935. Intraday and interday coefficients of variation were 1.51 and 1.35%, respectively.

Reference
Mittal S, Alexander KS, Dollimore D. A high-performance liquid chromatography assay for yohimbine HCl analysis. *Drug Dev Ind Pharm*. 2000; 26: 1059–65.

Zafirlukast

Chemical Name

Cyclopentyl 3-[2-methoxy-4-[(*o*-tolylsulfonyl)carbamoyl]benzyl]-1-methylindole-5-carbamate

Other Name

Accolate

Form

Zafirlukast

Molecular Formula

$C_{31}H_{33}N_3O_6S$

MW

575.7

CAS

107753-78-6

Appearance

Zafirlukast is a white to pale yellow fine amorphous powder.

Solubility

Zafirlukast is practically insoluble in water and has a solubility of 0.9 mg/mL in alcohol at 20 °C.

Method

Radhakrishna et al. reported an HPLC method for the determination of zafirlukast in bulk drug and its pharmaceutical formulation. The liquid chromatograph consisted of a Waters 510 pump, a Rheodyne 10-μL loop injector, and a Waters 996 photodiode-array detector. The stationary phase was a Waters Symmetry Shield RP18 column (250 × 4.6 mm, 5-μm particle size). The mobile phase consisted of 300 mL of 0.01 M monobasic potassium phosphate aqueous buffer and 700 mL of acetonitrile; it was adjusted to pH 3.5 with phosphoric acid. The flow rate was 1.0 mL/min. 5-Methyl 2-nitrophenol was used as an internal standard. UV detection was performed at 223 nm. The sample injection volume was 10 μL. Under these conditions, the retention times of the internal standard and zafirlukast were about 5.1 and 11.2 minutes, respectively (estimated from the published chromatogram).

Twenty zafirlukast tablets were ground and a portion of the powder equivalent to 20 mg of zafirlukast was accurately weighed, extracted with acetonitrile, and centrifuged. The supernatant was appropriately diluted with mobile phase and injected.

In order to verify the stability-indicating ability of the method, zafirlukast samples were refluxed separately with 0.1 N sodium hydroxide solution and 0.1 N hydrochloric acid at 60 °C for 12 hours. Zafirlukast solutions were also exposed to ultraviolet light (254 nm) for 24 hours or kept at 60 °C for 12 hours. In all cases, the degradation products were well resolved from the zafirlukast peak.

A calibration curve was constructed from 30 to 175 μg/mL. The correlation coefficient was 0.999. The coefficient of variation of the analysis was 0.3%.

Reference
Radhakrishna T, Satyanarayana J, Satyanarayana A. Determination of zafirlukast by stability indicating LC and derivative spectrophotometry. *J Pharm Biomed Anal.* 2002; 30: 695–703.

Zidovudine

Chemical Name
3'-Azido-3'-deoxythymidine

Other Names
AZT, Retrovir

Form	Molecular Formula	MW	CAS
Zidovudine	$C_{10}H_{13}N_5O_4$	267.2	30516-87-1

Appearance
Zidovudine is an odorless white to off-white crystalline solid.

Solubility
Zidovudine has solubilities of 20 mg/mL in water and 71 mg/mL in alcohol at 25 °C.

Method 1
Santoro et al. reported an assay method for the quantitative determination of zidovudine in capsules. The HPLC system consisted of a Varian model 5000 solvent-delivery pump, a model 4000 variable UV detector, a model 4400 integrator, and a Rheodyne model 7125 injector. The stationary phase was a LiChrospher 100 RP-18 column (125 x 4.0 mm, 5-μm particle size). The mobile phase was a mixture of water and methanol (75:25, vol/vol). The flow rate was 1.0 mL/min. UV detection was carried out at 265 nm.

A sample equivalent to 100.0 mg of zidovudine was accurately weighed, transferred to a 100-mL volumetric flask, mixed with 50 mL of water, sonicated for 30 minutes, and diluted to the mark with water. The solution was further diluted with mobile phase and filtered before analysis. The injection volume was 20 μL. Under these conditions, the retention time of zidovudine was about 5.7 minutes.

The drug products were incubated at different temperatures and humidity conditions (25 °C/70%, 40 °C/75%, and 50 °C/90%) for 90 days. Under these conditions, the peak of the intact drug was resolved from that of its degradation product, thymine (1.7 minutes).

A standard curve of zidovudine was obtained from 70.0 to 150.0 µg/mL. The correlation coefficient was 0.9995. The limit of detection and the limit of quantitation were 2.80 and 9.34 µg/mL, respectively. The coefficient of variation for the analysis of zidovudine was 0.79%.

Reference

Santoro MI, Taborianski AM, Kedor-Hackmann AKS. Stability-indicating methods for quantitative determination of zidovudine and stavudine in capsules. *Quim Nova.* 2006; 29: 240–4.

Method 2

Caufield and Stewart reported a method for the simultaneous determination of zidovudine and chlordiazepoxide. The Hewlett-Packard series 1090 system included a pump, an autosampler with a 25-µL loop, and a Gilson model 117 variable-wavelength UV detector or a Waters model 996 photodiode-array detector. The stationary phase was a Supelco Discovery RP-Amide C_{16} column (250 × 4.6 mm, 5-µm particle size). The mobile phase consisted of 25 mM monobasic sodium phosphate monohydrate (pH adjusted to 3.0 with 0.1 M phosphoric acid) and acetonitrile (80:20, vol/vol). The flow rate was 1.0 mL/min. Solutions of zidovudine and chlordiazepoxide were prepared with an aqueous acetonitrile diluent matching the mobile phase compositions. UV detection was performed at 265 nm. Under these conditions, retention times of zidovudine and chlordiazepoxide were 6.2 and 14.7 minutes, respectively.

To demonstrate that the method was stability indicating, solutions of zidovudine were subjected to acid hydrolysis (6 M hydrochloric acid), base hydrolysis (6 M sodium hydroxide solution), oxidation (0.3% hydrogen peroxide), heat (90 °C), and radiation (254 nm). Solutions of chlordiazepoxide were also subjected to acid hydrolysis (1 M hydrochloric acid), base hydrolysis (1 M sodium hydroxide solution), oxidation (0.3% hydrogen peroxide), heat (60 °C), and radiation (254 nm). In all cases, zidovudine and chlordiazepoxide were separated from their degradation products on their chromatograms.

A standard curve of zidovudine was constructed from 52.5 to 210 µg/mL. The correlation coefficient was greater than 0.9999. The intraday and interday coefficients of variation were 0.17 and 0.86%, respectively.

A similar method was reported by Musami et al.

References

Caufield WV, Stewart JT. HPLC separations of zidovudine and selected pharmaceuticals using a hexadecylsilane amide column. *Chromatographia.* 2001; 54: 561–8.

Musami P, Stewart JT, Taylor EW. Stability of zidovudine and ranitidine in 0.9% sodium chloride and 5% dextrose injections stored at ambient temperature (23 ± 2 °C) and 4 °C in 50-mL polyvinylchloride bags up to 24 hours. *Int J Pharm Compound.* 2004; 8: 236–9.

Method 3

Lam et al. investigated the stability of zidovudine in 5% dextrose injection and 0.9% sodium chloride injection. The HPLC system consisted of a Beckman model 110A isocratic pump, a Kratos model GM770R variable-wavelength UV detector, a Rheodyne model 7125 injector, and a Perkin-Elmer model LCI-100 integrator. The stationary phase was a DuPont Zorbax C_8 column (250 × 4.6 mm, 5-µm particle size). The mobile phase was a mixture of 20% acetonitrile and 80% 0.025 M monobasic sodium phosphate at pH 4.6. The flow rate was 1 mL/min. UV detection was performed at 266 nm.

Samples were diluted 1:100 with double-distilled water. The retention time of zidovudine was 7.5 minutes.

The stability-indicating nature of the assay was demonstrated by an accelerated degradation study. One milliliter of the standard zidovudine reference was mixed with 1 mL of 0.1 N hydrochloric acid or 0.1 N sodium hydroxide and heated at 100° C for 24 hours. Degradation product peaks did not interfere with the zidovudine peak.

A calibration curve for zidovudine was constructed from 12.7 to 25.4 µg/mL. The correlation coefficient was greater than 0.999. The intraday and interday variabilities were 0.5 and 1.6%, respectively.

Reference

Lam NP, Kennedy PE, Jarosinski PF, et al. Stability of zidovudine in 5% dextrose injection and 0.9% sodium chloride injection. *Am J Hosp Pharm.* 1991; 48: 280–2.

Ziprasidone

Chemical Name

5-[2-[4-(1,2-Benzisothiazol-3-yl)-1-piperazinyl]ethyl]-6-chloro-1,3-dihydro-2*H*-indol-2-one

Other Names

Geodon, Zeldox

Form	Molecular Formula	MW	CAS
Ziprasidone	$C_{21}H_{21}ClN_4OS$	412.9	146939-27-7
Ziprasidone hydrochloride	$C_{21}H_{21}ClN_4OS.HCl.H_2O$	467.4	138982-67-9
Ziprasidone mesylate	$C_{21}H_{21}ClN_4OS.CH_4O_3S.3H_2O$	563.1	199191-69-0

Method

El-Sherif et al. developed a chromatographic method for the analysis of ziprasidone in bulk materials and in pharmaceutical formulations. The Hewlett-Packard series 1100 chromatograph was equipped with a quaternary pump, a photodiode-array detector, and a manual injector with a 20-µL loop. The stationary phase was a LiChrosorb RP-18 column

(250 × 4 mm, 10-μm particle size). The mobile phase consisted of water, acetonitrile, and phosphoric acid (76:24:0.5, vol/vol/vol). The flow rate was 1.5 mL/min. UV detection was performed at 229 nm.

A portion of the powder from 20 capsules equivalent to 25 mg of ziprasidone was weighed into a 50-mL volumetric flask, half filled with methanol, shaken for 15 minutes, filled to the mark with methanol, further diluted to a nominal concentration of 0.25 mg of ziprasidone with methanol, and assayed. The injection volume was 20 μL. Under these conditions, the retention time of ziprasidone was 19.1 minutes.

This method was demonstrated to be stability indicating by assaying a mixture containing the drug and its degradation products. The degradation products eluted at retention times of 4.1, 31.7, and 42.2 minutes.

A calibration curve for ziprasidone was constructed from 10 to 500 μg/mL. The correlation coefficient was 0.999. The intraday and interday coefficients of variation were less than 2.1 and 1.2%, respectively. The limit of detection and the limit of quantitation were 0.77 and 3.02 μg/mL.

Reference

El-Sherif ZA, El-Zeany B, El-Houssini OM, et al. Stability indicating reversed-phase high-performance liquid chromatographic and thin layer densitometric methods for the determination of ziprasidone in bulk powder and in pharmaceutical formulations. *Biomed Chromatogr.* 2004; 18: 143–9.

Zoledronic Acid

Chemical Name

(1-Hydroxy-2-imidazol-1-yl-phosphonoethyl)phosphonic acid monohydrate

Other Name

Zometa

Form	Molecular Formula	MW	CAS
Zoledronic acid	$C_5H_{10}N_2O_7P_2$	272.1	118072-93-8
Zoledronate disodium	$C_5H_8N_2Na_2O_7P_2 \cdot 4H_2O$	388.1	165800-07-7
Zoledronate trisodium	$C_5H_7N_2Na_3O_7P_2 \cdot 2\frac{1}{2}H_2O$	383.1	165800-08-8

Appearance
Zoledronic acid occurs as a white crystalline powder.

Method
Rao et al. developed a method for the determination of zoledronic acid. The Agilent 1100 series liquid chromatographic system included a photodiode-array detector. The stationary phase was a Waters XTerra RP$_{18}$ column (250 × 4.6 mm, 5-μm particle size). The mobile phase consisted of methanol and a phosphate buffer (5:95, vol/vol); the phosphate aqueous buffer contained 8 mM monobasic potassium phosphate, 2 mM dibasic sodium phosphate, and 7 mM tetra-n-butyl ammonium hydrogen sulfate, pH 6.6. The flow rate was 0.7 mL/min. UV detection was carried out at 215 nm.

The drug was dissolved in mobile phase. The injection volume was 10 μL. The total run time was 15 minutes. Under these conditions, the retention time of zoledronic acid was 8.8 minutes.

The analytical method was demonstrated to be stability indicating by accelerated degradation studies. The drug substance was subjected to heat (60 °C), light, acid hydrolysis (0.5 N hydrochloric acid), base hydrolysis (0.1 N sodium hydroxide solution), water hydrolysis, and oxidation (10% hydrogen peroxide at 25 and 60 °C). Zoledronic acid was separated from its impurities and degradation products on the chromatograms.

A standard curve for zoledronic acid was constructed from 50 to 300 μg/mL. The correlation coefficient was greater than 0.999. The coefficient of variation of the assay was less than 1%.

Reference
Rao BM, Srinivasu MK, Rani CP, et al. A validated stability indicating ion-pair RP-LC method for zoledronic acid. *J Pharm Biomed Anal.* 2005; 39: 781–90.

Zolmitriptan

Chemical Name
4(S)-4-[3-(2-Dimethylaminoethyl)-1H-5-indolylmethyl]-1,3-oxazolan-2-one

Other Name
Zomig

Form	Molecular Formula	MW	CAS
Zolmitriptan	$C_{16}H_{21}N_3O_2$	287.4	139264-17-8

Appearance
Zolmitriptan occurs as a white to almost white powder.

Solubility
Zolmitriptan is readily soluble in water.

Method
Rao et al. developed a method for the analysis of zolmitriptan. The Agilent 1100 series liquid chromatograph was equipped with a photodiode-array detector. The stationary phase was a Waters Xterra RP$_{18}$ column (250 × 4.6 mm, 5-μm particle size), which was maintained at 30 °C. The mobile phase consisted of two solutions, A and B, delivered gradiently. Solution A was a mixture of 10 mM monobasic ammonium phosphate (pH adjusted to 9.85 with ammonia solution), methanol, and acetonitrile (70:20:10, vol/vol/vol). Solution B was a mixture of 10 mM monobasic ammonium phosphate (pH adjusted to 9.85 with ammonia solution) and acetonitrile (30:70, vol/vol). The mobile phase was delivered at 1.0 mL/min in a gradient mode as follows:

Time, minutes	Solution A, %	Solution B, %
0	100	0
10	100	0
30	45	55
35	45	55
36	100	0
46	100	0

UV detection was performed at 225 nm. Zolmitriptan solutions were prepared by dissolving the drug in a mixture of water and acetonitrile (1:1, vol/vol). The injection volume was 10 μL. Under these conditions, the retention time of zolmitriptan was 14.0 minutes.

The drug was subjected to light, heat (60 °C), acid hydrolysis (0.1 N hydrochloric acid), base hydrolysis (0.1 N sodium hydroxide solution), water hydrolysis, and oxidation (0.01% hydrogen peroxide). In all these cases, zolmitriptan was well separated from its degradation products on its chromatogram, demonstrating the stability-indicating nature of the analytical method.

A calibration curve for zolmitriptan was constructed from 25 to 150 μg/mL. The correlation coefficient was greater than 0.999. The coefficient of variation of the assay was less than 1%.

Reference
Rao BM, Srinivasu MK, Sridhar G, et al. A stability indicating LC method for zolmitriptan. *J Pharm Biomed Anal.* 2005; 39: 503–9.

Zomepirac Sodium

Chemical Name
Sodium 5-(*p*-chlorobenzoyl)-1,4-dimethylpyrrole-2-acetate dihydrate

Other Names
Zomax, Zomaxin, Zopirac

Form	Molecular Formula	MW	CAS
Zomepirac sodium	$C_{15}H_{13}ClNNaO_3 \cdot 2H_2O$	349.8	64092-49-5

Appearance
Zomepirac sodium occurs as crystals.

Method
Chen et al. developed an HPLC method for the determination of zomepirac in the presence of its degradation products. The liquid chromatographic system consisted of an Alcott 760 pump system, a Jasco 875 UV detector, and a CSW 1.7 integator. The stationary phase was an Inertsil ODS-3V column (250 × 4.6 mm, 5-μm particle size) with an Inertsil guard column (50 × 4.6 mm, 7-μm particle size). The mobile phase consisted of acetonitrile, methanol, and 1% acetic acid aqueous solution (2:64:34, vol/vol/vol). The flow rate was 1 mL/min. UV detection was performed at 254 nm. Butylparaben was used as an internal standard. The injection volume was 20 μL. Under these conditions, retention times of butylparaben and zomepirac were about 15.9 and 17.8 minutes, respectively.

The stability-indicating nature of this method was confirmed by accelerated degradation of zomepirac. Zomepirac in methanol was treated with 0.2 N hydrochloric acid or 0.2 N sodium hydroxide solution and incubated at 60 °C for 3 days. A zomepirac methanolic solution was also irradiated under a Hanovia 200-W high-pressure mercury lamp at a distance of 25 cm for 40 hours. The peak of intact zomepirac was well separated from the peaks of its degradation products.

A standard curve for zomepirac in methanol was constructed from 5.0 to 100 μM. The correlation coefficient was greater than 0.999. The intraday and interday coefficients of variation of the analysis were less than 2.2 and 3.9%, respectively.

Reference
Chen C-Y, Chen F-A, Chen C-J, et al. Stability-indicating HPLC assay method of zomepirac. *J Food Drug Anal.* 2003; 11: 87–91.

Zopiclone

Chemical Name
6-(5-Chloro-2-pyridinyl)-6,7-dihydro-7-oxo-5H-pyrrolo[3,4-b]pyrazin-5-yl 4-methylpiper-azine-1-carboxylate

Other Names
Imovane, Zileze

Form	Molecular Formula	MW	CAS
Zopiclone	$C_{17}H_{17}ClN_6O_3$	388.8	43200-80-2

Appearance
Zopiclone is a white or slightly yellow powder.

Solubility
Zopiclone is practically insoluble in water and in alcohol.

pK_a
Zopiclone has a pK_a of 6.7.

Method
Bounine et al. described the HPLC determination of zopiclone in tablets. The liquid chromatographic system consisted of a Kontron model 420 pump, a model 465 auto-injector, a Milton-Roy LDC model 3100 UV detector, and a Prolabo Stabitherm column oven. The stationary phase was a Merck LiChrospher-60 RP Select B column (125 × 4.0 mm, 5-μm particle size). The mobile phase was a mixture of a buffer, aceto-nitrile, and tetrahydrofuran (81:18:1, vol/vol/vol), where the buffer was prepared by dissolving 3.4 g of monosodium hexanesulfonate and 7.0 g of monobasic sodium phosphate dihydrate in 1 L of water. The flow rate was 1.5 mL/min. UV detection was performed at 303 nm.

A portion of powder from ground tablets equivalent to 5 mg of zopiclone was weighed into a 50-mL volumetric flask, mixed with 40 mL of 0.1 M hydrochloric acid, ultrasonicated for 15 minutes, diluted to volume with 0.1 M hydrochloric acid, and filtered through a 1.6-μm glass-fiber filter. The injection volume was 20 μL. Under these conditions, the retention time of zopiclone was about 7.3 minutes (estimated from the published chromatogram).

The stability-indicating ability of the method was demonstrated by assaying a spiked standard solution containing impurities and potential degradation products. No interference with the determination of zopiclone was observed.

A standard curve for zopiclone was constructed from 51.1 to 153.3 μg/mL. The correlation coefficient was 1.000.

Reference
Bounine JP, Tardif B, Beltran P, et al. High-performance liquid chromatographic stability-indicating determination of zopiclone in tablets. *J Chromatogr A*. 1994; 677: 87–93.

Zorubicin

Chemical Name

(2S-cis)-Benzoic acid [1-[4-[(3-amino-2,3,6-trideoxy-α-L-*lyxo*-hexopyranosyl)oxy]-1,2,3,4,6,11-hexahydro-2,5,12-trihydroxy-7-methoxy-6,11-dioxo-2-naphthacenyl]-ethylidene]hydrazide

Other Name

Rubidazone

Form

Form	Molecular Formula	MW	CAS
Zorubicin	$C_{34}H_{35}N_3O_{10}$	645.7	54083-22-6
Zorubicin hydrochloride	$C_{34}H_{35}N_3O_{10}$.HCl	682.1	36508-71-1

Appearance

Zorubicin hydrochloride occurs as a red-orange crystalline powder.

Method

Benaji et al. evaluated the stability and compatibility of zorubicin in intravenous fluids and polyvinyl chloride infusion bags. A Hewlett-Packard 1090M HPLC system was equipped with a variable-volume injector, an automatic sampling system, a Hewlett-Packard 7994 linear photodiode-array UV detector, and a Hewlett-Packard ThinkJet printer. The stationary phase was a Phenomenex Ultramex C_{18} ODS column (150 × 4.6 mm, 5-μm particle size). The mobile phase consisted of sodium dihydrogen phosphate buffer (0.03 M) with 0.1% tetraethylammonium, acetonitrile, and tetrahydrofuran (65:33:1, vol/vol/vol). The phosphate buffer was prepared in water adjusted to pH 4.5. The flow rate was 2 mL/min. UV detection was performed at 254 nm. Epirubicin was used as an internal standard.

Samples were diluted in phosphate buffer. The injection volume was 10 μL. The retention times for epirubicin, daunorubicin (a degradation product), and zorubicin were 1.52, 2.28, and 3.09 minutes, respectively.

The stability-indicating capability of the assay was demonstrated by accelerated degradation of the drug. Samples of zorubicin were mixed with 0.1 N hydrochloric acid, 0.1 N sodium hydroxide, and hydrogen peroxide. Degradation product peaks did not interfere with the intact zorubicin peak.

Calibration curves for zorubicin were constructed from 5 to 25 μg/mL. The correlation coefficient of the calibration curves was 0.9998. Intraday and interday coefficients of variation for the assay were less than 2.99 and 2.71%, respectively.

Reference

Benaji B, Dine T, Luyckx M, et al. Stability and compatibility studies of zorubicin in intravenous fluids and PVC infusion bags. *J Pharm Biomed Anal.* 1996; 14: 695–705.

INDEX

A

Boldface entries are monograph titles. All other entries are brand names or synonyms.